ELECTRON TRANSFER—FROM ISOLATED MOLECULES TO BIOMOLECULES

Part 2

ADVANCES IN CHEMICAL PHYSICS

VOLUME 107

ELECTRON TRANSFER—
FROM ISOLATED
MOLECULES TO
BIOMOLECULES
Part 2

Edited by

JOSHUA JORTNER

School of Chemistry
Tel Aviv University
Tel Aviv, Israel

and

M. BIXON

School of Chemistry
Tel Aviv University
Tel Aviv, Israel

ADVANCES IN CHEMICAL PHYSICS
VOLUME 107

Series Editors

I. PRIGOGINE

Center for Studies in Statistical Mechanics
and Complex Systems
The University of Texas
Austin, Texas
and
International Solvay Institutes
Université Libre de Bruxelles
Brussels, Belgium

STUART A. RICE

Department of Chemistry
and
The James Franck Institute
The University of Chicago
Chicago, Illinois

AN INTERSCIENCE® PUBLICATION

JOHN WILEY & SONS, INC.

NEW YORK • CHICHESTER • WEINHEIM • BRISBANE • SINGAPORE • TORONTO

Library of Congress Catalog Number 58-9935

ISBN 0-471-25291-3

Printed in the United States of America.

10 9 8 7 6 5 4 3 2 1

CONTRIBUTORS TO VOLUME 106

ISRAELA BECKER, School of Chemistry, Tel Aviv University, Ramat Aviv, Tel Aviv, Israel

DAVID N. BERATAN, Department of Chemistry, University of Pittsburgh, Pittsburgh, Pennslyvania

M. BIXON, School of Chemistry, Tel Aviv University, Ramat Aviv, Tel Aviv, Israel

H. L. CARRELL, Institute of Cancer Research, Fox Chase Cancer Center, Philadelphia, Pennsylvania

PINGYUN CHEN, Department of Chemistry, The University of North Carolina at Chapel Hill, Chapel Hill, North Carolina

VLADIMIR CHERNYAK, Department of Chemistry, University of Rochester, Rochester, New York

ORI CHESHNOVSKY, School of Chemistry, Tel Aviv University, Ramat Aviv, Tel Aviv, Israel

MATTHIJS P. DE HAAS, IRI, Delft University of Technology, Mekelweg 15, 2629JB Delft, The Netherlands

CAROLINE E. H. DESSENT, Sterling Chemistry Laboratory, Yale University, New Haven, Connecticut

CHRISTOPHER DURNELL, Department of Chemistry, University of Chicago, Chicago, Illinois

G. R. FLEMING, Department of Chemistry, University of Chicago, Chicago, Illinois

R. MICHAEL GARAVITO, Department of Biochemistry, Michigan State University, East Lansing, Michigan

STEVEN GOODMAN, Department of Chemistry, University of Chicago, Chicago, Illinois

GITTE IVERSEN, Department of Chemistry, Technical University of Denmark, Building 207, Lyngby, Denmark

MARK A. JOHNSON, Sterling Chemistry Laboratory, Yale University, New Haven, Connecticut

JOSHUA JORTNER, School of Chemistry, Tel Aviv University, Ramat Aviv, Tel Aviv, Israel

YURIS I. KHARKATS, The A.N. Frumkin Institute of Electrochemistry, Russian Academy of Sciences, Leninskij Prospect 31, Moscow, Russia

ALEKSANDR M. KUZNETSOV, The A.N. Frumkin Institute of Electrochemistry, Russian Academy of Sciences, Leninskij Prospect 31, Moscow, Russia

MATTHEW J. LANG, Department of Chemistry, University of Chicago, Chicago, Illinois

DONALD H. LEVY, Department of Chemistry and The James Franck Institute, University of Chicago, Chicago, Illinois

R. A. MARCUS, Noyes Laboratory of Chemical Physics 127-72, California Institute of Technology, Pasadena, California

THOMAS J. MEYER, Department of Chemistry, The University of North Carolina at Chapel Hill, Chapel Hill, North Carolina

SHAUL MUKAMEL, Department of Chemistry, University of Rochester, Rochester, New York

MARSHALL D. NEWTON, Department of Chemistry, Brookhaven National Laboratory, Box 5000, Upton, New York

AKIRA OKADA, Department of Chemistry, University of Rochester, Rochester, New York

KRISTIN M. OMBERG, Department of Chemistry, The University of North Carolina at Chapel Hill, Chapel Hill, North Carolina

MICHAEL N. PADDON-ROW, School of Chemistry, University of New South Wales, Sydney, Australia

SPIROS S. SKOURTIS, Department of Natural Sciences, University of Cyprus, 1678 Nicosia, Cyprus

NORMAN SUTIN, Chemistry Department, Brookhaven National Laboratory, P.O. Box 5000, Upton, New York

JENS ULSTRUP, Technical University of Denmark, Department of Chemistry, Building 207, Lyngby, Denmark

JAN W. VERHOEVEN, Laboratory of Organic Chemistry, University of Amsterdam, Nieuwe Achtergracht 129, Amsterdam, The Netherlands

JOHN M. WARMAN, IRI, Delft University of Technology, Mekelweg 15, Delft, The Netherlands

BAS WEGEWIJS, Laboratory of Organic Chemistry, University of Amsterdam, Nieuwe Achtergracht 129, Amsterdam, The Netherlands

NIEN-CHU C. YANG, Department of Chemistry, University of Chicago, Chicago, Illinois

SONG-LEI ZHANG, Department of Chemistry, University of Chicago, Chicago, Illinois

CONTRIBUTORS TO VOLUME 107

BIMAN BAGCHI, Solid State and Structural Chemistry Unit, Indian Institute of Science, Bangalore, India

PAUL F. BARBARA, Department of Chemistry, University of Minnesota, Minneapolis, Minnesota,

MINHAENG CHO, Department of Chemistry, Korea University, Seoul, Korea

W. DAVIS, Department of Chemistry and Materials Research, Nortwestern University, Evanston, Illinois

B. D. FAINBERG, School of Chemistry, Tel aviv University, Ramat Aviv, Tel Aviv, Israel

OLE FARVER, Institute of Analytical and Pharmaceutical Chemistry, Royal Danish School of Pharmacy 2 Universitetsparken, Copenhagen, Denmark

GRAHAM R. FLEMING, Department of Chemistry, University of Chicago, Chicago, Illinois

HAROLD L. FRIEDMAN, Department of Chemistry, State University of New York at Stony Brook, Stony Brook, New York

N. GAYATHRI, Solid State and Structural Chemistry Unit, Indian Institute of Science, Bangalore, India

SHELBY HATCH, Department of Chemistry, University of Princeton, Princeton, New Jersey

ROBIN M. HOCHSTRASSER, Department of Chemistry, University of Pennsylvania, 231 South 34th Street, Philadelphia, Pennsylvania

D. HUPPERT, School of Chemistry, Tel Aviv University, Ramat Aviv, Tel Aviv, Israel

M. KEMP, Department of Chemistry and Materials Research, Northwestern University, Evanston, Illinois

SONJA KOMAR-PANICUCCI, Peptide, Inc. Cambridge, Massachusettes

Y. MAO, Department of Chemistry and Materials Research, Northwestern University, Evanston, Illinois

NOBORU MATAGA, Institute for Laser Technology, 1-8-4 Utsubo-Honmachi, Nishi-Ku, Osaka, Japan

GEORGE MCLENDON, Department of Chemistry, Princeton University, Princeton, New Jersey

HIROSHI MIYASAKA, Department of Polymer Engineering and Science, Kyoto Institute of Technology, Matsugasaki, Sakyo, Kyoto, Japan

V. MUJICA, Universidad Central de Venezuela, Facultad De Clenclas, Escuela de Quimica, Apartado, Caracas, Venezuela

A. NITZAN, School of Chemistry, Tel Aviv University, Ramat Aviv, Tel Aviv, Israel

ERIC J. C. OLSON, Department of Chemistry, University of Minnesota, Minneapolis, Minnesota

J. N. ONUCHIC, Department of Physics, University of California at San Diego, La Jolla, California

ISRAEL PECHT, Department of Immunology, The Weizmann Institute of Science, Rehovot, Israel

FERNANDO O. RAINERI, Department of Chemistry, State University of New York at Stony Brook, Stony Brook, New York

M. A. RATNER, Department of Chemistry and Materials Research, North-Western University, Evanston, Illinois

J. J. REGAN, Beckman Institute, California Institute of Technology, Pasadena, California

A. ROITBERG, National Institute of Standards and Technology, Bio-technology Division, Building 222, A-353, Gaithersburg, Maryland

HITOSHI SUMI, Institute of Materials Science, University of Tsukuba, Tsukuba, Ibaraki Japan

KLAAS WYNNE, Femtosecond Research Centre, Department of Physics and Applied Physics, University of Strathclyde, 107 Rottenrow, Glasgow, United Kingdom

KEITARO YOSHIHARA, Institute for Molecular Science, Nyodall, Okazaki, Japan

INTRODUCTION

Few of us can any longer keep up with the flood of scientific literature, even in specialized subfields. Any attempt to do more and be broadly educated with respect to a large domain of science has the appearance of tilting at windmills. Yet the synthesis of ideas drawn from different subjects into new, powerful, general concepts is as valuable as ever, and the desire to remain educated persists in all scientists. This series, *Advances in Chemical Physics*, is devoted to helping the reader obtain general information about a wide variety of topics in chemical physics, a field that we interpret very broadly. Our intent is to have experts present comprehenisve analyses of subjets of interest and to encourage the expression of individual points of view. We hope that this approach to the presentation of an overview of a subject will both stimulate new research and serve as a personalized learning text for beginners in a field.

I. Prigogine
Stuart A. Rice

.

CONTENTS TO VOLUME 106

CONTENTS TO VOLUME 107

PREFACE

Remarkable progress has been made in the elucidation of the processes of energy acquisition, storage, and disposal in large molecules, clusters, condensed-phase, and biophysical systems, as explored from the microscopic point of view. The broad area of nonradiative dynamics, from isolated molecules to biomolecules, plays an important role in the development of modern chemistry. Electron transfer processes constitute a landmark example for intramolecular, condensed-phase, and biophysical nonradiative dynamics. Nonradiative electron transfer phenomena encompass electron transfer and hole transfer between localized states, involving intramolecular, intracluster, condensed phase, interfacial, and protein medium charge separation, migration, recombination, and localization, as well as electron and hole transport between a large number of constituents. Radiative charge transfer absorption, fluorescence, and resonance Raman processes in Mulliken charge-transfer complexes and donor–acceptor molecules are concurrently of considerable interest. The sweep and grandeur of electron transfer phenomena are reflected in the broad spectrum of diverse systems, encompassing a multitude of scientific disciplines, which involve isolated solvent-free supermolecules in supersonic jets; elemental, polar, ionic, and metal clusters in cluster beams; ions, complexes, organic and inorganic supermolecules, and solvated electrons in polar and in nonpolar solvents; metal, semiconductor, and superconductor electrodes in solution; surfaces and interfaces involving thin films, adsorbants and surface states; crystalline and amorphous semiconductors, molecular crystals, polymers, and biopolymers; and biological systems pertaining to respiratory and enzymatic protein systems, DNA repair, and the primary charge separation proceses in photosynthesis, which ensure the efficiency and robustness of the central life-sustaining processes on Earth.

Exploration of the ubiquitous electron transfer processes in chemistry, physics, and biology constitutes an interdisciplinary research area, blending concepts and experimental techniques from a wide variety of fields. The very considerable recent advances stem from concurrent progress in experiment and in theory. On the experimental front, considerable progress was made with the advent of femtosecond lasers, allowing for real-time interrogation of intramolecular and intermolecular electron transfer dynamics on the time scale of nuclear motion. Major impact was exerted by modern experimental approaches for the preparation and characterization of elemental isolated molecule and cluster systems, by advances in chemical synthesis of donor–acceptor supermolecules, and by the growing sophistication of biochemical synthesis with the application of genetic engineering methods and of chemical engineering for the preparation of well-characterized biophysical systems. The theoretical arsenal rests on the seminal Marcus electron transfer theory with the incorporation of electronic and nuclear quantum effects, the theory and simulations of radiationless processes, wave-packet dynamics, coherence effects, cluster dynamic size effects, nonadiabatic condensed-phase dynamics and nonlinear optical effects, which provide the conceptual framework for intramolecular, cluster, condensed-phase, and biophysical electron transfer dynamics. The chapters assembled in these volumes, which describe many of the experimental and theoretical results now in hand, promote the goal of establishing a unified description of the broad spectrum of chemical, physical, and biological electron transfer phenomena. In spite of tremendous progress and impact, the field may not yet be mature enough to permit a compilation of a definite treatise. Instead, the authoritative contributions assembled in these volumes reflect, in our opinion, the kind of information that will underline the development of an integrated approach to electron transfer phenomena, from isolated molecules to biomolecules, on time scales from those appropriate for "conventional" chemical processes (i.e., seconds to picoseconds) to ultrafast (femtoseconds) processes, transcendenting the time scale for nuclear motion.

The first two chapters are intended to provide an overview of the historical development of the entire field, from the seminal theoretical concepts of Franck and Libby in the late 1940s, which provided the theoretical cornerstone for condensed-phase electron transfer, up to the present time. We are greatly indebted to the pioneers of electron transfer science, Professor Rudoph A. Marcus and Professor Norman Sutin, for contributing these overview chapters. The third introductory chapter is meant to review some of the concepts, problems, ideas, experiments, and the theoretical arsenal in the field. The chapters have been organized in topical groups, the general material being as follows:

1. Overview
2. Isolated Molecules
3. Clusters
4. General Theory
5. Spectra and Electron Transfer Kinetics in Bridged Compounds
6. Solvent Control

7. Ultrafast Electron Transfer and Coherence Effects
8. Molecular Electronics
9. Electron Transfer and Chemistry
10. Biomolecules

We have followed the general policy of the *Advances in Chemical Physics* that the authors are given complete freedom as to the size, scope, and format of their contribution. Our point of view is that the person who pioneered the topic is the best judge of the appropriate mode of its presentation. We very much hope that these volumes will offer an overview of the entire field, reflecting the forefront of current research efforts and exploring the perspectives and future of electron transfer research.

We are grateful to numerous colleagues and friends whose lively and probing discussions at scientific meetings and during scientific encounters convinced us of the merits of the project. We thank the authors for their willingness to contribute to this endeavor and for their adherence to the timetable. Thanks are due to the editor of *Advances in Chemical Physics*, Stuart A. Rice, for welcoming and supporting this project. We thank Ms. C. A. Fjerstad and the editorial staff of Wiley Interscience for support of the publication of these volumes. The wide range of subjects touched on in these volumes bears witness of the scope and quality of modern electron transfer research and of the enthusiasm of its practitioners.

JOSHUA JORTNER
M. BIXON
Tel Aviv, Israel

ELECTRON TRANSFER—FROM ISOLATED MOLECULES TO BIOMOLECULES

Part 2

ADVANCES IN CHEMICAL PHYSICS

VOLUME 107

INTERPLAY BETWEEN ULTRAFAST POLAR SOLVATION AND VIBRATIONAL DYNAMICS IN ELECTRON TRANSFER REACTIONS: ROLE OF HIGH-FREQUENCY VIBRATIONAL MODES

BIMAN BAGCHI

Solid State and Structural Chemistry Unit, Indian Institute of Science, Bangalore 560012, India
Jawaharlal Nehru Center for Advanced Scientific Research, Bangalore

N. GAYATHRI

Solid State and Structural Chemistry Unit, Indian Institute of Science, Bangalore 560012, India

CONTENTS

Electron Transfer: From Isolated Molecules to Biomolecules, Part Two, edited by Joshua Jortner and M. Bixon. Advances in Chemical Physics Series, Volume 107, series editors I. Prigogine and Stuart A. Rice.
ISBN 0-471-25291-3   © 1999 John Wiley & Sons, Inc.

I. INTRODUCTION

Electron transfer reactions in solution are often coupled to both intramolecular vibrational relaxation and polar solvation dynamics [1–4]. The coupling to solvent relaxation is easy to understand, especially for electron transfer in a polar liquid, as the electric field of the charge itself is strongly coupled to solvent polarization. This solvent polarization plays a central role in the definition of both the reaction coordinate and the reaction energy surface of an outer-sphere electron transfer reaction. On the other hand, the realization that the intramolecular vibrational modes and their dynamics can play an important role in the electron transfer reactions came somewhat later. Interestingly, the currently held consensus that the vibrational energy relaxation plays a greater role in many photoinduced electron transfer reactions than the solvent polarization relaxation is exactly opposite to what was believed even a decade ago. Although our understanding of the details of electron transfer reaction remains imperfect in most cases, it is becoming clear that both vibrational energy and solvent polarization relaxations are important. Another new development in this field is the discovery that polarization relaxation in many common dipolar liquids contains an ultrafast component with time constant on the order of 100 fs or even less. The amplitude of this ultrafast component is significant. This raises the interesting question regarding the role of this component in the electron transfer reaction, especially in the presence of the participation of vibrational modes.

The objective of the present chapter is to articulate the recent theoretical and experimental advances in understanding the dynamics of electron transfer in polar liquids. The emphasis is on understanding the rich dynamical behavior that can emerge from the interplay between ultrafast solvation dynamics and the intramolecular vibrational modes in photoinduced electron transfer reactions.

In the conventional one-dimensional form of Marcus theory [1], the reaction coordinate of an electron transfer is the energy gap between the electronic surfaces of the reactant and product. For a symmetric electron transfer, this coordinate is essentially the difference in the solvation energy between the reactant and product states. Therefore, motion along the reaction coordinate can, in principle, be strongly coupled to the solvent polarization fluctuations. In the extreme limit of very slow solvent relaxation, the rate can even be controlled by the rate of solvent polarization relaxation (provided that alternative reaction channels are not available). The first clear theoretical prediction of a dynamic solvent effect in electron transfer was made by Zusman [5], who showed that for slow solvent relaxation the rate of an adiabatic electron transfer reaction is inversely proportional to the longitudinal polarization relaxation time, τ_L, of the polar solvent. Since the pioneering work of Zusman, there have been a large number of theoretical studies devoted to dynamic solvent effects, and most of the early studies predicted a strong influence of solvent dynamics on electron transfer [6,17–19]. The strong solvent relaxation dependence was predicted not only for the reactions in the normal region but even for the barrierless electron transfer. There are two reasons for this. First, solvent relaxation was assumed to be overdamped, and the dielectric relaxation of the medium was assumed to be Debye-like. These approximations lead to the prediction that solvation dynamics of a charge is single exponential, with the time constant equal to the longitudinal relaxation time, τ_L. Second, only motion along the classical polarization coordinate was assumed to be relevant in electron transfer. It is now known that both of these assumptions may have only limited validity [20–26].

Initial experimental studies to find the predicted solvent relaxation dependence of the electron transfer rate were made only in mid-1980s. These studies seemed to verify the predicted strong dependence of electron transfer rate on τ_L. Notable among the experimental studies in this phase are the work of Kosower, Huppert and co-workers on napthalene-N,N-dimethylamide and related derivatives [7–9], of McManis and Weaver on self-exchange reactions in metallocenes [10,38], and of Barbara et al. on bianthryl [11,12]. All these studies found solvent relaxation dependence of the electron transfer rate to varying degrees.

In contrast to this older picture, more recent systematic studies on relatively clean photoinduced electron transfer reactions have revealed rather a different picture. Experimental results from Yoshihara's group have shown that in some cases of low (or zero)-barrier electron transfer, the rate can be 50–100 times larger than $1/\tau_L$ [49]. Experiments by Barbara and co-workers found that the electron transfer rate in betaine-30 in glycerol triacetate (GTA) decouples almost completely from solvent polarization relaxation [43]. It should be pointed out here that betaine-30 is deep into the Marcus inverted regime [43,44,72].

The foregoing experimental studies, in turn, led to interesting theoretical developments, particularly on the role of intramolecular vibrational modes in electron transfer. Sumi and Marcus presented an elegant formulation [14] which showed that if relaxation along the vibrational coordinate (Q) is much faster than relaxation along the polarization coordinate (X), the effects of this coordinate can be included via a position-dependent intrinsic reaction rate of electron transfer. Sumi et al. [15,16] showed that this model can give rise to rich dynamical behavior, such as nonexponential kinetics and fractional τ_L dependence, which is sometimes observed in experiments. In the treatment of Sumi and Marcus, the vibrational coordinate was treated classically, as it was envisaged as a low-frequency mode, such as torsional motion.

Subsequently, Jortner and Bixon [73,20–22] developed a detailed treatment of the role of intramolecular vibrational modes in photoinduced electron transfer. These authors pointed out that the high-frequency modes of the product surface can effectively open up new reaction channels via Franck–Condon overlap with the ground and/or excited vibronic levels of the reactant. Thus an electron transfer reaction that is classically in the Marcus inverted regime can occur as a barrierless reaction with much greater rate than was possible otherwise.

In the next phase of development, the back electron transfer reaction in betaine-30 played an important role. It was observed that in this case the electron transfer rate was 10^6 times larger than that predicted by Sumi–Marcus theory. Naturally, the role of the high-frequency mode was invoked to explain this result. However, the difficulty persisted because no existing theory could explain the crossover from the solvent-dependent relaxation at low viscosities to solvent independence at high viscosities. To explain this rather anomalous result, Barbara et al. proposed that a minimal model of electron transfer should consist of a solvent polarization mode (X), a low-frequency classical vibrational mode (Q), and a high-frequency vibrational mode [43]. This model was termed the *hybrid model*, as it combines the ideas of Sumi and Marcus and those of Jortner and Bixon. The theoretical analysis of this model will form a major part of this chapter.

Theoretical studies carried out in our group indicate that both the ultrafast solvation and vibrational modes can combine to give rise to highly interesting dynamical behavior in photoinduced electron transfer reactions in solution [41,77,78]. An important new outcome is that the dynamic solvent effects in electron transfer are weakened not only because of the participation of the high-frequency modes but also because ultrafast solvation in polar liquids makes the reaction channels from the high-frequency vibrational modes easily accessible.

The ultrafast solvation can also significantly alter earlier conclusions on solvent dynamic effects on high-barrier adiabatic reaction. Recent studies have shown that ultrafast solvation can dramatically enhance the rate of barrier crossing over the rate expected if these modes were absent. In fact, one again finds that the ultrafast solvation leads to a weakening of the effects of solvent relaxation on adiabatic electron transfer reaction, and in most cases a significant ultrafast component can lead to the transition-state theory result.

In fact, the main theme of this chapter is that the presence of the ultrafast component in solvation can lead to a significant enhancement of the electron transfer rate and can, in addition, lead to a marked weakening of dependence on solvent polarization relaxation time τ_L. It appears that this ultrafast component serves a twin purpose. First, even a modest ultrafast component is enough to trigger the electron transfer. Second, this component reduces the frictional resistance significantly. Both of these aspects are discussed in detail in this review.

The organization of the remainder of the chapter is as follows. In the next section we describe predictions from the early theoretical models and compare them with experimental results. In Section III we discuss the role of intramolecular vibrational modes in electron transfer reactions. This section includes a discussion of the Sumi–Marcus and Jortner–Bixon theories. In Section IV we present the general theoretical formulation necessary to describe electron transfer reactions in a multidimensional potential energy surface. In Sections V and VI we discuss the results of solvent effects on the back electron transfer reaction in excited betaines in aprotic and protic solvents, respectively. In Section VII we discuss the free energy gap dependence of photoinduced electron transfer reactions, with particular emphasis on non-Marcus energy gap dependence. In Section VIII we present the results of theoretical studies on the effects of ultrafast solvation observed in water, acetonitrile, and methanol on a simple adiabatic electron transfer reaction. Section IX contains a review of recent work on nonexponentiality in the time evolution of electron transfer reactions in polar media. Section X contains a list of future problems that may be worth pursuing. Section XI concludes with a brief discussion.

II. EARLY THEORETICAL MODELS: PREDICTION OF STRONG DEPENDENCE OF ELECTRON TRANSFER RATE ON SOLVENT POLARIZATION RELAXATION

Outer-sphere electron transfer reactions are rather unique in the sense that a large activation energy is often involved, although no chemical bonds are broken or formed. The activation energy comes from interaction of the charge of the electron with solvent polarization as the former moves from one stable solvated position to another stable position. A key to the great success of Marcus theory is proper formulation of this solvent-dependent reaction coordinate. This reaction coordinate in Marcus theory is a collective coordinate. The nature of the reaction coordinate is best understood by considering a symmetric self-exchange reaction, such as $M^{2+} + M^{3+} \rightleftharpoons M^{3+} + M^{2+}$. The relevant change is in the solvent polarization around the reaction system. The reaction coordinate is defined by the expression

$$X = -\int d\mathbf{r}\, \Delta \mathbf{D}(\mathbf{r}) \cdot \mathbf{P}(\mathbf{r}) \qquad (2.1)$$

where $\Delta \mathbf{D}(\mathbf{r})$ is the change in the bare electric field (the electric displacement vector) of the reaction system due to charge transfer and $\mathbf{P}(\mathbf{r})$ is the solvent polarization. The free energy surface when plotted against this reaction coordinate shows parabolic dependence on X, one for each state. Note that X serves as an unambiguous reaction coordinate only when the activation barrier from solvent polarization is large and dominant. In many cases of photoinduced electron transfer reactions, the activation energy is nearly zero. In such cases, other coordinates, such as the intramolecular vibrational modes, also become relevant and one needs to think in terms of a multidimensional reaction energy surface.

When the initial state is neutral and the final state is a charge transfer state, the reaction coordinate is essentially the solvation energy of the charge transfer state. Even when the initial state is charged, the reaction coordinate is essentially the energy of interaction between the electric field of the ion–dipole with the solvent polarization. Therefore, the dynamics along the reaction coordinate is largely controlled by the same dynamics as probed by solvation dynamic experiments.

It was thus realized quite early that solvation dynamics of polar species can greatly influence and in some cases even dominate the rates of electron transfer reactions. Initial theoretical treatments analyzed the solvent influence on electron transfer with the reaction coordinate in outer-sphere electron transfer reactions as essentially the solvation energy of an ion [1–9].

The first important work was carried out by Zusman [5] and was based on the one-dimensional Kramers approach [58]. It was assumed that the reactant surface (1) and product surface (2) are both harmonic, and their harmonic frequencies are identical. The surfaces are then described as

$$V_1(X) = \frac{X^2}{4\lambda_X} \tag{2.2}$$

$$V_2(X) = \frac{(X - 2\lambda_X)^2}{4\lambda_X} + \Delta G \tag{2.3}$$

Here X denotes the solvent coordinate, λ_X is the solvent reorganization energy [1], and ΔG is the free energy change of the reaction. A simple schematic representation of the diabatic surfaces, $V_R(X)$ and $V_P(X)$, is shown in Figure 1.

In the initial studies, the electron transfer was only assumed to occur at the point where (according to the Fermi-Golden rule) the energies of the reactant and the product surfaces are equal. For a purely one-dimensional description, this condition is satisfied at $X_s = \Delta G + \lambda_X$, where the two

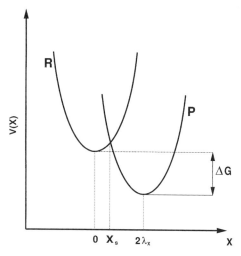

Figure 1.   General schematic representation of the reactant and product free energy surfaces as harmonic functions of the solvent reaction coordinate (X) for a one-dimensional Marcus model of electron transfer reaction. R and P represent the reactant and product states, respectively. ΔG is the difference between the free energy minima of R and P states. λ_x is the Marcus reorganization energy, and X_s is the point of intersection at which the electron transfer takes place.

surfaces intersect to form the only reactive point site, referred to as a *sink*. For symmetric reactions, $\Delta G = 0$, and as a result, $X_s = \lambda_X$ (see Figure 1 for notation).

According to the Zusman model, the electron transfer rate in the adiabatic limit is given by the equation

$$k_{\mathrm{ET}} = \frac{1}{\tau_L} \left(\frac{16\pi k_B T}{\lambda_X} \right)^{-1/2} \exp(-\Delta G^*/k_B T) \qquad (2.4)$$

where τ_L is the solvent longitudinal relaxation time, ΔG^* $(= (\lambda_X + \Delta G)^2/4\lambda_X)$ the free energy of activation, λ_X the solvent reorganization energy [1], and $k_B T$ the Boltzmann constant times the absolute temperature. τ_L is equal to $(\epsilon_\infty/\epsilon_0)\tau_D$, where τ_D is the dielectric relaxation time; ϵ_0 is the static and ϵ_∞ the infinite-frequency dielectric constant of the solvent. Expression (2.4) implies that the electron transfer rate cannot exceed the solvent relaxation rate $(1/\tau_L)$.

Since this pioneering work by Zusman, a large number of theoretical studies have been devoted to solvent effects on outer-sphere electron transfer reactions [6,17–37]. We can list only a few here. It was mainly to treat the zero- and low-barrier situations that Zusman's model was extended by others. Many of these treatments address related issues, and the results are often similar. All these studies employ essentially a one-dimensional description of the reaction potential energy surface with the solvation energy of an ion-pair as the reaction coordinate. They all lead to the prediction of strong solvent dependence of the electron transfer rate. Calef and Wolynes [6] studied in detail the dynamic solvent effects on an adiabatic electron transfer reaction and showed that under certain circumstances, the transfer of charge between two centers can be modelled as a one-dimensional diffusion over a barrier. The rate was found to be inversely proportional to the longitudinal relaxation time, τ_L. In an elegant paper by Efrima and Bixon [19], a stochastic solvent model was employed to describe the dynamics of the polarization fluctuations of the solvent. In this study, a careful analysis of the kinetics of an electron transfer reaction reveals that the effect of solvent dynamics can assume considerable significance only in the case of an adiabatic reaction. Later, an elegant extension of the formulation of Zusman has been presented by Hynes [18]. In this work, Hynes described the dynamical influence of slow, non-Debye polar solvent relaxation on the electron transfer rate in terms of frequency-dependent friction acting on the reaction coordinate and demonstrated that the Zusman model Markovian description along the reaction coordinate can be generalized to include non-Markovian dynamics. A notable feature of Hynes' work is the derivation of a general relation between frequency-dependent friction and the solvation time correlation function, the latter being determined by the time-dependent fluorescence Stokes shift. The extent of the solvent effect was found to vary with the degree of reaction adiabaticity. Also, the short-time solvent dynamics was shown to be more important than the long-time solvent relaxation.

In a detailed study, Rips and Jortner [17] modeled the outer-sphere electron transfer reaction as a two-surface problem. In their treatment, a real-time path integral formalism has been used to derive the influence functionals of the medium in the Gaussian approximation and for the electron transfer rate. The paper of Rips and Jortner represents an important work not only for its clarity but also because it considers the high- and low-barrier reactions within the same formalism. They arrived at the following simple expression for the high-barrier electron transfer rate:

$$k_{ET}^{-1} = \frac{\sqrt{4\pi k_B T \lambda_X}}{k_0} \left(1 + \frac{2k_0 \tau_L}{\lambda_X} \right) \exp(\beta \Delta G^*) \qquad (2.5)$$

where λ_X is the solvent reorganization energy, ΔG^* is the activation energy, k_0 refers to the intrinsic rate of surface crossing, and $\beta = (k_B T)^{-1}$. Expression (2.5) interpolates between the adiabatic and nonadiabatic limits. When $k_0 \gg \lambda_X/\tau_L$, we recover the expression for the adiabatic limit where the rate, k_{ET}, is inversely proportional to τ_L. In the opposite limit (i.e., $k_0 \ll \lambda_X/\tau_L$), we find the rate for the nonadiabatic case.

Several subsequent modifications of the Zusman model have led to the development of a more accurate and general description. Barbara, Fonseca, and co-workers presented a time-dependent solution of the Zusman model and applied it to the study of ultrafast charge separation in 4-(9-anthryl)-N,N'-dimethylaniline (ADMA) [23,43–47]. This study includes memory effects in the overdamped situation. The excited-state population was obtained by a Langevin dynamics simulation. Another detailed study of the extension of Zusman-type kinetic equations to model three-surface problems were carried out by Najbar and co-workers [24] using the Green's function technique of Zusman. Rips and Jortner [17] have also presented a numerical solution of the Zusman model for the low-barrier reaction. Roy and Bagchi have developed a time-dependent solution of a generalized Zusman model using a powerful Green's function technique [41]. A similar approach has been used by Tachiya and Murata to study the charge transfer behaviour in contact-ion pairs [69]. For symmetric exchange reactions, Roy and Bagchi derived an expression more general than the Rips–Jortner expression (2.5) for the average rate. This expression is valid for any barrier height and is given by the simple expression [41]

$$\frac{1}{k_I} = \frac{\sqrt{4\pi k_B T \lambda_X} \exp(\beta \Delta G^*)}{2k_0} + \frac{1}{D_X} \int_{X_0}^{X_s} dX \exp[\beta V_1(X)] \int_{-\infty}^{X} dX' \exp[-\beta V_1(X')]$$

(2.6)

where $V_1(X)$ is the reactant potential energy, D_X the solvation energy diffusion coefficient and X_0 the point on the reaction coordinate corresponding to the initial population. A particular feature of electron-exchange reactions is that they are processes with weak electronic coupling and a high activation barrier (on the order of 0.2 eV). In the high-barrier limit, the rate expression above reduces exactly to the rate expression of Rips and Jortner [Eq. (2.5)].

Early experimental studies seemed to reveal clear evidence of solvent-controlled adiabatic electron transfer. We discuss a few examples here. The charge transfer rates of napthalene-N,N-dimethylamide and related derivatives in a series of alkanols from methanol to decanol have been measured by Kosower, Huppert, and co-workers [7–9]. A characteristic feature of the dielectric response of alcohols is multiexponential behavior with three or four relaxation time constants. The average electron transfer time was found to be about the same as the longest dielectric relaxation time component. In another series of experiments, Kosower and Huppert et al. [7–9] conducted measurements on such systems in several pentadiols and found explicit time dependence of the charge separation on the solvents. Yet another system that was found to exhibit strong dependence on solvent relaxation is the photoelectron transfer reaction in the excited bianthryl [11,12]. This is one among a series of beautiful experiments carried out by Barbara et al. to determine both the solvent and intramolecular parameters.

Investigations of solvent effects in intermolecular electron exchange between molecules and their anions or cations, such as a redox couple of the type [38]

$$\text{Ox} + \text{e} \rightarrow \text{Red}$$

have also been made. Such reactions have been studied by Weaver, McManis, and co-workers on electron self-exchange (i.e., symmetrical) reactions in metal systems (e.g., CO^{2+}/CO^{3+}) [38] and have found the rate of exchange to be inversely proportional to the solvent relaxation time, as expected for a solvent-controlled adiabatic reaction.

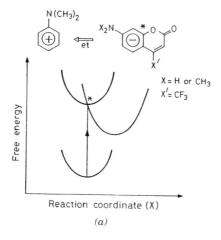

(a)

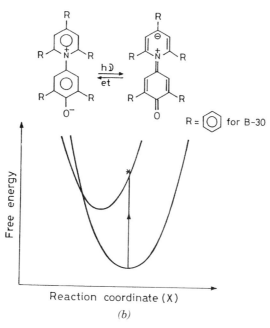

Reaction coordinate (X)

(b)

Figure 2. (a) Description of the electron transfer reaction in the dimethyl aniline–coumarine system. We have also shown a schematic diagram of the classical (Marcus) one-dimensional reaction free energy surface for this reaction, which is in the barrierless regime. This reaction occurs with a rate 50–100 times faster than the inverse of the longitudinal relaxation time, τ_L^{-1}. (b) Description of the back electron transfer reaction in the betaine-30 system. We have also shown a schematic diagram of the classical (Marcus) one-dimensional reaction free energy surface. This reaction is classically in the Marcus inverted regime. However, the rate of this reaction is about 10^6 times larger than the prediction of the Sumi–Marcus theory.

It is clear from the experimental studies cited above that the slow component of electron transfer dynamics is coupled to τ_L, the solvent polarization relaxation time. Zusman model–based theories could explain this solvent dependence to a good extent in several cases. However, the Zusman model is expected to be rigorously valid only in the limit of a high activation barrier where the effects of nonreactive modes are not important. Neither did the above-mentioned extended versions account for the effects of the intramolecular vibrational modes, which can couple directly to electron transfer. Another important point is that most of the experimental studies mentioned earlier were made with limited time resolutions that were available in the late 1980s. Thus a significant portion of the electron transfer dynamics might very well have been missed in those studies.

It was, therefore, not surprising that more recent results point to a rather different picture of dynamic solvent effects on photoinduced electron transfer reactions [43–54]. Kobayashi et al. [49] investigated quenching by electron transfer of fluorescence decays from electron acceptor dyes in electron donor solvents. They found fluorescence quenching as fast as 100 fs for coumarin in N,N-dimethylaniline (DMA). This is 50 times faster than the diffusional solvent polarization relaxation time. This is in sharp contrast to the predictions of the Zusman and related models. This result can be explained in the following way. No significant solvent reorganization is required for this reaction and the reaction occurs without an activation barrier, as shown in Figure 2(a). This picture is further substantiated by the fact that DMA is weakly polar. Another case that indicates the inadequacy of the earlier theories is the back electron transfer reaction in betaine-30, already mentioned. In Figure 2(b) the electron transfer reaction and the Marcus free energy diagram for betaine are depicted. This reaction is in the Marcus inverted regime. This reaction also shows solvent relaxation dependence at variance with the prediction of theoretical models. It has become clear from experiments on these two systems that the sink-broadening effects of nonsolvent modes can be as important as the polar solvent modes. Therefore, a description of electron transfer reaction with a one-dimensional reaction coordinate with a point reactive site is insufficient. Next, we describe the role of vibrational modes in electron transfer reactions in polar solutions.

III. ROLE OF VIBRATIONAL MODES IN WEAKENING SOLVENT DEPENDENCE

As already mentioned, despite the initial claims, clear evidence of solvent dynamic effects on photo-induced electron transfer has been surprisingly scarce [43,44,72,90], despite the large number of electron transfer reactions studied in polar media. Recent experiments on electron transfer reactions in femtosecond time scales seem to demonstrate that the electron transfer is often decoupled from solvent polarization. This fact has led to new thinking about the importance of intramolecular vibrational modes in the dynamics of electron transfer [14,20,43,44,72,73,90]. In slowly relaxing solvents, these vibrational modes can even bring about a crossover from solvent-controlled dynamics to solvent-independent vibrationally controlled dynamics. In the inverted region especially, contributions from the high-frequency modes may be so significant as to control the reaction rate.

Although predictions of solvent dynamic effects were based on a rather simple one-dimensional model of motion along the solvent reaction coordinate, understanding the absence of solvent effects has proved more challenging. Jortner and Bixon [73] have pointed out that an absence of solvent effects can be understood as arising from the presence of many reaction channels that can easily form from the high-frequency modes of the product surface. These channels can be broadened further by low-frequency vibrational modes, as envisaged by Sumi and Marcus [14].

A. Sumi–Marcus Theory: Role of Classical Intramolecular Vibrational Modes

To treat zero- or low-barrier electron transfer reactions, Sumi and Marcus [14] proposed a quasi-two-dimensional model where the electron transfer is coupled to the usual solvent polarization mode (X) and, in addition, to a fast relaxing "average" low-frequency mode (Q). The Q-mode can also be interpreted as a rapidly relaxing solvent mode that is different from the usual bulk solvent polarization mode. At the level of free energy, this Q-mode is treated at the same level as the polarization mode. However, the dynamics of the Q-mode is not treated in the Sumi–Marcus theory—it is eliminated by assuming that relaxation along the Q-coordinate is much faster than the relaxation along the X-coordinate. However, the Q-mode still profoundly influences the dynamics of electron transfer. First, the presence of the Q-mode broadens the reactive region, which is commonly referred to as the *sink* or *reaction window*. Now the reaction can occur over a wide region with an X-dependent rate. An extended sink near the potential minimum reduces the solvent influence on electron transfer because when solvent relaxation becomes slow, reactive trajectories can reach the reactive region by moving along the Q-coordinate. This is shown in Figure 3.

There are several far-reaching consequences of the low-frequency vibrational mode on electron transfer dynamics. First, as already discussed, it allows the reaction to be decoupled from solvent dynamics for slow liquids. Second, since the reaction can occur from various positions of the reactant potential energy surface with various intrinsic reaction rates, one may

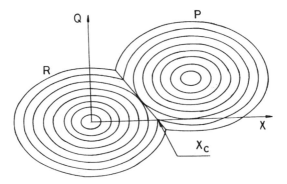

Figure 3. Free energy contour diagram of the two-dimensional reactant (R) and product (P) reaction surfaces, assumed to be harmonic functions of the solvent (X) and the vibrational coordinate (Q). This contour diagram is one of the main assumptions of the Sumi–Marcus theory of electron transfer reaction. At the origin is the minimum of the reactant surface, while the same for the product surface is at nonzero positive values of X and Q. The diagonal line is the equipotential energy line and represents the sink projection of the intersection between the two surfaces onto the $(X\text{--}Q)$ plane along which the transfer occurs.

observe pronounced *non-exponential kinetics* in low-barrier electron transfer reactions. In Sumi–Marcus theory, the features described above are mainly responsible for effecting a change from solvent control to vibrational control of the dynamics. In the Sumi–Marcus model, the dynamics along the vibrational mode (Q) was included by assuming the relaxation along this mode to be infinitely fast. The reaction was then described by assuming a one-dimensional Brownian diffusive motion along the solvent coordinate (X), while the reaction occurs at each X with a rate constant $k(X)$ during the diffusion. It is the solvent coordinate-dependent rate $k(X)$ that contains effects from the intramolecular mode through its dependence on the effective free energy of activation, $\Delta G(X)$, of the reaction along the solvent coordinate. The effects of the internal mode are, therefore, reflected indirectly in the solvent coordinate-dependent rate $k(X)$, which is given by

$$k(X) = \nu_q \exp\left[-\frac{\Delta G(X)}{k_B T}\right] \tag{3.1}$$

where ν_q is the frequency factor and $\Delta G(X)$ is the X-dependent free energy. $\Delta G(X)$ is given by the expression

$$\Delta G(X) = \frac{\lambda_X}{2\lambda_Q} (X - X_c)^2 \tag{3.2}$$

where X_c is the value of X corresponding to $Q = 0$ on the sink projection on the two-dimensional plane spanned by the solvent and vibrational coordinates, X and Q (see Figure 3). $\lambda_i (i = X, Q)$ is the reorganization energy for the ith coordinate.

The extent of influence of the internal modes on the dynamics is determined by the relative inner-shell and outer-shell reorganization energies, λ_Q and λ_X respectively. When $\lambda_X / \lambda_Q \ll 1$, the reaction window is broad in X and there is a weak X-dependence of the dynamics. When $\lambda_X / \lambda_Q \gg 1$, the sink reaction window is narrow in X and the dynamics becomes solvent-controlled, as predicted by the Zusman model.

To explicitly treat dynamic solvent effects, Sumi, Nadler, and Marcus [14–16] developed a detailed theoretical formalism for the dynamics of the barrierless electron transfer reactions based on a modified Smoluchowski equation for the time evolution of the probability distribution $[P_1(X, t)]$ on reactant surface 1 to describe the motion of the system on a harmonic reactant surface along the reaction coordinate X and with an X-position-dependent sink:

$$\frac{\partial P_1(X, t)}{\partial t} = D_X \frac{\partial^2 P_1(X, t)}{\partial X^2} + \frac{D_X}{k_B T} \frac{\partial}{\partial X}$$

$$\times \left[P_1(X, t) \frac{dV(X)}{dX} \right] - k(X) P_1(X, t) \tag{3.3}$$

The first term on the right accounts for the motion of the system on surface 1 in a potential $V(X)$, and the second accounts for decay resulting from the transfer of electrons to the product state. Note that a one-surface description was used here. In many low-barrier reactions, such as the ones studied experimentally by Yoshihara and co-workers [49–53], it is the full two-surface situation with significant back transfer of electron that is generally encountered. The one-surface description will be relevant in certain limiting conditions when the general two-surface problem reduces to a one-surface problem: for example, an asymmetric situation where the forward reaction from surface 1 to surface 2 is a barrierless process whereas the back transfer of electron from surface 2 to surface 1 is a high-barrier process.

An effective one-surface situation in which the reaction occurs with unit probability when the reactants arrive at the origin can be modelled by placing a pinhole sink at the origin of surface 1 [32]. In such a case the following is the expression for the survival probability on surface 1:

$$P_1(t) = \int_{-\infty}^{0} dX [P_0(X) + P_0(-X)] \, \text{erf} \, F(X, t) \tag{3.4}$$

where

$$F(X,t) = \frac{1}{\sqrt{2X_P}} \left\{ 2k_B T \left[1 - \exp\left(\frac{-2t}{\tau_s}\right) \right] \right\}^{-1/2} \exp\left(-\frac{t}{\tau_s}\right) \tag{3.5}$$

The error function, erf a, is defined as usual by

$$\text{erf } a = \frac{2}{\sqrt{\pi}} \int_0^a dy \exp(-y^2) \tag{3.6}$$

$P_0(X)$ is the initial probability distribution in the reactant well. If $P_0(X)$ is given by the equilibrium Boltzmann distribution, $P_1(t)$ reduces to the following much simpler form:

$$P_1(t) = \frac{2}{\pi} \sin^{-1} \left[\exp\left(-\frac{t}{\tau_s}\right) \right] \tag{3.7}$$

Equation (3.7) predicts that the decay of the reactant population is, in general, non-exponential in nature. The long-time decay, however, is still given by a single exponential with a rate constant equal to τ_s. Equation (3.7) is the expression first presented by Sumi and Marcus in the context of electron transfer reactions, but it was derived earlier by Szabo et al. in a different context [79].

Yoshihara et al. [53] reported good agreement between their experimental results and the theoretical simulations done according to the Sumi–Marcus model in their study of the temperature dependence of intermolecular electron transfer between oxazine 1 and an electron-donating solvent, aniline. But the predictions of the Sumi–Marcus model in the inverted region were found to be more than 10^6 times too slow in systems such as 4-(9-anthryl)-N,N-dimethylaniline (ADMA), bis-{N,N'-dimethylaminophenyl}-sulphone (DMAPS), and betaine-30, where the rates observed were found to exceed the solvent relaxation rate [43]. This discrepancy has been attributed to classical approximation of the intramolecular mode. Thus the reaction needs to occur near the crossing of the classical surfaces only. This limitation has been rectified by Jortner and Bixon in their work of considerable importance [20,22] and is discussed below.

B.   Jortner–Bixon Theory: Role of High Frequency Vibronic Reaction Channels

In the Jortner–Bixon treatment, the electron transfer process is considered isomorphous to nonradiative relaxation processes in large molecules.

In marked departure from the Sumi–Marcus theory, Jortner and Bixon argued that the relevant vibrational modes in an electron transfer reaction, particularly for reactions in the Marcus inverted regime, are the high-frequency modes, as they act in the electron transfer only via Franck–Condon overlap. The quantum "jumps" of a high-frequency vibrational mode can open several new reaction sinks for electron transfer, as shown in figure 4. Jortner and Bixon [73] therefore undertook a quantum mechanical treatment of the high-frequency vibrational coordinate in place of the classical one of Sumi–Marcus. Their treatment predicted rates that were close to experimental results in fast-relaxing polar aprotic solvents. The increase in the rate observed was because of the additional sinks that arise from intersection of the ground reactant surface with the multiple vibronic product surfaces. However, in this Jortner–Bixon treatment, the sinks were still point reactive sites, as it neglects the low-frequency vibrational mode. The microscopic rate from the

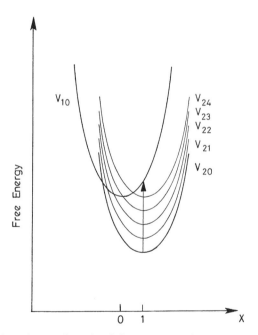

Figure 4. Schematic one-dimensional free energy surface representation for the vibronic ground state V_{10} and the manifold of vibrationally excited product states V_{2n}. This is a reduced description of the Jortner–Bixon scheme. The vibronic states are separated vertically by the quantum energy $h\nu_q$. This figure shows the additional quantum channels due to the vibronic levels of the product surface.

ground level of the reactant state to the nth vibronic level of the product state was assumed to be given by the relation

$$k^{0 \to n} = \frac{2\pi V_{el}^2}{\hbar(4\pi\lambda_{solv}k_B T)^{1/2}} \exp\left(-\frac{\Delta G_{0 \to n}}{k_B T}\right) \qquad (3.8)$$

where V_{el} is the effective electronic coupling [20,73] and $\Delta G_{0 \to n}$ is the effective activation energy for the $0 \to n$ channel given by the expression

$$\Delta G_{0 \to n} = \lambda_{solv} + \Delta G - 2X\lambda_{solv} + nh\nu_q \qquad (3.9)$$

Here λ_{solv} is the solvent reorganization energy and ν_q is the frequency of the quantum mode. The net rate is the sum of rates $k^{0 \to n}$ corresponding to the n vibronic levels of the product D^+A^- state. The merit of this model is that it correctly explains the great enhancement of the rate over the Sumi–Marcus prediction and, in fact, gives rise to rate values close to those observed experimentally with realistic values of the parameters. However, as already mentioned, in the original model the reaction channels were still assumed to be point sinks with no width, and as a result, this model failed to describe the crossover to solvent-independent behavior observed in experiments on betaine-30 in highly viscous and slowly relaxing solvents. This has led Barbara and co-workers to propose a hybrid model [43] which combines the ideas of Sumi and Marcus and of Jortner and Bixon to obtain a unified consistent description. We describe the hybrid model in Section III.C. However, even this hybrid model did not include the ultrafast component of solvation dynamics, primarily because it was not fully known at the time of formulation of the hybrid model.

Later, Bixon and Jortner analyzed the effects of medium-induced dynamics with model calculations to determine energy (E)-dependent rates, $k(E)$, for multimode harmonic systems with displaced potential surfaces using average Franck–Condon densities evaluated using quantum and classical formalisms [20,22]. For activationless electron transfer, an asymptotic analysis suggests that $k(E)$ is proportional to $\sqrt{E + n\epsilon}$, where $n\epsilon$ is the zero-point energy for n high-frequency modes. The above indicates a weak energy dependence. This in turn predicts weak solvent dependence, as the reaction can proceed without solvent relaxation. Thus the formulation of Jortner and Bixon seems to describe the activationless and inverted-region electron transfer as the experimental rates exhibit weak variation between limits of slow and fast medium-induced dynamics.

There are, however, several limitations to this approach. First, as already mentioned, the reaction sites remain point sinks. Thus one cannot rule out solvent independence. Moreover, the dynamics of motion on the potential surfaces was not considered. Some of these limitations were removed in the elegant hybrid model of Barbara *et al.*

C.   Hybrid Model of Barbara et al.: Crossover from Solvent to Vibrational Control

To reconcile the observed theoretical and experimental differences, Barbara et al. [43] proposed a combined Sumi–Marcus [14] and Jortner–Bixon [73] model. The authors pointed out that any model of electron transfer reaction in the inverted regime should *minimally* include a low-frequency solvent mode (classical), an intramolecular low-frequency mode (classical), and a high-frequency intramolecular mode (quantum mechanical). The role of the low-frequency vibrational mode is to broaden the point sinks formed by the high-frequency modes. If the dynamics along the low-frequency mode is assumed to be infinitely fast, the microscopic rate $k^{0 \to n}$ can be given by the form

$$k^{0 \to n} = \frac{2\pi V_{el}^2}{\hbar (4\pi \lambda_{cl,vib} k_B T)^{1/2}} \exp \left[-\frac{(\Delta G_{0 \to n} + \lambda_{cl,vib})^2}{4\lambda_{cl,vib} k_B T} \right] \qquad (3.10)$$

where $\lambda_{cl,vib}$ is the reorganization energy of the low-frequency, classical vibrational mode, which is not solvent dependent. Note that $\lambda_{cl,vib}$ is now in place of λ_{solv} in Eq. (3.9) to account for the low-frequency contribution. In the 'hybrid' model, therefore, $\lambda_{cl,vib}$, λ_{solv}, and ΔG are the three relevant parameters. However, the values of these parameters are often not known *a priori*. It is customary to obtain these relevant parameters by fitting the experimentally obtained absorption intensity profiles to line-shape models. Also, exact identification of the nature of the vibrational modes that are involved is often difficult. In the case of aromatic systems, skeletal ring vibrations in the high-frequency range (1000–2000 cm^{-1}) [42,75] have been observed. In the study of electron and proton transfer reactions involving aromatic rings, it is common practice to use a single "average" vibrational mode (usually in the 1500 cm^{-1} range) as an approximate representation of the skeletal vibrations of the molecular rings. The low-frequency mode, on the other hand, can be the large-amplitude torsional or bending motion (usually in the range 50–250 cm^{-1}). Also, intermolecular stretching modes in the intermediate-frequency range (250–1000 cm^{-1}) can couple to electron transfer [42,75]. Barbara and co-workers presented a numerical work which suggested that the minimal model is indeed capable of explaining some of the behavior observed in the inverted region. It should be pointed out that a similar description has recently been used by Syage to model proton transfer reactions in solution [74].

The 'hybrid model' predicted rates close to the experimental results of betaine-30 and *tert*-butylbetaine in a wide range of solvent environments [43]. However, they did not consider the dynamics of the low-frequency vibrational mode explicitly. In addition, in the original form of this model, only the Markovian response of the solvent mode was included. This is equivalent to assuming that the solvation dynamics is single exponential. However, generalization of this model is required in view of the observation that in most solvents solvation dynamics is nonexponential and seems to contain two vastly different time scales [83–91]. Although the origin of these different time scales is well understood in ultrafast liquids (e.g., water and acetonitrile), the same is certainly not clear for more complex liquids, such as formamides and higher alcohols. However, it appears to be certain that even in liquids commonly believed to be slow, there exists an ultrafast component of time constants two to three orders of magnitude smaller than the slow component commonly observed [90,91]. Theoretical study of the electron transfer rate in the presence of such diverse time scales poses an interesting challenge because the conventional approach to solving the reaction-diffusion equations soon becomes inefficient, especially in the presence of multiple delocalized sinks.

In the next section we discuss a theoretical formulation that is capable of describing chemical dynamics on a multidimensional potential energy surface and in the presence of both the non-Markovian dynamics and multiple delocalized reaction windows.

IV. GENERAL THEORETICAL FORMULATION OF MULTIDIMENSIONAL ELECTRON TRANSFER

We have recently developed a novel theoretical technique to solve the equation of motion for any reaction occurring on a multidimensional potential energy surface where the reaction occurs via a delocalized sink [41,77,78]. The merit of this scheme is that one can describe the situation where motion along the reaction coordinate is non-Markovian and the reaction window is broad. The importance of the first is clearly evident as almost all polar liquids seem to have a significant ultrafast component; thus one must include non-Markovian relaxation behavior. The second aspect is equally important, as treatment of the reaction in the presence of a delocalized sink is notoriously difficult, as most of the known techniques fail here.

A.   Model

In general, a photoinduced electron transfer reaction proceeds on a multi-dimensional potential energy surface where the coordinates consist of both polar solvent and nonsolvent modes. To keep the problem tractable, all the potentials are usually assumed to be harmonic. Let us assume an n-dimensional electron transfer process with a single classical solvent coordinate (X) and $(n-1)$ nonsolvent vibrational coordinates $(Q_1, Q_2, \ldots, Q_{n-1})$. The n-dimensional potential energy surfaces V_1 and V_2 for the reactant and product states can, therefore, be described by the following equations:

$$\text{Reactant:} \quad V_1(X, Q_1, Q_2, \ldots, Q_{n-1}) = \tfrac{1}{2} 2\lambda_X X^2 + \sum_{i=1}^{n-1} \tfrac{1}{2} 2\lambda_{Qi} Q_i^2 \tag{4.1}$$

$$\text{Product:} \quad V_2(X, Q_1, Q_2, \ldots Q_{n-1}) = \tfrac{1}{2} 2\lambda_X (X-1)^2 + \sum_{i=1}^{n-1} \tfrac{1}{2} 2\lambda_{Qi}(Q_i - 1)^2$$

$$+ \Delta G \tag{4.2}$$

where λ_X, $\lambda_{Qi}(i = 1, \ldots, n-1)$ are the corresponding reorganization energies of these modes, and ΔG is the free energy gap of the reaction (i.e., the vertical energy gap between the two surface minimums). Since we are interested primarily in the photoelectron transfer reactions, we shall often refer to reactant surface 1 as the locally excited state and the product surface P as the charge transfer state.

As mentioned earlier, the reaction occurs along the sink curve, obtained by the intersection of the two potential energy surfaces. For an n-dimensional potential energy surface, the sink surface is always an $(n-1)$-dimensional surface. For the sake of simplicity, we use $\mathbf{Q}$ to represent the set of the vibrational coordinates $Q_1, Q_2, \ldots, Q_{n-1}$ in the remainder of the chapter.

B.   Equation of Motion on the Reactant and Product Surfaces

A consequence of the many-body nature of the interaction between the rapidly fluctuating solvent environment and the reactant species is complex stochastic motion of the system, resembling Brownian dynamics. As mentioned earlier, this influence of the solvent dynamics on electron transfer was formulated as a Markovian stochastic process by Zusman [5]. This one-dimensional diffusive motion along the reactive mode X can be generalized to higher dimensions when more than one reactive mode is involved in the reaction process. As mentioned earlier, one of the strengths of the present treatment is that it is feasible to treat this diffusive relaxation as non-Markovian.

The time evolution of the probability distribution of the system on the locally excited surface $[P_1(X, \mathbf{Q}, t)]$ is assumed to be given by the following master equation [14,27,32,35]:

$$\frac{\partial P_1(X, \mathbf{Q}, t)}{\partial t} = (\mathscr{L}_X + \mathscr{L}_\mathbf{Q}) P_1(X, \mathbf{Q}, t)$$
$$- S(X, \mathbf{Q}) P_1(X, \mathbf{Q}, t) + S(X, \mathbf{Q}) P_2(X, \mathbf{Q}, t) \quad (4.3)$$

where $\mathscr{L}_\mathbf{Q} = \sum_{i=1}^{n-1} \mathscr{L}_\mathbf{Q}$. The first term simulates diffusion in a potential well $V_1(X, \mathbf{Q})$. The second and third terms take into account the actual transfer and back transfer along the sink curve. $S(X, \mathbf{Q})$ is the position-dependent sink function, which describes the path along which the electron transfer takes place between the locally excited and charge transfer surfaces. Equation (4.3) with P_1 and P_2 interchanged describes the motion in the charge transfer state on the potential energy surface $V_2(X, \mathbf{Q})$.

The operator $\mathscr{L}_\xi(\xi = X, Q_i)$ in Eq. (4.3) can be any general time evolution operator. Since we are interested in reactions in solution, it is convenient to assume that these operators are stochastic operators that describe the relaxation of any nonequilibrium population to an equilibrium population with well-defined rates supplied from outside. For relaxation along the solvent coordinate X, this rate should be related to the solvation rate, while for a given Q, this rate is the rate of vibrational population relaxation in solution. In choosing these operators, one needs to keep in mind that the relaxation of both X and Q are often nonexponential. In particular, the solvation time correlation function is often biphasic with a large difference between the two time constants. All the work that has been carried out to date approximates these operators by some kind of Smoluchowski operators. However, to describe the complex nonexponential kinetics of X (and even of Q), one needs to consider a non-Markovian operator. In the work we report below, the operator $\mathscr{L}_\xi(\xi = X, Q_i)$ is assumed to be a general Smoluchowski operator that is given in the form

$$\mathscr{L}_\xi = D_\xi(t) \left[\frac{\partial^2}{\partial \xi^2} + \frac{1}{k_B T} \frac{\partial}{\partial \xi} \frac{dV(\xi)}{d\xi} \right] \quad (4.4)$$

where $D_\xi(t)$ is the time-dependent diffusion coefficient of motion along the reaction coordinate. $D_\xi(t)$ is given by the relation [18]

$$D_\xi(t) = -k_B T \, d \ln \Delta_\xi(t)/dt \quad (4.5)$$

where $\Delta_\xi(t)$ is the time correlation function of the ξth reaction coordinate. It is generally assumed that the reaction time correlation function can be represented by $\sum_j w_j \exp(-t/t_j)$, where $\sum_j w_j = 1$.

The diffusion coefficient (D_ξ) is time-independent for a Markovian single exponential decay (as in the case of a Debye solvent) and is time dependent when the relaxation is characterized by a non-Markovian multiexponential time decay (as in the case of a non-Debye solvent). The effective relaxation time τ_{eff} is given by

$$\tau_{\text{eff}} = \int_0^t dt\, \Delta_\xi(t) \tag{4.6}$$

Following Hynes [18], the solvent reaction coordinate time correlation function, $\Delta_X(t)$, is defined as

$$\Delta_X(t) = \frac{\langle X(0)X(t)\rangle}{\langle X^2(0)\rangle} \tag{4.7}$$

where $\langle \cdots \rangle$ denotes the average over the solvent degrees of freedom in equilibrium with the reactant state. It is this quantity $\Delta_X(t)$ that reflects the dynamics of the solvent polarization fluctuations:

$$\hat{\Delta}_X(z) = \frac{z + \zeta_X}{z^2 + \Omega_X^2 + z\zeta_X} \tag{4.8}$$

where $\Omega^2 = (2\lambda_X\mu_X)^{-1}$. For homogeneous redox reactions, $\Delta_X(t)$ is the solvation time correlation function of an ion, while for the photoinduced electron transfer in bianthryl, $\Delta_X(t)$ is that of a dipole.

Equations (4.4) and (4.5) can be generalized to include different time evolution operators for the reactant and product states, as the frequencies of the two surfaces can be different. It is also possible to include inertial motion, as discussed in Section VIII.

C. Solution by Green's Function Technique

The powerful Green's function technique has been widely used to solve initial-value problems; a particularly well-known case is the Einstein–Langevin theory of Brownian motion [79–82]. Recently, a detailed time-dependent solution of a generalized one-dimensional Zusman model of an electron transfer reaction was carried out using this technique [41]. The main advantage of this technique is that an almost analytic solution can be obtained in the Laplace frequency plane. The time-dependent solution can subsequently be obtained by a simple Laplace inversion. As shown here, the analysis in Laplace coordinates and a description of the continuous sink in terms of delta-function point sinks affords a convenient solution by using the Green's function method. However, the important new development is that the calculation of the average rate can be reduced to a simple matrix inversion problem. This is particularly useful when solvent response is biphasic with widely different time scales.

By Laplace transformation, the dynamical equations for the two potential energy surfaces are converted into the following equations:

$$[z - (\mathscr{L}_X + \mathscr{L}_Q)]P_1(X, \mathbf{Q}, z) = P_1(X, \mathbf{Q}, t = 0)$$
$$- S(X, \mathbf{Q})P_1(X, \mathbf{Q}, z) + S(X, \mathbf{Q})P_2(X, \mathbf{Q}, z)$$
$$(4.9)$$

$$[z - (\mathscr{L}_X + \mathscr{L}_Q)]P_2(X, \mathbf{Q}, z) = P_2(X, \mathbf{Q}, t = 0)$$
$$- S(X, \mathbf{Q})P_2(X, \mathbf{Q}, z) + S(X, \mathbf{Q})P_1(X, \mathbf{Q}, z)$$
$$(4.10)$$

where the $P_i(X, \mathbf{Q}, t = 0)$ denote the initial equilibrium probability distribution on the ith surface and z is the (Laplace) frequency conjugate to the time t. By definition, the Green's function for the two surfaces follows the equations

$$[z - (\mathscr{L}_X + \mathscr{L}_Q)]G_1(X, \mathbf{Q}, z | \tilde{X}, \tilde{\mathbf{Q}}) = \delta(X - \tilde{X})\delta(Q - \tilde{Q}) \qquad (4.11)$$

$$[z - (\mathscr{L}_{X'} + \mathscr{L}_{Q'})]G_2(X', \mathbf{Q}', z | \tilde{X}', \tilde{\mathbf{Q}}') = \delta(X' - \tilde{X}')\delta(Q' - \tilde{Q}') \qquad (4.12)$$

where $X' = X - 1$ and $\delta(\mathbf{Q} - \mathbf{Q}_\xi)$ denotes $\prod_{i=1}^{n} \delta(Q_i - Q_{i\xi})$ in the foregoing and subsequent discussions. As in the case of X', Q'_ξ is equal to $Q_\xi - 1$.

Now, for a one-dimensional harmonic potential surface i described by the reaction coordinate A_i, the Green's function is given in the form [18]

$$G_i(A_i, t | \tilde{A}_i) = \frac{1}{\sqrt{2\pi\sigma_i^2[1 - \Delta_i(t)^2]}} \exp\left\{ -\frac{[A_i - \tilde{A}_i\Delta_i(t)]^2}{2\sigma_i^2[1 - \Delta_i(t)^2]} \right\} \qquad (4.13)$$

where $\sigma_i^2 = k_B T / 2\lambda_i$. Note that the Green's function for the surface P is given by $X' = X - 1$ and $Q' = Q - 1$.

The composite Green's function for an N-dimensional surface (in the absence of any coupling between the reaction coordinates) is given by the product of the Green's function for the N individual coordinates:

$$G_j(A_1, A_2, \ldots, A_N, t | \tilde{A}_1, \tilde{A}_2, \ldots, \tilde{A}_N) = \prod_{i=1}^{N} G_j(A_i, t | \tilde{A}_i) \qquad (4.14)$$

Note that the Green's function above is the solution of the dynamical equation in the absence of sinks; we need the solution in the presence of sinks. This is achieved by using the Green's function in a Laplace-transformed dynamical equation and subsequently obtaining the values for $P_1(X_s, \mathbf{Q}_s, z)$, where z is the Laplace frequency.

The solutions for $P_1(X, \mathbf{Q}, z)$ and $P_2(X, \mathbf{Q}, z)$ are given in terms of two coupled equations of the following form:

$$P_1(X, \mathbf{Q}, z) = \int d\tilde{\mathbf{Q}} \int d\tilde{X}\, G_1(X, \mathbf{Q}, z | \tilde{X}, \tilde{\mathbf{Q}})[P_1(\tilde{X}, \tilde{\mathbf{Q}}, 0)$$
$$- S(\tilde{X}, \tilde{\mathbf{Q}})P_1(\tilde{X}, \tilde{\mathbf{Q}}, z) + S(\tilde{X}, \tilde{\mathbf{Q}})P_2(\tilde{X}, \tilde{\mathbf{Q}}, z)] \qquad (4.15)$$

$$P_2(X, \mathbf{Q}, z) = \int d\tilde{\mathbf{Q}}' \int d\tilde{X}'\, G_2(X, \mathbf{Q}, z | \tilde{X}', \tilde{\mathbf{Q}}')[P_2(\tilde{X}', \tilde{\mathbf{Q}}, 0)$$
$$- S(\tilde{X}', \tilde{\mathbf{Q}}')P_2(\tilde{X}, \tilde{\mathbf{Q}}, z) + S(\tilde{X}', \tilde{\mathbf{Q}}')P_1(\tilde{X}', \tilde{\mathbf{Q}}', z)] \qquad (4.16)$$

For simplicity, the initial population excited on the reactant (locally excited) surface may be characterized as a delta-function source at (X_0, Q_0). Mathmatically, this is written as $P_i(X, \mathbf{Q}, t = 0) = \delta(X - X_0)\delta(\mathbf{Q} - \mathbf{Q}_0)\delta_{1i}$.

The sink function, $S(X, \mathbf{Q})$, can be written as $S(X, \mathbf{Q}) = \int d\tilde{\mathbf{Q}} \int d\tilde{X}\, S(\tilde{X}, \tilde{\mathbf{Q}})\delta(X - \tilde{X})\delta(\mathbf{Q} - \tilde{\mathbf{Q}})$. We next exploit this property to divide the continuous sink curve into a number of intervals. The sink function was assumed to be of the following form in the discretized representation:

$$S(\tilde{X}, \tilde{\mathbf{Q}}) = \sum_s k_s \delta(X - X_s)\delta(\mathbf{Q} - \mathbf{Q}_s) \qquad (4.17)$$

where k_s, is the strength of each interval. k_s is defined as $\int dX \int dQ\, S(X, Q)$. The determination of k_s is described in the next section. Each sink point acts as a delta-function sink. Use of the form of the sink function in Eqs. (4.16) and (4.17) gives the following expressions for the population densities. Note that this discretization is perfectly general and valid for any delocalized sink.

The merit of the present formulation is the development and application of a tractable and robust scheme to treat the extended sink.

$P_1(X, \mathbf{Q}, z)$ and $P_2(X, \mathbf{Q}, z)$ can then be expressed as

$$P_1(X, \mathbf{Q}, z) = G_1(X, \mathbf{Q}, z | X_0, \mathbf{Q}_0)$$
$$- \sum_s k_s G_1(X, \mathbf{Q}, z | X_s, \mathbf{Q}_s) P_1(X_s, \mathbf{Q}_s, z)$$
$$+ \sum_s k_s G_1(X, \mathbf{Q}, z | X_s, \mathbf{Q}_s) P_2(X_s, \mathbf{Q}_s, z) \qquad (4.18)$$

$$P_2(X, \mathbf{Q}, z) = G_2(X, \mathbf{Q}, z | X_0, \mathbf{Q}_0)$$
$$- \sum_s k_s G_2(X, \mathbf{Q}, z) | X_s, \mathbf{Q}_s) P_2(X_s, \mathbf{Q}_s, z)$$
$$+ \sum_s k_s G_2(X, \mathbf{Q}, z) | X_s, \mathbf{Q}_s) P_1(X_s, \mathbf{Q}_s, z) \qquad (4.19)$$

The sum is over the sink points where the populations are given by $P_i(X_s, \mathbf{Q}_s, z)$. By finding $P(X_s, \mathbf{Q}_s, z)$ and $P_2(X_s, \mathbf{Q}_s, z)$ using Eqs. (4.18) and (4.19), a set of linear equations are generated. These can be written in matrix form as

$$\mathbf{B} \cdot \mathbf{P} = \mathbf{G}_0 \qquad (4.20)$$

The elements of $\mathbf{B}$, $\mathbf{P}$, and $\mathbf{G}$ are given as

$$[B_{mm'}] = \delta_{mm'} + k_{m'} G_1(X_m, \mathbf{Q}_m, z | X_{m'}, \mathbf{Q}_{m'}) \qquad (4.21)$$

$$[B_{(m+k)m'}] = -k_{m'} G_2(X_k, \mathbf{Q}_k, z | X_{m'}, \mathbf{Q}_{m'}) \qquad (4.22)$$

$$[B_{m(m'+k')}] = -k_{k'} G_1(X_m, \mathbf{Q}_m, z | X_{k'}, \mathbf{Q}_{k'}) \qquad (4.23)$$

$$[B_{(m+k)(m'+k')}] = \delta_{(m+k),(m'+k')} - k_{s'} G_2(X_k, \mathbf{Q}_k, z | X_{k'}, \mathbf{Q}_{k'}) \qquad (4.24)$$

$$[P_m] = P_1(X_m, \mathbf{Q}_m, z) \qquad (4.25)$$

$$[P_{m+k}] = P_2(X_k, \mathbf{Q}_k, z) \qquad (4.26)$$

$$G_{0m} = G_1(X_m, \mathbf{Q}_m, z | X_0, \mathbf{Q}_0) \qquad (4.27)$$

$$[G_{0m+k}] = 0 \qquad (4.28)$$

The population on the reactant (locally excited) surface as a function of the Laplace frequency can be written in the form [41,59]

$$P_1(z) = \frac{1}{z} \left[1 - \sum_s k_s P_1(X_s, \mathbf{Q}_s, z) + \sum_s k_s P_2(X_s, \mathbf{Q}_s, z) \right] \qquad (4.29)$$

It is $P_1(z)$ that is obtained by the Green's function method. $P_1(t)$ is then obtained by the Laplace inversion of $P_1(z)$ [67] and can be used to obtain the average and the long-time rates of charge transfer.

The average rate of the reaction (k_I) is defined as

$$k_I^{-1} = \tau_a = \int_0^\infty dt\, P_1(t) = P_1(z=0) \qquad (4.30)$$

The long-time rate of the electron transfer reaction is given by the expression

$$k_L = -\lim_{t \to \infty} \frac{\partial \ln P_1(t)}{\partial t} \qquad (4.31)$$

As $P_1(t)$ is likely to be nonexponential in the low barrier reactions, an average survival time of a second kind (τ_b) is also used to characterize the dynamics. τ_b is given by the relation [14]

$$\tau_b = \int_0^\infty dt\, t P_1(t) \Big/ \int_0^\infty dt\, P_1(t) \qquad (4.32)$$

The main strengths of the Green's function scheme are its validity for any arbitrary sink and its generalization to a multidimensional case and to study the non-Markovian dynamics.

D.   Calculation of the Average Rate for a Delocalized Sink

To obtain the survival probability measured in fluorescence experiments, $P_1(z)$ is first obtained by numerically solving the system of equations (4.20)–(4.29) and then Laplace inverting $P_1(z)$ to obtain $P_1(t)$. The average rate k_I^{-1} can then be obtained according to Eq. (4.30). However, when the solvation time correlation function is biphasic with widely different time scales, this method is not robust because evaluation of $P_1(t)$ faces stability problems, as this procedure is computer intensive. Fortunately, there is a direct, almost analytical method of obtaining this average rate. This method uses the well-known Cramer's technique to solve the system of linear equations given by the matrix equation (4.20). In the zero-frequency limit, this solution gives the average rate in the form

$$k_I^{-1} = Lt_{z \to 0} \left(\det \mathbf{B} - \sum_s k_s \det \mathbf{B}^j \right) \Big/ \sum_s k_s \det \mathbf{B} \qquad (4.33)$$

where det represents the determinant of a matrix and the matrix elements of $\mathbf{B}$ are given [Eqs. (4.21)–(4.24)]. The elements of matrix $\mathbf{B}^j$ are obtained from matrix $\mathbf{B}$ by replacing the jth column by the column vector $\mathbf{G}$.

However, Eq. (4.33) becomes ill defined in the z-tending-to-zero limit as the Laplace transform of the Green's function diverges in that limit. Therefore, if a straightforward numerical evaluation of k_I^{-1} is attempted by using Eq. (4.30), this divergence makes the method useless. This problem is especially severe when the solvation dynamics is biphasic with a large difference between the rates of the two relaxation time components of $\Delta(t)$. Note that such a large difference (even to the extent of 100 or so) has been observed in recent solvation experiments. This difficulty can be circumvented by using a method suggested by Samanta and Ghosh [81], who developed a simple and robust scheme to avoid this problem.

Let us first define a quantity $\Delta G(X_s, \mathbf{Q}_s, z | X_{m'}, \mathbf{Q}_{m'})$ in the form

$$\Delta G(X_s, \mathbf{Q}_s, z | X_{m'}, \mathbf{Q}_{m'}) = Lt_{z \to 0} \int_0^\infty dt \exp(-zt)$$

$$\times [G_1(X_m, \mathbf{Q}_m, t | X_{m'}, \mathbf{Q}_{m'}) - G_{eq}(X_m, \mathbf{Q}_m)] \qquad (4.34)$$

where $G_{eq}(X_m, \mathbf{Q}_m) = G_1(X_m, \mathbf{Q}_m, t = \infty)$ is the equilibrium distribution. Note that this ΔG (not to be confused with the free energy) is well defined in the $z \to 0$ limit. A simple substitution of $\Delta G(X_s, \mathbf{Q}_s, z | X_{m'}, \mathbf{Q}_{m'})$ in place of $G_1(X_m, \mathbf{Q}_m, z | X_{m'}, \mathbf{Q}_{m'})$ in Eq. (4.33) gives the average rate in the form

$$k_I^{-1} = \det \mathbf{C} - \sum_s k_s \det \mathbf{C}^j \Big/ \sum_s k_s \det \mathbf{C}^{j'} \qquad (4.35)$$

where the matrix elements of $\mathbf{C}$ are of the form

$$[c_{mn}] = \delta_{mn} + k_n \Delta G(X_m, \mathbf{Q}_m, z | X_n, \mathbf{Q}_n) \qquad (4.36)$$

The elements of matrix $\mathbf{C}^j$ are obtained from matrix $\mathbf{C}$ by replacing the elements c_{ij} in the jth column with the elements $\Delta G(X_i, \mathbf{Q}_i, z | X_0, \mathbf{Q}_0)$. Similarly, matrix $\mathbf{C}^{j'}$ is obtained by replacing the elements of the jth column with the equilibrium values $G_{eq}(X_i, \mathbf{Q}_i)$.

The scheme above is valid for any arbitrary sink. The usefulness of the scheme lies in its generalization to a multidimensional potential energy surface and its ability to provide solutions even for non-Markovian dynamics [71,76].

E. High-Frequency Modes in the Dynamics of Electron Transfer

Here we describe how to include the effects of high-frequency vibrational modes which are needed to explain the high rate of electron transfer as in a case such as betaine-30 in the Marcus inverted region. In this case the system is minimally modeled by a low-frequency harmonic and classical solvent mode, a similar low-frequency vibrational mode, and a high-frequency, harmonic mode. The high-frequency mode is to be treated quantum mechanically. For simplicity, normalized coordinates are often used such that the bottom of the reactant surface is at 0, and that of the product at 1. The potentials for the reactant and product states, therefore, become

$$V_{1n}(X,Q) = \tfrac{1}{2}2\lambda_X X^2 + \tfrac{1}{2}2\lambda_Q Q^2 + nh\nu_q \tag{4.37}$$

$$V_{2n}(X,Q) = \tfrac{1}{2}2\lambda_X(X-1)^2 + \tfrac{1}{2}2\lambda_Q(Q-1)^2$$
$$+ nh\nu_q + \lambda_q + \Delta G \tag{4.38}$$

where V_{1n} and V_{2n} denote the reactant and product states, respectively, arising from the nth vibrational level of the high-frequency quantum mode. ΔG is the free energy gap of the reaction. X represents the solvent coordinate, while Q and q are the low- and high-frequency vibrational coordinates, respectively. λ_X, λ_Q, and λ_q are the corresponding energies of reorganization of these modes. ν_q is the frequency of the quantum mode.

Quantum treatment of the high-frequency mode is introduced as a change in the effective free energy gap $-\Delta G_n (= nh\nu_q + \lambda_q + \Delta G)$ between two-dimensional reactant and product surfaces. One usually assumes that relaxation of the high-frequency mode is much faster than any relevant process, so the three-mode problem is reduced to a two-mode multisurface problem. This is described in Figure 5. This approach can easily be generalized to an m-mode case where more than one high-frequency mode is involved.

As the system is excited onto the ground vibronic level of the reactant surface, that is, from V_{10}, and the higher vibronic levels are not involved (V_{1n}, $n = 1, 2, \ldots$), it is sufficient to consider the electron transfer reactive sites (sinks) that are present only along intersection of the V_{10} surface with product surfaces.

A classical sink is said to be positioned in the normal region or in the inverted region, depending on whether λ, the sum of the reorganization

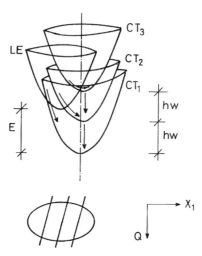

Figure 5. General two-dimensional multisurface schematic representation for reaction in the Marcus inverted region. LE (or V_{10}) and CT_1, CT_2, (or $V_{2n,n=0,1,2,...}$) are the effective potential energy surfaces for the ground reactant and the vibronic product states, respectively. E (or ΔG_n) is the difference between the potential energy heights between the reactant and the product states. w (or ν_q) is the quantum frequency. This is the reaction surface suggested in the hybrid model proposed by Barbara and co-workers.

energies λ_X and Δ_Q, is greater or less than $-\Delta G_n$, respectively. If the free energy gap $-\Delta G$ is sufficiently large as to result in an inverted-region case, the classical Marcus picture predicts the electron transfer reaction to be deeply in the inverted region and the reaction rate to be very small. When $\lambda = \Delta G$, the barrierless case results.

The sink curves can be obtained by equating Eqs. (4.37) and (4.38) and are given in the form

$$X = c + mQ \tag{4.39}$$

where $m = -\lambda_Q/\lambda_X$ and $c = \lambda_X + \lambda_Q + \Delta G_n$. The straight-line equation (4.39) represents the projection of the sink curves on the X–Q plane when the locally excited and charge transfer surfaces are both harmonic. Note that the gradient (m) and intercept (c) of the projected line are controlled by the potential parameters. These projections have been shown in Figure 5.

A simple geometrical interpretation of the two-dimensional model helps in the intuitive physical interpretation of several different situations. This is depicted in Figure 6. In the absence of the high-frequency mode, only one sink curve is present. A section of the two-dimensional potential energy surface, cut horizontally at height h on the V-axis and projected onto the

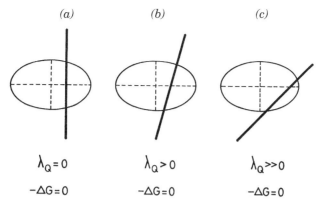

Figure 6. These figures provide a geometrical interpretation of the Sumi–Marcus–Bixon–Jortner model of electron transfer. The projections of the sink (bold line) on the X–Q plane showing the changes in sink orientation and its intercept with the X-axis with change in shift along the vibrational coordinate (λ_Q). The ellipse is the projection of a section of the locally excited surface cut at a potential height h. The horizontal and vertical dashed lines represent the X and Q axes, respectively. The position of the zero potential (0,0) is the intersection of the dotted lines. (a) For $\lambda_Q = 0$, the sink is parallel to the Q-axis; (b) for $\lambda_Q > 0$, the sink is oriented more toward the X-axis; (c) when $\lambda_Q \Delta$ is large, the sink is tilted almost parallel to the X-axis.

X–Q plane, would appear as an ellipse and the projected sink curve as a straight line.

When $-\Delta G = 0$ and there is no shift of the charge transfer surface along the vibrational coordinate (i.e., $\lambda_Q = 0$), $m = 0$, so the sink curve is parallel to the Q-axis. Under this condition, as shown in Figure 6(a), the sink curve is always in the normal region, as the intercept on the X-axis is positive. The intercept becomes more positive with an increase in λ_Q. As a result, the sink curve is shifted farther toward higher positive values of X. Also, the sink curve is tilted more toward the X-axis as the gradient becomes more negative with an increase in λ_Q, giving rise to a broader sink when projected along X.

When $-\Delta G = 0$, the intercept (c) is always positive. Sink lies on the normal side. As $-\Delta G$ is gradually increased from zero, the intercept c becomes more negative. This shifts the sink curve from the normal region toward the inverted region, as shown in Figure 7. This crossover from the normal to the inverted region passes through the barrierless region.

In the presence of a high-frequency mode, multiple sinks are present. The projection would appear as shown in Figure 5. However, the sinks are of differing strengths as the sink transfer rate corresponding to the $0 \rightarrow n$ transition involving the high-frequency mode is $(2\pi V_{el}^2/\hbar)|\langle 0, n\rangle|^2$, where V_{el} is the electronic coupling and $|\langle 0, n\rangle|^2$ is the Franck–Condon overlap of the

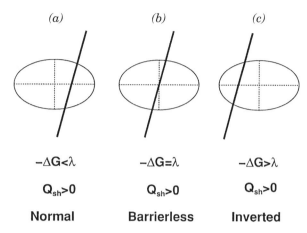

Figure 7. Marcus free energy gap dependence, shown by the sink projections (bold line) on the X–Q plane. The X and Q axes are represented by the horizontal and vertical dashed lines. (a) The sink lies in the normal region when $-\Delta G < \lambda$ (includes the case $-\Delta G = 0$). (b) A barrierless case results when $-\Delta G = \lambda$, as the position of the sink minimum is (0,0). (c) In the inverted region, $-\Delta G > \lambda$.

nuclear wavefunctions of the ground reactant and nth product states. The Franck–Condon factor between the initial n^1 and the final n^2 is given by the relation

$$|\langle n^1, n^2\rangle|^2 = \exp\left(-\frac{\delta^2}{2}\right) n^1! \, n^2! \left[\sum_{r=0}^{\min(n^1, n^2)} \frac{(-1)^{n^1-r}(\delta^2/\sqrt{2})^{n^1+n^2-2r}}{r! \, (n^1-r)! \, (n^2-r)!}\right]^2 \quad (4.40)$$

where $\delta^2 = 2\lambda_q/h\nu_q$ is the coupling parameter. As already discussed, the rates can still be high in the inverted region because of the additional reaction channels opened up due to the high-frequency quantum levels; some of these channels may not only be efficient but may also be located near the barrierless region (or even in the normal region). But the main constraint is that the Franck–Condon overlap decreases rapidly with higher quantum levels of the change transfer states. As a result, most of the pertinent charge transfer states that can contribute significantly to the transfer arise only from the lower quantum levels. These have large energy gaps with respect to the locally excited state and the transfer from the locally excited to charge transfer state can, therefore, be assumed to be negligible. The inverted-region case can then be treated as a single-surface (V_{10}) problem with multiple-sink windows. However, the foregoing assumption may not be correct for electron transfer in the normal regime.

The time evolution of the probability distribution $P_{10}(X,Q,t)$ of the system on the V_{10} potential energy surface is then given by the equation

$$\frac{\partial P_{10}(X,Q,t)}{\partial t} = (\mathcal{L}_X + \mathcal{L}_Q)P_{10}(X,Q,t) - S(X,Q)P_{10}(X,Q,t) \qquad (4.41)$$

The first term describes the relaxation in the $V_{10}(X,Q)$ potential. The second term accounts for the actual transfer of the electron to the various charge transfer states along the sink windows through the sink function $S(X,Q)$ defined as in Eq. (4.17). As already mentioned, the intrinsic sink transfer rate (k_0) corresponding to the $0 \rightarrow n$ transition involving the high-frequency mode is $(2\pi V_{el}^2/\hbar)|\langle 0,n \rangle|^2$.

The solution is obtained using the technique described in Section IV. Equation (4.41) leads to a set of linear equations that are represented in the matrix form as

$$\mathbf{B} \cdot \mathbf{P} = \mathbf{G}_0 \qquad (4.42)$$

where the elements of matrices of $\mathbf{B}$, $\mathbf{P}$ and $\mathbf{G}_0$ are, respectively,

$$[b_{mn}] = \delta_{mn} + k_n G_{10}(X_m, Q_m, z | X_n, Q_n) \qquad (4.43)$$

$$[p_m] = P_{10}(X_m, Q_m, z) \qquad (4.44)$$

$$[g_{0m}] = G_{10}(X_m, Q_m, z | X_0, Q_0) \qquad (4.45)$$

(X_0, Q_0) is the point of excitation on the locally excited surface or reactant surface. For simplicity, the initial distribution at (X_0, Q_0) has been characterized in this chapter as a delta-function source. Whereas in betaine, (X_0, Q_0) defines the minimum of the product surface, there are systems [e.g., 4-(9-anthryl)-N,N'-dimethyaniline (ADMA)] wherein three surfaces are involved and (X_0, Q_0) does not coincide with the coordinate of the minimum of the product surface. These systems are characterized by a ground surface from which the population is excited, the reactant surface, and the product surface [46].

The values of $P_{10}(X_s, Q_s, z)$ obtained by the Green's function method described above are then used in the following relation to find the population $P_{10}(z)$ in the Laplace plane:

$$P_{10}(z) = \frac{1}{z}\left[1 - \sum_s k_s P_{10}(X_s, Q_s, z)\right] \qquad (4.46)$$

V. COMPARISON OF THEORY WITH EXPERIMENT: BETAINE
IN APROTIC POLAR SOLVENTS

In the presence of high-frequency quantum modes, sinks of differing strengths are involved and the dynamics is highly complex. However, some simplifications occur because even though many sink channels are present in principle, only a few effectively participate in the charge transfer process. The choice of sinks is determined by the following factors: (1) the closeness of the relevant sink to the point of excitation [43,44,69,72]. (2) the barrier height of the sinks from the reactant potential surface minimum; (3) the solvent relaxation and the low-frequency vibrational mode relaxation times, τ_X and τ_Q, respectively; (4) the intrinsic rate of the sink channels; and (5) the width of the initial distribution of the excited population.

In Figure 8 we show the time evolution of the population decay from the reactant surface as the number of sinks is increased. These sinks are added in order from the inverted side. In this particular case, the full two-dimensional problem has been solved to obtain $P_{10}(t)$. This was a nontrivial exercise, even though we have assumed (in this model calculation) that the relaxations along both the solvent and vibrational coordinate are Markovian. Figure 8 shows the dynamics to become faster with the addition of sink lines near the barrier region and finally, to saturate. The figure clearly demonstrates the dramatic effects of the additional reaction windows arising from the high-frequency modes, which were neglected in the treatment of Sumi and Marcus. In this calculation the excitation point was chosen close to the barrierless region to resemble situations encountered in systems such as ADMA [46].

However, the situation is quite different in systems such as betaines. As mentioned earlier, the betaine-30 reaction system studied in a wide range of solvents shows a characteristic deep inverted region. Even so, the reaction occurs under highly nonequilibrium conditions, as the excited-state population, located far above the potential surface minimum, reacts through the normal sinks as it relaxes down toward the barrierless region. However, only a few effectively participate, as both the solvent relaxation rate and the intrinsic rate of the sinks compete to determine the choice of these sinks in the charge transfer process. The proximity of the efficient sinks to the excitation point and the solvent relaxation rate are, therefore, the two crucial factors that determine the extent of the reaction that can take place during the relaxation process itself. If a sink of infinite rate of charge transfer were to be present on the normal side, the entire relaxing population would fully react even before the potential minimum was reached. This, predictably, would give rise to a rate that would decrease with increasing solvent relaxation time. Such a model has been considered by Tachiya and

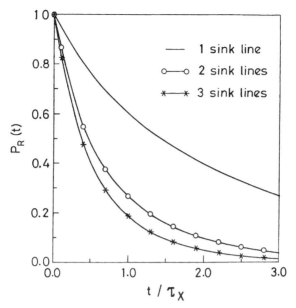

Figure 8. Time evolution of the survival probability $P_{10}(t)$ of the reactant population as the sinks from the high-frequency modes become progressively available. The energy parameters (scaled in $1000\,\text{cm}^{-1}$) are $\lambda_X = 2.0$, $\lambda_Q = 1.0$, $h\nu_q = 1.5$, $-\Delta G = 4.5$. These parameters have been so chosen as to mimic a reaction near the barrierless region. The excitation point is also chosen close to the barrierless region ($X_0 = 0.0, Q_0 = 0.4$). The first sink lies in the inverted region, the second in the barrierless region, and the third in the normal region. The sinks are assumed to be of equal strength with an intrinsic rate at all points along the sink k_0 equal to 10.0, scaled in the solvent relaxation time (τ_X). The relaxation time along $Q(\tau_Q)$ is $0.04\,\tau_X$. The dynamics is found to become faster steadily with the addition of another sink line close to the barrierless region and finally to saturate. Most of the reaction is complete within τ_X as observed in typical barrierless reactions.

Murata [69] to explain the solvent relaxation rate dependence of the average rate in the case of charge recombination reactions in contact ion pairs. In the present case, however, the sinks are not of infinite strength. In fact, the reactant population on its sojurn from the initial excitation point faces several inefficient reaction windows, each becoming more efficient than the preceding one. The relevance of each of these sinks is determined not only by its strength but also by the rate of the population relaxation. Therefore, both the relaxation rate along the solvent coordinate and the intrinsic rate of the sinks involved are important. This can give rise to highly nonexponential decay and a wide range of interesting dynamic behavior [69,82]. The situation becomes even more complex when relaxation along the reaction coordinates is highly nonexponential.

As mentioned earlier, calculations based on the hybrid model seemed to reveal complete solvent independence of the rate in the case of very slowly relaxing solvents. However, proper implementation of this model revealed that this independence is present only over a much shorter range of solvent relaxation rates than that obtained by Walker et al. [43]. The discrepancy in the earlier calculations seems to arise from use of a Markovian solvent dynamics with an average solvent relaxation rate. Let us elaborate on this. In the case of rapidly relaxing solvents, the diffusion rate of the reactant population toward the zero-barrier region is fast. The charge transfer rate will depend on the intrinsic transfer rates from the sinks that are close to the barrierless region, as the sinks on the normal region may not be sufficiently efficient to trap the fast-relaxation population. On the other hand, for slowly relaxing solvents, the rate may still follow a dependence close to $1/\tau_X$. For betaines, the initial population is placed at approximately $15–20k_BT$ (for T around 292 K) above the minimum on the reactant surface, and the intrinsic time constants of the sinks near the barrierless region are approximately in the range 0.1–10 ps [43,44,72,90]. The sinks on the normal region are much weaker because of the reduced Franck–Condon overlaps. Efficient high-frequency sinks are, therefore, present only near $0–2k_BT$. Thus a slow solvent relaxation (say, on the order of 10–1000 ps) as in the case of alcohols may turn out to be the rate-determining step if the population is excited at large potential heights. Then how could the earlier calculations have obtained a solvation-independent rate for slow solvents? It appears to us that Akesson et al. [44] and Walker et al. [43] analyzed the reaction dynamics with an incorrect estimation of the potential height of the initial excitation point, which was reported only to be in the range of $2k_BT$ [44], as against the correct value of around $15k_BT$. Incidentally, in betaines, efficient high-frequency sinks are also present near $2k_BT$. This explains the non-observance of the effects of slow solvent relaxation in the theoretical predictions of the hybrid model as given by Walker et al.

Note that even when relaxation along the low-frequency vibrational mode (Q) is much faster than that along the solvent mode, downward relaxation of the relaxing population will still be slow, as it will experience a high level of friction along the solvent coordinate, as there is no coupling between the X and Q modes in the hybrid model. Thus the solvent relaxation cannot be avoided.

In slow solvents, the weaker sinks in the normal region are likely to be more utilized than in the case of rapidly relaxing solvents. This can easily be understood from Figure 9, where electron transfer times are plotted as the upper limit of the X-sink position. The results were obtained by deliberately neglecting the reactive sites above the indicated upper limit value of the X-sink-coordinate position to study the extent of sink utility for two different solvation relaxations, of time constants 1 ps and 100 ps, in the Markovian limit. The other parameters were so chosen as to simulate the above-mentioned energy and sink conditions in betaines. Let us note here that the $X = 1$ position corresponds to the excitation point of the system (see Figure 4). This means that the contributions of the reactive sites on the normal side were added progressively as the upper limit of the X-sink-position value was increased to 1. This type of analysis merits the present formulation as a flexible and robust scheme for a selective study of various portions of the extended sink(s). The results shown in Figure 9 clearly indicate the normal reactive portion, especially the higher region, to be highly ineffective in the case of the faster relaxation time of 1 ps, as the charge transfer times do not change for values beyond an X-sink limit equal to 0.25. Obviously, the reaction is confined primarily to the barrierless region in this case. On the other hand, for the slower relaxation of 100 ps, the reaction times decrease progressively as the X-sink upper limit is increased. This clearly indicates that the normal sinks on the higher-potential side can be efficient enough for slow relaxations. Note that in both cases, the reaction times are large in the range X-sink upper limit < 0, as the reaction proceeds mainly by overcoming the activation barrier at the reactive sites in the inverted region. This still does not help to explain the high rate observed experimentally.

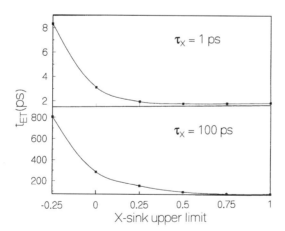

Figure 9. Plot of theoretical electron transfer times (t_{ET}) for two different solvent relaxation times, τ_X, equal to 1 and 100 ps, against the upper limit of the X-sink coordinate position. The results, obtained in the Markovian limit, clearly show the usefulness of the normal sinks only in the case of the slowly relaxing solvent. The potential parameter values used were of betaine-30 in butanol at 292 K as reported by Reid and Barbara [90] and with an electronic coupling (V_{el}) equal to 2800 cm^{-1}.

However, recent ultrafast experiments seem to indicate the presence of an ultrafast component preceding a slower-relaxing component in nearly all liquids [83–88,91]. For example, relaxation dynamics in acetonitrile shows an initial ultrafast Gaussian component with a relaxation time constant on the order of 100 fs, followed by a slower one in the 400 fs range [83]. That the faster component accounts for nearly 70% of the total relaxation clearly indicates that this component may play a very important role. The first, faster component comes from an inertial Gaussian response and the second, slower one from diffusional relaxation with an average time constant τ_s. The relative contributions of these components strongly affect the course of the reaction, as is evident from observations of betaine-30 in polar and nonpolar solvents. Whereas in the case of some polar solvents, the inertial solvation can be more than 50%, in alcohols (especially the higher ones) it appears to be even less than 20% [90]. However, recent experiments [87] seem to indicate that the presence of the ultrafast contribution is nearly universal to all liquids. If this is so, this component may have a significant influence on the rates of many chemical reactions. This appears to be the most likely candidate to explain the anomalous behavior observed in betaines.

Therefore, the hybrid model has been generalized to include the effects of non-Markovian solvent dynamics, which is now assumed to be characterized by a multiexponential solvation time correlation function $\Delta(t)$, given by

$$\Delta(t) = \sum_i w_i \exp(-t/\tau_i) \tag{5.1}$$

where the $\tau_{i=1,2,3,...}$ represent the relaxation times and w_i their corresponding weights. A similar biexponential theoretical treatment has been used by Nagasawa et al. [53,54,89] to study electron transfer behavior in oxazine 1/dimethylaniline and oxazine 1/aniline systems. The Markovian limit can be obtained by assuming a single exponential relaxation. Note that multiexponential behavior has been observed in many liquids [83–88,91]. Even when the first component is Gaussian, the main conclusions arrived at here using an exponential fit remain unchanged. Also, as the determination of the rate involves rather extensive numerical computation, it may be easier to study relaxation behavior using a multi-exponential fit.

In Figure 10 the time evolution of the reactant population under both Markovian and non-Markovian conditions is shown. The potential parameters have been chosen in accordance with the experimental values obtained in betaines as reported by Barbara and co-workers [44,43]. In this calculation, we have taken $w_1 = 0.65$, $\tau_i = 0.5$ ps, $w_2 = 0.35$, and

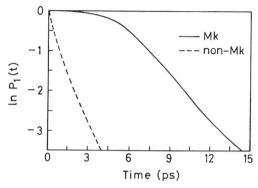

Figure 10. Semilog plot showing the dependence of the time evolution of the reactant survival probability $P_1(t)$ on the nature of solvent polarization relaxation. The solvent time correlation functions in the Markovian and non-Markovian cases are $\Delta_{Mk}(t) = \exp(-t/20.0)$ and $\Delta_{non-Mk}(t) = 0.65 \exp(-t/0.5) + 0.35 \exp(-t/20.0)$, respectively. Times are scaled in picoseconds. The values of the energy parameters (in $1000 \, cm^1$): $\lambda_X = 2.5$, $\lambda_Q = 1.5$, $\lambda_q = 1.0$, $\hbar\omega_q = 1.8$, $-\Delta G = 10.5$, and $V_{el} = 2.8$. Values are from [43] and are believed to be sensible for the back electron transfer in polar aprotic solvents.

$\tau_1 = 20 \, ps$. Figure 10 clearly shows the strong influence of the ultrafast component on the decay profile. The dynamics is nonexponential in both cases.

In Figure 11 the calculated average electron transfer time t_{ET} has been plotted against the effective (i.e., average) solvent relaxation time τ_{eff}. These results are obtained using the generalized hybrid model, which includes a single classical solvent mode (with biphasic response), a single classical low-frequency vibrational mode, and a single high-frequency quantum mode. All the parameter values, including the values of the frequencies and the respective organization energies used in this calculation, were taken from papers of Barbara and co-workers, who obtained these values by fitting them to the absorption spectrum of betaine-30 in polar and nonpolar solvents using Hush approximation. In Figure 11 we have shown experimental results from the Barbara group. It is apparent from the figure that a semiquantitative agreement with the results of Barbara et al. can be obtained for a rather wide range of solvent relaxation parameters. Note the strong dependence of the calculated rate on the amplitude of the ultrafast component when its value is small. The reason for this strong dependence is discussed below. It is, however, found that an extrapolation to larger τ_X gives rates which are slower than those observed experimentally. The reason for this is not known at present.

As noted earlier, the average electron transfer rate can be obtained from Eq. (4.30). On the other hand, the average solvent relaxation time (τ_{eff}) can be obtained from the solvent relaxation time components (τ_i) and their

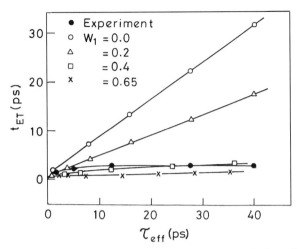

Figure 11. Theoretically obtained electron transfer times against effective solvent relaxation time t_{eff} ($= w_1\tau_1 + w_2\tau_2$). The results show a marked increase in rate with an increase in the weight w_1 of the first component τ_1 ($= 0.5$ ps). The values of the energy parameters are the same as those used in Figure 10. This figure can be used to understand the solvent dynamic effects on the back electron transfer in betaine-30 on polar aprotic solvents. The experimental results are also shown for comparison. Note that excellent agreement with the experimental result is obtained if the ultrafast component contributes 40–60%.

relative weights (w_i). An increase in the weight of the ultrafast component results in faster shifting of the average position of the relaxing population toward the equilibrium value. Thus, for a reaction to occur with the observed 3- to 4-ps time constants (where average solvation time is in the hundreds of picoseconds range), the reactant must avail itself of the efficient sinks near the barrierless region, which means that the system must relax from the initial high-energy state to a configuration where such sinks are present. However, if only the slow component is active, this is not possible. On the other hand, it is evident from Figure 11 that if only 30–40% of the total solvation energy correlation function relaxes via an ultrafast component, a significant part of reactant population can access the efficient sinks. This is because the population relaxation in a harmonic surface follows the simple equation

$$P(t) = [1.0 - \Delta^2(t)]^{-1/2} \exp\{-[X - X_0\Delta(t)]^2\}/[1.0 - \Delta^2(t)]^2 \quad (5.2)$$

where X is the solvation coordinate, X_0 the initial excitation point, and $\Delta(t)$ the normalized solvation time correlation function. If $\Delta(t)$ contains (say) 50% ultrafast component, $P(t)$ relaxes to about $0.5X_0$ in a very short time. However, in terms of reaction free energy, for betaines this means a relaxation of around $12k_B T$. This nonlinear dependence of the mean position of the population on the fast component is responsible for bringing the system down to the place where efficient sinks (or reaction windows) are present. Hence the larger the contribution of the fast component to the solvation time correlation function, the faster the overall electron transfer dynamics and, therefore, the higher the average rate. This mechanism explains the strong dependence on the fast component when its value is small and also the independence of the rate when the amplitude becomes larger than a minimum value (as is clear from Figure 11). The situation, however, is entirely different for the Markovian case, which neglects the fast component, or when the weight of the ultrafast component is small, as may be the case for alcohols. Thus we expect a situation where the electron transfer rate becomes dependent on the average solvation rate. What is really interesting here is the dominance of the faster component in the electron transfer.

VI.   COMPARISON OF THEORY WITH EXPERIMENT: BETAINE IN PROTIC SOLVENTS

In Figure 12, the average time constants of the betaine-30 reaction in alcohols at a temperature of 292 K are shown for V_{el} values of 1800 and 2800 cm^{-1}. The potential parameters have been chosen in accordance with the experimental values obtained for betaines in normal alcohols as reported by Reid and Barbara [90]. They used the average solvation time constant in their calculations. The calculations shown in Figure 12 on the other hand, were performed using the solvation time correlation function reported by Horng et al. [91]. These authors used coumarin as a probe and found that the solvation time correlation function can be represented by a sum of exponentials. Coumarin is known to be a "good" solute that does not form specific hydrogen bonds in alcohols, as is the case with betaines. The average solvation time, τ_{eff}, derives a significant contribution from the slower and weightier component in alcohols, and this explains the strong solvent dependence of the rate in Figure 12 for low values of τ_{eff}. The reaction here proceeds mainly through the slow channels. The results, however, are only qualitatively correct, as they do not match well with values observed experimentally; clearly, much work is still required to understand this problem. However, it is satisfactory to note the near independence of the rate calculated at higher values of the average relaxation time. This is similar to the pattern observed in experiments and is again due to the combined effects of the fast components in the solvation dynamics and

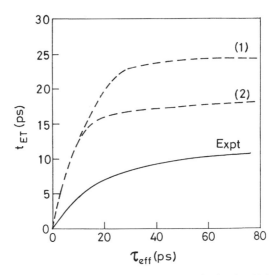

Figure 12. Theoretically obtained electron transfer times for betaine-30 in various alcohols, at 292 K for V_{el} equal to $1800\,cm^{-1}$ [dashed curve (a)] and $2800\,cm^{-1}$ [dashed curve (b)]. The theoretical results show more effective vibrational contributions from enhanced coupling strength. The experimental results are shown by the solid line. The values of the energy parameters used are as reported by Reid and Barbara [90]. The solvent relaxation data are obtained from the reported experimental values of coumarin in alcohols by Horng et al. [91].

the additional reaction windows from high-frequency quantum mode solvation dynamics.

Reid and Barbara had earlier suggested a modification to account for the solvent–solute hydrogen-bonding interactions by allowing a 30% ultrafast component that relaxes faster than the solvent relaxation time constant τ_s to the solvent relaxation dynamics in alcohols. Despite the inclusion of an ultrafast component (of 0.5 ps), the results of a study done to analyze the temperature dependence of reaction transfer of betaine-30 in alcohols using parameters reported by them shows a strong dependence on the slower component but the rates are still not close to the values observed experimentally (Table I). But within the framework of the model and the choice of parameters for the system, the theoretical results appear sensible. The reason for the much slower theoretical rates is again as follows. The weight of the ultrafast component is much smaller than the slower component in the case of alcohols. Note that even in the case of our study on betaine-30 in fast relaxing solvents, with potential parameters comparable to those used here for betaine-30 in alcohols, the results showed solvent independence over a rather wide range of solvent relaxation time parameters only for a large amount of the ultrafast component (as is usually the case in polar solvents)

TABLE I
Theoretically Obtained Electron Transfer Times for Betaine-30 in Various Alcohols[a]

Solvent	$T(K)$	τ_i (ps)	τ_2 (ps)	τ_{eff} (ps)	t_{ET} (ps)
Ethanol	292	0.5	15.0	10.65	12.891
	275	0.5	50.0	35.15	34.452
	250	0.5	110.0	77.15	61.668
	213	0.5	320.0	224.15	150.676
1-Propanol	292	0.5	30.0	21.15	21.795
	265	0.5	100.0	70.15	50.595
	240	0.5	380.0	266.15	137.095
	210	0.5	1100.0	770.15	382.142
	180	0.5	3400.0	2380.15	531.434
1-Butanol	292	0.5	45.0	31.65	28.177
	265	0.5	150.0	105.15	65.650
	255	0.5	410.0	287.15	128.742
	238	0.5	900.0	630.15	255.729
	218	0.5	2400.0	1680.15	503.082
1-Pentanol	292	0.5	75.0	52.65	29.431
	278	0.5	250.0	175.15	65.042
	250	0.5	1000.0	700.15	221.235
	221	0.5	4200.0	2940.15	624.311

[a] Obtained over a wide range of temperatures for V_{el} equal to $1800\,cm^{-1}$ and using a biexponential fit [30% ultrafast componenc of 0.5 ps (τ_1) and 70% slow component with a time constant (τ_2) equal to the average time constant τ_s] for the solvent relaxation. The values of the energy parameters and solvation time constants (τ_2) used are as reported by Reid and Barbara [47]. The results show a strong dependence on the slower relaxation time constant, especially in slow solvents.

[77]. Even so, we found that an extrapolation of the results to a higher τ_X (or τ_{eff}) value still gives rates which are slower than those observed experimentally. If any noticeable differences between the other parameters of the two systems are worthy of mention, they are mainly the following [43,90]:

1. The free energy gap is larger in the case of alcohols, which means that the reaction system here is in a deeper inverted region than in the case of polar solvents.

2. The reported value of quantum high frequency obtained from line-shape fits is $2350\,cm^{-1}$ in the case of alcohols and $1500-1800\,cm^{-1}$ in the case of polar solvents. The sinks are, therefore, less densely spaced in the case of alcohols, so the system may be expected to be sensitive to changes in the free energy gap in the inverted region, as the relevant high-frequency sinks are likely to be shifted in a direction from the normal region to the inverted region with increase in free energy gap

[77]. This effect could be more important in the low-temperature region, as the free energy gap becomes larger as the temperature is lowered with increased stabilization of the product surface(s). To study the sensitivity of the electron transfer rate on the frequency of the quantum mode, a calculation has been carried out by varying this frequency. The results are depicted in Figure 13, where the calculated average electron transfer time (inverse of the rate) is plotted against the free energy gap for three values of the frequency. While the results converge in the normal region (lower free energy gap), there is a significant dependence of the reaction time on frequency. The quantifies the point made above.

3. The value of electronic coupling (V_{el}) used in the case of alcohols is only $1800 \, cm^{-1}$, as against $2500 \, cm^{-1}$ in the case of polar solvents. This means that the sinks are generally more weak for betaine-30 in alcohols.

These three factors and the dominating slower relaxation in alcohols cause the theoretical rates to be much lower than those observed experi-

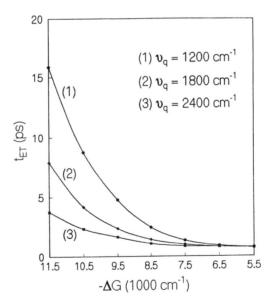

Figure 13. Average electron transfer time, t_{ET}, plotted as a function of the free energy change $(-\Delta G)$ showing the extent of changes observed in the rate on changing the frequency of the quantum mode over a range of $-\Delta G$ values. The results are found to converge in the lower free energy gap range. The value of V_{el} is $1200 \, cm^{-1}$. The values of the other fixed parameters are the same as in Figure 10.

mentally. If one were to ask what could be wrong or missing in the formulation, the answer, perhaps, could be one or more of the following:

1. Estimation of the electronic coupling strength through Hush approximation could be wrong. The reported values for betaine-30 in alcohols are only in the range $1650 \pm 250 \, \text{cm}^{-1}$ [90]. This accusation is probably not very justified. But it is easy to imagine (and explain the discrepancy to some extent) that had the coupling strength been higher (say, around $2800 \, \text{cm}^{-1}$), the predicted rates could have been larger.

2. Theoretical modeling so far has involved only an average high-frequency mode to represent the overall contributions from high-frequency factors [75,42]. Perhaps more than one high-frequency mode needs to be explicitly included in the parameter fitting of line-shape profiles and in the formalism which would indicate more densely spaced sink channels. However, this does not necessarily ensure an increase in the rate if the coupling strength is not sufficiently strong (as is the case here) to render the sinks efficient enough for the various relaxing systems, despite the weakening in their strengths due to the product of Franck–Condon factors of the high-frequency modes. The rates could then be slower than those observed for a single high-frequency mode.

3. The assumed weight of 30% for the ultrafast temperature-independent factors (hydrogen-bonding interactions) should perhaps have been more, as the difference between experimental and theoretical values is larger in the low-temperature range. For the solvation-dependent factor is so slow in the sufficiently low-temperature region that the faster hydrogen-bond interactions (and other fast low-temperature influences, if any) should have been the major contributor. Also, the reported λ_{sol} (or λ_X) value for alcohols obtained from line-shape fits shows an increase as the temperature is lowered [90]. As $\lambda_i \, (i = X, Q, \ldots)$ values are the force constants in quadratic equations (4.1) and (4.2), which determine the shape of the harmonic surfaces, a higher λ_{sol} value gives rise to steeper potential surfaces. There could be less friction in the relaxation process on steeper surfaces than on shallower and broader ones. Theoretically, this aspect can be manifested in the form of a faster ultrafast component and/or with a higher weight. One should, perhaps, examine more deeply the details of low-temperature processes, as it is not very clear whether the extent of hydrogen-bonding interaction contributions is strictly temperature independent.

TABLE II

Theoretically Obtained Electron Transfer Times for Betaine-30 in Ethannl and Pentanol[a]

Solvent	T(K)	w_1	τ_1 (ps)	w_2	τ_2 (ps)	τ_{eff} (ps)	t_{ET} (ps)
Ethanol	292	0.400	0.2	0.600	15.0	9.08	8.73
	275	0.425	0.2	0.575	50.0	28.835	21.05
	250	0.450	0.2	0.550	110.0	60.595	33.14
	213	0.475	0.2	0.525	320.0	168.095	62.69
1-Pentanol	292	0.400	0.2	0.600	75.0	45.08	17.50
	278	0.425	0.2	0.575	250.0	143.835	32.40
	250	0.450	0.2	0.550	1000.0	550.09	66.06
	221	0.475	0.2	0.525	4200.0	2205.095	329.56

[a]Obtained over a range of temperatures using V_{el} equal to $2800\,\mathrm{cm}^{-1}$ and a biexponential solvent relaxation, with time constants and weights as shown below. The results show closer agreement to experimental values in comparison with the results in Table I. The values of the energy parameters and solvation time constants (τ_2) used are as reported in [21].

In view of the above-mentioned factors, a test study was attempted with suitable changes to some of the parameters made deliberately (Table II). The time constant of the ultrafast component was decreased from 0.5 ps to 0.2 ps. Its weight was itself increased from 30% to 40% for a temperature of 292 K and increased further as the temperature was lowered from 292 K. The results were obtained for $V_{\mathrm{el}} = 2800$ cm^{-1} and show considerable improvement over previous results, especially in the case of slow solvation times and lower temperatures (Table I).

VII. NON-MARCUS FREE ENERGY GAP DEPENDENCE

The simple relation between the rate of electron transfer and the free energy gap (ΔG), which gives rise to the famous Marcus inverted parabola, is now well established for reactions where either the nonequilibrium effects or the multidimensional aspects of the reaction surface are not important [68]. However, deviations from the Marcus parabola are known to occur in some cases, especially in photoinduced electron transfer reactions [55–57]. In fact, one sometimes finds a nearly exponential dependence of the rate on the free energy gap, which is more common in nonradiative relaxation processes. This exponential-like free energy gap has been a subject of much discussion recently, and various mechanisms have been proposed to explain the apparent breakdown of the Marcus parabolic dependence. In the following we briefly review some of our work, which serves to emphasize the possible role of nonequilibrium effects coupled with ultrafast solvation in giving rise to non-Marcus energy gap dependence.

The results discussed in Sections V and VI demonstrate that the high-frequency quantum modes can greatly enhance the rate of an electron transfer reaction in the Marcus inverted region which would classically possess a very low rate. This alone can lead to a complete breakdown of the Marcus parabolic dependence of rate on the free energy gap for large gaps. In fact, Jortner et al. have already pointed out that in the presence of a large number of participating high-frequency modes, one recovers an almost linear dependence of the logarithmic of the rate on the free energy gap [22].

The results discussed in previous sections also demonstrate an aspect that has not been discussed in the literature in the past—that ultrafast modes can also help in accelerating the rate significantly. This is an important point because many experiments are carried out in liquid acetonitrile, which possesses a significant ultrafast component. Thus one can indeed obtain high rates in the Marcus inverted regime.

In some reactions these high-frequency quantum modes can play an important role in the normal Marcus region, too, and this is perhaps even more interesting [70]. This can indeed happen for those photoinduced reactions where the reactant and product surfaces are so displaced with respect to each other that the initial preparation places the reactant at a position above the classical crossing point (as in the case of betaine-30 and in some ion-pair formations). Here the reaction can occur as the system relaxes toward the potential minimum of the reactant surface. Thus, as the system relaxes, it has to pass through the sinks or reaction windows created by the high-frequency modes. In such a situation the reaction can be much faster than the rate predicted by the Marcus theory. Note that this is a purely nonequilibrium effect, not anticipated in the Marcus theory.

In fact, the scenario described above for both the normal and inverted regions can together give rise to a marked departure from the Marcus parabolic dependence of the logarithm of the rate. We discuss below the fact that numerical calculations bore out these expectations.

Figure 14 shows the free energy gap dependence of the average rate calculated for betaine-like system. By the latter we mean that all the necessary parameters are taken from that of betaine-30 in polar aprotic solvents. Note the almost linear dependence above the barrierless region in the plot of $\ln k_{ET}$ versus ΔG. We would like to remind the reader that this quasilinear dependence has been obtained by using the same generalized hybrid model discussed in previous sections, and the breakdown of Marcus parabolic dependence in the inverted region is a direct consequence of the intervention of the high-frequency quantum vibrational mode, which dramatically enhances the rate by providing extra reaction channels, as discussed above. This linear dependence is similar to that observed in the charge recombination of cyclophane-derived intimate radical ion pairs and in

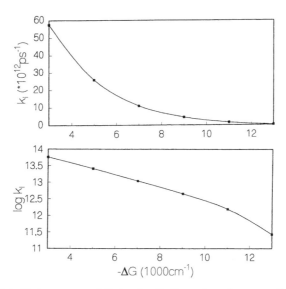

Figure 14. Plot of $\ln k_{ET}$ versus $-\Delta G$, showing the linear dependence near the barrierless and the inverted region. However, the results are expected to show the nonlinearity in the high-barrier normal regime. The calculation in this region has not been done yet, as a proper calculation should also account for the back transfers from the product surfaces.

berrelene-based supermolecules. The Jortner and Bixon model calculation of the multimode energy-dependent rate (Section III.B) could provide an explanation for this type of behavior. However, as mentioned earlier, this description did not account for the dynamics of motion on potential energy surfaces.

Note that in Figure 14 the rates in the Marcus normal region are also considerably higher than that expected from a Marcus parabola. However, the rates start to decrease faster than the linear dependence as we approach $\Delta G = 0$. This is because a part of the reactant population that escapes the reaction window during the system's downward sojourn can react only by the usual activated process. Since the activation energy is very high when the free energy gap is nearly zero, the latter rate is very slow. This slow component in the decay of the reactant survival probability makes the average rate low. Thus in the normal regime one may find a strong biphasic dependence in the decay of the reactant population. The fast-decaying part is due to the reaction under nonequilibrium conditions from the sinks in the normal regime, while the slowly decaying part is that due to decay by the usual Marcus mechanism. This is rather interesting.

Actually, Tachiya and Murata [69] earlier provided a one-dimensional semiquantitative explanation for the nonobservance of Marcus free energy gap dependence for recombination reactions in a series of contact ion pairs (CIPs) which is rather similar to the explanation provided above. These experiments were carried out in a series of donor–acceptor systems by Asahi and Mataga [70]. Our reason for dwelling on this problem is that it provides a very simple example of the nonequilibrium effects discussed above, especially biphasic decay of the reactant population in the normal free energy gap regime. In the one-dimensional model of Tachiya and Murata with solvent polarization as the reaction coordinate, the main parameters are the coupling constant V_{el} and the solvent relaxation time (τ_X). Tachiya and Murata used the former as a fitting parameter with reasonable values for both the CIP and the SSIP (solvent-separated ion pair). The important observation to be made here is that the value of V_{el} should be much larger for CIP than for SSIP. This is because V_{el} is known to obey a simple exponential dependence on the distance of separation r between the ion pair given by $V_{el} = J \exp(-r/r_0)$, where $r_0 = 1$ Å provides a good fit to the experimental data. J, the exchange integral, is chosen to have the typical value.

In Figure 15 we show the calculated average rate plotted against the free energy gap in a standard Marcus plot. The figure should be compared with

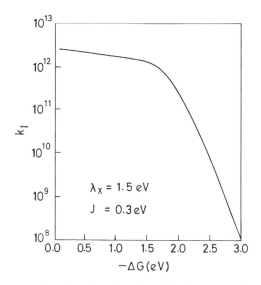

Figure 15. Average rate, k_I, plotted as a function of the free energy change $(-\Delta G)$ showing the non-Marcus energy gap dependence for CIP ($J = 0.3\,eV$). The small gap shown in the upper left of the graph is deliberate. As discussed in the text, for high activation energies, there can be a long, slow time decay component in $P_1(t)$ which may make k_1 dip near $-\Delta G = 0$. It is unlikely that experiments detect this slow decay, as it is predicted to occur in the same time scale as the electron transfer in the solvent-separated ion pair.

Figure 2 of Tachiya and Murata [69]. There are two main differences. We have used an appropriate initial distribution, which is the equilibrium distribution in the ground state. Second, the rates are obtained numerically. However, the main features are quite similar, especially the total breakdown of the Marcus parabola that occurs in the normal region.

In this case the time dependence of the reactant population shows interesting features because a part of the reaction population can escape capture during relaxation and would then subsequently react with a high activation barrier. The probability of escape is determined by the value of both the exchange integral and the relaxation rate. The calculated results are shown in Figures 16 and 17 for two different values of the exchange integral J. The following comments are in order. First, in both the normal and zero-barrier regions, the decay is strongly nonexponential for $J = 0.3$ (shown in Figure 16), which is the value reported by Tachiya and Murata for charge recombination in CIP. When the value of J is made smaller, the initial fast nonexponential decay is followed by a slow exponential-like decay, as shown in

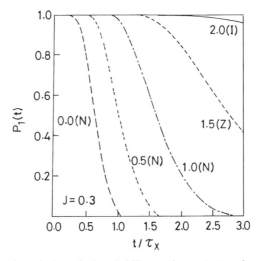

Figure 16. Time-dependent survival probability on the reactant surface, $P_1(t)$, for electron transfer from a contact ion pair for the Tachiya–Murata model (see the discussion in Sec. VII). Here $P_1(t)$ is plotted as a function of scaled time, showing the changes in the dynamics that are observed with changes in the free energy ΔG. The times are scaled by the solvent relaxation time τ_X. Energies are in units of eV. The values of other parameters are $\lambda_X = 1.5$ eV and $J = 0.3$ eV, where the former is the solvent reorganization energy and the latter is the exchange integral, respectively. N, Z, and I represent the normal, zero-barrier, and inverted regions, respectively. Note that the initial distribution is a Gaussian at $X = 0$ $(X = X_0)$ with the width of the equilibrium distribution in the ground-state surface.

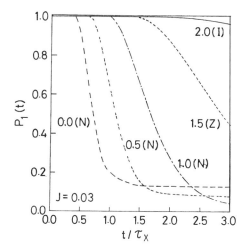

Figure 17. Time evolution of the reactant population $P_1(t)$ for the exchange integral $J = 0.03$ eV. This value of the exchange integral is appropriate for the electron transfer in the solvent-separated ion pair (SSIP). The dynamics is much slower and less nonexponential in all the cases than the results obtained for $J = 0.3$ eV (see Figure 16). The values of the rest of the parameters are as used to obtain the results shown in Figure 16.

Figure 17. The decay becomes mostly exponential when J is made even smaller, as in the case of charge recombination in SSIP.

The very slow decay for small J values is due to the obvious fact that in this case a part of the reactant population remains unreacted in its downward passage through the sink region. The subsequent reaction of this population occurs via the standard activated mechanism, with a rate much slower than the initial rate. Thus the decay is strongly biphasic. When J is very small, which might be the case for charge recombination in SSIP, most of electron transfer occurs via the activated mechanism in the normal region and one recovers the Marcus parabolic dependence. Thus the time dependence of the reactant population is very sensitive to the value of J. This can be useful in detecting the nature of an electron transfer reaction.

Up to now we have discussed only relaxation and quantum effects as probable reasons for the nonobservance of Marcus parabolic dependence. Apart from the above-mentioned effects, there can be other factors that may influence the dynamics substantially. Substituent and isotope effects on the donor–acceptor pair are known to affect significantly the free energy gap of the reaction and/or the frequency of the vibrational modes [54,75]. For example, the stretching frequency of the C—H bond decreases on deuteration. If the high-frequency quantum mode is related to this frequency, lowering of the frequency gives rise to more closely spaced sink channels. This, however, does not immediately ensure an increase in the reaction rate because a major constraint is the rapidly decreasing Franck–Condon overlap with higher quantum levels of the product state. For this reason, when large energy gaps are involved, the rates are found to decrease on deuteration as the sinks close to the barrierless region become weaker. This dependence has already been shown in Figure 13, where the calculated average relaxation time is plotted against the free energy gap in the presence of different frequencies of the high-frequency mode. The charge transfer rate is found to be more sensitive to changes in the high-frequency mode when the reaction is in the inverted region than in the normal region. The degree of this sensitivity depends, of course, on several other factors which collectively determine the effectiveness of a sink. In particular, a large electronic coupling V_{el} value can make the barrierless and normal sinks more useful even when the free energy gap is very large. In the case of weak couplings, it is obvious that the behavior would result in a bell-shaped Marcus dependence, as seen in the case of charge recombination in weakly coupled RIPs. This has been satisfactorily explained in terms of one-dimensional nonadiabatic electron transfer theory. It is quite possible that high-frequency mode contributions were present in these systems but were not evident, as the couplings were already too weak. So perhaps the high-frequency model could provide a generic explanation of the wide-ranging behavior in electron transfer.

VIII.   EFFECTS OF ULTRAFAST SOLVATION ON THE RATE OF ADIABATIC OUTER-SPHERE ELECTRON TRANSFER REACTIONS IN WATER, ACETONITRILE, AND METHANOL

Recent experimental studies on polar solvation dynamics have demonstrated that many common dipolar liquids (e.g., water, acetonitrile, lower member of normal alcohols) contain a significant ultrafast Gaussian component in the solvation time correlation function [83–86,104]. For example, water seems to possess an ultrafast component whose time constant is about 50 fs and which contributes more than 60% to the relaxation of the total solvation energy. We have already seen in previous sections that such an ultrafast component can have significant effect on the rate of an electron transfer reaction in either the Marcus inverted or barrierless region. Here we explore the effects of these ultrafast modes on the rate of an outer-sphere electron transfer reaction in the Marcus normal region which occurs under the usual equilibrium conditions. In this case the reaction is activated, and as the reaction is assumed to be adiabatic, the usual one-dimensional description with the solvent polarization as the sole reaction coordinate appears to be valid. We discuss the results of several studies that seem to provide an interesting insight into the role of solvent dynamics on electron transfer reactions.

When the electron transfer reaction is modeled as a Marcus two-surface problem, the activation barrier can be rather sharp, even cusplike for a weakly adiabatic reaction. In such a situation, the act of electron transfer across the barrier can be coupled primarily to the high-frequency polar response of the liquid. The recent discovery of the ultrafast component in solvation dynamics has shown that this high-frequency response can be considerably different from that at zero frequency; earlier theoretical studies focused only on the latter. This alone might have led to erroneous conclusions. We should mention that computer simulation studies of Hynes and co-workers were perhaps the first studies that emphasized this point.

How should we study the effects of ultrafast solvent modes on an adiabatic electron transfer reaction? One simple approach would be to study the effects of these modes on the friction that acts on the reaction system as it moves along the reaction coordinate. Such a formulation has been developed by Hynes [18]. In this theory an electron transfer reaction comprises two distinct steps: (1) the energy diffusion step to reach the barrier region, and (2) the crossing of the barrier. The friction itself can have rather different effects on these two steps. The high-frequency librational and inter-molecular vibrational modes, probed during the barrier crossing, may offer only a small friction of the motion along the remainder of the reaction coordinate. Therefore, these underdamped modes may lead to significantly less damping of the rate of barrier crossing than would appear from the value of the zero frequency (i.e., the macroscopic) friction. On the other hand, the lower value of the friction reduces the rate of energy diffusion. In such a situation, the following interesting question arises. Is coupling with solvent underdamped modes sufficient to result in an energy-diffusion-controlled electron transfer reaction? The latter is expected when the friction falls below the critical value required to attain Kramers' turnover region and the rate is essentially limited by the energy diffusion step. In this chapter we investigate the possibility of the existence of a region where energy diffusion is important. This is expected to throw some light on the fundamental question as to why the Marcus transition-state theory predictions of the rate of outer-sphere electron transfer reactions in water agree so well with experiments.

The consequences of nonexponential decay of the solvation time correlation function in electron transfer reactions have been discussed in several earlier studies [38,92,93,96,97]. McManis and Weaver [38] used the theoretical formulation of Hynes [18] to study the effects of non-Debye relaxation of the solvent on the rate of an outer-sphere electron transfer reaction. The presence of a fast-relaxation component in the multiple Debye dispersion of the solvent was found to result into a substantial enhancement of the adiabatic barrier-crossing frequency. This closely resembles the marked increase in the rate resulting from the solvent librational and inertial modes studied in this work. However, McManis and Weaver did not consider the competition between the barrier-crossing dynamics and the energy diffusion. This has been addressed in a recent molecular dynamics simulation study by Smith et al. [92] on model adiabatic electron transfer reactions in a strongly overdamped solvent. It has been found that even in the limit of sharp barrier reactions, the adiabatic transmission coefficient is determined by the solvent dynamics probed at the barrier top rather than that in the well. Furthermore, the transmission coefficient was found to be well described by Grote–Hynes theory [93] and reflected the importance of an initial Gaussian component in the solvation dynamics rather than its long-time behavior. In another detailed study, Yan et al. [96] used unified Liouville space theory to obtain a general connection between the rate processes and solvation dynamics in dipolar solvents. In this formulation, the dynamic solvent effects are described in terms of the dielectric relaxation of the solvent. However, the general model was used for solvents represented by overdamped oscillators and the frequency-dependent dielectric function was obtained from a Debye or Cole–Davidson model. An interesting outcome of this study is that the rate of electron transfer is found to be in Kramers' turnover region for high values of the adiabaticity parameter and the longitudinal relaxation time of the solvent.

Since the relaxation of solvent polarization is biphasic in ultrafast solvents, the solvent response probed is non-Markovian in nature. A key input in the Hynes formulation is the reaction coordinate time correlation function (RCTCF), which for the model studied here is the normalized solvation energy time correlation function, $S(t)$. The latter can be obtained directly from experiments or from computer simulation results. We described earlier non-Markovian extended molecular hydrodynamic theory (EMH), which provides excellent agreement with experiments and the simulation results in model Stockmayer liquid and also in acetonitrile, water, and methanol. The advantage of using this theory to evaluate the RCTCF is that it allows an almost analytic approach to the solution of the entire problem. In particular, it enables us to explore the dependence of the rate of electron transfer reaction on the relevant parameters, such as the solvent librational modes. It should also be noted that because of the excellent agreement between theory and experiments and computer simulation studies, the results presented here will be the same even if one starts from experimental or simulation data instead of molecular hydrodynamic theory.

The theoretical studies have revealed several interesting results. First, in the presence of the ultrafast solvent modes, the rate of barrier-crossing increases dramatically compared to cases where these modes are absent. For weakly adiabatic reactions, the librational and inertial response lead essentially to the transition-state result; the rates calculated are virtually indistinguishable from k^{TST} even at large zero-frequency frictions (k^{TST} is the rate predicted by transition-state theory). The latter result seem to be in excellent agreement with computer simulation results of Zichi et al. [94]. In the case of water, the total rate of electron transfer reaction is dominated by the barrier-crossing dynamics. Further investigation of nonpolarizable waterlike solvents reveals that because of the apparent overdamping of the high-frequency intermolecular vibrational mode in water, the friction may never become sufficiently small for the rate of electron transfer reaction to be limited by the energy diffusion process. The rate of electron transfer reaction in acetonitrile, on the other hand, is determined by the energy diffusion process, which is somewhat slower than the barrier-crossing rate. Therefore, *the electron transfer reaction in acetonitrile may furnish a rather unique example of a reaction that could be the energy-diffusion-controlled regime.* In methanol, the rate of an adiabatic reaction is found to be controlled essentially by the barrier-crossing dynamics, although the reaction may become energy diffusion controlled if the solvent librational modes are highly underdamped.

A.   Model

We shall consider a simple model electron transfer reaction that involves a redox couple of the type [38]

$$Ox + e \rightarrow Red$$

This model was studied earlier by McManis and Weaver [10,38] to investigate the dynamic solvent effects on outer-sphere electron reactions in dipolar liquids; they did not however, consider the inertial and viscoelastic response of the liquid. We shall assume that the acceptor and the donor are in contact and that they are spheres of equal size. The free energy of activation, $\Delta G^{\#}$, is the intrinsic outer-sphere part of the barrier energy, which in this case is equal to a solvation energy of $e/2$ amount of charge, where e is the charge of an electron. Initially, the reactant is located at the minimum of the reactant well and the charge distribution of the reactant creates an inhomogeneous polarization in the solvent. The polarization fluctuates around its equilibrium value because of the thermal motion of the solvent molecules. It is the coupling of the charge distribution with the thermal motion of the solvent dipoles that leads to the necessary activation and subsequent relaxation in the product well. In the one-dimensional (Marcus) model of electron transfer, the reaction coordinate is the fluctuating energy gap between the two equilibrium surfaces and is given by the expression [14,15,39]

$$\Delta E(t) = \int d\mathbf{r} \, [\mathbf{E}_0^P(\mathbf{r}) - \mathbf{E}_0^R(\mathbf{r})] \cdot \mathbf{P}(\mathbf{r}, t) \tag{8.1}$$

where $\mathbf{E}_0^R$ and $\mathbf{E}_0^P$ are the bare electric fields of the reactants and products, respectively, and $\mathbf{P}(\mathbf{r}, t)$ is the time-dependent polarization at position $\mathbf{r}$ and time t. Note that $\Delta E(t)$ is a *collective quantity*, as it derives contributions from all the solvent molecules. This is a fundamental feature of electron transfer reactions. Therefore, the rate depends significantly on the dynamics of the solvent, especially in the outer-sphere electron transfer reactions. As in the conventional Marcus picture, there is a high activation barrier for motion along the reaction coordinate. It is further assumed that both the reactant and product potential energy surfaces are harmonic along the reaction coordinate. This enables us to use simple analytical expressions for calculating the rate of the electron transfer reaction. The validity of the assumption of harmonicity of the potential energy surfaces is discussed later. For the model of outer-sphere electron transfer reaction described above, we shall now discuss the elegant formulation developed by Hynes [18] to treat the effects of solvent dynamics on the rate of reaction.

B.   Theoretical Formulation

The rate of a high-barrier adiabatic outer-sphere electron transfer reaction is, in general, determined by both the rate of barrier crossing and the rate of energy diffusion. The combined rate may be given by the following simple, albeit approximate expression [18]:

$$\frac{1}{k_{ET}} = \frac{1}{k_b} + \frac{2}{k_{ED}} \tag{8.2}$$

where k_b is the rate of crossing of the activation barrier and k_{ED} is the rate of energy diffusion due to solvent polarization relaxation in the reactant potential energy well. In the determination of both k_b and k_{ED}, a critical role is played by the reaction time correlation function $\Delta_R(t)$, which is defined by

$$\Delta_R(t) = \frac{\langle \Delta E(0) \, \Delta E(t) \rangle}{\langle \Delta E(0) \, \Delta E(0) \rangle} \tag{8.3}$$

where the average is over the solvent degrees of freedom of the reactant state at equilibrium and $\Delta E(t)$ is defined by Eq. (8.1). We next summarize the results of Hynes' theory, which expresses both rate constants in terms of the reaction time correlation function, $\Delta_R(t)$.

1. Barrier-Crossing Rate

The barrier-crossing rate constant, k_b, for a parabolic barrier top is given by the well-known Grote–Hynes formula [93]

$$k_b = k^{TST} \frac{\lambda_R}{\omega_b} \tag{8.4}$$

where ω_b is the barrier frequency and k^{TST} is the transition-state rate. The latter is obtained from the familiar expression

$$k^{TST} = \frac{\omega_0}{2\pi} \exp(-\beta \Delta G^{\#}) \tag{8.5}$$

where ω_0 stands for the harmonic frequency of the reactant well, $\Delta G^{\#}$ is the free energy of activation and $\beta = (k_B T)^{-1}$. Another characteristic frequency that appears in Eq. (8.4) is λ_R. This is termed the reactive frequency and may be obtained by solving the following equation self-consistently:

$$\lambda_R = \frac{\omega_b^2}{\lambda_R + \hat{\zeta}(\lambda_R)} \tag{8.6}$$

Calculation of the barrier-crossing rate therefore requires as a crucial input the frequency (z)-dependent friction, $\hat{\zeta}(z)$, acting along the reaction coordinate. This is obtained using the following route. We first write the generalized Langevin equation for motion along the reaction coordinate:

$$\mu \frac{\partial^2}{\partial t^2} \Delta E(t) = -\frac{dV(E)}{dE} - \int_0^t dt' \, \zeta(t - t') \, \Delta E(t') + f(t) \tag{8.7}$$

where μ is the reactive mass, ζ the frequency-dependent friction, $V(E)$ the potential energy surface on which the reaction occurs and $f(t)$ the random noise term, related to friction by the second fluctuation–dissipation theorem. Equation (8.7) leads to the following expression for the reaction coordinate time correlation function:

$$\hat{\Delta}_R(z) = \frac{z + \hat{\zeta}(z)}{z^2 + z\hat{\zeta}(z) + \omega_0^2} \tag{8.8}$$

where z is the Laplace frequency conjugate to time t and

$$\hat{\Delta}_R(z) = \int_0^\infty dt \exp(-zt)\, \Delta_R(t) \tag{8.9}$$

is the Laplace transform of the normalized reaction coordinate time correlation function, $\Delta_R(t)$. Let us recall that the acceptor and donor ions are chosen to be of the same size. Then it follows from Eq. (8.1) that within the assumption of linear response of the solvent, the reaction coordinate time correlation function is the same as the solvation time correlation function, $S(t)$, of an ion (of the same size as the acceptor) on which one unit of electronic charge has been created. Under these conditions, the frequency-dependent friction $\hat{\zeta}(z)$ is given uniquely by the solvation time correlation function, $S(t)$. Therefore, Eq. (8.8) provides the important bridging relation between the electron transfer reaction and the solvation dynamics, as pointed out by Hynes [18].

To obtain the rate, we need to specify the reactant well and the barrier frequencies and also the effective mass. The well frequency and the effective mass are obtained using the expressions [18]

$$\langle |\Delta E(0)|^2 \rangle = (\beta \mu \omega_0^2)^{-1} \tag{8.10}$$

$$\langle |\Delta \dot{E}(0)|^2 \rangle = (\beta \mu)^{-1} \tag{8.11}$$

Following Hynes [18], the barrier frequency is treated as an input parameter under the condition that the ratio ω_b/ω_0 is nearly equal to 1 for strongly adiabatic reactions, while it is significantly greater than 1 for weakly adiabatic reactions. The latter would then correspond to a high barrier frequency, that is, a very sharp, cusplike barrier top.

2. Rate of Energy Diffusion

In another elegant formulation, Grote and Hynes [93] provided an expression for the rate of energy-diffusion-controlled reactions in solution which is given by

$$\frac{1}{k_{\mathrm{ED}}} = \int_0^{E_a} dE\, \frac{1}{D(E)} \exp(\beta E) \int_0^E dE'\, \frac{1}{\omega(E')} \exp(-\beta E') \tag{8.12}$$

Here $\omega(E)^{-1} \exp(-\beta E)$ is the (unnormalized) equilibrium energy distribution, $D(E)^{-1} \exp(\beta E)$ corresponds to the resistance to energy flow and E_a is the activation energy. Note that for a harmonic potential energy surface, $\omega(E) = \omega_0/2\pi$ and the expression (8.12) simplifies to the form

$$\frac{1}{k_{ED}} = \frac{2\pi}{\omega_0} \int_0^{E_a} dE \, \frac{1}{D(E)} \exp(\beta E) \int_0^E dE' \exp(-\beta E') \tag{8.13}$$

The energy diffusion coefficient, $D(E)$, can be expressed as follows in terms of the frequency-dependent friction acting along the reaction coordinate [98]:

$$D(E) = \frac{E}{\beta\omega_0} \int_0^\infty dt \, \zeta(t) \, \cos[\omega_0(t)] = \frac{E}{\beta\omega_0} \, \mathrm{Re}\, \tilde{\zeta}(\omega_0) \tag{8.14}$$

Here also, an important prerequisite for the calculation of the rate is the frequency-dependent friction, $\tilde{\zeta}(\omega)$. The only difference from the preceding case is that now we need to obtain the friction in the Laplace–Fourier frequency ω. This is again calculated using the generalized Langevin equation with $z = -i\omega$ in Eq. (8.9), and the final expression for the real part of the frequency-dependent friction is as follows:

$$\mathrm{Re}\, \tilde{\zeta}(\omega) = \frac{\omega_0^2 S'(\omega)}{[1 - \omega S''(\omega)]^2 + [\omega S'(\omega)]^2} \tag{8.15}$$

In the rest of the treatment, we represent the real part of the frequency-dependent friction by $\tilde{\zeta}(\omega)$. In expression (8.15), $S'(\omega)$ and $S''(\omega)$ are the real and imaginary parts of the frequency (ω)-dependent RCTCF $\tilde{S}(\omega)$, respectively, and are defined by the equations

$$S'(\omega) = \int_0^\infty dt \, \cos \omega t \, S(t) \tag{8.16}$$

$$S''(\omega) = \int_0^\infty dt \, \sin \omega t \, S(t) \tag{8.17}$$

Therefore, the critical input in this calculation is again the RCTCF, which can now be estimated from the extended molecular hydrodynamic theory. The molecular hydrodynamic theory can also be used to derive an important relationship between the time constant τ_G of the initial Gaussian decay and the reactant well frequency, ω_0. We may recall that the latter is an important input required for calculation of the rate of electron transfer reaction. At very short times, the high-frequency response of the solvent probed will experience almost zero friction. Correspondingly, one may obtain the following equation for the solvation time correlation function at ultrashort times [40]:

$$S(t) = \exp[-(t/\tau_G)^2] \tag{8.18}$$

where

$$\frac{2}{\tau_G^2} = \omega_0^2 = \frac{2f(110; k = 0)}{\tau_I^2} \tag{8.19}$$

where $f_{110}(k)$ is essentially the force constant of the longitudinal polarization fluctuation [40] and is related to the static dielectric constant via the exact relation [40,41]

$$1 - 1/\epsilon(k) = 3Y/f_{110}(k) \tag{8.20}$$

where $3Y$ is the usual dimensionless dipolar interaction energy parameter, equal to $\rho\mu^2/3k_BT$. Here ρ, μ k_B, and T are the solvent density, dipole moment of a solvent molecule, Boltzmann constant, and temperature, respectively.

From Eq. (8.19) it clearly follows that the reactant well frequency, ω_0, depends critically on two system parameters, $f(110; k = 0)$ and τ_I. A high value of $f(110; k = 0)$ or a small value of τ_I would result in a very fast Gaussian relaxation and a high reactant well frequency. Furthermore, this would imply that the frequency-dependent friction along the reaction coordinate increases as the latter is proportional to ω_0^2 [see Eq. (8.15)]. Therefore, it appears that although the presence of the solvent ultrafast modes makes the initial Gaussian decay faster, the friction may not be sufficiently low as to lead to a energy-diffusion-controlled regime. We explore this aspect quantitatively in VIII.C.

Clearly, an important quantity required for estimation of the energy diffusion rate is the activation energy of the forward reaction. This can be evaluated using a molecular theory and is given by [39]

$$\beta E_a \simeq \beta \Delta G^{\#} = \frac{\beta e^2}{4\pi\sigma} \int_0^\infty dk \left[1 - \frac{1}{\epsilon_L(k)}\right] \left(\frac{\sin kr_c}{kr_c}\right)^2 \tag{8.21}$$

where the entropy effects have been ignored. This is also the expression used to calculate the transition-state rate needed to understand the extent of dynamic solvent effects on electron transfer reactions.

C. Numerical Calculations

Among the many data that are required to find the electron transfer rate, the most elusive has been the reaction time correlation function, $\Delta(t)$. Fortunately, this is now available experimentally and also theoretically for several fast solvents, most notably for water, acetonitrile, and methanol. Next we present results of a recent study of the effects of these ultrafast modes on a model adiabatic reaction in these three solvents.

1. Calculation of the Rate in Water

Since the reaction time correlation function is assumed to be the same as the solvation time correlation function, this quantity can be obtained directly from published experimental results or from the theory, as the latter provides an almost quantitative agreement with the former. However, we have used the theoretical result for the main reason that it gives us much better accuracy in the numerical work. Presentation of the details of the theory of solvation dynamics is beyond the scope of this review; we mention here the bare essentials. The theory we use is based on molecular hydrodynamic considerations and needs both static and dynamic two-particle correlation functions of the pure solvent. The static quantity we need is the wavevector-dependent dielectric function $\epsilon(k)$, which is expressed in terms of the long-itudinal orientational pair correlation function. The dynamic quantities we need are essentially the rotational and translational memory functions. Fortunately, accurate estimate of all these quantities are available for water [100,101]. Below we provide a brief list of the calculational procedure [40,41].

The static parameters used for water in the numerical calculations are given in Table III. In this calculation, time is scaled in units of $\tau_I = 0.1$ ps.

TABLE III

Parameters Used for Calculation of Solvation Dynamics and Electron Transfer Reaction in Water

μ (D)	ρ_0 ($\mathring{A}^{-3}$)	$3Y$	ϵ_0	σ ($\mathring{A}$)	τ_I (ps)	τ_0 (ps)	ω_0 (ps^{-1})
1.8	0.033	11.5	79.2	2.8	0.1	0.75	48.4

[a]$3Y = (4\pi/3)\beta\mu^2\rho_0$, where μ is the dipole moment; ρ_0 is the average number density; $\beta = (k_B T)^{-1}$, the Boltzmann constant times the absolute temperature; ϵ is the static dielectric constant; $\tau_0 = (m\sigma^2/k_B T)^{1/2}$ for solvent molecules having mass m and diameter σ; and ω_0 is the harmonic reactant well frequency.

As mentioned, we discuss here only the bare essentials of calculation of the reaction time correlation function.

1. $\epsilon_L(k)$ is calculated using the results of Raineri et al. [99] and then $f(110; k)$ is estimated using Eq. (8.20).

2. The rotational friction necessary to obtain $\Delta(t)$ is calculated from the frequency-dependent dielectric function, $\epsilon(z)$, of water. The dielectric relaxation data used in this calculation are summarized in Table IV.

3. The translational friction has been calculated using contributions from the two intermolecular vibrational peaks at 44 and $215\,\mathrm{cm}^{-1}$.

4. A solute to solvent size ratio of 3 is used. Both have been assumed to be spherical.

We should mention that in calculating the Grote–Hynes rate, we need the reaction time correlation function in the Laplace frequency plane; fortunately, theory gives this directly in the Laplace plane. The $\hat{S}(z)$ thus obtained can subsequently be used to calculate the quantity $z + \hat{\zeta}(z)$. Let us recall that in the calculation of the reactive frequency, λ_R, the solvation dynamics is essentially probed via $z + \hat{\zeta}(z)$. Variation of the latter with frequency is shown in Figure 18. Note that there is a marked decrease in $z + \hat{\zeta}(z)$ near frequencies equal to the reactant well frequency $\omega_0 = 4.84$. The latter is calculated from the second derivative of the calculated reaction time correlation function at time $t = 0$. At higher frequencies, $\hat{\zeta}(z)$ is zero for all practical purposes and $z + \hat{\zeta}(z)$ varies as z. The latter accounts for the saturation of the rate of barrier crossing to the transition-state theory rate for high barrier frequencies. In Table V the rates calculated have been

TABLE IV
Dielectric Relaxation Data for Water

A. Low-frequency Debye dispersion[a] [115]

ϵ_0	τ_D (ps)	ϵ_∞
79.2	9.33	4.86

B. High-frequency vibrational and librational relaxation[b] [100,116]

n_1^2	Ω_1 (cm^{-1})	n_2^2	Ω_2 (cm^{-1})	n_3^2
4.86	199	2.1	650	1.77

[a] ϵ_0 is the static dielectric constant and ϵ_∞ is the infinite-frequency dielectric constant, obtained by fitting the low-frequency relaxation to a Debye form with relaxation time τ_D.

[b] $n_1^2 = \epsilon_\infty$; $n_3^2 =$ optical dielectric constant. The relative amplitudes of the two high-frequency modes have been obtained experimental data given in Figure 11 of [116].

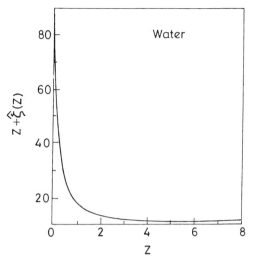

Figure 18. Frequency-dependent combination, $z + \zeta(z)$, plotted against the Laplace frequency z for an outer-sphere electron transfer reaction in water. The function in the ordinate plays an important role in determining the rate of barrier-crossing in Grote–Hynes theory. Here the frequency is scaled by $10\,ps^{-1}$. In the calculation of the ordinate, it is assumed that the solute–solvent size ratio is 3. The details of the calculation are plotted in Section VIII.

compared with the situation where the ultrafast solvent modes are absent and the dielectric relaxation has only a slow Debye dispersion, with the Debye parameters remaining the same as in Table IV. The solvent high-frequency modes are found to enhance the rate of the barrier crossing dramatically.

For calculation of the rate of energy diffusion, we again use the same representation of the frequency-dependent dielectric function, now in the Laplace–Fourier frequency plane with $z = -i\omega$ in Eq. 8.8. Equation 8.21

TABLE V
Calculated Rate of Barrier crossing in Water[a]

	ω_b/ω_0	λ_R	κ	ν_b
High-frequency	1.0	1.5337	0.3169	0.2241
modes of solvent	5.0	22.9895	0.9500	0.7318
present	6.0	28.9319	0.9875	0.9963
No high-frequency	1.0	0.1510	0.0243	0.024
solvent	5.0	3.4063	0.1098	0.1084
modes	9.0	12.6009	0.2030	0.2005

[a] The frequencies are scaled by $\tau_l = 0.1$ ps.

TABLE VI
Calculated Rate of Energy Diffusion and Rate of Electron Transfer in Water[a]

ω_0	$\tilde{\zeta}(\omega_0)$	βE_a	ν_{ED}	ν_b^{TST}	ν_{ET}
4.84	4.1897	12.9	49.54	0.7703	0.7471

[a]The frequencies are scaled by $\tau_i = 0.1$ ps.

provides an activation energy of 12.9, in the units of $k_B T$, for an ion/solvent size ratio of 3. This has been used to calculate the rate of energy diffusion, and the results are shown in Table VI. It is found that *despite the presence of the high-frequency solvent modes, energy diffusion is still efficient in water, and the rate of electron transfer reaction is determined primarily by the barrier-crossing rate.*

2. Calculation of the Rate in Acetonitrile

The rate of an outer-sphere electron transfer reaction in acetonitrile is calculated using the parameters presented in Table VII. The other quantities required to evaluate the RCTCF are obtained as follows.

1. $\epsilon_L(k)$ is obtained from the mean spherical approximation (MSA), correcting the $k \to 0$ and $k \to \infty$ using the results of Raineri et al. [99].
2. The rotational kernel, $\Gamma_R(z)$, is obtained using the Kerr relaxation data of acetonitrile given by McMorrow and Lotshaw [102].
3. The translational friction is approximated by its $z = 0$ limiting value, calculated using the Einstein relation.
4. The solute was chosen to be a sphere of radius three times that of the solvent, also assumed to be spherical.

The results of our calculation in acetonitrile are presented in Table VIII. Here frequencies are scaled in the units of τ_I. Note that in this solvent also,

TABLE VII
Parameters Used for Calculation of Solvation Dynamics and Electron Transfer Reaction in Acetonitrile[a]

μ (D)	ρ_0 (Å^{-3})	$3Y$	ϵ_0	σ (Å)	τ_I (ps)	τ_0 (ps)	ω_0 (ps^{-1})
3.5	0.0115	18.2	36.2	4.48	0.45	1.82	13.59

[a]$3Y = (4\pi/3)\beta\mu^2\rho_0$, where μ is the dipole moment; ρ_0 is the average number density; $\beta = (k_B T)^{-1}$, the Boltzmann constant times the absolute temperature; ϵ_0 is the static dielectric constant; $\tau_I = (I/k_B T)^{1/2}$ and $\tau = (m\sigma^2/k_B T)^{1/2}$ for solvent molecules having mass m, average moment of inertia I, and diameter σ; and ω_0 is the harmonic reactant well frequency.

TABLE VIII
Calculated Rate of Barrier Crossing in Acetonitrile[a]

ω_b/ω_0	λ_R	κ	ν_b
0.5	1.1193	0.3658	0.3563
1.0	4.4834	0.7326	0.7136
1.5	8.8985	0.9694	0.9441
1.8	11.0149	0.9998	0.9739

[a]The frequencies are scaled by $\tau_l = 0.45$ ps.

TABLE IX
Calculated Rate of Energy Diffusion and Rate of Electron Transfer in Acetonitrile[a]

ω_0	$\tilde{\zeta}(\omega_0)$	βE_a	ν_{ED}	ν_b^{TST}	ν_{ET}
6.1195	0.01448	9.88	0.1295	0.5618	0.0581

[a] The frequencies are scaled by $\tau_l = 0.45$ ps.

the transition-state result is attained rather quickly. Here $z + \hat{\zeta}(z)$ is significantly different from that of water (which is given in Figure 18). Similarly $\tilde{\zeta}(\omega)$ has a frequency dependence also dramatically different from that of water. Here also, $\tilde{\zeta}(\omega)$ is biphasic due to the biphasic nature of both $S'(\omega)$ and $S''(\omega)$. But the rapid oscillations present in water due to the high-frequency collective motion of the hydrogen-bonded network is absent in this nonassociated solvent.

The most important outcome of the present calculation (presented in Table IX) is that *the energy diffusion may become the rate-determining step in acetonitrile*. It is also clear that the barrier-crossing rate will contribute significantly (about 20%) to the total rate of electron transfer reaction. Note that here also $\tilde{\zeta}(\omega_0) = 0.0148$ is markedly less than $\zeta_0 = 11.4252$.

3. Calculation of the Rate in Methanol

The parameters used to characterize methanol are summarized in Table X. The molecular diameter, σ, of methanol is known [103] to be equal to 4.1 Å, and this gives $\tau_0 = 1.47$ ps at $T = 298$ K. The reactant well frequency, ω_0, is obtained from the short-term relaxation of the solvation time correlation function, $S(t)$ [17], by noting that in the inertial regime, the relaxation of $S(t)$ is given by $S(t) = \exp(-\frac{1}{2}\omega_0^2 t^2)$. We have used the following steps to evaluate the RCTCF, $S(t)$, in methanol.

1. $\epsilon_L(k)$ is obtained using the XRISM results of Raineri et al. [105].
2. The rotational dissipative kernel has been calculated starting from the complex multimodal dielectric relaxation data of methanol, as shown in Table X.

TABLE X

Parameters Used for Calculation of Solvation Dynamics and Electron Transfer Reaction in Methanol[a]

μ (D)	ρ_0 ($\mathring{A}^{-3}$)	$3Y$	ϵ_0	σ ($\mathring{A}$)	τ_I (ps)	τ_0 (ps)	ω_0 (ps^{-1})
1.7	0.015	4.33	32.5	4.1	0.057	1.47	50.41

[a] $3Y = (4\pi/3)\beta\mu^2\rho_0$, where μ is the dipole moment; ρ_0 is the average number density; $\beta = (k_BT)^{-1}$, the Boltzmann constant times the absolute temperature; ϵ_0 is the static dielectric constant; $\tau_I = (I/k_BT)^{1/2}$ and $\tau_0 = (m\sigma^2/k_BT)^{1/2}$ for solvent molecules having mass m, average moment of inertia I, and diameter σ; and ω_0 is the harmonic reactant well frequency.

3. The translational dissipative kernel is replaced by its single-particle limit and is subsequently estimated using the Stokes–Einstein relation.

4. The solute/solvent size ratio is chosen to be equal to 3.0.

The results of the calculations are presented in Tables XI and XII showing the barrier-crossing frequency, ν_b, the frequency of energy diffusion, ν_{ED}, and the frequency of electron transfer reaction, ν_{ET}. Here, as before, $\nu_j = k_j \exp(\beta E_a)$. Note that the results obtained for methanol are qualita-

.

TABLE XI

Calculated Rates of Barrier Crossing[a]

	ω_b/ω_0	λ_R	κ	ν_b
High-frequency	1.0	2.59	0.05	0.41
solvent modes	2.0	19.89	0.26	2.11
present	3.0	130.73	0.86	6.94
	4.0	196.74	0.98	7.83
No high-frequency	1.0	0.35	0.007	0.05
solvent modes	2.0	2.48	0.02	0.19
	3.0	6.33	0.04	0.34
	4.0	11.75	0.06	0.47

TABLE XII

Calculated Rates of Energy Diffusion and Rates of Electron Transfer[a]

ω_0	ν_b^{TST}	βE_a	γ_i/Ω_i ($i = 1, 2, 3$)	$\tilde{\zeta}(\omega_0)$	ν_{ED}	ν_{ET}
50.41	8.02	10.40	0.5, 0.1, 0.1	27.1519	251.30	7.54
			0.1, 0.01, 0.01	3.2528	30.11	5.32
			0.01, 0.001, 0.001	0.6846	6.34	2.27

[a] The frequencies and the rates are in ps^{-1}. $\nu_{ED} = k_{ED}\exp(\beta E_a) =$ frequency of energy diffusion, and $\nu_{ET} = k_{ET}\exp(\beta E_a) =$ net frequency of electron transfer reaction.

tively similar to those obtained previously for water. In particular, in both cases, the effects of the high-frequency librational modes are indeed remarkable. This has been demonstrated in Table XII, where we have presented the barrier frequencies calculated in the absence of high-frequency librational modes and the dielectric relaxation of the solvent, comprised of three Debye dispersions. The other parameters are kept the same as in methanol. Clearly, as in water, presence of the high-frequency solvent inertial modes has two distinct effects. First, *the barrier-crossing frequency increases almost by an order of magnitude*. Second, the transition-state rate sets in much faster in the limit of weakly adiabatic reactions. Another important prediction for adiabatic reactions in methanol is shown in Table XII. First, the presence of underdamped solvent inertial modes results in a much smaller value of $\tilde{\zeta}(\omega_0)$ than that of the zero-frequency friction. The latter, when calculated from the zero-frequency limit of Eq. (8.8) [or equivalently, from Eq. (8.9)] is found to be equal to $6.18 \times 10^3 \, \text{ps}^{-1}$. Also, the rate of the electron transfer reaction, unlike the rate of solvation energy relaxation, is found to be strongly dependent on the damping constant of these librational modes, the latter being treated as a parameter here. In addition, with increased underdamping of the high-frequency modes, the rate of reaction changes from being controlled by barrier-crossing dynamics to being determined by the rate of energy diffusion in the reactant well.

The frequencies presented in Tables V, VI, IX and XII also reveal the following interesting result. *Despite the rather different values of τ_L for water and methanol, the net frequency of the electron transfer is found to be almost the same in these two liquids*. As shown in the tables, $\nu_{ET} = 7.54 \, \text{ps}^{-1}$ in methanol is about $7.47 \, \text{ps}^{-1}$ in water. This shows that the dynamics of electron transfer in the ultrafast solvents are controlled primarily by the solvent inertial modes. We have already shown that in the presence of the latter, the barrier-crossing frequency is essentially given by the transition-state theory in the limit of weakly adiabatic reactions. This may explain the apparent lack of dependence of the rate on the solvent longitudinal relaxation time, which is a measure of the macroscopic polarization relaxation only of the dipolar liquid.

With the foregoing results at hand, let us now try to understand why the solvent inertial modes exhibit rather different effects on the rate of electron transfer reaction in water, acetonitrile, and methanol. A particularly useful way to look into this problem would be to study the frequency dependence of the friction, $\tilde{\zeta}(\omega)$, acting along the reaction coordinate. This is shown in Figure 19. Here the solute/solvent size ratio in each case is assumed to be 3.0. The frequencies are scaled by the inertial time constant, τ_I, in all three solvents. τ_I of methanol has been calculated from ω_0 using the relation $\omega_0^2 \tau_I^2 = 2f(110; k = 0)$ and is found to be $0.057 \, \text{ps}$. Also, we have scaled

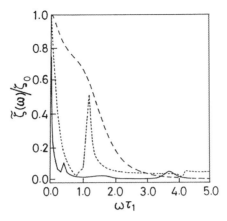

Figure 19. Real part of the frequency-dependent friction, $\zeta(\omega)$, acting along the electron transfer reaction coordinate (X), plotted against the Laplace–Fourier frequency, ω, for an outersphere electron transfer reaction for the three solvents considered here: solid line, methanol; short-dashed line, water; long-dashed line, acetonitrile. The ordinate is scaled by the zero-frequency friction (ζ_0) for each solvent. The value of ζ_0 is quite different for different solvents. For methanol, $\zeta_0 = 6.2 \times 10^{15}\,\text{s}^{-1}$, for water $\zeta_0 = 1.05 \times 10^{15}\,\text{s}^{-1}$, and for acetonitrile the zero-frequency friction has a much smaller value of $2.54 \times 10^{13}\,\text{s}^{-1}$. The abscissa in each graph is multiplied by the rotational inertial time in each solvent, equal to 0.06, 0.1, and 0.45 ps for methanol, water, and acetonitrile, respectively. The solute/solvent size ratio is assumed to equal to 3.0 in all the solvents.

the frictions by the respective zero-frequency values, which can be vastly different: in fact, $\hat{\zeta} = 6.18 \times 10^{15}\,\text{s}^{-1}$ for methanol, $1.05 \times 10^{15}\,\text{s}^{-1}$ for water, and $2.54 \times 10^{13}\,\text{s}^{-1}$ for acetonitrile. Note that $\tilde{\zeta}(\omega)$ in methanol shows features similar to those in water. This is because the *high-frequency behavior* of the memory kernel is determined primarily by the relative importance of the librational modes in the solvation dynamics. These are important in water and methanol because of the hydrogen bonding present in these solvents but not in acetonitrile. It can be seen from Figure 1 that the residual friction at ω_0 is smaller in methanol than in water. This is because of the lower weight carried by the Gaussian component in the solvation energy relaxation in methanol. Therefore, unlike the case of water, the rate of the reaction in methanol can be marginally into Kramers' turnover region [95] if the high-frequency solvent modes are significantly underdamped. In acetonitrile, on the other hand, the librational modes are not important. The energy diffusion control arises primarily from the following two factors. First, if we compare the dynamics of water and acetonitrile, it is found that solvation proceeds at comparable rates in both of these solvents. But being the less polar of the two, the friction along the reaction coordinate is expected to be less in acetonitrile. Thus the zero-frequency friction, $\zeta_0 = 11.4$ of acetonitrile, is much lower than that of water ($\zeta_0 = 89.2$). Second, the single-particle orientation in acetonitrile is also much faster than that in water because of the hydrogen bonding present in the latter. All these facts combine to make the friction in acetonitrile intrinsically small. The value of $\zeta(\omega_0)$ is then smaller than the critical value required to attain Kramers' turnover region.

D. Conclusions

Let us first summarize the main results of this section. We have presented a detailed investigation of the effect of ultrafast solvation of charged species on the rate of an outer-sphere adiabatic electron transfer reaction. The theoretical studies presented here were based on the simple and elegant formulation developed by Hynes to treat the effects of solvent dynamics on electron transfer reactions. Our investigations have revealed that the ultrafast dynamics of solvation can have several important consequences in electron transfer reaction. First, the rate of barrier-crossing can increase, even by more than an order of magnitude, when the ultrafast solvation modes are included. The resulting *barrier-crossing rate* is nearly equal to the transition-state theory result originally given by Marcus. This is especially true for weakly adiabatic limits for all the systems considered. Second, the energy diffusion necessary to reach the activation barrier is the rate-determining step for liquid acetonitrile. The situation is more complex for water and methanol, where the rate is essentially controlled by the barrier-crossing dynamics. It is also found that the electron transfer rate in water and methanol depends on the relative importance of the high-frequency modes in dielectric relaxation and also on the damping constants of these high-frequency modes. Therefore, in the traditional Kramers turnover figure [97,99], acetonitrile will fall well below the maximum, while water and methanol will be near the maximum, but on the higher-friction side. However, because of the intrinsic non-Markovian nature of the problem, such a figure may be of little use in augmenting our understanding.

Let us next comment on the limitations of the present work. First, it is based on the theory of Hynes, which perhaps provides a little too simplified description of the outer-sphere electron transfer reaction. In real systems, the potential energy surface can hardly be one-dimensional. This can have important consequences for the energy diffusion rate. This is because if the reaction is coupled to any high-frequency (may be intramolecular) modes, the required energy may flow from these modes, and as a result, the role of energy diffusion will be minimized. Therefore, a scenario similar to that in the viscosity dependence of the rate of cyclohexane inversion may arise. Another limitation of Hynes' (and hence, of our) formulation is that determination of the friction at high frequencies is somewhat ill defined. This is for the simple mathematical reason that $\hat{\zeta}(z)$ (a small quantity at large z) is extracted from a large quantity, as what comes out from Eq. (8.8) is $z + \hat{\zeta}(z)$. The situation is further complicated because $\hat{\Delta}_R(z)$ itself is very small at large z. Therefore, even a small error in $\hat{\Delta}_R(z)$ can give rise to a very large error in $\hat{\zeta}(z)$. The discussion above leads us to conclude that it is more appropriate to obtain $\hat{\zeta}(z)$ at large z from some other theoretical route. For instance, we believe that the approach of Cho and coworkers [95], which addresses the behavior of instantaneous normal modes in liquids, can provide important insight into the problem.

IX. NONEXPONENTIALITY IN ELECTRON TRANSFER KINETICS

Electron transfer reactions often show nonexponential kinetics [106]. The probable cause for this nonexponentiality has been a subject of considerable discussion in the literature in the recent past. For biological electron transfer reactions, this nonexponentiality is usually attributed to the existence of multiple configurations in the biosystem, each with different activation energy and solvent environment. Understanding of this nonexponentiality is therefore highly nontrivial and requires the elucidation of reaction parameters for the various configurations. We shall not discuss this further. Electron transfer reactions are known to be strongly nonexponential even in a homogeneous phase, such as in liquids and glasses. Actually, the Sumi–Marcus theory was first formulated to explain such observed nonexponentiality in low- or zero-barrier photoelectron transfer reactions. As discussed in Section III, the nonexponentiality in Sumi–Marcus theory comes from widening of the reaction zone due to the participation of a low-frequency intramolecular vibration in the electron transfer. The extent of this intrinsic nonexponetiality depends on the width of the reaction window determined by the ratio of the reorganization energies of the solvent and vibrational modes. There can, however, be another source of nonexponentiality in homogeneous systems. This is the non-Debye dielectric relaxation of the medium. In this section we discuss this issue briefly.

Attention to the role of non-Debye dielectric relaxation in the dynamics of electron transfer was triggered by an interesting article by McGuire and McLendon, who found that electron transfer reaction in glassy glycerol was highly nonexponential. In particular, it was found that at short times, the electron transfer kinetics exhibits an interesting fractional power-law dependence on the dielectric relaxation time of the medium. At long times, the usual linear dependence was observed. These experimental results, in turn, motivated several interesting theoretical studies aimed at understanding the kinetics of electron transfer in a non-Debye medium.

Rips and Jortner [108] analyzed the electron transfer reaction in rigid glycerol by assuming a continuous distribution of relaxation times. They considered a medium characterized by the Davidson–Cole (DC) dielectric spectrum with the complex dielectric susceptibility function [107]

$$\epsilon(\omega) = \epsilon_\infty + \epsilon_0 - \epsilon_\infty/(1 + i\omega\tau_{DC})^B \tag{9.1}$$

where τ_{DC} is the characteristic relaxation time and the exponent β is less than unity. For glycerol, β has a nearly temperature-independent value of about 0.6. The Cole–Davidson expression provides a good fit for a large number of polar solvents, particularly at low temperatures, where collective effects give rise to a marked deviation from the Debye spectrum.

There are now two ways of looking at the origin of the Davidson–Cole expression. First, it arises from cooperativity in the collective orientational relaxation due to the intermolecular interactions. The second approach is to consider the Davidson–Cole expression to arise from a distribution of relaxation times whose probability distribution is, fortunately, known exactly and is given by

$$g_{DC}(\tau_{DC}) = \begin{cases} \dfrac{\sin \pi\beta}{\pi\tau}\left(\dfrac{\tau}{\tau_{DC} - \tau}\right)^\beta & \tau \leq 0 \\ 0 & \tau \geq 0 \end{cases} \tag{9.2}$$

Rips and Jortner evaluated the average rate by averaging the rate for a given T_{DC} over the distribution function of Eq. (9.2). Two important conclusions emerged from their analysis. First, the average electron transfer rate displays a fractional dependence of the rate on τ_{DC}—the exponent being equal to β, the same exponent that determines the distribution of inhomogeneously distributed relaxation times. Second, the dependence of the adiabatic rate on electronic coupling gets modified to $V_{el}^{2(1-\beta)}$. Since $\beta = 0.6$, the dependence of the rate on V_{el} is considerably weaker than in the nonadiabatic limit.

Rips and Jortner did not investigate the time dependence of the reactant population. Such a calculation was carried out by Nadler and Marcus [109]. The decay is nonexponential, particularly at short times, where the rate of Nadler and Marcus agrees with that of Rips and Jortner and predicts the same fractional power-law dependence on τ_{DC}. In contrast, a different behavior holds for the long-time rate, which is proportional to τ_{DC}. The long-time decay is also nonexponential, but markedly less than that at shorter times.

It is interesting to note that one can formulate a dynamical disorder model of the problem above [111], where one need not assume a static inhomogeneous distribution of relaxation times, although the dielectric function can still be given by a highly non-Debye form such as the Davidson–Cole. This model is particularly simple for low-barrier reactions. Here the equation of motion is given by the one-dimensional version of Eq. (4.3), with the time evolution operator given by Eq. (4.4). Here the diffusion constant is again time dependent because of the non-Debye dielectric function. $D(t)$ is related to the solvation time correlation function $[\Delta(t)]$ by Eq. (4.5). In a barrierless photoelectron transfer reaction, the short-time dynamics is controlled by the time dependence of $D(t)$. Since the latter is determined by the solvation time correlation function, we have a simple way to determine the short-time dynamics. Since the liquid is non-Debye, $\Delta(t)$ is also nonexponential and can be fitted to a Kohlrausch–William–Watt (KWW) stretched exponential form

$$\Delta(t) = \exp(-t/\tau_S)^\alpha \qquad (9.3)$$

where the solvation time τ_S and α are determined by many factors. It has been found that in general the exponent α is closely related to the Davidson–Cole exponent β, nearly equal in value in some cases [110]. Thus the results of this dynamical disorder model [111] are in many ways similar to the results of Rips and Jortner and of Nadler and Marcus.

Thus one can identify the following causes of nonexponentiality in electron transfer kinetics:

1. The existence of a reaction system in multiple static configurations. This is common in biological systems.

2. A broad reaction window and a reaction that can take place with vastly different rates from different positions on the reaction coordinate.

3. The Rips–Jortner/Nadler–Marcus scenario, where the reaction occurs in a system characterized by a distribution of relaxation times.

4. A dynamically disordered system.

There may be other mechanisms as well. It is therefore not simple to identify the root of the observed nonexponentiality in a straightforward manner.

X. FUTURE PROBLEMS

In the preceding sections we have discussed several aspects of dynamic solvent effects on outer-sphere electron transfer reactions. The main theme of this review has been the interplay between the recently discovered ultrafast solvation in dipolar liquids and the intramolecular vibrational modes of the reactant–product system. It has been shown that *both* these factors may be important for often-observed apparently contradictory observations reported in the literature. Notable among them is dramatic weakening of the τ_L dependence of the rate if the ultrafast component becomes highly dominant.

The formalism presented here is rather general, as it allows one to consider the non-Markovian dynamics on the one hand and the multidimensional reaction potential energy surface on the other. Therefore, it can be applied to related problems of chemical dynamics. Some of them are listed below.

1. The twisted intramolecular charge transfer (TICT) state formation results from internal twisting of molecular subsystems toward a new geometrical arrangement. For example, in the case of organic compounds such as dimethyl aminobenzonitrile (DMABN), the twist changes the conformation from planar to orthogonal [60]. Rettig et al. have discussed in detail the various theoretical and experimental viewpoints on formation of the TICT state [60]. The large intramolecular charge separation that arises from the differences in electronegativity of the different functional groups of the molecule results in the TICT state becoming highly polar. As the TICT excited state carries a large dipole moment, it is sensitive to the polarity of the solvent environment. By sufficient stabilization from the surrounding solvent molecules, the TICT formation can become barrierless. In this way, the solvent can strongly influence the kinetics by affecting the excited-state surface. As discussed by Miller, the solvent dielectric friction should be considered as another mode that couples to the reactive motion (the twisting of the bulky groups about a body fixed axis) [61]. Thus this is an example of a two-dimensional barrierless reaction where friction is substantially different along the two coordinates, as are the force constants.

2. The dynamics of intramolecular reactions inside protein macromolecules are of multidimensional type. The recombination dynamics of the binding of ligands such as CO or O_2 to iron inside heme proteins is an example [62,63]. In this case the reaction coordinate is the motion of CO or O_2 through a reaction center (protein) referred to as a *gate* as it should be open wide enough to allow passage of the ligand. The reaction center geometry is subjected to slow random fluctuations from the surrounding molecules. This motion, which is perpendicular to the reaction coordinate, was therefore assumed to be viscosity dependent, while the reaction coordinate was assumed to be viscosity

independent. Agmon and Hopfield treated it as a one-dimensional problem with a position-dependent sink term [62]. But as the frictional dependence of the two degrees of freedom is different, it can give rise to complex dynamics, and a full two-dimensional treatment is necessary.

3. In many conventional barrierless reactions such as isomerization of the excited-state *cis*-stilbene, a two-dimensional approach may help to explain the experimentally observed deviations from Kramers' predictions [63–66]. Unlike stiff stilbene, where the sole reaction coordinate is rotational motion about the ethylene double bond, *cis*-stilbene has in addition to ethylenic torsion, rotational motions of the carbon–phenyl bonds, which can complicate the dynamics substantially. As it involves rotational motion of the bulky group along the reaction coordinate and less phenyl group motion along a perpendicular coordinate, a viscosity-dependent reaction coordinate and a viscosity-independent perpendicular coordinate may therefore be assumed.

On the theoretical formulation part, several major problems remain to be addressed. Some of them are listed below.

1. The inertial oscillations along both the solvent and vibrational coordinates will require the use of generalized Fokker–Planck operators in place of the Smoluchowski operators that have been used in this study. This may lead to a large increase in the computational effort but will certainly be a worthwhile exercise. The inertial modes can significantly enhance the rate because the system may be able to access more efficient sinks that lie on the inverted regime.

2. We have totally ignored the relaxation of the high-frequency vibrational modes which are bound to be populated at the time of preparation of the reactant. It is assumed here that the population relaxation of these modes is extremely fast. But if one is probing the dynamics with 20–50 fs time resolution (as would be done in the near future), one must also consider the relaxation of energy in the high-frequency modes. In some cases it can be rather slow [112,113].

3. In the study of the effects of ultrafast solvation on adiabatic electron transfer reactions, a better treatment of the rate of energy diffusion is required. Here several improvements are possible, including the prescription of Pollak, Grabert, and Hängii [114] to obtain the rate. We have faced serious problems in the implementation of this scheme, particularly at high frequencies.

4. Finally, one should also look into the effects of ultrafast solvation on low, barrier electron transfer reactions. This is an important problem that we hope to address in the future.

XI.   CONCLUSIONS

The dynamics of electron transfer reactions in photoexcited molecules are often different from the ground-state reactions. These reactions are often much faster, occurring in the subpicosecond time domain. The reason for this rapidity of reaction is that these reactions essentially occur without any significant activation barrier. With some reactions, such as oxazine 1 in dimethyl aniline, the lack of activation can be understood directly from the Marcus free energy construction. In other cases it might be the participation of the high-frequency vibrational modes that renders the reaction barrierless. Whereas the former scenario is straightforward, the latter is certainly not and is still a matter of debate and research. However, it is certainly clear that the intramolecular vibrational modes are critically important in the barrierless scenario. This is simply because in the absence of a significant solvation activation energy, the (Marcus) solvent reaction coordinate is no longer the only important coordinate. Low-frequency vibrations of the solute, which can effectively couple to the electron transfer, now have an important role to play in determining the dynamics of the reaction. This was the lesson we learned from the Sumi–Marcus theory. However, this is not enough to explain the reaction in systems such as betaine-30, which occurs with great rapidity, although deep in the Marcus inverted regime. Here the role of the high-frequency modes is the rate-determining factor. All these are relatively recent developments in the field of electron transfer.

The discovery that the solvation dynamics in several common dipolar liquids (water, acetonitrile, and methanol) can progress at an exceedingly fast rate makes the scenario above even more interesting, in addition to making the old theories of solvent dynamic effects essentially obsolete. Now, relaxation along the reaction coordinate has a biphasic character with vastly different time constants. Clearly, in such cases the ultrafast component will dominate if the reaction occurs under nonequilibrium conditions. On the other hand, if full relaxation of the reaction coordinate is required, the slow component may form a bottleneck. In the former case we may not see any solvent dynamic effects, whereas in the latter case, a strong solvent relaxation dependence is, of course, predicted.

In this review we have discussed some of the issues that arise because of the interplay between ultrafast solvation and the vibrational modes of the reactant. The theoretical formalism used in this study is rather general. It is particularly efficient to find the average rate of electron transfer from a wide reaction window. The generalized hybrid model, which includes both the solvent and the vibrational dynamics, seems capable of explaining many of the results observed, although in alcohols it falls short of explaining the rates quantitatively. There is still much work left to be done on these problems. But the studies discussed here certainly suggest that the ultrafast component can play a significant role in electron transfer kinetics, especially in understanding the lack of solvent dependence that has been observed experimentally.

We have discussed a small list of problems that need to be addressed in the future. They are certainly subjective and by no means exhaustive. Needless to say, that the field of electron transfer reaction in solution will remain an active area of research for many years to come and we can certainly look forward to much better understanding in the coming years.

ACKNOWLEDGMENTS

We thank Srabani Roy for collaboration and discussions. We thank P. F. Barbara, J. Jortner, V. Jarzeba, K. Yoshihara, K. Tominaga, H. Heitele, A. Nitzan, K. L. Sebastian, M. Tachiya, S. Ruhman, S. Ghosh, S. Umapathy, and V. Krishnan for help and discussions during the course of this work. The work was supported in part by the Department of Science and Technology, Council of Scientific and Industrial Research, and the Indo-French Center for Promotion of Advanced Research.

REFERENCES

1. R. A. Marcus, *J. Chem. Phys.* **24**, 966, 979 (1956); *Annu. Rev. Phys. Chem.* **15**, 155 (1964).

2. H. Heitele, *Angew. Chem. Int. Ed. Engl.* **32**, 359 (1993).

3. P. F. Barbara and W. Jarzeba, *Adv. Photochem.* **15**, 1 (1990).

4. G. R. Fleming and P. G. Wolynes, *Phys. Today*, **43**, 36 (1990).

5. L .D. Zusman, *Chem. Phys.* **49**, 295 (1980); **80**, 29 (1983); **119**, 51 (1988); **144**, 1 (1990).

6. D. F. Calef and P. G. Wolynes, *J. Chem. Phys.* **78**, 4145 (1983).

7. E. M. Kosower and D. Huppert, *Annu. Rev. Phys. Chem.* **37**, 127 (1986).

8. D. Huppert, V. Ittah, and E. M. Kosower, *Chem. Phys. Lett.* **144**, 15, (1988).

9. D. Huppert, V. Ittah, A. Masad, and E. M. Kosower, *Chem. Phys. Lett.*, **150**, 349 (1988).

10. M. J. Weaver and G. E. McManis III, *Acc. Chem. Res.* **23**, 294 (1990).

11. T. J. Kang, G. C. Walker, P. F. Barbara, and T. Fonseca, *Chem. Phys.* **149**, 81 (1990).

12. M. A. Kahlow, W. Jarceba, T. J. Kang, and P. F. Barbara, *J. Chem. Phys.* **90**, 151 (1989).

13. J. Najbar and W. Jarzeba, *Chem. Phys. Lett.* **196**, 504 (1992).

14. H. Sumi and R. A. Marcus, *J. Chem. Phys.* **84**, 4272 (1986).

15. W. Nadler and R. A. Marcus, *J. Chem. Phys.* **86**, 3096 (1987).

16. W. Nadler and R. A. Marcus, *Chem. Phys. Lett.* **144**, 24 (1988).

17. I. Rips and J. Jortner, *J. Chem. Phys.* **87**, 2090, 6513 (1987); **88**, 818 (1988).

18. J. T. Hynes, *J. Phys. Chem.* **90**, 3701 (1986).

19. S. Efrima and M. Bixon, *J. Chem. Phys.* **70**, 3531 (1979).

20. M. Bixon and J. Jortner, *Chem. Phys.* **176**, 467 (1993).

21. M. Bixon, J. Jortner, and J. W. Verhoeven, *J. Am. Chem. Soc.* **116**, 7349 (1994).

22. J. Jortner and M. Bixon, *Ber. Bunsenges. Phys. Chem.* **99**, 296 (1995); J. Jortner, M. Bixon, B. Wegewijs, J. W. Verhoven, and R. P. H. Rettschnick, *Chem. Phys. Lett.* **205**, 451 (1995); J. Jortner, M. Bixon, H. Heitele, and M. E. Michel-Beyerle, *Chem. Phys. Lett.* **197**, 131 (1992); J. Jortner and M. Bixon, *J. Photochem. Photobiol. A* **82**, 5 (1994).

23. T. Fonseca, *J. Chem. Phys.* **91**, 2869 (1989).

24. J. Najbar and W. Jarzeba, *Chem. Phys. Lett.* **196**, 504 (1992).

25. R. Zwanzig, *Proc. Natl. Acad. Sci. USA* **87**, 5856 (1990).

26. B. Bagchi and A. Chandra, *Adv. Chem. Phys.* **80**, 1 (1991).

27. R. Zwanzig, *Acc. Chem. Res.* **23**, 148 (1990).

28. K. S. Singwi and A. Sjolander, *Phys. Rev.* **119**, 863 (1960).

29. A. Szabo, D. Shoup, S. H. Northrup, and J. A. McCammon, *J. Chem. Phys.* **77**, 4484 (1984).

30. A. K. Harrison and R. Zwanzig, *Phys. Rev. A* **32**, 1072 (1985).

31. A. Szabo, *J. Chem. Phys.* **81**, 150 (1984).

32. B. Bagchi, G. R. Fleming, and D. Oxtoby, *J. Chem. Phys.* **78**, 7375 (1983).

33. W. Magnus, F. Oberhettinger, and R. P. Soni, *Formulas and Theorems for the Special Functions* (Springer-Verlag, New York, 1966); H. Bateman, *Higher Transcendental Functions*, (McGraw Hill. New York, 1953).

34. M. Tachiya, *J. Phys. Chem.* **97**, 5911 (1993).

35. B. Bagchi and G. R. Fleming, *J. Phys. Chem.* **94**, 9 (1990).

36. G. Oster and N. Nishijima, *J. Am. Chem. Soc.* **78**, 1581 (1956).

37. M. Tachiya, *Chem. Phys. Lett.* **159**, 505 (1989); *J. Phys. Chem.* **93**, 7050 (1989).

38. G. E. McManis and M. J. Weaver, *J. Chem. Phys.* **91**, 1720 (1989); G. E. McManis, A. Gochev, and M. J. Weaver, *Chem. Phys.* **152**, 107 (1991); M. J. Weaver, *Chem. Rev.* **92**, 463 (1993).

39. B. Bagchi, A. Chandra, and G. R. Fleming, *J. Phys. Chem.* **94**, 5197 (1990).

40. S. Roy and B. Bagchi, *J. Phys. Chem.* **98**, 9207 (1994).

41. S. Roy and B. Bagchi, *J. Chem. Phys.* **100**, 8802 (1994); **102**, 7937 (1995); **102**, 6719 (1995).

42. Y. I. Dakhnovoskii, R. Doolen, and J. D. Simon, *J. Chem. Phys.* **101**, 6640 (1994).

43. G. C. Walker, E. Akesson, A. E. Johnson, N. E. Levinger, and P. F. Barbara, *J. Phys. Chem.* **96**, 3728 (1992).

44. E. Akesson, A. E. Johnson, N. E. Levinger, G. C. Walker, T. P. DuBruil, and P. F. Barbara, *J. Chem. Phys.* **96**, 7859 (1992).

45. K. Tominaga, G. C. Walker, T. J. Kang, P. F. Barbara, and T. Fonseca, *J. Phys. Chem.* **95**, 10485 (1991).

46. K. Tominaga, G. C. Walker, W. Jarzeba, and P. F. Barbara, *J. Phys. Chem.* **95**, 10475 (1991).

47. P. J. Reid, S. Alex, W. Jarzeba, R. E. Schlief, A. E. Johnson, and P. F. Barbara, *Chem. Phys. Lett.* **229**, 93 (1994).

48. F. Pöllinger, H. Heitele, M. E. Michel-Beyerle, C. Anders, M. Futscher, and H. A. Staab, *Chem. Phys. Lett.* **198**, 645 (1992).

49. T. Kobayashi, Y. Takagi, H. Kandori, K. Kemnitz, and K. Yoshihara, *Chem. Phys. Lett.* **180**, 416 (1991).

50. H. Kandori, K. Kemnitz, and K. Yoshihara, *J. Phys. Chem.* **96**, 8042 (1992).

51. A. P. Yartsev, Y. Nagasawa, and K. Yoshihara, *Chem. Phys. Lett.* **102**, 546 (1993).

52. Y. Nagasawa,. A. P. Yartsev, K. Tominaga, A. E. Johnson, and K. Yoshihara, *J. Am. Chem. Soc.* **115**, 7922 (1993).

53. Y. Nagasawa, A. P. Yartsev, K. Tominaga, A. E. Johnson, and K. Yoshihara, *J. Chem. Phys.* **101**, 5717 (1994).

54. Y. Nagasawa, Ph.D. thesis, Graduate University for Advanced Studies, Japan, 1993, and references therein.

55. D. Rehm and A. Weller, *Isr. J. Chem.* **8**, 259 (1970).

56. M. Tachiya and S. Murata, *J. Phys. Chem.* **96**, 8441 (1992).

57. F. Markel, N. S. Ferris, I. R. Gould, and A. B. Myers, *J. Am. Chem. Soc.* **114**, 6208 (1992).

58. H. A. Kramers, *Physica,* **7**, 284 (1940).

59. C. S. Poornimadevi and B. Bagchi, *Chem. Phys. Lett.* **149**, 411 (1988); K. L. Sebastian, *Phys. Rev. A* **46**, 1732 (1992).

60. E. Lippert, W. Rettig, Bonacic-Kouterky, *Adv. Phys. Chem.* **68**, 1 (1987).

61. W. H. Miller, *J. Chem. Phys.* **61**, 1923 (1974).

62. N. Agmon and J. Hopfield, *J. Chem. Phys.* **78**, 6947 (1983).

63. N. Agmon and R. Kosloff, *J. Phys. Chem.* **91**, 1988 (1987); N. Agmon and S. Rabinovich, *J. Phys. Chem.* **97**, 7270 (1992); S. Rabinovich and N. Agmon, *Phys. Rev. E* **47** 3717 (1993).

64. D. C. Todd et al., *J. Chem. Phys.* **93**, 8658 (1990).

65. R. M. Hochstrasser, *Chem. Phys. Lett.* **118**, 1 (1985).

66. D. F. Calef and J. M. Deutch, *Annu. Rev. Phys. Chem.* **34**, 493 (1983).

67. H. Stehfest, *Commun. ACM* **13**, 624 (1970).

68. J. R. Miller, L. T. Calcaterra, and G. L. Closs, *J. Am. Chem. Soc.* **106**, 3047 (1984).

69. M. Tachiya and S. Murata, *J. Am. Chem. Soc.* **116**, 2424 (1994).

70. T. Asahi and N. Mataga, *J. Phys. Chem.* **95**, 1961 (1991).

71. J. Zhu and J. C. Rasaiah, *J. Chem. Phys.* **95**, 3325 (1991); **96**, 1435 (1992); J. C. Rasaiah and J. Zhu, *J. Chem. Phys.* **98**, 1213 (1993).

72. A. E. Johnson, N. E. Levinger, W. Jarzeba, R. E. Schleif, D. A. V. Kliner, and P. F. Barbara, *Chem. Phys.* **176**, 555 (1993).

73. J. Jortner and M. Bixon, *J. Chem. Phys.* **88**, 167. (1988)

74. (a) J. A. Syage, in *Fast Elementary Processes in Chemical and Biological Systems*, AIP Press: New York, 1995; (b) J. A. Syage, *Faraday Discuss.* **97**, 401 (1994).

75. R. Doolen and J. D. Simon, *J. Am. Chem. Soc.* **116**, 1155 (1994).

76. N. Gayathri, S. Roy, and B. Bagchi, to be published.

77. N. Gayathri and B. Bagchi, *J. Phys. Chem.* **100**, 3056 (1996).

78. N. Gayathri and B. Bagchi, *Chim. Phys.* **93**, 1652 (1996).

79. K. Schulten, Z. Schulten, and A. Szabo, *Physica A* **100**, 599 (1980).

80. A. Szabo, G. Lamm, and G. H. Weiss, *J. Stat. Phys.* **34**, 225 (1984).

81. A. Samanta and S. K. Ghosh, *Phys. Rev. E* **47**, 4568 (1993).

82. N. Gayathri and B. Bagchi, *J. Mol. Struct. (Theochem)*, **361**, 117 (1996).

83. S. J. Rosenthal, X. Xie, M. Du, and G. R. Fleming, *J. Chem. Phys.* **95**, 4715 (1991).

84. M. Cho, S. J. Rosenthal, N. F. Scherer, L. D. Ziegler, and G. R. Fleming, *J. Chem. Phys.* **96**, 5033 (1992).

85. R. Jimenez, G. R. Fleming, P. V. Kumar, and M. Maroncelli, *Nature* **369**, 471 (1994).

86. S. J. Rosenthal, R. Jimenez, G. R. Fleming, P. V. and M. Maroncelli, *J. Mol. Liq.* **60**, 25 (1994).

87. C. F. Chapman, R. S. Fee, and M. Maroncelli, *J. Phys. Chem.* **94**, 4929; (1990); R. S. Fee and M. Maroncelli, *Chem. Phys.* **183**, 235 (1994).

88. Y. J. Chang and E. W. Castner Jr., *J. Chem. Phys.* **99**, 7289 (1993); S. Palese, L. Schilling, R. J. D. Miller, P. R. Staver, and W. T. Lotshaw, *J. Phys. Chem.* **98**, 6308 (1993).

89. Y. Nagssawa, A. P. Yartsev, K. Tominaga, P. B. Bisht, A. E. Johnson, and K. Yoshihara, *J. Phys. Chem.* **99**, 653 (1995).

90. P. J. Reid and P. F. Barbara, *J. Phys. Chem.* **99**, 17311 (1995).

91. M. L. Horng, J. A. Gardecki, A. Papazyan, and M. Maroncelli, *J. Phys. Chem.* **99**, 17311 (1995).

92. B. B. Smith, A. Staib, and J. T. Hynes *Chem. Phys.* **176**, 521 (1993).

93. R. F. Grote and J. T. Hynes, *J. Chem. Phys.* **77**, 3786 (1982).

94. D. A. Zichi, G. Ciccotti, J. T. Hynes, and M. Ferrario, *J. Phys. Chem.* **93**, 6261 (1989).

95. M. Cho, G. R. Fleming, S. Saito, I. Ohmine, and R. M. Stratt, *J. Chem. Phys.* **100**, 6672 (1994).

96. Y. J. Yan, M. Sparpaglione, and S. Mukamel, *J. Phys. Chem.* **92**, 4842 (1988).

97. G. van der Zwan and J. T. Hynes, *J. Phys. Chem.* **89**, 4181 (1985).

98. B. Carmeli and A. Nitzan, *Phys. Rev. A* **29**, 1481 (1984).

99. F. O. Raineri, H. Resat, and H. L. Friedman, *J. Chem. Phys.* **96**, 3058 (1992); F. O. Raineri, H. Resat, B.-C. Perng, F. Hirata, and H. L. Friedman, *J. Chem. Phys.* **100**, 1477 (1994).

100. J. B. Hasted, S. K. Husain, F. A. M. Frescura, and R. Birch, *Chem. Phys. Lett.* **118**, 622 (1985).

101. P. Colonomos and P. G. Wolynes, *J. Chem. Phys.* **71**, 2644 (1979).

102. D. McMorrow and W. T. Lotshaw, *J. Phys. Chem.* **95**, 10395 (1991).

103. P. Colonomos and P. G. Wolynes, *J. Chem. Phys.* **71**, 2644 (1979).

104. D. Bingemann and N. P. Ernstring, *J. Chem. Phys.* **102**, 2691 (1995).

105. H. L. Friedman, F. O. Raineri, and H. Resat, in *Molecular Liquids, NATO-ASI Series,* J. Teixeira-Diaz, ed., Kluwer, Dordrecht, The Netherlands, 1992; F. O. Raineri, H. Resat, and H. L. Friedman, *J. Chem. Phys.* **96**, 3068 (1992); H. Resat, F. O. Raineri, and H. L. Friedman, *J. Chem. Phys.* **97**, 2618 (1992); **98**, 7277 (1993); F. O. Raineri, H. Resat, B.-C. Perng, F. Hirata, and H. L. Friedman, *J. Chem. Phys.* **100**, 1477 (1994); F. O. Raineri, B.-C. Perng and H. L. Friedman, *Chem. Phys.* **183**, 187 (1994).

106. M. McGuire and G. McLendon, *J. Phys. Chem.* **90**, 2549 (1986).

107. D. W. Davidson and R. H. Cole, *J. Chem. Phys.* **19**, 1484 (1951); **18**, 1417 (1950).

108. I. Rips and J. Jortner, *Chem. Phys. Lett.* **133**, 411 (1987).
109. W. Nadler and R. A. Marcus, *Chem. Phys. Lett.* **144**, 24 (1988).
110. R. Kohlrausch, *Pogg. Ann.* **12**, 393 (1847); G. Williams and D. C. Watts, *Trans. Faraday Soc.* **66**, 80 (1966).
111. B. Bagchi, unpublished work.
112. D. W. Oxtoby, *Adv. Chem. Phys.* **40**, 1 (1979).
113. D. W. Oxtoby, *Annu. Rev. Phys. Chem.,* **32**, 77 (1981).
114. E. Pollak, H. Grabert, and P. Hängi, *J. Chem. Phys.* **91**, 4073 (1989).
115. J. Barthel, K. Bachhuber, R. Buchner, and H. Hetzenauer, *Chem. Phys. Lett.* **165**, 369 (1990).
116. D. Bertolini and A. Tani, *Mol. Phys.* **75**, 1065 (1992).

SOLVENT CONTROL OF ELECTRON TRANSFER REACTIONS

FERNANDO O. RAINERI AND HAROLD L. FRIEDMAN

Department of Chemistry, State University of New York at Stony Brook, Stony Brook, NY 11794-3400

CONTENTS

Electron Transfer: From Isolated Molecules to Biomolecules, Part Two, edited by Joshua Jortner and M. Bixon. Advances in Chemical Physics Series, Volume 107, series editors I. Prigogine and Stuart A. Rice.
ISBN 0-471-25291-3   © 1999 John Wiley & Sons, Inc.

I. INTRODUCTION

There is enormous interest in understanding the basic aspects (structure, energetics, and dynamics) of elementary solvation processes in the liquid phase and, in particular, the dynamical participation of the solvent in the events that follow a sudden redistribution of the charges of a solute molecule. Traditionally, the solvent has been described as a dielectric continuum, especially in relation to its influence on the energetics of charge transfer (CT) reactions [1,2]. Although very useful because of its simplicity of implementation [3], the continuum dielectric formulation has important limitations. An interesting issue that has not received sufficient attention until recently [4–8] is the experimental finding that a solvent with vanishing molecular dipole moment can show unmistakable polarity, as reflected by its influence on the energetics of CT reactions [9–12].

Our goal in this work is to discuss a unified *molecular* theory of equilibrium and nonequilibrium solvation that is ideally suited to the study of the structural, energetic, and dynamical aspects of the solvation process relevant to CT reactions in solution. The theory is based on a simple *renormalized* linear response development that incorporates nonlinear aspects of equilibrium solvation. The renormalization is carried over the solute–solvent potential energy of interaction; the outcome is a statistical mechanical theory based on a *surrogate* (renormalized) Hamiltonian. We refer to this approach as a surrogate Hamiltonian (SH) theory of solvation [4–6]. An important feature of the theory is that both the solute and solvent molecules are represented by interaction site models (ISMs) of the sort used primarily in molecular dynamics and Monte Carlo simulation studies of solution processes, that is, models in which the potential energy of interaction between two molecules is a sum of pairwise additive site–site terms, including, in addition to Lennard-Jones (LJ) or similar short-range interactions, Coulombic interactions between partial charges located at the molecular sites. In this way the SH theory incorporates the symmetries of the molecular charge distributions and the harsh repulsive forces (i.e., the "size" and "shape" of the molecules). The statistical mechanical averages are calculated by integral equation methods of liquid-state theory rather than by simulation. As we shall see, in the context of the energetics of CT reactions and solvatochromic effects, the surrogate Hamiltonian theory of solvation applies equally well to dipolar and nondipolar solvents [4–6], whereas with regard to nonequilibrium solvation, the SH theory has only been applied to cases in which the solvent is dipolar.

II. COMMENTS ON THE CONVENTIONAL THEORY OF CHARGE TRANSFER ENERGETICS

In Marcus theory [1–3] the solvent is represented by a *linear* dielectric continuum without spatial dispersion; the response is *local*: the induced electric polarization vector $\mathbf{P}(\mathbf{x})$ at point $\mathbf{x}$ is proportional to the electric field $\mathbf{E}(\mathbf{x})$ at that point. This theory leads to parabolic diabatic free energy profiles $F^P(\eta)$ and $F^S(\eta)$ with the same curvature. Here and in the following, P and S refer, respectively, to the initial or *precursor* and the final or *successor* electronic states of a generic intra- or intermolecular donor–acceptor *solute* complex. For electron transfer reactions (and related processes) the P and S states of the solute are distinguished by their different charge distributions. The profiles as functions of the reaction coordinate η are completely characterized by two quantities: the reaction free energy $\Delta\mathscr{A}$ (also referred to as

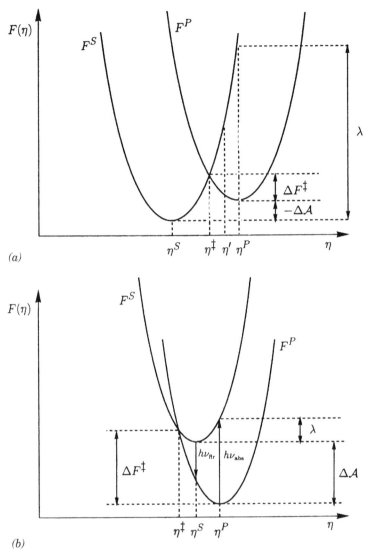

Figure 1. Free energy profiles for a CT process in solution, corresponding, respectively, to the "normal" (a) and "inverted" (b) regime. In (a), the $P \to S$ process could be thermal or optical, while a reverse ($S \to P$) optical process is also possible. In (b), vertical absorption ($P \to S$) and emission ($S \to P$) are illustrated, corresponding, respectively, to transition energies $h\nu_{abs}$ and $h\nu_{flr}$. The constrained free energy profiles $F^D(\eta)$ [$F^P(\eta)$ or $F^S(\eta)$] are functions of the reaction coordinate η. The figure serves to define the parameters (λ, $\Delta\mathscr{A}$) that control the profiles in the linear regime (i.e., their curvature and relative horizontal and vertical displacements). At an arbitrary value of η [e.g., η' in (a)], the vertical energy gap is $F^S(\eta') - F^P(\eta') = \Delta E + \eta'$ [see Eq. (4.5)].

the reaction driving force) and the reorganization (free) energy λ. These parameters determine the other key quantities that control thermal and optical electron transfer (see Fig. 1 for notation), discussed later in the chapter.

When internal (or inner-shell) reorganization can be neglected, λ denotes the *solvent* reorganization energy. Except for the presentation in Section VIII.H, only this case is considered in this chapter. In the simple two-sphere implementation of Marcus theory [1–3] the energy parameters relevant to the $P \rightarrow S$ transition are estimated on the basis of a model in which the electron donor (d) and acceptor (a) moieties occupy nonoverlapping spherical cavities in a dielectric continuum, yielding the following expression for the solvent reorganization energy [1–3]:

$$\lambda_{\text{cont}} = \frac{(\Delta Q)^2 \mathscr{C}}{2} \left(\frac{1}{b_d} + \frac{1}{b_a} - \frac{2}{L} \right) \qquad (2.1)$$

Here ΔQ is the charge transferred in the $P \rightarrow S$ process, while

$$\mathscr{C} = 1/\epsilon_\infty - 1/\epsilon_0 \qquad (2.2)$$

is the Pekar factor [13], which is determined by the $\omega = 0$ (ϵ_0) and $\omega = \infty$ (ϵ_∞) values of the solvent dielectric function. Furthermore, b_d and b_a are the radii of the two spheres, and L is the separation of their centers, generally restricted to $L \geq b_d + b_a$. An essential feature of Eq. (2.1) is the simple relationship between the solvent reorganization energy and the dielectric parameters of the solvent. Another characteristic feature is the simple Coulombic L^{-1} dependence of λ on the donor–acceptor separation L. The functional dependence of the solvent reorganization energy on the Pekar factor $\mathscr{C}$ and on the distance L has often been confirmed [14].

The estimation of solvent effects by a *local* continuum dielectric theory is not sufficient in general. In particular, the continuum dielectric theory completely fails to capture the dielectric response of a nondipolar solvent to a change in the charge distribution of a solute [4–6,12]. By *non-dipolar* we refer to solvents comprising molecules with permanent dipole moment of small or zero magnitude but with finite higher-order multipoles [15]. These contribute significantly to the intermolecular forces at small distances. Examples of such solvents are benzene, toluene, and 1,4-dioxane. Charge transfer experiments in some nondipolar solvents indicate that λ for such solvents is much larger than the values that one would obtain by using the local dielectric theory [Eqs. (2.1) and (2.2)] [9–12].

Another limitation of the continuum dielectric formulation is that it ignores the local density variation of the real solvent around the solute, compared with the uniform density of the solvent in the model leading to Eq. (2.1).

III. THEORY OF THE LONGITUDINAL DIELECTRIC RESPONSE IN MOLECULAR LIQUIDS

As a preamble to the molecular theory of charge transfer energetics and nonequilibrium solvation, we first discuss some aspects of a general formulation of the dielectric properties of molecular fluids. In this approach dipolar and nondipolar solvents are treated on the same basis. For a more complete discussion we refer the reader to a previous report [16].

A. Dielectric Response in Dipolar and Nondipolar Liquids

Before considering the dielectric formulation per se, it is instructive to analyze in some detail why the formulation of the dielectric response of nondipolar solvents (and more generally, of any molecular solvent) to highly nonuniform fields (such as the electric field generated by a solute molecule) requires the consideration of spatial dispersion. For the sake of simplicity, we shall ignore for a while the electronic polarizability of the solvent molecules.

The dielectric fluctuation formula for the static wavevector-dependent longitudinal dielectric function $\epsilon_L(k)$ that one commonly finds in the literature for a nonpolarizable solvent (i.e., with $\epsilon_\infty = 1$) is [17–19]

$$1 - \frac{1}{\epsilon_L(k)} = 4\pi\beta\left\{\frac{1}{V}\langle \hat{M}_L(\mathbf{k})\hat{M}_L(-\mathbf{k})\rangle\right\}_\infty \tag{3.1a}$$

where

$$\hat{M}_L(\mathbf{k}) = \mathbf{e}_k \cdot \sum_e \boldsymbol{\mu}_a e^{i\mathbf{k}\cdot\mathbf{x}_a} \tag{3.1b}$$

is the longitudinal part of the microscopic *dipole* polarization density. In Eq. (3.1b) $\boldsymbol{\mu}_a$ is the (nonfluctuating) dipole moment of molecule a, while $\mathbf{x}_a$ is the position of the center of mass of the molecule. Furthermore, $\mathbf{e}_k = \mathbf{k}/k$ is the unit vector along the wavevector $\mathbf{k}$. Equation (3.1a) is appropriate for dipolar solvents for not too large values of k (see [19] and [20] and below).

On the other hand, when the solvent molecules are nondipolar ($\boldsymbol{\mu}_a = \mathbf{0}$), the microscopic dipole polarization density $\hat{M}_L(\mathbf{k})$ density is obviously zero, and according to Eq. (3.1a), we would have $\epsilon_L(k) = 1$ for every value of k. This has the unreasonable implication that an applied electric field $\mathbf{E}^0(\mathbf{x})$, no matter how rapidly it may vary in space, would fail to polarize the fluid.

Clearly, to describe with sufficient generality the polarization response in any molecular fluid (dipolar or nondipolar) we must extend the definition of the polarization density vector beyond the dipole approximation. Macroscopically, we may write [5,16,19,21]

$$\mathbf{P}(\mathbf{x}) = \mathbf{P}^d(\mathbf{x}) + \mathbf{P}^q(\mathbf{x}) + \mathbf{P}^0(\mathbf{x}) + \cdots \tag{3.2}$$

where the vectors $\mathbf{P}^d(\mathbf{x})$, $\mathbf{P}^q(\mathbf{x})$, and $\mathbf{P}^0(\mathbf{x})$ represent, respectively, the dipole, quadrupole, and octupole polarization densities.

Naturally, we also need a general constitutive relation, capable of describing the response of each multipolar component of $\mathbf{P}(\mathbf{x})$ to the external electric field. A *qualitative* argument seems most instructive at this point. Consider, for instance, the quadrupole polarization density $\mathbf{P}^q(\mathbf{x})$. Quadrupoles are not affected by a uniform electric field; they couple instead to the gradient of the field. Similarly, octupoles couple to the gradient–gradient of the field, and so on. This simple argument suggests an extended constitutive equation of the form

$$\mathbf{P}(\mathbf{x}) = \chi_0^0 \cdot \mathbf{E}^0(\mathbf{x}) + \chi_1^0 \cdot \nabla\mathbf{E}^0(\mathbf{x}) + \chi_2^0 \cdot \nabla\nabla\mathbf{E}^0(\mathbf{x}) + \cdots \tag{3.3}$$

(where the dot in an expression like $\chi_1^0 \cdot \nabla\mathbf{E}^0(\mathbf{x})$ must be understood as the appropriate tensor contraction). A little consideration indicates, however, that we may equally well write this expression in the more succinct form

$$\mathbf{P}(\mathbf{x}) = \int d^3\mathbf{x}' \chi^0(\mathbf{x} - \mathbf{x}') \cdot \mathbf{E}^0(\mathbf{x}') \tag{3.4}$$

[as one can easily verify by substituting the Taylor expansion $\mathbf{E}^0(\mathbf{x}') = \mathbf{E}^0(\mathbf{x}) + (\mathbf{x}' - \mathbf{x}) \cdot \nabla\mathbf{E}^0(\mathbf{x}) + \cdots$ in the integral and then integrating term by term]. We recognize the integral form as the statement of a *nonlocal* constitutive equation, involving an external dielectric susceptibility tensor $\chi^0(\mathbf{x} - \mathbf{x}')$ with spatial dispersion [16–20,22].

From this argument we conclude that any attempt to describe the dielectric response of a nondipolar solvent must necessarily take into account the nonlocality of the response (spatial dispersion) [23]. The same conclusion applies to the case of dipolar solvents when we are interested in capturing the contribution to the dielectric response from molecular multipoles of higher order.

At this point it is appropriate to note that the preceding argument is somewhat incomplete. It ignores the fact that, for example, most configurations of a system of dipolar particles will couple not only to the spatial average of the electric field, but also to the spatial derivatives of the field.

This is, in fact, the origin of the k-dependence of $\epsilon_L(k)$ in a formula like Eq. (3.1), which only considers the dipolar contribution to the polarization density. It is clear, however, that this simplification in our argument does not alter in any way the conclusion of the preceding paragraph; it only implies that each of the terms on the right-hand side of Eq. (3.3) must be replaced by a more general (nonlocal) form.

Finally, it is also clear that including electronic polarizability does not change these conclusions. Electronic polarizability involves *additional* (high-frequency) degrees of freedom that couple in their own way to the external electric field, but this does not change the fact that the quadrupoles and molecular permanent multipoles of higher order do not couple to a uniform electric field.

B. Molecular Theory of the Dielectric Response Including Electronic Polarizability

1. Frequency- and Wavevector-Dependent Longitudinal Dielectric Function

We consider an infinite body of a molecular solvent in the field of an *external* (i.e., controllable) charge distribution $n^0(\mathbf{x}, t)$. The external electrostatic potential $\varphi^0(\mathbf{x}, t)$ associated with $n^0(\mathbf{x}, t)$ is given by Poisson's equation

$$\nabla^2 \varphi^0(\mathbf{x}, t) = -4\pi n^0(\mathbf{x}, t) \tag{3.5}$$

where $\nabla = \partial/\partial\mathbf{x}$. As it changes in time, $\varphi^0(\mathbf{x}, t)$ polarizes the solvent, in-ducing a charge distribution $\rho_{pol}(\mathbf{x}, t)$. The *screened* potential $\varphi(\mathbf{x}, t)$ is defined by the relation

$$\nabla^2 \varphi^0(\mathbf{x}, t) = -4\pi \rho_{tot}(\mathbf{x}, t) \tag{3.6}$$

where $\rho_{tot}(\mathbf{x}, t)$ is the total charge density at point $\mathbf{x}$ at time t (i.e., the sum of the external and the solvent-induced charge distributions):

$$\rho_{tot}(\mathbf{x}, t) = n^0(\mathbf{x}, t) + \rho_{pol}(\mathbf{x}, t) \tag{3.7}$$

We emphasize that $\rho_{pol}(\mathbf{x}, t)$ is a macroscopic quantity; we discuss its mol-ecular interpretation in the following subsection.

Equations (3.5) and (3.6) may be written in the equivalent forms

$$\varphi^0(\mathbf{k}, \omega) = v(k) n^0(\mathbf{k}, \omega) \tag{3.8}$$

and

$$\varphi(\mathbf{k}, \omega) = v(k) \rho_{tot}(\mathbf{k}, \omega) \tag{3.9}$$

where we introduce the notation $v(k) \equiv 4\pi/k^2$, and where the space- and time-dependent functions $f(\mathbf{x}, t)$ [e.g., $\rho_{tot}(\mathbf{x}, t)$] are expressed in terms of their Fourier transforms

$$f(\mathbf{k}, \omega) = \int_{-\infty}^{\infty} dt \int d^3\mathbf{x} \, e^{-i(\omega t - \mathbf{k} \cdot \mathbf{x})} f(\mathbf{x}, t) \tag{3.10}$$

The induced polarization charge density $\rho_{pol}(\mathbf{k}, \omega)$ comprises two components [5,16,24]:

$$\rho_{pol}(\mathbf{k}, \omega) = \rho_e(\mathbf{k}, \omega) + \rho_\mu(\mathbf{k}, \omega) \tag{3.11}$$

The fast component, $\rho_e(\mathbf{k}, \omega)$, is due to the response of the electronic charge of the solvent molecules, while the slow component, $\rho_\mu(\mathbf{k}, \omega)$, has its origin in the vibrational, orientational, and translational motions of the solvent molecules.

The frequency- and wavevector-dependent longitudinal dielectric function $\epsilon_L(k, \omega)$ is defined by the constitutive relation [16,24,25]

$$\rho_{tot}(\mathbf{k}, \omega) = n^0(\mathbf{k}, \omega)/\epsilon_L(k, \omega) \tag{3.12a}$$

or alternatively, using Eq. (3.7),

$$\rho_{pol}(\mathbf{k}, \omega) = \left(\frac{1}{\epsilon_L(k, \omega)} - 1 \right) n^0(\mathbf{k}, \omega) \tag{3.12b}$$

In Eq. (3.12b), $\rho_{pol}(\mathbf{k}, \omega)$ refers to all of the induced charge density [cf. Eq. (3.11)]. For the description of the fast component, $\rho_e(\mathbf{k}, \omega)$, it is customary to define a high-frequency or *optical* dielectric constant ϵ_∞ according to the relation [5,16,24]

$$\rho_e(\mathbf{k}, \omega) = \left(\frac{1}{\epsilon_\infty} - 1 \right) n^0(\mathbf{k}, \omega) \tag{3.13a}$$

It then follows that the relation between $\rho_\mu(\mathbf{k}, \omega)$ and $n^0(\mathbf{k}, \omega)$ is

$$\rho_\mu(\mathbf{k}, \omega) = \left(\frac{1}{\epsilon_L(k, \omega)} - \frac{1}{\epsilon_\infty} \right) n^0(\mathbf{k}, \omega) \tag{3.13b}$$

It is clear that Eqs. (3.12) incorporate both the temporal and spatial dispersion of the dielectric response. Strictly speaking, ϵ_∞ in Eqs. (3.13) should also be interpreted as a wavevector-dependent longitudinal optical dielectric function $\epsilon_\infty(k)$. Its dependence on k, however, becomes manifest only when the wavelength $2\pi/k$ of the external potential $\varphi^0(\mathbf{k}, \omega)$ is comparable with the spatial extension of the charge distribution of a solvent molecule. For the purpose of this paper it will be sufficient to regard ϵ_∞ as independent of k [5,6].

Equations (3.12b) and (3.13) describe the solvent dielectric response (either ρ_{pol} or its fast and slow components) induced by the external charge distribution $n^0(\mathbf{k}, \omega)$. An alternative way of formulating the response of the slow component $\rho_\mu(\mathbf{k}, \omega)$ employs the *external* charge susceptibility $\chi_\mu^0(k, \omega)$ [24,26]:

$$\rho_\mu(\mathbf{k}, \omega) = \chi_\mu^0(k, \omega)\varphi^0(\mathbf{k}, \omega) \tag{3.14}$$

In view of Eqs. (3.8) and (3.13b), the dielectric function $\epsilon_L(k, \omega)$ and the external charge susceptibility satisfy the relation

$$\frac{1}{\epsilon_L(k, \omega)} - \frac{1}{\epsilon_\infty} = v(k)\chi_\mu^0(k, \omega) \tag{3.15}$$

This completes the phenomenological description of the dielectric response of a molecular solvent to an external charge distribution. The formulation is general, without restrictions on the features of the solvent molecules. While it applies to the dipolar solvents, it also describes the dielectric behavior of the nondipolar solvents. In the following section we consider the molecular interpretation of the external charge susceptibility.

2. Polarizable Interaction Site Models

To be useful for the purposes discussed in Sections I and II, the microscopic formulation of the dielectric theory should be capable of handling solvent models comprising molecules with arbitrary charge distributions. Furthermore, the models must also incorporate the electronic polarizability of the molecules.

These requirements are satisfied by representing the solvent molecules by polarizable interaction site models (pol-ISM) [5,16]. In these models, which are generalizations of the standard (nonpolarizable) ISM [27], the interaction sites are endowed with *fluctuating* partial charges [5,28–30]. More specifically [5,16], the partial charge $\hat{z}_{aj}$ at site j of solvent molecule a is expressed as

$$\hat{z}_{aj} = q_{aj}^0 + \delta\hat{z}_{aj} \tag{3.16}$$

where q_{aj}^0 is the charge of the site when the molecule is isolated, while $\delta\hat{z}_{aj}$ is the fluctuating part due to the interactions with other solvent molecules and any external field.

For a given configuration of the solvent molecules, the fluctuating part $\delta\hat{z}_{aj}$ of the charge at site j in molecule a is related to the instantaneous electrostatic potential at the various sites (located at points $\mathbf{x}_{am}$) of the given molecule by [5,16]

$$\delta\hat{z}_{aj} = -\sum_{m\in a} \nu_{a,jm}\left(\varphi_{am}^0 + \sum_{bl} \hat{t}_{aj,bl}\hat{z}_{bl}\right) \tag{3.17}$$

where $\nu_{a,jm}$ is the jm element of the charge-susceptibility matrix $\mathbf{v}_a$ of molecule a. The quantity inside the parentheses is the sum of the external electrostatic potential $\varphi_{am}^0 \equiv \varphi^0(\mathbf{x}_{am})$ at the position $\mathbf{x}_{am}$ of site am, and an additional contribution due to the partial charges of the other molecules; $\hat{t}_{am,bl} \ (= |\mathbf{x}_{bl} - \mathbf{x}_{am}|^{-1}$ for $a \neq b$ and 0 otherwise) is the electrostatic Green's function for the interaction between two point charges separated by the distance $|\mathbf{x}_{bl} - \mathbf{x}_{am}|$ in vacuum.

In matrix form Eqs. (3.16) and (3.17) read

$$\hat{\mathbf{z}} = \mathbf{q}_0 - \nu(\boldsymbol{\varphi}^0 + \hat{\mathbf{t}}\hat{\mathbf{z}}) \tag{3.18}$$

where $\hat{\mathbf{z}}, \mathbf{q}^0$, and $\boldsymbol{\varphi}^0$ are column vectors with elements $\hat{z}_{aj}, q_{aj}^0$, and φ_{aj}^0, respectively, while ν and $\hat{\mathbf{t}}$ are square matrices with elements $\delta_{ab}\nu_{a,jm}$ and $\hat{t}_{am,bl}$. Notice that the subscripts a and b refer to all solvent molecules, while the subscripts j, l, and m refer to the interaction sites within the molecules.

For a given configuration, Eq. (3.18) may be solved for the instantaneous partial charges $\hat{z}_{aj}$ at the interaction sites:

$$\hat{z}_{aj} = \hat{q}_{aj} + \hat{e}_{aj} \tag{3.19}$$

where

$$\hat{q}_{aj} = [(1 + \mathbf{v}\hat{\mathbf{t}})^{-1}\mathbf{q}^0]_{aj} \tag{3.20a}$$

is the contribution to the instantaneous charge at the interaction site aj when $\varphi^0(\mathbf{x}) = 0$, while

$$\hat{e}_{aj} = -[\hat{\mathbf{J}}\boldsymbol{\varphi}^0]_{aj} \equiv -[(1 + \mathbf{v}\hat{\mathbf{t}})^{-1}\mathbf{v}\boldsymbol{\varphi}^0]_{aj} \tag{3.20b}$$

is the additional modification caused by the external potential $\varphi^0(\mathbf{x})$. Notice that the second equality in Eq. (3.20b) defines the matrix $\hat{\mathbf{J}}$, of elements $\hat{J}_{aj,bl}$, that will play a role in the molecular expression of the optical dielectric constant ϵ_∞.

It is clear that a pol-ISM model, having permanent partial charges q_{aj}^0 assigned to the interaction sites, can represent molecular charge distributions of quite general character, including of course "nondipolar" species with nonzero multipole moments of higher order. At the same time, the fluctuating contribution $\delta\hat{z}_{aj}$, which depends on the instantaneous liquid configuration as in Eq. (3.17), allows for incorporation of the electronic polarizability effects.

3. Molecular Expression for the External Charge Susceptibility

We consider now the microscopic formulation of the linear dielectric response of the solvent when acted upon by an external time-varying electrostatic potential $\varphi^0(\mathbf{x}, t)$. The time-dependent Hamiltonian is given by

$$H(t) = H_w + H'(t) \tag{3.21}$$

where H_w is the uniform solvent Hamiltonian [i.e., when $\varphi^0(\mathbf{x}, t) = 0$]. A detailed expression of H_w will not be required in this work; it may be found in [5]. The perturbation $H'(t)$ is given by the expression [5]

$$H'(t) = \int d^3\mathbf{x}\, \hat{\rho}_\mu(\mathbf{x})\varphi^0(\mathbf{x}, t) - \frac{1}{2}\int d^3\mathbf{x}\, d^3\mathbf{x}'\, \hat{\mathscr{J}}(\mathbf{x}, \mathbf{x}')\varphi^0(\mathbf{x}, t)\varphi^0(\mathbf{x}', t) \tag{3.22}$$

where we have introduced the solvent dynamical variables [cf. Eqs. (3.20)]

$$\hat{\rho}_\mu(\mathbf{x}) = \sum_{aj} \hat{q}_{aj}\delta(\mathbf{x} - \mathbf{x}_{aj}) \tag{3.23a}$$

$$\hat{\mathscr{J}}(\mathbf{x}, \mathbf{x}') = \sum_{aj,bl} \hat{J}_{aj,bl}\delta(\mathbf{x} - \mathbf{x}_{aj})\delta(\mathbf{x}' - \mathbf{x}_{bl}) \tag{3.23b}$$

Notice the different dependence of the first and second terms of $H'(t)$ on $\varphi^0(\mathbf{x}, t)$ (first and second order, respectively).

To derive the molecular expressions for the external susceptibility, we focus on $\rho_\mu(\mathbf{k}, t)$ and $\rho_e(\mathbf{k}, t)$, the spatial Fourier transforms of the macroscopic charge densities $\rho_\mu(\mathbf{x}, t)$ and $\rho_e(\mathbf{x}, t)$ of Section III.B.1. They are calculated as the statistical mechanical averages

$$\rho_\mu(\mathbf{k}, t) = \langle \hat{\rho}_\mu(\mathbf{k}); t \rangle \tag{3.24a}$$

$$\rho_e(\mathbf{k}, t) = \langle \hat{\rho}_e(\mathbf{k}); t \rangle \tag{3.24b}$$

of the Fourier components of the microscopic charge densities

$$\hat{\rho}_\mu(\mathbf{k}) = \sum_{aj} \hat{q}_{aj} e^{i\mathbf{k}\cdot\mathbf{x}_{aj}} \tag{3.25a}$$

$$\hat{\rho}_e(\mathbf{k}) = \sum_{aj} \hat{e}_{aj} e^{i\mathbf{k}\cdot\mathbf{x}_{aj}} \tag{3.25b}$$

over the nonequilibrium distribution function that describes the state of the solvent at time t under the action of the electrostatic potential $\varphi^0(\mathbf{x}, t)$. The fluctuating partial charges $\hat{q}_{aj}$ and $\hat{e}_{aj}$ were introduced in Eqs. (3.20).

Straightforward application of the theory of linear response to mechanical perturbations [31] gives the following molecular formula for the external charge susceptibility [16]:

$$\chi^0_\mu(k, \omega) = -\beta S_\mu(k)[1 - i\omega\Phi_\mu(k, \omega)] \tag{3.26}$$

where $\Phi_\mu(k, \omega)$ is the Fourier–Laplace transform

$$\Phi_\mu(k, \omega) = \int_0^\infty dt\, e^{-i\omega t} \Phi_\mu(k, t) \tag{3.27}$$

of the normalized time correlation function of the fluctuating slow part of the solvent polarization charge density

$$\Phi_\mu(k, t) = \left\{ \frac{\langle \hat{\rho}_\mu(\mathbf{k}, t)\hat{\rho}_\mu(-\mathbf{k})\rangle}{\langle \hat{\rho}_\mu(\mathbf{k})\hat{\rho}_\mu(-\mathbf{k})\rangle} \right\}_\infty \tag{3.28}$$

Furthermore,

$$S_\mu(k) = \left\{ \frac{1}{V} \langle \hat{\rho}_\mu(\mathbf{k})\hat{\rho}_\mu(-\mathbf{k})\rangle \right\}_\infty \tag{3.29}$$

is the associated static structure factor. The symbol $\{\cdots\}_\infty$ in Eqs. (3.28) and (3.29) specifies the thermodynamic limit of the enclosed expression, and V is the volume of the system. Furthermore, $\langle\cdots\rangle$ denotes an ensemble average over the equilibrium distribution function of the solvent (governed by the Hamiltonian H_w). In Eq. (3.26), $\beta = (k_B T)^{-1}$ is the inverse of the temperature expressed in energy units (k_B is the Boltzmann constant and T the temperature).

Equations (3.15) and (3.26) thus give

$$\frac{1}{\epsilon_L(k, \omega)} - \frac{1}{\epsilon_\infty} = -\beta v(k) S_\mu(k)[1 - i\omega\Phi_\mu(k, \omega)] \tag{3.30}$$

for the molecular expression of the frequency- and wavevector-dependent longitudinal dielecric function $\epsilon_L(k, \omega)$. For the optical dielectric function, linear response theory gives [5,16]

$$\frac{1}{\epsilon_\infty} - 1 = -v(k)\left\{\frac{1}{V}\left\langle\sum_{aj,bl} \hat{J}_{aj,bl} e^{i\mathbf{k}\cdot(\mathbf{x}_{aj}-\mathbf{x}_{bl})}\right\rangle\right\}_\infty \tag{3.31}$$

The dielectric fluctuation formulas (3.26), (3.30), and (3.31) provide the bridge between the phenomenological description of the dispersive dielectric response of the solvent (Sec. III.B.1) and the molecular Hamiltonian of the liquid. Equations (3.25)–(3.30) are the generalization to pol-ISM models of the corresponding dielectric formula reported previously [24,32] for a non-polarizable ISM model.

When the external charge distribution n^0 is independent of time, Eq. (3.14) simplifies to

$$\rho_\mu(\mathbf{k}) = \chi_\mu^0(k)\varphi^0(\mathbf{k}) \tag{3.32}$$

where $\chi_\mu^0(k) \equiv \chi_\mu^0(k, \omega = 0) = -\beta S_\mu(k)$ is the static external charge susceptibility. Correspondingly, Eq. (3.30) at $\omega = 0$ gives the relation [5,16]

$$\mathscr{C}(k) \equiv \frac{1}{\epsilon_\infty} - \frac{1}{\epsilon_L(k)} = \beta v(k) S_\mu(k) \tag{3.33}$$

between the wavevector-dependent Pekar factor [5,16] $\mathscr{C}(k)$ and the polarization charge structure factor $S_\mu(k)$ introduced in Eq. (3.29).

We conclude this section by noting that Eq. (3.4) is completely equivalent to Eq. (3.32). This follows from the relations [21] $\rho_\mu(\mathbf{x}) = -\nabla \cdot \mathbf{P}_\mu(\mathbf{x})$ and $\mathbf{E}^0(\mathbf{x}) = -\nabla\varphi^0(\mathbf{x})$ and the fact that in Section III.A the electronic polarizability of the solvent molecules was ignored [i.e., $\mathbf{P}(\mathbf{x})$ in Eqs. (3.2) and (3.4) equals $\mathbf{P}_\mu(\mathbf{x})$]. Hence the external charge susceptibility $\chi_\mu^0(k)$ [cf. Eq. (3.32)] is connected to the Fourier transform $\boldsymbol{\chi}^0(k)$ of the external polarization susceptibility tensor [cf. Eq. (3.4)] by the expression $\chi_\mu^0(k) = -\mathbf{k} \cdot \boldsymbol{\chi}^0(k) \cdot \mathbf{k}$.

C.   Results for the Wavevector-Dependent Pekar Factor $\mathscr{C}(k)$

In this section we report the wavevector dependence of the Pekar factor $\mathscr{C}(k)$ for several dipolar and nondipolar solvents. They are calculated from an approximation to Eq. (3.33) using integral equation techniques [5,6,16], specifically the Reference interaction site method (RISM) of Chandler and Andersen [33] with the hypernetted chain closure (XRISM [34]). We refer the reader to a recent publication [6] for a detailed explanation of the calculational procedure; only the results are discussed here. Nevertheless, we note that the computations are actually performed with nonpolarizable ISM models of the solvents (of the OPLS [35] or similar type [6,16,36]), with the results corrected a posteriori for the solvent electronic polarizability.

Figure 2(*a*) presents the wavevector-dependent Pekar factor for nine representative dipolar solvents: methyl acetate, acetone, dimethyl sulfoxide, tetrahydrofuran, chloroform, benzonitrile, acetonitrile, methanol, and water. In Figure 2(*b*) we report the corresponding results for six representative nondipolar solvents: tetrachloroethylene, carbon tetrachloride, 1,4-difluorobenzene, toluene, benzene, and 1,3,5-trifluorobenzene. (Literature references for the ISM models of most of the solvents are provided in Sec. VI.D.1.) Notice that we include toluene in the nondipolar group, since the contribution of its small dipole moment is expected to be minor relative to that from higher moments. [The results for carbon tetrachloride, toluene, and benzene were reported in a recent publication (Fig. 12 of [6]), but unfortunately in each of them the vertical scale needs to be multiplied by a simple numerical factor (0.457, 0.432, and 0.441 for carbon tetrachloride, toluene, and benzene, respectively). Only the data reported in Figure 12 of [6] are affected by the error. The corrected results were presented in [16] and in Figure 2(*b*).]

Figure 2(*a*) and (*b*) clearly illustrate that there are no profound differences in the global features of the wavevector-dependent Pekar factor of dipolar and nondipolar solvents, although the peak amplitudes vary considerably from solvent to solvent. The shape of $\mathscr{C}(k)$ is due entirely to the k-dependence of the static longitudinal dielectric function $\epsilon_L(k)$. The peaks of $\mathscr{C}(k)$ reflect the pronounced excursions of $\epsilon_L(k)$ to negative values, and the figures show that this occurs in all the solvents considered.

That $\epsilon_L(k)$ is negative at intermediate k in dipolar solvents is by now well known (see, e.g., [19] and the references to previous work reported there; for a more recent account, see [37]). It follows from Eq. (3.33) and the fact that in a stable system the structure factor $S_\mu(k)$ is positive, that $\mathscr{C}(k)$ satisfies the inequality $\mathscr{C}(k) \geq 0$. With $\epsilon_\infty > 0$, this inequality will be satisfied as long as either

$$\epsilon_L(k) \geq \epsilon_\infty \qquad \text{or} \qquad \epsilon_L(k) < 0 \qquad (3.34)$$

indicating that negative values of $\epsilon_L(k)$ are indeed allowed. However, the observation that $\epsilon_L(k)$ is negative at intermediate k also for nondipolar solvents is very recent [6,16] and reinforces the concept that except for the special point $k = 0$ (at which $\mathscr{C} \simeq 0$ because the permanent molecular dipole moment of the solvent molecules is zero), the intermolecular spatial correla-

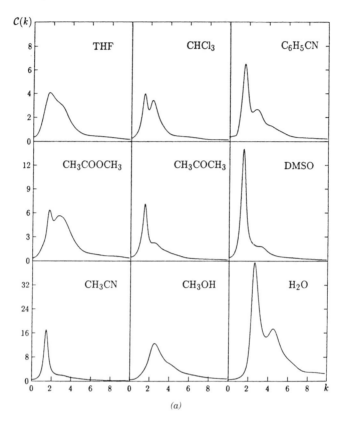

Figure 2. Wavevector-dependent Pekar factor $\mathscr{C}(k)$, calculated for polarizable ISM models of (a) representative dipolar solvents, and (b) representative solvents with very small or vanishing dipole moments. Notice that each of the rows in (a) has a different vertical scale. In (b) the data for tetrachloroethylene and 1,3,5-trifluorobenzene have been multiplied, respectively, by the scale factors 20 and 0.25. The abscissa is the wavevector k in units of Å^{-1}.

tions responsible for the features of $\epsilon_L(k)$ are not dissimilar in dipolar and nondipolar solvents.

IV. BASIC DESCRIPTION OF THE ENERGETICS OF CHARGE TRANSFER REACTIONS IN SOLUTION

A. Diabatic Free Energy Profiles

We consider a charge transfer process that carries the solute molecule (or solute complex) from a *precursor* (*P*) electronic state to a *successor* (*S*)

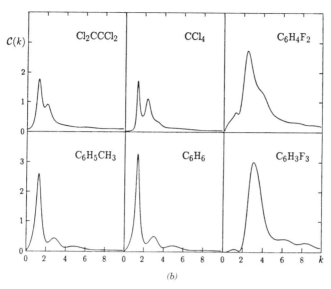

Figure 2. (*Continued*)

electronic state. When the solute is represented by an ISM model (with the interaction sites labeled with the subscript λ), the P and S states are characterized, respectively, by the sets of partial charges $\{Q_\lambda^P\}$ and $\{Q_\lambda^S\}$ at the interaction sites.

The precursor $F^P(\eta)$ and successor $F^S(\eta)$ diabatic free energy profiles are given by the formula [4–6,38–45]

$$-\beta F^D(\eta) = \ln\left[\theta \int d\Omega \int d\Gamma\, e^{-\beta \mathscr{H}^D}\delta(\eta - \hat{\mathscr{U}})\right] \qquad D = P, S \qquad (4.1)$$

as functions of the reaction coordinate η. The parameter θ has dimensions of energy and is independent of D, the solute electronic state [5]. By Ω we indicate the set of Euler angles that describe the orientation of the solute with respect to the laboratory reference frame, while Γ indicates the set of coordinates and momenta of the solvent interaction sites in the same reference frame.

The basic [5,6] Hamiltonian $\mathscr{H}^D$ of the system when the solute is in electronic state D is

$$\mathscr{H}^D = E^D + H^D = E^D + H_w + \hat{\Psi}^D \qquad D = P, S \qquad (4.2)$$

where H_w is the Hamiltonian of the homogeneous solvent in the absence of the solute, while E^D and $\hat{\Psi}^D$ are, respectively, the energy of the solute in the electronic state D in vacuum and the potential energy of interaction between the solute in electronic state D and the solvent molecules. Clearly, H^D is the Hamiltonian of the solvent in the field of the solute in state D. For a given configuration of the solvent around the solute, the energy difference between the S and P states of the system is

$$\mathscr{H}^S - \mathscr{H}^P = \Delta E + \hat{\mathscr{U}} \qquad (4.3)$$

where $\Delta E \equiv E^S - E^P$, while

$$\hat{\mathscr{U}} = H^S - H^P = \hat{\Psi}^S - \hat{\Psi}^P \qquad (4.4)$$

is the solvent contribution to the vertical energy gap. Notice that the Dirac delta function $\delta(\eta - \hat{\mathscr{U}})$ in Eq. (4.1) implicitly conveys the physical meaning of the reaction coordinate.

It is important to note that an exact relation [5,38,39b–d,40] follows from the definitions of $F^P(\eta)$ and $F^S(\eta)$: namely, the difference between the two free energy profiles at a given value of the reaction coordinate is

$$\Delta F(\eta) = F^S(\eta) - F^P(\eta) = \Delta E + \eta \qquad (4.5)$$

It follows [cf. Fig. 1(a)] that the two free energy profiles cross at $\eta^{\ddagger} = -\Delta E$. Also from Eq. (4.5), and making reference to Figure 1(b), we find that the absorption (ν_{abs}) and fluorescence (ν_{flr}) frequencies of the solute are given by the relations

$$h\nu_{abs} = \Delta F(\eta^P) = \Delta E + \eta^P \qquad (4.6a)$$

$$h\nu_{flr} = \Delta F(\eta^S) = \Delta E + \eta^S \qquad (4.6b)$$

where η^P and η^S are defined in Figure 1. For the Stokes shift we obtain

$$E_{ss} \equiv h\nu_{abs} - h\nu_{flr} = -(\eta^S - \eta^P) \equiv -\Delta\eta \qquad (4.7)$$

B. Generalized Characteristic Function

Equation (4.1) may be written in the alternative more convenient form [5]

$$e^{-\beta F^D(\eta)} = \theta e^{-\beta \mathscr{A}^D} p^D(\eta) \qquad D = P, S \qquad (4.8)$$

where

$$-\beta \mathscr{A}^D = \ln\left(\int d\Omega \int d\Gamma e^{-\beta \mathscr{H}^D}\right) \tag{4.9}$$

is the Helmholtz free energy of the system when the solute is in the electronic state D, while

$$p^D(\eta) = \langle \delta(\eta - \hat{\mathscr{u}}) \rangle^D \tag{4.10}$$

is the probability density that the solvent contribution $\hat{\mathscr{u}}$ to the vertical energy gap has the numerical value η. In this equation $\langle \cdots \rangle^D$ denotes an equilibrium ensemble average under the distribution function $f^D(\Gamma) \sim e^{-\beta \mathscr{H}^D}$ of the system when the solute is in the D-state.

To calculate $p^D(\eta)$ from a molecular model it is useful [5] to introduce the generalized characteristic function $g^D(z)$ of the complex variable $z = \alpha + i\zeta$ (where α and ζ are real numbers)

$$g^D(z) \equiv \langle e^{iz\hat{\mathscr{u}}} \rangle^D \qquad D = P, S \tag{4.11}$$

The probability density $p^D(\eta)$ can then be recovered from $g^D(z)$ by the Fourier integral

$$p^D(\eta) = \frac{1}{2\pi} \int_{-\infty}^{\infty} d\alpha e^{-i\alpha\eta} g^D(\alpha) \qquad D = P, S \tag{4.12}$$

where the integration is carried along the real line $z = \alpha$. It follows from Eqs. (4.8) and (4.12) that $g^D(z = \alpha)$ contains all the information required to calculate the diabatic free energy profile $F^D(\eta)$. Furthermore, using Eqs. (4.2) to (4.4), (4.9), and (4.11), it is simple to express the reaction free energy $\Delta \mathscr{A} \equiv \mathscr{A}^S - \mathscr{A}^P$ in terms of $g^P(z)$ or $g^S(z)$, with z purely imaginary. We find that [5]

$$\Delta \mathscr{A} = \Delta E + \Delta A \tag{4.13a}$$

where the solvation free energy change in the reaction is given by

$$-\beta \Delta A = \ln g^P(z = i\beta) = -\ln g^S(z = -i\beta) \tag{4.13b}$$

The relation of the generalized characteristic function $g^D(z)$ to the free energy profile $F^D(\eta)$ [cf. Eqs. (4.8) and (4.12)] and to the reaction free energy $\Delta \mathscr{A}$ [cf. Eqs. (4.13)] indicates the convenience of addressing the calculation of the energetics of charge transfer reactions in terms of $g^D(z)$. The advantage is that an approximation made at the level of the characteristic function automatically leads to consistent approximations for the shapes of the free energy profiles $[F^D(\eta)]$ and for the energetics $[\Delta\mathscr{A}]$. Furthermore, as long as the approximate characteristic functions satisfy the relation $g^S(\alpha)/g^S(-i\beta) = g^P(\alpha + i\beta)$, Eq. (4.5) is satisfied.

C.   Complex-Valued Hamiltonian Route

It is possible to formulate a rather elegant approach for calculation of the generalized characteristic functions $g^D(z)$ [46]. Although the method takes into account the *nonlinear* aspects of the problem (at the level of the hyper-netted-chain approximation), presently it is limited to model systems in which the electronic polarizability of the solvent molecules is ignored. This shortcoming may be ameliorated with simple corrections based on the surrogate Hamiltonian theory of solvation [5,6,46].

The method is based on the introduction of complex-valued diabatic Hamiltonians [47]

$$\mathscr{H}^D(z) = \mathscr{H}^D - i(z/\beta)\hat{\mathscr{U}}   D = P, S \qquad (4.14a)$$

[$\mathscr{H}^D$ and $\hat{\mathscr{U}}$ were defined, respectively, in Eqs. (4.2) and (4.4)] and the associated canonical partition function

$$Z^D(z) = \int d\Gamma e^{-\beta\mathscr{H}^D(z)} \qquad (4.14b)$$

and complex-valued Helmholtz free energy function

$$\mathscr{A}^D(z) = -\beta^{-1}\ln Z^D(z) \qquad (4.14c)$$

With these definitions it is straightforward to express the generalized characteristic function $g^D(z)$ in the concise form [46]

$$g^D(z) = Z^D(z)/Z^D(0) = e^{-\beta[\mathscr{A}^D(z) - \mathscr{A}^D(0)]}$$

$$= e^{-\beta[\tilde{\mathscr{A}}^D(z) - \tilde{\mathscr{A}}^D(0)]} \qquad (4.15)$$

$$= e^{-\beta[\tilde{\mu}^D(z) - \tilde{\mu}^D(0)]}$$

in which $\mathscr{Z}^D(0)$ and $\mathscr{A}^D(0)$ are, respectively, the canonical partition function and Helmholtz free energy corresponding to the Hamiltonian $\mathscr{H}^D(z = 0) = \mathscr{H}^D$. The third equality expresses $g^D(z)$ in terms of the excess complex-valued free energy $\tilde{\mathscr{A}}^D(z) \equiv \mathscr{A}^D(z) - \mathscr{A}^{D,\text{id}}$, where $\mathscr{A}^{D,\text{id}} = E^D + A^{\text{id}}$ is the free energy of the noninteracting system when the solute is in the electronic state D. The fourth equality of Eq. (4.15) gives, for our purposes, the most convenient expression for $g^D(z)$, in terms of the complex-valued excess chemical potential of the solute $\tilde{\mu}^D(z) \equiv \tilde{A}^D(z) - \tilde{A}_w$, where $\tilde{A}^D(z)$ is the solvation part of the complex-valued excess free energy, and $\tilde{A}_w \equiv A_w - A_w^{\text{id}}$ is the excess free energy of the pure solvent.

Equation (4.15) indicates that the generalized characteristic function $g^D(z)$ is related to the excess complex-valued chemical potential $\tilde{\mu}^D(z)$ of a fictitious solute that interacts with the solvent molecules through the complex-valued potential energy of interaction [cf. Eqs. (4.2) and (4.14a)]

$$\hat{\Psi}^D(z) = \hat{\Psi}^D - i(z/\beta)\hat{\mathscr{U}} = \sum_\lambda \sum_{aj} \left[u_{\lambda j}^D(r_{\lambda,aj}) - i(z/\beta)\frac{\Delta Q_\lambda q_j}{r_{\lambda,aj}} \right] \qquad (4.16)$$

where $r_{\lambda,aj} \equiv |\mathbf{x}_{aj} - \mathbf{x}_\lambda|$ is the distance between site j of the solvent molecule a (located at $\mathbf{x}_{aj}$) and site λ of the solute molecule (located at $\mathbf{x}_\lambda$). Furthermore, $u_{\lambda j}^D(r)$ is the *basic* (i.e., actual) potential energy of interaction between the solute site λ and a solvent interaction site of type j when the solute is in the electronic state D. The explicit form of the imaginary part of the interaction in Eq. (4.16) follows from the interpretation

$$\hat{\mathscr{U}} = \sum_\lambda \sum_{aj} \frac{\Delta Q_\lambda q_j}{r_{\lambda,aj}} \qquad (4.17)$$

for the solvent contribution to the vertical energy gap. Here $\Delta Q_\lambda \equiv Q_\lambda^S - Q_\lambda^P$ is the change in the partial charge of the solute site λ in the $P \to S$ transition, while q_j is the (nonfluctuating) partial charge of a solvent site of type j. It is important to realize that Eq. (4.17) applies only in the case that the model solvent is *nonpolarizable*, and that only the Coulombic part of the solute–solvent potential energy of interaction changes in the $P \to S$ transition.

We have recently shown [46] that the complex-valued chemical potential $\tilde{\mu}^D(z)$ can be calculated for systems comprising solute and solvent molecules represented by nonpolarizable ISM models. The implementation is based on a simple extension to complex solute–solvent interaction potentials of the reference interaction site method (RISM) of Chandler and Andersen [33] using the hypernetted-chain (HNC) closure. The RISM–HNC integral equation as a practical methodology to calculate the equilibrium structure functions of liquids and solutions comprising polyatomic polar molecules was first proposed and developed extensively by Rossky and co-workers [34]; they referred to the procedure as the extended reference interaction site method (XRISM). The complex-valued Hamiltonian method exploits the Singer-Chandler [48] RISM–HNC formula for the excess chemical potential (generalized to complex-valued functions) of a solute at infinite dilution in a molecular solvent. For more details we refer to the original report [46].

The complex-valued Hamiltonian method is useful to investigate non-linear effects in the diabatic free energy profiles of charge transfer reactions in solution. An alternative real-valued route to the calculation of the diabatic free energy profiles with the RISM–HNC integral equation has been developed by Hirata and co-workers [49].

D. Example of the Complex-Valued Hamiltonian Route

We now apply the complex-valued Hamiltonian theory to a relatively complicated chemical problem, the energetics of the ferrocene–ferrocenium electron exchange reaction (Cp = cyclopentadienyl)

$$FeCp_2 + FeCp_2^+ \rightarrow FeCp_2^+ + FeCp_2 \tag{4.18}$$

in acetonitrile.

We represent the "solute" by a disjoint 22-site ISM model, comprising in the P state the $FeCp_2$ and $FeCp_2^+$ moieties in D_{5h} symmetry, as shown in Figure 3. In the S state the $FeCp_2$ and $FeCp_2^+$ moieties are interchanged. The distance between the two metallic centers is 6.75 Å. Each CH group in the Cp rings is represented by an interaction site, with Lennard-Jones parameters $\varepsilon_{CH}/k_B = 55.36$ K and $\sigma_{CH} = 3.75$ Å taken from the OPLS force field [50]. For the iron atom we use $\varepsilon_{Fe}/k_B = 6.542$ K and $\sigma_{Fe} = 5.7$ Å for

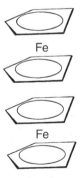

Figure 3. Structure of ferrocene–ferrocenium intermolecular encounter complex used as the solute in the calculation of the free energy profiles for the reaction in Eq. (4.18) in acetonitrile.

both oxidation states. The solute partial charges and the intramolecular distances are taken from the work of Newton et al. [51]. For the solvent acetonitrile we use the three-site ISM model developed by Edwards *et al.* [52].

In Figure 4 we show the diabatic free energy profiles calculated with the complex-valued Hamiltonian method [46]. For comparison we also show the free energy profiles calculated with the SH-RDT approximation (discussed in Secs. V.B and V.C.1). The SH-RDT profiles are exactly parabolic; the figure illustrates that for this system the deviations of the profiles calculated under the complex-valued Hamiltonian method from parabolic shape is quite small.

The free energy profiles calculated with the complex-valued Hamiltonian method in Figure 4 lead to an activation free energy $F_{np}^{\ddagger} = 8.87\,\text{kcal mol}^{-1}$ for Eq. (4.18). As it stands, this number cannot be compared with experiment, the reason being that the ISM model of the solvent and the theoretical method ignore the electronic polarizability of the solvent molecules. (This

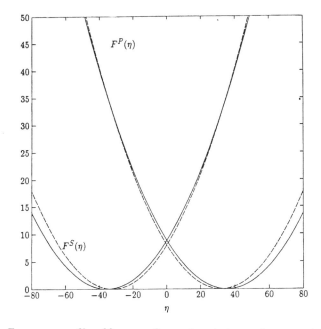

Figure 4.   Free energy profiles of ferrocene–ferrocenium electron exchange reaction [Eq. (4.18)] in acetonitrile at $T = 307\,\text{K}$. Solid curve, calculated with the complex-valued Hamiltonian method [Eqs. (4.8), (4.12), and (4.15)]; dashed curve, calculated with SH-RDT theory [Eqs. (5.14) and (5.22)]. Free energy profiles $F^D(\eta)$ and reaction coordinate η are in units of kcal mol^{-1}.

explains the subscript np = nonpolarizable in $F_{np}^{\ddagger}$). For a meaningful comparison with the experimental data it is very important to treat the solvent electronic polarizability in a consistent way, as discussed at some length in [5] and [6]. The resulting estimate of the activation free energy of the exchange reaction (4.18) after correction for electronic polarizability [6,46] is $F^{\ddagger} = 6.07\,\text{kcal mol}^{-1}$, which should be compared with the experimental estimate [53] $F_{exp}^{\ddagger} \simeq 5\text{–}6\,\text{kcal mol}^{-1}$. The SH-RDT value for the activation free energy is $F^{\ddagger} = 5.22\,\text{kcal mol}^{-1}$ [46]. Thus, after correction for electronic polarizability both the complex-valued Hamiltonian and SH-RDT methods compare reasonably well with experiment.

V. SURROGATE HAMILTONIAN THEORY OF CHARGE TRANSFER ENERGETICS

A. Surrogate Hamiltonian Description

We describe [4–6] the system in terms of surrogate (subscript Σ) diabatic Hamiltonians

$$\mathscr{H}_{\Sigma}^{D} = E^{D} + H_{w} + \hat{\Psi}_{\Sigma}^{D} \qquad D = P, S \tag{5.1}$$

in which the solute–solvent potential energy of interaction $\hat{\Psi}^{D}$ in Eq. (4.2) is replaced by a *renormalized* form $\hat{\Psi}_{\Sigma}^{D}$ that needs to be specified. For a given configuration of the solvent, the energy difference between the P and S states of the system is given by

$$\mathscr{H}_{\Sigma}^{S} - \mathscr{H}_{\Sigma}^{P} = \Delta E + \hat{\mathscr{E}} \tag{5.2}$$

where

$$\hat{\mathscr{E}} = \hat{\Psi}_{\Sigma}^{S} - \hat{\Psi}_{\Sigma}^{P} \tag{5.3}$$

is the solvent contribution to the vertical energy gap in the surrogate system.

In contrast with the basic description, where the equilibrium distribution function $f^{D}(\Gamma) \sim e^{-\beta\mathscr{H}^{D}}$ subsumes all the information about the likelihood of a particular solvent configuration around the solute in state D, the (time-independent) SH theory assumes [5,6] that the form of the equilibrium distribution function is

$$f_{\Sigma}^{D}(\Gamma) = \frac{1}{\Omega} f^{w}(\Gamma)[1 - \beta\delta(\hat{\Psi}_{\Sigma}^{D})] \tag{5.4}$$

where $f^w(\Gamma) \sim e^{-\beta H_w}$ is the equilibrium distribution function of the homogeneous solvent, $1/\Omega = (\int d\Omega)^{-1}$ is the probability density for a particular orientation of the solute, and $\delta(\hat{\Psi}_\Sigma^D) \equiv \hat{\Psi}_\Sigma^D - \langle \hat{\Psi}_\Sigma^D \rangle$ is the part of $\hat{\Psi}_\Sigma^D$ that fluctuates with respect to its average value $\langle \hat{\Psi}_\Sigma^D \rangle$ under the pure solvent distribution function $f^w(\Gamma)$. The renormalized potential energy of interaction $\hat{\Psi}_\Sigma^D$ in Eqs. (5.1) and (5.4) is fixed by requiring that

$$\langle\!\langle \hat{\rho}(\mathbf{r}) \rangle\!\rangle^D = \langle \hat{\rho}(\mathbf{r}) \rangle^D \tag{5.5}$$

for a certain solvent density field $\hat{\rho}(\mathbf{r})$ (where $\mathbf{r}$ locates a point in the solvent relative to the solute). Here $\langle \cdots \rangle^D$ and $\langle\!\langle \cdots \rangle\!\rangle^D$ denote ensemble averages under $f^D(\Gamma)$ and $f_\Sigma^D(\Gamma)$, respectively. Condition (5.5) implies that the solvation structure [as described by $\hat{\rho}(\mathbf{r})$] is the same in the basic and surrogate descriptions of the system, an obvious requirement if we expect the SH theory to be accurate. An important simplification of the surrogate formulation is that $f_\Sigma^D(\Gamma)$ is *linear* in the renormalized solute–solvent interactions $\hat{\Psi}_\Sigma^D$.

In the surrogate description Eq. (4.8) becomes [5]

$$e^{-\beta F_\Sigma^D(\eta)} = \theta e^{-\beta \mathscr{A}_\Sigma^D} p_\Sigma^D(\eta) \qquad D = P, S \tag{5.6}$$

where the free energy profiles $F^D(\eta)$ are expressed in terms of the surrogate estimates $p_\Sigma^D(\eta)$ and $\mathscr{A}_\Sigma^D$ of the probability density $p^D(\eta)$ and free energy $\mathscr{A}^D$. To calculate them we focus [5] on the surrogate version of Eq. (4.11):

$$g_\Sigma^D(z) \equiv \langle\!\langle e^{iz\hat{\mathscr{E}}} \rangle\!\rangle^D \tag{5.7}$$

in which the generalized characteristic function is calculated with the surrogates of the distribution function $f_\Sigma^D(\Gamma)$ and vertical energy gap $\hat{\mathscr{E}}$. Consistency between Eqs. (5.4) and (5.5) demands [5,6] that the cumulant expansion of $\langle\!\langle e^{iz\hat{\mathscr{E}}} \rangle\!\rangle^D$ terminates at second order, that is,

$$g_\Sigma^D(z) = \exp\left(iz\langle\!\langle \hat{\mathscr{E}} \rangle\!\rangle_c^D - \frac{z^2}{2} \langle\!\langle \hat{\mathscr{E}}^2 \rangle\!\rangle_c^D \right) \qquad D = P, S \tag{5.8}$$

Furthermore, in evaluating the first $\langle\!\langle \hat{\mathscr{E}} \rangle\!\rangle_c^D$ and second $\langle\!\langle \hat{\mathscr{E}}^2 \rangle\!\rangle_c^D$ cumulants of $\hat{\mathscr{E}}$ under $f_\Sigma^D(\Gamma)$, only terms that are at most quadratic in the renormalized solute–solvent interactions are retained [5]. This is because of the renormalization; including higher–order terms would amount to overcounting the solute–solvent interactions. We calculate [5]

$$\eta_\Sigma^{P,S} \equiv \langle\!\langle \hat{\mathcal{E}} \rangle\!\rangle_c^{P,S} = \langle\!\langle \hat{\mathcal{E}} \rangle\!\rangle^{P,S} = \langle \hat{\mathcal{E}} \rangle - \beta \langle \delta\hat{\mathcal{E}}\delta.\hat{\mathcal{N}} \rangle \pm \frac{\beta}{2}\langle \delta\hat{\mathcal{E}}\delta\hat{\mathcal{E}} \rangle \qquad (5.9)$$

$$\frac{1}{\kappa_\Sigma} \equiv \beta\langle\!\langle \hat{\mathcal{E}}^2 \rangle\!\rangle_c^P = \beta\langle\!\langle \hat{\mathcal{E}}^2 \rangle\!\rangle_c^S = \beta\langle \delta\hat{\mathcal{E}}\,\delta\hat{\mathcal{E}} \rangle \qquad (5.10)$$

where we have introduced the dynamical variables [5]

$$\delta\hat{\mathcal{E}} \equiv \hat{\mathcal{E}} - \langle \hat{\mathcal{E}} \rangle \qquad (5.11a)$$

and

$$\hat{\mathcal{N}} \equiv (\hat{\Psi}_\Sigma^P + \hat{\Psi}_\Sigma^S)/2 \qquad \delta\hat{\mathcal{N}} \equiv \hat{\mathcal{N}} - \langle \hat{\mathcal{N}} \rangle \qquad (5.11b)$$

and where to calculate the second-order cumulants we have used the relation $\langle\!\langle \hat{\mathcal{E}}^2 \rangle\!\rangle_c^D = \langle\!\langle \hat{\mathcal{E}}^2 \rangle\!\rangle^D - (\langle\!\langle \hat{\mathcal{E}} \rangle\!\rangle^D)^2$. It is important to recall that $\langle \cdots \rangle$ represents an expectation value over the distribution function $f^w(\Gamma)$ of the *homogeneous* solvent (in the absence of the solute molecule); in the SH approach all the details and nonlinear aspects of the solute–solvent interactions are relegated to the *renormalized* character of the solute–solvent potential energy of interaction $\hat{\Psi}_\Sigma^D$. We discuss the specific aspects of the renormalization procedure in Section V.C.

Based on the expression (5.8) for the surrogate characteristic function, we calculate with the surrogate analogs of Eqs. (4.12) and (4.13) the surrogate estimates of $p_\Sigma^D(\eta)$ and of the reaction free energy $\Delta\mathcal{A}_\Sigma$. Using Eq. (5.6), we then obtain the surrogate free energy profile $F^D(\eta)$.

B.   Diabatic Free Energy Profiles and Solvation Coefficients

We begin with the surrogate estimate of the free energy change $\Delta\mathcal{A}_\Sigma$, which we calculate with the analog of Eqs. (4.13). Using Eqs. (5.8)–(5.10) we get

$$\Delta\mathcal{A}_\Sigma = \Delta E + \Delta A_\Sigma \qquad (5.12a)$$

where

$$\Delta A_\Sigma = -\beta^{-1} \ln g_\Sigma^P(z = i\beta) = \langle \hat{\mathcal{E}} \rangle - \beta\langle \delta\hat{\mathcal{E}}\,\delta\hat{\mathcal{N}} \rangle \qquad (5.12b)$$

With regard to the surrogate estimate of the vertical energy gap probability density $p_\Sigma^D(\eta)$, Eqs. (4.12) and (5.8) give a Gaussian

$$p_\Sigma^D(\eta) = (\beta\kappa_\Sigma/2\pi)^{1/2} \exp[-\beta\kappa_\Sigma(\eta - \eta_\Sigma^D)^2/2] \qquad (5.13)$$

where η_Σ^D and κ_Σ were defined, respectively, in Eqs. (5.9) and (5.10).

Finally, we consider the free energy profiles, using Eqs. (5.6), (5.12), and (5.13). As expected from a (renormalized) linear response formulation, the SH theory gives harmonic diabatic free energy profiles with the same curvatures [4–6]:

$$F_\Sigma^P(\eta) = \frac{1}{4\lambda_\Sigma}(\eta - \eta_\Sigma^P)^2 \tag{5.14a}$$

$$F_\Sigma^S(\eta) = (\Delta E + \Delta A_\Sigma) + \frac{1}{4\lambda_\Sigma}(\eta - \eta_\Sigma^S)^2 \tag{5.14b}$$

where η_Σ^P and η_Σ^S [cf. Eq. (5.9)] are, respectively, the values of the reaction coordinate at which $F_\Sigma^P(\eta)$ and $F_\Sigma^S(\eta)$ have their minima (Fig. 1), while λ_Σ is related to the curvature κ_Σ [cf. Eq. (5.10)] of the profiles as $\lambda_\Sigma = (2\kappa_\Sigma)^{-1}$. From Eqs. (5.9), (5.10), and (5.12) then follow the important relations

$$\eta_\Sigma^P = \Delta A_\Sigma + \lambda_\Sigma \qquad \eta_\Sigma^S = \Delta A_\Sigma - \lambda_\Sigma \tag{5.15}$$

Equations (5.14) correspond to the choice $\theta = (\beta/4\pi\lambda_\Sigma)^{-1/2}e^{\beta\mathscr{A}_\Sigma^P}$ for the scaling energy parameter, which makes $F_\Sigma^P(\eta)$ vanish at the minimum [i.e., $F_\Sigma^P(\eta_\Sigma^P) = 0$]. As is clear from Eqs. (5.14) and (5.15), $F_\Sigma^P(\eta)$ and $F_\Sigma^S(\eta)$ depend on the surrogate estimates of two solvation coefficients: the solvent reorganization energy λ_Σ and the change in solvation free energy ΔA_Σ in the $P \rightarrow S$ transition of the solute. The former is defined by either of the first two equalities in Eq. (5.16) (valid for parabolic free energy profiles [5]):

$$\lambda_\Sigma = F_\Sigma^S(\eta_\Sigma^P) - F_\Sigma^S(\eta_\Sigma^S) = F_\Sigma^P(\eta_\Sigma^S) - F_\Sigma^P(\eta_\Sigma^P) = \frac{\beta}{2}\langle\delta\hat{\mathscr{e}}\delta\hat{\mathscr{e}}\rangle \tag{5.16}$$

The last equality gives the expression of λ_Σ in terms of the fluctuations of the surrogate vertical energy gap [5]. It is straightforward to show that $F_\Sigma^P(\eta)$ and $F_\Sigma^S(\eta)$ satisfy Eq. (4.5).

C. Three Implementations of the Surrogate Hamiltonian Theory

When the equilibrium solvation profile $\langle\hat{\rho}(\mathbf{r})\rangle^D$ in Eq. (5.5) is described by the appropriate solute site λ–solvent site j correlation functions $h_{\lambda j}^D(r)$ (or some combination of these functions), Eqs. (5.4) and (5.5) lead to simple expressions for the renormalized solute–solvent potential energy of interaction $\hat{\Psi}_\Sigma^D$ in terms of the RISM[33] solute–solvent site–site direct correlation functions $c_{\lambda j}^D(r)$. This identification involves, however, the assumption that the renormalized solute–solvent interactions in the polarizable ISM description of the solvent are the same as the renormalized solute–solvent interactions in the *equivalent* [5,6] nonpolarizable ISM model of the solvent. We refer to this assumption as the mean field polarizability approximation [5,6]. This simple mean field approximation allows us to group together [using the dielectric fluctuation formula, (3.31)] all terms that depend on the molecular charge susceptibilities $\nu_{a,jm}$ in terms of the optical dielectric constant ϵ_∞ of the solvent. For a full discussion of this development and its implications, we refer the reader to the original reports [5,6].

1. Surrogate Hamiltonian–Renormalized Dielectric Theory

In the SH–renormalized dielectric theory [4–6,16] (SH-RDT) we choose the solvent polarization charge density

$$\hat{\rho}_{\lambda,\mu}(\mathbf{r}) = \sum_{aj} \hat{q}_{aj}\delta(\mathbf{r} - \mathbf{r}_{\lambda,aj}) \tag{5.17}$$

at position $\mathbf{r}$ relative to each solute interaction site λ (i.e., $\mathbf{r}_{\lambda,aj} \equiv \mathbf{x}_{aj} - \mathbf{x}_\lambda$) as the solvent density field $\hat{\rho}(\mathbf{r})$ in Eq. (5.5); that is,

$$\langle\!\langle \hat{\rho}_{\lambda,\mu}(\mathbf{r}) \rangle\!\rangle^D = \langle \hat{\rho}_{\lambda,\mu}(\mathbf{r}) \rangle^D \tag{5.18}$$

Moreover, we assume that the renormalized solute–solvent potential energy of interaction $\hat{\Psi}_\Sigma^D \equiv \hat{\Psi}_{RDT}^D$ when the solute is in the electronic state D has the form [5] [compare with Eq. (3.22)]

$$\hat{\Psi}_{RDT}^D = \sum_\lambda \int d^3\mathbf{r}\, \hat{\rho}_{\lambda,\mu}(\mathbf{r})\varphi_{\Sigma,\lambda}^D(\mathbf{r}) - \frac{1}{2}\sum_{\lambda\lambda'} \int d^3\mathbf{r}\, d^3\mathbf{r}'\, \hat{\mathscr{J}}_{\lambda\lambda'}(\mathbf{r},\mathbf{r}')\Phi_{\Sigma,\lambda\lambda'}^D(\mathbf{r},\mathbf{r}') \tag{5.19}$$

where [cf. Eq. (3.23b)]

$$\hat{\mathscr{J}}_{\lambda\lambda'}(\mathbf{r},\mathbf{r}') = \sum_{aj,bl} \hat{J}_{aj,bl}\delta(\mathbf{r} - \mathbf{r}_{\lambda,aj})\delta(\mathbf{r}' - \mathbf{r}_{\lambda',bl}) \tag{5.20a}$$

$$\Phi_{\Sigma,\lambda\lambda'}^D(\mathbf{r},\mathbf{r}') = \varphi_{\Sigma,\lambda}^D(\mathbf{r})\varphi_{\Sigma,\lambda'}^D(\mathbf{r}') \tag{5.20b}$$

Clearly, the first and second terms in Eq. (5.19) correspond, respectively, to the solvent *slow* (configurational) and *fast* (electronic) contributions to the renormalized solute–solvent interactions [5]. To complete the specification of the SH-RDT theory we need to determine the renormalized electrostatic potentials $\varphi_{\Sigma,\lambda}^D(\mathbf{r})$ so that Eq. (5.18) is satisfied. This condition, together with the mean field polarizability approximation mentioned before, identifies the renormalized electrostatic potential $\varphi_{\Sigma,\lambda}^{D}(\mathbf{r})$ as a sum of convolutions of the RISM direct correlation functions $c_{\lambda j}^{D}(r)$ with certain functions of the pure solvent [introduced in Eq. (6.17)] that characterize its density response to an external nonuniform electrostatic potential [4–6,16,54–56].

Having thus specified $\hat{\Psi}_{RDT}^{D}$, the expression for the surrogate vertical energy gap $\hat{\mathscr{E}}$ and the auxiliary dynamical variable $\delta\hat{\mathcal{N}}$ follow automatically from Eqs. (5.3) and (5.11). It remains to calculate the solvent expectation values in Eqs. (5.12) and (5.16), for which we refer the reader to the original report [5]. Here we only explain the nature of the results.

In the SH-RDT theory any solvation coefficient I_{RDT} for the $P \rightarrow S$ electronic transition of the solute (solvent reorganization energy λ_{RDT}, change in solvation free energy ΔA_{RDT}) factors into the generic form

$$I_{RDT} \sim \int_0^\infty dk \, \psi_\mu(k) \Upsilon_{RDT}(k) \tag{5.21}$$

The function $\psi_\mu(k)$, which is a property of the pure solvent, is related to the equilibrium fluctuations of the solvent polarization charge density. The second factor $\Upsilon_{RDT}(k)$ is associated with the solute–solvent interactions, which in the SH theory are described in terms of the site–site intermolecular potentials of interaction instead of a cavity carved out of a continuum dielectric medium.

An example of Eq. 5.21 is the formula for the solvent reorganization energy [5,6,36],

$$\lambda_{RDT} = \frac{1}{\pi} \int_0^\infty dk \, \mathscr{C}(k) \Gamma_{RDT}(k) \tag{5.22}$$

where $\mathscr{C}(k)$ [cf. Eq. (3.33)] is the wavevector-dependent Pekar factor, [5,6,16] defined in terms of the optical dielectric constant ϵ_∞ and the wavevector-dependent static longitudinal dielectric function $\epsilon_L(k)$ of the solvent. Correspondingly,

$$\Gamma_{RDT}(k) = \sum_{\lambda\lambda'} \omega_{\lambda\lambda'}(k) \, \Delta n_{\Sigma,\lambda}(k) \, \Delta n_{\Sigma,\lambda'}(k) \tag{5.23}$$

is the SH-RDT solute–solvent coupling function for the solvent reorganization energy. In Eq. (5.23), $\omega_{\lambda\lambda'}(k)$ is the Fourier transform of the solute site–site intramolecular correlation function [33], while $\Delta n_{\Sigma,\lambda}(k) \equiv n_{\Sigma,\lambda}^{S}(k) - n_{\Sigma,\lambda}^{P}(k)$, represents the Fourier component of the change in the *effective* charge distribution $n_{\Sigma,\lambda}(k) = (4\pi/k^2)^{-1} \varphi_{\Sigma,\lambda}(k)$ of the solute site λ in the $P \rightarrow S$ electronic transition. We emphasize the distinction between

$\Delta n_{\Sigma,\lambda}(k)$ and $\Delta n_{0,\lambda}(\mathbf{k}) = \Delta Q_\lambda e^{i\mathbf{k}\cdot\mathbf{x}_\lambda}$, the corresponding change of the bare or basic charge density. The difference arises from the renormalization of the solute–solvent interactions, which is the principal feature of the SH theory. We refer the reader to the original report [5] for the SH-RDT formula of the change in solvation free energy ΔA_{RDT} in the $P \rightarrow S$ transition.

2. Surrogate Hamiltonian–Renormalized Site Density Theory

A different implementation of the SH method, the renormalized site density theory (SH-RST), is obtained when we choose for the solvent density field $\hat{\rho}(\mathbf{r})$ in Eq. (5.5) each of the microscopic solute–solvent pair densities

$$\hat{n}_{\lambda,j}(\mathbf{r}) = \sum_a \delta(\mathbf{r} - \mathbf{r}_{\lambda,aj}) \tag{5.24}$$

These microscopic densities are the dynamical variables associated with the local density of solvent sites of a given type at a point $\mathbf{r}$ measured relative to the solute interaction site λ.

In the SH-RST theory we also assume that the renormalized solute–solvent potential energy of interaction $\hat{\Psi}_\Sigma^D \equiv \hat{\Psi}_{RST}^D$ when the solute is in the electronic state D is of the form [5,56]

$$\hat{\Psi}_{RST}^D = \sum_\lambda \sum_j \int d^3\mathbf{r}\, \hat{n}_{\lambda,j}(\mathbf{r}) u_{\Sigma,\lambda j}^D(\mathbf{r}) - \frac{1}{2}\sum_{\lambda\lambda'} \int d^3\mathbf{r}\, d^3\mathbf{r}'\, \hat{\mathscr{J}}_{\lambda\lambda'}(\mathbf{r},\mathbf{r}')\Phi_{\Sigma,\lambda\lambda'}^D(\mathbf{r},\mathbf{r}')$$

$$\tag{5.25}$$

where $u_{\Sigma,\lambda j}^D(r)$ is a renormalized (solute site λ)–(solvent site j) potential energy of interaction that needs to be specified. On the other hand, for the solvent fast contribution to $\hat{\Psi}_{RST}^D$ we use the same form as in the SH-RDT theory [cf. Eqs. (5.19) and (5.20)], with an identical interpretation of the renormalized electrostatic potential $\varphi_{\Sigma,\lambda}^D(\mathbf{r})$ [5].

To specify the renormalized potentials $u_{\Sigma,\lambda j}^D(r)$ we impose condition (5.5) in the form $\langle\langle \hat{n}_{\lambda,j}(\mathbf{r})\rangle\rangle^D = \langle \hat{n}_{\lambda,j}(\mathbf{r})\rangle^D$ and invoke the mean field polarizability approximation. The outcome is the interpretation [5,56]

$$u_{\Sigma,\lambda j}^D(r) = -k_B T c_{\lambda j}^D(r) \tag{5.26}$$

of the renormalized potential in terms of the RISM direct correlation function. This interpretation of $u_{\Sigma,\lambda j}^D(r)$ in the SH-RST theory coincides with a similar interpretation for the renormalized solute–solvent interactions in the Gaussian bath approximation of Chandler and co-workers [57].

The SH-RST formulas for $\hat{\mathscr{E}}$ and $\delta\hat{\mathscr{N}}$ follow from Eqs. (5.3), (5.11), and (5.25). We refer the reader to the original report [5] for calculation of the solvent expectation values in Eqs. (5.12) and (5.16).

The solvation coefficients I_{RST} for the $P \to S$ transition under the SH-RST theory also factor, as in Eq. (5.21), into the product of structure functions of the pure solvent times functions associated with the solute–solvent interactions. The factorization is more complicated, however, involving the summation over each solvent–solvent pair of interaction sites. In particular, the SH-RST formula for the solvent reorganization energy is [5,6,36]

$$\lambda_{\text{RST}} = \frac{\beta}{2(2\pi)^3} \int d^3\mathbf{k} \sum_{jl} S'_{jl}(k)\mathscr{B}_{\text{RST},jl}(k) \tag{5.27}$$

where $S'_{jl}(k)$ is a modified partial structure factor of the pure solvent, and $\mathscr{B}_{\text{BST},jl}(k)$ takes care of the solute–solvent coupling. The difference between $S'_{jl}(k)$ and the standard structure factors $S_{jl}(k) = \rho[\omega_{jl}(k) + \rho h_{jl}(k)]$ is necessary to correct for the fact that in our calculations we actually represent the solvent by a nonpolarizable interaction site model [6]. The SH-RST solute–solvent coupling function

$$\mathscr{B}_{\text{RST},jl}(k) = \sum_{\lambda\lambda'} \omega_{\lambda\lambda'}(k)\,\Delta u_{\Sigma,\lambda j}(k)\,\Delta u_{\Sigma,\lambda' l}(k) \tag{5.28}$$

is expressed in terms of the Fourier transform $\Delta u_{\Sigma,\lambda j}(k)$ of the change $\Delta u_{\Sigma,\lambda j}(r) \equiv u^S_{\Sigma,\lambda j}(r) - u^P_{\Sigma,\lambda j}(r)$ in the renormalized solute–solvent site–site pair potentials of the SH-RST theory [cf. Eq. (5.26)]

We refer the reader to the original reports [5,6] for the SH-RST for the formula of the change in solvation free energy ΔA_{RST} in the $P \to S$ transition.

3. Surrogate Hamiltonian–Harmonic XRISM Approximation

The harmonic XRISM approximation [6] (SH-HXA) is derived by a simple approximation [6] to the energetics of charge transfer reactions calculated with the RISM-HNC (or extended-RISM, XRISM [34]) integral equation. We recall (Sec. IV.C) that one can *exactly* derive the free energy profiles within the context of the XRISM method as long as one is willing to deal with a fictitious solute that couples to the solvent by means of complex-valued interactions [cf. Eq. (4.16)]. (For an alternative using real-valued interactions, see [49].) If we confine ourselves to using the basic real-valued interactions, the XRISM method only gives straightforward access to the solvation free energy change ΔA_{XRISM} in the $P \to S$ transition (as calculated with the Singer–Chandler formula [48]), and to the solvent contributions to the vertical energy gaps [6,54] when the solvent is in unconstrained equilibrium under the field of the solute in the P and S states (see Fig. 1):

$$\eta_{XRISM}^{D} = \sum_{\lambda} \Delta Q_{\lambda} \int d^3\mathbf{r}\, \frac{1}{r} \left[\rho \sum_{j} q_j h_{\lambda j}^{D}(r)\right] \qquad D = P, S \qquad (5.29)$$

in terms of the solute–solvent site–site correlation functions $h_{\lambda j}^{D}(r)$ calculated under XRISM.

To completely reconstruct the solvation contribution to the CT energetics from the fragmented information ΔA_{XRISM}, η_{XRISM}^{P}, and η_{XRISM}^{S}, we invoke in the SH-HXA theory the same harmonic expressions, Eqs. (5.14), for the diabatic free energy profiles

$$F_{HXA}^{P}(\eta) = \frac{1}{4\lambda_{HXA}}(\eta - \eta_{HXA}^{P})^2 \qquad (5.30a)$$

$$F_{HXA}^{S}(\eta) = (\Delta E + \Delta A_{HXA}) + \frac{1}{4\lambda_{HXA}}(\eta - \eta_{HXA}^{S})^2 \qquad (5.30b)$$

The SH-HXA parameters η_{HXA}^{P}, η_{HXA}^{S}, ΔA_{HXA}, and λ_{HXA} are estimated with the help of the XRISM expressions for η_{XRISM}^{P}, η_{XRISM}^{S}, and ΔA_{XRISM}. This is done with the identifications [6]

$$\Delta A_{HXA} \equiv \Delta A_{XRISM} \qquad (5.31)$$

$$\lambda_{HXA} \equiv \frac{\eta_{XRISM}^{P} - \eta_{XRISM}^{S}}{2} = -\frac{1}{2}\sum_{\lambda}\Delta Q_{\lambda}\int d^3\mathbf{r}\,\frac{1}{r}\left[\sum_{j} q_j \Delta h_{\lambda j}(r)\right] \qquad (5.32)$$

We note [cf. Eqs. (4.6) and Fig. 1] that $\eta_{XRISM}^{P} - \eta_{XRISM}^{S}$ is the estimate of the steady-state Stokes shift E_{ss} in the optical $P \to S$ transition according to the XRISM approximation. Consistency with the harmonic dependence on the reaction coordinate in Eqs. (5.30) requires that the HXA vertical energy gaps be given by the equations [6] [cf. Eq. (5.15)]

$$\eta_{HXA}^{P} = \Delta A_{HXA} + \lambda_{HXA} \qquad \eta_{HXA}^{S} = \Delta A_{HXA} - \lambda_{HXA} \qquad (5.33)$$

It is important to realize that the vertical energy gaps η_{HXA}^{D} need not be equal to the XRISM energy gaps η_{XRISM}^{D}, Eq. (5.29). This is because the definition of λ_{HXA}, Eq. (5.32), is just a reasonable device in the case when the profiles are not quadratic.

The final SH-HXA formula for the solvent reorganization energy closely resembles the corresponding expression in the SH-RST theory [6]:

$$\lambda_{\text{HXA}} = \frac{\beta}{2(2\pi)^3} \int d^3\mathbf{k} \sum_{jl} S'_{jl}(k) \mathscr{B}_{\text{HXA},jl}(k) \tag{5.34}$$

in terms of the same corrected solvent partial structure factors $S'_{jl}(k)$ that appear in Eq. (5.27), but with different solute–solvent coupling functions

$$\mathscr{B}_{\text{HXA},jl}(k) = \sum_{\lambda\lambda'} \omega_{\lambda\lambda'}(k) \, \Delta u^{\text{coul}}_{\lambda j}(k) \, \Delta u_{\Sigma,\lambda'l}(k) \tag{5.35}$$

Here $\Delta u_{\Sigma,\lambda j}(k)$ is the same solute–solvent renormalized pair potential as in the SH-RST approximation, while $\Delta u^{\text{coul}}_{\lambda j}(k)$ is the Fourier transform of $\Delta Q_\lambda q_j/r$: namely, the change in the long-range contribution to the *basic* site–site potential energy of interaction between solute site λ and solvent site j.

D.   Example of Diabatic Free Energy Profiles

The purpose of this section is to examine the performance of the polarizability corrections mentioned at the beginning of Section V.C. The benchmarks for the test are the simulation results reported by King and Warshel [40] for a charge separation process in two different (three-site) ISM models of water, dubbed by the authors WAT1 and WAT2. Model WAT1 is nonpolarizable, with effective partial charges q_j at the interaction sites. In contrast, WAT2 is a polarizable model; each interaction site is assigned a permanent charge q_j^0 (with $|q_j^0| < |q_j|$) and an isotropic dipole polarizability α_j. (The two models have slightly different LJ size parameters). The parameters of the models are given in [40]. From the point of view of this work, we consider WAT1 as the nonpolarizable ISM model that corresponds to the polarizable ISM model WAT2.

The solute comprises two interaction sites (d and a), separated by a distance $L = 5\,\text{Å}$ [larger than $(\sigma_d + \sigma_a)/2 = 3.395\,\text{Å}$]. In the P state the solute sites are neutral, while in the S state $Q_d^S = -Q_a^S = 1e$. The LJ parameters of the solute may be found in [40].

In Figure 5(a) we compare the SH-RDT and SH-HXA free energy profiles (calculated without polarizability correction) against the simulated results [40] in the WATI model of water. The energy parameters λ and ΔA are compared in Table I. The figure and the table show that the agreement between the theories and simulation in the nonpolarizable model is very satisfactory. In Figure 5(b) the theoretical and simulation results are

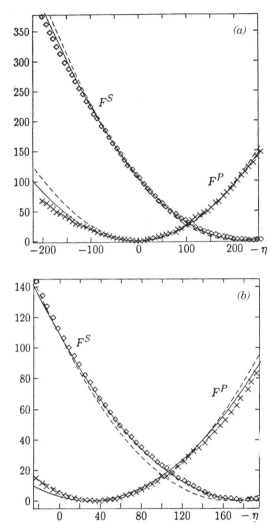

Figure 5.   Polarizability corrections to the diabatic free energy profiles. The solute is a two-site ISM solute that is nonpolar in the P state and undergoes charge separation in water. (a) Free energy profiles in model nonpolarizable water; (b) free energy profiles in a polarizable model of water. See the text and [40] for further details. Crosses and diamonds are the free energy profiles calculated by King and Warshel from an MD simulation [40]. The solid and dashed curves in (a) are the free energy profiles calculated, respectively, in the SH-HXA and SH-RDT approximations with the nonpolarizable water model. The solid and dashed curves in (b) have the same meaning as in (a) and were calculated by applying the correction for electronic polarizability [6] to the solvent reorganization energies associated with the free energy profiles of (a). Free energy and reaction coordinate are in units of kcal mol^{-}1. Notice the difference in the energy scales in (a) and (b).

TABLE I

Comparison Between Theory and Simulation for the Solvation Free Energy Change ΔA and Solvent Reorganization Energy λ of Charge Separation Reactions in water[a]

System	ΔA				λ			
	[40]	HXA	RST	RDT	[40]	HXA	RST	RDT
NP water[b]	−106.3	−105.75	−110.02	−102.36	105.6	112.55	104.50	99.32
P water[c]	−108.03	−105.75	−110.02	−102.36	70.0	77.48	68.42	63.25

[a]For details of the calculation, see the text and the caption of Figure 5.
[b]Charge separation in the nonpolarizable ISM model of water (WAT1) of King and Warshel [40].
[c]Charge separation in the polarizable ISM model of water (WAT2) of King and Warshel [40].

compared for the process in the polarizable WAT2 solvent model. The SH-RDT and SH-HXA free energy profiles in WAT2 were calculated from the respective profiles in WAT1 by applying the polarizability correction [6] to the solvent reorganization energies appearing in the expressions for the free energy profiles of Figure 5(*a*). The corresponding results for λ and ΔA are reported in Table I. The comparison in Figure 5(*b*) and Table I shows that the performance of our method of accounting for the electronic polarizability of the solvent molecules is satisfactory for this system. Unfortunately, at present we cannot judge the accuracy of the polarizability corrections when applied to CT processes in nondipolar solvents.

VI. SOLVENT REORGANIZATION ENERGY

A. Dependence of the Solvent Reorganization Energy with the Donor–Acceptor Distance: Dipolar and Nondipolar Solvents

It is of interest to analyze the difference in the solvent reorganization energy for a given CT process in dipolar and nondipolar solvents. For dipolar solvents the classical Marcus result [1], Eq. (2.1), exhibits an L^{-1} dependence with negative slope, where L is the distance that separates the donor and acceptor groups of the solute. The question naturally arises as to what is the L-dependence of the solvent reorganization energy in a nondipolar solvent, where Eq. (2.1) no longer applies (as discussed in Sec. III.A). In this section we examine this important issue under the SH-RDT theory.

We note that the SH-RDT formulas for λ_{RDT}, Eqs. (5.22) and (5.23), apply immediately to both classes of solvents. The reason for this is that from the point of view of a molecular-level theory (formulated, like the present one, in terms of interaction site models), there is little special about nondipolar fluids. The only "remarkable" consequence associated with the vanishing of the permanent dipole moment of the molecules of a nondipolar solvent is the smallness [58] of the Pekar factor at zero wave-vector. There is no obvious reason for expecting substantial qualitative differences in the behavior of $\mathscr{C}(k)$ at finite values of k for dipolar and nondipolar solvents, and this is clearly demonstrated in the results reported in Figure 2(a) and (b).

For simplicity here we consider a two-site donor–acceptor complex solute, the distance between the donor (d) and acceptor (a) centers being L. For such model the coupling function $\Gamma_{RDT}(k)$ [Eq. (5.23)] is given by

$$\Gamma_{RDT}(k) = |\Delta n_{\Sigma,d}(k)|^2 + |\Delta n_{\Sigma,a}(k)|^2 + 2j_0(kL)\,\Delta n_{\Sigma,d}(k)\,\Delta n_{\Sigma,a}(k) \quad (6.1)$$

where the spherical Bessel function $j_0(kL)$ in the third term is the donor–acceptor intermolecular correlation function of the solute. To extract the L-dependence of λ_{RDT}, we substitute Eq. (6.1) into Eq. (5.22). We obtain

$$\lambda_{RDT}(L) = \lambda_{RDT}^* + \lambda_{RDT}^{\times}(L) \quad (6.2)$$

where the *self term*

$$\lambda_{RDT}^* = \lambda_{RDT}(L = \infty) = \frac{1}{\pi}\int_0^{\infty} dk\,\mathscr{C}(k)[|\Delta n_{\Sigma,d}(k)|^2 + |\Delta n_{\Sigma,a}(k)|^2] \quad (6.3)$$

is independent of L. The L-dependence is confined to the *cross term*

$$\lambda_{RDT}^{\times}(L) = \frac{2}{\pi}\int_0^{\infty} dk\,\mathscr{C}(k)j_0(kL)\,\Delta n_{\Sigma,d}(k)\,\Delta n_{\Sigma,a}(k) \quad (6.4)$$

Without performing calculations there is little that one can say about the magnitude of the self term. This is because λ_{RDT}^* is *essentially* dependent on the effective solute charge densities $\Delta n_{\Sigma,\lambda}(k)$. The case for the cross term is more favorable. In [5] we have shown that when L is larger than the sum $b_d + b_a$ of the "effective" radii of the donor (b_d) and acceptor (b_a) groups, and also under certain mild conditions on the effective charge densities, the cross part $\lambda_{RDT}^{\times}(L)$ for a $P \to S$ process in which $\Delta Q_d = -\Delta Q_a \equiv \Delta Q$ may be expressed in the form

$$\lambda_{RDT}^{\times}(L) = -\frac{(\Delta Q)^2 \mathscr{C}}{L} + R(L) \quad (6.5)$$

The explicit (leading) term is of the classical Marcus–Hush form [1,2], involving the factor $\mathscr{C}/L$, and is clearly independent of the details of the $\Delta n_{\Sigma,\lambda}(k)$. In contrast [5], the remainder

$$R(L) = 4 \sum_{p} (G'_p \cos k_p L - G''_p \sin k_p L) \, \frac{e^{-\nu_p L}}{L} \qquad (6.6)$$

is strongly dependent on the features of the effective charge distributions, very much like the self term λ^*_{RDT}, through the real and imaginary part of the residues, G'_p and G''_p, respectively, of the function $G(\xi) \equiv \mathscr{C}(\xi) \, \Delta n_{\Sigma,d}(\xi) \, \Delta n_{\Sigma,a}(\xi)$ at the poles $\xi_p = k_p + i\nu_p$ in the first quadrant $(k_p > 0, \nu_p > 0)$ of the $\xi = k + i\nu$ complex plane. In principle G'_p, G''_p, k_p, and ν_p could depend mildly on the donor–acceptor distance L. This is because the $\Delta n_{\Sigma,\lambda}(k)$ are determined by the RISM solute–solvent direct correlation functions $\Delta c_{\lambda j}(k)$, which could depend on L. The results of numerical calculations of λ discussed in the example below, as well as those reported in [6], suggest that this residual dependence on L is negligible.

We emphasize that the derivation [5] of Eq. (6.5) does not depend in any sense on whether the solvent is dipolar or nondipolar. This leads to the following observations [6]:

1. For dipolar fluids, for which $\mathscr{C} \neq 0$, Eq. (6.5) would give the familiar L^{-1} dependence for λ_{RDT} if the leading term of Eq. (6.5) dominates the remainder $R(L)$.

2. In contrast, $\mathscr{C} \simeq 0$ for a nondipolar solvent, so that $\lambda^\times_{\mathrm{RDT}}(L)$ is entirely determined by the remainder $R(L)$. There is the possibility, however, that $\lambda^*_{\mathrm{RDT}} \gg \lambda^\times_{\mathrm{RDT}}(L) \simeq R(L)$, in which case the solvent reorganization energy would be practically independent of L.

In model calculations reported in [6] for a series of dipolar and nondipolar solvents we find that the L-dependence of $\lambda_{\mathrm{RDT}}(L)$ implied above in observations 1 and 2 is very well realized. Furthermore, exactly the same results for the L-dependence is calculated for $\lambda_{\mathrm{RST}}(L)$ and $\lambda_{\mathrm{HXA}}(L)$ [6]. As another illustration of the L-dependence of λ in the SH-RDT theory, we examine the solvent reorganization energy for the intramolecular charge transfer reaction

$$\mathrm{DB}_n\mathrm{A} \rightarrow \mathrm{D}^{+1/2}\mathrm{B}_n\mathrm{A}^{-1/2} \qquad (6.7)$$

for an homologous series of model solutes in acetonitrile and benzene. The solutes comprise $n + 2$ interaction sites in a linear arrangement. The solutes have in common the donor D (subscript d) and the acceptor A (subscript a)

groups, but they differ in the number n of bridge interaction sites B. We consider the cases $n = 0, 1, 2, 3$ (when $n = 0$ there is no bridge group connecting sites D and A). All the interaction sites (D, A, and B) are assigned the Lennard-Jones parameters $\sigma = 4.58$ Å and $\varepsilon/k_B = 38.0$ K; these parameters remain unchanged in the $P \rightarrow S$ transition. In each solute molecule the distance between consecutive sites is $l = 3\sigma/4 = 3.435$ Å, so that the distance between the donor and acceptor centers in the solute with n bridge sites is given by the formula $L = (n + 1)l$.

We identify DB_nA and $D^{+1/2}B_nA^{-1/2}$ in Eq. (6.7) with the P and S states of the solute with n bridge sites. The D and A sites are neutral in the P state ($Q_d^P = Q_a^P = 0$), while they acquire the partial charges $Q_d^S = -Q_a^S = e/2$ in the S state (e is the protonic charge). The bridge sites are neutral in both electronic states. Hence the charge transferred in the $P \rightarrow S$ transition is $\Delta Q = \Delta Q_d = -\Delta Q_a = e/2$.

In Figure 6 we report the results for the solvent reorganization energy of the solutes in acetonitrile (solid circles) and in benzene (open circles) as functions of the inverse $1/L$ of the donor–acceptor distance. For acetonitrile we used the three-site ISM model of Edwards et al. [52], while for benzene we used the 12-site model of Jorgensen and Severance [59]. Here we merely note that our calculations take into account the electronic polarizability of the solvent molecules; for more details we refer the reader to [6] and [16].

Acetonitrile and benzene are examples of, respectively, dipolar and non-dipolar solvents. To analyze these results it is convenient to recall the dielectric continuum result, Eq. (2.1), of Marcus [1] and Hush [2] for the solvent reorganization energy. It is important to note that Eq. (2.1) is derived under the condition $L \geq b_d + b_a$. This excludes the $n = 0$ solute, for which the donor and acceptor Lennard-Jones spheres overlap.

Figure 6 shows that in acetonitrile the $1/L$ dependence of the solvent reorganization energy predicted by Eq. (2.1) is well obeyed. This is in agreement with the observation 1 following Eq. (6.6), as we discuss now in more quantitative terms. The continuum model [cf. Eq. (2.1)] gives a slope $d\lambda_{cont}/d(1/L) = -(\Delta Q)^2 \mathscr{C} = -42.8$ kcal mol^{-1} Å, while the slope according to the SH-RDT (calculated by a linear least-squares fit to the data of the $n = 1, 2, 3$ solutes) is $d\lambda_{RDT}/d(1/L) = -42.0$ kcal mol^{-1} Å. Thus the slopes of the solvent reorganization energy versus $1/L$ for the charge transfer reaction in acetonitrile (or, more generally, in a *dipolar* solvent) predicted by the continuum and molecular theories are practically the same.

To compare the magnitudes of λ_{cont} and λ_{RDT} in acetonitrile or any other solvent we need to know the cavity radii b_d and b_a of the donor and acceptor groups. These depend on the solvent, an issue examined in [6].

The results for the reorganization energy in benzene are also in agreement with observation 2 following Eq. (6.6). We first note that the Pekar factor

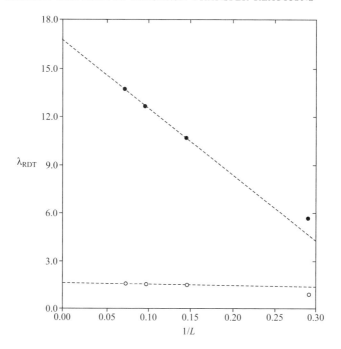

Figure 6. Solvent reorganization energy λ_{RDT} [see Eq. (5.22)] for the model charge transfer reactions $DB_nA \rightarrow D^{+1/2}B_nA^{1/2}$ ($n = 0, 1, 2, 3$) in acetonitrile (solid circles) and benzene (open circles). Dependence of λ_{RDT} on the inverse of L, the donor–acceptor distance. The solute models are straight chains of $n + 2$ overlapping interaction sites. The Lennard-Jones parameters, which are the same for all $n + 2$ solute sites, are given in the text. The dashed lines correspond to least squares fits to the data points of the solutes with $n = 1, 2$, and 3 bridge interaction sites. λ_{RDT} is in units of kcal mol^{-1}; L is in units of Å.

[cf. Eq.(2.2)] for benzene is $\mathscr{C} \simeq 0.007$. Then, using the naive estimate $b_d = b_a \simeq \sigma/2 = 2.29$ Å, we find with Eq. (2.1) that $\lambda_{cont} \simeq 0.2$ kcal mol^{-1} (or approximately 0.009 eV) for the DB_2A solute. The SH-RDT results in Figure 6 are significantly different. The figure shows that $\lambda_{RDT} \simeq 1.5$ kcal mol^{-1} (or $\simeq 0.07$ eV) for the $n = 1, 2, 3$ solutes, which is about an order of magnitude larger than predicted with Eq. (2.1). To put these results in perspective, we note that the charge transferred in Eq. (6.7) is $\Delta Q = e/2$. The solvent reorganization energy for the transfer of a whole electron would be approximately $4 \times \lambda_{RDT} \sim 0.28$ eV, which is in the range of values reported in the literature for electron transfer reactions in benzene and other nondipolar solvents [4–6,9–12]. Thus, although small, the solvent reorganization energy in benzene is much larger than the prediction of the continuum dielectric formula, and certainly not negligible.

Figure 6 also shows that in contrast with the results in acetonitrile, the L-dependence of λ_{RDT} is practically negligible in benzene. The two observations, $\lambda_{RDT} \neq 0$ and $d\lambda_{RDT}/d(1/L) \approx 0$, together with Eqs. (6.2) and (6.5), indicate that the situation indicated in observation 2 following Eq. (6.6) is fully realized in this example.

B. Spatial Resolution of the Solvent Reorganization Energy

In this section we analyze the behavior of several structural functions associated with the solvent reorganization energy, with emphasis on the comparison between the molecular SH-RDT and SH-HXA theories on one hand, and a nonlocal dielectric cavity theory (DCT) theory on the other. (We do not discuss the results of the SH-RST theory in this section; they are very similar to those of the SH-HXA theory.)

The DCT theory is based on a simple cavity model [6,54,60]

$$\Delta c_{\lambda j}(r) = \begin{cases} -\beta \, \Delta Q_\lambda q_j/r & \text{for } r > b_\lambda \\ -\beta \, \Delta Q_\lambda q_j/b_\lambda & \text{for } r < b_\lambda \end{cases} \tag{6.8}$$

for the solute–solvent RISM direct correlation functions. Here b_λ is an effective radius of solute site λ and is taken to be independent of the type of solvent site j. In [6] and [54] we showed that Eqs. (6.8) give a very simple expression for the effective charge density [6,54]

$$\Delta n_{DCT,\lambda}(k) = \Delta Q_\lambda j_0(kb_\lambda) \tag{6.9}$$

where $j_0(kb_\lambda)$ is the spherical Bessel function of zero order. The subscript DCT (dielectric cavity theory) emphasizes that the size parameters b_λ are not derived from a molecular theory. Based on $\Delta n_{DCT,\lambda}(k)$, Eq. (5.23) gives the corresponding coupling function $\Gamma_{DCT}(k)$.

We compare the SH-RDT, SH-HXA, and DCT theories for a simple CT transfer process, in which a monoatomic solute (a single interaction site, index α) undergoes ionization from the P state, with charge $Q_\alpha^P = 0$ to the S state with charge $Q_\alpha^S = e/2$. We refer to this CT process as the $P \to S$ *transition of the α-solute*. We study the CT process in acetonitrile and benzene, which we regard as useful paradigms of dipolar and nondipolar solvents, respectively. To calculate the solvent reorganization energy with the SH-RDT and SH-HXA theories we require a representation of the solute in terms of the solute–solvent site–site intermolecular potentials $u_{\alpha j}(r)$. We choose the solute LJ parameters $\varepsilon_{\alpha\alpha}/k_B = 38.0\,\text{K}$ and $\sigma_{\alpha\alpha} = 3.6\,\text{Å}$. In addition, to calculate the solvent reorganization energy with the DCT we need a cavity representation of the solute, in terms of its effective cavity radius b_α.

In this example the cavity radii b_α of the α-solute in acetonitrile and in benzene were selected so that in each solvent λ_{DCT} approximately matched the corresponding SH-HXA solvent reorganization energy λ_{HXA}. The values obtained are $b_\alpha = 3.68$ Å in acetonitrile and $b_\alpha = 4.12$ Å in benzene, thus illustrating the dependence of the cavity radius on the solvent.

We now compare the SH-HXA, SH-RDT, and DCT theories in terms of the extent to which different regions of the solvent around the solute contribute to the solvent reorganization energy. To examine this issue it is convenient to introduce r-dependent "integrand" functions $\lambda(r)$, one for each of the three approximations, defined so that $\lambda = \int_0^\infty dr\, \lambda(r)$, where λ is the *global* solvent reorganization energy. From Appendix A of [6] we have

$$\lambda(r) = -2\pi r^2 \, \Delta\rho_{\alpha,\mu}(r)\, \Delta\varphi_\alpha(r) \qquad (6.10)$$

where $\Delta\rho_{\alpha,\mu}(r)$ represents the change of the average solvent polarization charge density at a distance r from the solute [6]. Moreover, a different interpretation of the change in solute electrostatic potential $\Delta\varphi_\alpha(r)$ is required for each of the approximations [6]. These are (1) $\Delta\varphi_\alpha(r) = \Delta\varphi_{0,\alpha}(r) = \Delta Q_\alpha/r$ for the SH-HXA approximation; (2) $\Delta\varphi_\alpha(r) = \Delta\varphi_{\Sigma,\alpha}(r)$, the change in the renormalized electrostatic potential in the SH-RDT theory; and (3) $\Delta\varphi_\alpha(r) = \Delta\varphi_{DCT,\alpha}(r) = \Delta Q_\alpha/r$ when $r > b_\alpha$ and $\Delta Q_\alpha/b_\alpha$ when $r < b_\alpha$ in the DCT theory.

Also of interest is the cumulative solvent reorganization energy function, one for each of the approximations, defined by the equation

$$\lambda^{cum}(r) = \int_0^r ds\, \lambda(s) \qquad (6.11)$$

Notice that $\lambda^{cum}(r) \to \lambda$ when $r \to \infty$.

In Figure 7 we present the $\lambda(r)$ and $\lambda^{cum}(r)$ for the $P \to S$ transition of the α-solute in acetonitrile, while in Figure 8 we present the same functions when the solvent is benzene. It is clear from Figure 7 that in acetonitrile the SH-HXA and SH-RDT theories are in very good agreement with each other, signaling that the nondielectric contribution (see the next section) in λ_{HXA} is small for this process. Even in benzene (Figure 8), where the differences between the SH-HXA and SH-RDT are more pronounced, $\lambda_{RDT}(r)$ is always in phase with $\lambda_{HXA}(r)$. Furthermore, we observe that for a given theory, the features of $\lambda(r)$ and $\lambda^{cum}(r)$ in both solvents are qualitatively similar.

It is interesting to note that for the three approximations, the integrand function $\lambda(r)$ can have negative values in some ranges of r. Despite this, the cumulative function $\lambda^{cum}(r)$ is, under the SH-HXA and SH-RDT theories,

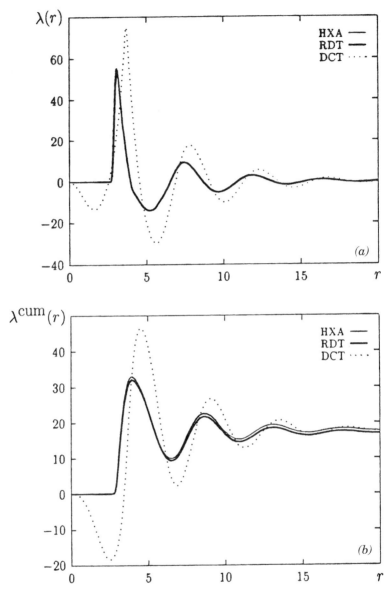

Figure 7. Integrand function $\lambda(r)$ [Eq. (6.10)] and cumulative solvent reorganization energy function $\lambda^{\text{cum}}(r)$ [Eq. (6.11)] for the $P \to S$ transition of the α-solute in acetonitrile: comparison of SH-HXA, SH-RDT, and DCT theories. (a) Integrand function $\lambda(r)$; (b) cumulative solvent reorganization energy function $\lambda^{\text{cum}}(r)$. Thick solid curves, SH-HXA; thin solid curves, SH-RDT; dotted curves, DCT. $\lambda(r)$ is in units of kcal mol^{-1} Å^{-1}; $\lambda^{\text{cum}}(r)$ in units of kcal mol^{-1}; r in units of Å.

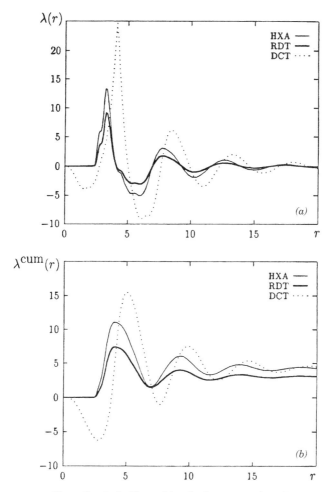

Figure 8. As in Figure 7 but for benzene as the solvent.

positive at every r. This property is closely connected with the fact, evident in the figures, that $\lambda(r)$ is zero when r is smaller than a certain distance, say r^*, indicating that both molecular theories comply with the physical requirement that the contribution to λ from configurations in which the solute and solvent molecules overlap is zero. This feature is not shared by the DCT, for which Figures 7(a) and 8(a) show that $\lambda_{\text{DCT}}(r)$ is negative for $r < r^*$. Hence, although the effective cavity radius b_a has been selected such that $\lambda_{\text{DCT}} \simeq \lambda_{\text{HXA}}$, the solvation structure is not correctly accounted for by the DCT theory.

Figures 7(*b*) and 8(*b*) demonstrate that both in dipolar and nondipolar solvents the largest contribution to λ comes from the shell of solvent closest to the solute. For dipolar solvents this observation is consistent with the results of recent MD simulations [61].

C. Dielectric and Nondielectric Contributions to the Solvent Reorganization Energy

In this section we show that the SH-RDT is embedded in the SH-RST, contributing with what one would generally identify as the dielectric part of the solvation response induced by the $P \rightarrow S$ transition of the solute. To derive the SH-RDT from the SH-RST it is convenient to introduce the projection operator [56]

$$\mathscr{P}_{\mu}^{\mathbf{k}}(\cdots) = \langle (\cdots) \hat{\rho}_{\lambda,\mu}(-\mathbf{k}) \rangle \langle \hat{\rho}_{\lambda,\mu}(\mathbf{k}) \hat{\rho}_{\lambda,\mu}(-\mathbf{k}) \rangle^{-1} \hat{\rho}_{\lambda,\mu}(\mathbf{k}) \qquad (6.12)$$

and its complement $\mathscr{Q}_{\mu}^{\mathbf{k}} = 1 - \mathscr{P}_{\mu}^{\mathbf{k}}$; we recall that $\langle \cdots \rangle$ represents an average with the equilibrium distribution function $f^{w} \sim e^{-\beta H_{w}}$ of the homogeneous solvent. Furthermore,

$$\hat{\rho}_{\lambda,\mu}(\mathbf{k}) = \sum_{aj} \hat{q}_{aj} e^{i\mathbf{k}\cdot\mathbf{r}_{\lambda,aj}} = e^{-i\mathbf{k}\cdot\mathbf{x}_{\lambda}} \hat{\rho}_{\mu}(\mathbf{k}) \qquad (6.13)$$

is the Fourier component of the slow part of the solvent polarization charge density $\hat{\rho}_{\lambda,\mu}(\mathbf{r})$ (relative to the solute interaction site λ) introduced in Eq. (5.17). The second equality follows from substituting $\mathbf{r}_{\lambda,aj} \equiv \mathbf{x}_{aj} - \mathbf{x}_{\lambda}$ and taking into account Eq. (3.25a).

With the help of the projection operators $\mathscr{P}_{\mu}^{\mathbf{k}}$ and $\mathscr{Q}_{\mu}^{\mathbf{k}}$ we can dissect the Fourier component

$$\hat{n}_{\lambda,j}(\mathbf{k}) = \sum_{a} e^{i\mathbf{k}\cdot\mathbf{r}_{\lambda,aj}} = e^{-i\mathbf{k}\cdot\mathbf{x}_{\lambda}} \sum_{a} e^{i\mathbf{k}\cdot\mathbf{x}_{aj}} = e^{-i\mathbf{k}\cdot\mathbf{x}_{\lambda}} \hat{n}_{j}(\mathbf{k}) \qquad (6.14a)$$

$$\delta\hat{n}_{j}(\mathbf{k}) \equiv \hat{n}_{j}(\mathbf{k}) - \langle \hat{n}_{j}(\mathbf{k}) \rangle \qquad (6.14b)$$

of each of the solvent site number densities $\hat{n}_{\lambda,j}(\mathbf{r})$ [cf. Eq. (5.24)] in two orthogonal components. We have

$$\hat{n}_{\lambda,j}(\mathbf{k}) = (\mathscr{P}_{\mu}^{\mathbf{k}} + \mathscr{Q}_{\mu}^{\mathbf{k}})\hat{n}_{\lambda,j}(\mathbf{k}) = \vartheta_{j}(k)\hat{\rho}_{\lambda,\mu}(\mathbf{k}) + \hat{\zeta}_{\lambda,j}(\mathbf{k}) \qquad (6.15)$$

where [5,54–56]

$$\hat{\zeta}_{\lambda,j}(\mathbf{k}) \equiv \hat{n}_{\lambda,j}(\mathbf{k}) - \vartheta_j(k)\hat{\rho}_{\lambda,\mu}(\mathbf{k}) = e^{-i\mathbf{k}\cdot\mathbf{x}_\lambda}\hat{\zeta}_j(\mathbf{k}) \qquad (6.16a)$$

$$\delta\hat{\zeta}_j(\mathbf{k}) \equiv \hat{\zeta}_j(\mathbf{k}) - \langle\hat{\zeta}_j(\mathbf{k})\rangle \qquad (16.6b)$$

is the Fourier component of the part of $\hat{n}_{\lambda,j}(\mathbf{k})$ that is *orthogonal* to the solvent polarization charge density fluctuation $\hat{\rho}_{\lambda,\mu}(\mathbf{k})$, while

$$\vartheta_j(k) \equiv \frac{\langle\delta\hat{n}_j(\mathbf{k})\hat{\rho}_\mu(-\mathbf{k})\rangle}{\langle\hat{\rho}_\mu(\mathbf{k})\hat{\rho}_\mu(-\mathbf{k})\rangle} \qquad (6.17)$$

is the number/charge response ratio, which describes the density response of the solvent (as expressed by the induced density of type j interaction site) to a nonuniform external electrostatic potential [5].

Based on the decomposition (6.15) together with Eqs. (5.19) and (5.25), it is clear that

$$\hat{\Psi}^D_{\text{RST}} = \hat{\Psi}^D_{\text{RDT}} + \frac{1}{(2\pi)^3}\int d^3k \sum_\lambda \sum_j \hat{\zeta}_{\lambda,j}(\mathbf{k})u^D_{\Sigma,\lambda j}(-\mathbf{k}) \qquad (6.18)$$

that is, the SH-RDT renormalized solute–solvent potential energy of interaction is the *dielectric* contribution to $\hat{\Psi}^D_{\text{RST}}$. The name derives from the fact that $\hat{\Psi}^D_{\text{RDT}}$ is expressed in terms of the solvent polarization charge density responsible for the dielectric properties of the solvent. We associate the remaining contribution to $\hat{\Psi}^D_{\text{RST}}$, which is linear in the solvent densities $\hat{\zeta}_{\lambda,j}\mathbf{k})$, as the *nondielectric* part; it originates from the *neutral* solvent densities $\hat{\zeta}_{\lambda,j}(\mathbf{k})$, which are *orthogonal* [5,56]

$$\langle\hat{\zeta}_{\lambda,j}(\mathbf{k})\hat{\rho}_{\lambda',\mu}(-\mathbf{k})\rangle = 0 \qquad (6.19)$$

with respect to the average $\langle\cdots\rangle$ under the equilibrium distribution f^w of the homogeneous solvent. Correspondingly, a similar decomposition applies for the surrogate energy variables

$$(\hat{\mathscr{E}})_{\text{RST}} = (\hat{\mathscr{E}})_{\text{RDT}} + (\hat{\mathscr{E}})_\zeta \qquad (6.20a)$$

$$(\hat{\mathscr{N}})_{\text{RST}} = (\hat{\mathscr{N}})_{\text{RDT}} + (\hat{\mathscr{N}})_\zeta \qquad (6.20b)$$

where $(\hat{\mathscr{E}})_\zeta$ and $(\hat{\mathscr{N}})_\zeta$ represent the nondielectric parts.

It then follows, as a result of the orthogonality property (6.19), that λ_{RST} and ΔA_{RST} separate into *dielectric* (corresponding to the SH-RDT approximation) and *nondielectric* (subscript ζ) components [5,6]:

$$\lambda_{RST} = \lambda_{RDT} + \lambda_\zeta \qquad (6.21a)$$

$$\Delta A_{RST} = \Delta A_{RDT} + \Delta A_\zeta \qquad (6.21b)$$

For example, the nondielectric component of λ_{RST} is given by the expression

$$\lambda_\zeta = \frac{\beta}{2(2\pi)^3} \int d^3\mathbf{k} \sum_{ij} S_{\zeta,jl}(k) \mathscr{B}_{RST,jl}(k) \qquad (6.22)$$

where

$$S_{\zeta,jl}(k) = \left\{ \frac{1}{V} \langle \delta\hat{\zeta}_j(\mathbf{k}) \delta\hat{\zeta}_l(-\mathbf{k}) \rangle \right\}_\infty \qquad (6.23)$$

is the static structure factor of the neutral part of the solvent site densities. The corresponding expression for ΔA_ζ may be found in [4] and [5].

The dissection of SH-RST estimates of the solvent reorganization energy and solvation free energy change into two contributions [Eqs. (6.21)] is quite interesting. The first piece (λ_{RDT} or ΔA_{RDT}) obviously has a dielectric flavor, as indicated by the Pekar factor in Eq. (5.22) for λ_{RDT}. Because of the orthogonality of the μ and ζ components achieved with the projection operators $\mathscr{P}_\mu^{\mathbf{k}}$ and $\mathscr{Q}_\mu^{\mathbf{k}}$, it is tempting to interpret the remaining piece as a measure of nondielectric effects, associated with the solvent neutral density redistribution in the neighborhood of the solute.

An analogous decomposition of λ into polarization (λ_p) and density (λ_d) parts has recently been reported by Matyushov [45]. His development is based on a model solvent comprising hard spheres with embedded point dipoles, so that the theory applies only to dipolar solvents. Inspection of Matyushov's equations shows a clear correspondence between his λ_p and our λ_{RDT}, both of which depend on the wavevector-dependent Pekar factor $\mathscr{C}(k)$ and (quadratically) on the change in the charge distribution of the solute in the $P \to S$ transition. There are, however, some interesting differences between λ_{RDT} and Matyushov's λ_p. The SH-RDT result applies to arbitrary ISM models of the solvent (and is not limited to a two-site model of the solute, as in the applications of Matyushov [45]). Furthermore, the wavevector-dependent Pekar factor associated with the SH-RDT theory captures the contribution of the spatial dispersion of the solvent dielectric response generated by the extended charge distribution of the solvent molecules. Finally, unlike Matyushov's λ_p, our λ_{RDT} also incorporates a contribution from the angle-averaged solute–solvent spatial correlations. Interestingly, this contribution originates from the nonvanishing spatial extent of the molecular charge distribution of the solvent molecules in our theory (in contrast with the point dipole representation of the charge distribution of the solvent molecules in Matyushov's theory).

D.   Examples of Complex Systems

1. $E_T(30)$ Solvent Polarity Scale [6]

Our first application considers the solvatochromic effect on the absorption transition energy of a solute probe in a series of solvents. The magnitude of the solvatochromic effect on the UV/visible $P \to S$ transition of a solute probe offers a very useful scheme for ranking the effective polarity of different solvents [62,63].

In this section we consider the ordering of nine molecular solvents according to the magnitude of the solvent contribution to the average vertical energy gap η^P (absorption) in the $P \to S$ transition of a simple ISM model of the betaine-30 dye [2,6-diphenyl-4-(2,4,6-triphenyl-1-pyridinio)-1-phenolate] shown in Figure 9(a). The pronounced negative solvatochromism of this betaine dye is the basis of the $E_T(30)$ empirical solvent polarity scale of Reichardt [63,64]. The solvatochromic effect stems from the stabilization of the highly dipolar zwiterionic ground state (the P state, with $\mu^P \simeq 15\,\mathrm{D}$) relative to the less dipolar excited state (the S state, with $\mu^S \simeq 6\,\mathrm{D}$) [63].

The solvents considered in this study are carbon tetrachloride [65], toluene [66], benzene [59], tetrahydrofuran [67], chloroform [68], benzonitrile [66], acetonitrile [52], methanol [69], and water [70]. The first three pertain to the class of nondipolar solvents. An immediate problem is how to represent the P and S states of the dye by reasonable ISM models. Consideration of the task of solving the XRISM integral equation for the solute–solvent correlation functions $h_{\lambda j}^D(r)$ and $c_{\lambda j}^D(r)$ for each of the nine solvents suggests that we seek a simple model of the dye involving only a small number of interaction sites. This view is reinforced by the limited information available on the P and S charge distributions of the molecule. We have therefore chosen to represent the solute by a very simple ISM, requiring from the model: (1) consistency with the magnitude of the dye's dipole moment in the ground and excited states (as reported in the literature [63,71]), and (2) intersite distances and volumes in qualitatively agreement with typical values of bond distances and with the molecular volume of the dye. It should be noticed that the dipole moments $\boldsymbol{\mu}^P$ and $\boldsymbol{\mu}^S$ of the ground

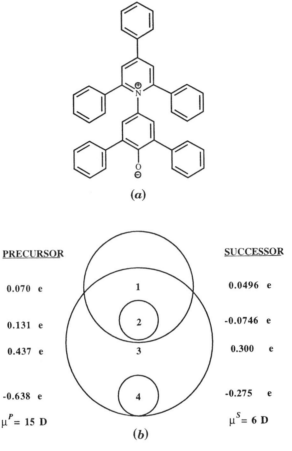

Figure 9. (*a*) 2,6-Diphenyl-4-(2,4,6-triphenyl-1-pyridinio)-1-phenolate betaine-30 dye probe molecule, based on which the $E_T(30)$ empirical solvent polarity scale is constructed; (*b*) interaction site model representation of betaine-30, displaying the partial charges (in units of *e*) of the *P* and *S* states. The Lennard-Jones parameters of the model are given in the text.

and excited states have nearly the same direction (pointing from the oxygen to the nitrogen atom) [63,71].

We represent the *P* and *S* states of the betaine dye with the linear four-site ISM model shown schematically in Figure 9(*b*); the partial charges at the interaction sites in both electronic states are tabulated in the figure. The rationale for the parameters chosen for the model may be found in the original report [6]; here we merely note that sites 2 and 4 represent, respectively, the pyridinium nitrogen and the phenolate oxygen atoms, while sites 1 and 3 correspond, respectively, to the carbon atom in the position para to the nitrogen atom and the carbon bound to the nitrogen atom. Based on these assignments, we choose the site LJ well-depth parameters that carbon, nitrogen, and oxygen have in Jorgensen's OPLS force field: $\varepsilon_{11}/k_B = \varepsilon_{33}/k_B = 55.36$ K (the united-atom representation of the CH group in pyridine [68]); $\varepsilon_{22}/k_B = 85.55$ K (the N atom in pyridine [68]), and $\varepsilon_{44}/k_B = 105.69$ K (the oxygen atom in the carbonyl group [72]), while for the intramolecular distances we choose $l_{12} = 2.805$ Å, $l_{23} = 1.418$ Å, and $l_{34} = 4.127$ Å. The LJ size parameters of sites 1 and 3 are chosen to roughly accommodate the betaine molecule: $\sigma_{11} = 10.0$ Å and $\sigma_{33} = 11.36$ Å; the nitrogen (site 2) and oxygen (site 4) are completely buried in the molecule. The LJ diameter of the nitrogen site has the OPLS value for nitrogen in pyridine $\sigma_{22} = 3.25$ Å, while for σ_{44} we choose $\sigma_{44} = 3.11$ Å so that spheres 3 and 4 are in contact in the phenolate end of the molecule.

Our results are shown in Figure 10, where we compare the calculated and experimental vertical transition energy of the solute in the nine solvents [6]. Represented on the vertical axis is the theoretical estimate (SH-RST, SH-RDT, SH-HXA) of the difference between the vertical transition energy of the solute in a particular solvent, say solvent y, $(h\nu_{\text{abs}})_y = \Delta E + (\eta^P)_y$, and the vertical transition energy in a reference *model* solvent $(h\nu_{\text{abs}})_{\text{ref}} = \Delta E + (\eta^P)_{\text{ref}}$. Hence the transition energy in vacuum ΔE cancels and we are left with $(\eta^P)_y - (\eta^P)_{\text{ref}}$. Correspondingly, represented in the horizontal axis is the experimental absorption energy of betaine-30 in solvent y relative to the absorption energy in the reference solvent tetramethylsilane [the solvent of lowest polarity in the $E_T(30)$ scale; we expect $(\lambda)_{\text{ref}} \simeq 0$ but $(\Delta A)_{\text{ref}} \neq 0$ because of electronic polarizability]. Consistent with the choice of the saturated solvent tetramethylsilane as the experimental reference, we adopt as a model reference a solvent in which the partial charges are set to zero [for such a solvent the theories of Sec. V.C give $(\eta^P)_{\text{ref}} = (\lambda)_{\text{ref}} = (\Delta A)_{\text{ref}} = 0$].

The generally good correlation shown in Figure 10 indicates that the ISM model of betaine-30 together with the SH-RST, SH-RDT, and SH-HXA approximations give a sensible representation of the differential solvation of the P and S states of the dye in a group of solvents that covers a considerable range of polarity, including both dipolar and nondipolar solvents. In particular, we see the correct positions of the nondipolar solvents CCl_4, benzene, and toluene.

In Table II we report the results for the average vertical energy gaps η^P, and the corresponding free energy variation ΔA and solvent reorganization energy λ, all of them relative to the completely nonpolar reference model solute. The results reported for each of the molecular theories incorporate

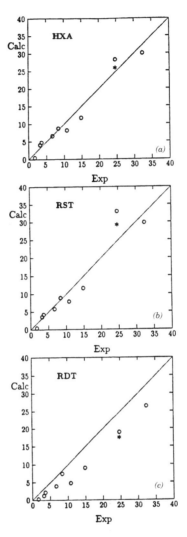

Figure 10. $E_T(30)$ solvent polarity scale: comparison between theoretical (ordinates) and experimental (abscisas) absorption transition energy of the betaine-30 dye in the nine solvents examined in this study. The theoretical transition energies $h\nu_{abs}$ are referred to the transition energy of a reference solvent model in which the partial charges are all zero [which, according to the theories discussed in this paper, is $(h\nu_{abs})_{ref} = \Delta E$]. The experimental transition energies are expressed relative to the transition energy of the betaine-30 dye in tetramethylsilane. (a) SH-HXA, (b) SH-RST, and (c) SH-RDT approximations. The solvents are, in increasing order of the abscissas: CCl_4, toluene, benzene, tetrahydrofuran, chloroform, benzonitrile, acetonitrile, methanol, and water. The circles and asterisks representation for methanol correspond to two different sets of LJ parameters for the hydroxyl H in the ISM model; the parameters are reported in Table II. Transition energies are in units of kcal mol^{-1}.

TABLE II

Solvent Contributions to the Absorption Transition Energy (η^P), Solvation Free Energy (ΔA), and Reorganization Energy (λ) of Betaine-30 (kcal mol^{-1})

Solvent	η^P				λ			ΔA		
	EXP^a	XRISM^{b,c}	RST^b	RDT^b	HXA^b	RST^b	RDT^b	HXA^b	RST^b	RDT^b
CCl$_4$	1.7	0.42	0.39	0.22	0.12	0.09	0.05	0.30	0.29	0.17
Toluene	3.2	3.95	3.48	1.16	1.08	0.82	0.20	2.86	2.66	0.96
Benzene	3.6	4.66	4.17	2.08	1.21	0.91	0.34	3.41	3.28	1.73
THF	6.7	6.56	5.72	3.90	1.45	1.09	0.71	5.10	4.63	3.18
Chloroform	8.4	8.69	8.90	7.43	1.82	1.63	1.40	6.80	7.27	6.03
Benzonitrile	10.8	8.16	7.86	4.78	1.75	1.24	0.72	6.38	6.62	4.05
Acetonitrile	14.9	11.69	11.59	9.05	2.68	2.26	1.91	9.04	9.34	7.14
Methanol^{d,e}	24.7	28.19	33.00	18.93	7.99	9.64	4.65	19.81	23.32	14.28
Methanol^{d,f}	24.7	25.82	29.18	17.45	7.32	8.38	4.28	18.13	20.81	13.18
Water^g	32.4	29.99	29.78	26.39	7.80	7.16	6.16	21.67	22.58	20.26

^{a}Value of $(\eta^P)_{EXP}$ relative to the value in tetramethylsilane.

^{b}Expressed relative to the corresponding value in a model reference solvent with partial charges set equal to zero.

^{c}Calculated with Eq. (5.29) and polarizability corrections; as a useful check for the harmonic nature of the diabatic free energy profiles, the results reported in this column should be compared with $[\lambda_{HXA}$ (column 6) $+ \Delta A_{HXA}$ (column 9)].

^{d}We report results for two variants of the methanol ISM model of [69]; they differ in the LJ interaction parameters of the hydroxyl hydrogen. For both models only the repulsive part of the LJ interaction for the hydroxyl H is considered to calculate the solute–solvent interactions.

^{e}Model represented by circles Figure 10; hydroxyl hydrogen LJ parameters are $\sigma_{HH} = 1.0$ Å, $\varepsilon_{HH} = 27.7$ K for both solute–solvent and solvent–solvent interactions.

^{f}Model represented by asterisk in Figure 10; hydroxyl hydrogen LJ parameters are $\varepsilon_{HH} = 1.0$ Å for solvent–solvent interactions, $\sigma_{HH} = 1.1928$ Å for solute–solvent interactions, and $\sigma_{HH} = 27.7$ K in all cases.

^{g}Only the repulsive part of the LJ interaction for the hydroxyl H is considered to calculate the solute–solvent interactions. Hydroxyl hydrogen LJ parameters are $\sigma_{HH} = 1.251$ Å, $\varepsilon_{HH} = 23.15$ K.

the correction for the solvent electronic polarizability mentioned in Section V.C.

According to Eq. (5.33), the combination $\Delta A_{\text{HXA}} + \lambda_{\text{HXA}}$ corresponds to the precursor vertical energy gap η_{HXA}^P in the SH-HXA theory. The results for η_{HXA}^P should be contrasted with the XRISM results, η_{XRISM}^P [Eq. (5.29)], that are also reported in Table II. This comparison can give us an indication of a departure from a quadratic dependence of the diabatic free energy profiles $F^D(\eta)$ with the reaction coordinate η. From the excellent agreement between η_{XRISM}^P (column 3 in Table II) and $\eta_{\text{HXA}}^P = \Delta A_{\text{HXA}} + \lambda_{\text{HXA}}$ [(column 6) + (column 9)] we conclude that the harmonic dependence on the reaction coordinate is a reasonable approximation to the fully XRISM free energy profiles for the betaine-30 $P \rightarrow S$ transition in the nine solvents studied in this work.

Inspection of Table II shows that $\lambda_{\text{RST}} \geq \lambda_{\text{RDT}}$ in every case. This is a necessary condition that follows from the fact that the RDT theory gives only the dielectric part of the more general RST theory (Sec. VI.C). In general, the SH-HXA and SH-RST theories give comparable results for all the energetic quantities η^P, ΔA, and λ. On the other hand, the RDT gives values for the same energy quantities that are always smaller in magnitude than the results of the other two molecular theories. Notice, however, that the solvent polarity order predicted by the three approximations is the same.

Possible reasons for the reversal of the polarity ranking of chloroform and benzonitrile, as calculated by the three theories, were discussed in [6]. Also in this reference we discuss the difficulties with the ISM model of methanol.

2. Stokes Shifts of Coumarin-153 in Solvents of Different Polarity [36]

In this section we compare the predictions of the surrogate Hamiltonian theories (SH-RDT, SH-RST, and SH-HXA) with the experimental result [12] for the Stokes shifts E_{ss} of the dye coumarin-153 in a series of solvents covering a wide range of polarity. In view of the (albeit renormalized) linear response character of the SH theories, the Stokes shift is calculated as twice the solvent reorganization energy.

We focus on coumarin-153 (C153) because for this solute an extensive set of measurements of E_{ss} on a number of solvents has been reported [12]. Furthermore, Kumar and Maroncelli [73] have developed ISM models for C153 in both the P and S states, which we use as the basis for developing the simplified ISM models for C153 used in our calculations.

We constructed a 19-site model (Fig. 11 and Table III) that is a united-atom reduced version of a 36-site model used by Kumar and Maroncelli [73]. Auxiliary calculations of E_{ss} of C153 in MeCN and a few other solvents

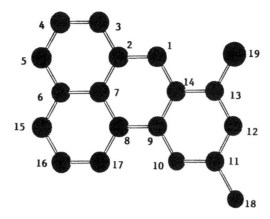

Figure 11. Ball-and-stick representation of the interaction site model of C153. The ISM model has 19 interaction sites and is constructed on the basis of the more detailed model with 36 interaction sites developed by Kumar and Maroncelli [73]. The code numbers for the interaction sites and the respective partial charges in the ground (*P*) and excited (*S*) electronic states are given in Table III, with the Lennard-Jones parameters of the model (the same in both electronic states).

TABLE III
Model Parameters of C153[a]

Site Label λ	Group	Q_λ^P/e	Q_λ^S/e	σ (Å)	ε/k_B (K)
1	CH	−0.058237	−0.229037	3.75	55.3543
2	C	−0.016544	0.1544560	3.5	40.27
3	CH_2	0.085098	0.0752980	3.905	59.38
4	CH_2	0.022271	0.037371	3.905	59.38
5	CH_2	0.039630	0.037830	3.905	59.38
6	N	−0.104904	0.075496	3.25	85.62
7	C	0.170811	0.064811	3.5	40.27
8	C	−0.260409	−0.19771	3.5	40.27
9	C	0.432737	0.414537	3.5	40.27
10	O	−0.456873	0.456373	2.96	105.8
11	C	0.970242	0.925642	3.5	40.27
12	CH	−0.398171	−0.530371	3.75	55.354
13	C	0.357045	0.0425450	3.5	40.27
14	C	−0.320389	0.010711	3.5	40.27
15	CH_2	0.054793	0.062293	3.905	59.38
16	CH_2	0.001805	0.018405	3.905	59.38
17	CH_2	0.171611	0.161511	3.905	59.38
18	O	−0.609718	−0.597918	2.96	105.8
19	CF_3	−0.080795	−0.0688950	4.5	52.8

[a]See Figure 11 for the site labels.

showed that our calculated Stokes shifts do not depend significantly on whether we use the 19- or 36-site C153 model. Since it takes much longer to perform the calculations for the 36-site model, here we only report results obtained with the 19-site model.

The 14 solvents considered in this study are listed in the caption of Figure 12; six of them are nondipolar. Their ISM models were either taken directly from the literature (see references in Sec. VI.D.1) or were constructed from the values recommended in the OPLS force field.

In Figure 12 we represent E_{ss} calculated under the SH-RDT, SH-RST, and SH-HXA approximations against the corresponding experimental data in 14 solvents. The figure shows that all three approximations basically reproduce the experimental data, with the best results given by the SH-HXA approximation. We note in particular that E_{ss} in the nondipolar solvents 1,4-difluorobenzene, benzene, and toluene is comparable to the Stokes shift in chloroform; this interesting feature is well reproduced by the SH theories. The slope $dE_{ss,\Sigma}/dE_{ss,exp}$ is basically unity for each of the SH approximations; closer inspection reveals, however, some minor reversals. It is not clear whether these are due to imperfections of the theory or to inadequacy of the ISM models.

It is evident that in each solvent E_{ss} under SH-RDT is consistently smaller than under the other two SH approximations. This is a manifestation of the fact that λ_{RDT} only derives from the dielectric contribution to the solvent reorganization energy (Sec. VI.C). The contributions due to the solvent number density redistribution around the solute induced by the $P \rightarrow S$ transition are underestimated in the SH-RDT approximation (Sec. VI.C and [5] and [6]).

However, the overall results for the SH-RST and SH-HXA are very encouraging, considering the degree of complexity of the solute and the fact that a considerable number of solvents have large numbers of interaction sites. In view of the complexity of the solute and solvent molecules, the good agreement between the SH calculations and experiment suggests that the SH theory might well be accurate enough to study the influence of the solvent on the energetics of electron transfer reactions in solution.

3.   Complex Donor-Acceptor Solutes: Porphyrin–Quinone Dyads [6]

In a recent paper, Mataga et al. [11a] studied the effect of solvent polarity on the photo-induced intramolecular charge separation (CS) and on the charge recombination (CR) of the charge-separated product for a series of fixed-distance donor–acceptor dyads. Through the use of donor–acceptor combinations among metallated (Zn) or free-base (H_2) porphyrin donors, and among several quinone acceptor groups, the CS reaction was studied over a wide range of driving force in benzene, butyronitrile (BuCN), and

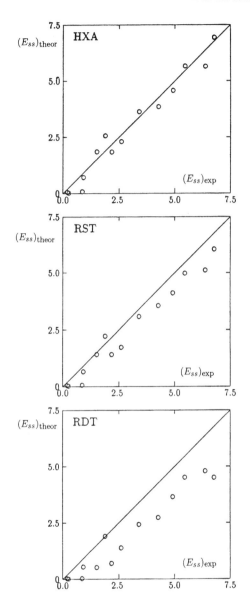

Figure 12. Solvent dependence of the steady-state Stokes shift E_{ss} of C153: theoretical (ordinates) versus the experimental results of [12]. The solvents are in order of increasing abscissas: (1) tetrachloroethylene, (2) carbon disulfide, (3) carbon tetrachloride, (4) 1,3,5-trifluorobenzene, (5) toluene, (6) chloroform, (7) benzene, (8) 1,4-difluorobenzene, (9) tetrahydrofuran, (10) methyl acetate, (11) acetone, (12) dimethyl sulfoxide, (13) acetonitrile, and (14) methanol. E_{ss} is in units of kcal mol^{-1}.

tetrahydrofuran (THF). In every case the CS reaction is found in the normal regime. The authors reported the values of the solvent reorganization energy in benzene, BuCN, and THF extracted from the analysis of the CS rate constant k_{CS} according to the single-mode semiclassical approximation [11a].

To apply the molecular theories to the calculation of λ in the various solvents we need suitable ISM models for the PQ dyads. We present here the results obtained with two and three-site ISM representations of the solutes. In our models we make no distinction between the metallated or free-base porphyrin donor group, nor between the various quinones. Although it is difficult to judge the validity of this approximation [4,74], we note that the same assumption is implicitly made in [11a] to extract λ for each solvent. Given the assumptions associated with the use of the single-mode semiclassical approximation in [11a], it does not seem warranted to use more realistic ISM models for the PQ systems.

Our ISM models are represented schematically in Figure 13. We consider two very simple ISM representations of the PQ dyads, referred to as models 1 and 2, that comprise two and three interaction sites, respectively. Subscripts d and a label the donor (porphyrin) and acceptor (quinone) interaction sites.

The choice of features for our model 1 was guided by the two-sphere cavity model proposed in [11a]. That model comprised spheres of radii

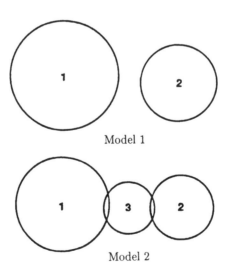

Figure 13.   Two-site (model 1) and three-site (model 2) ISM for the PQ dyads considered in the calculations. The parameters of the models are given in the text.

$b_d = 5$ Å and $b_a = 3.5$ Å, but did not report the donor–acceptor distance l_{da}. We find that $l_{da} = 11$ Å is consistent with the driving forces for CS that can be read on Figure 10 of [11a] for the HQ1M dyad in THF and BuCN. We adopt this value of l_{da}, while for the LJ size parameters we make the simplest possible choice, namely $\sigma_{dd} = 2b_d$ and $\sigma_{aa} = 2b_a$. For the LJ energy parameters we take $\varepsilon_{dd}/k_B = \varepsilon_{aa}/k_B = 75.0$ K (corresponding to the OPLS well-depth ε_{CC} of the carbon in the nitrile group [66,75]). The model is completed with the specification of the partial charges at the d and a sites in the P and S states. We consider the CS reaction, $Q_d^P = Q_a^P = 0 \rightarrow Q_d^S = -Q_a^S = 1e$.

For model 2, which comprises three collinear sites with the donor (d) and acceptor (a) sites at the ends, we also assume the donor–acceptor distance $l_{da} = 11$ Å. A bridge site (subscript b) is included in the model to represent the volume inaccessible to the solvent. As in model 1, we choose $\varepsilon_{dd}/k_B = \varepsilon_{aa}/k_B = \varepsilon_{bb}/k_B = 75.0$ K, while the size LJ parameters of the model are $\sigma_{dd} = 9.0$ Å, $\sigma_{aa} = 6.0$ Å, and $\sigma_b = 5.0$ Å. For the d and a sites the charges in the P and S states are chosen as in model 1; for the b site we choose $Q_b^P = Q_b^S = 0$. The LJ size parameters of this model were chosen so that the result for the solvent reorganization energy λ_{HXA} in THF under the HXA approximation agrees with the experimental result reported in [11a].

In Table IV for models 1 and 2, we compare the theoretical results (SH-HXA, SH-RST, and SH-RDT) with the experimental results [11a] (λ_{exp}) for the solvent reorganization energy in the photoinduced CS reaction of the PQ dyads in various solvents. Table IV also includes, for completeness, the theoretical predictions for the corresponding solvation free energy change ΔA. As mentioned before, comparison with the results of [11a] is complicated by the fact that in that work λ was obtained from the analysis of the dependence of k_{CS} on the driving force $-\Delta \mathcal{A}_{CS}$ according to the single-mode semiclassical approximation. Further uncertainty in the comparison arises from the fact that in [11a] the driving forces $-\Delta \mathcal{A}_{CS}$ in THF and BuCN were obtained from the half-wave potentials for the one-electron oxidation of porphyrins and one-electron reduction of quinones in CH_2Cl_2, after applying a solvation free energy correction for the difference in solvents. The latter correction was calculated [11a] with a continuum (local) dielectric formula that requires specification of a cavity model for the solute. To calculate the $-\Delta G_{CS}$ in benzene, a different approximation was made [11a].

Inspection of Table IV reveals that except for the SH-HXA and SH-RST results in benzene and toluene, the theoretical values of λ for model 1 are smaller than the results of [11a]. For example, λ_{HXA} in THF is about 4 kcal mol^{-1} smaller than λ_{exp}. On the other hand, the results for model 2 of the solute, which was parametrized so that $\lambda_{HXA} \simeq \lambda_{exp}$ in THF, are larger than

TABLE IV
Reorganization Energy and Free Energy Change of PQ Models (kcal mol^{-1})

Solvent	λ				ΔA		
	HXA	RST	RDT	EXP[a]	HXA	RST	RDT
		Two-Site Model					
Toluene	6.32	5.47	1.77		−7.54	−7.40	−3.07
Benzene	5.93	5.19	2.24	4.15	−8.03	−7.66	−4.52
THF	16.35	14.48	11.90	20.16	−27.58	−27.74	−23.02
Benzonitrile	17.28	16.02	12.04		−32.87	−41.37	−27.29
Butyronitrile[b]	25.64	24.72	23.48	26.52			
Acetonitrile	27.30	26.38	24.15		−42.80	−46.41	−40.68
		Three-Site Model					
Toluene	8.07	7.15	2.31		−9.46	−9.46	−3.69
Benzene	7.84	6.92	3.00	4.15	−10.15	−9.92	−5.77
THF	19.60	17.30	14.07	20.06	−32.52	−32.75	−26.98
Benzonitrile	20.52	18.91	14.07		−38.74	−48.20	−31.82
Butyronitrile[b]	30.34	29.18	27.57	26.52			
Acetonitrile	32.29	31.13	26.52		−50.57	−54.42	−48.89

[a]From [11a].
[b]See [6] for details of the calculations with this solvent.

the corresponding values of λ for model 1 and compare better with λ_{exp} in the case of the dipolar solvents. For the nondipolar solvents the solvent reorganization energies calculated for model 2 are almost twice the value of λ_{exp} for the SH-HXA and SH-RST approximations. In addition to the concern with the specific values of λ_{exp} pointed out in the preceding paragraph, it is clear that the modeling of the PQ dyads requires refinement. We note that the conformational flexibility of the BuCN molecules is a feature that is beyond the scope of the integral equation techniques used in our studies; the simplified procedure used to obtain the results presented here is described in [6].

In Table IV, for each theory (SH-RST, SH-RDT, and SH-HXA) the values for λ in benzene and toluene are very similar and not negligible. These qualitative conclusions are entirely in agreement with the results of [11a] (see Fig. 10 of that reference) and therefore may be not too dependent of the detailed features of the model representation of the PQ dyads.

To conclude, we note that for all the solvents reported in Table IV the values calculated for λ_{HXA} and λ_{RST} are rather similar, whereas λ_{RDT} is in most cases considerably smaller. This is a clear indicator of the importance of the nondielectric contribution to λ that is present in the RST and HXA results.

VII.   ADIABATIC ELECTRON TRANSFER REACTIONS

A.   Adiabatic Free Energy Profiles

When the electronic coupling J between the P and S diabatic states of the solute is large, the reaction is better described in terms of the ground (G) and excited (E) adiabatic free energy profiles [76,77]:

$$-\beta F^{G,E}(\eta) = \ln\left[\theta \int d\Omega \int d\Gamma \, e^{-\beta \mathscr{H}^{G,E}} \delta(\eta - \hat{\mathscr{U}})\right] \qquad (7.1)$$

where the ground- and excited-state nuclear Hamiltonians $\mathscr{H}^G$ and $\mathscr{H}^E$ are given by the formula [78]

$$\mathscr{H}^{G,E} = E^N + H_w + \hat{\Psi}^N \mp \tfrac{1}{2}[(\Delta E + \hat{\mathscr{U}})^2 + 4J^2]^{1/2} \qquad (7.2)$$

(The minus and plus signs in this equation correspond, respectively, to the ground G and excited E adiabatic states.) For convenience we have introduced the notation

$$E^N \equiv (E^P + E^S)/2 \qquad (7.3a)$$

$$\hat{\Psi}^N \equiv (\hat{\Psi}^P + \hat{\Psi}^S)/2 \qquad (7.3b)$$

The superscript N emphasizes that E^N and $\hat{\Psi}^N$ may be interpreted, respectively, as the solute electronic energy (in vacuum) and the *basic* solute–solvent potential energy of interaction when the solute is in (an apparent) electronic state N.

From the definitions, Eqs. (4.1) and (7.1), of the diabatic and adiabatic free energy profiles, it is straightforward to show the relation [76,77]

$$F^{G,E}(\eta) = F^N(\eta) \mp \tfrac{1}{2}[(\Delta E + \eta)^2 + 4J^2]^{1/2} \qquad (7.4)$$

where, in analogy with Eqs. (7.3), we introduced the auxiliary free energy profile of the diabatic N-electronic state

$$F^N(\eta) \equiv [F^P(\eta) + F^S(\eta)]/2 \qquad (7.5)$$

To derive Eq. (7.4), we have made use of the exact relation Eq. (4.5). It is important to note that Eq. (7.4), relating the adiabatic and diabatic free energy profiles, follows directly from the definitions, without approximations.

In the surrogate description of the SH theory of solvation, Eq. (7.4) for the adiabatic free energy profiles becomes

$$F_{\Sigma}^{G,E}(\eta) = F_{\Sigma}^{N}(\eta) \mp \tfrac{1}{2}[(\Delta E + \eta)^2 + 4J^2]^{1/2} \tag{7.6}$$

which expresses $F_{\Sigma}^{G,E}(\eta)$ in terms of the combination $F_{\Sigma}^{N}(\eta) = [F_{\Sigma}^{P}(\eta) + F_{\Sigma}^{S}(\eta)]/2$ of the diabatic free energy profiles. With Eqs. (5.14) for the latter we calculate

$$F_{\Sigma}^{G,E}(\eta) = \frac{\Delta E}{2} \mp \tfrac{1}{2}[(\Delta E + \eta)^2 + 4J^2]^{1/2}$$

$$+ \frac{2\eta_{\Sigma}^{P} - \lambda_{\Sigma}}{4} + \frac{1}{4\lambda_{\Sigma}}(\eta - \Delta A_{\Sigma})^2 \tag{7.7}$$

which is a central result of this work. It gives the adiabatic free energy profiles in terms of the solvation coefficients λ_{Σ} [Eq. (5.16)], ΔA_{Σ} [Eq. (5.12b)], and η_{Σ}^{P} [Eq. (5.15)]. These coefficients may be calculated with any of the implementations of the SH theory: SH-RST, SH-RDT, and SH-HXA.

B.   Example of a Model Symmetric Charge Transfer Reaction

We consider the intramolecular model symmetric electron transfer reaction

$$A^{-1/2}B^{1/2} \rightarrow A^{1/2}B^{-1/2} \tag{7.8}$$

(with $A = B$) in three model solvents representative of dipolar aprotic solvents. These systems have been studied by molecular dynamics (MD) simulation by Smith et al. [77]; they reported the corresponding adiabatic free energy profiles and activation free energies. The three solvents, dubbed in [77] "standard" (SD), "low charge" (LC), and "high charge" (HC), comprise dipolar diatomic molecules with fixed intersite separation of 2 Å and have the same number density 0.012 Å^{-3} at $T = 250$ K. The solvent LJ parameters (the same for both sites) are $\sigma = 2.5$ Å and $\varepsilon/k_B = 200$ K. The partial charges, $q = q_+ = -q_-$, at the solvent interaction sites are $q(SD) = e/4$, $q(LC) = e/8$, and $q(HC) = e/2$ (where e is the protonic charge); the corresponding dipole moments are $\mu_w(SD) = 2.4$ D, $\mu_w(LC) = 1.2$ D, and $\mu_w(HC) = 4.8$ D.

The solute particle is also a rigid dumbbell, with the interaction sites separated by a distance of 3 Å. Both sites have the same LJ interaction parameters as the solvent sites. The charge distributions of the P and S states [respectively, the left and right sides of Eq. (7.8)] are

$Q_A^P = -Q_B^P = -e/2$ and $Q_A^S = -Q_B^S = e/2$. In all cases the electronic coupling is $J = 1\,\mathrm{kcal\,mol^{-1}}$.

Because of the symmetry of the models, we have $\Delta E = \Delta A = 0$, and the P and S diabatic free energy profiles satisfy $F^S(\eta) = F^P(-\eta)$ and intersect at $\eta^{\ddagger} = 0$ (the value of the reaction coordinate at the transition state). The adiabatic profiles satisfy $F^{G,E}(\eta) = F^{G,E}(-\eta)$. Another consequence of the symmetry of the models is that the SH-RDT and SH-RST theories give exactly the same results for the free energy profiles. Accordingly, the subscript Σ in the remainder of this section refers to both SH-RDT and SH-RST.

In Figure 14 we represent the SH theory results for the adiabatic ground $F_{\Sigma}^G(\eta)$- and excited $F_{\Sigma}^E(\eta)$-state free energy profiles for reaction in the SD, LC, and HC solvents. The ground-state profiles should be compared with the profiles reported in Figure 5 of [77]. To facilitate this comparison, we note that the minima of the simulated $F^G(\eta)$ occur at (approximately) $\eta \simeq \pm 30$ (SD); ± 17 (LC); ± 47 (HC) $\mathrm{kcal\,mol^{-1}}$. The corresponding minima of the SH profiles occur at ± 29.9 (SD); ± 16.1 (LC); and ± 43.8 (HC) $\mathrm{kcal\,mol^{-1}}$.

In Table V we compare the SH and MD results for the nonadiabatic ($F_{\mathrm{na}}^{\ddagger}$) and adiabatic ($F_{\mathrm{ad}}^{\ddagger}$) activation free energies. These were calculated as follows. For the symmetric exchange reaction under consideration the nonadiabatic activation free energy [calculated with the diabatic profile $F_{\Sigma}^P(\eta)$] is given by the formula

$$F_{\mathrm{na}}^{\ddagger} \equiv F_{\Sigma}^P(\eta = 0) - F_{\Sigma}^P(\eta = \lambda_{\Sigma}) = \frac{\lambda_{\Sigma}}{4} \qquad (7.9)$$

where to derive the second equality we took into account Eq. (5.14a) and that the minimum of $F_{\Sigma}^P(\eta)$ occurs at $\eta_{\Sigma}^P = \lambda_{\Sigma}$.

For the adiabatic activation free energy we have

$$F_{\mathrm{ad}}^{\ddagger} = F_{\Sigma}^G(\eta = 0) - F_{\Sigma}^G[\eta = (\lambda_{\Sigma}^2 - 4J^2)^{1/2}] = \frac{(\lambda_{\Sigma} - 2J)^2}{4\lambda_{\Sigma}} \qquad (7.10)$$

where $\eta = (\lambda_{\Sigma}^2 - 4J^2)^{1/2}$ is the value of the reaction coordinate at which the solvent is equilibrated to the reactants; the second equality follows after taking into account Eq. (7.7).

Figure 14 and Table V show that the adiabatic free energy profiles calculated with the SH theories (SH-RDT and SH-RST) compare fairly well with the MD results, to within 10–15%.

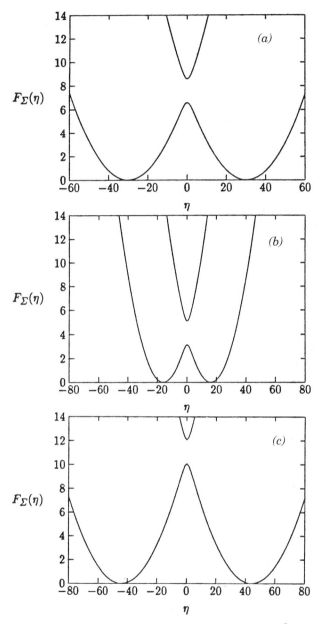

Figure 14.   Ground- and excited-state adiabatic free energy profiles $F_\Sigma^G(\eta)$ and $F_\Sigma^E(\eta)$ for the symmetric exchange reaction (7.8) in dipolar diatomic solvents: (*a*) SD solvent; (*b*) LC solvent; (*c*) HC solvent. The free energy profiles $F_\Sigma^{G,E}(\eta)$ and reaction coordinate η are in units of kcal mol^{-1}.

TABLE V
Activation Free Energy $F^{\ddagger}$ for the Symmetric Exchange Reaction (7.8) (kcal mol^{-1})

Solvent	$\mu_w{}^a$	$F_{na}^{\ddagger}(MD)^b$	$F_{na}^{\ddagger}(SH)^c$	$F_{ad}^{\ddagger}(MD)^b$	$F_{ad}^{\ddagger}(SH)^d$
SD	2.4	8.6	7.55	6.9	6.59
LC	1.2	4.9	4 06	3.2	3.12
HC	4.8	12.5	11.05	11.3	10.07

^{a}Dipole moment of the solvent molecules in units of debye.
^{b}MD results from Table 3 of [77].
^{c}Calculated with Eq. (7.9).
^{d}Calculated with Eq. (7.10).

VIII.   SURROGATE HAMILTONIAN THEORY OF NONEQUILIBRIUM SOLVATION

A.   Solvation Dynamics

The possibility of a dynamical solvent influence on the rate of electron transfer reactions in solution has drawn considerable attention in recent years [79–96]. At the same time, and intimately related, the last decade has witnessed a vast experimental [3d,97–99] and theoretical [100] effort directed to uncover the molecular details of the dynamical solvent response triggered by a jump in the charge distribution of a solute molecule. Typically, the charge jump (by photoexcitation) carries the solute from its ground (or precursor P) state to its first electronic excited (or successor S) state. The solvent, which is initially equilibrated under the field of the P-state solute, is abruptly destabilized by the field of the newly created S-state solute charge distribution. In suitable cases the solute fluoresces, and the ensuing process of nonequilibrium solvation is monitored by the evolution of the solvation time correlation function [3d,97,99,100].

$$\mathscr{L}(t) = \frac{\nu_{flr}(t) - \nu_{flr}(\infty)}{\nu_{flr}(0) - \nu_{flr}(\infty)} \tag{8.1}$$

where $\nu_{flr}(t)$ is the fluorescence frequency of the solute at time t after the charge jump. This description corresponds to the time-dependent fluorescence Stokes shift experiments [3d,97,99,100]. Establishing the detailed molecular mechanism leading to $\mathscr{L}(t)$ is crucially important for understanding the role of nonequilibrium solvation in the rate constants of electron and charge transfer reactions in solution [79–96].

A useful discussion of the wide variety of theoretical methods that have been applied to analyze the solvent dynamics responsible for the features of

$\mathcal{Z}(t)$ has recently appeared [100]. Because of the amount of detailed information that they provide, molecular dynamics computer simulation studies of solvation dynamics have been important aids in the development of additional approximate theories [44b,61,73,77,101–112]. The most recent simulation effort has been directed toward the study of systems using realistic representations of the solute and solvent molecules. Among the more recent approximate treatments are the instantaneous normal-mode analysis [113] (which is also capable of detailed representation of solute and solvent molecules); semianalytical methods that focus on the solvent response and regard the solute as the source of an external potential [114], and a novel continuum dielectric formulation of the solvent that satisfies the dielectric boundary conditions for arbitrary shapes of the solute [115].

In the remainder of this review we describe the generalization of the surrogate Hamiltonian theory of solvation to explain the solvent response triggered by the sudden change of the charge distribution of a solute complex from the precursor P to the successor S electronic state. As in the static case, our goal is a formulation that applies to molecular interaction site models (ISM) of the solute and solvent species [54–56]. Invoking the surrogate Hamiltonian description, the dynamical SH theory generates an approximate nonequilibrium distribution function that describes the state of the solvent as it adapts to a change in the distribution of charges on the solute molecule. The surrogate Hamiltonian is expressed in terms of renormalized solute–solvent interactions, a feature that leads to a simple and natural linear response description of the solvent dynamics in the vicinity of the solute. Unlike the static case, our presentation of the dynamical theory will not consider the electronic degrees of freedom of the solvent molecules. Hence in the following the solvent contribution to the vertical energy gap $\hat{\mathcal{U}}$ will always be given by Eq. (4.17).

B.   Surrogate Nonequilibrium Distribution Function

1.   Basic Description

The photoexcitation of the solute in the time-dependent fluorescent Stokes shift experiment carries the molecule from the precursor (P) or ground electronic state to the successor (S) or excited electronic state. The subsequent solvent response is conveniently monitored by evolution of the solvation time correlation function (TCF) $\mathcal{Z}(t)$ defined in Eq. (8.1). The fluorescence frequency at time t is given by [61b]

$$h\nu_{\text{flr}}(t) = \Delta E + \langle \hat{\mathcal{U}}; t \rangle \tag{8.2}$$

where $\langle \hat{\mathcal{U}}; t \rangle$ is the expectation value of the vertical energy gap under the nonequilibrium distribution function $f(t)$ at time t after the instantaneous $P \to S$ transition. From Eqs. (8.1) and (8.2) it follows that the solvation TCF has the statistical mechanical formulation [54–56]

$$\mathcal{L}(t) = \frac{\langle \hat{\mathcal{U}}; t \rangle - \langle \hat{\mathcal{U}}; \infty \rangle}{\langle \hat{\mathcal{U}}; 0 \rangle - \langle \hat{\mathcal{U}}; \infty \rangle} \tag{8.3}$$

To calculate the right-hand side of Eq. (8.3), we need to consider the nonequilibrium distribution $f(t)$ that describes the state of the solvent at every instant. A convenient approach is to regard the solute as the source of an external field that is fixed in space; in this way we can concentrate on the solvent dynamics driven by this field. Only those microscopic properties that depend on the coordinates and momenta of the solvent interaction sites will be considered as dynamical variables.

The time-dependent Hamiltonian of the system (the solvent in the field of the solute) is

$$H(t) = H_w + \hat{\Psi}(t) \tag{8.4}$$

where

$$\hat{\Psi}(t) = [1 - \theta(t)]\hat{\Psi}^P + \theta(t)\hat{\Psi}^S \tag{8.5}$$

describes the sudden change at $t = 0$ of solute–solvent potential energy of interaction due to the $P \to S$ transition; $\theta(t)$ is the unit step function.

The nonequilibrium distribution function $f(t)$, that describes the state of the solvent at instant t under the field $\hat{\Psi}(t)$ of the solute, satisfies the Liouville equation

$$\partial_t f(t) = -\mathcal{L}(t)f(t) \tag{8.6}$$

where $\mathcal{L}(t)$ is the Liouville operator associated with the time-dependent Hamiltonian $H(t)$ [Eq. (8.4)]. Equation (8.6) can be formally solved by separating $H(t)$ into a time-independent reference part H_0 and a time-dependent perturbation $H_1(t)$. In general, H_0 will comprise the Hamiltonian H_w of the homogeneous solvent and those parts of the solute–solvent potential energy of interaction that do not change in the $P \to S$ transition. We perform the corresponding separation of the Liouville operator into a time-independent reference part and a time-dependent perturbation; that is,

$$\mathcal{L}(t) = \mathcal{L}_0 + \mathcal{L}_1(t) \tag{8.7}$$

Then, with Eqs. (8.6) and (8.7), we can formally express the nonequilibrium distribution function $f(t)$ in terms of its initial value $f(0)$:

$$f(t) = e^{-t\mathcal{L}_0} \exp_+ \left[- \int_0^t dt' \, \tilde{\mathcal{L}}_1(t') \right] f(0) \qquad (8.8)$$

where the ordered exponential is defined by the expansion [116]

$$\exp_+ \left[- \int_0^t dt' \, \tilde{\mathcal{L}}_1(t') \right] = 1 - \int_0^t dt_1 \, \tilde{\mathcal{L}}_1(t_1) + \int_0^t dt_1 \, \tilde{\mathcal{L}}_1(t_1) \int_0^{t_1} dt_2 \, \tilde{\mathcal{L}}_1(t_2) + \cdots$$

$$(8.9)$$

and where $\tilde{\mathcal{L}}_1(t)$ is the perturbation Liouville operator in the interaction representation

$$\tilde{\mathcal{L}}_1(t) = e^{t\mathcal{L}_0} \mathcal{L}_1(t) e^{-t\mathcal{L}_0} \qquad (8.10)$$

In the time-dependent fluorescence Stokes shift experiment the initial condition is $f(0) = f^P$, which corresponds to the solvent in equilibrium under the solute in the electronic state P.

Linearization of Eq. (8.8) with respect of the perturbation $\mathcal{L}_1(t)$ leads to the conventional linear response estimate of $f(t)$. This approach is very useful; it provides the theoretical foundation for the linear response estimates of $\mathcal{X}(t)$ [cf. Eq. (8.39) below] calculated in molecular dynamics equilibrium computer simulations. Unfortunately, the two-body additive nature of the perturbation [$\hat{\mathcal{U}}$ given by Eq. (4.17)] and the fact that the reference operator $e^{t\mathcal{L}_0}$ dictates the time evolution of the solvent in the field of the unperturbed solute makes progress very difficult [54] if the goal is a semi-analytical theory. To overcome this problem we turn to the surrogate Hamiltonian description of the problem.

2. Surrogate Description

In the SH theory of solvation dynamics the system is described by a time-dependent surrogate (subscript Σ) Hamiltonian $H_\Sigma(t)$ of the same form as the basic Hamiltonian $H(t)$ [54–56]. The surrogate Hamiltonian $H_\Sigma(t)$, however, is formulated in terms of renormalized solute–solvent interactions $\hat{\Psi}_\Sigma(t)$, so that a simple linear response treatment is sufficient to describe the equilibrium solvation structure around the solute in the electronic states P (at $t = 0$) and S (at $t = \infty$) [54,56].

In analogy with Eq. (8.6), the surrogate estimate $f_\Sigma(t)$ of the nonequilibrium distribution function satisfies the Liouville equation

$$\partial_t f_\Sigma(t) = -\mathcal{L}_\Sigma(t) f_\Sigma(t) \qquad (8.11)$$

where the renormalized Liouville operator $\mathscr{L}_\Sigma(t)$ is defined in terms of the surrogate Hamiltonian

$$H_\Sigma(t) = H_w + \hat{\Psi}_\Sigma(t) \qquad (8.12)$$

where the time-dependent solute–solvent renormalized potential energy of interaction is the surrogate analog of Eq. (8.5):

$$\hat{\Psi}_\Sigma(t) = [1 - \theta(t)]\hat{\Psi}_\Sigma^P + \theta(t)\hat{\Psi}_\Sigma^S \qquad (8.13)$$

As in Eq. (8.7), we separate the surrogate Liouville operator into two parts:

$$\mathscr{L}_\Sigma(t) = \mathscr{L}_w + \mathscr{L}_{\Sigma,\Psi}(t) \qquad (8.14)$$

corresponding to a partition of $H(t)$ into reference $H_0 = H_w$ and perturbation $H_1(t) = \hat{\Psi}_\Sigma(t)$ parts. Notice that unlike the partition made in the basic description, all the solute–solvent potential energy of interaction is now included in the perturbation. The explicit expression of the operator $\mathscr{L}_\Sigma(t)$ may be found in [55] and [56]. The formal solution of Eq. (8.11) is again given by Eq. (8.8) but with the new interpretation of the reference $\mathscr{L}_0 = \mathscr{L}_w$ and the perturbation $\mathscr{L}_1(t) = \mathscr{L}_{\Sigma,\Psi}(t)$ Liouville operators:

$$f_\Sigma(t) = e^{-t\mathscr{L}_w} \exp_+\left[-\int_0^t dt'\, \tilde{\mathscr{L}}_{\Sigma,\Psi}(t')\right] f_\Sigma(0) \qquad (8.15a)$$

Furthermore, the initial condition is now the one appropriate for the surrogate Hamiltonian description; that is,

$$f_\Sigma(0) = f_\Sigma^P \qquad (8.15b)$$

with the surrogate distribution function f_Σ^P of the solvent in equilibrium with the solute in the P state given by Eq. (5.4).

The renormalized perturbation Liouville operator in the interaction representation is given by $\tilde{\mathscr{L}}_{\Sigma,\Psi}(t) = e^{t\mathscr{L}_w}\mathscr{L}_{\Sigma,\Psi}(t)e^{-t\mathscr{L}_w}$. However, we note that on account of Eq. (8.13), following the photoexcitation $(t \geq 0)$ we have $\mathscr{L}_{\Sigma,\Psi}(t) = \mathscr{L}_{\Sigma,\Psi}^S$, so that $\tilde{\mathscr{L}}_{\Sigma,\Psi}(t) = e^{t\mathscr{L}_w}\mathscr{L}_{\Sigma,\Psi}^S e^{-t\mathscr{L}_w}$, and the dynamics is governed by the Liouville operator of the solvent in the field of the solute in the S state.

To proceed we take into account that due to the renormalization of the solute–solvent interactions (that will be performed in a way similar to the time-independent SH theory), $f_\Sigma(t)$ cannot have a nonlinear dependence on $\delta\hat{\Psi}_\Sigma^D$. With f_Σ^P given by Eq. (5.4), we immediately derive

$$f_\Sigma(t) = e^{-t\mathscr{L}_w}\left[1 - \int_0^t dt'\, \tilde{\mathscr{L}}_{\Sigma,\Psi}(t')\right] f^w - \beta f^w e^{-t\mathscr{L}_w}\delta\hat{\Psi}_\Sigma^P \qquad (8.16)$$

which is of first order in the renormalized solute–solvent interactions.

The final expression for the surrogate nonequilibrium distribution function is, after integration by parts [55,56]

$$f_\Sigma(t) = f_\Sigma^S + \beta f^w e^{-t\mathscr{L}_w} \delta\hat{\mathscr{E}} = f_\Sigma^S + \beta f^w \delta\hat{\mathscr{E}}(-t) \tag{8.17}$$

where $f_\Sigma(\infty) = f_\Sigma^S$ is the surrogate distribution function, Eq. (5.4), of the solvent in equilibrium under the field of the solute in the S electronic state. In view of the definition of $\delta\hat{\mathscr{E}}$ [Eq. (5.11a)], it is simple to verify that the initial condition, Eq. (8.15b), is satisfied.

An important feature of the approximate $f_\Sigma(t)$ given by Eq. (8.17) is that its time dependence is dictated by the Liouville operator $\mathscr{L}_w$ of the homogeneous solvent. In this approximation the solute–solvent coupling is accounted for (in a renormalized way) by the surrogate vertical energy gap dynamical variable $\hat{\mathscr{E}}$.

C. Solvation Time Correlation Function

1. Derivation of the Formula for the Solvation Time Correlation Function

We may calculate $\mathscr{L}(t)$ by using the surrogate analog of Eq. (8.3); that is,

$$\mathscr{L}_\Sigma(t) = \frac{\langle\!\langle\hat{\mathscr{E}}; t\rangle\!\rangle - \langle\!\langle\hat{\mathscr{E}}; \infty\rangle\!\rangle}{\langle\!\langle\hat{\mathscr{E}}; 0\rangle\!\rangle - \langle\!\langle\hat{\mathscr{E}}; \infty\rangle\!\rangle} \tag{8.18}$$

where $\langle\!\langle\hat{\mathscr{E}}; t\rangle\!\rangle$ represents the expectation value of the surrogate vertical energy gap $\hat{\mathscr{E}}$ under the distribution function $f_\Sigma(t)$. With Eq. (8.17) we calculate

$$\langle\!\langle\hat{\mathscr{E}}; t\rangle\!\rangle = \int d\Gamma\, \hat{\mathscr{E}} f_\Sigma(t) = \langle\!\langle\hat{\mathscr{E}}\rangle\!\rangle^S + \beta\langle\delta\hat{\mathscr{E}}(t)\delta\hat{\mathscr{E}}\rangle \tag{8.19}$$

where, as in Section V.A, the symbols $\langle\!\langle\cdots\rangle\!\rangle^S$ and $\langle\cdots\rangle$ indicate, respectively, expectation values with the equilibrium distribution functions f_Σ^S [Eq. (5.4)] and f^w of the homogeneous solvent. Furthermore, we emphasize again that according to the SH dynamical theory, the time evolution in the TCF $\langle\delta\hat{\mathscr{E}}(t)\delta\hat{\mathscr{E}}\rangle$ is governed by the Liouville operator $\mathscr{L}_w$ of the homogeneous solvent.

Combining Eqs. (8.18) and (8.19), we obtain [54–56]

$$\mathscr{L}_\Sigma(t) = \frac{\langle\delta\hat{\mathscr{E}}(t)\delta\hat{\mathscr{E}}\rangle}{\langle\delta\hat{\mathscr{E}}\,\delta\hat{\mathscr{E}}\rangle} \tag{8.20}$$

which is a central result of the SH nonequilibrium theory.

2. Other Observables

The expectation value of a dynamical variable $\hat{\mathscr{G}}$ under $f_\Sigma(t)$ [Eq. (8.17)] is given by the analog of Eq. (8.19):

$$\langle\langle\hat{\mathscr{G}};t\rangle\rangle = \langle\langle\hat{\mathscr{G}}\rangle\rangle^S + \beta\langle\delta\hat{\mathscr{G}}(t)\,\delta\hat{\mathscr{E}}\rangle$$

$$= \langle\langle\hat{\mathscr{G}}\rangle\rangle^S + \beta\langle\delta\hat{\mathscr{G}}\,\delta\hat{\mathscr{E}}\rangle\Phi_{\mathscr{G}}(t) \qquad (8.21a)$$

where we have introduced the normalized equilibrium TCF

$$\Phi_{\mathscr{G}}(t) = \frac{\langle\delta\hat{\mathscr{G}}(t)\,\delta\hat{\mathscr{E}}\rangle}{\langle\delta\hat{\mathscr{G}}\,\delta\hat{\mathscr{E}}\rangle} \qquad (8.21b)$$

Therefore, depending on our ability to calculate the equilibrium TCFs $\Phi_{\mathscr{G}}(t)$, the SH theory of nonequilibrium solvation provides a systematic way of calculating other observables of interest for the characterization of the solvation mechanism triggered by the $P \rightarrow S$ transition of the solute. By simple extension of the formulation discussed in [55], together with certain approximations for calculating time correlation functions of collective dynamical variables (summarized in Appendixes A and B), it is possible to analyze the responses $\langle\langle\hat{\mathscr{G}};t\rangle\rangle$ for cases in which $\hat{\mathscr{G}}$ is of the generic form

$$\hat{\mathscr{G}} = \sum_{\lambda'l} \int d^3\mathbf{r}\,\hat{n}_{\lambda',l}(\mathbf{r})G_{\lambda'l}(\mathbf{r}) \qquad (8.22)$$

The functions $G_{\lambda'l}(\mathbf{r})$ determine the physical meaning of $\hat{\mathscr{G}}$ [55]. With suitable choices of the functions $G_{\lambda'l}(\mathbf{r})$, we can investigate the spatial resolution of the solvent response to the sudden transition (Sec. VIII.F.2) [55].

D. Two Implementations of the Dynamical Surrogate Hamiltonian Theory

As in the case of the static SH theory of solvation, we generate different approximations to the surrogate estimate of the solvation TCF $\mathscr{L}_\Sigma(t)$ by choosing different forms of the renormalized solute–solvent potential energies of interaction $\hat{\Psi}_\Sigma^P$ and $\hat{\Psi}_\Sigma^S$. The surrogate procedure [54–56] requires that at $t = 0$ and at $t = \infty$, which correspond to the situation in which the solvent is in equilibrium with the solute in the P and S electronic states, respectively, the solvation structure [measured as in Eq. (5.5) by a set of solute–solvent pair densities $\hat{\rho}(\mathbf{r})$] be described correctly. In other words, we establish the identity of $\hat{\Psi}_\Sigma^P$ and $\hat{\Psi}_\Sigma^S$ by the requirements

$$\langle\!\langle \hat{\rho}(\mathbf{r}); 0 \rangle\!\rangle = \langle\!\langle \hat{\rho}(\mathbf{r}) \rangle\!\rangle^P = \langle \hat{\rho}(\mathbf{r}); 0 \rangle = \langle \hat{\rho}(\mathbf{r}) \rangle^P \qquad (8.23a)$$

$$\langle\!\langle \hat{\rho}(\mathbf{r}); \infty \rangle\!\rangle = \langle\!\langle \hat{\rho}(\mathbf{r}) \rangle\!\rangle^S = \langle \hat{\rho}(\mathbf{r}); \infty \rangle = \langle \hat{\rho}(\mathbf{r}) \rangle^S \qquad (8.23b)$$

Notice that $\langle \cdots ; t \rangle$ and $\langle\!\langle \cdots ; t \rangle\!\rangle$ are nonequilibrium averages [with, respectively, $f(t)$ and $f_\Sigma(t)$] at $t = 0$ and $t = \infty$. On the other hand, $\langle \cdots \rangle^D$ and $\langle\!\langle \cdots \rangle\!\rangle^D$ are equilibrium averages with $f^D \sim \exp[-\beta(H_w + \hat{\Psi}^D)]$ and f_Σ^D [Eq. (5.4)], which correspond to the equilibrium situations at $t = 0$ ($D = P$) and at $t = \infty$ ($D = S$).

1. Surrogate Hamiltonian–Renormalized Dielectric Theory

In the SH-RDT theory the renormalized solute–solvent potential energy of interaction has the form of Eq. (5.19) without the second term associated with the solvent fast degrees of freedom. In Fourier space [54–56]

$$\hat{\Psi}_{\text{RDT}}^D = \frac{1}{(2\pi)^3} \int d^3k \sum_\lambda \hat{\rho}_{\lambda,\mu}(\mathbf{k}) \varphi_{\Sigma,\lambda}^D(-\mathbf{k}) \qquad D = P, S \qquad (8.24)$$

where $\hat{\rho}_{\lambda,\mu}(\mathbf{k})$ is the Fourier component of the solvent polarization charge density $\hat{\rho}_{\lambda,\mu}(\mathbf{r})$ at position $\mathbf{r}$ relative to the solute site λ [cf. Eq. (5.17)], while $\varphi_{\Sigma,\lambda}^D(\mathbf{k})$ is the Fourier transform of a renormalized electrostatic potential $\varphi_{\Sigma,\lambda}^D(\mathbf{r})$ associated with the interaction site λ of the solute in the D-electronic state. We recall that the SH-RDT theory gives an expression for $\varphi_{\Sigma,\lambda}^D(\mathbf{k})$ in terms of the site–site solute–solvent RISM direct correlation functions [33] when the solute is in the D-electronic state and certain response functions of the pure solvent [specifically the number/charge response ratios $\vartheta_j(k)$ introduced in Eq. (6.17)]. We refer the reader to [54–56] for more details.

For the surrogate vertical energy gap we have

$$(\hat{\mathscr{E}})_{\text{RDT}} = \hat{\Psi}_{\text{RDT}}^S - \hat{\Psi}_{\text{RDT}}^P = \frac{1}{(2\pi)^3} \int d^3k \sum_\lambda \hat{\rho}_{\lambda,\mu}(\mathbf{k}) \Delta\varphi_{\Sigma,\lambda}^D(-\mathbf{k}) \qquad (8.25)$$

where $\Delta\varphi_{\Sigma,\lambda}(\mathbf{k}) \equiv \varphi_{\Sigma,\lambda}^S(\mathbf{k}) - \varphi_{\Sigma,\lambda}^{P(\mathbf{k})}$ is the Fourier transform of the change in the *renormalized* electrostatic potential due to the solute charge redistribution in the $P \to S$ transition.

From our earlier work [54–56,117] it may be shown that Eq. (8.20) can be rearranged in the form [using Eq. (8.25) for the surrogate vertical energy gap]

$$\mathscr{L}_{\text{RDT}}(t) = \int_0^\infty dk \, \Phi_\mu(k, t) W_{\text{RDT}}(k) \qquad (8.26)$$

where the TCF $\Phi_\mu(k, t)$ was introduced in Eq. (3.28). We recall that $\Phi_\mu(k, t)$ is the TCF that determines the frequency- and wavevector-dependent long-itudinal dielectric function $\epsilon_L(k, \omega)$ of the pure solvent [16,24,118]. Further-more, the weight function [117]

$$W_{\mathrm{RDT}}(k) = \mathscr{C}(k)\Gamma_{\mathrm{RDT}}(k) \bigg/ \int_0^\infty dk \mathscr{C}(k)\Gamma_{\mathrm{RDT}}(k) \qquad (8.27)$$

may be interpreted as a normalized distribution of solute–solvent interaction coupling. The functions $\mathscr{C}(k)$ and $\Gamma_{\mathrm{RDT}}(k)$ were introduced in Eqs. (3.33) and (5.23), respectively.

We notice that Eq. (8.26) has the same general form as Eq. (5.21), with the solvent function ψ_μ dependent not only on the wavevector k but also on time. According to Eq. (8.26), it is the dielectric dynamics of the homogeneous solvent, as expressed in $\Phi_\mu(k, t)$, that is the source of the time dependence of the SH-RDT estimate of $\mathscr{L}(t)$. In this approximation the effect of the solute–solvent interactions is carried by the static weight function $W_{\mathrm{RDT}}(k)$. This factorization (to a function of the homogeneous solvent dynamics times a function of the static solute–solvent structure) is an important approximation and is a distinctive feature of the SH theory of solvation. The renormalized character of $W_{\mathrm{RDT}}(k)$ allows us to bypass the two-time many-point correlation functions that would necessarily appear in a more complete dynamical theory that addresses the inhomogeneity of the solvent in the field of the solute particle.

To implement Eq. (8.26) we require estimates for the pure solvent dynamics and structure, as represented by $\Phi_\mu(k, t)$ and the Pekar factor $\mathscr{C}(k)$. The calculation of $\Phi_\mu(k, t)$ is conveniently performed with the reference memory function approximation [24,118,119] implemented with collective reference dynamical variables. A brief summary of the RFMA methodology is given in Appendix A.

2. Surrogate Hamiltonian–Renormalized Site Density Theory

By its nature, the SH-RDT theory cannot be applied to the study of non-equilibrium solvation in nondipolar solvents (i.e., solvents like benzene or 1,3,5-trifluorobenzene), comprising molecules that lack a permanent dipole moment (or like toluene, for which the dipole moment is very small). As demonstrated experimentally by Maroncelli and co-workers [12], and confirmed theoretically by our work on the energetics of charge transfer reactions [4–6], these solvents are quite capable of responding to changes in the charge distribution of a solute molecule. This is because, although dipole-less, the solvent molecules still have nonzero electric multipoles of higher order (a consequence of having a charge distribution of finite size) that can couple with the *inhomogeneous* (nonuniform) electric field of the P and S electronic states of the solute. Although this type of coupling is captured by the SH-RDT, the results of Sections VI.C and D [6,36] indicate that the SH-RDT is not sufficiently accurate to explain the energetic aspects of the solvation response in this class of solvents; the neutral-density solvation response (contributing primarily to the nondielectric part of the solvation coefficients) turns out to be quite substantial (approximately 40% of the energetics).

There are also other interesting nonequilibrium solvation problems involving dipolar solvents in which the SH-RDT is not adequate. Important examples are the experiments of Berg and co-workers [120] for nonpolar solvation in polar solvents. In these experiments the charge distribution of the P and S electronic states of the solute is the same; what changes are other contributions to the solute–solvent interaction potentials. Recent theoretical work on this problem has been reported by Ladanyi and Stratt [113] and also by Bagchi [121].

To study these problems within the SH theory of nonequilibrium solvation the renormalized site density formulation (SH-RST) is more appropriate [56,122]. As discussed in Section V.C.2, in the SH-RST we formulate the renormalized solute–solvent potential energy of interaction $\hat{\Psi}_{RST}^D$ as the coupling of the individual solvent site densities $\hat{n}_{\lambda,j}(\mathbf{r})$ [Eq. (5.24)] to renormalized solute–solvent site–site interaction potentials $u_{\Sigma,\lambda j}^D(r)$ [Eq. (5.26)].

As in the dynamical formulation of the SH-RDT theory, presently the dynamical SH-RST is limited to nonpolarizable ISM models of the solute and solvent molecules. Therefore, the renormalized solute–solvent potential energies of interaction $\hat{\Psi}_{RST}^P$ and $\hat{\Psi}_{RST}^S$ in Eq. (8.13) are formulated as in Eq. (5.25) but without the second term. In terms of Fourier components

$$\hat{\Psi}_{RST}^D = \frac{1}{(2\pi)^3} \int d^3\mathbf{k} \sum_\lambda \sum_j \hat{n}_{\lambda,j}(\mathbf{k}) u_{\Sigma,\lambda j}^D(-\mathbf{k}) \qquad (8.28a)$$

where $\hat{n}_{\lambda,j}(\mathbf{k})$ was defined in Eq. (6.14) and $u_{\Sigma,\lambda j}^D(\mathbf{k})$ is the Fourier transform of the renormalized SH-RST pair potential introduced in Eq. (5.26). For the SH-RST vertical energy gap, we have

$$(\hat{\mathcal{E}})_{RST} = \hat{\Psi}_{RST}^S - \hat{\Psi}_{RST}^P = \frac{1}{(2\pi)^3} \int d^3\mathbf{k} \sum_\lambda \sum_j \hat{n}_{\lambda,j}(\mathbf{k}) \, \Delta u_{\Sigma,\lambda j}(-\mathbf{k}) \quad (8.28b)$$

where $\Delta u_{\Sigma,\lambda j}(\mathbf{k}) \equiv u_{\Sigma,\lambda j}^S(\mathbf{k}) - u_{\Sigma,\lambda j}^P(\mathbf{k})$.

Using Eq. (8.28b) in $\mathscr{L}_\Sigma(t)$ given by Eq. (8.20) gives the SH-RST formula for the solvation TCF as [56,122]

$$\mathscr{L}_{RST}(t) = \int_0^\infty dk \, \mathrm{tr}[\boldsymbol{\Phi}_n(k, t) \cdot \mathbf{W}_{RST}(k)] \tag{8.29}$$

As in the SH-RDT formula, $\mathscr{L}_{RST}(t)$ "factorizes" into the product of a solvent dynamical matrix $\boldsymbol{\Phi}_n(k, t)$ and a time-independent matrix $\mathbf{W}_{RST}(t)$ that reflects the change in the renormalized solute–solvent interactions in the $P \to S$ transition. The former is defined as

$$\boldsymbol{\Phi}_n(k, t) \equiv \mathbf{F}(k, t) \cdot \mathbf{S}(k)^{-1} \tag{8.30}$$

where $\mathbf{F}(k, t)$ is the matrix of the solvent site–site dynamic intermediate scattering functions

$$F_{jl}(k, t) = \left\{ \frac{1}{V} \langle \delta\hat{n}_j(\mathbf{k}, t) \delta\hat{n}_l(-\mathbf{k}) \rangle \right\}_\infty \tag{8.31}$$

and $\mathbf{S}(k)$ is the corresponding matrix of static structure factors $S_{jl}(k) = F_{jl}(k, t = 0)$. The solute–solvent coupling is expressed at the molecular level in terms of the matrix

$$\mathbf{W}_{RST}(k) = k^2 \mathbf{S}(k) \cdot \mathbf{B}_{RST}(k) \Big/ \int_0^\infty dk \, \mathrm{tr}[k^2 \mathbf{S}(k) \cdot \mathbf{B}_{RST}(k)] \tag{8.30}$$

where $\mathbf{B}_{RST}(k)$ is the matrix of elements $\mathscr{B}_{RST,jl}(k)$ defined in Eq. (5.28).

It is straightforward to calculate $\mathbf{W}_{RST}(k)$ once the change in the solute–solvent direct correlation functions is known. The difficult part of the dynamical SH-RST theory is calculation of the solvent dynamical matrix $\boldsymbol{\Phi}_n(k, t)$. At present only the simplest approximation for $\mathbf{F}(k, t)$ has been considered [122]; it is based on the site–site Smoluchowski–Vlasov (SSSV) theory initially proposed by Hirata [123]. The SSSV is basically an extension to ISM models of the Smoluchowski–Vlasov theory of Calef and Wolynes [124]. A more flexible reformulation of this theory using a matrix version of the reference memory function approximation discussed in Appendix A (with single-particle reference dynamics) gives the time evolution of the $F_{jl}(k, t)$ in terms of their single-molecule (self) counterparts $F_{jl}^s(k, t)$ [125]. We believe, however, that ultimately, a method relying on reference dynamical variables of collective character is required to explain nonequilibrium solvation dynamics at long times. This is because any method based on single-particle reference dynamics leads to an incorrect description of the primary correlation functions in the hydrodynamic stage.

E. Simpler Power-Law Approximation [117]

There has also been some effort in obtaining friendlier expressions for the solvation TCF that would be convenient to implement. For example, we reported [126] a simple new method for estimating time correlation functions of collective dynamical variables, the reference frequency modulation approximation (RFMA). As reviewed briefly in Appendix B, the RFMA leads to the approximate relation [126]

$$\Phi_Y(t) = \Phi_X(t)^{\mathscr{R}_{YX}} \qquad (8.33a)$$

between the normalized TCFs $\Phi_Y(t)$ and $\Phi_X(t)$ [[Eqs. (A.2) and (A.3)] of a pair of dynamical variables Y and X. The exponent

$$\mathscr{R}_{YX} = \mathscr{K}_Y / \mathscr{K}_X \qquad (8.33b)$$

is defined in terms of the ratio of the initial values $\mathscr{K}_Y$ and $\mathscr{K}_X$ of the first memory functions of the variables Y and X [[Eq. (A.5)]. The basis of Eq. (8.33) is the convolutionless generalized Langevin equation reported by Tokuyama and Mori [127]. By applying the RFMA approximation to a simplified version of the expression for the solvation TCF given by the SH-RDT, Eq. (8.26), we recovered the power-law relation

$$\mathscr{L}(t) \simeq \Phi_1(t)^{\alpha} \qquad (8.34)$$

found empirically by Maroncelli and co-workers [128,129]. Both the pure-solvent single-dipole TCF $\Phi_1(t) = \langle \mathbf{\mu}(t) \cdot \mathbf{\mu} \rangle / \langle \mathbf{\mu} \cdot \mathbf{\mu} \rangle$ and the exponent α on the right-hand side of this equation depend only on the solvent. Unfortunately, the approximations that lead to Eq. (8.34) are so drastic that any dependence of $\mathscr{L}(t)$ on the solute features is lost. In contrast, experimental results [130] as well as detailed studies by computer simulation [73] and various approximate theories [56,113] have revealed an important dependence of the solvation TCF on the nature of the solute, especially with respect to the multipolar order of the charge jump.

However, the RFMA result, Eq. (8.33), suggests a simple procedure for estimating the solvation TCF $\mathscr{L}_Y(t)$ of a solute Y in some solvent if the solvation TCF of another solute in the same solvent is known [117]. More specifically, we consider the calculation of $\mathscr{L}_Y(t)$ for the $P \to S$ transition of a solute Y in a given solvent in terms of the (presumed) known solvation TCF $\mathscr{L}_R(t)$ for the $P' \to S'$ transition of a reference solute R in the same solvent. Thus, since the SH estimate of $\mathscr{L}(t)$ [[Eq. (8.20)] is a normalized TCF, we identify the primary TCF $\Phi_Y(t)$ of Eq. (8.33a) with $\mathscr{L}_Y(t)$ and, correspondingly, the reference TCF $\Phi_X(t)$ with $\mathscr{L}_R(t)$. In this way Eq. (8.33a) becomes

$$\mathscr{L}_Y(t) = \mathscr{L}_R(t)^{\mathscr{R}_{YR}} \tag{8.35}$$

From the derivation of the RFMA in Appendix B it follows that Eq. (8.35) corresponds to the choice of the SH vertical energy gaps $(\hat{\mathscr{E}})_Y$ and $(\hat{\mathscr{E}})_R$ [cf. Eq. (5.3)], respectively, as the primary and reference dynamical variables Y and X. Therefore, the renormalization factor $\mathscr{R}_{YR}$ in Eq. (8.35) is given by the ratio of the initial values of the memory functions $(\mathscr{K}_{\mathscr{E}})_Y$ and $(\mathscr{K}_{\mathscr{E}})_R$ of $\mathscr{L}_Y(t)$ and $\mathscr{L}_R(t)$, respectively:

$$\mathscr{R}_{YR} = \frac{(\mathscr{K}_{\mathscr{E}})_Y}{(\mathscr{K}_{\mathscr{E}})_R} = \frac{(\langle|\hat{\mathscr{E}}|^2\rangle/\langle|\hat{\mathscr{E}}|^2\rangle)_Y}{(\langle|\hat{\mathscr{E}}|^2\rangle/\langle|\hat{\mathscr{E}}|^2\rangle)_R} \tag{8.36}$$

Using the formula [54,131,132]

$$(\mathscr{K}_{\mathscr{E}})_Y = -(d^2\mathscr{L}_Y(t)/dt^2)_{t=0} \tag{8.37}$$

[and a similar expression for $(\mathscr{K}_{\mathscr{E}})_R$] together with Eq. (8.26) for the SH-RDT estimate of the solvation time correlation functions, we calculate

$$\mathscr{R}_{YR} = \int_0^\infty dk\, \mathscr{K}_\mu(k) W_{\mathrm{RDT}}^{(Y)}(k) \Big/ \int_0^\infty dk\, \mathscr{K}_\mu(k) W_{\mathrm{RDT}}^{(R)}(k) \tag{8.38}$$

where $W_{\mathrm{RDT}}^{(Y)}(k)$ is the solute–solvent coupling weight function for the solute Y [Eq. (8.27), and a similar interpretation for $W_{\mathrm{RDT}}^{(R)}(k)$], while $\mathscr{K}_\mu(k) = -(d^2\Phi_\mu(k,t)/dt^2)_{t=0}$ is the initial value of the memory function of the polarization charge density TCF $\Phi_\mu(k,t)$. An analogous relation may be derived for the implementation of Eq. (8.35) under the SH-RST theory.

F. Examples

1. General Features of the Solvation Time Correlation Function

Before we present some applications of the SH dynamical theory, it is useful to discuss the types of results that can be generated from molecular dynamics (MD) computer simulations. By $\mathscr{L}_{\mathrm{NEMD}}(t)$ we represent the solvation TCF calculated with the right-hand side of Eq. (8.3), with the brakets representing an average over nonequilibrium trajectories obtained by molecular dynamics. By $\mathscr{L}^D(t)$ (with $D = P$ or S) we indicate the linear response estimate

$$\mathscr{L}^D(t) = \left\{ \frac{\langle \delta \hat{u}(t)\, \delta \hat{u} \rangle}{\langle \delta \hat{u}\, \delta \hat{u} \rangle} \right\}^D \tag{8.39}$$

The symbol $\{\cdots\}^D$ emphasizes the dependence of the *equilibrium* TCF $\mathscr{L}^D(t)$ on the basic Hamiltonian $H^D = H_w + \hat{\Psi}^D$ of the solvent in the field of the solute in the electronic state D. This dependence occurs through (1) the equilibrium distribution function $f^D \sim e^{-\beta H^D}$; (2) the time displacement operator $e^{t\mathscr{L}^D}$, where $\mathscr{L}^D$ is the Liouville operator associated with H^D; and (3) the dynamical variable $\delta \hat{u}^D \equiv \hat{u} - \langle \hat{u} \rangle^D$, with $\langle \hat{u} \rangle^D$ being the expectation value of $\hat{u}$ under f^D. The results to be presented in the figures below under $\mathscr{L}^P(t)$ or $\mathscr{L}^S(t)$ correspond to simulation results by other groups, calculated by *equilibrium* MD simulations of the solvent in the field of the solute in the P or S state.

For another comparison, we also report in some figures an estimate of the solvation TCF $\mathscr{L}(t)$, which corresponds to the dynamical extension of the dielectric cavity theory (DCT) discussed in Section VI.B [24,54]. The DCT estimate of the solvation TCF is given by the formula [54]

$$\mathscr{L}_{\mathrm{DCT}}(t) = \int_0^\infty dk\, \Phi_\mu(k, t) W_{\mathrm{DCT}}(k) \tag{8.40}$$

which should be contrasted with Eq. (8.26) for $\mathscr{L}_{\mathrm{RDT}}(t)$. The solute–solvent coupling weight function

$$W_{\mathrm{DCT}}(k) = \mathscr{C}(k)\Gamma_{\mathrm{DCT}}(k) \Big/ \int_0^\infty dk\, \mathscr{C}(k)\Gamma_{\mathrm{DCT}}(k) \tag{8.41}$$

is defined in terms of

$$\Gamma_{\mathrm{DCT}}(k) = \sum_{\lambda\lambda'} \omega_{\lambda\lambda'}(k)\, \Delta n_{\mathrm{DCT},\lambda}(k)\, \Delta n_{\mathrm{DCT},\lambda'}(k) \tag{8.42}$$

where the change in the DCT effective solute charge density $\Delta n_{\mathrm{DCT},\lambda}(k)$ was introduced in Eq. (6.9). The DCT approximation for the effective charge density, $\Delta n_{\mathrm{DCT},\lambda}(k)$, is expressed in terms of a spherical cavity model (an excluded volume for the solvent molecules at the surface of which electrostatic boundary conditions are met), with the radius b_λ of the cavity appearing as the only vestige of the molecular aspects of the solute–solvent interactions.

We begin with a simple system, to point out the salient features of the solvation TCF. We compare the SH-RDT estimate $\mathcal{L}_{RDT}(t)$ with the MD results for the dynamics of ion solvation in water. The simulation results are taken from the MD study by Maroncelli and Fleming [101], who used the ST2 water model of Stillinger and Rahman [133]. On the other hand our calculations use the transferable intermolecular potential model TIP4P of water due to Jorgensen et al. [70]. The motivation for choosing TIP4P is that a detailed characterization of the frequency-dependent dielectric function ϵ_ω by MD applied to this water model has been given by Neumann [134]. It provides our input for the calculation of $\Phi_\mu(k,t)$ under the RMFA approximation (cf. Appendix A). A further reason is our ability [54,118] to calculate the structure functions $h_{jl}(r)$ of this model under XRISM.

In Figure 15 we compare the predictions of the surrogate theory and the MD results of Maroncelli and Fleming [101] for two of their systems: Figure 15(a) $S0 \rightarrow S+$ (small neutral → small cation), and Figure 15(b) $L0 \rightarrow L+$ (large neutral → large cation). In both cases $Q^P/e = 0$ and $Q^S/e = 1$. The solute LJ parameters are $\epsilon/k_B = 38$ K, $\sigma = 3.1$ Å and $\epsilon/k_B = 2668.6$ K, $\sigma = 6.975$ Å, for the small and large solute, respectively. For $\mathcal{L}_{DCT}(t)$ the radii of the cavities (small solute: $b = 2.5$ Å; large solute $b = 4.5$ Å) were selected as the distance of closest approach between the uncharged solute and the oxygen site of the TIP4P water molecules (determined from the XRISM pair correlation functions).

Figure 15 shows that the surrogate theory agrees very well with the MD results. In this case the difference between the MD curves is not large, indicating that nonlinear effects, if present, are very mild. The SH-RST theory successfully reproduces the initial fast decay of the solvation TCF in water. Furthermore, the important oscillatory features of the MD curves, due to the librational motions of the solvent [101,118,135], appear to be accurately reproduced by the surrogate theory. The overall agreement between theory and simulation is perhaps somewhat surprising, since they refer to different models of the solvent. With respect to the DCT results, for both systems the shape of $\mathcal{L}_{DCT}(t)$ closely resembles that of $\mathcal{L}_{RDT}(t)$ because both calculations are based on the same solvent dynamics $\Phi_\mu(k,t)$. However, for the chosen cavity sizes b the initial decay of the solvation TCF is small compared with the MD and surrogate theories. It is worthwhile pointing out that the cavity results depend strongly on the size b of the cavity.

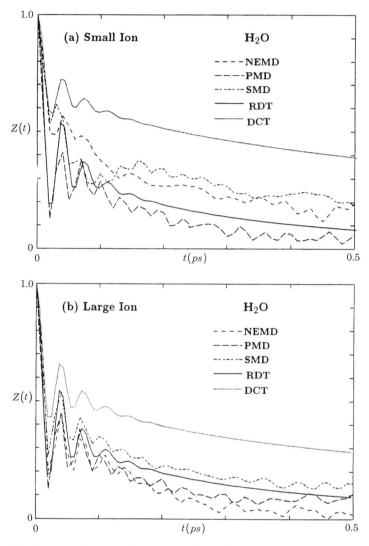

Figure 15. Solvation time correlation functions for an ion in water: (*a*) small ion; (*b*) large ion. The curves NEMD, PMD, and SMD are from MD simulations by Maroncelli and Fleming [101]. They correspond to an ion in the ST2 model water. The theoretical results RDT and DCT use the TIP4P model of water. NEMD, solvation TCF $\mathscr{Z}(t)$ [Eq. (8.3)] calculated from nonequilibrium MD trajectories corresponding to the $P \rightarrow S$ transition; PMD, solvation TCF $\mathscr{Z}^P(t)$ calculated by equilibrium MD [Eq. (8.39)] with the solvent in the field of the uncharged solute; SMD, solvation TCF $\mathscr{Z}^S(t)$ calculated by equilibrium MD [Eq. (8.39)] with the solvent in the field of the charged solute; RDT, solvation TCF $\mathscr{Z}_{RDT}(t)$ calculated with Eq. (8.26); DCT, solvation TCF, calculated with Eq. (8.40). The LJ parameters of the solutes are given in the text. t is in units of ps.

2. Resolution of Global Responses into Spatial Profiles

Here we compare the results of the SH-RDT theory with MD results of Maroncelli [61a] for ion solvation dynamics in the three-site ISM model of acetonitrile developed by Edwards et al. [52]. The $P \rightarrow S$ transition considered is the one labeled $S^0 \rightarrow S^+$ in Maroncelli's work; the solute is spherical with LJ parameters $\varepsilon/k_B = 38$ K and $\sigma = 3.1$ Å, and undergoes an ionization from $Q^P/e = 0$ to $Q^S/e = 1$. As input for the calculation of $\Phi_\mu(k, t)$ with the RMFA methodology (App. A) we use the TCF $\Phi_{M,L}(t)$ [Eq. (A.11)] reported in [52]. In addition to the solvation TCF $\mathscr{Z}(t)$, we also consider other dynamical solvation responses of global and local character (Sec. VIII.C).

In Figure 16 we confront the theoretical estimates $\mathscr{Z}_{RDT}(t)$ and $\mathscr{Z}_{DCT}(t)$ of the solvation TCF with simulation data. We first notice the differences between the three MD curves: $\mathscr{Z}_{NEMD}(t)$ (nonequilibrium trajectories) and $\mathscr{Z}^P(t)$ and $\mathscr{Z}^S(t)$ [equilibrium simulations, Eq. (8.39)]. As Figure 16 shows,

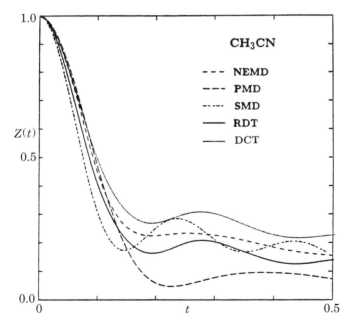

Figure 16. Solvation time correlation functions for ion solvation in acetonitrile. The curves NEMD, PMD, and SMD are from MD simulations by Maroncelli [61a], while the theoretical curves RDT and DCT are based on Eqs. (8.26) and (8.40), respectively. See the caption of Figure 15 for details. The solute LJ parameters of the solute are given in the text. t is in units of ps.

the surrogate estimate $\mathscr{L}_{\text{RDT}}(t)$ agrees surprisingly well with the NEMD response, particularly with regard to the very fast initial decay (ca. 80% within the first 150 fs). In common with $\mathscr{L}_{\text{NEMD}}(t)$, the SH-RDT estimate lies closer to $\mathscr{L}^P(t)$ at short times, whereas at longer times it lies closer to $\mathscr{L}^S(t)$. Furthermore, the small-amplitude oscillatory features of $\mathscr{L}_{\text{RDT}}$ resemble those of $\mathscr{L}_{\text{NEMD}}(t)$ rather than the much more pronounced oscillations shown by $\mathscr{L}^S(t)$ (the oscillations are out of phase, however). For the curve $\mathscr{L}_{\text{DCT}}(t)$ presented in Figure 16, the cavity radius is $b = 2.4$ Å, which corresponds to the distance of closest approach (as inferred from the XRISM pair correlation function) between the uncharged solute and the nitrogen site of acetonitrile. For this choice of cavity radius $\mathscr{L}_{\text{DCT}}(t)$ compares relatively well with the $\mathscr{L}_{\text{RDT}}(t)$ and the solvation TCF's calculated from simulation.

Next we focus on the spatial and temporal aspects of the solvation response, in particular on the contributions $\langle\!\langle [\hat{V}]_n; t \rangle\!\rangle$ and $\langle\!\langle [\hat{\mathscr{E}}]_n; t \rangle\!\rangle$ from solvent region n (defined below) to the nonequilibrium electrostatic reaction potential $\langle\!\langle \hat{V}; t \rangle\!\rangle$ and the surrogate energy gap $\langle\!\langle \hat{\mathscr{E}}; t \rangle\!\rangle$. The dynamical variable associated with the reaction potential is

$$\hat{V} = \int d^3\mathbf{r}\, \frac{\hat{\rho}_{\lambda,\mu}(\mathbf{r})}{r} \tag{8.43}$$

with $\hat{\rho}_{\lambda,\mu}(\mathbf{r})$ given by Eq. (5.17) and r is measured relative to the center of the solute molecule. [Note that in Eq. (8.43) we use the subscript λ to indicate the only site of the solute.]

We recognize that $\hat{V}$ is of the form of Eq. (8.22), with the solute site function [taking into account from Eqs. (5.17) and (5.24) that $\hat{\rho}_{\lambda,\mu}(\mathbf{r}) = \sum_l q_l \hat{n}_{\lambda,\mu}(\mathbf{r})$] given by $G_{\lambda'l}(r) = q_l \delta_{\lambda\lambda'}/r$.

We note that $e\langle\!\langle \hat{V}; t \rangle\!\rangle$ is equivalent to the nonequilibrium average under $f_\Sigma(t)$ of the *basic* long range part $\sum_{aj} eq_j/r_{\lambda,aj}$ of the solute–solvent coupling. On the other hand, $\langle\!\langle \hat{\mathscr{E}}; t \rangle\!\rangle$ gives the nonequilibrium average under $f_\Sigma(t)$ of the difference $\hat{\Psi}_\Sigma^S - \hat{\Psi}_\Sigma^P$ between the *renormalized* solute–solvent couplings. Both expectation values should agree if the short-range part of the solute–solvent coupling, included in $\hat{\Psi}_\Sigma^P$ and $\hat{\Psi}_\Sigma^S$, cancels when taking their difference. The local responses $\langle\!\langle [\hat{V}]_n; t \rangle\!\rangle$ and $\langle\!\langle [\hat{\mathscr{E}}]_n; t \rangle\!\rangle$ are calculated with more complicated expressions for the functions $G_{\lambda'l}(r)$; for a detailed account we refer the reader to the original report [55].

In Figure 17 we show the zonal nonequilibrium contributions $e\langle\!\langle [\hat{V}]_n; t \rangle\!\rangle$ and $\langle\!\langle [\hat{\mathscr{E}}]_n; t \rangle\!\rangle$ from different solvent regions around the solute as functions of time. Also displayed are the global responses $e\langle\!\langle \hat{V}; t \rangle\!\rangle$ and $\langle\!\langle \hat{\mathscr{E}}; t \rangle\!\rangle$.

The solvent regions $n = 1, 2, 3$ are defined as follows (for details, see [55]): zone 1, 2.2 Å $\leq r \leq 6.17$ Å; zone 2, 6.17 Å $\leq r \leq 10.6$ Å; zone 3,

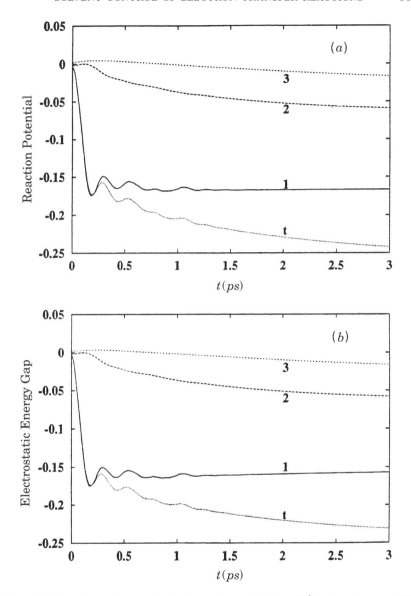

Figure 17. Ion solvation in acetonitrile: (*a*) zonal contributions $e\langle\!\langle[\hat{V}]_n;t\rangle\!\rangle$ to the nonequilibrium reaction potential; (*b*) zonal contributions $\langle\!\langle[\hat{\mathscr{E}}]_n;t\rangle\!\rangle$ to the nonequilibrium energy gap, as functions of time. The solvent regions $n = 1, 2, 3$ are defined as follows (for details, see [55]): zone 1; $2.2\,\text{Å} \leq r \leq 6.17\,\text{Å}$; zone 2; $6.17\,\text{Å} \leq r \leq 10.6\,\text{Å}$; zone 3, $10.6\,\text{Å} \leq r \leq \infty$. The curves labeled **t** in (*a*) and (*b*) are, respectively, the global nonequilibrium reaction potential $e\langle\!\langle\hat{V};t\rangle\!\rangle$ and energy gap $\langle\!\langle\hat{\mathscr{E}};t\rangle\!\rangle$. Reaction potential and energy gap are in units of hartree. t is in units of ps.

$10.6\,\text{Å} \leq r \leq \infty$. Zones 1 and 2 contain 13 and 46 solvent molecules, respectively, as calculated with the succesor pair correlation function $h_{\lambda j}^S(r)$ of the solute with the methyl group of the acetonitrile molecule.

The results in Figure 17(a) should be compared with Maroncelli's MD simulation results for the local contributions to the reaction potential, reported in Figure 16a of [61a]. For our zonal division, clearly region 1 is primarily responsible for the behavior of global response $\langle\langle \hat{V}; t \rangle\rangle$, especially for the initial very fast relaxation. Furthermore, the close agreement between Figure 17(a) and (b) indicates that at least for ion solvation in acetonitrile, $\hat{\mathscr{E}}$ is indeed a good surrogate of the factual energy gap $\hat{\mathscr{U}}$.

We note, however, that the contributions from the various zones depend strongly on their definition. Thus, although the comparison of Figure 17 with the simulation results is very encouraging, it appears that the simulation zones 1 and 2 are smaller (zones 1 and 2 contain 6 and 49 solvent molecules, respectively [61a]) than the zones in Figure 17. For a detailed discussion of this effect, see [55].

In Figure 18 we present the temporal evolution of the solvent polarization charge profile $\langle\langle \hat{\rho}_{\lambda,\mu}(\mathbf{r}); t \rangle\rangle$ around the spherical solute; each curve corresponds to a different time. The two continuous curves correspond to $\langle \hat{\rho}_{\lambda,\mu}(\mathbf{r}) \rangle^P = \rho \sum_j q_j h_{\lambda j}^P(r)$ and $\langle \hat{\rho}_{\lambda,\mu}(\mathbf{r}) \rangle^S = \rho \sum_j q_j h_{\lambda j}^S(r)$. The curves at intermediate times are displayed only under the peaks and were calculated with Eq. (8.21) and $\hat{\mathscr{G}} = \hat{\rho}_{\lambda,\mu}(\mathbf{r})$. There is a small but nonvanishing contribution at intermediate times (not displayed) for $r \leq 2.2\,\text{Å}$, an unphysical feature of the SH approximation.

The behavior of $\langle\langle \hat{\rho}_{\lambda,\mu}(\mathbf{r}); t \rangle\rangle$ in Figure 18 is very interesting. At $t = 0$ the polarization charge profile reflects the equilibrium solvation of the uncharged solute; the (negative) nitrogen site of the solvent molecules in the first solvation shell is only slightly favored compared with the (positive) methyl and carbon (of the nitrile group) sites. In other words, when the solute molecule is in the P state, the solvent molecules are arranged such that the partial charges of any molecule are approximately at the same distance from the solute, and therefore they almost "neutralize" themselves in their contribution to $\langle \hat{\rho}_{\lambda,\mu}(\mathbf{r}) \rangle^S$.

As Figure 18 shows, the buildup of negative charge close to the solute ($r \simeq 2.7\,\text{Å}$) following the $P \to S$ transition is very fast, with practically 70–80% of the first (negative) peak formed within 100 fs. Moreover, this peak shifts only slightly toward the solute with time. After this initial stage, the peak grows at a much slower rate (within picoseconds). Also shown in the figure is an equally fast buildup of positive charge at $r \simeq 3.2\,\text{Å}$ (maximum height at $t \approx 100\,\text{fs}$) which disappears at later times. Finally, there is also a slow buildup of positive charge at $r \simeq 5\,\text{Å}$.

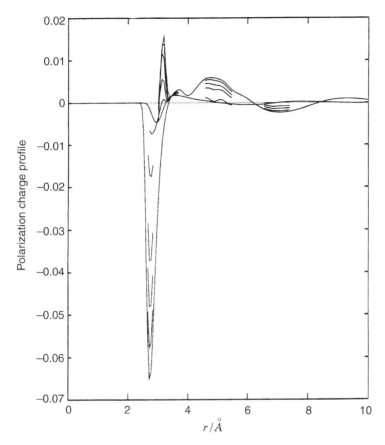

Figure 18. Ion solvation in acetonitrile: nonequilibrium solvent polarization charge profile $\langle\!\langle \hat{\rho}_{\lambda,\mu}(r); t \rangle\!\rangle$ at different times. From top to bottom the various curves at $r \simeq 2.7$ Å are the charge profiles at $t = 0$ (solid curve), 16 fs, 32 fs, 64 fs, 112 fs, 1 ps, and $t = \infty$ (solid curve). From bottom to top the curves at $r \simeq 5$ Å are the charge profiles at $t = 0$ (solid curve), 112 fs, 1 ps, 2 ps, 4 ps, and $t = \infty$ (solid curve). The curves at $r \simeq 7$ Å are from top to bottom as those at $r \simeq 5$ Å from bottom to top. Notice how much the peak near 2.7 Å changes in 112 fs and how little the maximum near 5 Å changes in the same interval. $\langle\!\langle \hat{\rho}_{\lambda,\mu}(\mathbf{r}); t \rangle\!\rangle$ is in units of $e \times \text{Å}^{-3}$, where e is the protonic charge; r is in units of Å.

These features of $\langle\!\langle \hat{\rho}_{\lambda,\mu}(\mathbf{r}); t \rangle\!\rangle$ provide an indication of the response mechanism of the first solvation shell. A plausible explanation is that these solvent molecules rotate, with the nitrogen sites almost preserving their distance to the solute while the methyl sites move toward the bulk. The first stages of this reorganization should occur by free rotation (within ≈ 100 fs); although the amplitude of this motion is small, it is sufficient to break the almost self neutralizing contribution of each molecule to $\langle\langle\hat{\rho}_{\lambda,\mu}(\mathbf{r}); t\rangle\rangle$ at $r \simeq 2.7\,\text{Å}$. This gives rise to the fast charge separation in the radial direction (the nitrogen sites at $r \simeq 2.7\,\text{Å}$ and the methyl sites at $r \simeq 3.2\,\text{Å}$). Afterward, the reorientation of the solvent molecules in the first shell has a slower and diffusive character. In concert with the slow final buildup of the negative peak at $r \simeq 2.7\,\text{Å}$, the disappearance of the $r \simeq 3.2\,\text{Å}$ peak and buildup of the $r \simeq 3.8\,\text{Å}$ peak (positive, corresponding to the carbon atom of the nitrile group) and $r \simeq 5\,\text{Å}$ peak (methyl group) take place.

From the discussion above it follows that the primary mechanism of response by the first shell of solvent molecules (particularly inertially at the initial stages) is rotational motion. This conclusion is consistent with the MD results of Maroncelli [61a].

3. Dependence of $\mathscr{L}_\Sigma(t)$ on the Charge Distribution of the Solute

In this section we test the ability of the RFMA procedure described in Section VIII.E [Eqs. (8.35) and (8.38)] to reproduce the results for $\mathscr{L}_{\text{RDT}}(t)$ calculated as in the previous two examples. With this aim in mind, we examine the dependence of the solvation TCF on the features of the solute charge distribution. The solvents considered are acetonitrile (MeCN) and methanol (MeOH).

We consider the solvation dynamics of a family of four benzene-like model solutes in the solvents MeCN and MeOH. These are the solute models already studied by Kumar and Maroncelli by molecular dynamics computer simulations [73]; study of a related set of solute models was reported elsewhere [56].

The solutes have 12 interaction sites arranged in the same geometry as the atomic sites of a benzene molecule (cf. Fig. 19). For convenience, we refer to the ring interaction sites as the "carbon atoms" (C) and to the remaining interaction sites as the "hydrogen atoms" (H). The relevant intramolecular distances of the models are $\varepsilon_{CC} = 1.41\,\text{Å}$; $l_{CH} = 1.09\,\text{Å}$, and the Lennard-Jones potential parameters are $\sigma_{CC} = 3.50\,\text{Å}$, $\varepsilon_{CC}/k_B = 40.3\,\text{K}$; $\sigma_{HH} = 2.50\,\text{Å}$; and $\varepsilon_{HH}/k_B = 25.18\,\text{K}$.

The partial charges of all the solutes in the electronic state P are: $Q_C^P = -0.135e$ for all the carbon sites and $Q_H^P = 0.135e$ for all the hydrogen sites. The *change* in the charge distribution of each of the solutes is represented in Figure 19: a $+$ (or $-$) sign attached to the solute interaction site λ indicates that in the electronic state S the partial charge at the site is $Q_\lambda^S = Q_\lambda^P + e$ (or $Q_\lambda^S = Q_\lambda^P - e$). The partial charges of the interaction sites not marked with a $+$ or a $-$ sign in the figure do not change in the $P \to S$ transition (i.e., $Q_\lambda^S = Q_\lambda^P$).

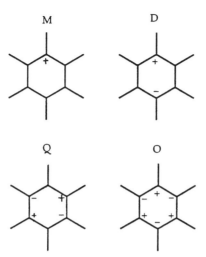

Figure 19. Benzene-like solutes with 12 interaction sites. The labels M, D, Q, and O correspond to the leading electric multipole moment of the change $\Delta n_0(\mathbf{x})$ in the charge distribution of a solute in the $P \rightarrow S$ transition. See the text for details.

Figure 19 shows that the solute species are labeled after the leading electric multipole moment of the change $\Delta n_0(\mathbf{x}) \equiv n_0^S(\mathbf{x}) - n_0^P(\mathbf{x})$ in the charge distribution in the $P \rightarrow S$ transition: monopole (M), dipole (D), quadrupole (Q), and octopole (O). Here $n_0^D(\mathbf{x}) = \sum_\lambda Q_\lambda^D \delta(\mathbf{x} - \mathbf{x}_\lambda)$ is the bare charge distribution of a solute in the electronic state D. Notice that because the molecular charge distributions have finite spatial extension, the higher-order multipoles of $\Delta n_0(\mathbf{x})$ are nonzero.

We now turn to the results for the solvation TCFs of the M, D, Q, and O solutes in MeCN and MeOH, as calculated using the RMFA and RFMA methods; the results are shown in Figures 20 and 21. The solid lines in the figures correspond to $\mathscr{L}_{\mathrm{RDT}}(t)$ calculated with the RFMA approximation [Eqs. (8.26), (8.27), and (A.13)]. The simulation data for the reference TCF $\Phi_{M,L}(t)$ used in the calculations are taken from [52] and [136] for MeCN and MeOH, respectively. It is evident from the figures that the dependence of the solvation TCF on the order, L, of the leading multipole moment of $\Delta n_0(\mathbf{x})$ is quite pronounced in either solvent. This feature was reported by Kumar and Maroncelli [73] and in our previous report [56]. It should be noted that the solid curves [$\mathscr{L}_{\mathrm{RDT}}(t)$ with RMFA] are in very good agreement with the simulation results of Kumar and Maroncelli [73]. This comparison is important, since both studies use the same ISM models for the solute and solvent molecules.

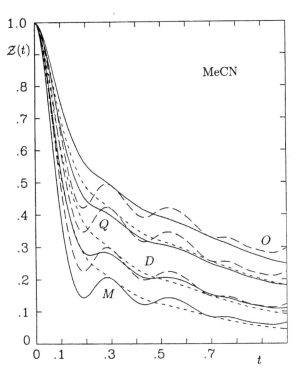

Figure 20. Solvation TCFs $\mathscr{Z}(t)$ for the benzenelike solutes of Figure 19 in acetonitrile. Solid curves, $\mathscr{Z}_{RDT}(t)$ calculated with Eq. (8.26) and the RMFA theory (App. A); long-dashed curves, calculated with the RFMA theory using $\mathscr{Z}_{RDT}(t)$ for the monopole solute as the reference in Eq. (8.35); short-dashed curves, calculated with the RFMA theory using $\mathscr{Z}_{RDT}(t)$ for the octopole solute as the reference in Eq. (8.35). t is in units of ps.

The two other sets of curves in Figures 20 and 21 (long- and short-dashed lines) correspond to calculations of $\mathscr{Z}(t)$ with the new RFMA power-law formula [Eqs. (8.35) and (8.38)] using two different reference solutes R. For the *reference* solvation TCF [$\mathscr{Z}_R(t)$ in Eq. (8.35)] we take $\mathscr{Z}_{RDT}(t)$ for one of the benzenelike solutes calculated with the RMFA method (i.e., one of the solid curves in Figs. 20 and 21). Thus the long-dashed curves in the figures correspond to the solvation TCFs $\mathscr{Z}_Y(t)$ for the benzenelike solutes $Y = D, Q, O$ taking $\mathscr{Z}_M(t)$ as the reference TCF (i.e., the reference solute is the monopole $R = M$ and $\mathscr{Z}_M(t)$ is calculated with the RMFA). The short-dashed curves are the solvation TCFs $\mathscr{Z}_Y(t)$ for the benzenelike solutes $Y = M, D, Q$ taking $\mathscr{Z}_O(t)$ as the reference TCF (i.e., the reference solute is the octopole $R = O$). The $\mathscr{Z}_Y(t)$ results with the RFMA should be compared with the corresponding solid lines in Figures 20 and 21.

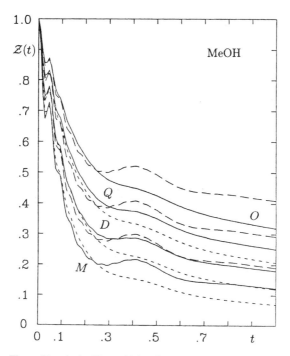

Figure 21.   As in Figure 20 but for the solutes in methanol.

Figure 20 shows that the new RFMA power-law approximation gives very reasonable results in acetonitrile. Figure 21 shows that in methanol the agreement between the RMFA and RFMA estimates is poorer, but still qualitatively correct. It is clear from Eq. (8.35) that the distinctive features of the reference TCF $\mathscr{Z}_R(t)$ will be reproduced in the RFMA estimate $\mathscr{Z}_Y(t)$. Therefore, it is not surprising that $\mathscr{Z}_D(t)$, $\mathscr{Z}_Q(t)$, and $\mathscr{Z}_O(t)$, calculated with the RMFA reference $\mathscr{Z}_M(t)$, will show pronounced oscillatory features at intermediate times. In contrast, the oscillatory component is missing in the RFMA estimates $\mathscr{Z}_M(t)$, $\mathscr{Z}_D(t)$, and $\mathscr{Z}_Q(t)$ calculated with the RMFA TCF $\mathscr{Z}_O(t)$ as the reference.

The figures illustrate that the RFMA TCFs $\mathscr{Z}_Y(t)$ based on either of the alternative RMFA reference functions $\mathscr{Z}_M(t)$ or $\mathscr{Z}_O(t)$ are remarkably consistent with each other and with the corresponding RMFA results for times less that 300 fs. We recall [126] that the generic RFMA power-law formula (8.33) is exact at very short times, in the sense that the initial curvature $(d^2\Phi_Y(t)/dt^2)_{t=0}$ of the primary TCF calculated with Eq. (8.33) has the correct value.

It is also noteworthy that the correct ordering of the TCFs calculated with the RFMA power law is always maintained at long times, although the deviations with the corresponding RMFA curves are larger when methanol is the solvent. At long times we expect [126] that $\mathscr{Z}_R(t) \sim e^{-1/T_R}$ [i.e., T_R is the longest decay time of $\mathscr{Z}_R(t)$]. Equation (8.35) then implies [126] that $\mathscr{Z}_Y(t) \sim e^{-t/T_Y}$, with $1/T_Y = \mathscr{R}_{YR}/T_R$. The results shown in Figures 20 and 21 indicate that the new RFMA power law proposed here is a good first approximation for relating the behavior of the solvation TCF of different solutes in the same solvent.

To summarize the results of Section VIII.E and this subsection, we have explored a simplified implementation of the SH-RDT formula for the solvation TCF. The resulting simplified procedure gives $\mathscr{Z}(t)$ in reasonable agreement with more complete implementation of the SH-RDT theory under the RMFA approximation. We note that we have not conditioned the application of Eq. (8.35) to reference (R) and target (Y) solute pairs that have the same symmetry, the same geometry, or even the same number of interaction sites.

The simpler RFMA approach discussed in this chapter should be useful in cases where the more robust [54–56] RMFA (Sec. VIII.D) procedure is very difficult to implement. This situation arises when the required dielectric information of the solvent is not available, especially with regards to the far-infrared contribution to the frequency-dependent dielectric constant ϵ_ω. In such cases, if the solvation TCF of another solute R has already been measured, the RFMA method can be very helpful. A case in point is the very interesting paper by Barbara et al. [137], where the required $\mathscr{Z}(t)$ of bianthryl was assumed to be equal to the solvation TCF of coumarin probes. We believe that Eq. (8.35) will improve on this type of approximation.

4. Solvation Dynamics of Coumarin-153 in Acetonitrile

For a more realistic application of Eq. (8.26) we compare (Fig. 22) the prediction $\mathscr{Z}_{\mathrm{RDT}}(t)$ of the SH-RDT theory with both experiment and molecular dynamics simulation for the solvation TCF of the dye coumarin-153 (C153) in acetonitrile. In Figure 22 the curve labeled EXP is the result of ultrafast time-dependent fluorescence Stokes shift experiments performed by Maroncelli and co-workers [99]. The curve labeled MD is $\mathscr{Z}(t)$ calculated by molecular dynamics (from the equilibrium fluctuations of the vertical energy gap $\hat{\mathscr{U}}$ when the solute is in the P state) by Kumar and Maroncelli [73]. Finally, the curve labeled RDT corresponds to the solvation TCF calculated with Eq. (8.26). The dielectric TCF $\Phi_\mu(k, t)$ is calculated with the RMFA approximation discussed in Appendix A; the reference TCF $\Phi_{M,L}(t)$ in Eq. (A.13) is taken from [52]. The MD and SH-RDT calculations use a comparable nonpolarizable ISM models for the solute [36 interaction sites in [73];

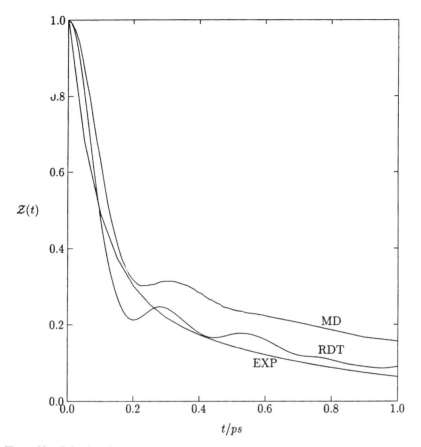

Figure 22. Solvation time correlation function $\mathcal{Z}(t)$ for the solvation dynamics of the dye C153 in acetonitrile. EXP, experimental result from [99]; MD, molecular dynamics simulation result with a 36-site ISM model of C153 taken from [73]; RDT; the result of the SH-RDT theory [Eq. (8.26)] applied to a 19-site ISM model of C153. t is in units of ps.

19 sites (Fig. 11 and Table III) in the SH-RDT calculations] and the non-polarizable ISM model of [52] to represent the solvent.

The inertial stage of the solvation dynamics is evident in the SH-RDT and MD curves on Figure 22 at the earliest times, where the two curves are in perfect agreement with each other. The time resolution of the experiment, however, is not sufficient to uncover the expected behavior of $\mathcal{Z}(t)$ at this early stage [99]. Overall, the three curves display a similar very fast initial decay (approximately 60% of the total decay within 0.2 ps). This ultrafast stage is followed by a slower decay, in which the MD and SH-RDT curves also display oscillatory features due to solvent motion of librational character. The SH-RDT result shows surprisingly good agreement with the experimental result for $t > 0.7$ ps, which might reflect the influence of non-linear effects in the solvation response that are captured by the theory.

5. Transient Absorption [138]

We consider now a model solute with three electronic states P, S_I, and S_{II}, which represent, respectively, the ground, first, and second excited states. We assume that at $t = 0$ the solute undergoes a $P \rightarrow S_I$ transition by photon absorption.

As was discussed in Section VIII.C, with $f_\Sigma(t)$ we may calculate the expectation value $\langle\langle \hat{\mathscr{G}}; t \rangle\rangle$ of any observable $\mathscr{G}$ of the form given in Eq. (8.22). We have shown [55] that $\langle\langle \hat{\mathscr{G}}; t \rangle\rangle$ may be expressed in terms of the $t = 0$ and $t = \infty$ expectation values (equilibrium averages) of $\hat{\mathscr{G}}$, and the equilibrium time correlation function $\Phi_{\mathscr{G}}(t)$ [cf. Eq. (8.21b)].

As an example we consider an anilinelike solute that comprises seven interaction sites, as indicated in Figure 23(a). All seven sites have the same LJ parameters $\sigma = 3.75$ Å and $\epsilon/k_B = 55.36$ K, while all the bond lengths are 1.4 Å.

The surrogate vertical energy gaps $\hat{\mathscr{E}} \equiv \hat{\Psi}_\Sigma^I - \hat{\Psi}_\Sigma^P$ and $\hat{\mathscr{E}}' = \hat{\Psi}_\Sigma^{II} - \hat{\Psi}_\Sigma^I$ are the dynamical variables relevant, respectively, to time-dependent fluorescence ($\mathscr{G} = \hat{\mathscr{E}}$) and to transient absorption experiments ($\hat{\mathscr{G}} = \hat{\mathscr{E}}'$). In Figure 23(b) and (c) we illustrate the temporal evolution of the solvent contributions $\langle\langle \hat{\mathscr{E}}; t \rangle\rangle$ and $\langle\langle \hat{\mathscr{E}}'; t \rangle\rangle$ to the time-dependent fluorescence and transient absorption experiments.

6. Time-Dependent Fluorescence Line Shape

The time evolution of the line shape of the normalized fluorescence spectrum can be analyzed in terms of the probability density $S(\nu', t)$ that the solvent contribution to the P–S energy gap $\hat{\mathscr{U}}$ has the numerical value $\nu' \equiv \nu - \Delta E/h$ at time t after the photoexcitation. Here ΔE is the energy difference between the S and P states in vacuum (cf. Section IV.A) and h is the Planck constant. Mathematically,

$$S(\nu', t) = \langle \delta(\nu' - \hat{\mathscr{U}}/h); t \rangle \tag{8.44}$$

where $\langle \cdots; t \rangle$ indicates the average under the nonequilibrium distribution function $f(t)$.

According to the dynamical SH theory, we obtain an estimate of $S(\nu', t)$ by calculating the surrogate analog of Eq. (8.44):

$$S_\Sigma(\nu', t) = \langle\langle \delta(\nu' - \hat{\mathscr{E}}/h); t \rangle\rangle \tag{8.45}$$

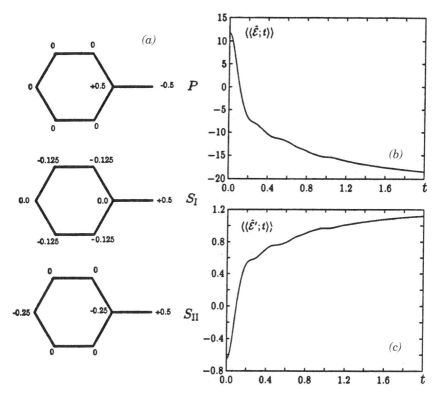

Figure 23. Solvation dynamics of an anilinelike model solute in acetonitrile: (*a*) charge distributions of the model solute; (*b*) average time evolution of the S_I–P vertical energy gap $\hat{\mathcal{E}}$ (fluorescence); (*c*) average time evolution of the S_{II}–S_I vertical energy gap $\hat{\mathcal{E}}'$ (transient absorption). P, S_I, and S_{II} represent, respectively, the ground, first, and second excited states. $P \rightarrow S_I$ photoexcitation takes place at $t = 0$. Partial charges are in units of e; vertical energy gaps in units of kcal mol^{-1}; time in units of ps.

where $\langle\langle \cdots ; t \rangle\rangle$ is the expectation value under the surrogate nonequilibrium distribution function $f_\Sigma(t)$ given by Eq. (8.17). Using the integral representation of the delta function we obtain

$$S_\Sigma(\nu', t) = \frac{1}{2\pi} \int_{-\infty}^{\infty} d\alpha\, e^{-i\alpha\nu'} \langle\langle e^{i\alpha\hat{\mathcal{E}}/h}; t \rangle\rangle \qquad (8.46)$$

where $\langle\langle e^{i\alpha\hat{\mathcal{E}}/h}; t \rangle\rangle$ is the time-dependent characteristic function associated with the probability density $S_\Sigma(\nu', t)$. As in the calculation of the generalized characteristic functions $g_\Sigma^D(\alpha)$ in Section V.A, we evaluate $\langle\langle e^{i\alpha\hat{\mathcal{E}}/h}; t \rangle\rangle$ by means of the cumulant expansion truncated at second order (the justification being the same as in Sec. V.A.). We have

$$\langle\langle e^{i\alpha\hat{\mathscr{E}}/h}; t\rangle\rangle = \exp\left(\frac{i\alpha}{h}\langle\langle\hat{\mathscr{E}}; t\rangle\rangle_c - \frac{\alpha^2}{2h^2}\langle\langle\hat{\mathscr{E}}^2; t\rangle\rangle_c\right) \tag{8.47}$$

The cumulant averages in this expression have the following meaning:

$$\langle\langle\hat{\mathscr{E}}; t\rangle\rangle_c = \langle\langle\hat{\mathscr{E}}; t\rangle\rangle = \eta_\Sigma(t) = \eta_\Sigma^S + 2\lambda_\Sigma\mathscr{L}_\Sigma(t) \tag{8.48a}$$

$$\langle\langle\hat{\mathscr{E}}^2; t\rangle\rangle_c = \langle\langle\hat{\mathscr{E}}^2; t\rangle\rangle - \langle\langle\hat{\mathscr{E}}; t\rangle\rangle^2 = (2/\beta)\lambda_\Sigma \tag{8.48b}$$

The first cumulant $\langle\langle\hat{\mathscr{E}}; t\rangle\rangle_c \equiv \eta_\Sigma(t)$ corresponds to the first moment $\overline{\nu'(t)}$ of the solvation contribution to the fluorescence spectrum. Equation (8.48a) shows that its time dependence is dictated by the surrogate estimate $\mathscr{L}_\Sigma(t)$ of the solvation TCF. The other parameters in this equation are the solvent contribution η_Σ^S to the steady-state fluorescence frequency (i.e., the fluorescence after the solvent has equilibrated to the field of the solute in the excited state) and the solvent reorganization energy λ_Σ.

The second cumulant average [Eq. (8.48b)] is related to the width of the fluorescence spectrum as measured by the centered second moment $\overline{(\nu'(t) - \overline{\nu'(t)})^2}$. Equation (8.48b) shows that at the level of approximation of the dynamical SH theory, the width is independent of time and directly proportional to the solvent reorganization energy. This prediction of the SH theory is not quite correct, as the width of the fluorescence spectrum has been found to vary with time [61b,99,139].

Finally, by substituting the expression for $\langle\langle e^{i\alpha\hat{\mathscr{E}}/h}; t\rangle\rangle$ into Eq. (8.46), we obtain for the normalized fluorescence spectrum

$$S_\Sigma(\nu', t) = h\left(\frac{\beta}{4\pi\lambda_\Sigma}\right)^{1/2}\exp\left\{-\frac{\beta[h\nu' - \eta_\Sigma(t)]^2}{4\lambda_\Sigma}\right\} \tag{8.49}$$

where $\eta_\Sigma(t)$ is defined in Eq. (8.48a).

G. Solvent Reaction Coordinate Dynamics on the Diabatic Wells

In this and following sections we establish the relation between the dynamical SH theory of solvation and some of the theories already proposed to explain dynamical solvent effects in the rate constant for charge transfer reactions in solution. We first calculate the surrogate estimate, $P_\Sigma^D(\eta, t|\eta_0)$, of the *conditional* probability density that the value of the reaction coordinate at time t is η given that its value at $t = 0$ was η_0. We note that $P_\Sigma^D(\eta, t|\eta_0)$ concerns the time-dependent fluctuations of the reaction coordinate η when the solute is in the electronic state D. Based on the definition of conditional probability density, we write [132,140]

$$P_\Sigma^D(\eta, t|\eta_0) = P_\Sigma^D(\eta, t; \eta_0)/p_\Sigma^D(\eta_0) \tag{8.50}$$

where $p_\Sigma^D(\eta_0)$ (the probability density that the reaction coordinate has the numerical value η_0) was calculated in Section V.B [Eq. (5.13)], while

$$P_\Sigma^D(\eta, t; \eta_0) = \langle\!\langle \delta(\eta - \hat{\mathscr{E}}(t)) \, \delta(\eta_0 - \hat{\mathscr{E}}) \rangle\!\rangle^D \tag{8.51}$$

is the *joint* probability [132,140] that the reaction coordinate takes the values η_0 at $t = 0$ and η at time t. The expectation value $\langle\!\langle \cdots \rangle\!\rangle^D$ is taken with the surrogate equilibrium distribution function f_Σ^D [Eq. (5.4)] of the solvent in equilibrium with the solute in the electronic state D. We note that the time dependence of $\hat{\mathscr{E}}(t)$ in Eq. (8.51) is dictated by the renormalized Liouville operator $\mathscr{L}_\Sigma^D$ associated with the surrogate diabatic Hamiltonian $H_\Sigma^D = H_w + \hat{\Psi}_\Sigma^D$.

To evaluate $P_\Sigma^D(\eta, t; \eta_0)$ we transform the right-hand side of Eq. (8.51) using the integral representations of the delta functions. We obtain

$$P_\Sigma^D(\eta, t; \eta_0) = \frac{1}{(2\pi)^2} \int_{-\infty}^{\infty} d\alpha \int_{-\infty}^{\infty} d\alpha_0 \, e^{-i\alpha\eta - i\alpha_0\eta_0} e^{K_\Sigma^D(\alpha, \alpha_0)} \tag{8.52}$$

where we have introduced the cumulant generating function $K_\Sigma^D(\alpha, \alpha_0)$ by the definition

$$e^{K_\Sigma^D(\alpha, \alpha_0)} \equiv \langle\!\langle e^{i\alpha\hat{\mathscr{E}}(t) + i\alpha_0\hat{\mathscr{E}}} \rangle\!\rangle^D \tag{8.53}$$

Proceeding as in Sections V.A and VIII.F.6 with the cumulant expansion truncated at second order, we obtain

$$K_\Sigma^D(\alpha, \alpha_0) = i\alpha\langle\!\langle \hat{\mathscr{E}}(t) \rangle\!\rangle_c^D + i\alpha_0\langle\!\langle \hat{\mathscr{E}} \rangle\!\rangle_c^D - \frac{\alpha^2}{2}\langle\!\langle [\hat{\mathscr{E}}(t)]^2 \rangle\!\rangle_c^D$$

$$- \frac{\alpha_0^2}{2}\langle\!\langle [\hat{\mathscr{E}}]^2 \rangle\!\rangle_c^D - \alpha\alpha_0\langle\!\langle \hat{\mathscr{E}}(t)\hat{\mathscr{E}} \rangle\!\rangle_c^D \tag{8.54}$$

with the following expressions for the cumulant averages:

$$\langle\!\langle \hat{\mathscr{E}}(t) \rangle\!\rangle_c^D = \langle\!\langle \hat{\mathscr{E}} \rangle\!\rangle_c^D = \eta_\Sigma^D \tag{8.55a}$$

$$\langle\!\langle [\hat{\mathscr{E}}(t)]^2 \rangle\!\rangle_c^D = \langle\!\langle \hat{\mathscr{E}}^2 \rangle\!\rangle_c^D = (2/\beta)\lambda_\Sigma \tag{8.55b}$$

$$\langle\!\langle \hat{\mathscr{E}}(t)\hat{\mathscr{E}} \rangle\!\rangle_c^D = \langle \delta\hat{\mathscr{E}}(t)\delta\hat{\mathscr{E}} \rangle = 2\lambda_\Sigma \mathscr{L}_\Sigma(t) \tag{8.55c}$$

The first equality in Eq. (8.55a) follows from the stationarity property $\langle\!\langle e^{t\mathscr{L}_\Sigma^D}\hat{\mathscr{E}} \rangle\!\rangle^D = \langle\!\langle \hat{\mathscr{E}} \rangle\!\rangle^D$, which holds up to terms that are of second order in the renormalized solute–solvent interactions. The second equality in Eq. (8.55a) identifies $\langle\!\langle \hat{\mathscr{E}} \rangle\!\rangle_c^D = \langle\!\langle \hat{\mathscr{E}} \rangle\!\rangle^D$ with the value η_Σ^D of the reaction coordinate at the minimum of the diabatic free energy profile $F_\Sigma^D(\eta)$ [Eq. (5.9)]. Equations (8.55b) and (8.55c) result from evaluating the cumulant averages to second order in the renormalized interactions; they correspond to Eqs. (5.10) and (5.16) [for Eq. (8.55b)] and to Eqs. (5.10), (5.16), and Eq. (8.20) [for Eq. (8.55C)].

Now combining Eqs. (8.50), (8.52), (8.54), and (8.55), we finally derive the expression

$$P_\Sigma^D(\eta, t|\eta_0) = \{\beta/4\pi\lambda_\Sigma[1 - \mathscr{L}_\Sigma^2(t)]\}^{1/2} \exp\left\{ \frac{-\beta[(\eta - \eta_\Sigma^D) - (\eta_0 - \eta_\Sigma^D)\mathscr{L}_\Sigma(t)]^2}{4\lambda_\Sigma[1 - \mathscr{L}_\Sigma^2(t)]} \right\} \tag{8.56}$$

It is clear that $P(\eta, t|\eta_0)$ is the conditional probability density for a stationary, non-Markovian, Gaussian stochastic process [83,88,141]. It may be shown that $P_\Sigma(\eta, t|\eta_0)$ given by Eq. (8.56) satisfies the differential equation [83,88,141]

$$\frac{\partial P_\Sigma^D(\eta, t|\eta_0)}{\partial t} = L_\Sigma^D(t) P_\Sigma^D(\eta, t|\eta_0) \tag{8.57}$$

where the operator $L_\Sigma^D(t)$ is defined by the equation

$$L_\Sigma^D(t) = \mathscr{D}_\Sigma(t) \frac{\partial}{\partial\eta} \left[\frac{\partial}{\partial\eta} + \frac{\partial(\beta F_\Sigma^D(\eta))}{\partial\eta} \right] \tag{8.58}$$

where $F_\Sigma^D(\eta)$ is the SH estimate of the diabatic free energy profile [Eq. (5.14)], and the time-dependent diffusion function is defined by the expression

$$\mathscr{D}_\Sigma(t) = -\langle \delta\hat{\mathscr{E}}^2 \rangle \frac{\dot{\mathscr{L}}_\Sigma(t)}{\mathscr{L}_\Sigma(t)} \tag{8.59}$$

Notice that $L_\Sigma^D(t)$ depends on the electronic state of the solute only through the force term $\partial(\beta F_\Sigma^D(\eta))/\partial\eta$; the diffusion function $\mathscr{D}_\Sigma(t)$ is the same in both electronic states.

Of course, the Gaussian character of $P_\Sigma^D(\eta, t|\eta_0)$ is not surprising [83]: it is a consequence of having evaluated the cumulant generating function $K_\Sigma^D(\alpha, \alpha_0)$ only up to terms that are of second order in the renormalized solute–solvent interactions $\hat{\Psi}_\Sigma^D$. Our justification for truncating the cumulant expansion at second order is the same as in Section V.A: because of their renormalized character, inclusion in $K_\Sigma^D(\alpha, \alpha_0)$ of terms that are of higher order in $\hat{\Psi}_\Sigma^D$ would result in the overcount of the solute–solvent interactions.

We notice that under the global approximation [126] $[\Phi_\mu(k, t) \simeq \Phi_\mu(k = 0, t)]$ the SH-RDT estimate of the solvation TCF [Eq. (8.26)] reduces to $\mathscr{L}_{\mathrm{RDT}}(t) = \Phi_\mu(k = 0, t)$. Furthermore, for a Debye solvent $\Phi_\mu(k = 0, t) = e^{-t/\tau_L}$, where τ_L is the longitudinal relaxation time. Under these circumstances the diffusion function is independent of time and equals $\mathscr{D} = \langle\delta\hat{e}^2\rangle/\tau_L$. When Eq. (8.57) is incorporated in the framework of a theory of dynamical solvent effects on electron transfer reactions, this global-Debye solvent approximation leads to a dependence with τ_L^{-1} for the rate constant of an adiabatic electron transfer reaction.

To conclude this section we would like to emphasize that all the solvation functions that occur in Eq. (8.56) or, equivalently, in Eqs. (8.57) to (8.59), are derived from a single *unified* molecular theory. Thus the solvent reorganization energy λ_Σ [Eq. (5.16)], the solvent contribution to the D-state vertical energy gap η^D [Eq. (5.15)], and the solvation TCF $\mathscr{L}_\Sigma(t)$ [Eq. (8.20)] are all calculated under any one of the implementations, SH-RDT or SH-RST, of the surrogate Hamiltonian theory of solvation. We think that this feature is important, in the sense that the approximations involved in the static and dynamic implementations of the SH theory of solvation are mutually consistent.

H. Connection with Electron Transfer Theory

Equation (8.57) describes the time-dependent equilibrium fluctuations of the solvent reaction coordinate η when the solute is in the D electronic state. This equation, which incorporates many of the SH theory results for equilibrium and nonequilibrium solvation, provides the basis for a straightforward connection with the theory of Sumi and Marcus [82] and subsequent extensions [87,88,92f] for the influence of solvation dynamics on the rate constant of electron transfer reactions in solution.

The Sumi–Marcus theory and its extensions are formulated in terms of diabatic free energy surfaces $\mathscr{F}^P$ and $\mathscr{F}^S$ that are functions of the solvation coordinate η and of the internal vibrational coordinates of the solute complex. In the following we consider that the vacuum energies E^P and E^S of the P and S diabatic states [cf. Eq. (4.2)] are functions of a single harmonic internal coordinate q:

$$E^D(q) = E_0^D + \tfrac{1}{2}a(q - q^D)^2 \qquad D = P, S \qquad (8.60)$$

where we assume that the force constant a is independent of the electronic state D, while q^D is the value of the vibrational coordinate q at the equilibrium configuration of the D electronic state in vacuum. The inner or vibrational reorganization energy is

$$\lambda_q \equiv E^S(q^P) - E^S(q^S) = E^P(q^S) - E^P(q^P) = \tfrac{1}{2}a(\Delta q)^2 \qquad (8.61)$$

where in the last equality we have introduced the notation $\Delta q \equiv q^S - q^P$.

When we take into account the q-dependence of the electronic energies E^D, Eqs. (5.14) for the surrogate estimates of the diabatic free energy profiles become

$$\mathcal{F}_\Sigma^P(x, y) = \frac{1}{4\lambda_\Sigma}x^2 + \frac{1}{2}ay^2 \qquad (8.62a)$$

$$\mathcal{F}_\Sigma^P(x, y) = \Delta\mathcal{A}_\Sigma + \frac{1}{4\lambda_\Sigma}(x - x_0)^2 + \frac{1}{2}a(y - y_0)^2 \qquad (8.62b)$$

where, for convenience, we have introduced the new variables x and y for, respectively, the solvent and vibrational coordinates; the variables (x, y) and (η, q) are related by the equations

$$x \equiv -(\eta - \eta_\Sigma^P) \qquad y \equiv q - q^P \qquad (8.63)$$

The constants x_0 and y_0 in Eq. (8.62b) are given by

$$x_0 = -(\eta_\Sigma^S - \eta_\Sigma^P) = -\Delta\eta_\Sigma = 2\lambda_\Sigma \qquad (8.64a)$$

$$y_0 = \Delta q = q^S - q^P \qquad (8.64b)$$

where in Eq. (8.64a) we have taken into account that $\eta_\Sigma^P - \eta_\Sigma^S$ is the surrogate estimate of the solvent contribution to the steady-state Stokes shift $E_{\Sigma,ss} = 2\lambda_\Sigma$ [cf. Eq. (4.7)]. Finally, the surrogate estimate $\Delta\mathcal{A}_\Sigma$ of the reaction free energy in $\mathcal{F}_\Sigma^S(x, y)$ is given by Eq. (5.12). From Eqs. (8.62) it follows that the energy gap between the P and S diabatic free energy surfaces is for any values of x and y,

$$\Delta\mathcal{F}_\Sigma(x, y) \equiv \mathcal{F}_\Sigma^S(x, y) - \mathcal{F}_\Sigma^P(x, y) = (\Delta\mathcal{A}_\Sigma + \lambda_\Sigma + \lambda_q) - x - 2(\lambda_q/y_0)y \qquad (8.65)$$

where $\lambda_\Sigma + \lambda_q$ is the total reorganization energy of the $P \to S$ electron transfer process.

Sumi and Marcus [82] assume that for any value of the solvent coordinate x, a fast reorganization of the inner coordinate y can make the electron transfer possible, taking the solute–solvent system from the P (or reactant) well to the S (or product) well. Zhu and Rasaiah [88] have extended this idea to the case of reversible reactions. The rate constants $k_f(x)$ and $k_b(x)$ (at a given value x of the solvent coordinate) for, respectively, the forward ($P \to S$, subscript f) and backward ($S \to P$, subscript b) electron transfer have the form [82,87,88,92]

$$k_i(x) = \nu_q e^{-\beta \mathscr{F}_i^*(x)} \qquad i = f, b \tag{8.66}$$

where the explicit expression of the frequency factor ν_q depends on whether the electron transfer reaction facilitated by the y-coordinate is adiabatic or nonadiabatic. The activation free energies $\mathscr{F}_f^*(x)$ and $\mathscr{F}_b^*(x)$ for the forward and backward reactions are defined by the equations

$$\mathscr{F}_f^*(x) \equiv \mathscr{F}_\Sigma^P(x, y^*(x)) - \mathscr{F}_\Sigma^P(x, 0) \tag{8.67a}$$

$$\mathscr{F}_b^*(x) \equiv \mathscr{F}_\Sigma^S(x, y^*(x)) - \mathscr{F}_\Sigma^S(x, y_0) \tag{8.67b}$$

Here $y^*(x)$ is the value of the y-coordinate at which, for a given value of x,

$$\Delta\mathscr{F}_\Sigma(x, y^*(x)) = 0 \tag{8.68a}$$

Using Eq. (8.65), we calculate

$$y^*(x) = -y_0 \frac{x - (\Delta\mathscr{A}_\Sigma + \lambda_\Sigma + \lambda_q)}{2\lambda_q} \tag{8.68b}$$

Substituting this result on Eqs. (8.67), we obtain

$$\mathscr{F}_f^*(x) = \frac{[x - (\Delta\mathscr{A}_\Sigma + \lambda_\Sigma + \lambda_q)]^2}{4\lambda_q} \tag{8.69a}$$

$$\mathscr{F}_b^*(x) = \frac{[x - (\Delta\mathscr{A}_\Sigma + \lambda_\Sigma + \lambda_q)]^2}{4\lambda_q} \tag{8.69b}$$

Except for a trivial scale factor because of our use of the variable x instead of $x/(2\lambda_\Sigma)^{1/2}$, Eqs. (8.69) have been obtained by many authors [82,87,88,92].

We now consider the dynamical problem. Equation (8.57) describes the relaxation dynamics of the solvent reaction coordinate when the solute is in the electronic state D; the possibility of a change in the electronic state due to electron transfer is not included. To account for the electronic transition facilitated by the inner coordinate, Eq. (8.57) is generalized to a pair of coupled relaxation-reaction equations

$$\frac{\partial P_\Sigma^P(x,t)}{\partial t} = L_\Sigma^P(x,t)P_\Sigma^P(x,t) - k_f(x)P_\Sigma^P(x,t) + k_b(x)P_\Sigma^P(x,t) \qquad (8.70a)$$

$$\frac{\partial P_\Sigma^S(x,t)}{\partial t} = L_\Sigma^S(x,t)P_\Sigma^S(x,t) - k_b(x)P_\Sigma^S(x,t) + k_f(x)P_\Sigma^S(x,t) \qquad (8.70b)$$

formulated in terms of the solvent coordinate x. Notice that we have simplified the notation of the conditional probability density $P_\Sigma^D(x,t|x_0)$ to read $P_\Sigma^D(x,t)$. The operators $L_\Sigma^D(x,t)$ are defined by the equation

$$L_\Sigma^D(x,t) = \mathscr{D}_\Sigma(t)\frac{\partial}{\partial x}\left[\frac{\partial}{\partial x} + \frac{\partial(\beta F_\Sigma^D(x))}{\partial x}\right] \qquad (8.71)$$

with the diffusion function $\mathscr{D}_\Sigma(t)$ still given by Eq. (8.59).

Recently, a convenient procedure has been proposed [88] for calculating [from Eqs. (8.70)] the survival probabilities for electron transfer between thermally equilibrated reactants or from a nonequilibrium initial state. The method is formulated in terms of an integral equation that can be straightforwardly solved numerically. The kernel of the integral equation is determined by the solvent dynamics $[\mathscr{D}_\Sigma(t)]$, the relative contributions $(\lambda_q/\lambda_\Sigma)$ of the inner and solvent reorganization energies, and the barrier heights for electron transfer. Quantum effects for the q-coordinate have also been incorporated in this formulation [87,88,92]. We refer the reader to [87,88,92] for a more complete account of this methodology applied to the solvent dynamical effects on electron transfer reactions in solution.

IX. CONCLUSIONS

We have presented a *molecular* theory of solvation energetics and dynamics governing CT processes, with special emphasis on the free energy profiles that characterize the precursor and successor states along the solvent reaction coordinate. Either a static or dynamical *renormalized* linear response formulation replaces the *basic* diabatic Hamiltonians by surrogates defined in terms of suitably renormalized solute–solvent interactions; the resulting surrogate Hamiltonian theory of solvation applies to quite realistic models of the solute and solvent molecules.

With regard to the energetics of CT reactions, three implementations of the theory (SHRST, SH-RDT, and SH-HXA) were discussed; the SH-RST and SH-RDT predict parabolic diabatic free energy profiles with the same curvature, whereas in the case of HXA these features apply by definition. These formulations take into account the electronic polarizability of the solvent molecules in a mean-field way. A fully nonlinear theory (at the level of the RISM-HNC integral equation) formulated in terms of a complex-valued Hamiltonian was also introduced. Presently, however, this theory is limited to nonpolarizable interaction site models for the solvent molecules.

We believe that the SH theory of solvation provides a reasonably accurate account of the energetics of charge transfer reactions involving relatively complicated solutes in solvents covering a wide range of polarity. The dynamical counterpart of the SH theory also holds promise for tackling more complicated systems, as demonstrated by the SH-RDT results for the solvation TCF of C153 (a 19-site ISM model) in acetonitrile and methanol.

ACKNOWLEDGMENTS

We greatly appreciate many collaborations and discussions with Marshall D. Newton, Fumio Hirata, Haluk Resat, Baw-Ching Perng, and Isabel Brandariz concerning the research on which we based this review. F.O.R. gratefully acknowledges a fellowship from the Consejo Nacional de Investigaciones Científicas y Técnicas (CONICET) de la República Argentina. This work was made possible by support provided by the National Science Foundation of the United States (CHE 9321963).

APPENDIX A: REFERENCE MEMORY FUNCTION APPROXIMATION

The reference memory function approximation (RMFA) [24,118,119] is a method for calculating, using the memory function equation [131,132]

$$d\Phi_Y(t)/dt = -\int_0^t dt' \, K_Y(t - t')\Phi_Y(t') \tag{A.1}$$

the normalized time correlation function

$$\Phi_Y(t) \equiv \frac{\langle Y(t) Y^* \rangle}{\langle YY^* \rangle} \tag{A.2}$$

of the desired *primary* dynamical variable Y. In this section the symbol $\langle \cdots \rangle$ denotes a general equilibrium ensemble average; a particular choice would be the ensemble average $\langle \cdots \rangle$ with the distribution function f^w of the pure solvent.

The memory function equation (A.1) is a consequence of the generalized Langevin equation (GLE$_1$) derived by Mori [131,132]. A distinctive feature of both the GLE$_1$ and the memory function equation is the convolution operation of the function of interest with the memory function $K_Y(t)$.

Rather than attempting the calculation of $K_Y(t)$, the RMFA method relies on the information contained in the TCF

$$\Phi_X(t) \equiv \frac{\langle X(t)X^* \rangle}{\langle XX^* \rangle} \tag{A.3}$$

of a related *reference* dynamical variable X. We assume that $\Phi_X(t)$ is already known, say from experiment, computer simulation, or by an independent theoretical approximation. The RMFA is based on the assumption that the normalized first memory functions of the primary and reference dynamical variables are equal at all times [24,118,119]:

$$K_Y(t)/\mathscr{K}_Y = K_X(t)/\mathscr{K}_X \tag{A.4}$$

Here [131,132]

$$\mathscr{K}_Y \equiv K_Y(t=0) = \langle |\dot{Y}|^2 \rangle / \langle |Y|^2 \rangle \tag{A.5a}$$

$$\mathscr{K}_X \equiv K_X(t=0) = \langle |\dot{X}|^2 \rangle / \langle |X|^2 \rangle \tag{A.5b}$$

are, respectively, the initial values of the first memory functions of the dynamical variables Y and X.

With Eq. (A.4) and the memory function equations [cf. Eq. (A.1)] for $\Phi_Y(t)$ and $\Phi_X(t)$ it is possible to show [24,118,119] that the corresponding Fourier–Laplace transforms

$$\Phi_Y(\omega) = \int_0^\infty dt\, e^{-i\omega t} \Phi_Y(t) \tag{A.6}$$

and $\Phi_X(\omega)$ [defined as in Eq. (A.6)] are connected by the equation

$$\Phi_Y(\omega) = [\mathscr{R}_{YX} + i\omega\Phi_X(\omega)(1 - \mathscr{R}_{YX})]^{-1}\Phi_X(\omega) \tag{A.7}$$

where the renormalization factor $\mathscr{R}_{YX}$ is defined in terms of the initial values $\mathscr{K}_Y$ and $\mathscr{K}_X$ of the respective memory functions $K_Y(t)$ and $K_X(t)$:

$$\mathscr{R}_{YX} \equiv \mathscr{K}_Y / \mathscr{K}_X \tag{A.8}$$

In previous reports [118,119] we showed that $\Phi_\mu(k,t)$ in Eq. (8.26) can be evaluated reliably with the RMFA method as we review now. From Eq. (3.28) it is clear that the primary dynamical variable is $Y = \hat{\rho}_\mu(\mathbf{k})$. For the reference dynamical variable X we choose the total dipole polarization

$$X = \hat{M}_L = \sum_a \mu_{az} = \left(\frac{1}{ik} \sum_{aj} q_j e^{i\mathbf{k}\cdot\mathbf{x}_{aj}} \right)_{k=0} \tag{A.9}$$

(μ_{az} is the z-Cartesian component of the dipole moment $\boldsymbol{\mu}_a$ of the solvent molecule a in the laboratory reference frame; we assume that the wavevector $\mathbf{k}$ is oriented along the z-Cartesian axis). In this case $\Phi_Y(\omega)$ and $\Phi_X(\omega)$ in Eq. (A.7) correspond, respectively, to $\Phi_\mu(k,\omega)$ and

$$\Phi_{M,L}(\omega) = \int_0^\infty dt\, e^{-i\omega t} \Phi_{M,L}(t) \tag{A.10}$$

where

$$\Phi_{M,L}(t) = \frac{\langle \hat{M}_L(t)\hat{M}_L \rangle}{\langle \hat{M}_L \hat{M}_L \rangle} \tag{A.11}$$

is the normalized TCF of $\hat{M}_L$, that determines the frequency-dependent dielectric constant ϵ_ω [17,118,142]. The idea of using $\hat{M}_L$ as the reference dynamical variable X is due to Fried and Mukamel [143].

For the present choices of X and Y, the renormalization factor $\mathscr{R}_{YX}$ [which we now call $\mathscr{R}_\mu(k)$] becomes [24,118,119] [cf. Eqs. (A.5) and (A.8)]

$$\mathscr{R}_\mu(k) = \frac{\mathscr{R}_\mu^J(k)}{\mathscr{R}_\mu^S(k)} = \frac{\langle |\hat{\rho}_\mu(\mathbf{k})|^2 \rangle / \langle |\hat{M}_L(\mathbf{k})|^2 \rangle}{\langle |\hat{\rho}(\mathbf{k})|^2 \rangle / \langle |\hat{M}_L(\mathbf{k})|^2 \rangle} \tag{A.12}$$

The second equality defines, respectively, the kinetic $\mathscr{R}_\mu^J(k)$ and structural $\mathscr{R}_\mu^S(k)$ parts of the renormalization factor. Their calculation has been discussed at length in [24], [118], and [119].

With these choices of primary and reference dynamical variables, Eq. (A.7) becomes

$$\Phi_\mu(k,\omega) = \{\mathscr{R}_\mu(k) + i\omega\Phi_{M,L}(\omega)[1 - \mathscr{R}_\mu(k)]\}^{-1}\Phi_{M,L}(\omega) \tag{A.13}$$

As in most of our previous reports [54–56], the results for $\mathscr{L}_{RDT}(t)$ [Eq. (8.26)] presented in Section VIII were obtained with the "*sim*" implementation of Eq. (A.13), in which the reference TCF $\Phi_{M,L}(t)$ is obtained from computer *simulation* of the appropriate ISM model solvent.

APPENDIX B: REFERENCE FREQUENCY MODULATION APPROXIMATION

The memory function equation [Eq. (A.1)] of the GLE_1 is not the only possible effective equation of motion for the primary TCF $\Phi_Y(t)$. An alternative is the frequency modulation equation [126],

$$\frac{d^2}{dt^2} \ln \Phi_Y(t) = -\psi_Y(t) \tag{B.1}$$

which relates the normalized TCF $\Phi_Y(t)$ [Eq. (A.2)] to the frequency modulation function $\psi_Y(t)$. This equation is obtained from the convolutionless generalized Langevin equation (GLE_2) first derived by Tokuyama and Mori [127].

In a previous report [126] we showed that an approximate relation analogous to Eq. (A.4) could be invoked for the frequency modulation functions of the primary dynamical variable Y and an appropriate reference dynamical variable X. Thus in the RFMA we assume that

$$\psi_Y(t)/\psi_Y(0) = \psi_X(t)/\psi_X(0) \tag{B.2}$$

at every t. It is straightforward to show that this assumption, together with the frequency modulation equation (B.1) for $\Phi_Y(t)$ and $\Phi_X(t)$, leads to the useful relation [126]

$$\Phi_Y(t) = \Phi_X(t)^{\mathscr{R}_{YX}} \tag{B.3}$$

between the primary and reference time correlation functions. The renormalization factor $\mathscr{R}_{YX}$ in Eq. (B.3) is also given by Eq. (A.8) because of the relations $\mathscr{K}_Y = \psi_Y(0)$ and $\mathscr{K}_X = \psi_X(0)$ [126,127].

REFERENCES

1. (a) R. A. Marcus, *J. Chem. Phys.* **24**, 966, 979 (1956); (b) *Discuss. Faraday Soc.* **29**, 21 (1960); (c) R. A. Marcus and N. Sutin, *Biochim. Biophys. Acta* **811**, 265 (1985); (d) R. A. Marcus, *Angew. Chem. Int. Ed. Engl.* **32**, 1111 (1993).

2. (a) N. S. Hush, *Trans. Faraday Soc.* **57**, 557 (1961); (b) *Electrochim. Acta* **13**, 1005 (1968). See also (c) P. P. Schmidt, *Electrochemistry*, Chemical Society Special Periodic Reports, H. R. Thirsk, ed., Vol. 5, Chap. 2, Academic Press, London, 1977, p. 21.

3. Some relevant papers are: (a) A. Weller, *Z. Phys. Chem. N.F.* **133**, 93 (1982); (b) G. van der Zwan and J. T. Hynes, *J. Phys. Chem.* **89**, 4181 (1985); (c) M. D. Newton and H. L. Friedman, *J. Chem. Phys.* **88**, 4460 (1988); (d) P. F. Barbara and W. Jarzeba, *Adv. Photochem.* **15**, 1, (1990); (e) H. J. Kim and J. T. Hynes, *J. Chem. Phys.* **96**, 5088 (1992); (f) M. Maroncelli, *J. Chem. Phys.* **106**, 1545 (1997).

4. H. L. Friedman, F. O. Raineri, B.-C. Perng, and M. D. Newton, *J. Mol. Liq.* **65/66**, 7 (1995).

5. B.-C. Perng, M. D. Newton, F. O. Raineri, and H. L. Friedman, *J. Chem. Phys.* **104**, 7153 (1996).

6. B.-C. Perng, M. D. Newton, F. O. Raineri, and H. L. Friedman, *J. Chem. Phys.* **104**, 7177 (1996).

7. D. V. Matyushov, *Mol. Phys.* **84**, 533 (1995).

8. H. J. Kim, *J. Chem. Phys.* **105**, 6818 (1996); **105**, 6833 (1996).

9. A. D. Joran, B. A. Leland, P. M. Felker, A. H. Zewail, J. J. Hopfield, and P. B. Dervan, *Nature* **327**, 508 (1987).

10. H. Oevering, M. N. Paddon-Row, M. Heppener, A. M. Oliver, E. Cotsaris, J. Verhoeven, and N. S. Hush, *J. Am. Chem. Soc.* **109**, 3258 (1987); M. Antolovich, P. J. Keyte, A. M. Oliver, M. N. Paddon-Row, J. Kroon, J. W. Verhoeven, S. A. Jonker, and J. M. Warman, *J. Phys. Chem.* **95**, 1933 (1991); J. M. Warman, K. J. Smit, M. P. de Haas, S. A. Jonker, M. N. Paddon-Row, A. M. Oliver, J. Kroon, H. Oevering, and J. W. Verhoeven, *J. Phys. Chem.* **95**, 1979 (1991); J. Kroon, J. W. Verhoeven, M. N. Paddon-Row, and A. M. Oliver, *Angew. Chem. Int. Ed. Engl.* **30**, 1358 (1991); J. M. Warman, K. J. Smit, S. A. Jonker, J. W. Verhoeven, H. Oevering, J. Kroon, M. N. Paddon-Row, and A. M Oliver, *Chem. Phys.* **170**, 359 (1993).

11. (a) T. Asahi, M. Ohkohchi, R. Matsusaka, N. Mataga, R. P. Zhang, A. Osuka, and K. Maruyama, *J. Am. Chem. Soc.* **115**, 5665 (1993); (b) H. Heitele, F. Pollinger, T. Haberle, M. E. Michel-Beyerle, and H. A. Staab, *J. Phys. Chem.* **98**, 7402 (1994).

12. L. Reynolds, J. A. Gardecki, S. J. V. Frankland, M. L. Horng, and M. Maroncelli, *J. Phys. Chem.* **100**, 10337 (1996).

13. S. Pekar, *Investigations of the Electric Theory of Crystals*, Moscow, 1951; English translation: U.S. AEC document AEC-tr-5575. The dependence of λ on dielectric constants is given by $\mathscr{C}$ in general only when image effects are neglected. For example, the case of an embedded point dipole at the center of a spherical solute cavity with $\epsilon = 1$ yields an expression for λ with the dielectric factor $(\epsilon_0 - 1)/(2\epsilon_0 + 1) - (\epsilon_\infty - 1)/(2\epsilon_\infty + 1)$. See, for example, [3b] and B. S. Brunschwig, S. Ehrenson, and N. Sutin, *J. Phys. Chem.* **91**, 4714 (1987).

14. M. J. Powers and T. J. Meyer, *J. Am. Chem. Soc.* **100**, 4393 (1978); G. E. McMannis, A. Gochev, R. M. Nielson, and M. J. Weaver, *J. Phys. Chem.* **93**, 7733 (1989); R. L. Blackbourn and J. T. Hupp, *J. Phys. Chem.* **94**, 1788 (1990).

15. The nondipolar class of solvents would also include solvents with no permanent multipole moments (e. g., rare gas atoms) or moments of negligible magnitude (presumably most saturated hydrocarbons). We do not consider such species in the applications reported in this work; however, see [7].

16. F. O. Raineri, B.-C. Perng, and H. L. Friedman, *Electrochim. Acta* **42**, 2749 (1977).

17. P. Madden and D. Kivelson, *Adv. Chem. Phys* **56**, 467 (1984).

18. (a) R. F. Loring and S. Mukamel, *J. Chem. Phys.* **87**, 1272 (1987); (b) A. Chandra and B. Bagchi, *J. Chem. Phys.* **90**, 1832 (1989); **91**, 3056 (1989); (c) T. Fonseca and B. M. Ladanyi, *J. Chem. Phys.* **93**, 8148 (1990); P. Attard, D. Wei, and G. N. Patey, *Chem. Phys. Lett.* **172**, 69 (1990).

19. F. O. Raineri, H. Resat, and H. L. Friedman, *J. Chem. Phys.* **96**, 3068 (1992).

20. A. A. Kornyshev, D. A. Kossakowski, and M. A. Vorotyntsev, in *Condensed Matter Physics Aspects of Electrochemistry*, M. P. Tosi and A. A. Kornyshev, eds., World Scientific, Singapore, 1991.

21. (a) J. D. Jackson, *Classical Electrodynamics*, Wiley, New York, 1975, Sec. 6.7; (b) P. Mazur, *Adv. Chem. Phys.* **1**, 309 (1958).

22. A. A. Kornyshev, in *The Chemical Physics of Solvation*, Part A, R. R. Dogonadze, E. Kálmán, A. A. Kornyshev, and J. Ulstrup, eds., Elsevier, Amsterdam, 1985, Chap. 3.

23. For a recent macroscopic treatment, see S. M. Chitanvis, *J. Chem. Phys.* **104**, 9065 (1996).

24. F. O. Raineri, Y. Zhou, H. L. Friedman, and G. Stell, *Chem. Phys.* **152**, 201 (1991).

25. (a) O. V. Dolgov, D. A. Kirzhnitz, and E. G. Maksimov, *Rev. Mod. Phys.* **53**, 81 (1981); (b) D. A. Kirzhnitz, in *The Dielectric Function of Condensed Systems*, L. V. Keldysh, D. A. Kirzhnitz, and A. A. Maradudin, eds., North-Holland, Amsterdam 1989, Chap. 2; (c) O. V. Dolgov and E. G. Maksimov, *ibid.*, Chap. 4.

26. D. Pines and P. Nozières, *The Theory of Quantum Liquids*, Vol. 1, Addison-Wesley, Reading, Mass., 1989.

27. M. P. Allen and D. J. Tildesley, *Computer Simulation of Liquids*, Clarendon Press, Oxford, 1990 Sec. 1.3.3.

28. J. Applequist, *J. Phys. Chem.* **97**, 6016 (1993).

29. (a) M. Sprik and M. L. Klein, *J. Chem. Phys.* **89**, 7556 (1988); (b) M. Sprik, *J. Phys. Chem.* **95**, 2283 (1991); (c) *J. Chem. Phys.* **95**, 6762 (1991).

30. (a) S. W. Rick, S. J. Stuart, and B. J. Berne, *J. Chem. Phys.* **101**, 6141 (1994); (b) M. H. New and B. J. Berne, *J. Am. Chem. Soc.* **117**, 7172 (1995); (c) S. W. Rick and W. J. Berne, *J. Am. Chem. Soc.* **118**, 672 (1996); (d) J. S. Bader and B. J. Berne, *J. Chem. Phys.* **104**, 1293 (1996); (e) S. J. Stuart and B. J. Berne, *J. Phys. Chem.* **100**, 11934 (1996).

31. (a) R. Kubo, *J. Phys. Soc. Jpn.* **12**, 570 (1957); (b) S. W. Lovesey, *Condensed Matter Physics: Dynamic Correlations*, W. A. Benjamin, Redwood City, Calif., 1980; (c) R. Kubo, M. Toda, and N. Hashitsume, *Statistical Physics*, Vol. 2, Springer-Verlag, Berlin, 1985; (d) S. Datagupta, *Relaxation Phenomena in Condensed Matter Physics*, Academic Press, San Diego, Calif., 1987.

32. H. L. Lemberg and F. Stillinger, *J. Chem. Phys.* **62**, 1677 (1975). This paper gives a relation between $\epsilon_L(k, \omega)$ and the time correlation function of the charge current density $\hat{\mathbf{j}}_\mu(\mathbf{r})$, where $\hat{\rho}_\mu(\mathbf{r}) = -\nabla \cdot \hat{\mathbf{j}}_\mu(\mathbf{r})$.

33. (a) D. Chandler and H. C. Andersen, *J. Chem. Phys.* **57**, 1930 (1972); (b) D. Chandler, in *The Liquid State of Matter: Fluids, Simple and Complex*, E. W. Montroll and J. L. Lebowitz, eds., North-Holland, Amsterdam, 1982, p. 275.

34. (a) F. Hirata and P. J. Rossky, *Chem. Phys. Lett.* **83**, 329 (1981); (b) F. Hirata, B. M. Pettitt, and P. J. Rossky, *J. Chem. Phys.* **77**, 509 (1982); (c) P. J. Rossky, B. M. Pettitt, and G. Stell, *Mol. Phys.* **50**, 1263 (1983).

35. W. L. Jorgensen and T. B. Nguyen, *J. Comput. Chem.* **14**, 195 (1993).

36. F. O. Raineri, H. L. Friedman, I. Brandariz, and B.-C. Perng, in *Electron and Ion Transfer in Condensed Matter,* Proceedings of the 1996 Adriatico Research Conference, A. A. Kornyshev, M. P. Tosi, and J. Ulstrup, eds., World Scientific, Singapore, 1997.

37. (a) P. A. Bopp, A. A. Kornyshev, and G. Sutmann, *Phys. Rev. Lett.* **76**, 1280 (1996); (b) M. S. Skaf, *Mol. Phys.* **90**, 25 (1997).

38. A. Warshel, *J. Phys. Chem.* **86**, 2218 (1982); A. Warshel and J.-K. Hwang, *J. Chem. Phys.* **84**, 4938 (1986); J.-K. Hwang and A. Warshel, *J. Am. Chem. Soc.* **109**, 715 (1987).

39. (a) E. A. Carter and J. T. Hynes, *J. Phys. Chem.* **93**, 2184 (1989); *J. Chem. Phys.* **94**, 5961 (1991); (b) M. Tachiya, *J. Phys. Chem.* **93**, 7050 (1989); (c) A. Yoshimori, T. Kakitani, Y. Enomoto, and N. Mataga, *J. Phys. Chem.* **93**, 8316 (1989); T. Kakitani, N. Matsuda, A. Yoshimori, and N. Mataga, *Prog. React. Kinet.* **20**, 347 (1995); (d) H.-X. Zhou and A. Szabo, *J. Chem. Phys.* **103**, 3481 (1995).

40. G. King and A. Warshel, *J. Chem. Phys.* **93**, 8682 (1990).

41. T. Fonseca, B. M. Ladanyi, and J. T. Hynes, *J. Phys. Chem.* **96**, 4085 (1992).

42. Y. Enomoto, T. Kakitani, A. Yoshimori, Y. Hatano, and M. Saito, *Chem. Phys. Lett.* **178**, 235 (1991); Y. Enomoto, T. Kakitani, A. Yoshimori, and Y. Hatano, *Chem. Phys. Lett.* **186**, 366 (1991).

43. V. Pérez, J. M. Lluch, and J. Bertrán, *J. Am. Chem. Soc.* **116**, 10117 (1994).

44. (a) R. A. Kuharski, J. S. Bader, D. Chandler, M. Sprik, M. Klein, and R. Impey, *J. Chem. Phys.* **89**, 3248 (1988); (b) J. S. Bader and D. Chandler *Chem. Phys. Lett.* **157**, 50l (1989).

45. D. V. Matyushov, *Chem. Phys.* **174**, 199 (1993); *Mol. Phys.* **79**, 795 (1993); D. V. Matyushov and R. Schmid, *J. Phys. Chem.* **98**, 5152 (1994).

46. F. O. Raineri, B.-C. Perng, and H. L. Friedman, *Z. Phys. Chem.* **Bd. 204**, S.109 (1998).

47. This approach was first exploited by Loring and co-workers in the closely related problem of inhomogeneous broadening in electronic spectra in the liquid phase: (a) R. F. Loring, *J. Phys. Chem.* **94**, 513 (1990); N. E. Shemetulskis and R. F. Loring, *J. Chem. Phys.* **95**, 4756 (1991); B. M. Ladanyi, N. E. Shemetulskis, and R. F. Loring, *J. Chem. Phys.* **96**, 8637 (1992).

48. S. J. Singer and D. Chandler, *Mol. Phys.* **55**, 621 (1985).

49. S.-H. Chong, S.-I. Miura, G. Basu, and F. Hirata, *J. Phys. Chem.* **99**, 10526 (1995); S.-H. Chong and F. Hirata, *Mol. Simul.* **16**, 3, (1996).

50. W. L. Jorgensen, J. D. Madura, and C. J. Swenson, *J. Am. Chem. Soc.* **106**, 6638 (1984).

51. M. D. Newton, K. Ohta, and E. Zhong, *J. Phys. Chem.* **95**, 2317 (1991).

52. D. M. F. Edwards, P. A. Madden, and I. R. McDonald, *Mol. Phys.* **51**, 1151 (1984).

53. G. E. McManis, R. M. Nielson, A. Gochev, and M. J. Weaver, *J. Am. Chem. Soc.* **111**, 5533 (1989).

54. F. O. Raineri, H. Resat, B.-C. Perng, F. Hirata, and H. L. Friedman, *J. Chem. Phys.* **100**, 1477 (1994).

55. F. O. Raineri, B.-C. Perng and H. L. Friedman, *Chem. Phys.* **183**, 187 (1994).

56. H. L. Friedman, F. O. Raineri, F. Hirata, and B.-C. Perng, *J. Stat. Phys.* **78**, 239 (1995).

57. (a) D. Chandler, Y. Singh, and D. M. Richardson, *J. Chem. Phys.* **81**, 1975 (1984); (b) L. R. Pratt and D. Chandler, *Methods Enzymol.* **127**, 48 (1986).

58. The vibrational motion of the atoms within a molecule (atomic polarization) is responsible for the usually small difference between ϵ_0 and ϵ_∞ in nondipolar solvents. For a discussion, see (a) D. L. Mills, *Nonlinear Optics*, Springer-Verlag, Berlin, 1991, Chapter 2; and (b) Y.

Kita, K. Kiyohara, M. Oobatake, S. Hayashi, and K. Machida, *J. Chem. Phys.* **101**, 7828 (1994).

59. W. L. Jorgensen and D. L. Severance, *J. Am. Chem. Soc.* **112**, 4768 (1990).

60. B. Roux, H.-A. Yu, and M. Karplus, *J. Phys. Chem.* **94**, 4683 (1990).

61. (a) M. Maroncelli, *J. Chem. Phys.* **94**, 2084 (1991); (b) E. A. Carter and J. T. Hynes, *J. Chem. Phys.* **94**, 5961 (1991); (c) D. K. Phelps, M. J. Weaver, and B. M. Ladanyi, *Chem. Phys.* **176**, 575 (1993).

62. E. M. Kosower, *An Introduction to Physical Organic Chemistry*, Wiley, New York, 1968.

63. (a) C. Reichardt, *Solvents and Solvent Effects in Organic Chemistry*, 2nd ed., VCH Publishers, Weinheim, Germany, 1988; (b) *Chem. Rev.* **94**, 2319 (1994).

64. (a) C. Reichardt, *Angew. Chem. Int. Ed. Engl.* **18**, 98 (1979); (b) *Chem. Soc. Rev.* **14**, 147 (1992); (c) C. Reichardt, S. Asharin-Fard, A. Blum, M. Eschner, A.-M. Mehranpour, P. Milart, T. Niem, G. Schäfer, and M. Wilk, *Pure Appl. Chem.* **65**, 2593 (1993).

65. S. E. DeBolt and P. A. Kollman, *J. Am. Chem. Soc.* **112**, 7515 (1990).

66. W. L. Jorgensen, E. E. Laird, T. B. Nguyen, and J. Tirado-Rives, *J. Comput. Chem.* **14**, 206 (1993).

67. J. Chandrasekhar and W. L. Jorgensen, *J. Chem. Phys.* **77**, 5073 (1982).

68. W. L. Jorgensen, J. M. Briggs, and M. L. Contreras, *J. Phys. Chem.* **94**, 1683 (1990).

69. M. Haughney, M. Ferrario, and I. R. McDonald, *J. Phys. Chem.* **91**, 4934 (1987).

70. W. L. Jorgensen, J. Chandrasekhar, J. D. Madura, R. W. Impey, and M. L. Klein, *J. Chem. Phys.* **79**, 926 (1983).

71. (a) W. Liptay, *Z. Naturforsch. A* **20a**, 1441 (1965); **21a**, 1605 (1966); (b) R. Bicca de Alencastro, J. D. Da Motta Neto, and M. C. Zerner, *Int. J. Quant. Chem. Quant. Chem. Symp.* **28**, 361 (1994).

72. W. L. Jorgensen and C. J. Swenson, *J. Am. Chem. Soc.* **107**, 569 (1985).

73. P. V. Kumar and M. Maroncelli, *J. Chem. Phys.* **103**, 3038 (1995).

74. (a) I. R. Gould, D. Noukakis, L. Gomez-Jahn, J. L. Goodman, and S. Farid, *J. Am. Chem. Soc.* **115**, 4405 (1993); (b) R. Doolen, J. D. Simon, and K. K. Baldridge, *J. Phys. Chem.* **99**, 13938 (1995).

75. W. L. Jorgensen and J. M. Briggs, *Mol. Phys.* **63**, 547 (1988).

76. D. A. Zichi, G. Ciccotti, J. T. Hynes and M. Ferrario, *J. Phys. Chem.* **93**. (1989) 6261.

77. B. B. Smith, A. Staib, and J. T. Hynes, *Chem. Phys.* **176**, 521 (1993).

78. R. R. Dogonadze, A. M. Kuznetsov, and T. A. Marsagishvili, *Electrochim. Acta* **25**, 1 (1980).

79. (a) L. D. Zusman, *Chem. Phys.* **49**, 295 (1980); (b) **80**, 29 (1983); (c) **119**, 51 (1988); (d) **144**, 1 (1990); (f) A. B. Helman, *Chem. Phys.* **133**, 271 (1989).

80. B. L. Tembe, H. L. Friedman, and M. D. Newton, *J. Chem. Phys.* **76**, 1490 (1990).

81. (a) D. F. Calef and P. G. Wolynes, *J. Phys. Chem.* **87**, 3387 (1993); (b) D. F. Calef, in *Photoinduced Electron Transfer*, Part A, *Conceptual Basis*, M. A. Fox and M. Chanon, eds., Elsevier, Amsterdam, 1988.

82. (a) H. Sumi and R. A. Marcus, *J. Chem. Phys.* **84**, 4894 (1986); (b) W. Nadler and R. A. Marcus, *J. Chem. Phys.* **86**, 3906 (1987); (c) H. Sumi, *J. Phys. Chem.* **95**, 3334 (1991).

83. J. T. Hynes, *J. Phys. Chem.* **90**, 3701 (1986).

84. (a) I. Rips and J. Jortner, *J. Chem. Phys.* **87**, 2090 (1987); (b) J. Jortner and M. Bixon, *J. Chem. Phys.* **88**, 167 (1988); (c) I. Rips, J. Klafter, and J. Jortner, in *Photochemical Energy*

Conversion, J. R. Norris, Jr. and D. Meisel, eds., Elsevier, Amsterdam, 1988; (d) I. Rips, J. Klafter, and J. Jortner, *J. Phys. Chem.* **94**, 8557 (1990); (e) M. Bixon and J. Jortner, *Chem. Phys.* **176**, 467 (1993).

85. M. Sparpaglione and S. Mukamel, *J. Phys. Chem.* **91**, 3938 (1987); *J. Chem. Phys.* **88**, 3263 (1988); **88**, 4300 (1988); Y. J. Yang, M. Sparpaglione, and S. Mukamel, *J. Phys. Chem.* **92**, 4842 (1988); Y. J. Yang and S. Mukamel, *J. Phys. Chem.* **93**, 6991 (1989); *Acc. Chem. Res.* **22**, 301 (1989).

86. M. Maroncelli, J. MacInnis, and G. R. Fleming, *Science*, **243**, 1674 (1989).

87. (a) E. Akesson, G. C. Walker, and P. F. Barbara, *J. Chem. Phys.* **95**, 4188 (1991); (b) P. F. Barbara, G. C. Walker, and T. P. Smith, *Science*, **256**, 975 (1992); (c) G. C. Walker, E. Akesson, A. E. Johnson, N. E. Levinger, and P. F. Barbara, *J. Phys. Chem.* **96**, 3728 (1992); (d) A. E. Johnson, N. E. Levinger, W. Jarzeba, R. E. Schlief, D. A. V. Kliner, and P. F. Barbara, *Chem. Phys.* **176**, 555 (1993); (e) P. J. Reid and P. F. Barbara, *J. Chem. Phys.* **99**, 3554 (1995).

88. (a) J. Zhu and J. C. Rasaiah, *J. Chem. Phys.* **95**, 3325 (1991); (b) **96**, 1435 (1992); (c) J. C. Rasaiah and J. Zhu, *J. Chem. Phys.* **98**, 1213 (1993); (d) J. Zhu and J. C. Rasaiah, *J. Chem. Phys.* **101**, 9966 (1994).

89. (a) M. J. Weaver and G. E. McManis III, *Acc. Chem. Res.* **23**, 294 (1990); (b) M. J. Weaver, *Chem. Rev.* **92**, 463 (1992); (c) D. K. Phelps and M. J. Weaver, *J. Phys. Chem.* **96**, 7187 (1992); (d) M. J. Weaver, *J. Mol. Liq.* **60**, 57 (1994).

90. H. Heitele, *Angew. Chem. Int. Ed. Engl.* **32**, 359 (1993).

91. P. J. Rossky and J. D. Simon, *Nature* **370**, 263 (1994).

92. (a) S. Roy and B. Bagchi, *J. Chem. Phys.* **100**, 8802 (1994); (b) *J. Phys. Chem.* **98**, 9207 (1994); (c) *J. Chem. Phys.* **102**, 6719 (1995); (d) **102**, 7937 (1995); (e) N. Gayathri and B. Bagchi, *J. Mol. Struct. (Teochem)* **361**, 117 (1996); (f) *J. Phys. Chem.* **100**, 3056 (1996).

93. (a) Z. Wang, J. Tang, and J. R. Norris, *J. Chem. Phys.* **97**, 7251 (1992); (b) J. Tang, *J. Chem. Phys.* **104**, 9408 (1996).

94. A. Samanta and S. K. Gosh, *J. Chem. Phys.* **102**, 3172 (1995); *Chem. Phys.* **214**, 61 (1997).

95. P. F. Barbara, T. J. Meyer, and M. A. Ratner, *J. Phys. Chem.* **100**, 13148 (1996).

96. I. Benjamin and E. Pollak, *J. Chem. Phys.* **105**, 9093 (1996).

97. (a) M. Cho, S. J. Rosenthal, N. F. Scherer, L. D. Ziegler, and G. R. Fleming, *J. Chem. Phys.* **96**, 5033 (1992); (b) R. Jimenez, G. R. Fleming, P. V. Kumar, and M. Maroncelli, *Nature* **369**, (1994) 471; (c) S. J. Rosenthal, R. Jimenez, G. R. Fleming, P. V. Kumar, and M. Maroncelli, *J. Mol. Liq.* **60**, (1994) 25; (d) M. Cho, J.-Y. Yu, T. Joo, Y. Nagasawa, S. A. Passino, and G. R. Fleming, *J. Phys. Chem.* **100**, 11944 (1996); (e) G. R. Fleming and M. Cho, *Annu. Rev. Phys. Chem.* **47**, 109 (1996).

98. (a) S. Kinoshita, *J. Chem. Phys.* **91**, 5175 (1989); (b) H. Murakami, S. Kinoshita, Y. Hirata, T. Okada, and N. Mataga, *J. Chem. Phys.* **97**, 7881 (1992); (c) T. Lian, Y. Kholodenko, and R. M. Hochstrasser, *J. Phys. Chem.* **99**, 2546 (1995); (d) W. P. de Boeij, M. S. Pshenichnikov, and D. A. Wiersma, *Chem. Phys. Lett.* **247**, 264 (1995); (e) *J. Phys. Chem.* **100**, 11806 (1996); (f) *J. Chem. Phys.* **105**, 2953 (1996).

99. M. L. Horng, J. A. Gardecki, A. Papazyan, and M. Maroncelli, *J. Phys. Chem.* **99**, 17311 (1995).

100. For a recent general discussion, see R. M. Stratt and M. Maroncelli, *J. Phys. Chem.* **100**, 7 12981 (1996).

101. M. Maroncelli and G. R. Fleming, *J. Chem. Phys.* **89**, 5044 (1988).

102. O. A. Karim, A. D. J. Haymet, M. J. Banet, and J. D. Simon, *J. Phys. Chem.* **92**, 3391 (1988).

103. (a) R. M. Levy, D. B. Kitchen, J. T. Blair, and K. Krogh-Jespersen, *J. Phys. Chem.* **94**, 4470 (1990); (b) M. Belhadj, D. B. Kitchen, K. Krogh-Jespersen, and R. M. Levy, *J. Phys. Chem.* **95**, 1082 (1991).

104. T. Fonseca and B. M. Ladanyi, *J. Phys. Chem.* **95**, 2116 (1991).

105. M. Bruehl and J. T. Hynes, *J. Phys. Chem.* **96**, 4068 (1992).

106. (a) L. Perera and M. Berkowitz, *J. Chem. Phys.* **96**, 3092 (1992); (b) E. Neria and A. Nitzan, *J. Chem. Phys.* **96**, 5433 (1992).

107. M. Maroncelli, P. V. Kumar, A. Papazyan, M. L. Horng, S. J. Rosenthal, and G. R. Fleming, in *Ultrafast Reaction Dynamics and Solvent Effects*, Proceedings of the International Workshop held in Abbaye de Royaumont, France, May 1993.

108. T. Fonseca and B. M. Ladanyi, *J. Mol. Liq.* **60**, 1, (1994).

109. P. L. Muino and P. Callis, *J. Chem. Phys.* **100**, 4093 (1994).

110. R. Olender and A. Nitzan, *J. Chem. Phys.* **102**, 7180 (1995).

111. B. D. Bursulaya, D. A. Zichi, and H. J. Kim, *J. Phys. Chem.* **99**, 10069 (1995).

112. M. S. Skaf and B. M. Ladanyi, *J. Phys. Chem.* **100**, 18258 (1996).

113. B. M. Ladanyi and R. M. Stratt, *J. Phys. Chem.* **100**, 1266 (1996).

114. For recent references along these lines, see (a) R. Biswas and B. Bagchi, *J. Phys. Chem.* **100**, 1238 (1996); (b) A. Chandra, D. Wei, and G. N. Patey, *J. Chem. Phys.* **99**, 4926 (1993); and the references to previous work by these authors cited therein.

115. X. Song, D. Chandler, and R. A. Marcus, *J. Phys. Chem.* **100**, 11954 (1996).

116. S. Mukamel, *Principles of Nonlinear Optical Spectroscopy*, Oxford University Press, New York, 1995, Chap. 2.

117. F. O. Raineri, H. L. Friedman, and B.-C. Perng, *J. Mol. Liq.*, **73,74**, 419 (1997).

118. (a) H. Resat, F. O. Raineri, and H. L. Friedman, *J. Chem. Phys.* **97**, 2618 (1992); (b) **98**, 7277 (1993).

119. H. L. Friedman, F. O. Raineri, and H. Resat, in *Molecular Liquids: New Perspectives in Physics and Chemistry*, J. J. C. Teixeira-Dias, ed., Kluwer, Dotdrecht, The Netherlands, 1995.

120. J. Yu, T. J. Kang, and M. Berg, *J. Chem. Phys.* **94**, 5787 (1994); T. J. Kang, J. Yu, and M. Berg, *Chem. Phys. Lett.* **174**, 476 (1991); J. T. Fourkas and M. Berg, *J. Chem. Phys.* **98**, 7773 (1993); J. T. Fourkas, A. Benigno, and M. Berg, *J. Chem. Phys.* **99**, 8552 (1993).

121. B. Bagchi, *J. Chem. Phys.* **100**, 6658 (1994).

122. F. Hirata, T. Munakata, F. O. Raineri, and H. L. Friedman, *Mol. Liq.* **65/66**, 15 (1995).

123. F. Hirata, *J. Chem. Phys.* **96**, 4619 (1992).

124. D. F. Calef and P. G. Wolynes, *J. Chem. Phys.* **78**, 4145 (1983).

125. F. O. Raineri, F. Hirata, and H. L. Friedman, unpublished results.

126. F. O. Raineri and H. L. Friedman, *J. Chem. Phys.* **101**, 6111 (1994).

127. M. Tokuyama and H. Mori, *Prog. Theor. Phys.* **55**, 411 (1975).

128. M. Maroncelli, P. V. Kumar, and A. Papazyan, *J. Phys. Chem.* **97**, 13 (1993).

129. Another attempt at a justification of Eq. (8.34) may be found in S. Roy and B. Bagchi, *Chem. Phys.* **183**, 207 (1994).

130. C. F. Chapman, R. S. Fee, and M. Maroncelli, *J. Phys. Chem.* **99**, 4811 (1995).

131. H. Mori, *Prog. Theor. Phys.* **33**, 423 (1965).

132. B. J. Berne, in *Physical Chemistry: An Advanced Treatise*, D. Henderson, ed., Academic Press, San Diego, Calif., 1971.

133. F. H. Stillinger and A. Rahman, *J. Chem. Phys.* **60**, 1545 (1974).

134. M. Neumann, *J. Chem. Phys.* **85**, 1567 (1986).

135. M. Maroncelli, *J. Mol. Liq.* **57**, l, (1993).

136. M. Skaf, T. Fonseca, and B. M. Ladanyi, *J. Chem. Phys.* **98**, (1993) 8929.

137. T. J. Kang, W. Jarzeba, P. F. Barbara, and T. Fonseca, *Chem. Phys.* **149**, (1990), 81.

138. F. O. Raineri, H. L. Friedman, B.-C. Perng, and I. Brandariz, in *Fentochemistry: Ultrafast Chemical and Physical Processes in Molecular Systems*, M. Chergui, ed., World Scientific, Singapore, 1996.

139. D. Bingemann and N. P. Ernsting, *J. Chem. Phys.* **102**, 2691 (1995).

140. C. W. Gardiner, *Handbook of Stochastic Methods*, Springer-Verlag, Berlin, 1994.

141. R. F. Fox, *Phys. Rep.* **48**, 181 (1978).

142. R. L. Fulton, *Mol. Phys.* **29**, 405 (1975); *J. Chem. Phys.* **62**, 4355 (1975); **63**, 77 (1975).

143. L. Fried and S. Mukamel, *J. Chem. Phys.* **93**, 932 (1990).

THEORETICAL AND EXPERIMENTAL STUDY OF ULTRAFAST SOLVATION DYNAMICS BY TRANSIENT FOUR-PHOTON SPECTROSCOPY

B. D. FAINBERG

*Raymond and Beverly Sackler Faculty of Exact Sciences,
School of Chemistry, Tel Aviv University, Tel Aviv 69978, Israel
Physics Department, Center for Technological Education, Holon,
52 Golomb Street, Holon 58102, Israel*

D. HUPPERT

*Raymond and Beverly Sackler Faculty of Exact Sciences,
School of Chemistry, Tel Aviv University, Tel Aviv 69978, Israel*

CONTENTS

Electron Transfer: From Isolated Molecules to Biomolecules, Part Two, edited by Joshua Jortner and M. Bixon. Advances in Chemical Physics Series, Volume 107, series editors I. Prigogine and Stuart A. Rice.
ISBN 0-471-25291-3 © 1999 John Wiley & Sons, Inc.

I. INTRODUCTION

Time-resolved luminescence (TRL) and four-photon spectroscopy have been applied to probe the dynamics of electronic spectra of molecules in solutions (solvation dynamics) [1–21]. In TRL spectroscopy a fluorescent probe molecule is electronically excited and the fluorescence spectrum is monitored as a function of time. Relaxation of the solvent polarization around the newly created excited molecular state led to a time-dependent Stokes shift of the luminescence spectrum. Such investigations are aimed at studying the mechanism of solvation effects on electron transfer processes, proton transfer, and so on [1–3,6,7,9]. In this regard it is worth noting the works by Fleming's and Barbara's groups on the observation of ultrafast (subpicosecond) components in the solvation process [1,6,7,9,22,23] and systematic studies of solvation dynamics by Maroncelli and others [24,25]. The experimental efforts were supplemented by results of molecular dynamics simulations and theory by Maroncelli and Fleming [2,26], Neria and Nitzan [27], Fonseca and Ladanyi [28], Perera and Berkowitz [29], and Bagchi and others [30–32].

The four-photon experiments were carried out with both very short pump pulses (pulse duration $t_p \sim 10\,\text{fs}$) [11–17, 33] and with pulses long compared with the reciprocal bandwidth of the absorption spectrum and irreversible electronic dephasing $T'(t_p \sim 100\,\mathrm{fs})$ [18–21] (see also [34]). Photon echo measurements conducted with former pulses in Shank's, Wiersma's, and Fleming's groups, and by Vöhringer and Scherer, provided important information on solvation in the condensed phase [11–17]. For example, three pulse-stimulated photon echo experiments [8,16,17] showed that the echo peak shift, as a function of a delay between the second and third pulses, could give accurate information about solvation dynamics. A large contribution to the theory of four-photon spectroscopy has been made by Mukamel and co-authors (see [10] and references therein).

In the four-photon spectroscopy methods with pulses long compared with reciprocal bandwidth of the solvent contribution to the absorption spectrum ($t_p \gg \sigma_{2s}^{-1/2}$, where σ_{2s} is the solvent contribution to the central second moment of the absorption spectrum [18–21,35], pump pulses of frequency ω create light-induced changes in the sample under investigation, which are measured with a time-delayed probe pulse. Due to the condition $t_p \gg \sigma_{2s}^{-1/2}$, pump pulses have a relatively narrow bandwidth and therefore create a narrow hole in the initial thermal distribution with respect to a generalized solvation coordinate in the ground electronic state (Figure 1) and, simultaneously, a narrow spike in the excited electronic state. These distributions tend to the equilibrium point of the corresponding potentials over time.

By varying the excitation frequency ω, one can change the spike and the hole position on the corresponding potential. The rates of the spike and the

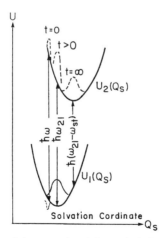

Figure 1.   Potential surfaces of the ground and the excited electronic states of a solute molecule in a liquid: one-dimensional potential surfaces as a function of a generalized solvent polarization coordinate.

hole movements depend on their position. The changes related to the spike and the hole are measured at the same or another frequency ω_1 by the delayed probe pulse. Therefore, one can control relative contribution of the ground state (a hole) and the excited state (a spike) to an observed signal. This property of the spectroscopy with pulses $t_p \gg \sigma_{2s}^{-1/2}$ can be used for the nonlinear solvation study, when the breakdown of linear response for solvation dynamics occurs [21,36,37] (see Sec. V).

Recently, an excellent review [38] has been published devoted to photon echo and fluorescence Stokes shift experiments. Therefore, we concern ourselves here with four-photon spectroscopy with pulses $t_p \gg \sigma_{2s}^{-1/2}$. We will call this spectroscopy resonance transient four-photon spectroscopy (RTFPS) with pulses long compared with the electronic dephasing, bearing in mind that the reciprocal bandwidth of the absorption spectrum is determined by the dephasing time of the corresponding electronic transition. We will mean only the solvent contribution to this bandwidth (σ_{2s}). It is worthy to note that in the case of inhomogeneously broadened transitions, one must distinguish two dephasing times [39]: reversible ($\sim \sigma_{2s}^{-1/2}$) and irreversible T' ($T' > \sigma_{2s}^{-1/2}$). Therefore, in the last case we will mean the fastest (reversible) dephasing time.

The outline of this chapter is as follows. In Section II we present a general theory for RTFPS with pulses long compared with the electronic dephasing. In Section III we describe our experimental results obtained by the heterodyne optical Kerr effect (HOKE) spectroscopy on ultrafast solvation dynamics study of two organic molecules: rhodamine 800 (R800) and DTTCB in different solvents. In Section IV we present the results concerning the quantum beats for cryptocyanine and DTTCI accompanying solvation dynamics. In Section V we discuss a prospect of spectroscopy with pulses longer than the reciprocal bandwidth of the absorption spectrum: nonlinear solvation study. In the appendixes we carry out auxiliary calculations.

II. THEORY OF RTFPS WITH PULSES LONG COMPARED WITH REVERSIBLE ELECTRONIC DEPHASING

A. Hamiltonian of Chromofore Molecule in Solvent and Basic Methods of Spectroscopy

Let us consider a molecule with two electronic states $n = 1$ and 2 in a solvent described by the Hamiltonian

$$H_0 = \sum_{n=1}^{2} |n\rangle [E_n - i\hbar\gamma_n + W_n(\mathbf{Q})]\langle n| \qquad (2.1)$$

where $E_2 > E_1$, E_n and $2\gamma_n$ are the energy and inverse lifetime of state n, and $W_n(\mathbf{Q})$ is the adiabatic Hamiltonian of reservoir R (the vibrational subsystems of a molecule and a solvent interacting with the two-level electron system under consideration in state n).

The molecule is affected by electromagnetic radiation of three beams:

$$\mathbf{E}(\mathbf{r}, t) = \mathbf{E}^+(\mathbf{r}, t) + \mathbf{E}^-(\mathbf{r}, t) = \tfrac{1}{2}\vec{E}(\mathbf{r}, t)\exp(-i\omega t) + \text{c.c.} \qquad (2.2)$$

where

$$\vec{E}(\mathbf{r}, t) = \sum_{m=1}^{3} \vec{\mathscr{E}}_m(t)\exp(i\mathbf{k}_m\mathbf{r})$$

Since we are interested in the solvent–solute intermolecular relaxation, we shall single out the solvent contribution to $W_n(\mathbf{Q})$: $W_n(\mathbf{Q}) = W_{nM} + W_{ns}$, where W_{ns} is the sum of the Hamiltonian governing the nuclear degrees of freedom of the solvent in the absence of the solute and the part that describes interactions between the solute and the nuclear degrees of freedom of the solvent; W_{nM} is the Hamiltonian representing the nuclear degrees of freedom of the solute molecule.

A signal in any method of nonlinear spectroscopy can be expressed by the nonlinear polarization $\mathbf{P}^{NL}$. We will consider the following methods of the RTFPS with pump pulses long compared with the electronic dephasing: the resonance transient grating spectroscopy (RTGS) [18–20], the transmission pump-probe experiment [40–42], the heterodyne optical Kerr effect (HOKE) spectroscopy [21,43–45], and time-resolved hole-burning experiments [40,46–51].

In the RTGS two pump pulses with wavevectors $\mathbf{k}_1$ and $\mathbf{k}_2$ and frequency ω create a light-induced grating in the sample under investigation with a wavevector $\mathbf{q} = \mathbf{k}_1 - \mathbf{k}_2$. The grating effectiveness is measured by the diffraction of a time-delayed probe pulse $\omega, \mathbf{k}_3$ with the generation of a signal with a new wavevector $\mathbf{k}_s = \mathbf{k}_3 \pm \mathbf{q}$. The signal power I_s in the direction $\mathbf{k}_s$ at time t is proportional to the square of the modulus of the corresponding positive-frequency component of the cubic polarization $\mathbf{P}^{(3)+}$: $I_s(t) \sim |\mathbf{P}^{(3)+}(\mathbf{r}, t)|^2$. In pulsed experiments, the dependence of the signal energy J_s is usually measured on the delay time τ of the probe pulse relative to the pump pulses:

$$J_s(\tau) \sim \int_{-\infty}^{\infty} |\mathbf{P}^{(3)+}(\mathbf{r}, t)|^2\, dt \qquad (2.3)$$

In the transmission pump–probe experiment [40–42], a second pulse (whose duration is the same as the pump pulse) probes the sample transmission ΔT at a delay τ. This dependence $\Delta T(\tau)$ is given by [52]

$$\Delta T(\tau) \sim -\omega \operatorname{Im} \int_{-\infty}^{\infty} \mathscr{E}_{pr}^*(t - \tau) \mathscr{P}^{NL+}(t) \, dt \qquad (2.4)$$

where $\mathscr{E}_{pr}$ and $\mathscr{P}^{NL+}(t)$ are the amplitudes of the positive-frequency component of the probe field and the nonlinear polarization, respectively.

In resonance HOKE spectroscopy [21,44,45] (see Figure 2), a linearly polarized pump pulse at frequency ω induces anisotropy in an isotropic sample. After the passage of the pump pulse through the sample, a linearly polarized probe pulse at $\pi/4$ rad from the pump field polarization, is incident on the sample. A polarization analyzer is placed after the sample oriented at approximately $\pi/2$ (but not exactly) with respect to the probe pulse polarization. A small portion of the probe pulse that is not related to the induced anisotropy plays the role of a local oscillator (LO) with a controlled magnitude and phase. More information concerning the HOKE spectroscopy is given in Section III.

The HOKE signal can be written in the form

$$J_{HET} \sim -\operatorname{Im} \int_{-\infty}^{\infty} \mathscr{E}_{LO}^*(t - \tau) \exp(i\psi) \mathscr{P}^{NL+}(t) \, dt \qquad (2.5)$$

where ψ is the phase of the LO. If $\psi = 0$, the resonance HOKE spectroscopy provides information similar to that of the transmission pump-probe spectroscopy [see Eq. (2.4)]. If $\psi = \pi/2$, the resonance HOKE spectroscopy provides information about the real part of the nonlinear susceptibility (the change in the index of refraction).

In the time-resolved hole-burning experiment [40,46,47,53–58], the sample is excited with a ~ 100-fs pump pulse, and the absorption spectrum is measured with a 10-fs probe pulse that is delayed relative to the pump pulse by a variable τ. In another variant of such an experiment, a delayed pump pulse, broadened up to a continuum, can play a role of a probe pulse. The difference in the absorption spectrum at $\omega' + \omega$ is determined by [47,52]

$$\Delta\alpha(\omega') \sim -\operatorname{Im}[\mathscr{P}^{NL}(\omega')/\mathscr{E}_{pr}(\omega')] \qquad (2.6)$$

where

$$\mathscr{P}^{NL}(\omega') = \int_{-\infty}^{\infty} \mathscr{P}^{NL+}(t) \exp(i\omega' t) \, dt \qquad (2.7)$$

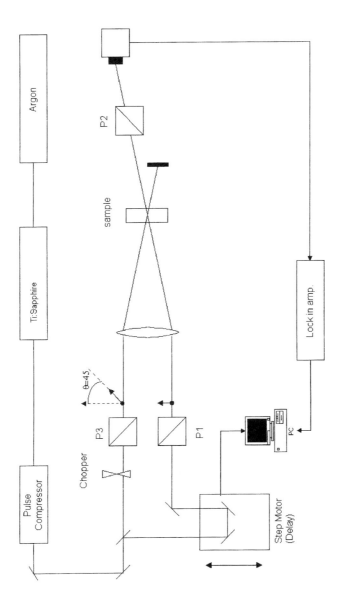

Figure 2. Schematic diagram of the experimental apparatus; P_1, P_2 comprise a crossed Glan-Thomson polarized pair. The laser system consists of a CW-mode locked Ti:sapphire laser pumped by a CW argon laser.

is the Fourier transform of the nonlinear polarization, and

$$\mathscr{E}_{\mathrm{pr}}(\omega') = \int_{-\infty}^{\infty} \mathscr{E}_{\mathrm{pr}}(t - \tau) \exp(i\omega' t) \, dt \qquad (2.8)$$

is the Fourier transform of the probe field amplitude.

B. Calculation of Nonlinear Polarization

The electromagnetic field (2.2) induces in the medium an optical polarization $\mathbf{P}(\mathbf{r}, t)$ which can be expanded in powers of $\mathbf{E}(\mathbf{r}, t)$ [39]. For cubic polarization of the system under investigation we obtain

$$\mathbf{P}^{(3)}(\mathbf{r}, t) = \mathbf{P}^{(3)+}(\mathbf{r}, t) + \text{c.c.} = NL^4 \langle \mathrm{Tr}_R(\mathbf{D}_{12}\rho_{21}^{(3)}(t)) + \text{c.c.} \rangle_{\mathrm{or}} \qquad (2.9)$$

where N is the density of particles in the system, L the Lorentz correction factor of the local field, $\mathbf{D}$ the dipole moment operator of a solute molecule, $\langle \cdots \rangle_{\mathrm{or}}$ denotes averaging over the different orientations of solute molecules, and $\rho^{(3)}$ is the density matrix of the system calculated in the third approximation with respect to $\mathbf{E}(\mathbf{r}, t)$.

The ath component $(a, b, c, d = x, y, z)$ of the amplitude of the positive-frequency component of the cubic polarization $\mathscr{P}_a^{(3)+}(t)$ describing the generation of a signal with a wave vector $\mathbf{k}_s = \mathbf{k}_{m'} + \mathbf{k}_{m''} - \mathbf{k}_{m''}(\mathbf{P}^{(3)+}(\mathbf{r}, t) = \bar{\mathscr{P}}^{(3)+}(t) \exp[i(\mathbf{k}_s\mathbf{r} - \omega t)])$ is given by the formula [59]

$$\mathscr{P}_a^{(3)+}(t) = \frac{-iNL^4}{8\hbar^3}$$

$$\times \sum_{mm'm''} \sum_{bcd} \int\!\!\int_0^{\infty}\!\!\int d\tau_1 \, d\tau_2 \, d\tau_3 \exp\{-[i(\omega_{21} - \omega) + \gamma]\tau_1$$

$$- T_1^{-1}\tau_2 - \gamma\tau_3\}\{\exp[i(\omega_{21} - \omega)\tau_3]F_{1abcd}(\tau_1, \tau_2, \tau_3)$$

$$\times \mathscr{E}_{m'c}(t - \tau_1 - \tau_2)\mathscr{E}_{md}^*(t - \tau_1 - \tau_2 - \tau_3)$$

$$+ \exp[-i(\omega_{21} - \omega)\tau_3]F_{2abcd}(\tau_1, \tau_2, \tau_3)$$

$$\times \mathscr{E}_{m'c}(t - \tau_1 - \tau_2 - \tau_3)\mathscr{E}_{md}^*(t - \tau_1 - \tau_2)\}\mathscr{E}_{m''b}(t - \tau_1) \qquad (2.10)$$

where $T_1 = (2\gamma_2)^{-1} \equiv (2\gamma)^{-1}$ is the lifetime of the excited state 2, $\omega_{21} = \omega_{\mathrm{el}} - \langle W_2 - W_1 \rangle/\hbar$ is the frequency of the Franck–Condon transition $1 \rightarrow 2$ (see the definition of W_2 and W_1 in Sec. II.A), and $\omega_{\mathrm{el}} = (E_2 - E_1)/\hbar$ is the frequency of purely electronic transition with corrections from the electronic degrees of freedom of the solvent [47,59]. The summation in Eq. (2.10) is carried out over all fields that satisfy the condition $\mathbf{k}_s = \mathbf{k}_{m'} + \mathbf{k}_{m''} - \mathbf{k}_m$. The functions $F_{1,2absd}(\tau_1, \tau_2, \tau_3)$ are sums of four-time correlations functions corresponding to the four photon character of light–matter interaction:

$$F_{1abcd}(\tau_1, \tau_2, \tau_3) = K_{dcab}(0, \tau_3, \tau_1 + \tau_2 + \tau_3, \tau_2 + \tau_3)$$

$$+ K_{dbac}(0, \tau_2 + \tau_3, \tau_1 + \tau_2 + \tau_3, \tau_3) \qquad (2.11)$$

$$F_{2abcd}(\tau_1, \tau_2, \tau_3) = K^*_{cdba}(0, \tau_3, \tau_2 + \tau_3, \tau_1 + \tau_2 + \tau_3)$$

$$+ K^*_{cabd}(0, \tau_1 + \tau_2 + \tau_3, \tau_2 + \tau_3, \tau_3) \qquad (2.12)$$

where

$$K_{abcd}(0, t_1, t_2, t_3) = \langle\langle D_{12}^a \exp(i\tilde{W}_2 t_1 / \hbar) D_{21}^b \exp(iW_1(t_2 - t_1) / \hbar)$$

$$\times D_{12}^c \exp(i\tilde{W}_2(t_3 - t_2) / \hbar) D_{21}^d \exp(-iW_1 t_3 / \hbar)\rangle\rangle_{\text{or}} \qquad (2.13)$$

are the tensor generalizations of the four-time correlation functions $K(0, t_1, t_2, t_3)$ which were introduced in four-photon spectroscopy by Mukamel [60,61]. Here $\langle\cdots\rangle \equiv \text{Tr}_R(\cdots \rho_R)$ denotes the operation of taking a trace over the reservoir variables, $\rho_R = \exp[-W_1/(kT)]/\text{Tr}_R \exp[-W_1/(kT)]$ is the density matrix of the reservoir in the state 1, and $\tilde{W}_2 = W_2 - \langle W_2 - W_1 \rangle$ is the adiabatic Hamiltonian in the excited state without the reservoir addition to the frequency of the Franck–Condon transition (the term $\langle W_2 - W_1 \rangle$).

It follows from Eqs. (2.10) to (2.13) that the nuclear response of any four-photon spectroscopy signal, generally speaking, depends on the polarizations of the excited beams because of the tensor character of the values $F_{1,2abcd}$ and $K_{abcd}(0, t_1, t_2, t_3)$.

The transient nonlinear optical response depends strongly on the relations between the intramolecular chromophore relaxation and solvation dynamics. Numerous experiments [16,18,22,25,62–64] show that the Franck–Condon molecular state achieved by an optical excitation, relaxes very fast and the relaxed intramolecular spectrum forms within 0.1 ps. Therefore, we shall consider that the intramolecular relaxation takes place within the pump pulse duration. Such a picture corresponds to a rather universal dynamical behavior of large polar chromophores in polar solvents, which may be represented by four well-separated time scales [16]: an intramolecular vibrational component, and intermolecular relaxation which consists of an ultrafast ($\sim$ 100 fs), 1–4 ps, and 10–100 ps decay components.

As to the interactions with the solvent, they satisfy the slow modulation limit [19,35,47,59,65] in the spirit of Kubo's theory of the stochastic modulation [66]:

$$\tau_s^2 \sigma_{2s} \gg 1 \qquad (2.14)$$

where $\sqrt{\sigma_{2s}}$ plays the role of the modulation amplitude and τ_s is the characteristic time of the attenuation of the solvation correlation function.

As a consequence of condition (2.14), times τ_1 and τ_3, become fast [67–71]. Therefore, we can integrate the right-hand side of Eq. (2.10) with respect to them if the exciting pulses are Gaussian [35]:

$$\mathscr{E}_m(t) = \mathscr{E}_0 \exp[-(\Delta^2/2)(t - t_m)^2 + i\omega t_m]$$

with pulse duration of

$$t_p = 1.665/\Delta \gg \sigma_{2s}^{-1/2} \qquad (2.15)$$

As a result, Eq. (2.10) is strongly simplified [35,59]

$$\mathscr{P}_a^{(3)+}(t) = \frac{1}{8} \sum_{mm'm''} \sum_{bcd} \int_0^\infty d\tau_2\, \chi_{abcd}^{(3)}(\omega, t, \tau_2) \mathscr{E}_{m''b}(t) \mathscr{E}_{m'c}(t - \tau_2) \mathscr{E}_{md}^*(t - \tau_2)$$

$$(2.16)$$

where $\chi_{abcd}^{(3)}(\omega, t, \tau_2)$ is the cubic susceptibility. It can be represented as a sum of products of Condon $\chi_{FC\alpha,\varphi}^{(3)}(\omega, t, \tau_2)$ and non-Condon $B_{abcd}^{HT\alpha,\varphi}(\tau_2)$ parts:

$$\chi_{abcd}^{(3)}(\omega, t, \tau_2) = \sum_{\alpha,\varphi} \chi_{FC\alpha,\varphi}^{(3)}(\omega, t, \tau_2) B_{abcd}^{HT\alpha,\varphi}(\tau_2) \qquad (2.17)$$

where indices α, φ of $\chi_{FC}^{(3)}$ and B_{abcd}^{HT} show that the corresponding values are related to nonequilibrium processes in the absorption (α) or emission (φ) (for more details, see below). The Condon factors $\chi_{FC\alpha,\varphi}^{(3)}(\omega, t, \tau_2)$ depend on the excitation frequency ω, t, and τ_2, but they do not depend on the polarization states of exciting beams. The non-Condon terms $B_{abcd}^{HT\alpha,\varphi}(\tau_2)$ do not depend on ω, but depend on τ_2 and the polarizations of the exciting beams. The origin of the non-Condon terms $B_{abcd}^{HT\alpha,\varphi}$ stems from the dependence of the dipole moment of the electronic transition on the nuclear coordinates $\mathbf{D}_{12}(\mathbf{Q})$. Such a dependence is explained by the Herzberg–Teller (HT) effect (i.e., mixing different electronic molecular states by nuclear motions).

We are interested mainly in non-Condon effects in solvation. There-fore, for simplicity we consider high-frequency intramolecular vibrations as Condon vibrations. We consider the translational and rotational motions of liquid molecules as nearly classical at room temperatures, since their characteristic frequencies are smaller than the thermal energy kT.

Here we do not consider the rotational motion of a solute molecule as a whole. The corresponding times are in the range of several hundreds of picoseconds for complex molecules and are not important for ultrafast investigations (< 10–100 ps). In the ultrafast range such effects are only important for small molecules. One can take into account the influence of the rotational motion of an impurity molecule on $\mathbf{P}^{(3)}$ by approach [72].

1. Condon Contributions to Cubic Susceptibility

At first let us consider a case of one optically active (OA) intramolecular vibration of frequency ω_0. Then the Condon contributions $\chi^{(3)}_{\mathrm{FC}\alpha,\varphi}(\omega, t, \tau_2)$ to the cubic susceptibility (2.17) can be written in the form [35]:

$$\chi^{(3)}_{\mathrm{FC}\alpha,\varphi}(\omega, t, \tau_2) = -i(2\pi^3)^{1/2} N L^4 \hbar^{-3} \exp(-\tau_2/T_1)(\sigma(\tau_2))^{-1/2}$$

$$\times \sum_{n,k=-\infty}^{\infty} I_n(S_0/\sinh\theta_0) I_k(S_0/\sinh\theta_0)$$

$$\times \exp[-2S_0 \coth\theta_0 + (n+k)\theta_0]$$

$$\times F^e_{n,s\alpha}(\omega - n\omega_0 - \omega_{\mathrm{el}}) w(z_{\alpha,\varphi}) \tag{2.18}$$

Here S_0 is the dimensionless parameter of the shift in the equilibrium point for the intramolecular vibration ω_0 under electronic excitation, $\theta_0 = \hbar\omega_0/(2kT)$, $I_n(x)$ is the modified Bessel function of the first kind [73], $F^e_{n,s\alpha}(\omega - n\omega_0 - \omega_{\mathrm{el}}) = (2\pi\sigma_{2s})^{-1/2} \exp[-(\omega - n\omega_0 - \omega_{\mathrm{el}} - \langle u_s \rangle/\hbar)^2/(2\sigma_{2s})]$ is the equilibrium absorption spectrum of a chromophore corresponding to an nth member of a progression with respect to the vibration $\omega_0, u_s = W_{2s} - W_{1s}$,

$$w(z) = \exp(-z^2)\left[1 + (2i/\sqrt{\pi})\int_0^z \exp(t^2)\, dt\right]$$

is the error function of the complex argument [73],

$$z_{\alpha,\varphi} = \{i\Delta^2[\tau_2(2+S(\tau_2)) - t(3+S(\tau_2)) + t_{m''} + t_{m'} + t_m(1+S(\tau_2))]$$

$$+ \omega - \omega_{\alpha,\varphi}(\tau_2) + \omega_0(\mp k + nS(\tau_2))\}/[2\sigma(\tau_2)]^{1/2} \qquad (2.19)$$

$$\sigma(\tau_2) = \sigma_{2s}\left\{1 - S^2(\tau_2) + \frac{\Delta^2}{\sigma_{2s}}[3 + 2S(\tau_2) + S^2(\tau_2)]\right\} \qquad (2.20)$$

is the time-dependent central second moment of the changes related to nonequilibrium processes in the absorption and emission spectra, at the active pulse frequency ω,

$$\omega_{\alpha,\varphi}(\tau_2) = \omega_{el} \pm \frac{\omega_{st}}{2} + S(\tau_2)\left[\omega - \left(\omega_{el} \pm \frac{\omega_{st}}{2}\right)\right] \qquad (2.21)$$

are the first moments related to the solvent contribution to transient absorption (α) and emission (φ) spectra, respectively, $\omega_{st} = 2\langle u_s \rangle$ is the solvent contribution to the Stokes shift between the equilibrium absorption and emission spectra, $\hbar^2\sigma_{2s}S(t) = \langle u_s(0)u_s(t)\rangle - \langle u_s\rangle^2$, $S(t)$ is the normalized solute–solvent correlation function, and $\sigma_{2s} = \hbar^{-2}(\langle u_s^2(0)\rangle - \langle u_s\rangle^2)$ is the solvent contribution to the second central moment of both the absorption and the luminescence spectra. The terms $w(z_{\alpha,\varphi})$ in the right-hand side of Eq. (2.18) describe contributions to the cubic polarizations of the nonequilibrium absorption and emission processes, respectively.

The third term on the right-hand side of Eq. (2.20), which is proportional to Δ^2/σ_{2s}, plays the role of the pulse-width correction to the hole or spike width. This term is important immediately after the optical excitation when $\tau_2 \approx 0$ and, therefore, $S(\tau_2) \approx 1$. The first term on the right-hand side of Eq. (2.19), which is proportional to $\Delta^2 \sim 1/t_p^2$, takes into account the contribution of the electronic transition coherence.

It is worthy to note that Eqs. (2.16) and (2.18) to (2.21) describe in a continuous fashion a transition from the time frame in which coherent effects such as photon echo exist to the time range where reversible dephasing disappears [35,65]. For acting pulse durations t_p satisfying the condition

$$\sigma_{2s}^{-1/2} \ll t_p \ll (\tau_s/\sigma_{2s})^{1/3} \equiv T' \qquad (2.22)$$

Eqs. (2.16) and (2.18) to (2.21) describe the effects of two- and three-pulse (stimulated) photon echo [35,65].

When

$$t_p \gg T' \qquad (2.23)$$

and the pump and the probe pulses do not overlap in time, one can ignore terms $\sim \Delta^2$ in Eqs. (2.19) and (2.20) [35,65]. In the last case Eq. (2.16) can be used for any pulse shape, and the cubic susceptibilities $\chi_{abcd}^{(3)}(\omega, t, \tau_2)$ and $\chi_{FC\alpha,\varphi}^{(3)}(\omega, t, \tau_2)$ in Eqs. (2.16) to (2.18) do not depend on time t (i.e., they convert to usual steady-state susceptibilities).

We used a notation T' for $(\tau_s/\sigma_{2s})^{1/3}$ since the last value plays a role of the irreversible dephasing time in the case under consideration [35,52,65].

2. Non-Markovian Model of an Optically Active Oscillator for the Optical Response

Theory presented in Section II.B.1 connects the nonlinear optical response of a molecule in a solution with the correlation function $S(t)$ of the optically active motions for which $u_s = W_{2s} - W_{1s} \neq 0$. The analytical form of $S(t)$ can be arbitrary in principal. Its calculation is an independent problem. We have developed a systematic way for the calculation of $S(t)$ of a optically active oscillator [20] which is closely connected with a continued-fraction representation of the time correlation functions, obtained by Mori with the projection operator formalism [74,75]. Our method is simpler, and therefore we hope that our derivation will be understandable for a broader audience.

In this subsection we do not limit our consideration by the classical (high-temperature) approximations. In the general (quantum) case, the correlation function $S(t)$ is complex and it is not an observable quantity. Therefore, it is difficult to treat its physical meaning. It is more convenient to deal with the relaxation function $\Phi_r(t)$, which describes the relaxation of a system after removal of the external disturbance [76]. Unlike the correlation function, $\Phi_r(t)$ is always a real observable function.

Let us introduce the Fourier transforms of $\hbar^2 \sigma_{2s} S(t)$ and $\Phi_r(t)$:

$$\begin{bmatrix} s(\omega) \\ \phi(\omega) \end{bmatrix} = \frac{1}{2\pi} \int_{-\infty}^{\infty} dt \, \exp(-i\omega t) \begin{bmatrix} \hbar^2 \sigma_{2s} S(t) \\ \Phi_r(t) \end{bmatrix}$$

$s(\omega)$ and $\phi(\omega)$ satisfy the relation [76,77]

$$s(\omega) = \{\hbar\omega/[1 - \exp(-\hbar\omega\beta)]\}\phi(\omega) \qquad (2.24)$$

where $\beta = 1/kT$. Using the inverse Fourier transformation, we obtain from Eq. (2.24):

$$S(t) = \frac{1}{2\hbar\sigma_{2s}} \left[2 \int_0^\infty \omega \coth \frac{\hbar\omega\beta}{2} \phi(\omega) \cos \omega t \, d\omega - i \frac{d\Phi_r}{dt} \right] \qquad (2.25)$$

The latter allows one to find the correlation function if the relaxation function is known. For the classical limit ($\hbar\omega\beta \ll 1$) $\hbar^2\sigma_{2s}S(\tau_2) = \beta^{-1}\Phi_r(\tau_2)$ [i.e., the normalized classical correlation function coincides with the normalized relaxation function $f_r(t) = \Phi_r(t)/\Phi_r(0)$] . In addition, for the classical case $\sigma_{2s} = \omega_{st}\hbar^{-1}\beta^{-1}$; therefore, $\Phi_{r_{cl}}(0) = \hbar\omega_{st}$.

Let us turn to the central magnitude u_s: $u_{cen} = \tilde{W}_{2s} - W_{1s}(\langle u_{cen}\rangle = 0)$. The magnitude $\Phi_r(t)$ can be written for our case in the form [76]

$$\Phi_r(t) = -\frac{i}{\hbar}\lim_{\varepsilon\to +0}\int_t^\infty \langle[u_{cen}, u_{cen}(t)]\rangle \exp(-\varepsilon t')\, dt' \qquad (2.26)$$

where $u_{cen}(t) = \exp[(i/\hbar)W_1 t]u_{cen}\exp[-(i/\hbar)W_1 t]$.

Consider the Laplace transform of the normalized relaxation $f_r(t) = \Phi_r(t)/\Phi_r(0)$:

$$\tilde{f}_r(p) = \int_0^\infty \exp(-pt)f_r(t)\, dt \qquad (2.27)$$

Suppose that $f_r(t)$ has only simple poles p_m. Then the function $f_r(t)$ can be represented in the form [78]

$$f_r(t) = \sum_m d_m \exp(p_m t) \qquad (2.28)$$

where d_m are the residues of the function $\tilde{f}_r(p)$ at $p = p_m$, and d_m are the coefficients at $(p - p_m)^{-1}$ in the Laurent series expansion of $f_r(p)$. The Laplace transformation $f_r(p)$ can be represented as

$$\tilde{f}_r(p) = \sum_m d_m(p - p_m)^{-1} \qquad (2.29)$$

Let us turn to the calculation of the preexponential factors in Eq. (2.28) for $f_r(t)$. Using Eq. (2.26), we obtain the power series expansion of $f_r(t)$ in t. Such an expansion will content only even powers of t, due to the evenness of $\Phi_r(t)$ [76] [$\Phi_r(t) = \Phi_r(-t)$]:

$$f_r(t) = 1 + \sum_{n=1}^\infty c_n \frac{t^{2n}}{(2n)!} \qquad (2.30)$$

where

$$c_n = -\frac{2}{\hbar\Phi_r(0)}\,\mathrm{Im}\langle u_{cen}(0)u_{cen}^{(2n-1)}(0)\rangle \qquad u_{cen}^{(2n-1)}(0) \equiv \frac{d^{2n-1}}{dt^{2n-1}}\,u_{cen}(t)|_{t=0}$$

$$(2.31)$$

For the high-temperature (classical) case .

$$c_n = (-1)^n \langle u_{\text{cen}}^{(n)}(0) u_{\text{cen}}^{(n)}(0) \rangle / \langle u_{\text{cen}}^2(0) \rangle \tag{2.32}$$

where we used the relation $\langle u_{\text{cen}}^{(2n)}(0) u_{\text{cen}}^{(0)}(0) \rangle = (-1)^n \langle u_{\text{cen}}^{(n)}(0) u_{\text{cen}}^{(n)}(0) \rangle$.

Comparing Eqs. (2.28) and (2.30), we find that the coefficients c_n must satisfy the relation

$$c_n = \sum_m d_m p_m^{2n} \tag{2.33}$$

and

$$\text{(a)} \ \sum_m d_m = 1 \qquad \text{(b)} \ \sum_m d_m p_m^{2n-1} = 0 \tag{2.34}$$

Let us construct some approximation to $f_r(t)$, taking into account successively two $(N = 2)$, three $(N = 3)$ (and so on) terms on the right-hand side of Eq. (2.29). Such an approach corresponds to the fitting of $f_r(t)$ by a two-, three- (and so on) pole formula. Apparently, such an approach has the character of an asymptotic (long-time) series expansion of $f_r(t)$ [78].

We shall write down the fractions obtained in such a way, in the form of continued fractions. Their terms can be calculated by Eqs. (2.33) and (2.34), aside from the last one, which will be an empirical constant. For $N = 2$ we have

$$\tilde{f}_r^{(2)}(p) = \frac{1}{p + (-c_1)/(p - (p_1 + p_2))} \tag{2.35}$$

where $p_1 p_2 = -c_1$ and $-(p_1 + p_2)$ is an empirical constant. The Laplace transform $\tilde{f}_r^{(2)}(p)$ coincides with the Laplace transform of the relaxation function of the Brownian oscillator (BO) with the frequency-independent attenuation γ_{osc} [68,79–81] for $\omega_{\text{osc}}^2 = -c_1$ and $\gamma_{\text{osc}} = -(p_1 + p_2)$. Thus the model of the optically active BO can be considered as a two-pole approximation to the relaxation function of the system under consideration, and its frequency $\omega_{\text{osc}} = \sqrt{-c_1}$.

For very large attenuation $\gamma_{\text{osc}} \gg \omega_{\text{osc}}$, $\tilde{f}_r(p)$ turns to the form

$$\tilde{f}_r(p) = 1/(p + \tau_c) \tag{2.36}$$

where $\tau_c = \omega_{\text{osc}}^2 / \gamma_{\text{osc}}$ and corresponds to Kubo's stochastic model [66].

For a three-pole case we obtain

$$\tilde{f}_r^{(3)}(p) = \cfrac{1}{p - \cfrac{c_1}{p + \cfrac{c_1 - c_2/c_1}{p - (p_1 + p_2 + p_3)}}} \tag{2.37}$$

$\tilde{f}_r^{(3)}(p)$ coincides with the Laplace transform of the relaxation function of the non-Markovian oscillator (NMO) with an exponential memory $\varphi(t) = \varphi(0) \exp(-\alpha|t|)$ [82,83] for $\omega_{osc}^2 = -c_1$, $\varphi(0) = c_1 - c_2/c_1$, and $\alpha = -(p_1 + p_2 + p_3)$ is an empirical constant. The function $\varphi(t)$ describes the memory effects in the relaxation process. Thus the model of an optically active NMO with an exponential memory function $\varphi(t)$ can be considered as a three-pole approximation to the relaxation function of the system under consideration. The normalized relaxation function for this oscillator is presented in Appendix B. The model of two such oscillators describes accurately various experimental and computer simulations data of ultrafast solvation dynamics [20,83]. In the case of solvation the parameters of a NMO can be expressed by the dielectric function of a solvent [20].

According to the results presented in this subsection, the optically active oscillator model can be considered as the corresponding N-pole approximation to the relaxation function. The analytical forms of classical and quantum relaxation functions are the same, and they are distinguished only by the formulas for the coefficients c_n [Eqs. (2.31) or (2.32)]. Thus in the quantum case one does not need to determine a correct quantum correlation function for an NMO itself, which is a rather complex problem [10,80].

3. Nonlinear Polarization in a Condon Case for Nonoverlapping Pump and Probe Pulses

The consideration of Section II.B.1 is confined by the Gaussian character of the value $u_s = W_{2s} - W_{1s}$. For nonoverlapping pump and probe pulses when condition (2.23) is satisfied, a nonlinear polarization in a Condon case can be expressed by the formula [20,83]

$$\mathbf{P}^{NL+}(\mathbf{r}, t) = \frac{\pi}{2\hbar} N \mathbf{D}_{12}(\mathbf{D}_{21}\vec{E}(\mathbf{r}, t))\{i[F_\alpha(\omega, \omega, t) - F_\varphi(\omega, \omega, t)]$$

$$+ [\Phi_\alpha(\omega, \omega, t) - \Phi_\varphi(\omega, \omega, t)]\} \tag{2.38}$$

for any u_s. Here

$$F_{\alpha,\varphi}(\omega_1, \omega, t) = \int_{-\infty}^{\infty} d\omega' F_{\alpha,\varphi M}(\omega') F_{\alpha,\varphi s}(\omega_1 - \omega_{el} - \omega', \omega, t) \tag{2.39}$$

are the spectra of the nonequilibrium absorption (α) or luminescence (φ) of a molecule in solution,

$$F_{\alpha,\varphi s}(\omega',\omega,t) = \frac{1}{2\pi} \int_{-\infty}^{\infty} d\tau_1 f_{\alpha,\varphi s}(\tau_1,t) \exp(-i\omega'\tau_1) \qquad (2.40)$$

and

$$F_{\alpha,\varphi M}(\omega') = \frac{1}{2\pi} \int_{-\infty}^{\infty} d\tau_1 f_{\alpha,\varphi M}(\tau_1) \exp(-i\omega'\tau_1) \qquad (2.41)$$

the corresponding intermolecular (s) and intramolecular (M) spectra:

$$\Phi_{\alpha,\varphi}(\omega_1,\omega,t) = \pi^{-1} P \int_{-\infty}^{\infty} d\omega' \frac{F_{\alpha,\varphi}(\omega_1,\omega,t)}{\omega'-\omega_1} \qquad (2.42)$$

are the nonequilibrium spectra of the refraction index, which are connected to the corresponding spectra $F_{\alpha,\varphi}(\omega_1,\omega,t)$ by the Kramers–Kronig formula, P is the symbol for the principal value.

$$f_{\alpha,\varphi M}(\tau_1) = \text{Tr}_M[\exp(\pm(i/\hbar)W_{2,1M}\tau_1)\exp(\mp(i/\hbar)W_{1,2M}\tau_1)\rho_{1,2M}] \qquad (2.43)$$

are the characteristic functions (the Fourier transforms) of the intramolecular absorption (α) or emission (φ) spectrum [84],

$$\rho_{1,2M} = \exp(-\beta W_{1,2M})/\text{Tr}_M \exp(-\beta W_{1,2M})$$

is the equilibrium density matrix of the solute molecule,

$$f_{\alpha,\varphi s}(\tau_1,t) = \text{Tr}_s[\exp((i/\hbar)u_s\tau_1)\rho_{1,2s}(t)] \qquad (2.44)$$

are the characteristic functions of the intermolecular absorption (α) or the emission (φ) spectra, and $\rho_{1,2s}(t)$ is the field-dependent density matrix of the system describing the evolution of the solvent nuclear degrees of freedom in the ground (1) or excited (2) electronic states. It can be calculated by using the method of successive approximations with respect to the light intensity [37,52,83].

The signals in the pump–probe and time-resolved hole-burning experiments are determined only by the nonequilibrium absorption and emission spectra [36,37,83]:

$$\Delta T(\tau) \sim -\omega[F_\alpha(\omega,\omega,\tau) - F_\varphi(\omega,\omega,\tau)] \qquad (2.45)$$

$$\Delta\alpha(\omega') \sim -[F_\alpha(\omega+\omega',\omega,\tau) - F_\varphi(\omega+\omega',\omega,\tau)] \qquad (2.46)$$

Equations (2.45) and (2.46) have been obtained for pump-pulse duration shorter than the solute–solvent relaxation time.

The formulas in this subsection are not limited by the four-photon approximation because they are based on the approach of [52] and [85], which has been developed for solving problems related to the interaction of vibronic transitions with strong fields.

4. Non-Condon Terms

Let us consider the non-Condon terms in equation (2.17) for $\chi^{(3)}_{abcd}(\omega, t, \tau_2)$. They have the following forms [59]:

$$B^{\mathrm{HT}(m)}_{abcd}(\tau_2) = \int\int d\vec{\mu}\, d\vec{\nu}\, \langle\tilde{\sigma}_{ab}(\vec{\nu})\tilde{\sigma}_{dc}(\vec{\mu})\rangle_{\mathrm{or}} \exp\left\{ -2\sum_j [\langle Q^2_{sj}(0)\rangle \right.$$

$$\left. \times (\mu_j^2 + \nu_j^2 + 2\mu_j\nu_j\Psi_{sj}(\tau_2)) + i\delta_{m\varphi}d_{sj}\nu_j(1 - \Psi_{sj}(\tau_2))] \right\} \quad (2.47)$$

where $m = \alpha, \varphi; \delta_{m\varphi}$ is the Kronecker delta,

$$\tilde{\sigma}_{ab}(\vec{\nu}) = \frac{1}{(2\pi)^M} \int d\mathbf{Q}_s \sigma_{ab}(\mathbf{Q}_s) \exp(-i\vec{\nu}\mathbf{Q}_s) \quad (2.48)$$

is the Fourier transformation of the tensor

$$\sigma_{ab}(\mathbf{Q}_s) = D^a_{12}(\mathbf{Q}_s/2)D^b_{21}(\mathbf{Q}_s/2) \quad (2.49)$$

M is the dimensionality of the vector $\mathbf{Q}_s$, $\Psi_{sj}(\tau_2) = \langle Q_{sj}(0)Q_{sj}(\tau_2)\rangle/\langle Q^2_{sj}(0)\rangle$ is the correlation function, corresponding to coordinate Q_{sj}. If this is an OA vibration, the solvation correlation function $S(\tau_2)$ is related to the correlation functions $\Psi_{sj}(\tau_2)$. In the classical case this relation is [44]

$$S(\tau_2) = \sum_j \omega_{\mathrm{st},j}\Psi_{sj}(\tau_2)/\omega_{\mathrm{st}} \quad (2.50)$$

where $\omega_{\mathrm{st},j}$ is the contribution of the jth intermolecular motion to the entire intermolecular Stokes shift ω_{st} ($\omega_{\mathrm{st}} = \sum_j \omega_{\mathrm{st},j}$). $S(\tau_2)$ can be considered as an average of the values $\Psi(\tau_2)$ distributed with the density $\omega_{\mathrm{st},j}/\omega_{\mathrm{st}}$. If the non-Condon contribution is due to a non-OA vibration that does not contribute to the Stokes shift ω_{st}, then $\Psi(\tau_2)$ is not related to $S(\tau_2)$.

The second addend in the square brackets in Eq. (2.47) describes the interference of the Franck–Condon and Herzberg–Teller contributions. The value of the parameter d_{sj} can be expressed by the equation [44]

$$|d_{sj}| = (\hbar\omega_{\mathrm{st},j})^{1/2}/\omega_{\mathrm{st}} \quad (2.51)$$

For freely orientating molecules, the orientational averages $\langle \tilde{\sigma}_{ab}(\vec{\nu}) \tilde{\sigma}_{dc}(\vec{\mu}) \rangle_{or}$ can be expressed by the tensor invariants $\tilde{\sigma}^0$, $\tilde{h}_s$ and $\tilde{h}_a$ [59] (see App. A). In the last case the values $B_{abcd}^{HT(m)}$ can be expressed by the values [44,59]

$$
B_{0,s,a}^{(m)}(\tau_2) = \int\int d\vec{\mu}\, d\vec{\nu} \left\{ -2 \sum_j [\langle Q_{sj}^2(0)\rangle (\mu_j^2 + \nu_j^2 + 2\mu_j\nu_j \Psi_{sj}(\tau_2)) \right.
$$

$$
\left. + i\delta_{m\varphi} d_{sj}\nu_j (1 - \Psi_{sj}(\tau_2))] \right\} \begin{cases} \tilde{\sigma}^0(\vec{\nu})\tilde{\sigma}^0(\vec{\mu}) \\ \tilde{h}_s(\vec{\mu},\vec{\nu}) \\ \tilde{h}_a(\vec{\mu},\vec{\nu}) \end{cases} \qquad (2.52)
$$

related to the tensor invariants.

If it will also be necessary to take the non-Condon effects into account for low-frequency ($< kT$) intramolecular vibrations (see Sec. IV), one can make it on the same basis as non-Condon effects in solvation.

III. EXPERIMENTAL STUDY OF ULTRAFAST SOLVATION DYNAMICS

A. Experimental Setup

The experimental system is shown in Figure 2. A passively continuous-wave (CW) mode locked Ti : sapphire laser (Coherent Mira 900F) is pumped by an argon ion laser (Coherent Innova 310) to produce a 76-MHz train of short pulses with 70–120 fs FWHM tunable in the optical range of 720–800 nm. The laser output beam of 6 nJ/pulse is split into pump and probe beams with an intensity ratio of 3 : 1, respectively.

In the time-resolved resonance HOKE experiment, the laser-induced anisotropy created by the resonant absorption of the pump-pulse photons is probed by a variably delayed, weak polarized probe pulse. The change in the polarization state of the probe beam is detected by the transmission through a crossed polarizer pair (P_1 and P_2 in Fig. 2) of the probe beam as a function of the time delay between the pump and probe pulses.

To amplify the optical Kerr signal and to avoid complexity due to the quadratic nature of the signal, we use heterodyne methods in the signal detection. A local oscillator is derived by a small rotation of the analyzer polarizer (P_2) by $< 1^0$ from the maximum extinction position part of the probe pulse, which is in phase and polarized orthogonal to the probe polarization. The magnitude of the local oscillator intensity is about 30 times that of the Kerr signal. The use of a local oscillator with field $\mathscr{E}_{LO}$ and light intensity I_{LO} to detect a signal with a field $\mathscr{E}_s$ and intensity I_s results in a detector response

$$I_{LO} + I_s + \frac{nc}{8\pi}(\mathscr{E}_s^*\mathscr{E}_{LO} + \mathscr{E}_{LO}^*\mathscr{E}_s)$$

The crossed term in parentheses is the heterodyne term.

The sample was measured in a rotating cell to avoid thermal contribution to the OKE signal. Unlike the nonresonant OKE measurement, which is used to measure the dynamics of liquids [43], the resonance measurements provide the solvation dynamics of probe molecules in solution.

B. Resonance Heterodyne Optical Kerr-Effect Spectroscopy of Solvation Dynamics in Water and D_2O

Recently, interesting results have been obtained concerning the ultrafast solvation dynamics in liquid water [9,22,32,86–88]. It was found, experimentally [9], by molecular dynamical simulations and theory [32,87,88] that the solvation of a solute molecule (or ion) in water is bimodal. The solvation correlation function is Gaussian at short times and exponential at long times. Solvation studies are of great importance, since the time response of solvent molecules to the electronic rearrangement of a solute has an essential influence on the rates of chemical reactions in liquid [9,89], particularly in liquid water.

A question arises when and if the solvation dynamics of a solute in deuterated water is similar to water [32]. The Debye relaxation time measured by dielectric relaxation technique for D_2O is slower than that of H_2O at the same temperature [90]. Deuterated water is a more ordered liquid with a stronger hydrogen bond compared to normal water [91]. It was predicted that significant isotope effect may be observed in ion solvation of normal and deuterated water in the (sub)picosecond range [32]. It was reported in [23] (see also [22]) that a small isotope effect exists in water for the longitudinal relaxation time.

Using the technique of the resonance heterodyne optical Kerr effect (HOKE) [21,43,92] we have studied the solvation dynamics of three organic molecules: rhodamine 800 (R800), 1,1'-diethyl-4,4'-carbocyanine iodide (cryptocyanine), and 3,3'-diethylthiatricarbocyanine bromide (DTTCB) in normal and deuterated water in the femto–and picosecond ranges [44,45]. We found a rather significant isotope effect in the picosecond range for R800 but not for DTTCB and cryptocyanine. We attribute the R800 results to a specific solvation in rhodamine 800 due to the formation (breaking) of an intermolecular solute–solvent hydrogen bond. Another important aspect of this study is that the solvation correlation function is bimodal with an ultrafast femtosecond component $<100\,\text{fs}$.

1. Calculation of HOKE Signal of R800 in Water and D_2O

Here we apply the general theory described in Section II.B to the calculation of the HOKE signal of R800 in water and D_2O. Let us consider the spectra of R800 in water and other solvents (Fig. 3). This molecule has a well-structured spectra which can be considered as a progression with respect to an optically active (OA) high-frequency vibration of about $1500\,\text{cm}^{-1}$ [93]. The members of this progression are well separated, and their amplitudes rapidly attenuate when the number of the progression member increases (practically, as one can see from Figure 3, the amplitude of the third component is rather small). Such behavior provides evidence of a small change of the molecular nuclear configuration on an electronic excitation. In other words, the Franck–Condon electron–vibrational interactions in rhodamine molecules are small. The resonance Raman scattering studies of rhodamine dyes [94,95] display intense lines in the range of about 1200–$1600\,\text{cm}^{-1}$ and the lowest-frequency one at $600\,\text{cm}^{-1}$ in both alcohol and water solutions. Therefore, one can assume that the intramolecular vibrational contribution to the line broadening of R800 in water in the range between the electronic transition frequency, ω_{el}, and the first maximum is minimal. In our experiments the excitation frequency corresponds to this range $(\omega = 13{,}986\,\text{cm}^{-1})$.

Let us discuss the interactions with the solvent. Bearing in mind our comments concerning the role of the intra- and inter-molecular interactions, we can assume that criterion (2.14) is correct for the first maxima in both absorption and luminescence spectra of R800 in water. In the last case σ_{2s} is the central second moment of the first maximum. Also, the criterion (2.15) is well realized in our experiments, since $t_p \sim 100\,\text{fs}$ (in the first series of our measurements, $t_p \approx 150\,\text{fs}$) and $\sigma_{2s}^{-1/2} \approx 14\,\text{fs}$.

Let us discuss the role of non-Condon effects for R800 in H_2O and D_2O. The absorption spectra of R800 in water and D_2O differ from the corresponding spectra in other solvents (Fig. 3). Solvents like H_2O and D_2O influence the relative intensities of spectral components in the absorption band. It can be described by the dependence of the dipole moment of the electronic transition $\mathbf{D}_{12}$ on a solvent coordinate $\mathbf{D}_{12}(\mathbf{Q}_s)$ [59] (i.e., by the non-Condon effect). Thus the electronic dipole moment dependence on a solvent coordinate must be a necessary component of our consideration. Moreover, the R800 absorption spectrum in D_2O differs from that of H_2O. The substitution of H by D influences the absorption spectrum shape. Thus one can assume that the dependence $\mathbf{D}_{12}(\mathbf{Q}_s)$ is determined by the solute–solvent hydrogen bond in water. The analytical form of the

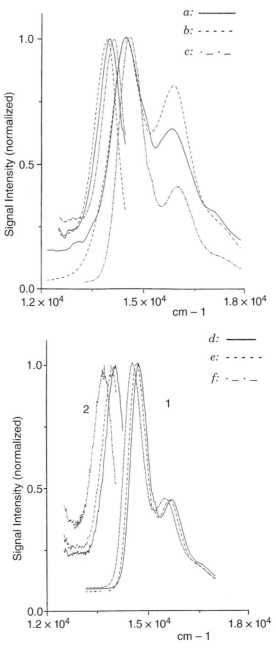

Figure 3. Absorption (1) and emission (2) spectra of R800 in water (*a*), D$_2$O (*b*), ethanol (*c*), acetone (*d*), propylene carbonate (*e*), and dimethyl sulfoxide (*f*).

$\mathbf{D}_{12}(\mathbf{Q}_s)$ dependence is determined by invoking a specific model for the interaction.

Let us consider the HOKE signal for the LO phase $\psi = 0$ [Eq. (2.5)]. Bearing in mind Eq. (2.18), we can write the imaginary part of the Condon contributions $\chi^{(3)}_{FC\alpha,\varphi}(\omega, t, \tau_2)$ in the form

$$\text{Im}\,\chi^{(3)}_{FC\alpha,\varphi}(\omega, t, \tau_2) = -(2\pi^3)^{1/2} N L^4 h^{-3} \exp(-\tau_2/T_1)[\sigma(\tau_2)]^{-1/2}$$

$$\times F^e_{0,s\alpha}(\omega - \omega_{el})\,\text{Re}\,w(z_{\alpha,\varphi}) \tag{3.1}$$

where we insert $t_m = 0$ and $t_{m''} + t_{m'} = \tau$ in Eq (2.19) for $z_{\alpha,\varphi}$.

The last equation corresponds to a case where only the first maxima of the absorption and the emission spectra are taken into consideration ($n = k = 0$). This simplification is justified due to the relative position of the excitation frequency ω with respect to the rhodamine's spectra.

The cubic polarization for the HOKE experiment (Y is the signal polarization axis, the probe pulse polarization is along the X axis, and the pump pulse is at $45°$ with respect to both X and Y) can be written in the form [see Eqs. (2.10), (2.17), (2.47), (2.52), and App. A]:

$$\mathscr{P}_y^{(3)+}(t) = \frac{1}{8}\sum_{\alpha,\varphi}\int_0^\infty d\tau_2\,\chi^{(3)}_{FC\alpha,\varphi}(\omega, t, \tau_2)\{\tfrac{1}{5}B_s^{\alpha,\varphi}(\tau_2)|\mathscr{E}_{2x}(t-\tau_2)|^2$$

$$\times \mathscr{E}_{3x}(t-\tau) + [B_0^{\alpha,\varphi}(\tau_2) + \tfrac{1}{30}B_s^{\alpha,\varphi}(\tau_2) - \tfrac{1}{6}B_a^{\alpha,\varphi}(\tau_2)]$$

$$\times \mathscr{E}_{1x}(t)\mathscr{E}_{3x}(t-\tau_2-\tau)\mathscr{E}_{2x}^*(t-\tau_2)\} \tag{3.2}$$

For subsequent calculations we ought to choose a concrete dependence of $\mathbf{D}(\mathbf{Q}_s)$.

When the dipole moment $\mathbf{D}_{12}(\mathbf{Q}_s)$ changes its direction but preserves its modulus [19, 59] [see Eqs. (A.11) to (A.14) below], the values $B_{0,s,a}^{\alpha,\varphi}$ are given by the following equations:

$$B_0^\alpha = B_0^\varphi = D_0^4/9;\ B_a^\alpha = B_a^\varphi = 0 \tag{3.3}$$

$$B_s^{(m)}(\tau_2) = (D_0^4/2)\left\{\tfrac{1}{3} + \exp\left[-\sum_j r_j^2(1 - \Psi_{sj}(\tau_2))\right]\right.$$

$$\left.\times \cos\left[\delta_{m\varphi}\sum_j r_j(1 - \Psi_{sj}(\tau_2))(\beta\hbar\omega_{st,j})^{1/2}\right]\right\} \tag{3.4}$$

where $r_j = 2\alpha_j\sqrt{\langle Q_{sj}^2(0)\rangle}$ are constants characterizing the correlations of the vector $\mathbf{D}_{21}$ with the jth intermolecular vibration, $D_0 = |D_{21}|$, $\omega_{s,j}$ is the contribution of the jth intermolecular motion to the total intermolecular Stokes shift ω_{st} ($\omega_{st} = \sum_j \omega_{st,j}$), and $\Psi_{sj}(\tau_2)$ is the normalized correlation function corresponding to the jth intermolecular vibration, which is related to the solvation correlation function $S(\tau_2)$ by Eq. (2.50). It is worth noting that the cosine term in the right-hand side of Eq. (3.4) for B_s^φ describes the interference of the Franck–Condon (dynamical Stokes shift) and the Herzberg–Teller relaxation dynamics.

2. Experimental Details

Rhodamine 800 was purchased from Exciton, 3,3-diethylthiatricarbocyanine bromide from Koch–Light, and they were used without further purification. $D_2O > 99.9\%$ was purchased from Aldrich.

3. Data Analysis

Our aim is to determine the solvation correlation function by resonance HOKE spectroscopy. According to Eqs. (2.5), (2.19) to (2.21), (3.1), and (3.2), we need to know, for this purpose, the following characteristics of the steady-state spectra: ω_{el} and the solvent contribution to the Stokes shift between the equilibrium absorption and emission spectra ω_{st}. The latter is related to the solvent's contribution to the second moment σ_{2s} by the relation $\omega_{st} = \hbar\beta\sigma_{2s}$. One can determine ω_{el} as the crossing point in the frequency scale of the equilibrium absorption and emission spectra of R800 ($\omega_{el} = 14{,}235\,cm^{-1}$ for water and is about the same for D_2O).

The solvent contribution to the central moment σ_{2s} can be determined by the relation $\delta\Omega = 2\sqrt{2\sigma_{2s}\ln 2}$, where $\delta\Omega$ is the half-width of the first absorption maximum. In order to exclude from our consideration the contribution of the second maximum and the optically active vibration of the frequency $\sim 600\,cm^{-1}$, we determined $\delta\Omega$ as twice the distance (in the frequency domain) between the luminescence maximum and the right-hand-side half maximum of the first luminescence maximum. Using this method, we obtain $\sigma_{2s} = 115{,}416\,cm^{-2}$ for the water solution and therefore $\omega_{st} = \hbar\beta\sigma_{2s} = 550\,cm^{-1}$, which conforms with the experimentally measured value. For D_2O the relation $\omega_{st} = \hbar\beta\sigma_{2s}$ is an approximate one; and in this case we used $\sigma_{2s} = 123{,}900\,cm^{-2}$.

Bearing this in mind, we fit our experimental data by Eqs. (2.5), (2.19) to (2.21), (2.50), (2.51), and (3.1) to (3.4). We present the correlation function $S(\tau_2)$ in the form of a sum of a Gaussian and one or two exponentials:

$$S(\tau_2) = a_f \exp[-(\tau_2/\tau_f)^2] + \sum_{i=1}^{2} a_i \exp(-\tau_2/\tau_{ei}) \qquad (3.5)$$

where $a_f + \sum_i a_i = 1$, τ_{e2} is the decay time of the slow (picosecond) exponential. We relate it to the solute–solvent H-bond, and therefore connect the correlation function for the non-Condon intermolecular motion on the right-hand side of Eq. (3.5) with this exponential:

$$\Psi_{sj}(\tau_2) = \exp(-\tau_2/\tau_{e2}) \tag{3.6}$$

Comparing Eqs. (3.5) and (2.50), we can express the value $\omega_{st,j}$ in Eq. (2.50) by the parameters a_f, a_2, and ω_{st}:

$$\omega_{st,j} = (1 - a_f - a_1)\omega_{st} \tag{3.7}$$

Correspondingly, the fitting parameters are $a_f, a_1, \tau_f, \tau_{e1}, \tau_{e2}$, and $r^2 \equiv r_j^2$.

The pulse duration t_p in our experiments is $t_p \approx 70 - 150\,\text{fs}$, depending on the laser excitation wavelength. In the case of ultrafast OKE experiments, the decay time T_1 in Eq. (3.1) is replaced by the orientation relaxa-

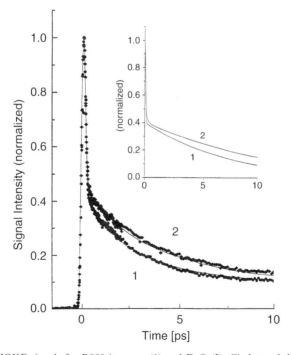

Figure 4. HOKE signals for R800 in water (1) and D_2O (2). Circles and diamonds, experimental data; solid curves, computer fit using Eqs. (2.5), (2.19) to (2.21), (2.50), (2.51), and (3.1) to (3.7) for $t_p = 150\,\text{fs}$, $r^2 = 2.5$, and other parameters given in Table II. Inset: solvation correlation functions for H_2O (1) and D_2O (2).

tion time τ_{or} of the solute molecules if the latter is shorter than T_1. For rhodamine dyes $T_1 \sim 1 - 2\,\text{ns} > \tau_{or} \sim 150\,\text{ps}$. We multiplied the experimental data by the factor $\exp(\tau/\tau_{or})$ and compared the theoretical and experimental data for delay times $\tau \ll \tau_{or} \approx 150\,\text{ps}$. Figure 4 shows the computer fit results of the experimental data of R800 in H_2O and D_2O. The fit of the theoretical calculations to the experimental curves is good. The inset in Figure 4 shows the solvation correlation functions $S(t)$ of R800 for H_2O and D_2O found by the computer fitting procedure.

We also carried out the corresponding measurements for R800 in water at different excitation frequencies ω [Fig. 5(a)]. Figure 5(b) shows theoretical spectra for different excitation conditions (i.e. ω and t_p for rhodamine 800 in water). We used the same parameter values of the previous fit (Fig. 4) for the curves shown in Figure 5(b). One can see that the theoretical curves reproduce all the fine details observed in the experimental data (in particular, the

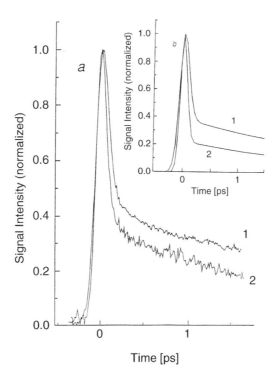

Figure 5. Experimental (*a*) and calculated (*b*) HOKE signals for R800 in water for various excitation frequencies ω and pulse duration t_p. (*a*) $\omega = 13{,}755\,\text{cm}^{-1}$, $t_p = 125\,\text{fs}$ (1); $\omega = 13{,}550\,\text{cm}^{-1}$, $t_p = 100\,\text{fs}$ (2). (*b*) $\omega = 13{,}831\,\text{cm}^{-1}$, $t_p = 130\,\text{fs}$ (1); $\omega = 13{,}441\,\text{cm}^{-1}$, $t_p = 90\,\text{fs}$ (2).

decrease in the amplitude of the slower signal component for "blue" excitations).

4. DISCUSSION

The correlation solvation functions for R800 in water and D_2O consist of two main components: an ultrafast Gaussian one with $\tau_f \sim 85\,fs < 100\,fs$, and a slow one with an exponential decay of a few picoseconds. Only a small part of the fast signal component can be explained by the coherent spike. The main contribution to it is due to the hole-burning effect.

The amplitude of the Gaussian component is about 60% for water, and the sum of a Gaussian and a fast exponential for D_2O is also 60%. This value is close to that observed by Fleming et al. ($\sim$50%) for coumarin-343 solvation in liquid water [9]. Its duration (85 fs) is about 1.7 times longer than that observed in [9]. The large difference can be explained as follows. The solvation, observed in [9], has been interpreted as an ion solvation [32]. According to [32], dipole solvation is slower than th ion solution. Therefore, if in the case of R800, the solvation is due to dipole or higher multipole interactions, its fast component is slower than in the case of ion solvation. The fast exponential of 146 fs for D_2O corresponds to that observed for a water solvation in [9] and [22].

Let us consider the slow components of the correlation functions for H_2O and D_2O [Eq. (3.5)] ($\tau_{e2} = 6.8\,ps$ for H_2O and $\tau_{e2} = 10\,ps$ for D_2O). They are close to the Debye relaxation times τ_D for these solvents (8.27 ps and 10.37 ps, respectively [90]). Such long components have not been observed in recent studies of solvation dynamics of other solutes in water [9,22]. We interpret our observations as a specific solvation related to formation (or breaking) of an intermolecular solute–solvent hydrogen bond between R800 and water molecules. The situation is similar to that observed by Berg and coauthors [96,97] on specific solvation dynamics of resorufin in alcohol solutions. In hydrogen-bonding solvents, the longest component of the Debye dielectric relaxation is assumed to be related to the rate of hydrogen-bond reorganization of the solvent [97–100]. According to [98], the time τ_D may reflect translation in water. In computer simulations the autocorrelation time of hydrogen bonds in water is 5–7 ps [97,100]. Thus the assumption that the slowest solvation is related to the reorganization of a hydrogen bond seems rather plausible. The experimental data for R800 show a significant isotope effect in water ($\sim$32% for times τ_{e2}), in contrast to study [96] in which an isotope effect in deuterated ethanol was not observed. It would be expected in view of the larger number of hydrogen bonds that water makes [102].

The hydrogen-bond formation (or breaking) assumption correlates with occurrence of non-Condon effects. The dependence $D(Q)$ is essential for a large change in Q. This is the case of hydrogen-bond formation (or breaking) where a large Q is accompanied by a large hopping distance (3.3 Å for water [98]) and a small activation energy. In Figure 6 the HOKE data for DTTCB solution in water and D_2O are shown. These data reflect only the fast dynamics of solvation (nonspecific one) and do not show any significant isotope effect.

In conclusion, using the technique of time-resolved HOKE, we have studied the ultrafast solvation dynamics of R800 and DTTCB in water and D_2O. According to our findings, the time dependence of the HOKE signal for R800 at the frequency domain under consideration is determined mainly by solute–solvent interactions. The significant change in the HOKE signal during the first $\sim100\,\mathrm{fs}$ is determined largely by the transient hole-burning effect. A biphasic behaviour of the solvation correlation function is essential for a good fit with the experimental data. The fast component of

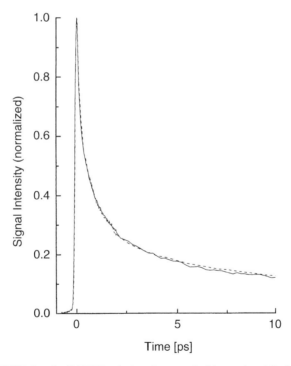

Figure 6. HOKE data for DTTCB solutions in water (solid curve) and D_2O (dashed curve); $t_p = 70\,\mathrm{fs}$, $\omega = 13{,}330\,\mathrm{cm}^{-1}$.

solvation dynamics for both R800 and DTTCB is determined by the non-specific solvation. The slowest component for R800 (which is close to the Debye relaxation time) is determined by a specific solvation related to formation (or breaking) of an intermolecular solute–solvent hydrogen bond. Correspondingly, we observe a significant isotope effect for the R800 solution and do not observe an isotope effect for DTTCB, which does not seem to form a solute–solvent hydrogen bond.

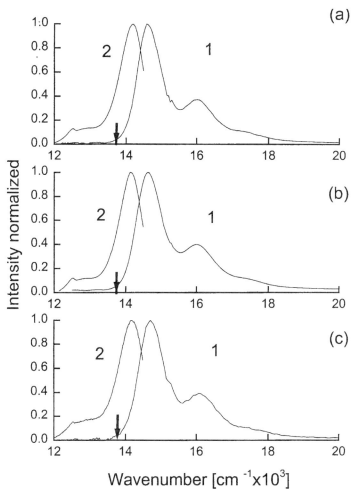

Figure 7. Absorption (1) and emission (2) spectra of R800 in methanol (*a*), butanol (*b*), and acetonitrile (*c*). The arrow shows the relative positions of excitation frequency ω.

C. Solvation Dynamics of R800 in Polar Solvents

To investigate the solvent influence on solvation dynamics, we carried out a comparison study of R800 solvation in various solvents. In addition to water and D_2O, we also studied its solvation in alcohols (methanol and butanol) and acetonitrile. The absorption and emission spectra of R800 in these solvents are shown in Fig. 7). We used laser pulses with a pulse duration of $t_p = 120\,\text{fs}$ at the frequency $\omega = 13,700\,\text{cm}^{-1}$. Figure 8 shows the computer fit to the experimental data of R800 in these solvents. The fit in the region 0–10 ps is carried out without inclusion of the non-Condon effects for these solvents. We used the same expression (3.5) for the solvation correlation functions, excluding acetonitrile and butanol. For butanol we added

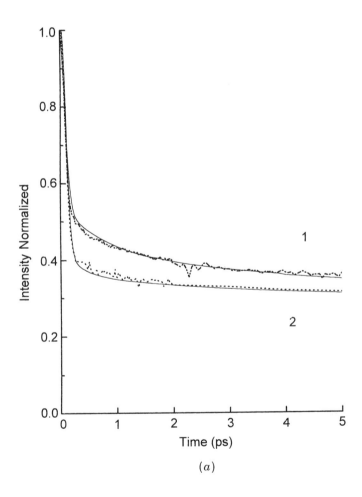

(a)

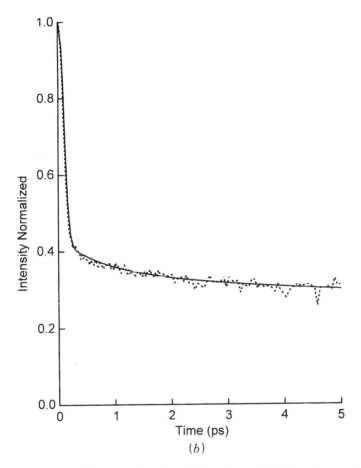

Figure 8. HOKE signals for R800 in methanol (1) and butanol (2) (*a*), and acetonitrile (*b*). Dots, experimental data; solid curves, computer fit.

one more exponential. For acetonitrile we used a cubic dependence for the ultrafast component: $a_f \exp[-(\tau_2/\tau_3)^3]$ instead of a Gaussian one, which better fitted the experimental data. The fit of the theoretical calculations to the experimental curves is good. Figure 9 shows the solvation correlation functions $S(t)$ of R800 for the last solvents found by the computer fitting procedure. Parameters σ_{2s} and ω_{el} and the best-fit parameter values for the solvation correlation function in all solvents are collected in Tables I and II, respectively.

It is worthy to note that we obtained a good fit for methanol only when a hot transition related to the intramolecular vibration of $600\,\mathrm{cm}^{-1}$ was taken

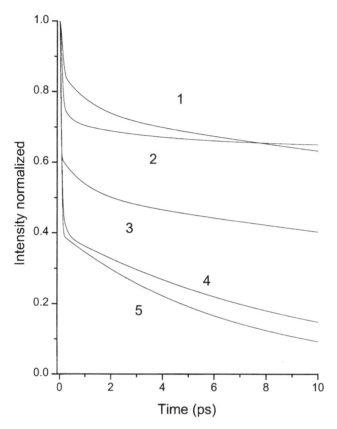

Figure 9. Solvation correlation functions for R800 in various solvents: methanol (1), butanol (2), acetonitrile (3), D_2O (4), and water (5).

TABLE I
Solvent Contribution to the Central Second Moment σ_{2s} and the Frequency of the Purely Electronic Transition ω_{el} of R800 for Various Solvents

Solvent	σ_{2s} (cm^{-2})	ω_{el} (cm^{-1})
Water	115,416	14,235
D_2O	123,900	14,235
Methanol	113,820	14,451
Butanol	79,630	14,406
Acetonitrile	104,712	14,372

TABLE II
Parameters of Solvation Correlation Functions for R800 in Various Solvents

Solvent	a_f	τ_f (ps)	a_1	τ_{e1} (ps)	a_2	τ_{e2} (ps)	a_3	τ_{e3} (ps)
Water	0.6	0.085	0	—	0.4	6.8	0	
D$_2$O	0.44	0.085	0.156	0.146	0.404	10	0	
Methanol	0.135	0.142	0.125	1.14	0.74	63.3	0	
Butanol	0.214	0.107	0.072	0.29	0.0635	3.13	0.651	556
Acetonitrile	0.3786	0.05	0.1147	1	0.507	42.2	0	

into account. This fact corresponds to the lowest-frequency excitation of R800 in methanol (see Table I and Fig. 7). In the last case we need to use the general expression [Eq (2.18)] for $n = -1$ and $k = 0$ instead of Eq. (3.1).

Let us confine ourselves and discuss the time dependence of the solvation correlation functions during the first 10 ps. Comparison of the ultrafast Gaussian components of the solvation correlation functions for methanol and water (Fig. 9 and Table II) shows that it is slower and has a smaller amplitude in methanol. This conclusion conforms qualitatively with the molecular dynamics simulations of solvation in methanol and water [102].

The amplitude of the ultrafast Gaussian component in methanol a_f is very close to that (0.14) observed with TRL of DASPI in the same solvent [103], although in our case this component is about twice faster. The times that are close to $\tau_{e1} = 1.14$ ps have also been observed for DASPI in methanol in [57] and [103]. This time corresponds to the fastest one (1.12 ps) for the multiple Debye description of $\epsilon(\omega)$ experimental data [24,104].

The value of the fastest component for R800 in acetonitrile (50 fs) conforms with TRL measurements of the ultrafast solvation component of LDS-750 in the same solvent (70 fs) [6] (which was assumed Gaussian), although in our case the fit with the cubic exponential gives a better result. The longer time behavior yields $\tau_{e1} = 1$ ps, which is close to that observed for DASPI in acetonitrile [103]. Acetonitrile is a dipolar aprotic solvent; therefore, it does not form a hydrogen bond. In this regard one can understand why the Condon description of the solvation dynamics in acetonitrile is satisfied.

IV.   QUANTUM BEATS ACCOMPANYING SOLVATION DYNAMICS

We observed in the HOKE signal low-frequency (about 100 cm^{-1}) beats during the solvation process. This issue is illustrated by Figures 10–21.

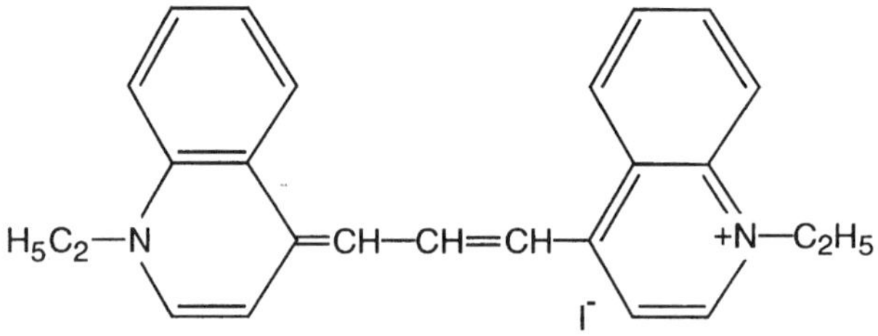

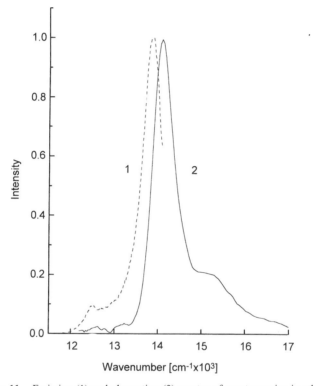

Figure 10.   Molecular structure of cryptocyanine.

Figure 11.   Emission (1) and absorption (2) spectra of cryptocyanine in ethanol.

Figures 12–13 and 20 show our results concerning the HOKE signal of crytocyanine and DTTCI (3,3′-diethylthiatricarbocyanine iodide), respectively, in alcohols. Previously, beats have been observed both on nonresonance [105–108] and on resonance excitation [16,17,41,42,109,110], in particular, in pump–probe spectroscopy [41,42] and in transient dynamics measurements of dichroism and birefringence [109]. An electron-vibrational

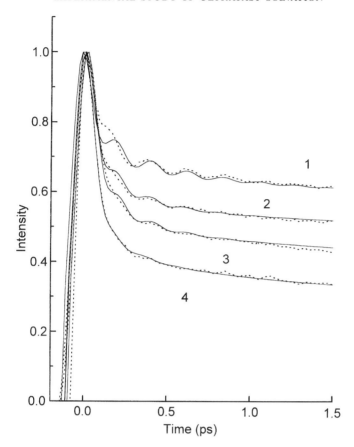

Figure 12. Short-time behavior of the HOKE signals for cryptocyanine in methanol, ethanol, butanol, and hexanol. Dots, experimental results; solid lines, computer fit using the theory presented in Sections II and III. The parameters used to fit the data are given in Tables IV and V.

theory of the beats on resonance excitation has been developed in [40], [109] and [111]. It has been pointed out in [111] that two mechanisms are plausible for the observed beats: HT (non-Condon) and Franck–Condon. The estimations of the relative contributions to the beats from these mechanisms [111] and whether they originate from either the excited or the ground electronic states [109,111] can be provided by the steady-state spectra of a chromophore in a particular solvent.

A specific property of our experiments is that the beats were observed *during* the solvation process. The beats' intensity and attenuation depended on the excitation frequency, the solvent, the solute, and the counter ion in

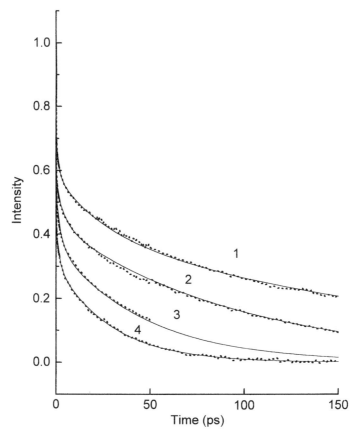

Figure 13. Long-time behavior of the HOKE signal for cryptocyanine in methanol, ethanol, butanol, and hexanol. Dots, experimental results; solid curves, computer fits to the experimental results. The parameters used to fit the long-time data are the same as those used for the short time shown in Figure 12.

ionic solutes. In this section we present our experimental results concerning the low-frequency beats, and give their interpretation, using the general theory developed in [59] and [111] (see Sec. II).

A. Simulation of Beats Accompanying Solvation Dynamics and Quantitative Estimations

Here we apply the theory described in Section II.B to calculate the signal of the low-frequency ($< kT$) beats. The last condition allows us to use a semi-classical theory of Sections II.B.1 and II.B.4. Let us first consider the solvation influence on the beat amplitude and phase. The correlation

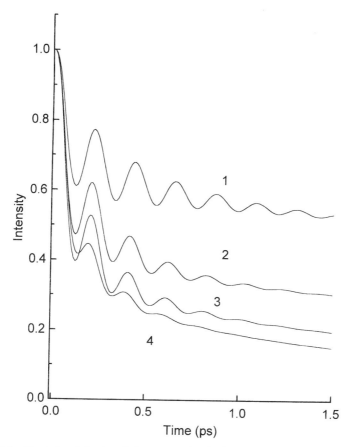

Figure 14. Short-time behavior of the computed solvation correlation functions for crypto-cyanine in methanol, ethanol, butanol, and hexanol.

function, describing both the solute–solvent interaction (solvation) and the low-frequency intramolecular vibrations (k), can be represented in the form

$$S(\tau) = \Psi_s(\tau) + \eta[\Psi_k(\tau) - \Psi_s(\tau)] \tag{4.1}$$

where $\eta = \omega_{st,k}/\omega_{st}$, ω_{st} is the Stokes shift determined by the intermolecular motions, $\omega_{st,k}$ is the contribution of the vibration k to the entire Stokes shift, and the index s denotes the solute–solvent contribution.

We will suppose that the parameter $\eta \ll 1$ [111] due to the fact that we observe only the fundamental frequency of the vibration which is responsible for the beats. Then using Eqs. (2.5), (2.16), (2.17), and (3.1)

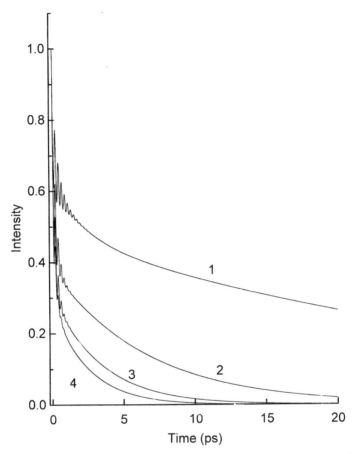

Figure 15. Long-time behavior of the computed solvation correlation functions for crypto-cyanine in methanol, ethanol, butanol, and hexanol.

for nonoverlapping and short pulses, we obtain in the first order with respect to η:

$$J_{\mathrm{HET}}(\tau) \sim \sum_{+,-} \frac{1}{\sqrt{1 - \Psi_s^2(\tau)}} \exp\left[-\frac{(\omega - \omega_{\mathrm{el}} \pm \omega_{\mathrm{st}}/2)^2}{\sigma_{2s}} \frac{1 - \Psi_s(\tau)}{1 + \Psi_s(\tau)} \right]$$
$$\times \left\{ 1 + [\Psi_k(\tau) - \Psi_s(\tau)] \left[\frac{2(\omega - \omega_{\mathrm{el}} \pm \omega_{\mathrm{st}}/2)^2}{\sigma_{2s}[1 + \Psi_s(\tau)]} + \frac{\Psi_s(\tau)}{1 - \Psi_s^2(\tau)} \right] \eta \right\}$$

$$(4.2)$$

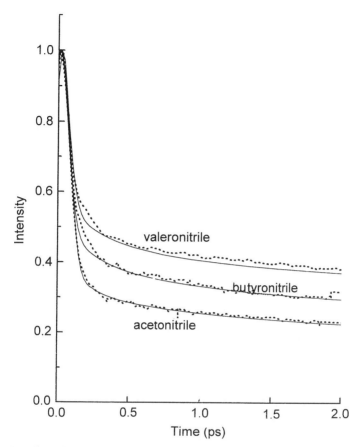

Figure 16. Short-time behavior of the HOKE signals for cryptocyanine in acetonitrile, butyronitrile, and valeronitrile. Dots, experimental results; solid curves, computer fit using the theory presented in Sections II and III. The parameters used to fit the data are given in Table IV.

One can directly see from Eq. (4.2) that the attenuation and phase of the beats are determined not only by the correlation function of the corresponding vibration $\Psi_k(\tau)$, but also by the solvation dynamics.

The maximum of the beats does not coincide with the maximum of the whole signal and appears later when the difference $\Psi_k(\tau) - \Psi_s(\tau)$ on the right-hand side of Eq. (4.2) differs from 0. Second, the factor preceding the term in braces on the right-hand side of Eq. (4.2) describes a proportional reduction of the beat amplitude which is related to solvation dynamics. Third, the term $[(\Psi_k(\tau) - \Psi_s(\tau))\Psi_s(\tau)/(1 - \Psi_s^2(\tau))]\eta$ on the right-hand side of Eq. (4.2), whose origin is the relaxation of the hole (or spike)

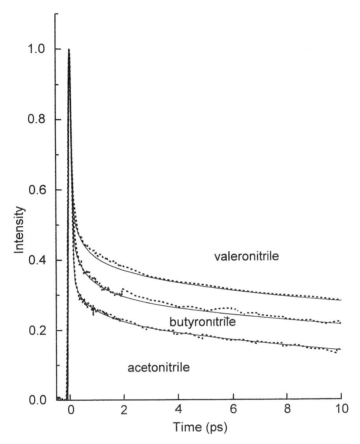

Figure 17. Middle-time behavior of the HOKE signals for cryptocyanine in acetonitrile, butyronitrile, and valeronitrile. Dots, experimental results; solid curves, computer fit using the theory presented in Sections II and III. The parameters used to fit the data are given in Table IV.

value, strongly diminishes upon the relaxation of solvation correlation function $\Psi_s(\tau)$. Thus, solvation dynamics accelerates the beats attenuation. The last two factors are conditioned by the hole-burning effect (the spike and the hole evolution). Therefore, one can say that hole burning in a solvation process increases the beats' contrast. We can expect that it is easier to observe beats in a slow-relaxing solvent that in a solvent where the Gaussian component decays very fast and has a large amplitude. The last conclusions are illustrated in Figures 12, 16, and 20.

Now let us discuss the question of how to differentiate whether the beats originate either from the excited or the ground electronic states. The

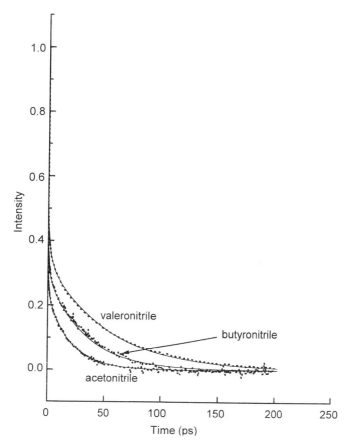

Figure 18.   Long-time behavior of the HOKE signals for cryptocyanine in acetonitrile, butyronitrile, and valeronitrile. Dots, experimental results; solid curves, computer fit using the theory presented in Sections II and III. The parameters used to fit the data are given in Table IV.

corresponding estimations for the solvating system differ from that for the steady-state spectra of a chromophore in a solvent [109,111]. We consider this issue by examples of a transmission pump–probe experiment [Eq. (2.4)] and HOKE spectroscopy [Eq. (2.5)] for $\psi = 0$.

Let us compare the contributions from the transient absorption $F_\alpha(\omega, \omega, \tau)$ and emission $F_\varphi(\omega, \omega, \tau)$ spectra related to the dynamics in the ground and excited electronic states, respectively. First we consider only the contribution of the intermolecular motion to $F_{\alpha,\varphi}(\omega, \omega, \tau)$. Then, using the four-photon approximation with respect to light–matter interaction, we obtain [37]

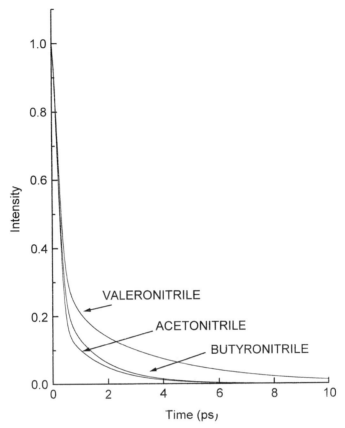

Figure 19. Computed solvation correlation functions for cryptocyanine in acetonitrile, butyronitrile, and valeronitrile.

$$\frac{|F_\varphi(\omega,\omega,\tau)|}{|F_\alpha(\omega,\omega,\tau)|} = \exp\left[\hbar\beta(\omega_{\text{el}} - \omega)\frac{1 - S(\tau)}{1 + S(\tau)}\right] \tag{4.3}$$

One can see that for the delay time $\tau = 0$ [$S(0) = 1$], the ratio (4.3) is equal to 1. If $\omega > \omega_{\text{el}}$, this ratio diminishes when τ increases [$S(\infty) = 0$] and approaches $\exp[-\hbar\beta(\omega - \omega_{\text{el}})]$, which is much smaller than 1 for $\hbar\beta(\omega - \omega_{\text{el}}) \gg 1$. This is explained by the fact that the spike relaxes much faster than the hole for $\hbar\beta(\omega - \omega_{\text{el}}) \gg 1$. It is related to the Franck–Condon principle; the sublevels of the excited electronic state achieved upon vertical optical transition correspond to a higher excitation level than the ones in the ground electronic states, and therefore the former relax faster [35].

Such a picture is qualitatively held or the case that includes intramolecular vibrations (see [37]). Thus, using different methods of spectroscopy with pulses long compared with the electronic dephasing at different excitation frequencies, we can investigate the beats in the ground electronic state or in the excited state separately. Simulations performed in Section V.B fully confirm this conclusion.

B. Experimental Results and Discussion

Figures 12, 13, 16, 17, 18, and 20 show our experimental HOKE data concerning beats for crytocyanine and DTTCI in alcohols and nitriles. We can see that the contrast of the beats is larger in the slow-relaxing solvents. The beats are almost invisible in solvents with a fast-relaxing Gaussian component of an appreciable amplitude such as nitriles and methanol. This fact can be explained by the hole-burning effect during a solvation process (see the discussion in Sec. IV.A.).

Let us consider the cryptocyanine data. The cryptocyanine molecule belonging to the polymethine dyes is shown in Figure 10. It consists of a conjugate system linked on both sides symmetrically to the same ring structure. Figure 11 shows the absorption and time-integrated emission of cryptocyanine in methanol. Both the absorption and emission exhibit a simple vibronic structure with a mirror symmetry. The Stokes shift is relatively small in comparison to R800. The Stokes shift, the second central moment, and the electronic origin of cryptocyanine are given in Table III. Using the fitting procedure described in the preceding section for rhodamine 800 yielded a good fit of the computed signal with the experimental one. The experimental data at long times (i.e., longer than 1.5 ps) was fitted to one or two exponents. The excited-state lifetime of cryptocyanine is relatively short. In low-viscosity liquids, the excited-state lifetime scales *linearly* with the

TABLE III
Solvent Contribution to the Central Second Moment σ_{2s}, the Frequency of the Purely Electronic Transition ω_{el}, and the Excited-State Lifetime of Cryptocyanine in Various Solvents

Solvent	ω_{el} (cm^{-1})	ω (cm^{-1})	σ_{2s} (cm^{-2})	τ_{ex} (ps)
Methanol	14,034	13,720	49,560	29
Ethanol	13,967	13,717	41,160	48
Butanol	13,908	13,705	44,730	100
Hexanol	13,873	13,690	44,310	213
Acetonitrile	14,027	13,610	62,370	21
Butyronitrile	13,934	13,730	53,340	33
Valeronitrile	13,907	13,620	52,920	53

solvent viscosity. The excited-state lifetime of cryptocyanine at room temperature was measured for various liquids using a time-correlated single-photon counting technique (TCSPC) and is given in Table III. The excited-state lifetimes measured by TCSPC are in agreement with the longest component of the OKE signals of the low-viscosity liquids. The long-time OKE signals with low time resolution for cryptocyanine in methanol, ethanol, butanol, and hexanol are shown in Figure 13. In addition to the excited-state lifetime, the HOKE signal consists of long-time components corresponding to the diffusional solvent response which contribute to the solvation correlation function. The data for cryptocyanine in methanol and ethanol were fitted with a single exponent of 2.5 ps and 3.5 ps, respectively (Table IV). These values are in reasonable agreement with other measurements of different probe molecules in the same liquids using other methods.

The short-time high-resolution HOKE signals of cryptocyanine up to 2 ps are shown in Figure 12 for the linear monoalcohol solvents. The signal for hexanol comprises a short component of 100 fs FWHM followed by several damped oscillations with a time period 250 fs. For butanol, the oscillations' amplitudes are much smaller than in hexanol solutions and the damping rate is much faster. In ethanol, the oscillations are barely observed, and in methanol solutions the oscillations are not observed in the OKE signal.

The short-time OKE signals of cryptocyanine in the alcohols were fitted by a Gaussian component and a model of an optically active NMO with exponential decaying memory function (see Sec. II.B.2). Using such a model may be explained as follows. The Markovian description of the dynamical system relaxation (a molecular vibration) is correct only for times which are much longer than the correlation time of the bath [66,112]. If one assumes that the beat attenuation is related to a solute–solvent interaction, both the beat attenuation time and the correlation time of the bath (solvent) τ_s will be

TABLE IV
Parameters of the Solvation Correlation Functions for Cryptocyanine in Various Solvents

Solvent	a_f	τ_f (ps)	a_1	τ_{e1} (ps)	a_2	τ_{e2} (ps)	η
Methanol	0.22	0.091	0.28	2.5			0.5
Ethanol	0.08	0.091	0.31	3.45			0.61
Butanol	0.06	0.091	0.41	6.66			0.53
Hexanol	0.05	0.091	0.15	2	0.48	33.33	0.32
Acetonitrile	0.6	0.091	0.2	0.22	0.2	1.43	
Butyronitrile	0.5	0.091	0.25	0.29	0.25	1.43	
Valeronitrile	0.45	0.091	0.3	0.4	0.25	3.33	

TABLE V
Parameters of the NMO for the Beat Description of Cryptocyanine in alcohols[a]

	$\gamma_{osc}(0)$ (cm^{-1})	ω_{osc} (cm^{-1})	α (cm^{-1})
Methanol	263	94	85
Ethanol	254	96	69
Butanol	301	95	61
Hexanol	284	107	45

[a] $\gamma_{osc}(0) = \varphi(0)/\alpha$ is the frequency-dependent attentuation $\gamma_{osc}(\omega) = $ Re $\int_0^\infty \varphi(t) \exp(-i\omega t)\, dt$ for $\omega = 0$.

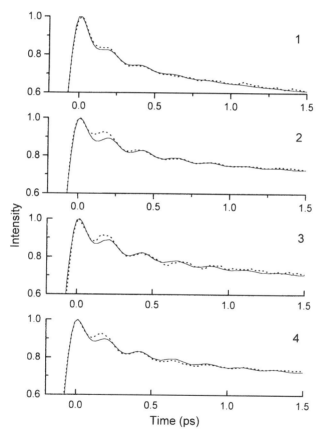

Figure 20. Short-time behavior of the HOKE signals for DTTCI in methanol (1), ethanol (2), propanol (3), and butanol (4). Dots, experimental results; solid curves, computer fit using the theory presented in Sections II and III. The parameters used to fit the data are given in Tables VII and VIII.

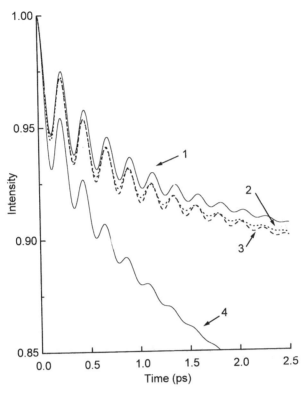

Figure 21. Short-time behavior of the computed solvation correlation functions for DTTCI in butanol (1), propanol (2), ethanol (3) and methanol (4).

of the same order, since τ_s is on the order of the solvation relaxation time. To describe such an effect, we must use at least a three-pole approximation to the corresponding correlation function, which corresponds to the NMO with an exponential memory function.

The fit of the experimental data by the NMO model is good for all solvents. The fitting parameters are given in Table V. From the fitting parameters given in Table V, it emerges that the main difference between solvents is the large difference in the value of α, the rate constant of the NMO memory function decay. Figures 20 and 21 show the HOKE signals for DTTCI in alcohols with the corresponding fits and the computed solvation correlation functions, respectively. Parameters of solvation correlation functions are given in Tables VI to VIII.

We have used the Condon theory to explain the main peculiarities of the beats observed. Figure 22, however, shows that the beats frequency is active

TABLE VI

Solvent Contribution to the Central Second Moment σ_{2s}, the Frequency of the Purely Electronic transition ω_{el}, and the Excited-State Lifetime of DTTCI in Alcohols

Solvent	ω_{el} (cm^{-1})	ω (cm^{-1})	σ_{2s} (cm^{-2})	τ_{ex} (ps)
Methanol	12,905	13,060	129,700	143
Ethanol	12,805	13,089	113,700	
Propanol	12,747	13,089	127,900	143
Butanol	12,720	13,089	104,712	143

TABLE VII

Parameters of the Solvation Correlation Functions for DTTCI in Alcohols

Solvent	a_f	τ_f (ps)	a_1	τ_{e1} (ps)	a_2	τ_{e2} (ps)	η
Methanol	0.005	0.091	0.912	25			0.083
Ethanol	0		0.005	0.29	0.925	100	0.07
Propanol	0		0.006	0.29	0.93	90.9	0.064
Butanol	0		0.02	0.29	0.92	100	0.06

TABLE VIII

Parameters of the NMO for the Beat Description of DTTCI in Alcohols.

	$\gamma_{osc}(\omega)$ (cm^{-1})	ω_{osc} (cm^{-1})	α (cm^{-1})
Methanol	436	86	37
Ethanol	548	87	27
Propanol	569	88	24
Butanol	672	89	21

in the infrared (IR) spectrum of the molecule DTTCI*. Therefore, generally speaking, one need also take into account the HT (non-Condon) mechanism of the beats, since only the HT mechanism is responsible for the presence of a definite frequency in an infrared spectrum [111,113]. We discuss this issue elsewhere.

---

* We thank Dr. V. Schreiber for measuring the IR spectrum of DTTCI.

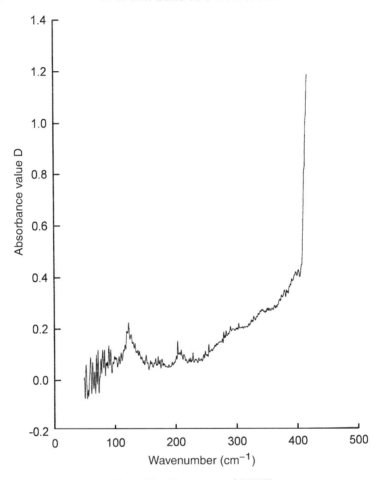

Figure 22.   IR spectrum of DTTCI.

V.   PROSPECT: SPECTROSCOPY OF NONLINEAR SOLVATION

In this section we discuss the advantages of transient four-photon spectro-scopy with pulses longer than the electronic transition dephasing. We show that it can be used for the nonlinear solvation study (i.e., when the linear response for the solvation dynamics breaks down).

As has already been intimated in Section I, one can control relative contribution of the ground state (a hole) and the excited state (a spike) to an observed signal by changing excitation frequency ω. This property of the spectroscopy with pulses long compared with electronic dephasing can be utilized for nonlinear solvation study. In the last few years, much attention has been given to the problem of nonlinear solvation [28,114–121]. In the case of linear solvation the spike and the hole motions can differ only by initial conditions of the excitation. However, in the case of nonlinear solvation, when the field created by a solute strongly changes during electronic excitation, relaxation of the solvent polarization occurs under conditions that are essentially different from those of the initial electronic state, and the spike and hole motions differ strongly, irrespective of the initial conditions of excitation.

A molecular dynamics simulation study of solvation dynamics in methanol [28] and polyethers [118] showed the breakdown of the linear response theory for this process. Solvation dynamics processes in the ground state of a solute differ from those of the excited electronic state [28]. In this section we show that spectroscopy methods considered in this review allow us to obtain separate information concerning solvation dynamics of a solute in the ground and excited electronic states, and therefore to study nonlinear solvation.

The aim of the theory is to relate the signal obtained in transient spectroscopy measurements with the solvation characteristics. There are four-time correlation functions in nonlinear spectroscopy where the system evolution is determined in the excited electronic state, and the thermal averaging is carried out in the ground electronic state [10,20,40,83,122,123]. Applying stochastic models to the calculation of such correlation functions results in missing any effects connecting with the situation when the chromophore affects the bath, in particular the dynamical Stokes shift (see, e.g., [123]). Indeed, the electronic excitation of a molecule results in a situation where the solvent configuration does not correspond to the upper electronic molecular state. The solvent (bath) is forced to relax to a new equilibrium configuration. It is precisely this reverse influence of the molecule on the solvent (bath) that cannot be taken into account in the limits of the stochastic approach. The latter makes the use of stochastic models in nonlinear spectroscopy of solvation dynamics meaningless, because the main effect of solvation is the Stokes shift. This is unfortunate since the stochastic models enable us to take into account in a simple way, many features of nuclear dynamics.

However, it is possible to overcome this difficulty if one takes into account the change in molecular electronic states before using stochastic approach [36,37]. One can express four-time correlation functions by the ones in which the thermal averaging is carried out in the same electronic state as the system evolution (equilibrium averages). In the last case, there is no change in the electronic molecular states while using the stochastic approach, and therefore applying the stochastic models will not result in missing the dynamical Stokes shift. We use such an approach in this section and investigate the influence of the intramolecular spectrum on the transient spectroscopy signal (TRL and RTFPS with pulses long compared with the electronic transition dephasing).

A. Four-Time Correlation Functions Related to Definite Electronic States

In the case of nonlinear solvation $u_s = W_{2s} - W_{1s}$ is not a Gaussian quantity, due to different solvation dynamics in the ground and in the excited electronic states. Therefore, we use a non-Gaussian formulation of the nonlinear polarization of Section II.B.3. We can represent the formula for the characteristic functions of the "intermolecular" spectra $f_{\alpha,\varphi s}(\tau_1, t)$ [see Eq. (2.44)] in the four-photon approximation in the form [36,37]

$$f_{js}(\tau_1, t) = \frac{(-1)^j}{4\hbar^2} \int_0^t d\tau_2 |\mathbf{D}_{21}\vec{E}(\mathbf{r}, t - \tau_2)|^2$$

$$\times \int_{-\infty}^{\infty} d\tau_3 f_{\alpha,M}^*(\tau_3) \exp[-\tau_2/T_1 + i\tau_3(\omega - \omega_{\text{el}})]M_j(\tau_1, \tau_2, \tau_3) \quad (5.1)$$

where $j = 1, 2; \alpha = 1, \varphi = 2;$

$$M_j(\tau_1, \tau_2, \tau_3) = \text{Tr}_s \left\{ \exp\left[\frac{i}{\hbar}(u_j(\tau_2)\tau_1 - u_j(0)\tau_3)\right] \rho_{1s}^e \right\} \quad (5.2)$$

The index e denotes the equilibrium state,

$$u_j(\tau_2) = \exp\left(\frac{i}{\hbar}W_j\tau_2\right) u_s \exp\left(-\frac{i}{\hbar}W_j\tau_2\right)$$

Averages [Eq. (5.2)] can be found using either classical or stochastic approaches. In the classical analysis [123] we can calculate $u_{1,2}(\tau_2)$ if we find the classical trajectories $\mathbf{Q}_{1,2}(\tau_2)$ in the ground ($\mathbf{Q}_1$) or excited ($\mathbf{Q}_2$) electronic states such that $u_{1,2}(\tau_2) \equiv u_s(\mathbf{Q}_{1,2}(\tau_2))$. In the last case the values $u_{1,2}(\tau_2) \equiv u_s(\mathbf{Q}_{1,2}(\tau_2))$ are C-numbers. Apparently, the value $u_1(\tau_2)$ is determined by a motion in the ground electronic state, and $u_2(\tau_2)$ by a motion in the excited electronic state. The thermal averaging in Eq. (5.2) is carried out with respect to the ground electronic state. However, applying stochastic models to the calculation of $M_2(\tau_1, \tau_2, \tau_3)$, where the system evolution is determined in the excited electronic state 2, and the thermal averaging is carried out in the ground electronic state 1 results in missing any bath effects on the chromophore, in particular, the dynamical Stokes shift. We can overcome such a difficulty if we will express $M_2(\tau_1, \tau_2, \tau_3)$, by a four-time correlation function related to a definite electronic state [36,37]:

$$M_j(\tau_1, \tau_2, \tau_3) = \frac{b_{js}}{b_{1s}} \exp\left[\frac{i}{\hbar} \langle u_s \rangle_j (\tau_1 - \tau_3 - i\delta_{2j}\beta\hbar)\right] \bar{M}_j(\tau_1, \tau_2, \tau_3) \qquad (5.3)$$

where $b_{js} = \text{Tr}_s \exp(-\beta W_{js})$,

$$\bar{M}_j(\tau_1, \tau_2, \tau_3) = \left\langle \exp\left\{\frac{i}{\hbar}[\bar{u}_j(\tau_2)\tau_1 - \bar{u}_j(0)(\tau_3 + i\delta_{2j}\beta\hbar)]\right\}\right\rangle_j \qquad (5.4)$$

are the central four-time correlation functions, $\bar{u}_j(\tau) = u_j(\tau) - \langle u(0) \rangle_j$ is the central value of $u_j(\tau_2)$; $\langle \cdots \rangle_j = \text{Tr}_s\{\cdots \rho_{js}^e\}$ denotes the average with respect to electronic state j; δ_{2j} is the Kronecker delta.

In Eq. (5.4) the averaging is carried out in the same electronic state as the classical trajectory calculations (equilibrium average), and the stochastic model can be used for the calculation of Eq. (5.4) and all the expressions resulting from it.

We expand the four-time correlation function by cumulants [124] (see the explanation of cumulant averages in Chapter 8 of [10]).

$$\bar{M}_j(\tau_1, \tau_2, \tau_3) = \exp\left\{\sum_{n=2}^{\infty} \left(\frac{i}{\hbar}\right)^n \frac{1}{n!} \langle[\bar{u}_j(\tau_2)\tau_1 - \bar{u}_j(0)(\tau_3 + i\delta_{2j}\beta\hbar)]^n\rangle_{cj}\right\} \qquad (5.5)$$

Here the suffix c means that $\langle \cdots \rangle_{cj}$ is defined as a cumulant average with respect to state j.

B. Simulation of Transient Four-Photon Spectroscopy Signals for Nonlinear Solvation

In this subsection we simulate by Eqs. (2.3), (2.38) to (2.43), (2.45), (2.46), (5.1), (5.3), (5.5), and (5.6) (see below) the signal for various methods of transient spectroscopy with pulses long compared with electronic dephasing. We use the solvation correlation functions of a nonlinear solvation calculated by Fonseca and Ladanyi for a dipole solute in methanol [28] (Fig. 23). Unfortunately, their simulations are limited by normalized correlations functions of the second order. Therefore, in our simulations we can use expansion equation (5.5) only up to the first term, thus confining our consideration up to the second-order cumulants. The corresponding formula for the cubic polarization differs from the one for linear solvation (Sec. II) by the presence of different solvation correlation functions describing the dynamics in the ground or excited electronic states.

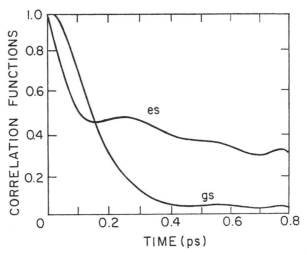

Figure 23.   Solvation correlation functions for the ground (gs) and excited (es) electronic states calculated in [28].

Figure 24(a) shows the calculation results for RTGS. For comparison we also show the corresponding signals when both ground- and excited-state correlation functions coincide (linear case) and are equal to either the correlation function of the excited state $S_{es}(t)$ (curve 2) or the ground state $S_{gs}(t)$ (curve 3). The equilibrium spectra of the molecule in solution $F_{\varphi}^{e}(\omega)$ (curve 4) and $F_{\alpha}^{e}(\omega)$ (curve 5) and the shapes of the intramolecular spectra $F_{\varphi M}(\omega)$ (curve 6) and the $F_{\alpha M}(\omega)$ (curve 7) (when the solvent contribution is absent) are shown in the insets to Figures 24 to 26. The arrows show the relative positions of the excitation frequency ω. One can see that for excitation at the frequency of the purely electronic transition, the signal provides combined information concerning the solvation dynamics in both states. But for the excitation at the maximum of the absorption band, the signal mainly reflects the solvation dynamics in the excited electronic state [Fig. 24(b)].

Figure 25 shows the calculation results using Eq. (2.45). It can be seen that for the excitation on the blue side of the absorption spectrum, the transmission pump–probe experiment provides information concerning the solvation dynamics in the ground electronic state. The same is true for the HOKE signal at $\psi = 0$ [Eq. (2.5)], since the right-hand side of Eq. (2.45) also describes a signal for the latter case (see above). Such behavior can be understood if we compare the contributions from the transient absorption and emission spectra related to the dynamics in the ground and in the excited electronic states, respectively (see Sec. IV.A).

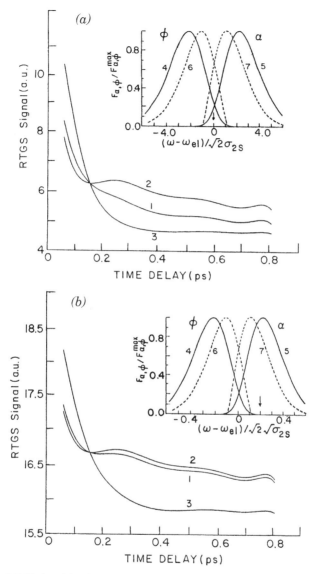

Figure 24. RTGS signal for the case of nonlinear solvation (1) calculated by the correlation functions of [28] (see Fig. 23); other parameters of the system are identical to the parameters used in the numerical calculations in [18] and [59]. Curves 2 and 3 in Figures 24 to 26 correspond to signals when ground- and excited-state correlation functions coincide (linear case) and are equal either to the correlation function of the excited state (2) or the ground state (3). Insets to Figures 24 to 26 show the equilibrium spectra of the molecule in solution (4, 5), and the shapes of the intramolecular spectra (6, 7). The arrows show the relative positions of the excitation frequency ω.

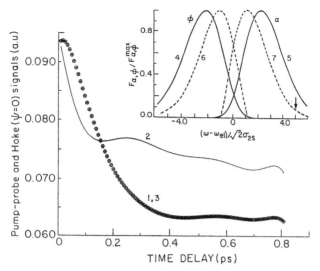

Figure 25. Transmission of the probe signal in pump-probe experiments (and the HOKE signal at $\psi = 0$) in the case of nonlinear solvation. The rest is the same as in Figure 24.

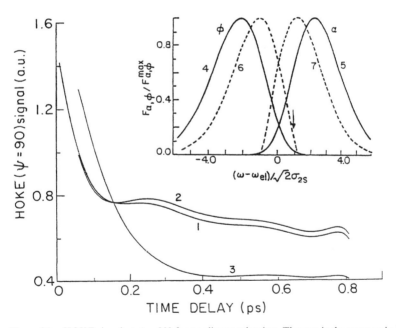

Figure 26. HOKE signal at $\psi = 90°$ for nonlinear solvation. The rest is the same as in Fig 24.

Now let us consider the HOKE signal when the LO phase $\psi = 90°$. Using Eqs. (2.5) and (2.38), we obtain

$$J_{\text{HET}}(\psi = 90°) \sim -[\Phi_\alpha(\omega, \omega, \tau) - \Phi_\varphi(\omega, \omega, \tau)] \qquad (5.6)$$

We can see that for the excitation as it is shown in Figure 26, the HOKE signal reflects mainly solvation in the excited electronic state.

Thus, using different methods of spectroscopy with pulses long compared with the electronic transition dephasing at different excitation frequencies, we can separately investigate the solvation dynamics in the ground electronic state or in the excited state, and this enables us to study the nonlinear solvation.

C. Spectral Moments of the Nonequilibrium Absorption and Luminescence of a Molecule in Solution

Our consideration in the previous subsection was of a preliminary nature since it was limited by the second-order cumulants. Here and in the following subsections we show how to characterize the nonlinear solvation (non-Gaussian) case correctly.

The nonlinear polarization and the transient spectroscopy signal can be expressed by the spectra of nonequilibrium absorption (α) and luminescence (φ) of a molecule in solution [Eqs. (2.38) to (2.43), (2.45), (2.46), (5.1), and (5.6)] which are convolutions of the inter- and intramolecular spectra [Eq. (2.39)]. Therefore, we calculate the normalized spectral moments $\langle \omega_{\alpha,\varphi s}^n(\tau) \rangle / \langle \omega_{\alpha,\varphi s}^0 \rangle$ of the intermolecular spectra $F_{\alpha,\varphi s}(\omega', \omega, \tau)$. Using them, one can easily find the spectral moments of the spectra $F_{\alpha,\varphi}(\omega_1, \omega, \tau)$.

For simplicity we consider that the pump pulses are shorter than the solute–solvent relaxation time and do not overlap with the probe pulses. The nth noncentral moment of nonequilibrium intermolecular spectrum $F_{\alpha,\varphi s}$ is determined by

$$\langle \omega_{\alpha,\varphi s}^n(\tau) \rangle = \int_{-\infty}^{\infty} \omega_1^n F_{\alpha,\varphi s}(\omega_1, \omega, \tau) \, d\omega_1 \qquad (5.7)$$

Using Eqs. (5.1), (5.3), and (5.5), one can obtain for the first moment [36,37]

$$\frac{\langle \omega_{\alpha,\varphi s}^n(\tau) \rangle}{\langle \omega_{\alpha,\varphi s}^0 \rangle} = \langle u_s \rangle_j / \hbar + \sigma_{1j}(\tau) \qquad (5.8)$$

where

$$\sigma_{1j}(\tau) = \frac{1}{F_\alpha^e(\omega)} \sum_{n=1}^{\infty} \frac{(-1)^n}{\hbar^{n+1}} \langle \bar{u}_j^n(0)\bar{u}_j(\tau)\rangle_{cj} \left\{ \frac{1}{n!} \frac{d^n F_\alpha^e(\omega)}{d\omega^n} \right.$$

$$\left. + \delta_{2j} \sum_{k=1}^{n} \frac{(-\hbar\beta)^k}{k!\,(n-k)!} \frac{d^{n-k} F_\alpha^e(\omega)}{d\omega^{n-k}} \right\} \tag{5.9}$$

The values $\sigma_{nj}(\tau)$ are the partial central moments:

$$\sigma_{nj}(\tau) = \int_{-\infty}^{\infty} (\omega_1 - \langle u_s\rangle_j/\hbar)^n \frac{F_{js}(\omega_1,\omega,\tau)}{\langle \omega_{js}^0 \rangle} \, d\omega_1 \tag{5.10}$$

where $F_{js}(\omega_1,\omega,\tau)$ are the intermolecular spectra of the nonequilibrium absorption $(j = \alpha)$ or luminescence $(j = \varphi)$ determined by Eq. (2.40).

Equation (5.9) relates the first moments of the nonequilibrium absorption $(j = 1)$ and emission $(j = 2)$ spectra to the correlation functions $\langle \bar{u}_j^n(0)\bar{u}_j(\tau)\rangle_{cj}$ of the solute–solvent interaction $u_s = W_{2s} - W_{1s}$. The coefficients of expansion are determined by the experimentally measurable values: derivatives of the equilibrium absorption spectrum $F_\alpha^e(\omega)$ of a solute molecule in a solution. Dependence of the first moments on $F_\alpha^e(\omega)$ reflects the fact that in the nonlinear case, the spectral dynamics depends on the excitation conditions. Equation (5.9) can be considered as a generalization of the fluctuation–dissipation theorem for the nonlinear solvation case [37].

In the particular case of the linear solvation when the magnitude u_s is Gaussian, the cumulants of order higher than the second are equal to zero, and expansion (5.9) comes abruptly to an end after the first term. In the last case Eq. (5.9) reflects the fluctuation–dissipation theorem [37].

Equation (5.9) expresses the first moments of nonequilibrium spectra by the derivatives of the equilibrium absorption spectrum of a solute molecule in solution $F_\alpha^e(\omega)$ which can be measured experimentally. The second partial central moment can be also presented in a form similar to Eq. (5.9). However, the formula for it is more complex and therefore is not presented here. In general, the corresponding formulas become complicated when the order of a moment increases. However, formulas for high-order moments can be written in a more compact form if one refrains from using $F_\alpha^e(\omega)$ [37]:

$$\sigma_{nj}(\tau) = \frac{1}{\hbar^n F_\alpha^e(\omega)} \frac{b_{js}}{b_{1s}} \langle \bar{u}_j^n(\tau) \exp(\delta_{2j}\beta u_s) F_{\alpha M}(\omega - \omega_{\mathrm{el}} - u_s/\hbar)\rangle_j \tag{5.11}$$

Equation (5.11) for $j = 2$ can be also presented in the form of a nonequilibrium average [37]:

$$\sigma_{n2}(\tau) = \frac{1}{\hbar^n F_\alpha^e(\omega)} \langle \bar{u}_2^n(\tau) F_{\alpha M}(\omega - \omega_{el} - u_s/\hbar) \rangle_1 \qquad (5.12)$$

Equations (5.11) and (5.12) express the partial central moments of nonequilibrium spectra by the intramolecular absorption spectrum $F_{\alpha M}$. It cannot be measured directly for a molecule that is in a polar solvent. However, the intramolecular spectrum $F_{\alpha M}(\omega)$ can be determined as the spectrum of the same solute in a nonpolar solvent [125].

D. Broad and Featureless Electronic Molecular Spectra

Let us consider the particular but very important and widely-distributed case of very broad and featureless electronic spectra of solute organic molecules in solutions. The examples are LDS-750 [18], phtalimides [126], and many others. For such molecules the square root of the second central moment of the equilibrium absorption spectrum is rather large $\sqrt{\sigma_\alpha} \sim 1700 \, \text{cm}^{-1}$. Because of this, the formulas for the partial central moments for the spectra of such molecules are strongly simplified [36,37]:

$$\sigma_{n2}(\tau) = \frac{b_{2s}}{b_{1s}} \hbar^{-n} \langle \bar{u}_2^n(\tau) \exp(\beta u_s) \rangle_2 \qquad (5.13)$$

Equation (5.13) can be also considered as a generalization of the fluctuation–dissipation theorem for the nonlinear solvation case [37].

We can also rewrite Eq. (5.13) in the form of the nonequilibrium averages [36,37]

$$\sigma_{n2}(\tau) = \hbar^{-n} \langle \bar{u}_2^n(\tau) \rangle_1 \qquad (5.14)$$

where $\bar{u}_2(\tau)$ is determined by the motion in excited electronic state 2; however, the averaging is carried out with respect to ground electronic state 1.

E. Time-Resolved Luminescence Spectroscopy

The time shift of the first moment of the luminescence spectrum is characterized by the equation [22,28]

$$C_\varphi(\tau) = \frac{\langle \omega_\varphi(\tau) \rangle - \langle \omega_\varphi(\infty) \rangle}{\langle \omega_\varphi(0) \rangle - \langle \omega_\varphi(\infty) \rangle} \qquad (5.15)$$

where $\langle \omega_{\alpha,\varphi}(\tau) \rangle = \int_{-\infty}^{\infty} \omega_1 F_{\alpha,\varphi}(\omega_1, \omega, \tau)\, d\omega_1$ is the first moment of the absorption (α) or emission (φ) spectrum.

The quantity $C_\varphi(\tau)$ can be presented in the form [36,37]

$$C_\varphi(\tau) = \sigma_{12}(\tau)/\sigma_{12}(0) \qquad (5.16)$$

where the normalized first moment of the TRL spectrum $\sigma_{12}(\tau)$ is determined in particular by Eq. (5.9) for $j = 2$. One can see that for the case of nonlinear solvation, the first moment of the TRL spectra $\sigma_{12}(\tau)$ is determined not only by the correlation function of the second order, but also by cumulant averages $\langle \bar{u}_2^n(0)\bar{u}_2(\tau) \rangle_{c2}$ higher than the second ($n > 1$).

Computer simulations [28] calculated the normalized correlation function in the ground electronic state ($\langle \bar{u}_1(0)\bar{u}_1(\tau) \rangle_1 / \langle \bar{u}_1^2(0) \rangle_1$) and in the excited electronic state ($\langle \bar{u}_2(0)\bar{u}_2(\tau) \rangle_2 / \langle \bar{u}_2^2(0) \rangle_2$), and the nonequilibrium response function $C(\tau)$ ($\sim \langle \bar{u}_2(\tau) \rangle_1$), which was assumed to correspond to the experimental measurements of $\sigma_{12}(\tau)$. One can see that the first normalized moment $\sigma_{12}(\tau)$ depends on the equilibrium absorption spectrum of the solute molecule in solution, $F_\varphi^e(\omega)$, and in general is not reduced to the nonequilibrium average $\langle \bar{u}_2(\tau) \rangle_1$. However, for the case of broad and featureless electron spectra and for excitation near the Franck–Condon frequency of the transition $1 \rightarrow 2$ [36,37], the partial central moments of the emission band $\sigma_{12}(\tau)$ almost do not depend on $F_\alpha^e(\omega)$ and are expressed by the nonequilibrium averages [Eq. (5.14)]. Thus the nonequilibrium average approximately describes the first moment of the TRL spectrum only in the case of broad and featureless electronic spectra of a solute molecule in solution for the excitation near the frequency of the Franck–Condon transition.

F. Time-Resolved Hole-Burning Study of Nonlinear Solvation

Let us consider the time-resolved hole-burning experiment [46]. Similar to TRL studies [22,28] [see Eq. (5.15)], one can characterize the time shift of the first moment of the difference absorption spectrum [Eq. (2.46)]

$$\langle \omega_{\Delta\alpha}(\tau) \rangle = \int_0^\infty \omega_1\, \Delta\alpha(\omega_1 - \omega)\, d\omega_1 \qquad (5.17)$$

by the equation

$$C_{\Delta\alpha}(\tau) = \frac{\langle \omega_{\Delta\alpha}(\tau) \rangle - \langle \omega_{\Delta\alpha}(\infty) \rangle}{\langle \omega_{\Delta\alpha}(0) \rangle - \langle \omega_{\Delta\alpha}(\infty) \rangle} \qquad (5.18)$$

where $\omega_1 = \omega' + \omega$. The quantity $\langle \omega_{\Delta\alpha}(\tau) \rangle$ can be expressed by the first moments of the nonequilibrium absorption (α) and luminescence (φ) spectra: $\langle \omega_{\Delta\alpha}(\tau) \rangle = \langle \omega_\varphi(\tau) \rangle - \langle \omega_\alpha(\tau) \rangle$.

The quantity $C_{\Delta\alpha}(\tau)$ can be presented by $\sigma_{1j}(\tau)$ [36,37]:

$$C_{\Delta\alpha}(\tau) = [\sigma_{12}(\tau) + \sigma_{11}(\tau)]/[\sigma_{12}(0) + \sigma_{11}(0)] \tag{5.19}$$

According to Eqs. (5.9) and (5.16), the term $C_\varphi(\tau)$ provides information on solvation dynamics in the excited electronic state, while $C_{\Delta\alpha}(\tau)$ provides both the solvation dynamics in the ground and excited electronic states. We shall show how the solvation dynamics in the ground electronic state can be found by the time-resolved hole-burning spectroscopy. Let us assume that we have determined both $\sigma_{12}(\tau)$ and $\sigma_{12}(0)$ by TRL spectroscopy. By measuring the dependence $C_{\Delta\alpha}(\tau)$, we can determine the function $\sigma_{11}(\tau)$, describing the dynamics in the ground electronic state. Using Eq. (5.19), one can show [36,37] that

$$\sigma_{11}(\tau) = C_{\Delta\alpha}(\tau)[2\sigma_{12}(0) - \omega_{st}] - \sigma_{12}(0)C_\varphi(\tau) \tag{5.20}$$

where $\omega_{st} = \hbar^{-1}(\langle u_s \rangle_1 - \langle u_s \rangle_2)$ is the solvent contribution to the Stokes shift between the equilibrium absorption and emission spectra.

G. Stochastic Approach to Transient Spectroscopy of Nonlinear Solvation Dynamics

In this subsection we show how to use a stochastic approach to calculation of the spectral moments of the nonequilibrium absorption and luminescence of a solvating molecule [37]. We consider $u_j(\tau)$ as a random function of a parameter τ. Equilibrium averages in formulas (5.9), (5.11), and (5.13) have the following form: $\langle \psi_1(u_j(\tau))\psi_2(u_j(0)) \rangle_j$, where $\psi_{1,2}(u_j)$ are given functions of u_j. Let us denote $u_j(\tau) = u_\tau$ and $u_j(0) = u_0$. Then the equilibrium averages under discussion can be presented in the form [127,128]

$$\langle \psi_1(u_\tau)\psi_2(u_0) \rangle_j = \int\int du_0 \, du_\tau \, w_j(u_0)v_j(u_\tau|\tau, u_0)\psi_1(u_\tau)\psi_2(u_0) \tag{5.21}$$

where $w_j(u_0)$ describes a law of probability in electronic state j, and $v_j(u_\tau|\tau, u_0)$ is the density of the conditional probability that u takes the value u_τ at time τ if it takes the value u_0 at time 0.

It is worth noting that u in Eq. (5.21) is a stationary random function, and therefore $w_j(u_0)$ does not depend on τ, and v_j depends only on one time variable. This is due to the fact that in our formulas the average is carried out with respect to the same electronic state as the determination of value u.

Thus, to calculate averages [Eq. (5.21)] we must know the corresponding conditional probability $v_j(u_\tau | \tau, u_0)$. It has been calculated for the rotational diffusion model in the case of nonlinear solvation [37].

Let us consider the nonequilibrium averages [Eq. (5.14)] which appear in, the theory of broad and featureless electronic molecular spectra (Sec. V.D). Using permutations under the trace operation, we can present Eq. (5.14) in the form

$$\sigma_{n2}(\tau) = \hbar^{-n} \langle \bar{u}_2^n(\tau) \rangle_1 = \hbar^{-n} \operatorname{Tr}_s[(u_s - \langle u_s \rangle_2)^n \tilde{\rho}_{2s}(\tau)] \qquad (5.22)$$

where $\tilde{\rho}_{2s}(\tau) = \exp(-(i/\hbar)W_2\tau)\rho_{1s}^e \exp((i/\hbar)W_2\tau)$ is the density matrix describing the evolution of the solvent nuclear degrees of freedom in the excited electronic state 2 for the specific initial condition: $\tilde{\rho}_{2s}(0) = \rho_{1s}^e$; that is, it coincides with ρ_{1s}^e for $\tau = 0$.

The classical analog of $\tilde{\rho}_{2s}(\tau)$ is a one-dimensional distribution $w_2(u_s, \tau)$ for a *nonstationary* random process for the initial condition

$$w_2(u_s, \tau) = w_1(u) \qquad (5.23)$$

where $w_1(u)$ describes the stationary probability in the ground electronic states.

Thus we can write the value $\sigma_{n2}(\tau)$ corresponding to Eq. (5.14), in the form

$$\sigma_{n2}(\tau) = \hbar^{-n} \int (u_s - \langle u_s \rangle_2)^n w_2(u_s, \tau) \, du_s \qquad (5.24)$$

where $w_2(u_s, t)$ must be determined for nonstationary conditions that correspond to the ground state for $t < 0$, and the excited one for $t \geq 0$. Sometimes, finding $w_2(u_s, t)$ for suitable initial conditions is an easier problem than finding the conditional density $v_2(u_\tau | \tau, u_0)$ for arbitrary u_0 [37].

The general formulas presented in Sections V.C, V.D, and V.G have been applied to the calculation of the spectral moments of a molecule in a model solvent [37]. According to Debye [129–131], the solvent was considered as composed of point dipoles **d**. Each dipole undergoes rotational Brownian motion as a result of interactions with a bath. The quantity u_s, the difference between interactions of the solvent with the excited-state solute and with the ground-state solute, can be represented in the form $u_s = \sum_n (W_{2s}^{(n)} - W_{1s}^{(n)})$, where $W_{2s}^{(n)}$ (or $W_{1s}^{(n)}$) denotes the interaction between the solvent molecule labeled n and the excited-state (or ground-state) solute. The value u_s for the interaction of the solute with a single solvent molecule is

$u' = -\mathbf{d} \cdot (\mathbf{E}^{(2)} - \mathbf{E}^{(1)})$, where $E^{(j)}$ is the electrical field created by a solute in the electronic state j.

To avoid nonprincipal complications, we considered that the electric field created by a solute in both electronic states 1 and 2 is directed along the same straight line, but can differ in its value or the direction with respect to this line. In this case one can write $u' = -\mathbf{d} \cdot (E^{(2)} - E^{(1)}) \cos \theta = -dE_{21} \cos \theta$, where θ is the angle between the dipole and the direction of field $\mathbf{E}^{(2)}$, and $E^{(1)}$ is the value of field $\mathbf{E}^{(1)}$ with a sign plus or minus depending on the orientation of $\mathbf{E}^{(1)}$ with respect to $\mathbf{E}^{(2)}$, and $E_{21} \equiv E^{(2)} - E^{(1)}$.

A long-time solution for the model under consideration in a strong electrical field has been obtained in [37]. Figure 27 illustrates nonlinear solvation behavior for the solute, which does not create a field in the ground electronic state $(E^{(1)} = 0, E_{21} = E^{(2)})$. Figure 28 shows the time-dependent first moment of the difference absorption spectrum for time-resolved hole-burning (HB) experiments. We can see that in the general case the signal is an intermediate one with respect to the dynamics in the ground and in the excited electronic states [Fig. 28(a)]. However, we can obtain a signal that

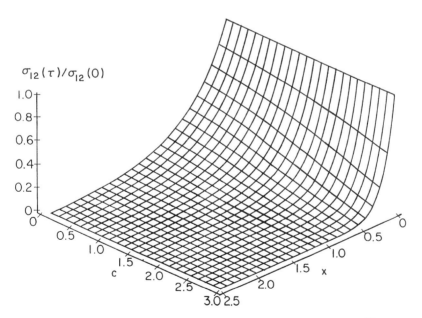

Figure 27.   TRL signal as a function of $x = D\tau$ for $y = 1$ and different $c = \beta dE^{(2)}$ where D is the rotational diffusion coefficient for the solvent and $y = \hbar(\omega_{el} - \omega)/(dE_{21})$.

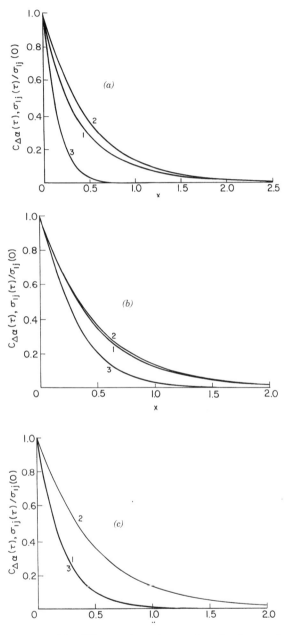

Figure 28. First moment of the difference absorption spectrum for time-resolved HB experiments (1) for $c = 3$ and $y = 1$ (a), $y = 0.75$ (b), and $y = 0.01$ (c). Transient first moments for the absorption (2) and the emission (3) are shown for comparison.

mimics either the excited- or ground-state dynamics by tuning the excitation frequency [(Fig. 28(b) and (c)].

H. Conclusion

In this section we have shown that by using different methods of spectroscopy with pulses long compared with electronic dephasing and tuning the excitation frequencies, one can investigate separately solvation dynamics in the ground or excited electronic states, and consequently, study nonlinear solvation.

We used the theory [20,83] developed for pulses long compared with electronic dephasing $(t_p \gg T')$. In this regard, the question arises whether the condition $t_p \gg T'$ is necessary for the separate study of solvation dynamics in the ground or in the excited electronic states. Following the considerations of this section, the pump pulses must be sufficiently long to provide a definite position of a spike and a hole on the potentials of the excited or ground electronic state, respectively. One can formulate the corresponding condition as follows: $t_p \gg \sigma_{2s}^{-1/2}$, where σ_{2s} is the contribution of solvation processes to the second central moment of an absorption (or emission) band $(\sigma_{2S}^{-1/2} \sim 10^{-14}\,\text{s}$ [19,20,59]). For some cases the criterion $t_p \gg \sigma_{2s}^{-1/2}$ is weaker than $t_p \gg T'$. However, the delay time τ between the pump and the probe pulses must be larger than irreversible dephasing time T'. In reality, short pump pulses $t_p \leq T'$ induce a polarization grating whose attenuation is determined by T' [i.e the relaxation of the nondiagonal (with respect to electronic indices) density matrix ρ_{21}]. The relaxation of ρ_{21} is determined by the evolution of the system in both the excited and ground electronic states. Therefore, the creation of a polarization grating is an unfavorable situation that interferes with the separate study of dynamics in the ground and the excited electronic states. The polarization grating relaxes for delay times $\tau > T'$. Thus the analysis conducted in this section is limited to experiments performed with pulses $t_p \gg 10^{-14}\,\text{s}$, and delay times between the pump and the probe pulses $\tau > T'$.

We have used a new approach for calculation of the four-time correlation functions in nonlinear spectroscopy of nonlinear solvation when the breakdown of the linear response of the solvation dynamics occurs. In this approach the thermal averaging is carried out in the same electronic state as the system evolution calculations (equilibrium averages).

This approach has a number of advantages. First, stochastic models can be used for the calculation of the corresponding averages in Eqs. (5.4), (5.5), (5.9), (5.11), and (5.13), while the information concerning the time-resolved Stokes shift is preserved. In contrast, the application of stochastic models to the calculation of $M_2(\tau_1, \tau_2, \tau_3)$ [Eq. (5.2)], where classical trajectories are determined in the excited electronic state and thermal averaging carried out in the ground electronic state cannot be used to model the time-resolved Stokes shift. Second, computer calculation of the equilibrium averages consumes less computing time than that of averages using classical trajectory calculations, and the averaging is carried out in different electronic states [132]. Third, Eqs. (5.4), (5.5), and (5.9) provide simple analytical expressions for the important particular case (Sec. V.D).

We would particularly like to note Eq. (5.11) which seems to us most perspective in terms of practical usage. It was used in [37].

We have investigated the correctness of the description of the TRL spectrum first moment by the nonequilibrium average which is commonly used for nonlinear solvation studies [28] (Sec. V.E). We have shown how solvation dynamics in the ground electronic state is obtained by time-resolved hole-burning spectroscopy (Sec. V.F).

We demonstrated the use of stochastic models for calculating the spectral moments of the non-equilibrium absorption and luminescence of a solvating molecule (Sec. V.G). We have formulated two approaches. The first is more general and corresponds to a model of a stationary random process. It can be realized with the four-time correlation functions related to definite electronic states. The second approach corresponds to a model of a nonstationary random process and can be used for calculation of the nonequilibrium averages which appear in the theory of broad and featureless electronic molecular spectra. We applied our results to the Debye model of a rotation diffusion for the case of nonlinear solvation in [37].

Finally, let us concern an experimental study of nonlinear solvation effects. Large nonlinear effects have been found in many kinds of solvents with hydrogen bonds by using molecular dynamics simulations [28,117,118]. The detailed studies of Ladanyi and co-workers [28,117] have shown that breakdown of the linear response occurs when substantial differences exist between the pattern of solute–solvent hydrogen bonding in the initial and final solute states. Kumar and Maroncelli have found nonlinear effects for the solvation of relatively small solutes of the size of benzene in methanol [133]. Their finding was consistent with measurements of the solute dependence of solvation dynamics in 1-propanol [134]. Simple aromatic amines (aniline, 1-aminonaphthalene, 2-aminoanthracene, 1-aminopyrene, and dimehtylaniline) showed behavior that was inconsistent with expectations based on nonspecific theories of solvation dynamics [134]. Taking into account the effects of the solute self-motion improves the agreement of a theory with experiment [135]. According to [134], the key features of these solutes that differentiate them are (1) that the hydrogen-bonding effect is localized to a single interaction and (2) that partly as a result of this localization, the perturbation caused by the $S_0 \rightarrow S_1$ transition causes the response driven in a nonlinear fashion.

Thus the simple aromatic amines under discussion can show nonlinear solvation behavior; therefore, they can be used in experiments concerning the nonlinear solvation study. Also, diatomic molecules show large nonlinear effects due to solute motion [136].

ACKNOWLEDGMENTS

We thank B. Zolotov and R. Davidi for the measurements of absorption and emission spectra and for four-photon measurement data handling. This work was supported by grants from the United States–Israel Binational Science Foundation (BSF) and the James Franck Binational German–Israel Program in Laser Matter Interaction.

APPENDIX A

Let us consider the case of freely orientating molecules. To calculate $\langle \tilde{\sigma}_{ab}(\vec{\nu})\tilde{\sigma}_{dc}(\vec{\mu})\rangle_{or}$, we shall expand the tensor $\tilde{\sigma}_{ab}(\vec{\nu})$ [or $\sigma_{ab}(\mathbf{Q}_s)$] by irreducible parts, (i.e., parts that transform only by themselves at any coordinate transformations):

$$\tilde{\sigma}_{ab}(\vec{\nu}) = \tilde{\sigma}^0(\vec{\nu})\delta_{ab} + \tilde{\sigma}_{ab}^s(\vec{\nu}) + \tilde{\sigma}_{ab}^a(\vec{\nu}) \tag{A.1}$$

where

$$\tilde{\sigma}^0(\vec{\nu}) = \tfrac{1}{3}\sum_a \tilde{\sigma}_{aa}(\vec{\nu}) = \tfrac{1}{3}\mathrm{Tr}\,\tilde{\sigma} \tag{A.2}$$

is a scalar,

$$\tilde{\sigma}_{ab}^s(\vec{\nu}) = \tfrac{1}{2}(\tilde{\sigma}_{ab}(\vec{\nu}) + \tilde{\sigma}_{ba}(\vec{\nu})) - \tilde{\sigma}^0(\vec{\nu})\delta_{ab} \tag{A.3}$$

is a symmetrical tensor, and

$$\tilde{\sigma}_{ab}^a(\vec{\nu}) = \tfrac{1}{2}[\tilde{\sigma}_{ab}(\vec{\nu}) - \tilde{\sigma}_{ba}(\vec{\nu})] \tag{A.4}$$

is an antisymmetric tensor.

One can show that the following values: $\tilde{\sigma}^0(\vec{\nu})$,

$$\tilde{h}_s(\vec{\nu}, \vec{\mu}) = \sum_{ab} \tilde{\sigma}_{ab}^s(\vec{\nu})\tilde{\sigma}_{ba}^s(\vec{\mu}) \tag{A.5}$$

$$\tilde{h}_a(\vec{\nu}, \vec{\mu}) = \sum_{ab} \tilde{\sigma}_{ab}^a(\vec{\nu})\tilde{\sigma}_{ba}^a(\vec{\mu}) \tag{A.6}$$

are invariants of the tensor $\tilde{\sigma}_{ab}(\vec{\nu})$ (i.e., values that are constants for tensors in any coordinate system). We can express any orientation average $\langle \tilde{\sigma}_{ab}(\vec{\nu})\tilde{\sigma}_{dc}(\vec{\mu})\rangle_{or}$ by the tensor invariants $\tilde{\sigma}^0$, $\tilde{h}_s$, and $\tilde{h}_a$:

$$\langle \tilde{\sigma}_{aa}(\vec{\nu})\tilde{\sigma}_{aa}(\vec{\mu})\rangle_{or} = \tilde{\sigma}^0(\vec{\nu})\tilde{\sigma}^0(\vec{\mu}) + \tfrac{2}{15}\tilde{h}_s(\vec{\nu},\vec{\mu}) \tag{A.7}$$

$$\langle \tilde{\sigma}_{aa}(\vec{\mu})\tilde{\sigma}_{bb}(\vec{\nu})\rangle_{or} = \tilde{\sigma}^0(\vec{\mu})\tilde{\sigma}^0(\vec{\nu}) - \tfrac{1}{15}\tilde{h}_s(\vec{\mu},\vec{\nu}) \qquad (a \neq b) \tag{A.8}$$

$$\langle \tilde{\sigma}_{ab}(\vec{\mu})\tilde{\sigma}_{ba}(\vec{\nu})\rangle_{or} = \tfrac{1}{10}\tilde{h}_s(\vec{\mu},\vec{\nu}) + \tfrac{1}{6}\tilde{h}_a(\vec{\mu},\vec{\nu}) \qquad (a \neq b) \tag{A.9}$$

$$\langle \tilde{\sigma}_{ab}(\vec{\mu})\tilde{\sigma}_{ab}(\vec{\nu})\rangle_{or} = \tfrac{1}{10}\tilde{h}_s(\vec{\mu},\vec{\nu}) - \tfrac{1}{6}\tilde{h}_a(\vec{\mu},\vec{\nu}) \qquad (a \neq b) \tag{A.10}$$

All the other averages are equal to zero.

As an example, let us consider a molecule, where the direction of its dipole moment depends on the excitation of some (intermolecular) motions [59]. In the molecular frame of references $(x'y'z')$

$$D_{x'}(\mathbf{Q}_s) = D_0 \cos(\vec{\alpha}\mathbf{Q}_s) = D_0 \cos\left(\sum_j \alpha_j Q_{sj}\right) \tag{A.11}$$

$$D_{y'}(\mathbf{Q}_s) = D_0 \sin(\vec{\alpha}\mathbf{Q}_s), \; D_{z'} = 0 \tag{A.12}$$

We obtain for this model

$$\tilde{\sigma}^0(\vec{\nu}) = (D_0^2/3)\delta(\vec{\nu}), \; \tilde{h}_a(\vec{\mu},\vec{\nu}) = 0 \tag{A.13}$$

$$\tilde{h}_s(\vec{\mu},\vec{\nu}) = (D_0^4/4)[\delta(\vec{\alpha}-\vec{\mu})\delta(\vec{\alpha}+\vec{\nu})$$
$$+ \delta(\vec{\alpha}+\vec{\mu})\delta(\vec{\alpha}-\vec{\nu}) + \tfrac{2}{3}\delta(\vec{\mu})\delta(\vec{\nu})] \tag{A.14}$$

where $\delta(\vec{\nu})$ is the δ-function of Dirac.

APPENDIX B

Let us calculate the relaxation function of the NMO with an exponential memory corresponding to three-pole approximation. We introduce the mean relaxation time of the oscillator τ_r by the formula $\tau_r = \int_0^\infty f_r(t)\,dt$. It is easy to show that $\tau_r = \tilde{f}_r(0) = \tilde{\varphi}(0)/\omega_{osc}^2$, where $\tilde{\varphi}(p)$ is the Laplace transform of the memory function $\varphi(t)$, and $\tilde{\varphi}(0) = \gamma(0)$ is the frequency-dependent attenuation $\gamma(\omega)$ for $\omega = 0$.

Passing to dimensionless variables $s = p/\alpha$, $q = \gamma(0)\tau_r$, and $z = \alpha\tau_r$, we obtain from Eq. (2.37)

$$\tilde{f}_r(p) = \frac{1}{\alpha} \frac{s^2 + s + q/z}{s^3 + s^2 + (q/z)(1 + z^{-1})s + q/z^2} \tag{B.1}$$

Computing the inverse Laplace transformation of Eq. (B.1), we obtain $f_r(t)$. For the underdamped regime it has the following form [20,83]:

$$f_r(t) = \frac{a^2 + w^2}{(3a + 1)^2 + w^2} \left\{ (2a + 1) \exp(aT) \left[(1 + d) \cos wT \right. \right.$$
$$\left. \left. - \frac{a}{w}(3 + d) \sin wT \right] - 2a \exp[-(2a + 1)T] \right\} \tag{B.2}$$

where $a \pm iw$ and $-(2a + 1)$ are the roots of the denominator in Eq. (B.1), $a < 0$, $2a + 1 > 0$, $T = \alpha t$, $d = (4a + 1)/(a^2 + w^2)$. One can see from Eq. (B.2) that unlike the BO, the NMO has a nonoscillating component even for the underdamped regime.

REFERENCES

1. M. A. Kahlow, W. Jarzeba, T. P. DuBruil, and P. F. Barbara, *Rev. Sci. Instrum.* **59**, 1098 (1988).

2. M. Maroncelli and G. R. Fleming, *J. Chem. Phys.* **89**, 875 (1988).

3. E. W. Castner, Jr., M. Maroncelli, and G. R. Fleming, *J. Chem. Phys.* **86**, 1090 (1987).

4. S. H. Lin, B. Fain, N. Hamer, and C. Y. Yeh, *Chem. Phys. Lett.* **162**, 73 (1989).

5. S. H. Lin, B. Fain, and C. Y. Yeh, *Phys. Rev. A* **41**, 2718 (1990).

6. S. J. Rosenthal, X. Xiaoliang, M. Du, and G. R. Fleming, *J. Chem. Phys.* **95**, 4715 (1991).

7. M. Cho et al., *J. Chem. Phys.* **96**, 5033 (1992).

8. M. Cho et al., *J. Phys. Chem.* **100**, 11944 (1996).

9. R. Jimenez, G. R. Fleming, P. V. Kumar, and M. Maroncelli, *Nature* **369**, 471 (1994).

10. S. Mukamel, *Principles of Nonlinear Optical Spectroscopy*, Oxford University Press, New York, 1995.

11. J.-Y. Bigot et al., *Phys. Rev. Lett.* **66**, 1138 (1991).

12. C. J. Bardeen and C. V. Shank, *Chem. Phys. Lett.* **226**, 310 (1994).

13. E. T. J. Nibbering, D. A. Wiersma, and K. Duppen, *Chem. Phys.* **183**, 167 (1994).

14. M. S. Pshenichnikov, K. Duppen, and D. A. Wiersma, *Phys. Rev. Lett.* **74**, 674 (1995).

15. P. Vöhringer et al., *J. Chem. Phys.* **102**, 4027 (1995).

16. T. Joo et al., *J. Chem. Phys.* **104**, 6089 (1996).

17. W. P. de Boeij, M. S. Pshenichnikov, and D. A. Wiersma, *J. Phys. Chem.* **100**, 11806 (1996).

18. S. Y. Goldberg et al., *Chem. Phys.* **183**, 217 (1994).

19. B. Fainberg, R. Richert, S. Y. Goldberg, and D. Huppert, *J. Lumin.* **60–61**, 709 (1994).

20. B. D. Fainberg and D. Huppert, *J. Mol. Liq.* **64**, 123 (1995); erratum, **68**, 281 (1996).

21. B. D. Fainberg et al., in Fast Elementary Processes in Chemical and Biological Systems, AIP Proc. Vol. 364, A. Tramer, ed., AIP Press, Woodbury, N.Y. 1996, p. 454.

22. W. Jarzeba et al., *J. Phys. Chem.* **92**, 7039 (1988).

23. Y. Kimura, J. C. Alfano, P. K. Walhout, and P. F. Barbara, *J. Phys. Chem.* **98**, 3450 (1994).

24. M. Maroncelli, *J. Mol. Liq.* **57**, 1 (1993).

25. M. L. Horng, J. Gardecki, A. Papazyan, and M. Maroncelli, *J. Phys. Chem.* **99**, 17311 (1995).

26. M. Maroncelli, *J. Chem. Phys.* **94**, 2084 (1991).

27. E. Neria and A. Nitzan, *J. Chem. Phys.* **96**, 5433 (1992).

28. T. Fonseca and B. M. Ladanyi, *J. Phys. Chem.* **95**, 2116 (1991).

29. L. Perera and M. Berkowitz, *J. Chem. Phys.* **96**, 3092 (1992).

30. B. Bagchi and A. Chandra, *J. Chem. Phys.* **97**, 5126 (1992).

31. S. Roy and B. Bagchi, *J. Chem. Phys.* **99**, 1310 (1993).

32. N. Nandi, S. Roy, and B. Bagchi, *J. Chem. Phys.*, **102**, 1390 (1995).

33. T. Joo and A. C. Albrecht, *Chem. Phys.* **176**, 233 (1993).

34. P. Cong, Y. J. Yan, H. P. Deuel, and J. D. Simon, *J. Chem. Phys.* **100**, 7855 (1994).

35. B. D. Fainberg, *Opt. Spectrosc.* **68**, 305 (1990) [*Opt. Spektrosk.* **68**, 525, (1990)].

36. B. D. Fainberg, B. Zolotov, and D. Huppert, *J. Nonlin. Opt. Phys. Mater.* **5**, 789 (1996).

37. B. D. Fainberg and B. Zolotov, *Chem. Phys.* **216**, 7 (1997).

38. G. R. Fleming and M. Cho, *Annu. Rev. Phys. Chem.* **47**, 109 (1996).

39. Y. R. Shen, *The Principles of Nonlinear Optics*, Wiley, New York, 1984.

40. Y. J. Yan and S. Mukamel, *Phys. Rev. A* **41**, 6485 (1990).

41. M. J. Rosker, F. W. Wise, and C. L. Tang, *Phys. Rev. Lett.* **57**, 321 (1986).

42. F. W. Wise, M. J. Rosker, and C. L. Tang, *J. Chem. Phys.* **86**, 2827 (1987).

43. D. McMorrow and W. T. Lotshaw, *J. Phys. Chem.* **95**, 10395 (1991).

44. B. Zolotov, A. Gan, B. D. Fainberg, and D. Huppert, *Chem. Phys. Lett.* **265**, 418 (1997).

45. B. Zolotov, A. Gan, B. D. Fainberg, and D. Huppert, *J. Lumin.* **72–74**, 842 (1997).

46. C. H. Brito-Cruz, R. L. Fork, W. H. Knox, and C. W. Shank, *Chem. Phys. Lett.* **132**, 341 (1986).

47. R. F. Loring, Y. J. Yan, and S. Mukamel, *J. Chem. Phys.* **87**, 5840 (1987).

48. M. D. Stephens, J. G. Saven, and J. L. Skinner, *J. Chem. Phys.* **106**, 2129 (1997).

49. B. Fain, S. H. Lin, and N. Hamer, *J. Chem. Phys.* **91**, 4485 (1989).

50. B. Fain and S. H. Lin, *Chem. Phys. Lett.* **207**, 287 (1993).

51. B. Fain, S. H. Lin, and V. Khidekel, *Phys. Rev. A* **47**, 3222 (1993).

52. B. D. Fainberg, *Chem. Phys.* **148**, 33 (1990).

53. J. Yu, T. J. Kang, and M. Berg, *J. Chem. Phys.* **94**, 5787 (1991).

54. J. T. Fourkas, A. Benigno, and M. Berg, *J. Chem. Phys.* **99**, 8552 (1993).

55. J. T. Fourkas and M. Berg, *J. Chem. Phys.* **98**, 7773 (1993).

56. J. Ma, D. V. Bout, and M. Berg, *J. Chem. Phys.* **103**, 9146 (1995).

57. D. Bingemann and N. P. Ernsting, *J. Chem. Phys.* **102**, 2691 (1995).

58. K. Nishiyama, Y. Asano, N. Hashimoto, and T. Okada, *J. Mol. Liq.*, **65/66**, 41 (1996).

59. B. Fainberg, *Isr. J. Chem.* **33**, 225 (1993).

60. S. Mukamel, *Phys. Rev. A* **28**, 3480 (1983).

61. S. Mukamel, *J. Phys. Chem.* **89**, 1077 (1985).

62. V. L. Bogdanov and V. P. Klochkov, *Opt. Spectrosc.* **44**, 412 (1978).

63. V. L. Bogdanov and V. P. Klochkov, *Opt. Spectrosc.* **45**, 51 (1978).

64. V. L. Bogdanov and V. P. Klochkov, *Opt. Spectrosc.* **52**, 41 (1982).

65. B. Fainberg, *Phys. Rev. A* **48**, 849 (1993).

66. R. Kubo, in *Relaxation, Fluctuation and Resonance in Magnetic Systems*, D. der Haar, ed., Oliver & Boyd, Edinburgh, 1962, p. 23.

67. B. D. Fainberg, *Opt. Spectrosc.* **58**, 323 (1985) [*Opt. Spektrosk.* **58**, 533 (1985)].

68. B. D. Fainberg and I. B. Neporent, *Opt., Spectrosc.* **61**, 31 (1986) [*Opt. Spektrosk.* **61**, 48 (1986)].

69. B. D. Fainberg, *Opt. Spectrosc.* **60**, 74 (1986) [*Opt. Spektrosk.* **60**, 120 (1986)].

70. B. D. Fainberg and I. N. Myakisheva, *Sov. J. Quant. Electron.* **17**, 1595 (1987) [*Kvantovaya Elektron.* (Moscow), **14**, 2509 (1987)].

71. B. D. Fainberg and I. N. Myakisheva, *Opt. Spectrosc.* **66**, 591 (1989) [*Opt. Spektrosk.* **66**, 1012 (1989)].

72. M. Cho, G. R. Fleming, and S. Mukamel, *J. Chem. Phys.* **98**, 5314 (1993).

73. M. Abramovitz and I. Stegun, *Handbook on Mathematical Functions*, Dover, New York, 1964.

74. H. Mori, *Prog. Theor. Phys.* **34**, 399 (1965).

75. B. J. Berne and G. D. Harp, *Adv. Chem. Phys.* **17**, 63 (1970).

76. R. Kubo, *J. Phys. Soc. Jpn.* **12**, 570 (1957).

77. Y. T. Mazurenko and V. A. Smirnov, *Opt. Spectrosc.* **45**, 12 (1978) [*Opt. Spekrosk.* **45**, 23 (1978)].

78. A. Ango, *Mathematics for Electro- and Radio Engineers*, Nauka, Moscow, 1965.

79. T. Takagahara, E. Hanamura, and R. Kubo, *J. Phys. Soc. Jpn.* **44**, 728 (1978).

80. Y. T. Mazurenko and V. A. Smirnov, *Opt. Spectrosc.* **47**, 262, 360 (1979) [*Opt. Spektrosk.* **47**, 471, 650 (1979)].

81. Y. J. Yan and S. Mukamel, *J. Chem. Phys.* **89**, 5160 (1988).

82. B. D. Fainberg, *Opt. Spectrosc.* **63**, 436 (1987) [*Opt. Spektrosk.* **63**, 738 (1987)].

83. B. D. Fainberg and D. Huppert, *Nonlin. Opt.* **11**, 329 (1995); erratum, **16**, 93 (1996).

84. M. Lax, *J. Chem. Phys.* **20**, 1752 (1952).

85. B. D. Fainberg, *Opt. Spectrosc.* **67**, 137 (1989) [*Opt. Spektrosk.* **67**, 241 (1989)].

86. A. Migus, Y. Gauduel, J. L. Martin, and A. Antonetti, *Phys. Rev. Lett.* **58**, 1559 (1987).

87. M. Maroncelli and G. R. Fleming, *J. Chem. Phys.* **89**, 5044 (1988).

88. L. E. Fried, N. Bernstein, and S. Mukamel, *Phys. Rev. Lett.* **68**, 1842 (1992).

89. J. T. Hynes, in *Ultrafast Dynamics of Chemical Systems*, J. D. Simon ed., Kluwer, Dordrecht, The Netherlands, 1994, pp. 345–381.

90. U. Kaatze, *Chem. Phys. Lett.* **203**, 1 (1993).

91. G. Nemethy and H. Scharagaa, *J. Chem. Phys.* **41**, 680 (1964).

92. M. Cho, N. F. Scherer, and G. R. Fleming, *J. Chem. Phys.* **96**, 5618 (1992).

93. B. D. Fainberg and B. S. Neporent, *Opt. Spectrosc.* **48**, 393 (1980) [*Opt. Spektrosk.*, **48**, 712 (1980)].

94. R. B. Andreev et al., *J. Appl. Spectrosc.* **25**, 1013 (1976) [*Zh. Prikl. Spektrosk.* **25**, 294 (1976)].

95. R. B. Andreev et al., *Opt., Spectrosc.* **41**, 462 (1976) [*Opt. Spektrosk.* **41**, 782 (1976)].

96. J. Yu and M. Berg, *Chem. Phys. Lett.* **208**, 315 (1995).

97. A. J. Berigno, E. Ahmed, and M. Berg, *J. Chem. Phys.* **109**, 7382 (1996).

98. N. Agmon, *J. Phys. Chem.* **100**, 1072 (1996).

99. D. Bertolini, M. Cassettari, and G. Salvetti, *J. Chem. Phys.* **76**, 3285 (1982).

100. S. K. Garg and C. P. Smyth, *J. Phys. Chem.* **69**, 1294 (1965).

101. D. A. Zichi and P. J. Rossky, *J. Chem. Phys.* **84**, 2814 (1986).

102. M. S. Skaf and B. M. Ladanyi, *J. Phys. Chem.* **100**, 18258 (1996).

103. A. M. Jonkman, P. der Meulen, H. Zang, and M. Glasbeek, *Chem. Phys. Lett.* **256**, 21 (1996).

104. J. Barnet, K. Bachhuber, R. Buchner, and H. Hetzenauer, *Chem. Phys. Lett.* **165**, 369 (1990).

105. Y. Yan and K. A. Nelson, *J. Chem. Phys.* **87**, 6240 (1987).

106. Y. Yan and K. A. Nelson, *J. Chem. Phys.* **87**, 6257 (1987).

107. S. Ruhman, A. G. Joly, and K. A. Nelson, *IEEE J. Quantum Electron.* **24**, 460 (1988).

108. S. Ruhman, B. Kohler, A. G. Joly, and K. A. Nelson, *IEEE J. Quantum Electron.* **24**, 470 (1988).

109. J. Chesnoy and A. Mokhtari, *Phys. Rev. A* **38**, 3566 (1988).

110. C. J. Bardeen, Q. Wang, and C. V. Shank, *Phys. Rev. Lett.* **75**, 3410 (1995).

111. B. D. Fainberg, *Opt. Spectrosc.* **65**, 722 (1988) [*Opt. Spektrosk.* **65**, 1223 (1988)].

112. V. M. Fain and Y. I. Khanin, Quantum Electronics, Vol. 1, *Basic Theory*, Pergamon Press, Braunschweig, Germany, 1969.

113. G. Herzberg, *Infrared and Raman Spectra of Polyatomic Molecules*, D. Van Nostrand, Princeton, N.J. 1954.

114. T. Kakitani and N. Mataga, *Chem. Phys.* **93**, 381 (1985).

115. T. Kakitani and N. Mataga, *J. Phys. Chem.* **89**, 8, 4752 (1985).

116. T. Fonseca, B. M. Ladanyi, and J. T. Hynes, *J. Phys. Chem.* **96**, 4085 (1992).

117. T. Fonseca and B. M. Ladanyi, *J. Mol. Liq.* **60**, 1 (1994).

118. R. Olender and A. Nitzan, *J. Chem. Phys.* **102**, 7180 (1995).

119. Y. Georgievskii, *J. Chem. Phys.* **104**, 5251 (1996).

120. H. L. Friedman, F. O. Raineri, F. Hirata, and B.-C. Perng, *J. Stat. Phys.* **78**, 239 (1995).

121. R. Biswas and B. Bagchi, *J. Phys. Chem.* **100**, 1238 (1996).

122. A. M. Walsh and R. F. Loring, *J. Chem. Phys.* **94**, 7575 (1991).

123. L. E. Fried and S. Mukamel, *Adv Chem. Phys.* **84**, 435 (1993).

124. R. Kubo, *J. Phys. Soc. Jpn.* **17**, 1100 (1962).

125. R. S. Fee and M. Maroncelli, *Chem. Phys.* **183**, 235 (1994).

126. T. V. Veselova et al., *Opt. Spectrosc.* **39**, 495 (1975) [*Opt. Spektrosk.* **39**, 870 (1975)].

127. S. M. Rytov, *Introduction to the Statistical Radiophysics*, Nauka, Moscow, 1966.

128. A. Abragam, *The Principles of Nuclear Magnetism*, Clarendon Press, Oxford, 1961.

129. P. Debye, Polar Molecules, Dover, New York, 1929.

130. H. Frölich, Theory of Dielectrics, Clarendon Press, Oxford, 1958.

131. C. J. F. Böttcher, *Theory of Electric Polarization*, Vol. 1, Elsevier, Amsterdam, 1973.

132. A. Nitzan, personal communication, 1996.

133. P. V. Kumar and M. Maroncelli, *J. Chem. Phys.* **103**, 3038 (1995).

134. C. F. Chapman, R. S. Fee, and M. Maroncelli, *J. Phys. Chem.* **99**, 4811 (1995).

135. R. Biswas and B. Bagchi, *J. Phys. Chem.* **100**, 4261 (1996).

136. S. J. Rosenthal et al., *J. Mol. Liq.* **60**, 25 (1994).

COHERENCE AND ADIABATICITY IN ULTRAFAST ELECTRON TRANSFER

KLAAS WYNNE

Department of Chemistry, University of Pennsylvania, 231 South 34th Street, Philadelphia, PA 19104
Femtosecond Research Centre, Department of Physics and Applied Physics, University of Strathclyde, 107 Rottenrow, Glasgow G4 0NG, UK

ROBIN M. HOCHSTRASSER

Department of Chemistry, University of Pennsylvania, 231 South 34th Street, Philadelphia, PA 19104

CONTENTS

Electron Transfer: From Isolated Molecules to Biomolecules, Part Two, edited by Joshua Jortner and M. Bixon. Advances in Chemical Physics Series, Volume 107, series editors I. Prigogine and Stuart A. Rice.
ISBN 0-471-25291-3   © 1999 John Wiley & Sons, Inc.

I. INTRODUCTION

The study of ultrafast photoinduced reactions such as electron transfer in the condensed phase and in biological systems has led to a wealth of information about the intrinsic factors that govern chemical reaction rates. Of particular interest here is the situation where the electron transfer reaction rate becomes comparable with the vibrational frequencies and intra- and intermolecular vibrational relaxation rates. In such cases a variety of optical spectroscopic techniques can be used to obtain structural and dynamic information. Absorption and resonance Raman scattering [1–22] spectroscopy can be used to study the earliest (ca. 10 fs) events after photoexcitation and to obtain information regarding the vibrational modes that are coupled to reactions. Ultrafast visible and near-infrared (IR) time-resolved pump–probe experiments have been used to study electron transfer in small inorganic complexes [19,23–25], electron donor–acceptor complexes [26–31], proteins that exhibit electron transfer between their cofactors [32–38] and other systems [39–45]. These experiments have measured electron transfer rates, energy distributions in the reaction products, and the dependence of initial excitation energy on the electron transfer rate. Studies using probe pulses in the IR [46–52] have an even higher sensitivity to measurement of vibrational distributions in the products. All this detailed experimental information now allows for stringent tests of theories of condensed phase reactions.

The rates of most chemical reactions including ultrafast electron transfer, can be described theoretically using a form of Fermi's golden rule [53–57], which calculates the rate as the product of an electronic coupling term and a vibrational overlap factor. Thus electron transfer rates could in principle be calculated exactly if enough were known about the size of the coupling, the nuclear structure changes that accompany the reaction, and the vibrational relaxation dynamics. The class of electron transfer systems that exhibit a charge transfer absorption band that can be excited directly is special in this regard. In these systems (Figure 1) there are often only two electronic states involved in all the relevant spectroscopy and reaction dynamics. The integrated absorption spectrum depends on the transition dipole moment between the two electronic states. The spectrum will contain an underlying progression of the vibrational modes that are displaced and hence Franck–Condon active in the transition. In electron transfer from the excited-state back to the ground state, those same vibrational modes that change frequency or equilibrium position in the absorption spectrum will be required to change during the reaction. One might, therefore, refer to these modes as also being Franck–Condon active in the electron transfer transition. One set of vibrational modes that will be Franck–Condon active in the reaction and

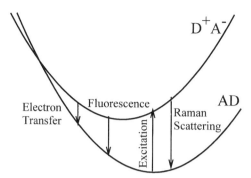

Figure 1. In direct electron transfer the reactive transition is the same as those of absorption, emission, and Raman scattering. Therefore, equilibrium spectroscopy can be used to obtain the parameters relevant for the calculation of absolute electron transfer rates.

in the absorption spectrum are those associated with the solvent or protein. In a typical experiment the solvent is initially at equilibrium with the ground electronic state of the system, but it will reorganize after the transition to the excited-state since this transition involves a change in the permanent dipole moment. The effect of this reorganization is a solvent-induced line broadening. When the coupling between molecule and solvent is sufficiently strong, the line shape approaches a Gaussian shape [58]. The spectral width is determined by the equilibrium distribution of solvent environments in the ensemble. One consequence of this solvent broadening is that it obscures the Franck–Condon progressions of the intramolecular vibrations.

Resonance Raman scattering experiments that measure the full Raman excitation profiles have been applied to analyze the Franck–Condon active vibrational modes in a variety of cases where broadening due to electronic dephasing obscures the Franck–Condon progressions associated with the electronic absorption spectrum [59–62]. Only recently have resonance Raman scattering studies become available that analyze the excitation profiles near charge transfer transitions [1–22]. The absorption spectrum and the electron transfer rate can be calculated using a simple golden-rule expression that incorporates the wavepacket description of these processes as described by many authors [15,20,63–72]. Since the calculation of these quantities with a dozen or so modes using fast Fourier transform methods takes less than a second on a Pentium PC, it is usually necessary to make only a few simplifying approximations regarding the Franck–Condon active modes if the resonance Raman scattering data are available. Resonance Raman spectroscopy cannot in most cases determine the frequency of the vibrational modes in the excited electronic state, nor is it very sensitive to Dushinsky rotation (in which the excited-state normal coordinates are a nontrivial linear combination of the ground-state normal coordinates [68]. Therefore, it is commonly assumed that the frequencies in the ground and excited-states are equal, that the modes are harmonic, and that Dushinsky rotation is absent. The electron transfer rate is determined by the vibrational overlap between low-energy vibrational states in the excited electronic state and high-energy vibrational states in the electronic ground-state. Resonance Raman scattering has not yet given any information about these high-energy vibrational states, and therefore anharmonic effects on the ground-state surface at high energies are usually neglected. Notwithstanding these assumptions, the data so obtained provide the best basis available for calculating absolute rates of electron transfer reactions.

Because the electron transfer is direct, the optical excitation only produces vibrational states that participate in the reaction; it does not produce inactive modes. This absence of *spectator modes* is a very important feature that can be used to great advantage in the experimental study of the role of vibrational coherence in electron transfer. In a recent publication [30] it was shown that the electron transfer reaction in the solution-phase electron donor–acceptor (EDA) complex between tetracyanoethylene (TCNE) and pyrene, which exhibits a charge transfer band and ultrafast electron transfer from the charge transfer state back to the ground state, is modulated by vibrational wavepacket motion induced by a femtosecond pump pulse on the excited-state surface. In this paper it will be shown that a relatively simple classical Monte Carlo model can account for the essential features of the experimental result. Since the electron transfer transition in TCNE–pyrene is coupled to a dozen or so Franck–Condon active vibrational modes (none of which could be excited coherently in the experiment because their frequency is too high), the effect of vibrational coherence in one of these modes on the electron transfer rate is relatively minor. This may be a feature that is common to many chemical reactions in polyatomic systems. Some implications of this analysis for biological electron transfer are discussed in this paper.

The electron transfer rate will typically be calculated with some form of the golden rule. A golden-rule rate expression is always valid as long as a set of basis states is chosen in which the coupling between these states is small. Virtually all known molecular systems that exhibit a charge transfer band (the Fe^{II}–Fe^{III} system in solution is a well-known exception [73]), and certainly those for which resonance Raman scattering data are available, have a strong absorption and a transition dipole moment on the order of a few debye. The Mulliken–Hush treatment [74,75] (see below) relates the strength of the transition dipole moment of a charge transfer band to the magnitude of the coupling between the diabatic neutral and charge separated states. It is these diabatic states that are typically used as a basis set in the perturbative, golden-rule treatment of electron transfer in systems with three electronic states involved. A transition dipole moment of 1 D in a charge transfer transition implies a diabatic coupling on the order of $1000 \, cm^{-1}$. This is much larger than many of the frequencies of the Franck–Condon active modes (including the solvent mode [76]), and consequently, the diabatic basis set cannot be used in a perturbative treatment. The adiabatic states diagonalize the Hamiltonian in the Born–Oppenheimer approximation and therefore the golden-rule treatment is valid in this basis set, with the nuclear kinetic energy operator as the Hamiltonian that gives rise to transitions. As we showed in a previous publication [28] knowledge of the Franck–Condon active modes obtained from resonance Raman scattering data can be used to calculate the magnitude of the non-Born–Oppenheimer coupling, from which absolute rate calculations are possible. In this article we present a new version of this non-Born–Oppenheimer coupling theory that expresses the rate in terms of wavepackets and momentum operators. We also show the relationship between the non-Born–Oppenheimer treatment in the adiabatic basis and the conventional treatment of electron transfer in the diabatic basis by means of the Landau–Zener expression for transition rates. This treatment will bring out the fact that our non-Born–Oppenheimer rate expressions are simply a quantum mechanical version of Landau–Zener theory in the strong-coupling limit.

Unfortunately, resonance Raman scattering data are not yet available for many systems of interest. We used [28] resonance Raman scattering data for the EDA complex between TCNE and hexamethylbenzene (HMB) in CCl_4 [13,16,17] to calculate absolute rates with the non-Born–Oppenheimer expressions and compared these predictions with the experimentally observed [28] rates in various polar and non-polar solvents. In the improved calculations given here the agreement between theory and experiment is good. To extend the usefulness of the resonance Raman scattering data for this particular EDA complex, the absorption spectra of two similar complexes, TCNE–pyrene and TCNE with tetracyanobenzene (TCB), were fitted assuming the same modes would be Franck–Condon active as in TCNE–HMB except with different Huang–Rhys factors. The electron transfer rates calculated using these sets of nuclear reorganization parameters are again in good agreement with experiment. It is concluded that the golden-rule treatment in the adiabatic basis in conjunction with nuclear reorganization parameters that are consistent with Raman data leads to correct predictions of the absolute rates that vary in the range between $(100 \, fs)^{-1} \, (1 \, ns)^{-1}$.

II.   FAST ELECTRON TRANSFER BETWEEN ADIABATIC STATES

A.   Diabatic States, Adiabatic States, and the Relevant Coupling

To calculate electron transfer rates it has to be clear what the relevant quantum states are. It is the experimental arrangement that determines which state is initially excited and which state is generated. In describing the quantum states of a molecule, one usually makes the Born–Oppenheimer approximation, in which the motion of the nuclei is adiabatic with respect to the motion of the electrons [56]. The adiabatic Born–Oppenheimer states are eigenstates of the total Hamiltonian of the system, with the nuclear coordinates taken as parameters and have the form $|\varphi(\mathbf{r}_e; \mathbf{x})\rangle|\chi(\mathbf{x})\rangle$, where $|\varphi(\mathbf{r}_e; \mathbf{x})\rangle$ is the electronic wavefunction that depends on the electronic coordinates $\mathbf{r}_e$ and parametrically on the nuclear coordinates $\mathbf{x}$, and $|\chi(\mathbf{x})\rangle$ is the vibrational wavefunction. In describing electron transfer (especially in EDA complexes) it is often convenient to choose the electronic states such that they are eigenfunctions of the Hamiltonian only for infinite separation between the donor and the acceptor. As the donor and acceptor are brought together, there will be an "electron transfer" coupling that couples the various states. These states, which are eigenstates only under special circumstances, are often referred to as *diabatic states*, and consequently the coupling between them is the diabatic coupling. This terminology is used here as well. More sophisticated ways of defining the diabatic states have been discussed before [77] but are difficult to apply in practice.

For a description of electron transfer processes it will suffice to consider only two diabatic electronic states: The covalent state $|c\rangle$, which corresponds to the state DA in which both donor and acceptor are neutral, and the ionic state $|i\rangle$, which corresponds to the state D^+A^-. In the Mulliken theory, electronic transitions occur from the ground covalent state to the ionic state only if these two diabatic states are admixed by some coupling. Therefore, the optical preparation step must create a mixed state. It is the composition of this state that we now consider. In the diabatic basis the electronic Hamiltonian is thus given by

$$\hat{H}_D = \begin{pmatrix} U_C(\mathbf{x}) & V(\mathbf{x}) \\ V(\mathbf{x}) & U_I(\mathbf{x}) \end{pmatrix}_D$$

$$= U_C(\mathbf{x})|c\rangle\langle c| + U_I(\mathbf{x})|i\rangle\langle i| + V(\mathbf{x})\{|c\rangle\langle i| + |i\rangle\langle c|\} \qquad (2.1)$$

where $U_{C/I}(\mathbf{x})$ is the diabatic vibrational potential in the covalent–ionic state and $V(\mathbf{x})$ is a (position-dependent) coupling between these states. The diagonal elements of $V(\mathbf{x})$ are contained in $U_{C/I}(\mathbf{x})$. In this picture, if the system is initially prepared in a pure diabatic state, (e.g., $|i\rangle$), the diabatic coupling $V(\mathbf{x})$ will induce transfer of population to the state $|c\rangle$.

The eigenvectors of the diabatic Hamiltonian $\hat{H}_D$ can be expressed in terms of $U_{C/I}(\mathbf{x})$ and $V(\mathbf{x})$; the eigenvector matrix $\hat{E}(\mathbf{x})$ in the diabatic frame, containing the adiabatic electronic eigenstates as column vectors, is given by [78]

$$\hat{E}(\mathbf{x}) = (|g\rangle, |e\rangle) = \begin{pmatrix} \cos\vartheta & \sin\vartheta \\ -\sin\vartheta & \cos\vartheta \end{pmatrix}_D, \qquad \tan 2\vartheta = \frac{2V(\mathbf{x})}{\Delta(\mathbf{x})},$$

$$\Delta(\mathbf{x}) = U_I(\mathbf{x}) - U_C(\mathbf{x}) \tag{2.2}$$

where $|g\rangle$ and $|e\rangle$ are the adiabatic ground- and excited-state kets (column vectors in the current representation). By diagonalizing the diabatic electronic Hamiltonian, Eq. (2.1), with the eigenvector matrix $\hat{E}(\mathbf{x})$, one obtains the adiabatic nuclear potential in the adiabatic frame. It is given by [79]

$$\hat{H}_A = \begin{pmatrix} U_C(\mathbf{x}) + V(\mathbf{x})\cot\vartheta & \\ & U_I(\mathbf{x}) - V(\mathbf{x})\cot\vartheta \end{pmatrix}_A \tag{2.3}$$

Thus the vibrational states $\{|\chi_g\rangle, |\chi_e\rangle\}$ are determined by

$$[U_C(\mathbf{x}) + V(\mathbf{x})\cot\vartheta + \hat{T}(\mathbf{x})]|\chi_g\rangle = E_g|\chi_g\rangle$$

and the vibronic, adiabatic states are $\{|g\rangle|\chi_g\rangle, |e\rangle|\chi_e\rangle\}$. If the system were initially prepared in an adiabatic state (e.g., $|e\rangle$), there could be no nonradiative transitions, such as electron transfer, to the state $|g\rangle$ induced by the adiabatic Hamiltonian, which by definition has no off-diagonal elements. However, the nuclear kinetic energy operator, $\hat{T}(\mathbf{x})$, gives rise to two off-diagonal electronic coupling terms [80–82]. We will consider only the one given by

$$(\hat{V}_{\text{NBO}}^k)_{ij} = -\frac{\hbar^2}{m_k} \left\langle \chi_i \left| \left\langle \varphi_i \left| \frac{\partial}{\partial x_k} \right| \varphi_j \right\rangle \frac{\partial}{\partial x_k} \right| \chi_j \right\rangle + \text{c.c.} \tag{2.4}$$

It is often assumed [80–82] that the term proportional to $\langle \varphi_i | \partial^2/\partial x_k^2 | \varphi_j \rangle$ can be ignored compared with the term given in Eq. (2.4), although this has not been explicitly demonstrated in electronic structure calculations as far as we know and would be difficult [77]. However, the second-order term has the same form as the diabatic coupling and could therefore be thought to be part of it. The kinetic energy operator also has diagonal elements that are not considered explicitly in the definition of the adiabatic states above. However, these diagonal elements can be considered to be an implicit part of the Hamiltonian, which implies that the shape of the potential surfaces will be modified by the kinetic operator [80].

In the expression above the non-Born–Oppenheimer coupling involves the promoting mode k, with normal mode coordinate x_k and reduced mass m_k. For explicit numerical calculations of the nuclear kinetic energy–induced coupling (see below), it will be convenient to assume that the diabatic states $|c\rangle$ and $|i\rangle$ do not depend on the nuclear coordinates $\mathbf{x}$. In general, this assumption will not be valid, and hence the implication

$$\int d\mathbf{x}\, \varphi_n^*(\mathbf{r}_e; \mathbf{x}) \frac{d}{dx_I} \varphi_m(\mathbf{r}_e; \mathbf{x}) = 0 \qquad (2.5)$$

where $\{n, m\} = \{c, i\}$ and l labels the promoter mode, is also not valid. The assumption that the diabatic states do not depend on the nuclear co-ordinates is equivalent to considering them as an expansion of states at just one internuclear separation. One convenient separation would be the equilibrium nuclear configuration of the ground state, which is also the Franck–Condon region. Clearly, the results thus obtained will be suspect at internuclear separations that differ greatly from this but it is not clear exactly under which condition the assumption will break down.

The Hellmann–Feynman theorem allows us to write part of the non-Born–Oppenheimer Hamiltonian as

$$\left\langle \varphi_i \left| \frac{\partial}{\partial x_k} \right| \varphi_j \right\rangle = \frac{\langle \varphi_i | (\partial \hat{H}_A)\mathbf{x})/\partial x_k) | \varphi_j \rangle}{U_j(\mathbf{x}) - U_i(\mathbf{x})} \qquad (2.6)$$

This form demonstrates that the Hamiltonian becomes particularly large in those regions where the potentials cross or nearly cross. For a large separation between two adiabatic surfaces, the small non-Born–Oppenheimer coupling can be treated with perturbation theory. Otherwise, the non-Born–Oppenheimer coupling becomes nonperturbative. Thus if two diabatic potentials cross at some crossing point $\mathbf{x}_c$ and the diabatic coupling V is large, this will lead to a large splitting of $2V$ between the adiabatic potentials at $\mathbf{x}_c$, and the non-Born–Oppenheimer coupling can be treated perturbatively. On the other hand, if the diabatic coupling is small, the splitting in the adiabatic frame is small and the non-Born–Oppenheimer effects will be large. However, in this case the diabatic coupling V is itself small and can be treated by perturbation theory in the diabatic frame. Hence the appropriate picture in which to treat electron transfer is wholly dependent on the magnitude of V. This issue is addressed again in Section II.C.

B. Stability of the Adiabatic States of Molecules in a Medium

In defining the adiabatic states as the appropriate basis for two-state electron transfer, no account was yet taken of the solvent. It is important to examine this point because a polar solvent might have a tendency to localize charge in the molecule and permit an adiabatic state to decay into a localized, diabatic, purely covalent or ionic state as a result of solvent interactions. Thereby an electron transfer process, which for the gas-phase molecule was determined by the non-Born–Oppenheimer coupling, might involve the diabatic coupling in solution. In the following we discuss two limiting situations. In the first, the effect of solvent is modeled as a stochastic modulation of the energy levels of the donor–acceptor complex. The second limit is when the solvent significantly modifies the structure of the charge transfer complex.

If the molecule undergoing electron transfer is coupled to the solvent, an important contribution to the energy will be from the coupling of the permanent dipole moment to the charges in the polar solvent. An appropriate picture to describe this coupling is therefore the diabatic picture. Consider the potentials V_c and V_i of the covalent and ionic states as a function of the reduced solvent coordinate q (Figure 2):

$$V_i(q) = \tfrac{1}{2}q^2 \qquad V_c(q) = \tfrac{1}{2}(q - q_0)^2 + \Delta G^0 \qquad \lambda_S = \tfrac{1}{2}q_0^2 \qquad (2.7)$$

where q_0 is the shift in equilibrium position between the two potentials, ΔG^0 is the driving force of the electron transfer reaction, and λ_S is the solvent reorganization energy. Coupling the system to the solvent will lead to forced Brownian motion along the solvent coordinate and consequently to a fluctuation of the energy difference between the covalent and the ionic states. If the position fluctuations are Gaussian [83], the energy fluctuations are an Ornstein–Uhlenbeck process with an average value of $-(\lambda_S + \Delta G^0)$ and variance [84] $2\lambda_S k_B T$. The Hamiltonian of the system can thus be written in two parts as

$$\hat{H}_\mathrm{D} = \begin{pmatrix} 0 & \beta \\ \beta & \Omega \end{pmatrix} \qquad \hat{G}_\mathrm{D}(t) = \begin{pmatrix} 0 & 0 \\ 0 & \xi(t) \end{pmatrix} \qquad (2.8)$$

where β is the diabatic coupling between the covalent and ionic states, Ω the average energy splitting between the two diabatic states [i.e., $-(\lambda_S + \Delta G^0)$ is absorbed in it: see Figure 2], and $\xi(t)$ the stochastic energy fluctuation

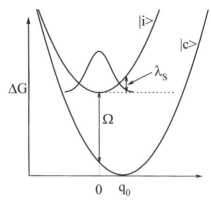

Figure 2. Diagram of the potentials described in Eq. (2.7). A molecule that undergoes electron transfer will be coupled to the solvent through the change in permanent dipole moment between the ionic and the covalent state. An initial distribution along the solvent coordinate in the ionic state will be centered at $q = 0$. Fluctuations in the ionic state potential well will lead to an average energy separation between the two diabatic states of Ω and energy fluctuations with a Gaussian distribution.

induced by the coupling of the solvent to the permanent dipole moment difference between the two diabatic states. We can transform again to a frame in which the Hamiltonian $\hat{H}_D$ is diagonal. For this case the adiabatic frame is the one that diagonalizes the entire Hamiltonian Eq. (2.8). However, the latter frame of reference is inconvenient for the calculations below because in it the fluctuation Hamiltonian, $\hat{G}_A$, has both diagonal and off-diagonal elements.

We may now regard the evolution of the system as being governed by a stochastic differential equation. The equation of motion can be transformed to an ordinary differential equation for the average properties using standard stochastic methods [83,85–87] or the equivalent Redfield theory [88,89]. If a vector of density matrix elements in the adiabatic frame is defined as $\mathbf{r} \equiv (\rho_{gg}, \rho_{ee}, \rho_{ge}, \rho_{eg})^T$, it can be shown that the equation of motion for this vector averaged over the stochastic process is given by [89]

$$\frac{d\langle\mathbf{r}\rangle}{dt} = \begin{bmatrix} -\Gamma & \Gamma & -\alpha(0) & -\alpha(0) \\ \Gamma & -\Gamma & \alpha(0) & \alpha(0) \\ -\alpha(-\omega) & \alpha(-\omega) & -\Gamma(\omega) - \gamma + i\omega & \Gamma(-\omega) \\ -\alpha(\omega) & \alpha(\omega) & \Gamma(\omega) & -\Gamma(\omega) - \gamma - i\omega \end{bmatrix} \cdot \langle\mathbf{r}\rangle \quad (2.9)$$

where

$$\omega = \sqrt{\Omega^2 + 4\beta^2} \tag{2.10}$$

$$\gamma \equiv \frac{\Omega^2}{\omega^2} \int d\tau \, \langle \xi(0)\xi(\tau) \rangle \tag{2.11}$$

$$\Gamma(\omega) \equiv \frac{2\beta^2}{\omega^2} \int d\tau \, \langle \xi(0)\xi(\tau) \rangle^{-i\omega\tau} \qquad \Gamma \equiv \tfrac{1}{2}[\Gamma(\omega) + \Gamma(-\omega)] \tag{2.12}$$

$$\alpha(\omega) \equiv \frac{-\beta\Omega}{\omega^2} \int d\tau \, \langle \xi(0)\xi(\tau) \rangle e^{-i\omega\tau} \tag{2.13}$$

In the expressions above, Γ is the rate of population exchange between adiabats, γ the rate of dephasing, and the $\alpha(\omega)$ are nonsecular terms. In deriving Eq. (2.9), it has been assumed that $\xi(t)$ is a classical stochastic process, and therefore this equation does not take into account detailed balance [88,90,91]. Although it would be straightforward to include detailed balance [87,89] in Eq. (2.9), it would have little effect since in the cases discussed here it is assumed that the splitting between the electronic states ω is large compared with $k_B T$. It can be seen that there is exchange of population between the adiabatic states with rate Γ, which could in principle lead to the "unmixing" of the adiabatic states into diabatic states. However, the rate of population exchange depends on the solvent power spectral density,

$$S(\omega) \equiv \int d\tau \, \langle \xi(0)\xi(\tau) \rangle e^{-i\omega\tau} \tag{2.14}$$

at frequency ω (i.e., the frequency corresponding to the energy separation between the two adiabatic states at $q = 0$) [see Eq. (2.10)]. In the cases that we focus on here, (i.e., systems with optical charge transfer bands), the frequency Ω, and therefore ω, is extremely high ($> 10,000 \, \text{cm}^{-1}$) and $S(\omega)$ is, for all practical purposes, zero. The nonsecular terms $\alpha(\omega)$ can be ignored in this case even if $S(\omega)$ were nonzero, since they contribute to the final result, $\langle \mathbf{r} \rangle(t)$, in the form of an integral over $\exp(i\omega t)$. This integral vanishes unless ω is close to zero. The rate of electronic dephasing depends on the solvent power spectral density at zero frequency as expected, which is nonzero. Therefore, superpositions between the two adiabatic electronic states will dephase rapidly with rate γ. It can thus be concluded from this model that solvent fluctuation will not localize the charge and will not drive the system to settle into diabatic states. Of course, this conclusion cannot be drawn in the case of (nearly) degenerate reactant and product states.

On the other hand, by making a strict separation between the charge transfer complex and the solvent bath, the model above omits a number of potentially important factors. It is well known that ions in solution can strongly bind to solvent molecules from the first solvation shell [92], so that some of the solvent molecules may have to be regarded as being an integral part of the "super molecular" charge transfer complex that is undergoing the electron transfer and subject to the bath fluctuations described above. Such a combined donor–acceptor–solvent structure may have electronic states that are indeed more localized and hence more diabatic. Electronic structure calculations that incorporate some of the solvent molecules explicitly would have to be performed to test this idea. Experimentally, it is known [93] that electron donor–acceptor complexes in the charge transfer state can fall apart into the component ions, a configuration that is often referred to as the solvent-separated radical ion pair (SSRIP). This process involves motion along a coordinate corresponding to the donor–acceptor separation accompanied by solvent molecules filling the space between the donor and acceptor. This "solvent coordinate" is different from the one discussed above, which does not significantly alter the equilibrium structure of the charge transfer complex. Polar solvents may lower the barrier for dissociation along this coordinate, and if the rate of electron transfer back to the ground-state is slow enough, this may lead to the formation of SSRIPs. For increasing donor–acceptor separations the diabatic coupling β will decrease [94,95], resulting in adiabatic states that have increasingly the same character as the diabatic states. The experimental observation of SSRIPs is no proof of localization in itself since the diabatic and adiabatic states coincide for infinite separation between donor and acceptor. It would be interesting to examine the effects on localization of SSRIP formation. Experimentally, it appears that SSRIPs are found only when the characteristic electron transfer time is longer than a few tens of nanoseconds [96], suggesting that it can be ignored in electron transfer processes that occur on the picosecond time scale.

C. Landau–Zener Approach

In this section we show the relationship between the matrix elements of the kinetic energy operator responsible for the coupling between the adiabatic states and the Landau–Zener effect that is commonly employed to evaluate curve-crossing dynamics. In addition, the requirements for curve crossing influences the dynamics, which in certain regimes of coupling is not properly predicted by a single golden-rule formula.

In a previous publication [28] the non-Born–Oppenheimer coupling was explicitly calculated for the TCNE–HMB electron donor–acceptor complex in the limit of large diabatic coupling. It was found that in this limit the electron transfer rate can be written as a golden-rule expression in the adiabatic frame with the non-Born–Oppenheimer matrix element as the relevant coupling parameter. In the limit of vanishing diabatic coupling, the non-Born–Oppenheimer coupling tends to infinity, but the rate can then be calculated with perturbation theory in the diabatic frame. A simple and exact expression that interpolates between these two limits does not exist. A model [97–99] that uses Redfield theory to model the nuclear dynamics and that treats the diabatic coupling exactly has been shown to interpolate properly between the two regimes. This model becomes computationally cumbersome for low-frequency modes or if more than a single mode is coupled to the reaction. It is possible, however, to find a simple expression if it is assumed that the nuclear motion is classical. For a single promoting mode, the non-Born–Oppenheimer Hamiltonian is given by

$$\hat{V}_{\text{NBO}} = -\frac{\hbar^2}{m}\,\hat{\pi}_{ge}\left\langle g\left|\frac{\partial}{\partial x}\right|e\right\rangle\frac{\partial}{\partial x} + \text{h.c.} \qquad (2.15)$$

where $|g\rangle$ and $|e\rangle$ are the adiabatic electronic wavefunctions, $\hat{\pi}_{ge} = |g\rangle\langle e|$ is the state flip operator, m is the reduced mass of the promoting mode, and x is its normal coordinate. The term $\langle g|\partial/\partial x|e\rangle$ can be calculated explicitly in a model that assumes that the potentials of the diabatic electronic states are harmonic, that is,

$$U_C = \tfrac{1}{2}m\omega^2 x^2 \qquad U_I = \tfrac{1}{2}m\omega^2(x-x_0)^2 + \Delta G^0 \qquad \lambda = \tfrac{1}{2}m\omega^2 x_0^2 \quad (2.16)$$

and coupled by a coupling V that, in general, may be a function of the coordinate x itself. Using Eq. (2.2) for the states $|g\rangle$ and $|e\rangle$ and the definitions from Section II.A, it is found that

$$\left\langle e\left|\frac{\partial}{\partial x}\right|g\right\rangle = \frac{V\Delta(x)' - \Delta(x)V'}{\Delta(x)^2 + 4V^2} = -\left\langle g\left|\frac{\partial}{\partial x}\right|e\right\rangle \qquad (2.17)$$

where the prime stands for a derivative with respect to the normal coordinate. It can thus be seen that the non-Born–Oppenheimer coupling in the position representation is a Lorentzian as a function of position that peaks at the crossing point of the diabatic potentials and has a width proportional to the diabatic coupling. In the limit of vanishing diabatic coupling, $V \to 0$, it tends to a Dirac delta function and becomes nonperturbative. This result can be used to justify the model described later to explain the transfer of coherence in the low frequency modes as a result of the electron transfer process.

If the diabatic potentials $U_{C/I}(x)$ are harmonic with normal mode frequency ω, the potential difference is linearly proportional to the normal-mode coordinate and can be written as

$$\Delta(x) = \varepsilon(x - x_C) \qquad \varepsilon \equiv \sqrt{2\lambda m\omega^2} \qquad (2.18)$$

where x_C is the point of intersection of the two potentials and $\lambda = Sh\omega$ is the reorganization energy of the vibrational mode, with S the Huang–Rhys factor. In addition, it can be assumed that the diabatic coupling is constant in the region around the crossing in the diabatic frame. The semiclassical Landau–Zener formalism assumes that a classical particle moves through the crossing region with constant speed v. The probability for jumping from one electronic surface to another is determined by quantum mechanics. The expression for the non-Born–Oppenheimer Hamiltonian can be put in a semiclassical form by replacing the momentum operator $(\hbar/i)(\partial/\partial x)$ with the classical momentum and the position operator by vt. The non-Born–Oppenheimer Hamiltonian can then be written in semiclassical form as

$$\hat{V}_{NBO}^{SC} = i\hbar \, \frac{V\,\Delta(x)'}{\Delta(x)^2 + 4V^2} \, \hat{\pi}_{ge} + \text{h.c.} \qquad (2.19)$$

where the prime now stands for a derivative with respect to time.

This result can be derived in an alternative fashion [100]. For a classical particle moving through the crossing region with a given speed, the semi-classical Hamiltonian in the diabatic picture is given by Eq. (2.1) with x replaced by vt. The adiabatic Hamiltonian can be found by transforming to the adiabatic frame with the eigenvector matrix $\hat{E}$, which will be of the form of Eq. (2.2) but time dependent. The relevant transformation is $\hat{H}_A = \hat{E}\hat{H}_D\hat{E}^{-1} - i\hat{E}(\partial/\partial t)\hat{E}^\dagger$. The second term in this transformation has off-diagonal elements that are identical to Eq. (2.19) given the assumptions made above (constant speed, constant diabatic coupling, and harmonic potentials).

The probability to jump from one adiabatic surface to another in a single pass of the nuclei through the crossing region due to the Hamiltonian in Eq. (2.19) has been calculated previously [100–103] and is given by

$$P = \exp\left(\frac{-4\pi V^2}{\hbar \varepsilon v}\right) \qquad (2.20)$$

The probability to jump from one surface to another taking into account multiple passes through the crossing region, P_m, depends on the path taken [56,104–107]. In the normal regime [in which the adiabatic potential has the shape of a double potential well and the path is from one well to the other; Figure 3(a)] this probability is given by $P_m = (1 - P)/(2 - P)$. In the inverted regime [in which the adiabatic potentials are nested and the path is from the top well to the bottom well; Figure 3(b)], this probability is given by $P_m = P(1 - P)$. To get a transition rate one can take the jump probability from Eq. (2.20), replace the speed by an average speed (which would be given by $\sqrt{2k_BT/\pi m}$ for a low-frequency classical mode), and scale it with the probability to reach the crossing point. The rate for a low-frequency classical mode is thus given by

$$k_{LZ} = P_m \frac{\omega}{2\pi} \exp\left[\frac{-(\Delta G^0 + \lambda)^2}{4\lambda k_B T}\right] \tag{2.21}$$

where the form of P_m depends on the regime as described above. Independent of the regime, in the limit of vanishing diabatic coupling, Eq. (2.21) reduces to

$$\lim_{V \to 0} k_{LZ} = \frac{2\pi V^2}{\hbar\sqrt{4\pi\lambda k_B T}} \exp\left[\frac{-(\Delta G^0 + \lambda)^2}{4\lambda k_B T}\right] \tag{2.22}$$

which is the well-known [53,55,56] golden-rule expression for the electron transfer rate. Figure 4 shows the rate of electron transfer as a function of the

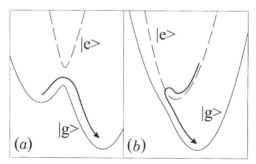

Figure 3. Strong coupling electron transfer in the normal and inverted regime. (a) In the normal regime, if the path taken is from one side of the double well to the other, the probability for a transition, taking into account recrossings, is given by $(1 - P)/(2 - P)$, where P is the probability for a single crossing. (b) In the inverted regime, if the path taken is from the top well to the bottom well, the probability for a transition is given by $P(1 - P)$.

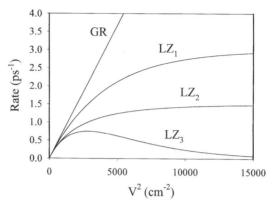

Figure 4.   Rate of electron transfer as a function of the coupling strength V according to the Landau–Zener (LZ) and the golden-rule (GR) equations. The parameters are $T = 298\,\mathrm{K}$, $\omega = 100\,\mathrm{cm}^{-1}$, $\Delta G^0 = -1000\,\mathrm{cm}^{-1}$, and $\lambda = 1000\,\mathrm{cm}^{-1}$. The three Landau–Zener curves correspond to: normal regime, single crossing (LZ_1); normal regime, multiple crossings (LZ_2); and inverted regime, multiple crossings (LZ_3).

diabatic coupling strength in the normal and inverted regime calculated using the golden-rule expression and the Landau–Zener expression. The rate predicted by the golden-rule expression keeps growing quadratically as a function of the diabatic coupling strength. The Landau–Zener expressions show that the rate saturates, or even declines in the inverted regime, for large diabatic coupling. For the particular parameters chosen, the rate predicted by the golden-rule expression is off by an order of magnitude for a diabatic coupling as small as $100\,\mathrm{cm}^{-1}$.

In the derivation above of the Landau–Zener expression for the electron transfer rate it was assumed that the promoting mode behaved classically. This does not imply that the derivation is valid only if the promoting mode is the solvent coordinate. In the case of EDA and some mixed-valence organometallic complexes in which a donor and acceptor are brought to close proximity giving rise to a charge transfer absorption band, many intramolecular vibrational modes change their equilibrium position on going from the $|g\rangle$ to the $|e\rangle$ state, as evidenced by the enhancement of these modes in the resonance Raman scattering spectrum [1–22]. All of these modes, along with the solvent mode, will act as promoter modes that can couple the $|g\rangle$ and $|e\rangle$ manifolds through a non-Born–Oppenheimer coupling. It is for this reason that electron transfer can take place in the absence of a solvent, in the gas phase [108]. To calculate the transition rate between adiabatic surfaces due to the non–Born–Oppenheimer coupling for a quantum vibrational mode, it appears one can use Redfield type calculations as discussed above [97-99]. However, such a calculation becomes unwieldy for more than one Franck–Condon active mode. In the limit of large diabatic coupling and a large number of active modes, the method described previously for the TCNE–HMB complex can be employed [28].

D. Wavepacket Description of NBO Transitions

The description of spectroscopic processes such as Raman scattering, absorption, and fluorescence by means of vibrational wavepacket motion has proved useful [15,20,63–72]. This concept can also be used to describe electron transfer, so rather than being viewed as dependent on Franck–Condon factors, the rate becomes determined by the time-dependent overlap of a vibrational wavepacket propagated on the final-state potential surface projecting onto specific modes of the initial surface. Just as in the other applications of this technique, a rapid, efficient method of calculating the rate emerges.

In the limit that the diabatic coupling becomes very large, one can apply perturbation theory in the adiabatic basis using the non-Born–Oppenheimer Hamiltonian as the coupling matrix element. Thus if the adiabatic basis set consists of the states $|1\rangle$ (reactant state) and $|2\rangle$ (product state) separated by the driving force $\Delta G^0 = \Omega_{21}$, the non-Born–Oppenheimer Hamiltonian can be written as [see Eqs. (2.17) and (2.18)]

$$\hat{V}_{\text{NBO}}(x) = \hat{L}(x) \frac{\partial}{\partial x} (\hat{\pi}_{12} - \hat{\pi}_{21}) \qquad \hat{L}(x) = \frac{\hbar^2}{m} \frac{V\varepsilon}{\Delta(x)^2 + 4V^2} \qquad (2.23)$$

Applying second-order perturbation theory in $\hat{V}_{\text{NBO}}$, one can derive the rate of change of the population of the product state (i.e., the rate of nonadiabatic transitions) as

$$\frac{d\rho_{22}}{dt} = \frac{-2}{\hbar^2} \text{Re} \int_0^\infty dt' \sum_{n=0}^\infty \rho_{nn} e^{-i(\Omega_{21} - n\omega_v)t'} \sum_{k,m} \left\langle n \left| \frac{\hat{L}(x_k)\partial}{\partial x_k} \hat{U}_2(t') \frac{\hat{L}(x_m)\partial}{\partial x_m} \right| n \right\rangle$$

$$(2.24)$$

where $\hat{U}_2(t)$ is the propagator for evolution on the surface of state $|2\rangle$ and ρ_{nn} is the initial distribution over vibrational states [e.g., a Boltzmann distribution $\rho_{nn} = [1 - \exp(-\hbar\omega/k_B T)]\exp(-n\hbar\hat{\omega}/k_B T)$]. In Eq. (2.24) the vibrational wavefunctions are written as $|n\rangle \equiv |n_1 n_2 \cdots\rangle$, where n_1 is the number of quanta in normal mode 1, and so on. If the various normal modes are not anharmonically coupled, as is assumed from here on, the wavefunctions can be written as $|n_1 n_2 \cdots\rangle = |n_1\rangle |n_2\rangle \cdots$. The sum over k and m in Eq. (2.24) is over all the promoting modes that give rise to non-Born–Oppenheimer transitions.

In the limit of very large diabatic coupling, the non–Born–Oppenheimer coupling can be approximated by its value at the crossing point [i.e., $L(x_i) \approx L(x_{i,C}) = \hbar^2 \varepsilon_i / 4m_i V$, where ε_i is defined for each normal mode by Eq. (2.18)]. The derivatives with respect to position can be written in terms of raising and lowering operators [109], and if the adiabatic potentials are approximated as harmonic, the vibrational correlation function can be rewritten for $k_B T \ll \hbar\omega$ as

$$\left\langle n \left| \frac{\partial}{\partial x_i} \hat{U}_2(t) \frac{\partial}{\partial x_j} \right| n \right\rangle = -\frac{\sqrt{m_i m_j \omega_i \omega_j}}{2\hbar}$$

$$\times \left[\delta_{ij} \langle 1_i | 1_i(t) \rangle \prod_{k \neq i} \langle 0_k | 0_k(t) \rangle + (1 - \delta_{ij}) \langle 1_i | 0_i(t) \rangle \langle 0_j | 1_j(t) \rangle \prod_{k=i,j} \langle 0_k | 0_k(t) \rangle \right]$$

$$(2.25)$$

where m_i is the reduced mass of normal coordinate i, ω_i its frequency, and δ_{ij} is the Kronecker delta function. The single-mode time-dependent vibrational overlap integrals are defined as

$$\langle n_i | m_i(t) \rangle = \langle n_i | \hat{U}_2(t) | m_i \rangle \tag{2.26}$$

where $\langle n_i | m_i(t) \rangle$ stands for the vibrational wavefunction with m quanta in mode i in the initial electronic state $|1\rangle$, propagated on the potential of final electronic state $|2\rangle$ and projected onto the vibrational state with n quanta in mode i in the initial electronic state $|1\rangle$. Explicit expressions for the vibrational overlap integrals will be given in Section II.F. It is straightforward to generalize Eq. (2.25) to be valid at any temperature. The single-mode vibrational correlation functions defined in Eq. (2.26) have been discussed extensively in the context of the calculation of absorption spectra and resonance Raman scattering excitation profiles [15,63–65,67,68,72,110]. A general form for the vibrational correlation function for harmonic vibrational potentials was derived by Yan and Mukamel [67]. From their expressions it is found that

$$\langle 1 | 0(t) \rangle = -\frac{\Delta}{\sqrt{2}} (1 - e^{-i\omega t}) \langle 0 | 0(t) \rangle$$

$$\langle 1 | 1(t) \rangle = \left[\frac{\Delta^2}{2} (1 - e^{-i\omega t})^2 + e^{-i\omega t} \right] \langle 0 | 0(t) \rangle \tag{2.27}$$

where Δ is the dimensionless displacement of the harmonic wells, which is related to the Huang–Rhys factor by $S = \frac{1}{2}\Delta^2$. A simple expression for $\langle 0|0(t)\rangle$, valid for $k_B T \ll \hbar\omega$ and equal ground and excited state frequencies is, [68]

$$\langle 0|0(t)\rangle = \exp[-S(1 - e^{-i\omega t}) - \frac{1}{2} i\omega t] \qquad (2.28)$$

A more general expression is given later. From Eqs. (2.25) and (2.27) it can thus be seen that in the "short-time approximation" or equivalently, in the regime of large driving force, the expression for the non-Born-Oppenheimer coupling-induced electron transfer rate assumes a normal, golden-rule form. The effective coupling in this approximation is

$$V^2_{\text{effective}} = \sum_i \frac{\lambda_i (\hbar\omega_i)^3}{16 V^2} \qquad (2.29)$$

where $\lambda_i = \frac{1}{2} \hbar\omega_i \Delta_i^2$ is the reorganization energy of mode i.

E. Coupling to the Bath: Solvent Reorganization Energy

The expression above for the electron transfer rate considers only the intramolecular vibrational modes, but it is also important to include the effect of bath modes. The bath modes, of which there must be about 10^{23}, obviously cannot be incorporated mode by mode on an equal footing with the internal modes. However, the usual strategy is to lump them together into a *single* harmonic mode that is usually referred to as *the* solvent mode. This solvent mode then can be included into the above expression (2.24) for the electron transfer rate as another normal mode of the system that is Franck–Condon active in the charge transfer transition. In the continuum Onsager cavity description the reorganization energy [111] of this solvent mode is given by [112]

$$\lambda_s = \frac{(\Delta\mu)^2}{4\pi\varepsilon_0 r^3} \left(\frac{\varepsilon_S - 1}{2\varepsilon_S + 1} - \frac{\varepsilon_\infty - 1}{2\varepsilon_\infty + 1} \right) \qquad (2.30)$$

where $\Delta\mu$ is the difference between the excited- and ground-state permanent dipole moments μ_e and μ_g r the cavity radius, ε_∞ the infinite frequency and ε_S the zero-frequency dielectric constant of the solvent. The interaction between the system and the bath will also lead to a stabilization energy, formally defined as the shift of the electronic transition energy of the solute on going from vacuum to solution, which is estimated from [112]

$$\Delta(\Delta G^0) = -\frac{\mu_e^2 - \mu_g^2}{4\pi\varepsilon_0 r^3} \frac{\varepsilon_S - 1}{2\varepsilon_S + 1} \tag{2.31}$$

Since the solvent mode will usually have a low frequency compared with $k_B T$ and a large reduced mass, a classical description of its dynamics is appropriate. If the solvent mode is described as a Brownian oscillator [113–115] in the overdamped Smoluchowski limit, the vibrational correlation function for the solvent mode can be written as [76,115]

$$\sum_{n_S} \rho_{n_S n_S} \langle n_S | n_S(t) \rangle = \exp[-g(t)] \tag{2.32}$$

where

$$g(t) = \left(\frac{\Delta_S}{\Lambda}\right)^2 (e^{-\Lambda t} + \Lambda t - 1) + i \frac{\lambda_S}{\Lambda} (e^{-\Lambda t} - 1) \tag{2.33}$$

$\Delta_S^2 = 2\lambda_S k_B T$, $\Lambda = 1/\tau_L$, where τ_L is the correlation time of the bath. To highlight that the solvent correlation function is of the same form as the intramolecular vibrational correlation functions discussed in the preceding section, Eq. (2.32) is written in terms of the time-dependent vibrational overlap integral of the solvent mode, where n_S is the number of quanta in this mode. The bath correlation time can be identified as the longitudinal dielectric relaxation time, which according to continuum theory [116] is given by $\tau_L \approx (\varepsilon_\infty/\varepsilon_S)\tau_D$, where τ_D is the Debye rotational diffusion time. More accurate values for τ_L can be obtained by experimentally measuring the rate of dynamic Stokes shifting of the emission from polar dyes [116–118]. Values for τ_L thus obtained are listed in Table I. The total Stokes shift between the absorption spectrum of the electron transfer system (the charge transfer band) and the emission spectrum is given by twice the total reorganization energy. (Expressions for the absorption and emission spectra are given in Section II.G.1.). If the intramolecular reorganization energy can be obtained from the resonance Raman profile analysis, the solvent reorganization energy λ_s can be obtained from the total Stokes shift or, with less accuracy, from the charge transfer absorption band alone. Therefore, in principle, all parameters occurring in Eq. (2.33) can be obtained from independent experiments.

F. Recipe for Calculating Electron Transfer Rates

The equations derived in previous sections can be combined into a recipe for calculating absolute electron transfer rates. By taking the basic rate

TABLE I
Electron Transfer Rates in TCNE–HMB[a]

Solvent	τ_{ET} (ps)	τ_L (ps)	τ_{NBO} (ps)
Acetonitrile	0.643(8)	0.56 [116]	1.5
Acetone	0.995(6)	0.83 [42]	2.8
Tetrahydrofuran	2.13(2)	1.42 [117]	5.6
Glyceroltriacetate	2.93(2)	125 [42]	3.4
CCl$_4$	11.2(1)	0.3 [17]	34

[a]Listed are the solvent, the experimentally observed electron transfer time (τ_ET), the experimental dielectric solvation time (τ_L), and the electron transfer rate calculated with the expressions involving the non-Born–Oppenheimer coupling [τ_{NBO}]; see Eg. (234)].

expression (2.24) in the limit of large diabatic coupling combined with expression (2.32) for the coupling to the solvent, one obtains the rate expression

$$k(\Delta G^0) = \sum_{i,j} \frac{\pi |\Delta_i \Delta_j| \omega_i^2 \omega_j^2}{16 V^2} \operatorname{Re} \frac{1}{\pi} \int_0^\infty dt' \, e^{-i\Delta G^0 t' - g(t')}$$

$$\times \left[\delta_{ij} \langle 1_i | 1_i(t) \rangle \prod_{k \neq i} \langle 0_k | 0_k(t) \rangle + (1 - \delta_{ij}) \langle 1_i | 0_i(t) \rangle \langle 0_j | 1_j(t) \rangle \right.$$

$$\left. \times \prod_{k i,j} \langle 0_k | 0_k(t) \rangle \right] \qquad (2.34)$$

where ΔG^0 is the driving force of the reaction, the sum is over all the Franck–Condon active (promoting) modes, Δ_i is the dimensionless shift of mode i, ω_i is the frequency of mode i, V is the diabatic coupling, and the bath correlation function is given in Eq. (2.33). Note that the rate expression above is written as a prefactor multiplied with the real part of a Fourier transform. The Fourier transform is normalized with respect to integration over ΔG^0. An alternative approximate form of the rate expression is obtained by replacing the non-Born–Oppenheimer coupling by an effective coupling

$$k^{\text{effective}}(\Delta G^0) = \frac{2\pi V_{\text{effective}}^2}{\hbar^2} \frac{1}{2\pi} \int_{-\infty}^\infty dt' \, \langle n | n(t') \rangle e^{-i\Delta G^0 t' - g(t')} \qquad (2.35)$$

where $V_{\text{effective}}$ is given by Eq. (2.29). Again this expression is written as a prefactor times a Fourier transform where the Fourier transform is normalized with respect to integration over ΔG^0.

Expressions for the intramolecular vibrational correlation functions have been given in the literature [15,63–65,67,68,72,110]. In the calculations in this paper we used the general expression given by Yan and Mukamel [67], which allows for different frequencies in the initial and final states. This expression is given by

$$\langle m|n(t)\rangle = \sigma_0(t)W_{mn}(t) \tag{2.36}$$

where

$$\sigma_0(t) = \frac{\exp[\Delta^2 f(t)]}{\sqrt{\psi(t)}}$$

$$W_{mn}(t) = \frac{\alpha(t)^{m+n}}{\sqrt{m!\,n!\,2^{m+n}}} \sum_{k=0}^{k^*} \frac{(2k)!}{k!}\,\eta_{mnk}\gamma(t)^k H_{m+n-2k}\frac{\lambda f(t)\Delta}{\alpha(t)} \tag{2.37}$$

In this expression H_n is the Hermite polynomial of order n, k^* is the integer part of $(m+n)/2$, and

$$\psi(t) = \frac{\omega_+^2}{4\omega'\omega''}\left[1 - \left(\frac{\omega_-}{\omega_+}\right)^2 \exp(-2i\omega't)\right] \qquad f(t) = -\frac{\omega''[1-\exp(-i\omega't)]}{\omega_+ - \omega_-\exp(-i\omega't)}$$

$$\alpha(t) = \sqrt{\frac{\omega_- - \omega_+\exp(-i\omega't)}{2\omega_+ - 2\omega_-\exp(-i\omega't)}} \qquad \gamma(t) = \frac{2\lambda^2 + i(\lambda^4-1)\sin\omega't'}{2\lambda^2 - i(\lambda^4-1)\sin\omega t}$$

$$\omega_\pm = \omega' \pm \omega'' \qquad \lambda = \sqrt{\frac{\omega'}{\omega''}} \qquad \eta_{mnk} = \sum_{q=0}^{2k}(-1)^q\binom{m}{2k-q}\binom{n}{q}$$

$$\binom{m}{l} = \begin{cases} \dfrac{m!}{l!(m-l)!}, & m \ge l \\[2ex] 0, & m < l, \end{cases} \tag{2.38}$$

In the foregoing, ω'' is the vibrational frequency in the initial state $|1\rangle$ and ω' is the frequency in the final state $|2\rangle$. In our calculations the rates were calculated using Eq. (2.34) or (2.35). The time correlation functions were calculated in a 512- to 8196-point array and numerically Fourier transformed using FFT [119] on a Pentium PC.

G.   Rate Calculations for Several Model Systems

In this section we show that the foregoing equations can be used to make reasonable predictions of electron transfer rates. All the examples chosen are Mulliken charge transfer complexes involving an aromatic molecule as the donor and tetracyanoethylene or tetracyanobenzene as the acceptor. The need for the theoretical description to incorporate all the relevant modes is made clear in Figure 5, which shows the electron transfer rate as a function of the driving force calculated with the full rate expression, Eq. (2.34), and in the "short-time approximation," Eq. (2.35). In Figure 5, curves A and B, it has been assumed that the reaction is coupled to a single intramolecular mode with a frequency of $\omega = 1500 \, \text{cm}^{-1}$ and a dimensionless shift of $\Delta = 1$ and a solvent mode with a reorganization energy of $\lambda_S = 500 \, \text{cm}^{-1}$ and a correlation time of $\tau_L = 300 \, \text{fs}$. It can be seen that the short-time approximation gives results identical to those of the full expression at large driving force, but for $\Delta G^0 \approx \lambda_{\text{total}}$ the short-time approximation can be off by as

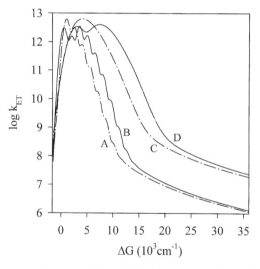

Figure 5.   Electron transfer rate calculated with the non-Born–Oppenheimer coupling. For all curves the diabatic coupling has been chosen as $V = 4000 \, \text{cm}^{-1}$ and the solvent correlation time as $\tau_L = 300 \, \text{fs}$. Curves A and C are calculated by approximating the non-Born–Oppenheimer coupling using the short-time approximation (2.35). Curves B and D are calculated taking into account the proper matrix elements of the momentum operator Eq. (2.34). Curves A and B are calculated for a transition coupled to a single vibrational mode with frequency $1500 \, \text{cm}^{-1}$, a dimensionless shift of $\Delta = 1$, and a solvent reorganization energy of $\lambda_s = 500 \, \text{cm}^{-1}$. Curves C and D are calculated using the parameters for TCNE–HMB in CCl_4 [i.e. vibrational frequencies (equal in ground and excited-states) and dimensionless shifts as given in Table II].

much as a factor of 6. Overall, the curves of the rates of electron transfer versus driving force predicted by our Eq. (2.34) do not appear very different from those based on the golden rule in the diabatic picture. *However, the rates predicted here decrease with increasing diabatic coupling rather than increase.*

Raman spectra taken on resonance with the charge transfer transition have been published for several compounds, but full sets of frequencies and dimensionless shifts have only been published for a few. In the following sections detailed calculations of the absolute rates of electron transfer using the non-Born–Oppenheimer theory are given for three of these systems.

1. Rate Calculation for TCNE–HMB

The tetracyanoethylene–hexamethylbenzene (TCNE–HMB) electron donor–acceptor complex has become the best-studied system that exhibits direct electron transfer. The x-ray crystal structure of the complex is known [120,121]. The complex has been studied in solutions [31,122–124], the crystalline state [125–129], and the gas phase [130,131] by a great variety of methods. Most important to our aims is that the TCNE–HMB complex was studied extensively by means of resonance Raman scattering spectroscopy[1,4,13–18]. Therefore, it is known which vibrational modes are Franck–Condon active in the charge transfer transition. We published [28] an ultrafast spectroscopy study of electron transfer in the TCNE–HMB complex in various polar solvents. The results of this study are summarized in Table I and it can be seen that the electron transfer rate varies between $(0.64 \text{ ps})^1$ in acetonitrile and $(11.2 \text{ ps})^{-1}$ in CCl_4. In the following the theory from previous sections is applied in conjunction with the resonance Raman scattering data, to calculate absolute rates for this system.

To calculate the absolute rates for electron transfer, the most recent resonance Raman scattering data from Myers and co-workers will be used [16,17]. Figure 6(a) shows the experimental static ground-state absorption spectrum of TCNE–HMB in CCl_4 solution and the absorption spectrum predicted theoretically. To calculate this absorption spectrum, the dimensionless shifts and vibrational mode frequencies as well as lineshape parameters are taken from the resonance Raman spectrum. The frequency-dependent extinction coefficient is then given by [68]

$$\varepsilon(\omega) = \frac{10 N_A \pi n \mu^2 \omega}{3\hbar c \varepsilon_0 \ln(10)} \sum_n P_n \text{Re} \int_0^\infty dt \langle n|n(t)\rangle \exp[i(\omega - \Delta G^0/\hbar + n\omega_n)t - g(t)]$$

$$(2.39)$$

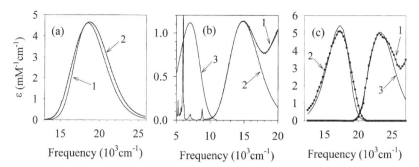

Figure 6. (*a*) Absorption spectrum of TCNE–HMB in CCl$_4$ (2), theoretical spectrum calculated using the parameters from the resonance Raman scattering data [13,16,17] (1), and the theoretical spectrum calculated by modifying the Raman scattering data as described in the text (2). (*b*) Absorption spectrum of TCNE–pyrene in benzonitrile (1) and the theoretical absorption (2) and fluorescence (3) spectrum calculated by modifying the Raman scattering data for TCNE–HMB data [13,16,17] as described in the text. (*c*) Absorption (1) and fluorescence (2) spectra of TCNB–HMB in CCl$_4$ and theoretical spectrum (3) obtained as in (*b*).

and the fluorescence spectrum by

$$\sigma(\omega) \propto \omega^3 \sum_n P_n \, \mathrm{Re} \int_0^\infty dt \langle n|n(t)\rangle \exp[-i(\omega - \Delta G^0/\hbar + n\omega_n)t - g(t)]$$

(2.40)

where $\langle n|n(t)\rangle$ is a multimode vibrational correlation function for which expressions are available in the literature [15,63–65,67,68,72,110] and $g(t)$ is the solvent correlation function defined in Eq. (2.33). The dimensionless shifts and vibrational frequencies used are listed in Table II and the parameters for the solvent correlation function in CCl$_4$ are $\lambda_S = 2450\,\mathrm{cm}^{-1}$ and $\tau_L = 300\,\mathrm{fs}$ as used in the original resonance Raman scattering study [16].

To calculate an electron transfer rate with Eq. (2.24), one needs the diabatic coupling which can be obtained from the experimental absorption spectrum using the Mulliken–Hush treatment [74,75], which expresses the diabatic coupling as

$$V = \frac{\mu_{12}\,\Delta E}{\Delta\mu}$$

(2.41)

where μ_{12} is the transition dipole moment of the charge transfer transition (found experimentally to be 3.5 D), ΔE is the Franck–Condon energy gap (approximately 18,000 cm^{-1}, and $\Delta\mu$ is the change in permanent dipole

TABLE II

Resonance Raman Scattering [16] Parameters for TCNE–HMB in CCl_4

Frequency (cm^{-1})	Δ	λ_i (cm^{-1})
165	3.80/0.80[a]	1191/250
450	1.33	398
542	0.54	79
600	0.57	97
968	0.42	85
1292	0.65	273
1389	0.33	76
1437	0.21	32
1551	0.94	685
1570	0.29	66
2222	0.45	225

[a]The dimensionless shift is 0.80 if the excited-state frequency is taken as $350\,cm^{-1}$ and 3.80 if this frequency is taken equal to that of the ground-state [13].

moment between the diabatic covalent and ionic states (estimated as 16 D). Using this expression one finds a diabatic coupling of approximately $4000\,cm^{-1}$. The diabatic coupling derived from the fluorescence spectrum [16] is significantly smaller, $1255\,cm^{-1}$. Since the uncertainties in determining the coupling this way are significantly larger, the former value of $4000\,cm^{-1}$ will be used. However, if the diabatic coupling is truly smaller in the region of vibrational coordinates where electron transfer takes place, all the rates calculated below will increase, not decrease. Application of the Mulliken–Hush treatment to TCNE–HMB has been discussed in more detail in a previous publication [28].

Using the same parameters as in the resonance Raman scattering study [16], a diabatic coupling of $4000\,cm^{-1}$ and Eqs. (2.24) and (2.32) to calculate the electron transfer rate, the rate is calculated to be $(16.1\,ps)^{-1}$. The experimentally observed rate in CCl_4 (see Table I) is $(11.2\,ps)^1$.

The parameters from the resonance Raman scattering data [16] lead to a predicted absorption spectrum that does not fit the experimental spectrum very well. One reason for this may be that it was assumed that the first mode (see Table II) has a ground-state frequency of $165\,cm^{-1}$ and an excited-state frequency of $350\,cm^{-1}$. This assumption was based on the idea that this mode corresponds mainly to the TCNE–HMB intercomplex stretch, which would attain a higher frequency in the (mainly ionic) excited-state due to increased coulombic attraction. In an experimental ultrafast spectroscopy study on the related complex TCNE–pyrene [30] it was found, however, that the ca. $165\,cm^{-1}$ mode has nearly identical ground- and excited-state frequencies. It was concluded from this study that the $165 \, \mathrm{cm}^{-1}$ mode is therefore not simply the DA stretch mode but a mixture of the DA stretch with perhaps TCNE out-of-plane modes whose frequencies would not be so sensitive to the charge separation. Therefore, in the following calculations the ground- and excited-state frequencies of this mode were set equal and the dimensionless shift was changed from its original value of 0.80 to 3.8, which has a minimal effect on the predicted resonance Raman scattering excitation profile [13,16,17] but has a significant effect on the predicted absorption spectrum. It is interesting to see if further fitting the absorption spectrum will lead to predicted electron transfer rates that are comparable. To this end the experimental absorption spectrum was fit, with the solvent reorganization energy λ_S and the driving force (0–0 transition energy) ΔG^0 as free parameters. Additionally, the dimensionless shifts were all scaled in the fit using a single scaling parameter S_C. When the experimental spectrum in CCl_4 is fit in this manner, an excellent fit is obtained (χ^2 close to 1) and one finds $\lambda_S = 200 \pm 100 \, \mathrm{cm}^{-1}$, $\Delta G^0 = 13{,}570 \pm 200 \, \mathrm{cm}^{-1}$, and $S_C = 1.31 \pm 0.01$. The solvent reorganization energy found this way is much closer to what might be expected for a nonpolar solvent such as CCl_4. The fluorescence spectrum predicted using these parameters peaks at ca. $10{,}000 \, \mathrm{cm}^{-1}$, (i.e., at slightly lower energy than experimentally observed [16,17]. Unfortunately, the increase of all the dimensionless shifts by 30% will lead to predicted resonance Raman scattering profiles that have intensities that are significantly too high ($< 25\%$ for the fundamentals, $< 40\%$ for overtones).

The rate in CCl_4 predicted using these parameters is $(33.9 \, \mathrm{ps})^{-1}$. The absorption spectra of four polar solvents were fit in an identical manner; the fit results are listed in Table III and the calculated rates in Table 1. The

TABLE III
Parameters Obtained from Fits to the Static Absorption Spectra[a]

| Solvent | λ_s (cm^{-1}) | | $-\Delta G^0$ (10^3 cm^{-1}) | $-\Delta\Delta G^0$ (10^3 cm^{-1}) | |
	Experimental	Theoretical[b]	Experimental	Experimental	Theoretical[b]
Acetonitrile	1770	1570	12.416	3.47	4.94
Acetone	1602	1460	12.903	3.00	4.79
Glycerol triacetate	1372	1020	12.882	3.00	4.15
Tetrahydrofuran	1270	1080	13.255	2.64	4.20
CCl$_4$	108±96	0	13.570 ± 200	2.32	2.32

[a]Driving force and classical reorganization energy from fits to the absorption spectra using all 11 modes listed in Table II. $\Delta\Delta G^0$ stands for ΔG^0 (solv)$-\Delta G^0$(CCl$_4$)
[b]From [182] and [183].

rates vary between $(1.5\,\text{ps})^{-1}$ in acetonitrile and $(5.6\,\text{ps})^{-1}$ in tetrahydrofuran (i.e., the variation with solvent of the theoretically predicted rate is not as large as experimentally observed. One reason for this deficiency is clear, in our calculations it is assumed that the electron transfer reaction starts with a complex that is thermally equilibrated in the reactant state. In the experiment, however, the reactant state is prepared by the laser pump pulse in a distinctly nonequilibrated state. The equilibration process along the solvent coordinate is slow in solvents with a long solvation time constant such as glycerol triacetate (GTA) and will lead to smaller electron transfer rates in these solvents. When this effect is included in the rate calculations, as discussed before [28,42,132,133], the theoretically predicted rates come into good agreement with the experiment. Also, fitting the charge transfer absorption spectra with the scaling of the dimensionless shifts, S_C, appears to be a stable method for obtaining the relevant reorganization parameters.

2. Extension to Similar EDA Complexes

a. TCNB–HMB. The properties of the Franck–Condon modes are known for only a few of the systems that exhibit direct electron transfer [1–22]. In the absence of complete information it has often been assumed that the charge transfer transition is coupled to the solvent and to a *single* intramolecular high-frequency mode with a frequency of about $1400\,\text{cm}^{-1}$. Under these extreme assumptions the absorption spectrum is given by

$$\varepsilon(\omega) \propto \omega \sum_{n=0}^{\infty} \frac{e^{-s}S^n}{n!} \exp\left[\frac{-(\hbar\omega - \lambda_S - n\hbar\omega_Q + \Delta G^0)^2}{4\lambda_S k_B T}\right] \qquad S = \frac{\lambda_Q}{\hbar\omega_Q} = \frac{1}{2}\Delta^2$$

$$(2.42)$$

where ω_Q is the frequency of the intramolecular vibrational mode and Δ its dimensionless shift. One cannot hope to get a useful estimate of the strength of the non-Born–Oppenheimer coupling starting from this oversimplified model (see below). It is therefore interesting to see if the data [13,16,17] regarding the Franck–Condon active modes of TCNE–HMB in CCl$_4$ can be used to predict the rates of electron transfer in other, similar, EDA complexes.

A good candidate for such a test is the EDA complex between tetracyanobenzene (TCNB) and HMB. First there is a substantial amount of data available regarding the rates of electron transfer and absorption and emission spectra [26,134–138]. Second, the donor in this complex is the same as in TCNE–HMB. In the resonance Raman spectrum of TCNE–HMB in CCl$_4$ [13], six out of 11 fundamental transitions are assigned to modes that are localized on the donor. Most of the remaining modes are related to C≡N and C=C motions of TCNE, which may be expected to be active in TCNB as well. The Huang–Rhys factors for the charge transfer transition in TCNB–HMB may be significantly different from those in TCNE–HMB. Therefore, to obtain appropriate parameters, the absorption and emission spectrum of TCNB–HMB can be fitted with the driving force, solvent reorganization energy, and a Huang–Rhys factor scaling parameter S_C as fitting parameters. Figure 6(c) shows the result obtained by simultaneously fitting the absorption and emission spectrum of TCNB–HMB in CCl_4 obtained from Figure 1 of [138]. The fit parameters are $\lambda_S = 10\,cm^{-1}$, $\Delta G^0 = 19.9 \times 10^3\,cm^{-1}$ and $S_C = 1.11$. Using Eqs. (2.24) and (2.32), a solvent dielectric relaxation time of $\tau_L = 300\,fs$ (as obtained from the resonance Raman study of TCNE–HMB in CCl_4 [13,16,17] and a diabatic coupling of $V = 860\,cm^{-1}$ (obtained from the emission strength of TCNB–HMB [138]), we calculate an electron transfer rate of $k_{NBO} = 12.8 \times 10^7 = (7.8\,ns)^{-1}$. This differs by only a factor of 4 from the experimentally observed rate [138] of $k_{EXP} = 3.8 \times 10^7$.

The calculated rate does not depend very strongly on the solvent dielectric correlation time. Changing it from $\tau_L = 300\,fs$ to $\tau_L = 10\,ns$ decreases the predicted rate to $k_{NBO} = 10.3 \times 10^7 = (9.7\,ns)^{-1}$. This is in marked contrast with the single-mode model that shows variations in the predicted rate over two orders of magnitude when the solvent dielectric relaxation time is varied in this way. This can be understood from the different partitioning of the reorganization energy in the one-mode and 11-mode models. Figure 5 shows that at large driving force the rate of electron transfer falls off more slowly with driving force. This effect originates from the fluctuations in the bath that are fast compared with τ_L. These fast fluctuations give the rate versus driving force curve a Lorentzian shape. Therefore, if a large part of the total reorganization energy is partitioned in the solvent reorganization energy, the predicted rate becomes very sensitive to the chosen value of the solvent dielectric relaxation time.

b. TCNE–Pyrene. A femtosecond spectroscopy study of TCNE–pyrene was recently published by us [30] and the results from this study are summarized in Table IV (see also below). No full-resonance Raman spectrum is available for the TCNE–pyrene complex, and therefore the absorption spectrum was again fitted to the 11-mode model derived from the TCNE–HMB resonance Raman scattering data [16,17]. The absorption spectrum of TCNE–pyrene in benzonitrile [see Figure 6(b)] has three overlapping bands that complicate obtaining a proper fit to the lowest charge transfer band. The spectrum was first fit with a function of the form of Eq. (2.42), simulating three charge transfer bands, which resulted in a good fit (χ^2 close to 1). In the fit function the amplitude of the low-energy charge transfer

TABLE IV
Electron Transfer Rates in TCNE–pyrene[a]

Solvent	τ_{ET} (ps)	τ_L (ps)	τ_{NBO} (ps)
Acetonitrile	0.291 ± 0.009	0.56 [116]	0.13^b
Benzonitrile	1.47 ± 0.02	4.76 [42]	1.3^c
Ethylacetate	1.51 ± 0.06	2.63 [42]	

[a]Listed are the solvent (abbreviations the same as in Table I), the experimentally observed electron transfer time τ_{ET}, the experimental dielectric solvation time (τ_L), and the electron transfer rate calculated with the expressions involving the non-Born–Oppenheimer coupling (τ_{NBO}, see the text).
[b]Rate calculated from the data obtained from the absorption spectrum in benzonitrile.
[c]Rate calculated as in note b but with the driving force set to $12\,500\,\mathrm{cm}^{-1}$ (energy of the pump photon).

band was then set to zero, after which the fit function was subtracted from the experimental data. This new data set consists of one, approximately Gaussian, band and some subtraction noise at high-energy that never exceeds 10% of the peak of the low-energy band. The new data set was fit with same 11-mode model as that was used for TCNE–HMB and TCNB–HMB above. The resulting fit parameters are $\lambda_S = 300 \pm 100\,\mathrm{cm}^{-1}$, $\Delta G^0 = 9990 \pm 1\,\mathrm{cm}^{-1}$, $\mu_{eg} = 1.631$ D, and $S_C = 1.25 \pm 0.02$. The indicated uncertainties in the parameters are only those due to noise in the fitted data set and "lack of fit" (the errors in the fit parameters are 68.3% joint confidence intervals [139]) and completely ignore the systematic errors introduced by subtracting off the high-energy absorption bands. The solvent reorganization energy obtained is significantly smaller than expected for this system. From the transition dipole moment and an estimated change in permanent dipole moment between the two diabatic states of 16 D, adiabatic coupling of $1500\,\mathrm{cm}^{-1}$ is calculated using Eq. (2.42).

Using the parameters from the fit to the absorption spectrum, and Eqs. (2.24) and (2.32), one calculates an electron transfer rate of $(128\,\mathrm{fs})^{-1}$. This rate is quite a bit larger than that experimentally observed in benzonitrile, however, it is very close to the rate observed in acetonitrile. The electron transfer rate is very fast compared with the solvation time of 4.76 ps of benzonitrile and is comparable with the solvation time in acetonitrile. Therefore, in benzonitrile, back electron transfer from the excited-state will take place with the solvent essentially "frozen" in the ground-state configuration. If the rate is calculated for a driving force of $12,500\,\mathrm{cm}^{-1}$ (the energy of one 800 nm pump photon), the result is $(1.3\,\mathrm{ps})^{-1}$, which is close to the measured value in benzonitrile. It thus appears that the rate equations derived in this paper in combination with the data obtained from the absorption spectrum can correctly predict both very small and extremely large rates of electron transfer.

III. COHERENCE IN ELECTRON TRANSFER

A. General Aspects of Coherence

The observation of vibrational coherence in the stimulated emission from the reactant state of the bacterial photosynthetic reaction center [35,140,141] prior to ultrafast primary electron transfer, sparked a whole series of experimental and theoretical studies of the role of coherence in chemical and biological reactions. Vibrational coherence related to such reactions has now been observed in several molecules of biological significance [142]. In recent experiments [143] wavepacket motion related to heme doming and iron–histidine motion was observed after the dissociation of NO from myoglobin–NO. Of considerable interest is coherent nuclear motion in the product state produced by the reaction, as was observed in the initial isomerization of rhodopsin [144–147] as well as bacteriorhodopsin [144]. Reactive motions are now known to generate coherent vibrational states for small molecules in solution, for example, in the photodissociation of I_3- into I_2- and I, the I_2-fragment is produced in a vibrationally coherent state [148–150] and similarly for the photodissociation of HgI_2 into coherently vibrating HgI and I [151,152]. Vibrational coherence was also observed in the cis-trans isomerization reaction of stilbene in solution [153]. The creation of coherence in products of reactions, known for some time in gases [154–157], may be a common feature of narrow-barrier crossing reactions and will be observed more frequently as more condensed phase reactions are studied with appropriate time-resolution. Electronic coherence, such as that induced by a short excitation pulse between two or more degenerate or nearly degenerate electronic states [86,87,158–160], can also influence the rate of a condensed-phase reaction if these states are short-lived transition states for a reaction [45].

The experimental observation of vibrational coherence in chemical and biological reactions has led to some theoretical studies. For example, Redfield and similar theories have been used to make predictions regarding barrierless coherent electron transfers [98,99,161–163] and other reactions [97,106,164–166]. However, in the condensed phase there are no experimental observations of vibrational coherence being introduced into the reactant state by a light pulse and then influencing the formation of the product of the reaction. For example, the reactant state vibrational coherence in the bacterial photosynthetic reaction center [35,140,141] is not known to have any effect on the rate of formation of the charge-separated product state.

It is important to define clearly the possible influences on a chemical reaction arising from having a coherent initial (reactant) state. There are two categories to consider:

1. The reaction products may exhibit a *modulation* at reactant vibrational frequencies because they can be generated only when the reactant-state wavepacket, propagating freely under the influence of the molecular Hamiltonian, contains nuclear coordinates that fall within a specific restricted range. For example, the wavepacket might have to be near a region of the surface where there is curve crossing. In this case the rates averaged over many cycles of the wavepacket motion are expected to be the same as those found for conventional nonimpulsive excitation. Examples of this effect are known from gas-phase [155] and solution-phase [148–153] experiments.

2. The time-averaged photochemistry of the wavepacket created by a pulse of light, as gauged by quantum yield or product-state distribution or nature of chemical products or average rate constant, may be *different* from that of a conventionally excited system. Examples of this behavior are the control experiments or other multiple-pulse or nonlinear optical methods in which pulse sequences are used that *interrupt* what we referred to as the free propagation of the wavepacket [167–170]. We do not consider these experimental configurations here. Nevertheless, the question is often asked [147] whether the chemistry in a coherent ensemble prepared conventionally by one short light pulse might differ from that of a system that has no coherence, simply because each molecule is in a superposition state? Clearly, the answer is negative if by chemistry we mean only the average populations of the states of the various species involved.

In electron transfer reactions, coherence can come in two guises: electronic or vibrational. Electronic coherence could occur if the excitation pulse created a superposition between the DA and D^+A^- (or their adiabatic counterparts) states. In direct electron transfer this is an unlikely occurrence since the two relevant states are separated by a very large energy gap (on the order of $10,000 \, cm^{-1}$ or larger). In indirect electron transfer (i.e., in a three-level system [40] consisting of a ground-state and DA and D^+A^- excited states), a short pulse could more readily excite an electronic coherent superposition. However, the superposition will get damped by coupling to the solvent. The time scale of this damping is obtained from the inverse of the FWHM of a typical charge transfer band. This suggests typical charge transfer transition dephasing times are on the order of 5 fs. Therefore, it is not expected that electronic coherence will have a significant effect on the reaction dynamics. To get a better understanding of the role that vibrational coherence might play in electron transfer, we should go back to Eq. (2.24) and similar equations that give the reaction rate. In the derivation of this equation it is assumed that the system is vibrationally at equilibrium in the initial electronic state $|1\rangle$ [Figure 7(a)]. This assumption leads to a decay of the initial state population that is essentially exponential with time, and hence a rate constant is well defined. However, if the initial state is prepared with a sufficiently short pulse, a superposition of vibrational states will be created. The generated wavepacket will then move on the surface of state $|1\rangle$. This wavepacket motion can lead to a modulation of the reaction rate as in category (1) above. It would be straightforward to derive an equation similar to Eq. (2.24) under the assumption that the excitation pulse is much shorter than all other time scales in the problem. This is, however, a situation that is hard to realize experimentally at present. In an experiment it is most likely that some of the lower-frequency modes are excited coherently, while the remaining high frequency modes are not impulsively driven because they undergo a few oscillations during the excitation. In the theoretical description these high frequency modes could be treated as described

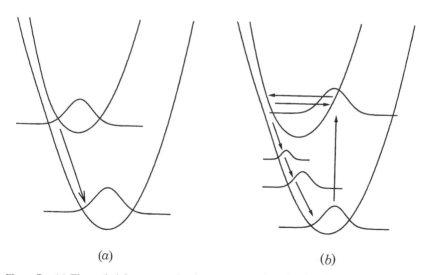

(a)   (b)

Figure 7. (a) The underlying assumption in most expressions for the electron transfer rate is that the initial state is at vibrational equilibrium. This results in a nearly exponential decay of the excited-state population. (b) If the initial state is prepared out of equilibrium by a short pulse, the rate may be modulated by wavepacket motion on the initial state surface.

by the methods given above and special "coherent" treatment given to the low-frequency modes. The need for this distinction between high- and low-frequency modes will disappear when tunable pulses of a few-femtoseconds duration become used routinely in experiments. An important additional point regarding the observation of coherent wavepacket motion is that it implies the existence of a nonequilibrium population distribution of vibrational states. This distribution will contain not only those modes that were excited impulsively but also the high-frequency modes having significant shifts. Furthermore, *the vibrational energy relaxation times of the high-frequency modes of most molecules will be long compared with the vibrational periods of the low-frequency modes* [171]. Thus electron transfer processes occurring on the time scale of the low-frequency internal motions *must necessarily be occurring from a nonequilibrium distribution that includes the lower-frequency modes whose coherent effects are being observed as well as any high-frequency modes that were excited.* It follows that in theoretical descriptions of picosecond or shorter time scale electron transfer kinetics that display quantum oscillations in either the reactant or product states the equilibrium approximation is not the proper initial condition.

B.   Model System Exhibiting Coherence in Electron Transfer: TCNE–Pyrene

Sometime ago [28] we stressed how useful it would be to have some model systems that would illustrate the principles needed to understand vibrational coherence transfer occurring concomitantly with an electron jumping from a donor to an acceptor. For the reasons given above, the Mulliken charge transfer complexes have turned out to be excellent models. Not only can they be essentially two- state systems, but the electron transfer is fast enough that the final states can be examined in spectroscopic experiments before the vibrational coherence has dephased.

In a recent paper [30] we presented the results of an experimental study of electron transfer in the EDA complex between TCNE and pyrene, with significantly better time resolution than previous studies [96,172,173]. The electron transfer times observed in this study are summarized in Table IV. The dynamics of electron transfer in TCNE–pyrene was observed by studying the charge transfer band bleaching recovery dynamics at 810 nm, transient absorption from the TCNE anion at 405 nm, and stimulated emission from the charge transfer state back to the ground state at 1.2 μm. The transients in benzonitrile solvent at 810 nm and 1.2 μm are shown in Figure 8. Coherent oscillations were seen, both in the ground-state absorption bleach signal and in the gain signal, from a vibrational mode of the TCNE–pyrene complex with a frequency of about 165 cm^{-1}. The phase of the oscillations observed in the bleaching signal depends on the solvent used,

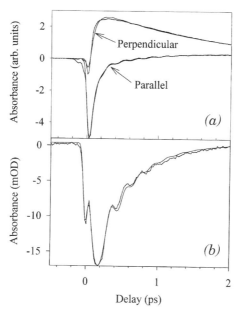

Figure 8. (*a*) First 2 ps of the transient ground-state absorption bleach of TCNE–pyrene in benzonitrile pumped and probed at 810 nm. The two curves correspond to the signals with the pump and probe pulses polarized parallel and perpendicular to each other. The data show an oscillation with a frequency of approximately 165 cm^{-1}. (*b*) As in (*a*) but probed at 1.2, at which the emission stimulated from the excited-state is observed. The oscillations in the data at approximately 165 cm^{-1} correspond to wavepacket motion on the excited-state surface.

and its value is inconsistent with impulsive stimulated Raman scattering [174]. The coherent oscillations in the stimulated emission show that the electron transfer reaction (from the excited charge transfer state back to the ground-state) is vibrationally coherent, and this suggests that the oscillations seen in the ground-state absorption bleach signal are caused by a coherent repopulation of the ground-state surface.

The coherent electron transfer in the TCNE–pyrene complex has been modeled with a simple classical Monte Carlo simulation, as described in the following section, that explicitly simulates the 165 cm^{-1} coordinate and incorporates the other vibrational modes in an appropriate quantum mechanical expression for the jump probability. This simulation, however simple, reproduces the essential features of the experimentally observed ground-state absorption bleach recovery signal, thereby corroborating the interpretation that the electron transfer reaction rate is modulated by vibrational coherence.

C.   Simulation of Reactive Coherence Transfer in Polyatomic Molecules

The vibrational coherence that is observed in the stimulated emission signal of TCNE–pyrene at 1.2 µm (see Fig. 8) indicates that the electron transfer reaction is in the coherent regime. The question is whether the oscillations that are seen in the bleach recovery dynamics observed at 810 nm are due to a modulation of the electron transfer rate as a result of this vibrational coherence. At first glance one would expect to see a stepwise increase in the ground-state population. As the vibrational wavepacket sloshes back and forth in the excited-state well, some population might "wash over the barrier" once every vibrational period and cause a sudden increase in the ground-state population. Experimentally (in the ground-state bleach recovery signal at 810 nm), one observes an approximately exponential increase in the ground-state population with a small modulation that appears to be sinusoidal rather than stepwise. This appears to be inconsistent with the interpretation of the electron transfer rate being modulated by the vibrational wavepacket motion. However, the simple picture given above of a wavepacket sloshing about and washing over the barrier will break down in any system larger than a diatomic. This can be seen quite easily in a model system with two harmonic modes. The barrier height and position along the coordinate of the first mode will clearly depend on the value of the coordinate of the second mode. In the inverted regime of electron transfer, quantum mechanical tunneling will take place that will effectively make the barrier height and position along the first coordinate dependent on the value of the second coordinate in the product state directly after the transition. Thus in multimode system the barrier along the coordinate of one mode can take on a whole range of values depending on the properties of the other Franck–Condon active modes.

A Monte Carlo simulation was performed to get a better understanding of vibrationally coherent electron transfer reactions. The ground- and excited-state potentials of TCNE–pyrene were approximated by harmonic wells (Fig. 9), with the driving force and reorganization energy chosen such that the electron transfer reaction is in the inverted regime [55] as it is in all the EDA complexes discussed in this chapter. A wavepacket of classical particles is set up with a distribution of positions and momenta such that the classical wavepacket oscillates back and forth in the excited (reactant) state well. During the simulation a function $P(q)$ (Fig. 9) determines the probability for each particle to jump to the ground-state surface, and the total population in the ground-state as a function of time is saved. The jump-probability function is chosen as

$$P(x) = \begin{cases} P_0 \exp[-\Delta(x)/C] & \Delta(x) \geq 0 \\ 0 & \Delta(x) < 0 \end{cases} \qquad (3.1)$$

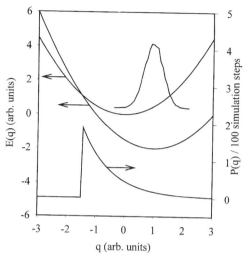

Figure 9. Model used in the Monte Carlo simulation described in the text. The parabolic curves are the free energy curves of the $|DA\rangle$ and $|D^+A^-\rangle$ states of an EDA complex as a function of the (reduced) solvation coordinate. The single-sided exponential curve is a plot of the jump-probability function described in the text. The bell-shaped curve is the initial position distribution used in the simulation.

where $\Delta(x)$ [see Eq. (2.2)] is the difference in energy between the two potentials. This jump-probability function, roughly based on the energy gap law, [175] has the general form expected for a nonadiabatic transition [54,55,175,176] and implicitly models the effect of a number of high-frequency accepting modes. In this model, if the parameter C is chosen small (i.e., if the jump-probability function decays quickly as a function of the energy gap), one indeed observes that the population of the ground-state increases in a stepwise manner. The TCNE–pyrene EDA complex is expected to have as many as ten accepting modes (in addition to the $165 \, cm^{-1}$ and the solvent mode), in analogy to the TCNE–HMB complex [13,15–17], and therefore C is expected to be rather large. From the slope of the curve of the electron transfer rate versus driving force in Figure 5, one can estimate the value of C as approximately $1000 \, cm^{-1}$. For large values of C the jump-probability function decays slowly as a function of the energy gap, in other words, the population can flow back to the ground-state at a nearly constant rate for all positions of the vibrational wavepacket. Consequently, the total population in the ground-state is found to increase approximately exponentially with small oscillations superimposed on this rise. Although these oscillations are formally "stepwise," they are not easily distinguished from sinusoidal in an experiment with a limited signal-to-noise

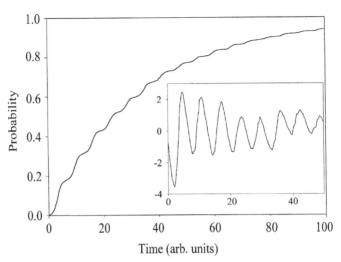

Figure 10. Result of a classical Monte Carlo simulation of vibrationally coherent electron transfer in the inverted regime (see text). Shown is the probability for the system to return to the ground state of the EDA complex as a function of time. Wavepacket motion in the excited state causes a modulation of the instantaneous electron transfer rate, resulting in a stepwise increase in the ground-state population. The inset shows the residuals ($100\times$) of an exponential fit to the population dynamics.

ratios. Figure 10 shows the result of a simulation with the parameters $C = 1.0$ (in reduced units) and $P_0 = 0.02/$(simulation step). It can be seen that the probability for the system to be in the ground-state rises exponentially as function of time and that this rise is modulated with a stepwise oscillation. The inset in Figure 10 shows the residuals of an exponential fit to the simulation result, which are nearly sinusoidal. It is found that the phase of the oscillation is a function of the driving force of the reaction. Thus it can be expected that a change of solvent will lead to a change of phase in the oscillation, as was experimentally observed [30]. The sinusoidal modulation observed in the ground-state absorption bleach signal of TCNE–pyrene at 810 nm is therefore seen to be consistent with a coherent modulation of the electron transfer rate.

D.   Coherent Effects in Electron Transfer: Qualitative Considerations

In Section III.C we discussed manifestations of vibrational coherence in electron transfer. However, the observation of the coherence provides information regarding a number of factors that would not be known if the coherent effects were not seen. The most obvious type of new information concerns the vibrational spectrum. In the solution phase at ambient temperatures electronic spectra provide very little information about the vibrational frequencies in the reactive, electronically excited-states. However, the electronic dephasing, which obliterates this information under linear absorption conditions, involves only the dephasing of states that are coupled directly by the optical fields. After the coupled electronic level pairs have completely dephased, there remains a vibrational coherence that normally will dephase on a much slower time scale. Therefore, in nonlinear absorption, where this coherence is probed by an additional field, the vibrational states may be examined with high accuracy even in the solution phase. More simply put, the excitation process generates an ensemble with a *polarizability* that oscillates at the vibrational frequencies because all the molecules began their motion at about the same region in nuclear configuration space. This polarizability will generate a detectable induced dipole when it is coupled to a probe optical field.

In large molecules the optical spectrum is usually relatively insensitive to the vibrational energy distribution in the initial electronic state. This is because there are often only small dimensionless shifts along the high-frequency modes with only the low-frequency modes forming significant progressions in the spectrum. When combined with large electronic dephasing, these conditions result in the spectra of the high-frequency modes being very insensitive to which vibrational quantum level forms the initial state. Therefore, the observation of coherence in pump-probe experiments provides otherwise unknown information regarding the vibrational energy distributions of the reactive states reached by optical excitation. For example, in the reaction center the oscillations [35] were found to persist for times comparable with the electron transfer process. This not only proves that the electron transfer is occurring from a nonthermal distribution of vibrational states but provides a method for the precise determination of that distribution. This vibrational *population* information would not be obtainable if the vibrations were not impulsively excited.

As mentioned earlier, the observation of vibrational coherence in the product state of a chemical reaction indicates how the coherence is transferred into the products by the forces acting on the nuclei as the system passes through the transition state. This is made particularly clear in the photodissociation of triatomic molecules, of which HgI_2 and I_3^- are examples, where one can see from the potential surface [177] how the coherence in the products is formed by the trajectories across the transition-state region. The time of crossing to the products (HgI or I_2^-) with their nuclei extended is close to half a vibrational period of the product molecule. Furthermore, the curvature of the surface forces the nuclei to maintain a relatively narrow distribution of internuclear distances while the system moves into the product channel. This ensures that the wavepacket that is formed is much narrower than the spread of bond lengths that would exist for a typical eigenstate at the mean energy of the vibrational distribution. In electron transfer processes the situation must be analogous. The vibrational coherence in the product state, which in our case is the charge recombined ground-state, will be observed through those modes for which a sufficiently narrow coordinate distribution remains immediately after the charge transfer. These distributions may be created in cyclic processes that are determined by the vibrational periods in the reactant state, and they are clearly dependent on the pulse width used in the experiment. The oscillations of modes having periods faster than the time resolution of the experiment are suppressed by the distribution of initial conditions created by the finite time scale of the excitation process. If we could observe this coherent distribution, including all the vibrational modes coupled to the electron transfer, it would represent a direct determination of the vibrational information needed to describe the electron transfer free from the caveats associated with the data from Raman spectroscopy. Of course, in this picture there is no electron transfer rate constant in the sense of some exponential process, because the system is prepared impulsively. The electron transfer occurs efficiently only when the coupled modes are optimally displaced on the reactant potential surface.

IV. CONCLUSIONS

The Marcus theory of electron transfer has been applied in many instances and has generally been successful. The rate versus driving force does have a normal and inverted regime [53], and this relation appears to be valid over as much as 13 orders of magnitude [178]. Given this success, it is now appropriate to try to calculate absolute rates of electron transfer based on separately obtainable parameters. This is what we have done here for the TCNE–HMB complex in polar and nonpolar solvents based on resonance Raman data of this complex in CCl_4 [13,16,17]. It was found [28] that if the diabatic Hush-type coupling, obtained from the charge transfer absorption spectrum using Eq. (2.41), is used directly to calculate the rate of electron transfer, the values obtained are orders of magnitude higher than those observed experimentally. We demonstrated here that by extracting the non-Born–Oppenheimer coupling between adiabatic states from the data, rates can be calculated that agree well with experiment. We found that the Huang–Rhys factors of TCNE–HMB were useful for the similar complexes TCNB–HMB and TCNE–pyrene. In this way we have been able to predict correctly electron transfer rates that vary over nearly six orders of magnitude. Two important caveats are evident from this work [28]. First, the resonance Raman excitation profiles are almost exclusively modeled under the assumption of harmonic potentials, and hence harmonic potentials are also used to calculated the electron transfer rates. Clearly, the assumption of harmonic potentials in the rate theories must break down for very large driving forces. Second, the Franck–Condon factors of high-frequency modes decay more slowly with driving force than those of low-frequency modes. Therefore, the high-frequency modes will be the dominating accepting modes at very large driving force (estimated [28] as $> 15 \times 10^3 \, cm^{-1}$), and special care has to be taken that the dimensionless shifts of these modes are measured accurately with resonance Raman scattering.

There has been a great deal of interest in the role of vibrational coherence in (chemical) reactions in the condensed phase. From our studies of vibrational wavepacket motion in TCNE–pyrene it can be concluded that vibrational coherence modulates the rate of back electron transfer but that the effect can be small for multimode electron transfer processes. From Monte Carlo simulations of a simple model we showed that a polyatomic system is likely to have a large number of Franck–Condon active modes that have a tendency to "smear out" the probability for a reactive jump as a function of the reaction coordinate. Of course, in theory, it is possible to have a reactive transition in a polyatomic system with only a single "dominating" mode. The creation of vibrational coherence is this dominating mode would then have a large effect, and the rate of electron transfer could be strongly modulated by this coherence. This is clearly the case with the photodissociation of diatomics such as the alkali halides [154,155].

This work has implications for the role of vibrational coherence in biological electron transfer (i.e., in the photosynthetic reaction center of *Rhodobacter. sphaeroides* [35,140,141]. It might be that biological electron transfer is somehow optimized to have only a single or a limited number of dominating modes. It has been suggested [179] that the reorganization energy for primary electron transfer in the photosynthetic reaction center of *Rhodobacter. capsulatus* is smaller than $500 \, cm^{-1}$, which might mean that there are only a small number of dominating modes. However, the way to prove that vibrational coherence is important (in the sense that it signals the dynamics in the transition-state region) in biological electron transfer is by monitoring the formation of the product state. Unfortunately, the visible excited-state absorption spectrum of the photosynthetic reaction center has overlapping absorptions from both the reactant and the product state. The near-infrared transient absorption spectrum [38] of the photosynthetic reaction center is less spectrally congested. For example, the triplet–doublet absorption in the product state at $8000 \, cm^{-1}$ ($1.2 \, \mu m$) [38,180,181] appears to be free from overlap with other transitions and could be used to monitor the formation of the product state. However in evaluating the results of such an experiment still another set of dimensionless shifts has to be considered.

ACKNOWLEDGMENTS

We would like to acknowledge the help of Chris Galli and Gavin D. Reid in taking some of the data presented here, Anne B. Myers for discussion of resonance Raman scattering and wavepacket analysis, and Joshua Jortner for discussions on the theory of reaction dynamics. This work was supported by NIH and NSF.

REFERENCES

1. K. H. Michaelian, K. E. Rieckhoff, and E. M. Voigt, *Chem. Phys. Lett.* **23**, 5 (1973).

2. B. I. Swanson, *Inorg. Chem.* **15**, 253 (1976).

3. K. H. Michaelian, K. E. Rieckhoff, and E. M. Voigt, *Chem. Phys. Lett.* **39**, 521 (1976).

4. M. L. Smith and J. L. McHale, *J. Phys. Chem.* **89**, 4002 (1985).

5. H. E. Toma, P. S. Santos, and A. B. P. Lever, *Inorg. Chem.* **27**, 3850 (1988).

6. H. E. Toma, E. Stadler, and P. S. Santos, *Spectrochim. Acta* **44A**, 1365 (1988).

7. S. K. Doorn and J. T. Hupp, *J. Am. Chem. Soc.* **111**, 1142 (1989).

8. S. K. Doorn and J. T. Hupp, *J. Am. Chem. Soc.* **111**, 4704 (1989).

9. J. L. McHale and M. J. Merriam, *J. Phys. Chem.* **93**, 526 (1989).

10. R. L. Blackbourn, C. S. Johnson, J. T. Hupp, M. A. Bryant, R. L, Sobocinski, and J. E., Pemberton, *J. Phys. Chem.* **95**, 10535 (1991).

11. S. K. Doorn, R. L. Blackbourn, C. S. Johnson, and J. T. Hupp, *Electrochim. Acta* **36**, 1775 (1991).

12. D. J. Stufkens, T. L. . Snoeck, W. Kaim, T. Roth, and B. Olbrich-Deussner, *J. Organomet. Chem.* **409**, 189 (1991).

13. F. Markel, N. S. Ferris, I. R. Gould, and A. B. Myers, *J. Am. Chem. Soc.* **114**, 6208 (1992).

14. B. M. Britt, H. B. Lueck, and J. L. McHale, *Chem. Phys. Lett.* **190**, 528 (1992).

15. A. B. Myers, *Chem. Phys.* **180**, 215 (1994).

16. K. Kulinowski, I. R. Gould, and A. B. Myers, *J. Phys. Chem.* **99**, 9017 (1995).

17. K. Kulinowski, I. R. Gould, N. S. Ferris, and A. B. Myers, *J. Phys. Chem.* **99**, 17715 (1995).

18. B. M. Britt, J. L. McHale, and D. M. Friedrich, *J. Phys. Chem.* **99**, 6347 (1995).

19. D. C. Arnett, P. Vohringer, and N. F. Scherer, *J. Am. Chem. Soc.* **117**, 12262 (1995).

20. J. L. Wootton and J. I. Zink, *J. Phys. Chem.* **99**, 7251 (1995).

21. H. Lu, V. Petrov, and J. Hupp, *Chem. Phys. Lett.* **235**, 521 (1995).

22. B. W. Pfennig, Y. Wu, R. Kumble, T. G. Spiro, and A. B. Bocarsly, *J. Phys. Chem.* **100**, 5745 (1996).

23. G. C. Walker, P. F. Barbara, S. K. Doorn, Y. Dong, and J. T. Hupp, *J. Phys. Chem.* **95**, 5712 (1991).

24. K. Tominaga, D. A. V. Kliner, A. E. Johnson, N. E. Levinger, and P. F. Barbara, *J. Chem. Phys.* **98**, 1228 (1993).

25. P. J. Reid, C. Silva, P. F. Barbara, L. Karki, and J. T. Hupp, *J. Phys. Chem.* **99**, 2609 (1995).

26. S. Ojima, H. Miyasaka, and N. Mataga, *J. Phys. Chem.* **94**, 4147 (1990).

27. N. Mataga, S. Nishikawa, T. Asahi, and T. Okada, *J. Phys. Chem.* **94**, 1443 (1990).

28. K. Wynne, C. Galli, and R. M. Hochstrasser, *J. Chem. Phys.* **100**, 4797 (1994).

29. W. Jarzeba, R. E. Schlief, and P. F. Barbara, *J. Phys. Chem.* **98**, 9102 (1994).

30. K. Wynne, G. D. Reid, and R. M. Hochstrasser, *J. Chem. Phys.* **105**, 2287 (1996).

31. (a) Y. Kimura, Y. Takebayashi, and N. Hirota, *Chem. Phys. Lett.* **257**, 429 (1996); (b) I. V. Rubtsov and K. Yoshihara, *J. Phys. Chem. A* **101**, 6138 (1997).

32. C. Kirmaier and D. Holten, *Proc. Natl. Acad Sci. USA* **87**, 3552 (1990).

33. C. K. Chan, T. J. DiMagno, L. X. Q. Chen, J. R. Norris, and G. R. Fleming, *Proc. Natl. Acad. Sci. USA,* **88**, 11202 (1991).

34. T. Arlt, S. Schmidt, W. Kaiser, C. Lauterwasser, M. Meyer, H. Scheer, and W. Zinth, *Proc. Natl. Acad. Sci. USA* **90**, 11757 (1993).

35. M. H. Vos, M. R. Jones, C. N. Hunter, J. Breton, J.-C. Lambry, and J.-L. Martin, *Biochemistry,* **33**, 6750 (1994).

36. G. C. Walker, S. Maiti, B. R. Cowen, C. C. Moser, P. L. Dutton, and R. M. Hochstrasser, *J. Phys. Chem.* **98**, 5778 (1994).

37. J. M. Peloquin, S. Lin, A. K. W. Taguchi, and N. Woodbury, *J. Phys. Chem.* **99**, 1349 (1995).

38. K. Wynne, G. Haran, G. D. Reid, C. C. Moser, P. L. Dutton, and R. M. Hochstrasser, *J. Phys. Chem.* **100**, 5140 (1996).

39. T. Kobayashi, Y. Takagi, H. Kandori, K. Kemnitz, and K. Yoshihara, *Chem. Phys. Lett.* **180**, 416 (1991).

40. K. Tominaga, G. C. Walker, W. Jarzeba, and P. F. Barbara, *J. Phys. Chem.* **95**, 10475 (1991).

41. R. J. Sension, A. Z. Szarka, G. R. Smith, and R. M. Hochstrasser, *Chem. Phys. Lett.* **185**, 179 (1991).

42. G. C. Walker, E. Åkeson, A. E. Johnson, N. E. Levinger, and P. F. Barbara, *J. Phys. Chem.* **96**, 3728 (1992).

43. A. Yartsev, Y. Nagasawa, A. Douhal, and K. Yoshihara, *Chem. Phys. Lett.* **207**, 546 (1993).

44. E. Tuite, J. M. Kelly, G. S. Beddard, and G. S. Reid, *Chem. Phys. Lett.* **226**, 517 (1994).

45. K. Wynne, S. M. LeCours, C. Galli, M. J. Therien, and R. M. Hochstrasser, *J. Am. Chem. Soc.* **117**, 3749 (1995).

46. S. K. Doorn, P. O. Stoutland, R. B. Dyer, and W. H. Woodruff, *J. Am. Chem. Soc.* **114**, 3133 (1992).

47. S. K. Doorn, R. B. Dyer, P. O. Stoutland, and W. H. Woodruff, *J. Am. Chem. Soc.* **115**, 6398 (1993).

48. S. Maiti, G. C. Walker, B. R. Cowen, R. Pippenger, C. C. Moser, P. L. Dutton, and R. M. Hochstrasser, *Proc. Natl. Acad. Sci. USA* **91**, 10360 (1994).

49. K. G. Spears, X. Wen, and S. M. Arrivo, *J. Phys. Chem.* **98**, 9693 (1994).

50. P. Hamm, M. Zurek, W. Mäntele, H. Scheer, and W. Zinth; *Proc. Natl. Acad. Sci. USA* **92**, 1826 (1995).

51. P. Hamm and W. Zinth, *J. Phys. Chem.* **99**, 13537 (1995).

52. K. Spears, X. Wen, and R. Zhang, *J. Phys. Chem.* **100**, 10206 (1996).

53. R. A. Marcus and N. Sutin, *Biochim. Biophys. Acta* **811**, 265 (1985).

54. H. Sumi and R. A. Marcus, *J. Chem. Phys.* **84**, 4894 (1986).

55. J. Jortner and M. Bixon, *J. Chem. Phys.* **88**, 167 (1988).

56. M. D. Newton and N. Sutin, *Annu. Rev. Phys. Chem.* **35**, 437 (1984).

57. P. F. Barbara, T. J. Meyer, and M. A. Ratner, *J. Phys. Chem.* **100**, 13148 (1996).

58. R. Kubo, in *Fluctuation, Relaxation and Resonance in Magnetic Systems*, D. ten Haar, Scottish Universities Summers School, Edinburgh, 1961, p. 23.

59. A. B. Myers, R. A. Mathies, D. J. Tannor, and E. J. Heller, *J. Chem. Phys.* **77**, 3857 (1982).

60. A. B. Myers, R. A. Harris, and R. A. Mathies, *J. Chem. Phys.* **79**, 603 (1983).

61. A. B. Myers and R. A. Mathies, *J. Chem. Phys.* **81**, 1552 (1984).

62. A. B. Myers, M. O. Trulson, and R. A. Mathies, *J. Chem. Phys.* **83**, 5000 (1985).

63. E. J. Heller, *J. Chem. Phys.* **62**, 1544 (1975).

64. E. J. Heller, R. L. Sundberg, and D. Tannor, *J. Phys. Chem.* **86**, 1822 (1982).

65. D. J. Tannor and E. J. Heller, *J. Chem. Phys.* **77**, 202 (1982).

66. R. D. Coalson and M. Karplus, *J. Chem. Phys.* **79**, 6150 (1983).

67. Y.-J. Yan and S. Mukamel, *J. Chem. Phys.* **85**, 5908 (1986).

68. A. B. Myers and R. A. Mathies, in *Biological Applications of Raman Spectroscopy*, Vol. 2, T. G. Spiro, ed., Wiley, New York, 1987, p. 1.

69. R. D. Coalson, *Chem. Phys. Lett.* **147**, 208 (1988).

70. M. V. R. Krishna and R. D. Coalson, *Chem. Phys.* **120**, 327 (1988).

71. E. Simoni, C. Reber, D. Talaga, and J. I. Zink, *J. Phys. Chem.* **97**, 12678 (1993).

72. A. B. Myers, in *Laser Techniques in Chemistry*, Vol. 23, A. B. Myers and T. R. Rizzo, eds., Wiley, New York, 1995, p. 325.

73. D. E. Khoshtariya, A. M. Kjaer, T. A. Marsagishvili, and J. Ulstrup, *J. Phys. Chem.* **95**, 8797 (1991).

74. N. S. Hush, *Electrochim. Acta* **13**, 1005 (1968).

75. R. J. Cave and M. D. Newton, *Chem. Phys. Lett.* **249**, 15 (1996).

76. B. Li, A. E. Johnson, S. Mukamel, and A. B. Myers, *J. Am. Chem. Soc.* **116**, 11039 (1994).

77. D. R. Yarkony, *J. Phys. Chem.* **100**, 18612 (1996).

78. C. Cohen-Tannoudji, B. Diu, and F. Laloe, *Quantum Mechanics*, Wiley, New York, 1977.

79. S. Larsson, *Theor. Chim. Acta* **60**, 111 (1981).

80. M. Born and K. Huang, *Dynamical Theory of Crystal Lattices*, Clarendon Press, Oxford, 1954.

81. A. Nitzan and J. Jortner, *J. Chem. Phys.* **56**, 3360 (1972).

82. K. F. Freed, in *Radiationless Processes in Molecules and Condensed Phases*, F. K. Fong, ed., Springer-Verlag, Berlin, 1976.

83. N. G. van Kampen, *Stochastic Processes in Physics and Chemistry*, North-Holland, Amsterdam, 1992.

84. J. W. Goodman, *Statistical Optics*, Wiley, New York, 1985.

85. H. Haken and P. Reineker, *Z. Physik*, **249**, 253 (1972).

86. K. Wynne and R. M. Hochstrasser, *Chem. Phys.* **171**, 179 (1993).

87. K. Wynne and R. M. Hochstrasser, *J. Raman Spectrosc.* **26**, 561 (1995).

88. A. G. Redfeld, *Adv. Magn. Reson.* **1**, 1 (1965).

89. R. Wertheimer and R. Silbey, *Chem. Phys. Lett.* **75**, 243 (1980).

90. F. E. Figueirido and R. M. Levy, *J. Chem. Phys.* **97**, 703 (1992).

91. J. S. Bader and B. J. Berne, *J. Chem. Phys.* **100**, 8359 (1994).

92. G. A. Krestov, N. P. Novosyolov, I. S. Perelygin, A. M. Kolker, L. P. Safonova, V. D. Ovchinnikova, and V. N. Trostin, *Ionic Solvation*, Ellis Horwood, New York, 1994.

93. I. R. Gould, R. H. Young, R. E. Moody, and S. Farid, *J. Phys. Chem.* **95**, 2068 (1991).

94. G. Briegleb, *Elektronen Donator Acceptor Komplexe*, Springer-Verlag, Berlin, 1961.

95. R. S. Mulliken and W. B. Person, *Molecular Complexes*, Wiley, New York, 1969.

96. T. Asahi and N. Mataga, *J. Phys. Chem.* **95**, 1956 (1991).

97. J. M. Jean, G. R. Fleming, and R. A. Friesner, *Ber. Bunsenges. Phys. Chem.* **95**, 253 (1991).

98. J. M. Jean, R. A. Friesner, and G. R. Fleming, *J. Chem. Phys.* **96**, 5827 (1992).

99. J. M. Jean and G. R. Fleming, *J. Chem. Phys.* **103**, 2092 (1995).

100. K.-A. Suominen, B. M. Garraway, and S. Stenholm, *Opt. Commun.* **82**, 260 (1991).

101. J. P. Davis and P. Pechukas, *J Chem. Phys.* **64**, 3129 (1976).

102. J.-T. Hwang and P. Pechukas, *J. Chem. Phys.* **67**, 4640 (1977).

103. B. W. Shore, *The Theory of Coherent Atomic Excitation*, Wiley, New York, 1990.

104. L. D. Landau and E. M. Lifshitz, *Quantum Mechanics (Non-relativistic Theory)*, 3rd ed., Pergamon Press, Oxford, 1977.

105. H. Sumi, *J. Phys. Soc. Jpn.* **49**, 1701 (1980).

106. J. N. Onuchic and P. G. Wolynes, *J. Phys. Chem.* **92**, 6495 (198

107. Y. Bu and C. Deng, *J. Phys. Chem.* **100**, 18093 (1996).

108. M. Chattoraj, S. L. Laursen, B. Paulson, D. D. Chung, G. L. Closs, and D. H. Levy, *J. Phys. Chem.* **96**, 8778 (1992).

109. W. H. Louisell, *Quantum Statistical Properties of Radiation*, Wiley, New York, 1973.

110. J. Sue, Y.-J. Yan, and S. Mukamel, *J. Chem. Phys.* **85**, 462 (1986).

111. R. A. Marcus and N. Sutin, *Biochim. Biophys. Acta* **811**, 265 (1986).

112. G. v. d. Zwan and J. T. Hynes, *J. Phys. Chem.* **89**, 4181 (1985).

113. S. Chandrasekhar, *Rev. Mod. Phys.* **15**, 1 (1943).

114. S. A. Adelman, *Adv. Chem. Phys.* **44**, 143 (1980).

115. Y.-J. Yan and S. Mukamel, *J. Chem. Phys.* **89**, 5160 (1988).

116. P. F. Barbara and W. Jarzeba, *Adv. Photochem.* **15**, 1 (1990).

117. J. D. Simon, *Acc. Chem. Res*, **21**, 128 (1988).

118. M. Maroncelli, J. MacInnis, and G. R. Fleming, *Science*, **243**, 1674 (1989).

119. W. H. Press, S. A. Teukolsky, W. T. Vetterling, and B. P. Flannery, *Numerical Recipes in C*, 2nd ed., Cambridge University Press, New York, 1995.

120. M. Saheki, H. Yamada, H. Yoshioka, and K. Nakatsu, *Acta Crystallogr. B* **32**, 662 (1976).

121. E. Maverick, K. N. Trueblood, and D. A. Bekoe, *Acta Crystallogr. B* **34**, 2777 (1978).

122. E. F. Caldin, J. E. Crooks, D. O Donnell, D. Smith, and S. Toner, *Chem. Soc. J. Faraday Trans. Pt. 1* **68**, 849 (1972).

123. M. Rossi, U. Buser, and E. Haselbach, *Helv. Chim. Acta* **59**, 1039 (1976).

124. W. Liptay, T. Rehm, D. Wehning, L. Schanne, W. Baumann, and W. Lang, *Z. Naturforsch. A* **37**, 1427 (1982).

125. B. Ware, K. Williamson, and J. P. Devlin, *J. Phys. Chem.* **72**, 3970 (1968).

126. C. J. Eckhardt and R. J. Hood, *J. Am. Chem. Soc.* **101**, 6170 (1979).

127. M. Tanaka, *Bull. Chem. Soc. Jpn.* **50**, 2881 (1977).

128. M. Rossi and E. Haselbach, *Helv. Chim. Acta,* **62**, 140 (1979).

129. C. W. Jurgensen, M. J. Peanasky, and H. G. Drickamer, *J. Chem. Phys.* **83**, 6108 (1985).

130. M. Kroll, *J. Am. Chem. Soc.* **90**, 1097 (1968).

131. J. Cioslowski, *Tetrahedron,* **42**, 735 (1986).

132. R. D. Coalson, D. G. Evans, and A. Nitzan, *J. Chem. Phys.* **101**, 436 (1994).

133. D. G. Evans and R. D. Coalson, *J. Chem. Phys.* **104**, 3598 (1996).

134. S. Ojima, H. Miyasaka, and N. Mataga, *J. Phys. Chem.* **94**, 5834 (1990).

135. S. Ojima, H. Miyasaka, and N. Mataga, *J. Phys. Chem.* **94**, 7534 (1990).

136. I. R. Gould, D. Noukakis, J. L. Goodman, R. H. Young, and S. Farid, *J. Am. Chem. Soc.* **115**, 3830 (1993).

137. I. R. Gould, D. Noukakis, L. Gomez-Jahn, J. L. Goodman, and S. Farid, *J. Am. Chem. Soc.* **115**, 4405 (1993).

138. I. R. Gould, D. Noukakis, L. Gomez-Jahn, R. H. Young, J. L. Goodman, and S. Farid, *Chem. Phys.* **176**, 439 (1993).

139. G. A. Seber and C. J. Wild, *Nonlinear Regression*, Wiley, New York, 1989.

140. M. H. Vos, J.-C. Lambry, S. J. Robles, D. C. Youvan, J. Breton, and J.-L. Martin, *Proc. Natl. Acad. Sci. USA* **88**, 8885 (1991).

141. M. H. Vos, F. Rappaport, J.-C. Lambry, J. Breton, and J.-L. Martin, *Nature,* **363**, 320 (1993).

142. K. Wynne, C. Galli, P. J. F. DeRege, M. J. Therien, and R. M. Hochstrasser, in *Ultrafast Phenomena VIII,* J.-L. Martin, A. Migus, G. A. Mourou, and A. H. Zewail, eds., Springer-Verlag, Berlin, 1993, p. 71.

143. L. Zhu, J. T. Sage, and P. M. Champion, *Science,* **266**, 629 (1994).

144. S. L. Dexheimer, Q. Wang, L. A. Peteanu, W. T. Pollard, R. A. Mathies, and C. V. Shank, *Chem. Phys. Lett.* **188**, 61 (1992).

145. W. T. Pollard, S. L. Dexheimer, Q. Wang, L. A. Peteanu, C. V. Shank, and R. A. Mathies, *J. Phys. Chem.* **96**, 6147 (1992).

146. L. A. Peteanu, R. W. Schoenlein, Q. Wang, R. A. Mathies, and C. V. Shank, *Proc. Natl. Acad. Sci. USA,* **90**, 11762 (1993).

147. Q. Wang, R. W. Schoenlein, L. A. Peteanu, R. A. Mathies, and C. V. Shank, *Science,* **266**, 422 (1994).

148. U. Banin, A. Waldman, and S. Ruhman, *J. Chem. Phys.* **96**, 2416 (1992).

149. U. Banin and S. Ruhman, *J. Chem. Phys.* **98**, 4391 (1993).

150. U. Banin, and S. Ruhman, *J. Chem. Phys.* **99**, 9318 (1993).

151. N. Pugliano, D. K. Palit, A. Z. Szarka, and R. M. Hochstrasser, *J. Chem. Phys.* **99**, 7273 (1993).

152. N. Pugliano, A. Z. Szarka, and R. M. Hochstrasser, *J. Chem. Phys.* **104**, 5062 (1996).

153. A. Z. Szarka, N. Pugliano, D. K. Palit, and R. M. Hochstrasser, *Chem. Phys. Lett.* **240**, 25 (1995).

154. T. S. Rose, M. J. Rosker, and A. H. Zewail, *J. Chem. Phys.* **88**, 6672 (1988).

155. M. J. Rosker, T. S. Rose, and A. H. Zewail, *Chem. Phys. Lett.* **146**, 175 (1988).

156. N. F. Scherer, R. J. Carlson, A. Matro, M. Du, A. J. Ruggiero, V. Romero-Rochin, J. A. Cina, G. R. Fleming, and S. A. Rice, *J. Chem. Phys.* **95**, 1487 (1991).

157. M. H. M. Janssen, M. Dantus, H. Guo, and A. H. Zewail, *Chem. Phys. Lett.* **214**, 281 (1993).

158. R. S. Knox and D. Gülen, *Photochem. Photobiol.* **57**, 40 (1993).

159. A. Matro and J. A. Cina, *J. Phys. Chem.* **99**, 2568 (1995).

160. C. K. Law, R. S. Knox, and J. H. Eberly, *Chem. Phys. Lett.* **258**, 352 (1996).

161. R. Friesner and R. Wertheimer, *Proc. Natl. Acad Sci. USA* **79**, 2138 (1982).

162. S. S. Skourtis, A. J. R. da Silva, W. Bialek, and J. N. Onuchic, *J. Phys. Chem.* **96**, 8034.

163. S. H. Lin, R. G. Alden, M. Hayashi, S. Suzuki, and H. A. Murchison, *J. Phys. Chem.* **97**, (1993).

164. F. Zhu, C. Galli, and R. M. Hochstrasser, *J. Chem. Phys.* **98**, 1042 (1993).

165. J. Manz, B. Reischl, T. Schroeder, F. Seyl, and B. Warmuth, *Chem. Phys. Lett.* **198**, 483 (1992).

166. M. Ben-Nun and R. D. Levine, *Chem. Phys. Lett.* **203**, 450 (1993).

167. B. Kohler, J. L. Krause, F. Raksi, K. R. Wilson, V. V. Yakovlev, R. M. Whitnell, and Y. Yan, *Acc. Chem. Res.* **28**, 133 (1995).

168. C. J. Bardeen, Q. Wang, and C. V. Shank, *Phys. Rev. Lett.* **75**, 3410 (1995).

169. B. Kohler, V. V. Yakovlev, J. Che, J. L. Krause, M. Messina, K. R. Wilson, N. Schwenter, R., and Y. Yan, *Phys. Rev. Lett.* **74**, 3360 (1995).

170. P. C. Planken, I. Brener, M. C. Nuss, M. S. C. Luo, and S. L. Chuang, *Phys. Rev. B* **48**, 4903 (1993).

171. J. C. Owrutsky, D. Raftery, and R. M. Hochstrasser, *Annu. Rev. Phys. Chem.* **45**, 519 (1994).

172. N. Mataga, Y. Kanda, and T. Okada, *J. Phys. Chem.* **90**, 3880 (1986).

173. T. Asahi and N. Mataga, *J. Phys. Chem.* **93**, 6575 (1989).

174. L. Dhar, J. A. Rogers, and K. A. Nelson, *Chem. Rev.* **94**, 157 (1994).

175. M. Bixon, J. Jormer, J. Cortes, H. Heitele, and M. E. Michel-Beyerle, *J. Phys. Chem.* **98**, 7289 (1994).

176. L. D. Zusman, *Chem. Phys.* **49**, 295 (1980).

177. G. A. Voth and R. M. Hochstrasser, *J. Phys. Chem.* **100**, 13034 (1996).

178. C. C. Moser, J. M. Keske, K. Warncke, R. S. Farid, and P. L. Dutton, *Nature,* **355**, 796 (1992).

179. Y. Jia, T. J. DiMagno, C.-K. Chan, Z. Wang, M. Du, D. K. Hanson, M. Schiffer, J. R. Norris, G. R. Fleming, and M. S. Popov, *J. Phys. Chem.* **97**, 13180 (1993).

180. J. R. Reimers and N. S. Hush, *J. Am. Chem. Soc.* **117**, 1302 (1995).

181. G. Haran, K. Wynne, C. C. Moser, P. L. Dutton, and R. M. Hochstrasser, *J. Phys. Chem.* **100**, 5562 (1996).

182. C. Reichardt, *Solvents and Solvent Effects in Organic Chemistry*, 2nd ed. Verlag Chemie, Weinheim, Germany, 1988.

183. *Handbook of Chemistry and Physics,* 57th ed., CRC Press, Boca Raton, Fla., 1977.

ELECTRON TRANSFER AND SOLVENT DYNAMICS IN TWO- AND THREE-STATE SYSTEMS

MINHAENG CHO

Department of Chemistry, Korea University, Seoul 136-701, Korea

GRAHAM R. FLEMING

Department of Chemistry, University of California, Berkeley, Berkeley CA 94720

CONTENTS

Electron Transfer: From Isolated Molecules to Biomolecules, Part Two, edited by Joshua Jortner and M. Bixon. Advances in Chemical Physics Series, Volume 107, series editors I. Prigogine and Stuart A. Rice.
ISBN 0-471-25291-3 © 1999 John Wiley & Sons, Inc.

I. INTRODUCTION

Our aim in this chapter is to incorporate the full range of time scales present in the system–bath interaction into the calculation of electron transfer rate kernels in a systematic way so that artificial distinctions between "fast" and "slow" variables can be avoided and the effect of temperature included in a natural way. The strengths of coupling and time-scale information is contained in a quantity we call the spectral density (i.e., the spectrum of frequencies present in the medium weighted by their influence on the observable of interest). In large measure through the work of Mukamel [1], the interrelations between various nonlinear optical spectroscopies are clarified and several techniques have been developed to obtain the spectral density for the optical transition frequency directly from experiment [2–8]. Both fluorescence Stokes shift and various types of photon echo spectroscopy have been brought to bear on this problem, as described in Section II, and for purposes of calculation in this article we use the optical transition frequency spectral density (also referred to as the spectroscopic spectral density) obtained by three-pulse echo peak shift studies of the dye IR144 in acetonitrile [5].

For such a solvation spectral density to be useful it must be transferable at least between situations in which the system–bath interaction is of the same general type [e.g., dominated by polar (Coulombic) effects]. In Section III we describe an approach to factoring the spectral density which allows us to obtain just such a transferable quantity, which we refer to as the bare solvent spectral density. In Sections IV and V, using the spectral density obtained via spectroscopic measurements, we present the procedure to calculate the electron transfer rate kernels of two- and three-state systems.

II. NONLINEAR SPECTROSCOPY AND SPECTRAL DENSITY

The dynamical aspects of the solvent degrees of freedom can be efficiently captured by the effective harmonic oscillator model in which the bath is assumed to be a collection of uncoupled harmonic oscillators [9–14]. There have been numerous attempts to formulate the nonlinear response functions containing information on the system–bath interaction in a nontrivial fashion [1]. As shown later, it is useful to recapitulate how to derive the corresponding response functions with a simple bath model (e.g., bosonic bath). Here we consider a chromophore with two electronic states and coupled to the bosonic bath linearly, that is,

$$
H = \begin{bmatrix} \varepsilon_g & 0 \\ 0 & \varepsilon_e \end{bmatrix} - \begin{bmatrix} 0 & \mu_{eg} \\ \mu_{eg}^* & 0 \end{bmatrix} E(\mathbf{r}, t)
$$

$$
+ \begin{bmatrix} \sum_\alpha \left(\dfrac{p_\alpha^2}{2m_\alpha} + \dfrac{1}{2} m_\alpha \omega_\alpha^2 q_\alpha^2 \right) & 0 \\ 0 & \sum_\alpha \left\{ \dfrac{p_\alpha^2}{2m_\alpha} + \dfrac{1}{2} m_\alpha \omega_\alpha^2 \left(q_\alpha - \dfrac{c_\alpha}{m_\alpha \omega_\alpha^2} \right)^2 \right\} \end{bmatrix}
$$

$$(2.1)$$

where ε_g and ε_e are the electronic energies of the ground and excited states, respectively. Invoking the electric dipole approximation and treating the external electric field classically, the field–chromophore interaction Hamiltonian is given by the second term in Eq. (2.1). μ_{eg} is the dipole matrix element. It will be assumed that the system is initially in the equilibrium state on the potential energy surface of the ground state. The nuclear Hamiltonian associated with the excited state consists of a set of displaced harmonic oscillators whose displacements are assumed to be $c_a/m_a\omega_a^2$. Here c_a denotes the coupling constant representing the strength of the interaction between the two-state chromophore and the ath harmonic model.

In particular, Mukamel and co-workers [1] developed the multimode Brownian oscillator picture to effectively model the system–bath interaction, where a few optically active harmonic oscillators are linearly coupled to a continuously distributed optically inactive harmonic modes. As proven in the Appendix, this model is essentially identical to the effective harmonic oscillator bath given in Eq. (2.1). One can understand this equality as follows. Since the optically active harmonic modes are coupled to the optically inactive modes, after performing the normal-mode transformation to find a new set of harmonic oscillators, one finds that the normal modes are now all optically active by sharing the oscillator strength in a way determined by the normal-mode-transformation eigenvector matrix elements. Therefore, despite the simplicity of the model Hamiltonian in Eq. (2.1), we find that a variety of spectroscopic phenomena arising from the system–bath interaction can be investigated by studying Eq. (2.1) in detail.

A. General Aspects of System–Bath Interaction: Fluorescence Stokes Shift and Solvation Dynamics

The time-dependent fluorescence Stokes shift [15] can be formulated directly from the model Hamiltonian, Eq. (2.1). First, we summarize some general background on fluctuation and relaxation in condensed phases. As a spectroscopic example of nonequilibrium relaxation phenomena, we describe the time-dependent fluorescence Stokes shift. We first assume that an impulsive perturbation by the external field induces a sudden electronic transition from the ground to excited states so that the solute charge distribution is changed in a stepwise fashion. Since initially the bath degrees of freedom are in equilibrium with the electronic ground-state charge distribution, the delta-function-approximated external field interaction produces a nonequilibrium state on the potential energy surface of the excited state. Then the nonequilibrium energy difference relaxes to a new equilibrium state, and this relaxation can be monitored by measuring the time-dependent fluorescence Stokes shift (FSS) function,

$$S(t) = \frac{\overline{\Delta E}(t) - \overline{\Delta E}(\infty)}{\overline{\Delta E}(0) - \overline{\Delta E}(\infty)} \qquad (2.2)$$

where the nonequilibrium energy difference between the excited and ground states, $\overline{\Delta E}(t)$, can be measured experimentally from the central frequency of the time-dependent fluorescence spectra [16,17]. The FSS function defined above reflects the relaxation of the nonequilibrium state; therefore, one can infer that it can be related to the fluctuation correlation function of the equilibrium state. The energy difference operator ΔE can be divided into the fluctuating and mean value components, that is,

$$\Delta E(t) = \langle \Delta E \rangle + \delta V_{SB}(t) \qquad (2.3)$$

By using linear response theory, the nonequilibrium quantity $\overline{\Delta E}(t)$ can be expressed as an integral over the response function, $G(t)$, so that the FSS function can be rewritten as

$$S(t) = \frac{\displaystyle\int_t^\infty d\tau\, G(\tau)}{\displaystyle\int_0^\infty d\tau\, G(\tau)} \qquad (2.4)$$

with

$$G(t) \equiv i\langle[\delta V_{SB}(0), \delta V_{SB}(t)]\rangle \tag{2.5}$$

Here the response function was defined as the mean value of the commutator of the Heisenberg operator $\delta V_{SB}(t)$. Thus the calculation of the FSS function is reduced to the calculation of the linear response function of the system–bath coupling potential. It turns out that it is very useful to define the spectral density as

$$\rho(\omega) \equiv \frac{\text{Im}[\tilde{G}(\omega)]}{2\pi\hbar^2\omega^2} \tag{2.6}$$

where $\tilde{G}(\omega)$ is the Fourier–Laplace transform [18]

$$\tilde{G}(\omega) = \int_0^\infty d\omega \, e^{i\omega t} G(t)$$

Depending on the composite system one is interested in, the spectral density defined above describes the optical broadening or the contribution of the system–bath interaction to the electron transfer rate.

We now use the model Hamiltonian given in Eq. (2.1) to find that

$$\langle\Delta E\rangle = \varepsilon_e - \varepsilon_g + \sum_\alpha \frac{c_\alpha^2}{2m_\alpha\omega_\alpha^2}$$

$$\delta V_{SB} = \sum_\alpha c_\alpha q_\alpha \tag{2.7}$$

Note that the fluctuating coupling potential is just a linear combination of the bath oscillator coordinates with c_α, the αth expansion coefficients. By inserting $\delta V_{SB}(t) = \sum_\alpha c_\alpha q_\alpha(t)$ into the definition of the response function and defining the spectral density $\rho_S(\omega)$ as

$$\rho_S(\omega) \equiv \frac{1}{2\hbar} \sum_\alpha \frac{c_\alpha^2}{m_\alpha\omega_\alpha^3} \delta(\omega - \omega_\alpha) \tag{2.8}$$

the FSS function can be expressed in terms of the spectroscopic spectral density,

$$S(t) = \frac{\hbar}{\lambda_S} \int_0^\infty d\omega \, \omega \rho_S(\omega) \cos \omega t \tag{2.9}$$

where the normalization constant λ_S is known as the solvation reorganization energy, defined as

$$\lambda_S \equiv \hbar \int_0^\infty d\omega \, \omega \rho_S(\omega) \qquad (2.10)$$

In $\rho_S(\omega)$ and λ_S, the subscript S emphasizes the fact that this spectral density and the solvation reorganization energy are associated with the spectroscopic phenomena described by the Hamiltonian equation (2.1). Later we introduce another spectral density associated with electron transfer processes in a common solvent. As shown in the following, this spectral density plays a central role in determining the dephasing and spectral diffusion processes in a variety of linear and nonlinear spectroscopies.

It should be mentioned that even though there are numerous ways to define the spectral density that give different functional forms for the reorganization energy and the FSS function, we find that Eq. (2.6) for the definition of the spectral density is most useful conceptually. That is because one can provide the following interpretations: (1) the solvation reorganization energy is identical to the first moment of the spectral density, $\rho_S(\omega)$, and (2) the one-sided cosine transformation of $\omega\rho_S(\omega)$ gives the time dependence of the FSS function. Then the average energy difference, $\langle \Delta E \rangle$, can be rewritten as $\varepsilon_e - \varepsilon_g + \lambda_S$.

The experimental determination of the time-dependent FSS function can be a direct method to obtain the spectral density $\rho_S(\omega)$. A practical limitation here is that the time resolution of the fluorescence up-conversion technique [16,17] is typically limited to 30–40 fs and even though ca. 100-fs components in $S(t)$ can readily be detected, it is difficult to obtain their functional form accurately with the spectral reconstruction technique employed to obtain $S(t)$ from the experimental data. Other techniques generally based on photon echo measurements possess substantially higher time resolution [4,15,19]. As mentioned in Section I, the spectral density, $\rho_S(\omega)$, can be approximately transformed into that associated with the electron transfer process in the same polar liquid. This is discussed in Section III.

There exist several other methods that can provide the spectral density. The solvation dynamics of a dye molecule in polar liquids can be studied by using classical molecular dynamics simulations by calculating $S(t)$ or the classical correlation function of the solvation energy fluctuation (note that the two contain the same information in the classical limit, that is, when $\beta\hbar\omega_a \ll 1$ for all ω_a) [20–23]. Then taking the frequency transformation of the solvation correlation function, one can obtain the spectral density. This procedure was used by Chandler and co-workers [14a] to evaluate the stationary-phase-approximated electron transfer rate constant. Theoretical studies based on the hydrodynamic approach to calculate the wavevector- and frequency-dependent dielectric function have been presented by numerous workers [24–26]. Instead of reviewing those works, we merely refer to their papers, since it is our primary goal to provide a systematic procedure to study electron transfer reaction in condensed phases based on spectroscopic measurements instead of other approaches mentioned above.

B. Nonlinear Response Functions and Spectral Density

In Section II.A the time-dependent fluorescence Stokes shift was described in terms of the spectral density defined in Eq. (2.8) when the optical spectroscopy in condensed phases is completely described by the model Hamiltonian equation (2.1). In this subsection, four-wave mixing (third-order) spectroscopy is discussed in terms of the spectral density defined in Eq. (2.8). Although the principal result, calculation of the nonlinear response functions for four-wave mixing spectroscopy, was presented previously and studied quite extensively [1], we believe that it is useful to summarize those results in this paper since the same procedure can be used for studying electron transfer reactions involving two or three states in condensed phases. As shown later, the two problems, optical spectroscopy of a two-state chromophore and electron transfer of a two-state system in condensed phases, are almost identical, except that the former involves a time-dependent perturbation, whereas the latter involves a time-independent coupling Hamiltonian. There is a distinct difference in the quantities to be measured. The former focuses on first- or third-order processes with respect to the system-field interaction to create a nonlinear electronic coherence state, which in turn produces the macroscopic polarization acting as a source in Maxwell's equations. On the other hand, for electron transfer we must consider even-order processes with respect to the nonadiabatic coupling matrix element, in order to calculate the time evolution of the populations of the two states. However the intermediate steps in investigating these two problems involve identical procedures, which is why we feel that it is helpful to recapitulate the formal derivation of the nonlinear response function associated with four-wave mixing spectroscopy, based on the Hamiltonian equation (2.1).

It turns out that it is useful to transform the Hamiltonian by using the unitary transformation operator,

$$u \equiv \exp\left\{ \frac{i}{\hbar} \sum_\alpha p_\alpha \begin{bmatrix} 0 & 0 \\ 0 & \dfrac{c_\alpha}{m_\alpha \omega_\alpha^2} \end{bmatrix} \right\} \tag{2.11}$$

Then the transformed Hamiltonian is given by

$$h = uHu^{-1} = h_0 + v(\mathbf{r}, t) \tag{2.12}$$

where

$$h_0 = h_s + h_b \tag{2.13}$$

with

$$h_s = \begin{bmatrix} \varepsilon_g & 0 \\ 0 & \varepsilon_e \end{bmatrix}$$

$$h_b = \sum_\alpha \left(\frac{p_\alpha^2}{2m_\alpha} + \frac{1}{2} m_\alpha \omega_\alpha^2 q_\alpha^2 \right)$$

$$v(\mathbf{r}, t) = - \begin{bmatrix} 0 & \mu_{eg} \exp(-i\hat{f}) \\ \mu_{eg}^* \exp(i\hat{f}) & 0 \end{bmatrix} E(\mathbf{r}, t) = -\hat{v} E(\mathbf{r}, t) \tag{2.14}$$

Here the operator $\hat{f}$ is defined as

$$\hat{f} \equiv \sum_\alpha \frac{c_\alpha}{\hbar m_\alpha \omega_\alpha^2} p_\alpha \tag{2.15}$$

and the dressed electric transition dipole operator $\hat{v}$ was naturally defined in Eq. (2.14). In comparison to the original Hamiltonian given in Eq. (2.1), the transformed Hamiltonian does not contain the linear system–bath coupling term, that is, the last term in Eq. (2.1). Instead, the field–system interaction term is dressed by the bath harmonic oscillators. Hereafter the field–system interaction Hamiltonian, $v(\mathbf{r}, t)$, will be treated as a perturbational Hamiltonian. It seems to be necessary to add and subtract the thermal average of $v(\mathbf{r}, t)$ from the total Hamiltonian to optimize the perturbational approach [10]. However, for a typical polar liquid with a typical chromophore, the system–bath interaction, whose magnitude is, roughly speaking, proportional to the fluorescence Stokes shift, is usually strong enough to ignore the thermal average of $v(\mathbf{r}, t)$.

Using the Hamiltonian in Eq. (2.12), we shall now calculate the third-order polarization that generates the signal field. To calculate the optical polarization, it is necessary to solve the quantum Liouville equation perturbatively

$$\frac{d\rho(\mathbf{r}, t)}{dt} = -\frac{i}{\hbar} [h_0, \rho(\mathbf{r}, t)] - \frac{i}{\hbar} [v(\mathbf{r}, t), \rho(\mathbf{r}, t)] \tag{2.16}$$

For the sake of notational simplicity, the Liouvillian operators are defined for an arbitrary operator A as

$$L_0 A = [h_0, A]$$

$$L_I A = [\hat{v}, A]$$

$$L_v A = [v(\mathbf{r}, t), A] = -L_I A E(\mathbf{r}, t) \qquad (2.17)$$

Then the time-dependent density operator can be expanded in powers of the system–bath interaction Hamiltonian as

$$\rho(t) = \sum_{n=0}^{\infty} \rho^{(n)}(t) \qquad (2.18)$$

where the nth-order term can be obtained by using the standard time-dependent perturbation theory,

$$\rho^{(n)}(\mathbf{r}, t) = \left(-\frac{i}{\hbar}\right)^n \int_{t_0}^{t} d\tau_n \int_{t_0}^{\tau_n} d\tau_{n-1} \cdots \int_{t_0}^{\tau_2} d\tau_1$$

$$\times e^{-(i/\hbar)L_0(t-\tau_n)} L_v(\tau_n) e^{-(i/\hbar)L_0(\tau_n-\tau_{n-1})} L_v(\tau_{n-1}) \cdots$$

$$\times e^{-(i/\hbar)L_0(\tau_2-\tau_1)} L_v(\tau_1) e^{-(i/\hbar)L_0(\tau_1-t_0)} \rho(t_0) \qquad (2.19)$$

where $\rho(t_0)$ is the initial canonical, equilibrium, density operator, given by

$$\rho(t_0) = |g\rangle \frac{\exp(-\beta h_b)}{\mathrm{Tr}[\exp(-\beta h_b)]} \langle g| \qquad (2.20)$$

The four-wave mixing spectroscopies are described completely by the third-order density operator, which can be obtained by inserting $L_v(\tau_j)$ into Eq. (2.19):

$$\rho^{(3)}(\mathbf{r}, t) = \left(\frac{i}{\hbar}\right)^3 \int_{t_0}^{t} d\tau_3 \int_{t_0}^{\tau_3} d\tau_2 \int_{t_0}^{\tau_2} d\tau_1 [e^{-(i/\hbar)L_0(t-\tau_3)} L_I e^{-(i/\hbar)L_0(\tau_3-\tau_2)} L_I$$

$$\times e^{-(i/\hbar)L_0(\tau_2-\tau_1)} L_I e^{-(i/\hbar)L_0(\tau_1-t_0)} \rho(t_0)] E(\mathbf{r}, \tau_3) E(\mathbf{r}, \tau_2) E(\mathbf{r}, \tau_1) \qquad (2.21)$$

The third-order polarization can be expressed as

$$P^{(3)}(\mathbf{r}, t) = \mathrm{Tr}[\hat{\mu} \rho^{(3)}(\mathbf{r}, t)] \qquad (2.22)$$

where the electric dipole operator $\hat{\mu}$ is defined as

$$\hat{\mu} = \begin{bmatrix} 0 & \mu_{eg} \\ \mu_{eg}^* & 0 \end{bmatrix} \tag{2.23}$$

Inserting the appropriate Liouville operators into Eq. (2.21) and fully expanding the commutators, we find that the third-order polarization can be written as

$$P^{(3)}(\mathbf{r}, t) = \text{Tr}[\hat{\mu}\rho^{(3)}(\mathbf{r}, t)]$$

$$= \left(\frac{i}{\hbar}\right)^3 \int_{t_0}^{t} d\tau_3 \int_{t_0}^{\tau_3} d\tau_2 \int_{t_0}^{\tau_2} d\tau_1 \text{Tr}\{\hat{\mu}(t)[\hat{v}(\tau_3), [\hat{v}(\tau_2), [\hat{v}(\tau_1), \rho(t_0)]]]\}$$

$$\times E(\mathbf{r}, \tau_3)E(\mathbf{r}, \tau_2)E(\mathbf{r}, \tau_1) \tag{2.24}$$

where the following interaction representations were used:

$$\hat{\mu}(t) \equiv \exp\left(\frac{i}{\hbar}h_0 t\right)\hat{\mu}\exp\left(-\frac{i}{\hbar}h_0 t\right) = \begin{bmatrix} 0 & \mu_{eg}e^{-i\varepsilon_{eg}t} \\ \mu_{eg}^*e^{i\varepsilon_{eg}t} & 0 \end{bmatrix}$$

$$\hat{v}(\tau_n) = \exp\left(\frac{i}{\hbar}h_0\tau_n\right)\hat{v}\exp\left(-\frac{i}{\hbar}h_0\tau_n\right)$$

$$= \begin{bmatrix} 0 & \mu_{eg}^*\exp\{-i\hat{f}(\tau_n) - i\varepsilon_{eg}\tau_n\} \\ \mu_{eg}^*\exp\{i\hat{f}(\tau_n) + i\varepsilon_{eg}\tau_n\} & 0 \end{bmatrix} \tag{2.25}$$

Since the integrand involves three commutators, the nonlinear response function consists of eight terms. As discussed by Mukamel [1], four terms among them are the Hermitian conjugates of the other four terms, so that calculation of four nonlinear response functions is sufficient to calculate the third-order nonlinear polarization in Eq. (2.24). Changing the integration variables to

$$t_1 = \tau_2 - \tau_1$$
$$t_2 = \tau_3 - \tau_2$$
$$t_3 = t - \tau_3$$

and inserting Eq. (2.25) into (2.24) and using Baker–Hausdorff operator identity repeatedly, one can rewrite the third-order polarization as

$$P^{(3)}(\mathbf{r}, t) = \int_0^\infty dt_3 \int_0^\infty dt_2 \int_0^\infty dt_1 \, S^{(3)}(t_3, t_2, t_1) E(\mathbf{r}, t - t_3)$$

$$\times E(\mathbf{r}, t - t_3 - t_2) E(\mathbf{r}, t - t_3 - t_2 - t_1) \qquad (2.26)$$

where the nonlinear response function is given by

$$S^{(3)}(t_3, t_2, t_1) = \left(\frac{i}{\hbar}\right)^3 \theta(t_3)\theta(t_2)\theta(t_1) \sum_{\alpha=1}^4 [R_\alpha(t_3, t_2, t_1) - R_\alpha^*(t_3, t_2, t_1)] \quad (2.27)$$

with

$$
\begin{aligned}
R_1(t_3, t_2, t_1) &= \exp(-i\bar{\omega}_{eg}t_1 - t\bar{\omega}_{eg}t_3) \exp\{-g^*(t_3) - g(t_1) \\
&\quad - g^*(t_2) + g^*(t_2 + t_3) + g(t_1 + t_2) - g(t_1 + t_2 + t_3)\} \\
R_2(t_3, t_2, t_1) &= \exp(i\bar{\omega}_{eg}t_1 - i\bar{\omega}_{eg}t_3) \exp\{-g^*(t_3) - g(t_1) \\
&\quad + g(t_2) - g(t_2 + t_3) - g^*(t_1 + t_2) + g^*(t_1 + t_2 + t_3)\} \\
R_3(t_3, t_2, t_1) &= \exp(i\bar{\omega}_{eg}t_1 - i\bar{\omega}_{eg}t_3) \exp[-g(t_3) - g^*(t_1) \\
&\quad + g^*(t_2) - g^*(t_2 + t_3) - g^*(t_1 + t_2) + g^*(t_1 + t_2 + t_3)] \\
R_4(t_3, t_2, t_1) &= \exp(-i\bar{\omega}_{eg}t_1 - i\bar{\omega}_{eg}t_3) \exp[-g(t_3) - g(t_1) \\
&\quad - g(t_2) + g(t_2 + t_3) + g(t_1 + t_2) - g(t_1 + t_2 + t_3)]
\end{aligned} \qquad (2.28)
$$

Here the average absorption frequency $\bar{\omega}_{eg}$ is equal to $(\varepsilon_e - \varepsilon_g + \lambda_S)/\hbar$. The lineshape function $g(t)$ is fully determined by the spectral density $\rho_S(\omega)$ via

$$g(t) \equiv -i\lambda_S t/\hbar + P(t) + iQ(t) \qquad (2.29)$$

where

$$P(t) \equiv \int d\omega \, \rho_S(\omega) \coth(\beta\hbar\omega/2)(1 - \cos\omega t)$$

$$Q(t) \equiv \int d\omega \, \rho_S(\omega) \sin\omega t \qquad (2.30)$$

Note that once the spectral density $\rho_S(\omega)$ is completely known, the linear and nonlinear spectroscopic signals can be calculated in principle.

Once the nonlinear response function is known from complete knowledge of the spectroscopic spectral density, any four-wave mixing spectroscopy of a two-state system can be predicted: for example, spontaneous and stimulated fluorescence signals, transient grating signals, photon echoes, and so on. In practice, experimental studies utilizing those nonlinear spectroscopic techniques aim at extraction of information on the spectral density. However, because the nonlinear response function given above depends on the spectral density in a highly complicated fashion, there does not exist any direct experimental method that is capable of providing unique, unbiased information on the spectroscopic spectral density. This is the major obstacle in interpreting the nonlinear spectroscopic signal. Still the time-dependent fluorescence Stokes shift measurement is clearly one of the most useful techniques for probing the system–bath-interaction-induced relaxation process in condensed phases. There are a few limitations of fluorescence Stokes shift experiments in comparison to other nonlinear spectroscopic techniques, such as the photon echo peak shift measurement [2,4–6,15], which we consider next.

C. Photon Echoes, Solvation Dynamics, and Spectral Density

Although the functional form of the nonlinear response function may look formidable, the third-order polarization in Eq. (2.26) can be greatly simplified by using a controllable external field configuration to probe interesting aspects of the system–bath interaction selectively. For instance, photon echo spectroscopy has been known as an excellent tool to measure the homogeneous contribution to optical line broadening. This statement is valid when the slow degrees of freedom are sufficiently sluggish so that they can be treated as static in essence. For instance, the distributed impurities in a glassy material give a distribution of the local structures around the chromophores, which is virtually static in nature, whereas the localized phonons can be considered as quickly fluctuating dynamical degrees of freedom [27]. The fluctuation of the latter degrees of freedom is therefore responsible for what is known as the *homogeneous dephasing process*. Unlike the glassy material, liquids do not have such a time-scale separation among the bath degrees of freedom, which is the primary reason why the conventional photon echo cannot be used to investigate the dynamical aspect of the liquids inducing line broadening. A few attempts to modify the conventional photon echo measurements have been presented. The echo signal is usually measured by integrating the echo intensity. Therefore, if there is no large static inhomogeneity, the integrated echo is, unfortunately, complicated by the undesirable additional integration process. To overcome this complication, Wiersma and co-workers [8] and Scherer and co-workers [7] applied an additional gate pulse to probe the echo field amplitude in time, thus to measure the echo intensity with a time resolution. In parallel with this type of echo experiment, Fleming and co-workers [2,5,6,19] and Cho et al. [15] showed, both experimentally and theoretically, that the echo peak shift as a function of the delay between the second and third pulses in the three-pulse photon echo experiment is a rich source of information on the solvation dynamics. Next, a theoretical description of how the integrated three-pulse photon echo can be used to probe the system–bath interaction represented by the spectral density is presented.

When the slowly varying amplitude approximation is invoked, the photon echo signal is proportional to the square of the third-order polarization, which in turn acts as a source term generating an electric field from the macroscopic sample. Since what is measured is the macroscopic polarization instead of the microscopic (single molecular) polarization, the ensemble average must be taken to give the phase-matching condition. When one introduces a sequence of the three pulses and assumes that the time profiles of the three pulses are all close to delta functions, the expression for the three-pulse photon echo signal can be simplified to the form [15]

$$S_{PE}(\tau, T) \propto \int_0^\infty dt\, I(t - \tau) \exp\{-2[P(\tau) - P(T) + P(t) + P(\tau + T)$$

$$+ P(T + t) - P(\tau + T + t)]\}$$

$$\times \cos^2[Q(T) + Q(t) - Q(T + t)] \tag{2.31}$$

where $P(t)$ and $Q(t)$ were defined in Eq. (2.30) and the static inhomogeneous contribution, if there is one, is represented by the Gaussian function,

$$I(t - \tau) = \exp[-\Delta_{\text{inh}}^2(t - \tau)^2] \tag{2.32}$$

Suppose that the inhomogeneous width is very large (i.e., Δ_{inh} is large); then the inhomogeneous function $I(t - \tau)$ approaches a delta function [when the normalization constant is assumed to be properly multiplied to Eq. (2.32)]. Then the echo signal will peak at $t = \tau$. This is the case when the conventional photon echo measurements can be used directly to eliminate the inhomogeneous contribution to the optical line broadening. However, as discussed before, there are few cases when this condition is met, so it is necessary to perform the integral in Eq. (2.31) to describe the three-pulse echo process completely.

Although the integrand involves highly complicated functions, since the time period during t is associated with the optical coherence period, the half-side integral in Eq. (2.31) can be carried out by expanding the integrand for short time t to find

$$S_{PE}(\tau, T) \propto \frac{1}{\sqrt{A(\tau, T)}} \exp\left[-2P(\tau) - \Delta_{inh}^2 \tau^2 + \frac{B(\tau, T)^2}{4A(\tau, T)}\right]$$

$$\times \left[1 + \text{erf}\left(\frac{B(\tau, T)}{2\sqrt{A(\tau, T)}}\right)\right] \tag{2.33}$$

where

$$A(\tau, T) = \overline{\omega^2 \coth(\beta\hbar\omega/2)} + \Delta_{inh}^2 + M(T) - M(\tau + T) + f(T)$$

$$B(\tau, T) \equiv 2\Delta_{inh}^2 \tau + 2\dot{P}(\tau + T) - 2\dot{P}(T) \tag{2.34}$$

Here the two auxiliary functions are defined as

$$M(T) \equiv \int d\omega\, \omega^2 \coth(\beta\hbar\omega/2)\rho_S(\omega)\cos\omega t$$

$$f(T) \equiv (\lambda_S/\hbar)^2[1 - M(T)/M(0)]^2 \tag{2.35}$$

The bar in $\overline{\omega^2 \coth(\beta\hbar\omega/2)}$ means that the average over the spectral density is taken, i.e.

$$\overline{\omega^2 \coth(\beta\hbar\omega/2)} = \int d\omega\, \omega^2 \coth(\beta\hbar\omega/2)\rho_S(\omega) \tag{2.36}$$

which corresponds to the mean-square fluctuation amplitude.

Figure 1 shows a series of three-pulse photon echo signals for different values of the population period, T, for IR144 in acetonitrile. The two signals shown correspond to the two time-reversed echo signals measured in the directions $\mathbf{k}_1 - \mathbf{k}_2 + \mathbf{k}_3$ and $-\mathbf{k}_1 + \mathbf{k}_2 + \mathbf{k}_3$. We define the peak shift $\tau^*(T)$ as half the distance between the two maxima. $\tau^*(T)$ can be determined very accurately (± 300 as) and clearly decreases as T increases. As we have discussed in detail elsewhere [15], $\tau^*(T)$ reflects the ability of the system to be rephased after the period T. Figure 2 shows $\tau^*(T)$ plotted versus T for two systems: IR144 in acetonitrile [5] and the same dye in a polymer glass PMMA at room temperature [6]. The initial portions of the curves are

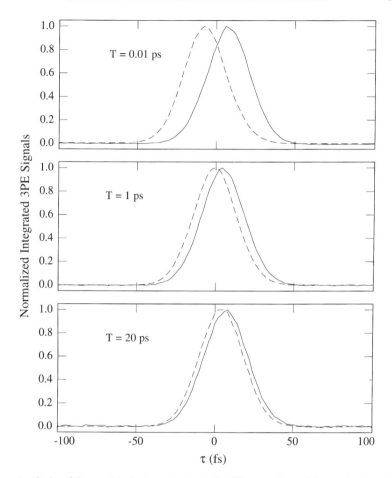

Figure 1. Series of three-pulse photon echo signals for different values of the population period T for IR144 in acetonitrile. The two curves correspond to the echo signals measured in the directions of $\mathbf{k}_1 - \mathbf{k}_2 + \mathbf{k}_3$ and $-\mathbf{k}_1 + \mathbf{k}_2 + \mathbf{k}_3$.

strikingly similar, but after about 200 fs, the data in PMMA become effectively constant, reflecting the static inhomogeneous distribution. By contrast, in the fluid solution, the peak shift continues to decay to a final value of zero.

In the cases of no and large inhomogeneous broadening limits, the traditional interpretation of the echo signal holds. However, when the inhomogeneous width is comparable to the root mean square of the fluctuation amplitude, the interpretation of the simple two pulse echo measurements is ambiguous. In this case, an approximate expression for the echo peak shift

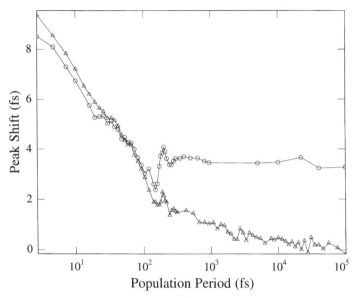

Figure 2. Three-pulse photon echo peak shift $\tau^*(T)$ versus log(population period, T). Circles, measured peak shift of IR144 in a polymer glass PMMA; triangles, data of IR44 in acetonitrile.

can be obtained by expanding Eq. (2.33) for short time τ and finding the maximum of the echo signal, and then the echo peak shift as a function of the second delay time T is

$$\tau^*(T) = \frac{\sqrt{\gamma + \Delta_{inh}^2 + f(T)}(\gamma S(T) + \Delta_{inh}^2)}{\sqrt{\pi}\,[\gamma(\gamma + 2\Delta_{inh}^2 + f(T)) + \Delta_{inh}^2 f(T)]} \qquad (2.37)$$

The asymptotic value of the echo peak shift is

$$\tau^*(T \to \infty) = \frac{\Delta_{inh}^2\sqrt{\gamma + \Delta_{inh}^2 + \lambda_S^2/\hbar^2}}{\sqrt{\pi}\,[\gamma(\gamma + 2\Delta_{inh}^2 + \lambda_S^2/\hbar^2) + \Delta_{inh}^2\lambda_S^2/\hbar^2]} \qquad (2.38)$$

where $\gamma \equiv 2\lambda_S k_B T/\hbar^2$. Although Eq. (2.37) does not fit the initial decaying part of the echo peak shift very well, it describes the longer time portion quantitatively. In particular, when the delay time is longer than the solvation correlation time, one can ignore the contribution from $f(T)$ in Eq. (2.37), so that the echo peak shift is found to be directly proportional to the fluorescence Stokes shift function or solvation correlation function,

$S(T)$. This is why the echo peak shift measurements can be used directly to investigate the solvation dynamics and in turn to obtain the spectroscopic spectral density by frequency Fourier transform. Furthermore, as can be seen from the asymptotic value of the echo peak shift, if there is no inhomogeneity (i.e., $\Delta_{inh} = 0$), the asymptotic value vanishes. Conversely, the observation of the finite asymptotic echo peak shift suggests that there is a finite inhomogeneous, truly static contribution to the optical broadening. This prediction has been tested experimentally. The data in Figure 2 confirm this directly. (For liquids such as ethanol, the echo peak shift decays to zero, whereas the echo peak shift of IR144 in a glassy material, PMMA, does not approach zero at long time T.)

By combining the information on the fluorescence Stokes shift magnitude obtained from the stationary absorption and fluorescence Stokes shift with the time dependence of the echo peak shift (i.e., approximately equal to the solvation correlation function) one can construct the spectroscopic spectral density. This will be used to calculate the electron transfer rate kernel in the same solvent.

III.   SOLVENT (BARE) SPECTRAL DENSITY

The spectroscopic spectral density containing information on the chromophore–bath interaction can be measured by using the spectroscopic methods described above. Unfortunately, the spectroscopic spectral density $\rho_S(\omega)$ cannot be used directly to calculate the electron transfer rate constant, because the electron transfer system–bath interaction may well be very different from the chromophore–bath interaction. For this reason, computer simulation studies have been used to calculate the corresponding spectral density for a composite system involving the electron transfer pair dissolved in a condensed medium [14]. The solvation energy correlation function in the classical limit is calculated, and using the fluctuation–dissipation theorem connecting the classical correlation function and response function, the electron transfer spectral density is obtained by using Eq. (2.9), where the spectroscopic spectral density is replaced with the electron transfer spectral density. Then the spectral density so obtained is put into the formal expression for the electron transfer rate constant (see Sec. IV). Experimental verification of the validity of the spectral density calculated from the computer simulation, however, has not been provided so far. In this section we develop a consistent procedure to extract the net solvent spectral density normalized by the solute-dependent part of the total spectral density. This solvent (bare) spectral density can then be used to study other processes, as long as they are completely specified by the appropriate spectral density. To establish this connection between the spectroscopic spectral density, $\rho_S(\omega)$, and the electron transfer spectral density, $\rho_{ET}(\omega)$, it is necessary to assume that the interaction between the system (either a chromophore or an electron transfer pair) and the bath is governed primarily by the electrostatic interactions. This assumption is acceptable when a polar solute or an electron transfer pair is dissolved in polar liquids. In this section we start with a discussion on the factorization of the spectroscopic spectral density into solute- and solvent-dependent parts, where the latter quantity will be used to construct the electron transfer spectral density.

A. Fluorescence Stokes Shift Magnitude

For a solute whose charge distributions of the ground and excited states are given arbitrarily, the electrostatic interaction energies between the solute and the solvent are given by

$$U_g(t) = -\int d\mathbf{r}\, \mathbf{P}(\mathbf{r}, t) \cdot \mathbf{E}_g(\mathbf{r})$$

$$U_e(t) = -\int d\mathbf{r}\, \mathbf{P}(\mathbf{r}, t) \cdot \mathbf{E}_e(\mathbf{r}) \qquad (3.1)$$

where $P(\mathbf{r}, t)$ is the polarization operator for the solvent, the contribution from the nuclear (orientational) degrees of freedom. Here the contribution from the electronic degrees of freedom of the solvent is not considered, since they can effectively be included in the average energy difference between the two states. $\mathbf{E}_g(\mathbf{r})$ and $\mathbf{E}_e(\mathbf{r})$ denote the electric fields arising from the solute molecule in the ground and excited states in the absence of the solvent.

When the charge distribution of a chromophore is described accurately by a permanent dipole, the fluorescence Stokes shift magnitude has been shown to be a useful tool for measuring the difference of the permanent dipole moments of the electronic ground and excited states of the solute. For instance, by using Onsager's reaction field method, Ooshika [28] and McRae [29] showed that the fluorescence Stokes shift magnitude, which is twice the solvation reorganization energy, can be estimated from the dielectric constant of the solvent medium,

$$\Delta E(0) - \Delta E(\infty) = 2\lambda_S \propto (1/a^3)|\boldsymbol{\mu}_e - \boldsymbol{\mu}_g|^2 \left(\frac{\varepsilon_0 - 1}{2\varepsilon_0 + 1} - \frac{\varepsilon_\infty - 1}{2\varepsilon_\infty + 1} \right) \quad (3.2)$$

where a is the effective Onsager radius of the solute and ε_0 and ε_∞ denote the zero- and high-frequency limits of the dielectric function $\varepsilon(\omega)$. $\boldsymbol{\mu}_g$ and $\boldsymbol{\mu}_e$ denote the ground- and excited-state dipole moments. Here it is assumed that the solute is represented by a point dipole at the center of a spherical cavity of radius a and the solvent is modeled by a dielectric continuum whose dielectric properties are described completely by $\varepsilon(\omega)$. To describe the time-dependent fluorescence Stokes shift, Mazurenko and Bakshiev [30], Bagchi et al. [31] and van der Zwan and Hynes [32] extended Onsager's cavity model and showed that the frequency-dependent dielectric function can be used directly to obtain the FSS function. Instead of a macroscopic approach such as Onsager's cavity model, Loring et al. [33] presented a microscopic description of the fluorescence Stokes shift in terms of the wave-vector and frequency-dependent dielectric function, and they found that

$$\Delta E(0) - \Delta E(\infty) = 2\lambda_S \propto (1/r^3)|\mathbf{\mu}_e - \mathbf{\mu}_g|^2 \left(\frac{1}{\varepsilon_\infty} - \frac{1}{\varepsilon_0}\right) \qquad (3.3)$$

when the continuum limit is taken. Here r is the effective solute radius given as a function of the integrated intensity of the difference electric fields of the ground- and excited-state charge distributions.

In any case, the fluorescence Stokes shift magnitude consists of two parts, the solute-dependent part and the bath-dependent part. The former is pro-portional to $(1/a^3)|\mathbf{\mu}_e - \mathbf{\mu}_g|^2$ in case of the Onsager's cavity model, or $(1/r^3)|\mathbf{\mu}_e - \mathbf{\mu}_g|^2$ in the microscopic model. The remaining part of the expressions for the Stokes shift magnitude is associated with the macroscopic (or microscopic) properties of the solvent system. This leads us to propose an approximate procedure to factorize the spectral density, which represents the dynamical time-scale distribution of the bath degrees of freedom as well as the strength of the system–bath interaction, into the purely solute- and bath-dependent parts.

B.   Factorization of the Spectral Density

From the definition of the system–bath interaction energy operators in Eq. (3.1), the fluctuating energy difference operator δV_{SB} is given by

$$\delta V_{SB}(t) = U_e - U_g - \langle U_e - U_g \rangle = -\int d\mathbf{r} \, \delta \mathbf{P}(\mathbf{r}, t) \cdot [\mathbf{E}_e(\mathbf{r}) - \mathbf{E}_g(\mathbf{r})] \qquad (3.4)$$

where the fluctuating solvent polarization operator in the Heisenberg repre-sentation is

$$\delta \mathbf{P}(\mathbf{r}, t) \equiv \mathbf{P}(\mathbf{r}, t) - \langle \mathbf{P}(\mathbf{r}, t) \rangle \qquad (3.5)$$

Inserting the fluctuating energy difference operator in Eq. (3.4) into the definition of the response function, Eq. (2.5), gives

$$G(t) = (2\pi)^{-3} \int d\mathbf{k} |\mathbf{E}_e(\mathbf{k}) - \mathbf{E}_g(\mathbf{k})|^2 G_{pp}(\mathbf{k}, t) \qquad (3.6)$$

Here the response function of the fluctuating solvent polarization is defined as

$$G_{pp}(\mathbf{k}, t) \equiv \int d\mathbf{r} \, e^{i\mathbf{k}\cdot\mathbf{r}} \langle [\delta\mathbf{P}(\mathbf{r}, 0) \cdot \hat{\mathbf{k}}, \delta\mathbf{P}(0, t) \cdot \hat{\mathbf{k}}] \rangle \tag{3.7}$$

Since the electric field created by a charge distribution with a finite extent is longitudinal, only the longitudinal component of the solvent polarization is considered.

From the definition of the spectral density in Eq. (2.6), when the fluctuating energy difference operator is given as Eq. (3.4), the spectroscopic spectral density is

$$\rho_S(\omega) = (2\pi)^{-3} \int d\mathbf{k} |\mathbf{E}_e(\mathbf{k}) - \mathbf{E}_g(\mathbf{k})|^2 \frac{\mathrm{Im}[G_{pp}(\mathbf{k}, \omega)]}{2\pi\hbar^2\omega^2} \tag{3.8}$$

where $G_{pp}(\mathbf{k}, \omega)$ is the Fourier–Laplace transform of the polarization response function. One can show that the spectral density defined above can be related to the wavevector- and frequency-dependent longitudinal dielectric function by using linear response theory. Then the polarization response function is equal to $(4\pi)^{-1}(\varepsilon(\mathbf{k}, \infty)^{-1} - \varepsilon(\mathbf{k}, \omega)^{-1})$ [26]. Furthermore, the relationship between the $\mathbf{k}$- and ω-dependent dielectric function and the correlation function of the longitudinal polarization can be used to obtain $\varepsilon(\mathbf{k}, \omega)$ from computer simulation studies. We shall not, however, pursue this procedure further in this review; instead, we will be content to use Eq. (3.8) to construct a bridge between the two spectral densities, $\rho_S(\omega)$ and $\rho_{\mathrm{ET}}(\omega)$.

To establish this connection, the polarization response function is Taylor expanded as

$$G_{pp}(\mathbf{k}, \omega) = \sum_{n=0} (\mathbf{k} \cdot \nabla_k)^n G_{pp}(\mathbf{k} = 0, \omega) \tag{3.9}$$

Thus the spectral density defined above can be rewritten by a series summation over the infinite Taylor expanded terms. For the sake of simplicity, we take only the first term in Eq. (3.9); that is, we assume that the $\mathbf{k}$- and ω-dependent polarization response function is a very slowly varying function with respect to k is comparison to the intensity of the difference electric field term. This amounts to the continuum limit,

$$G_{pp}(\mathbf{k}, \omega) \rightarrow G_{pp}(0, \omega) \tag{3.10}$$

Within this approximation, the spectroscopic spectral density can be written in a completely factorized form,

$$\rho_S(\omega) = \Gamma_S \sigma(\omega) \tag{3.11}$$

where

$$\Gamma_S \equiv (2\pi)^{-3} \int d\mathbf{k} |\mathbf{E}_e(\mathbf{k}) - \mathbf{E}_g(\mathbf{k})|^2$$

$$\sigma(\omega) \equiv \frac{\text{Im}[G_{pp}(\mathbf{k} = 0, \omega)]}{2\pi\hbar^2\omega^2} \tag{3.12}$$

Note that the factor Γ_S is determined purely by the solute charge distribution and not a function of the solvent degrees of freedom, whereas the spectra part, $\sigma(\Omega)$, represents the dynamics of the solvent degrees of freedom only. From the definitions of the spectral density, Eq. (3.11), the solvation reorganization energy representing the strength of the chromophore–bath interaction is given as

$$\lambda_S = \frac{\Gamma_S}{2\pi\hbar} \int_0^\infty d\omega \, \text{Im}[G_{pp}(0, \omega)]/\omega \tag{3.13}$$

that is, half of the fluorescence Stokes shift magnitude. In case of the high-temperature (classical) limit, the response function in the time domain, $G_{pp}(\mathbf{k}, t)$, is proportional to the time derivative of the classical correlation function of the solvent polarization [34]. Then using the classical fluctuation–dissipation relationship, one can obtain Eq. (3.3) from Eq. (3.13). Now it should be clear how from the spectroscopic spectral density one can extract the solvent (bare) spectral density that can be transferable from one experiment to another or to electron transfer process in the same solvent.

It is worthwhile emphasizing what types of approximations were used to factor a given spectral density into its solute and solvent parts as in Eq. (3.11). They are: (1) the electrostatic interaction given by Eq. (3.1) is dominant in the system–bath interaction, and (2) since the continuum limit was taken, the local structure of the solvent is ignored. However, a systematic improvement over the second approximation can be made by including higher-order Taylor expansion terms in Eq. (3.9). The consequence of the result in Eq. (3.11) is that the frequency distribution, without overall amplitudes, of the spectroscopic spectral density is universal as long as the two assumptions mentioned above are acceptable.

In practice, what should be calculated, assuming that the spectroscopic spectral density is known, is the charge distributions of the ground and excited states of the chromophore used in the experiment. One can use a semiempirical method to carry out this calculation or assume that the two states are well represented by permanent dipoles estimated from other types of experiments. Then it should be straightforward to calculate the scaling factor s and to obtain the solvent spectral density, $\sigma(\omega)$.

C. Electron Transfer Spectral Density

We next turn to the electron transfer process in the same solvent system. Instead of the charge distributions of the electronic ground and excited states, it is now necessary to consider the donor and acceptor states. We assume that the donor (acceptor) state has the transferring electron localized at the donor (acceptor) site. By denoting the two electric fields as $\mathbf{E}_D(\mathbf{r})$ and $\mathbf{E}_A(\mathbf{r})$, the corresponding scaling factor specifying the ET system is defined as

$$\Gamma_{ET} \equiv (2\pi)^{-3} \int d\mathbf{k} |\mathbf{E}_A(\mathbf{k}) - \mathbf{E}_D(\mathbf{k})|^2 \tag{3.14}$$

The electron transfer spectral density is

$$\rho_{ET}(\omega) = \Gamma_{ET}\sigma(\omega) \tag{3.15}$$

where the solvent (bare) spectral density $\sigma(\omega)$ was obtained from the spectroscopic measurements of the spectroscopic spectral density.

In this section we showed how the spectroscopic spectral density can be transformed into that associated with the electron transfer processes. Applications of this ET spectral density Eq. (3.15), are given in the following sections.

Before we close this section it is necessary to discuss the contribution from the high-frequency intramolecular vibrational modes to electron transfer processes. Although the bare solvent spectral density $\sigma(\omega)$ contains information on the dynamical aspects of solvent only, it is a straightforward procedure to include the intramolecular mode contribution to the electron transfer spectral density. Once the reorganization energies and frequencies of the relevant intramolecular modes are estimated, one can add them to $\rho_{ET}(\omega)$. The remaining procedure to calculate the electron transfer rate kernel is identical, regardless of the inclusion of the intramolecular modes.

IV. TWO-SITE ELECTRON TRANSFER

Electron transfer in condensed phases has been modeled by the spin-boson model, whose Hamiltonian is given by

$$
H = \begin{bmatrix} \varepsilon_D & 0 \\ 0 & \varepsilon_A \end{bmatrix} - \begin{bmatrix} 0 & \Delta \\ \Delta^* & 0 \end{bmatrix}
$$
$$
+ \begin{bmatrix} \displaystyle\sum_\alpha \left(\frac{p_\alpha^2}{2m_\alpha} + \frac{1}{2} m_\alpha \omega_\alpha^2 q_\alpha^2 \right) & 0 \\ 0 & \displaystyle\sum_\alpha \left\{ \frac{p_\alpha^2}{2m_\alpha} + \frac{1}{2} m_\alpha \omega_\alpha^2 \left(q_\alpha - \frac{c_\alpha}{m_\alpha \omega_\alpha^2} \right)^2 \right\} \end{bmatrix}
$$

$$(4.1)$$

where the electron exchange matrix element is denoted as Δ. ε_D and ε_A are the energies of the donor and acceptor states, respectively. Note that bath harmonic oscillators are displaced as the electron transits from the donor to acceptor states. This model Hamiltonian has also been found to be extremely useful in studying the role of the dissipative bath in the tunneling process at the low temperature [9,10]. Since there exists a vast amount of literature on this model Hamiltonian, we shall not attempt to redrive or present a complete summary of the well-known results. Instead, we present a numerical comparison between the two reduced equation-of-motion approaches, the partial ordering prescription (POP) and the chronological ordering prescription (COP) and the variational approach to the calculation of the generalized electron transfer rate kernel that is valid in the nonadiabatic as well as in the adiabatic limits.

In Section II, the model Hamiltonian, Eq. (2.1), for the spectroscopy was unitary transformed into Eq. (2.12), to include the system–bath interaction precisely. Similarly, it is useful to transform the electron transfer Hamiltonian (4.1) by using the unitary transformation operator,

$$
U \equiv \exp \left\{ \frac{i}{h} \sum_\alpha p_\alpha \begin{bmatrix} 0 & 0 \\ 0 & \dfrac{c_\alpha}{m_\alpha \omega_\alpha^2} \end{bmatrix} \right\}
\tag{4.2}
$$

The transformed Hamiltonian is given by

$$
H_{\mathrm{ET}} = U H U^{-1} = H_0 + V
\tag{4.3}
$$

where

$$H_0 = H_S + H_B \tag{4.4}$$

with

$$H_s = \begin{bmatrix} \varepsilon_D & 0 \\ 0 & \varepsilon_A \end{bmatrix}$$

$$H_B = \sum_\alpha \left(\frac{p_\alpha^2}{2m_\alpha} + \frac{1}{2} m_\alpha \omega_\alpha^2 q_\alpha^2 \right)$$

$$V = - \begin{bmatrix} 0 & \Delta \exp(-i\hat{F}) \\ \Delta^* \exp(i\hat{F}) & 0 \end{bmatrix} \tag{4.5}$$

Here the operator $\hat{F}$ is defined as

$$\hat{F} \equiv \sum_\alpha \frac{C_\alpha}{\hbar m_\alpha \omega_\alpha^2} p_\alpha \tag{4.6}$$

The transformed Hamiltonian does not contain the linear system–bath coupling term; instead, the bare electron exchange matrix element is dressed by the bath harmonic oscillators [i.e., $\Delta \exp(-i\hat{F})$]. Hereafter the dressed electron transfer operator V will be treated as a perturbation Hamiltonian. Since it is usually in the strong coupling limit (strong system–bath interaction) so that the thermal average of V over the bath degrees of freedom is negligibly small, it will not be necessary to add and subtract the thermal average of V from the total Hamiltonian to optimize the perturbational approach.

Before we present the formal derivation of the rate equation it is useful to discuss the initial condition of the electron transfer reaction. Usually, the initial state is assumed to be a thermal equilibrium state on the donor state. However, this is unlikely to be the case if the initial state is created by an ultrafast laser pulse, as in a photo-induced electron transfer. A small portion of the electronic ground-state population is excited to create the initial state on the electron donor (electronic excited state) surface, which has to be a nonequilibrium state on the potential energy surface of the donor state. This nonequilibrium state then relaxes to a quasi-equilibrium state on the donor surface as time progresses. During the relaxation process, continuous leakage of the donor population to the acceptor state occurs. Therefore, the complete description of the photo-induced electron transfer should include this nonequilibrium nature of the initial preparation. This was presented by one of the authors and Silbey in [35]. They found that the relaxation of the initial nonequilibrium state induces a time dependence to the electron transfer rate kernel. Since in this case, one has to consider the three states— electronic ground, electronic excited (or donor) state, and electron acceptor states—it was shown that detailed knowledge of the three potential energy surfaces in a two-dimensional solvation coordinate system was needed. Depending on the relative positions of the three potential energy surfaces, the nonequilibrium rate kernel shows nontrivial patterns and strongly deviates from the exponential decaying kinetics. However, in this review we rather focus on the case when the initial state is in equilibrium on the donor potential surface, even though it is straightforward to include the photo-induced nonequilibrium nature of the initial preparation step.

There exist two apparently different methods to obtain the reduced equations of motion. The first is to utilize the Mori–Zwanzig projection operator technique to find the generalized master equation [36,37],

$$\frac{d\mathbf{P}(t)}{dt} = \int_0^t d\tau \, \mathbf{K}_C(t - \tau)\mathbf{P}(\tau) \tag{4.7}$$

where $\mathbf{P}(t)$ denotes the time-dependent population vector and $\mathbf{K}_C(t)$ is the time-dependent rate kernel. Here the subscript C emphasizes that $\mathbf{K}_C(t)$ was obtained by using the *chronological ordering prescription* (COP) [38,39]. Note that the differential equation (4.7) involves a convolution integral. Thus it is convenient to use the Fourier–Laplace transform to evaluate the frequency-dependent populations and take its inverse transform to find its time-domain correspondence. In this procedure, one needs to calculate the frequency-dependent generalized rate kernel. Then the time-dependent population can be calculated as

$$\mathbf{P}(t) = -\text{IFL}\left[\frac{\mathbf{P}(t = 0)}{i\omega - \hat{\mathbf{K}}_C(\omega)}\right] \tag{4.8}$$

where IFI denotes the inverse Fourier–Laplace transform and $\tilde{\mathbf{K}}_C(\omega)$ is the Fourier–Laplace transform of the rate kernel, $\mathbf{K}_C(t)$. The second method is to use Kubo's generalized cumulant expansion to find a generalized master equation of the form [40]

$$\frac{d\mathbf{P}(t)}{dt} = \mathbf{K}_P(t)\mathbf{P}(t) \tag{4.9}$$

where $\mathbf{K}_P(t)$ is the generalized rate kernel expressed by using the partial ordering prescription (POP). The time-dependent population vector is, from Eq. (4.9),

$$\mathbf{P}(t) = \exp\left[\int_0^t d\tau\, \mathbf{K}_P(\tau)\right]\mathbf{P}(t=0) \qquad (4.10)$$

The two approaches are identical in two limiting cases: (1) when the rate kernels are calculated by the second-order perturbation theory in the Markovian limit, and (2) when the exact calculation of the rate kernels is performed by summing the infinite terms. Otherwise, the two approaches result in slightly different outcomes. The second procedure (POP) appears to be convenient in the calculation of the time-domain quantities, such as the calculation of the time-dependent population in the electron transfer reaction by applying the lowest-order perturbation theory (see Sec. IV.C). On the other hand, we find that the first approach (COP) is useful in studying the transition from nonadiabatic to adiabatic rate processes, since the frequency-dependent rate kernel can be systematically resumed by using the variational principle (see Sec. IV.D)

A.   Generalized Master Equation I: Chronological Ordering Prescription

The conventional projection operator method has been used extensively to obtain the reduced equation of motion after taking the average (or trace in quantum mechanical sense) of the density operator over the bath degrees of freedom. The Mori–Zwanzig procedure, particularly, gives a generalized master equation if the interesting variables are the populations of each state of the system. In this subsection we briefly outline the derivation, and some numerical calculations are given later. For the sake of simplicity, we assume that the electron exchange matrix element is not affected by the bath dynamics (i.e., Δ does not depend on the bath degrees of freedom). This is in spirit similar to the classical Condon approximation in the optical transition process where the electric dipole matrix element does not depend on the nuclear degrees of freedom.

In general, the rate equations can be obtained by solving the quantum Liouville equation

$$\frac{d\rho(t)}{dt} = -iL\rho(t) \qquad (4.11)$$

where the Liouville operators are defined as commutators,

$$LA \equiv [H_{ET}, A] = [H_0, A] + [V, A]$$

$$L_0 A \equiv [H_0, A] \tag{4.12}$$

$$L_V A \equiv [V, A]$$

For the sake of simplicity we shall denote the Liouville space vector by $|\cdots\rangle\rangle$, and the scalar product of two Hilbert space operators, A and B, is denoted as

$$\langle\langle A | B \rangle\rangle \equiv \text{Tr}[A^+ B] \tag{4.13}$$

Here the trace is over the complete Hilbert space, including the system and the solvent. By following Sparpaglione and Mukamel [41] closely, the Liouville space projection operator $\hat{P}$ is defined as

$$\hat{P} \equiv |\hat{D}\rho_d\rangle\rangle\langle\langle\hat{D}| + |\hat{A}\rho_a\rangle\rangle\langle\langle\hat{A}| \tag{4.14}$$

where the donor and acceptor operators in the Hilbert space are denoted by $\hat{D} \equiv |d\rangle\langle d|$ and $\hat{A} \equiv |a\rangle\langle a|$, respectively, and the two density operators associated with the donor and the acceptor, ρ_d and ρ_a, are, respectively,

$$\rho_d = \exp(-H_d/k_b T)/\text{Tr}[\exp(-H_d/k_b T)]$$
$$\rho_a = \exp(-H_a/k_b T)/\text{Tr}[\exp(-H_a/k_b T)] \tag{4.15}$$

The complementary operator $\hat{Q}$ is

$$\hat{Q} = 1 - \hat{P} \tag{4.16}$$

By using the standard projection operator techniques, the reduced equation of motion for the populations (*generalized master equation I*) can be obtained:

$$\frac{dp_d}{dt} = -\int_0^t d\tau [k_{da}(t - \tau)p_d(\tau) - k_{ad}(t - \tau)p_a(\tau)]$$
$$\frac{dp_a}{dt} = \int_0^t d\tau [k_{da}(t - \tau)p_d(\tau) - k_{ad}(t - \tau)p_a(\tau)] \tag{4.17}$$

where the populations of the donor and the acceptor are

$$p_d(t) = \text{Tr}[\hat{D}\rho(t)]$$
$$p_a(t) = \text{Tr}[\hat{A}\rho(t)] \tag{4.18}$$

The time-dependent rate kernel $k_{da}(t)(k_{ad}(t))$ describes the transition rate from donor(acceptor) to acceptor(donor) as given by

$$k_{da}(t) \equiv \langle\!\langle \hat{D}|L\exp(-i\hat{Q}Lt)\hat{Q}L|\hat{D}\rho_d\rangle\!\rangle$$
$$k_{ad}(t) \equiv \langle\!\langle \hat{A}|L\exp(-i\hat{Q}Lt)\hat{Q}L|\hat{A}\rho_a\rangle\!\rangle \tag{4.19}$$

The generalized master equation (4.17) with the time-dependent rate kernels, Eq. (4.19), is formally exact, since the full Liouville operator L instead of L_0 was used in Eq. (4.19).

It turns out that the Fourier–Laplace transform of the generalized rate equation is useful in the following derivation of the resumed rate kernel. By denoting the Fourier–Laplace transform of an arbitrary time-dependent function $f(t)$ as $F(\omega)$,

$$F(\omega) = \int_0^\infty dt \exp(i\omega t) f(t) \tag{4.20}$$

with its inverse transform

$$f(t) = (2\pi)^{-1} \int_{-\infty}^\infty d\omega \exp(-i\omega t) F(\omega) \tag{4.21}$$

the generalized rate equations (4.17) can be written as

$$i\omega \begin{bmatrix} P_d(\omega) \\ P_a(\omega) \end{bmatrix} + \begin{bmatrix} p_d(t=0) \\ p_a(t=0) \end{bmatrix} = \begin{bmatrix} K_{da}(\omega) & -K_{ad}(\omega) \\ -K_{da}(\omega) & K_{ad}(\omega) \end{bmatrix} \times \begin{bmatrix} P_d(\omega) \\ P_a(\omega) \end{bmatrix} \tag{4.22}$$

where $P(\omega)$ and $K(\omega)$ correspond to the Fourier–Laplace transforms of $p(t)$ and $k(t)$, respectively. Similarly, we define the Liouville space advanced Green functions as

$$G(\omega) = -i \int_0^\infty dt \exp(i\omega t) \exp(-iLt) = \frac{1}{\omega - L}$$
$$G_0(\omega) = -i \int_0^\infty dt \exp(i\omega t) \exp(-iL_0 t) = \frac{1}{\omega - L_0} \tag{4.23}$$

Thus the Fourier–Laplace transform of the rate kernel, for example, $K_{da}(\omega)$, is given by

$$K_{da}(\omega) = i\langle\!\langle \hat{D}|L_V \frac{1}{\omega - \hat{Q}L} \hat{Q}L_V|\hat{D}\rho_d\rangle\!\rangle \tag{4.24}$$

From now on we focus on $K_{da}(\omega)$ only, since the calculation of $K_{ad}(\omega)$ is precisely identical to that of $K_{da}(\omega)$. Using the formal relation

$$\hat{Q}L = L_0 + \hat{Q}L_V \tag{4.25}$$

and the operator identity

$$\frac{1}{\omega - \hat{Q}L} = \frac{1}{\omega - \hat{Q}L_0}\left(I + \frac{\hat{Q}L_V}{\omega - \hat{Q}L}\right)$$

$$= \frac{1}{\omega - \hat{Q}L_0}\sum_{n=0}^{\infty}\left[\hat{Q}L_V\frac{1}{\omega - \hat{Q}L_0}\right]^n \tag{4.26}$$

one can obtain the formally exact perturbative expansion of the rate kernel

$$K_{da}(\omega) = \sum_{n=1}^{\infty}K_{da}^{(2n)}(\omega) \tag{4.27}$$

where the $2n$th-order contribution to the rate kernel in Fourier–Laplace space is:

$$K_{da}^{(2n)}(\omega) = i\langle\langle\hat{D}|[L_V G_0(\omega)L_V G_0(\omega)\hat{Q}]^{(n-1)}L_V G_0(\omega)L_V|\hat{D}\rho_d\rangle\rangle \tag{4.28}$$

Note that the terms including odd numbers of actions of the L_1 operator; vanish, since the diagonal matrix elements are taken, and the contribution, $K_{da}^{(2n)}(\omega)$ is order $2n$ in the nonadiabatic coupling Δ. For example, the usual second-order Fermi golden-rule expression is the first term in Eq. (4.27), that is,

$$K_{da}^{(2)}(\omega) = i\langle\langle\hat{D}|L_V G_0(\omega)L_V|\hat{D}\rho_d\rangle\rangle \tag{4.29}$$

which is called the *nonadiabatic rate kernel* throughout this paper.

The calculation of the time-dependent population evolution is now reduced to that of the rate kernel in the Fourier–Laplace space if the Mori–Zwanzig projection operator technique is used. There are numerous cases where the second-order expression for the rate kernel is quantitatively accurate enough to predict the rate. However, there are two important cases where the nonadiabatic rate does not correctly represent the reaction rate. The first obvious case is when the electron exchange matrix element is sufficiently large so that one cannot ignore the higher-order terms in the perturbative expansion of the rate kernel in Eq. (4.27). The second is when the bath correlation time is very slow. In this case the solvation of the wavepacket created in the product state by the fluctuation in the system–bath interaction stays near the transition-state (reactive window or curve-crossing) region for a sufficient time. Consequently, multiple recrossings, associated with higher-order terms in Eq. (4.27), are likely to occur, and eventually the reaction rate becomes independent of the electron exchange matrix element. In Section IV.C, Schwinger's stationary variational principle will be used to provide a systematic procedure for obtaining an electron transfer rate kernel that is valid from the nonadiabatic to adiabatic limits.

B. Generalized Master Equation II: Partial Ordering Prescription

There exists another type of generalized master equation where the rate kernel is multiplicative in the differential equation of the population. Its derivation starts from the same quantum Liouville equation (4.11). Instead of applying the Mori–Zwanzig projection operator technique, by changing to the interaction representation with respect to the zeroth Liouville operator, Eq. (4.11) can be rewritten as

$$\frac{d\tilde{\rho}(t)}{dt} = -i\widetilde{L}_V(t)\tilde{\rho}(t) \tag{4.30}$$

where

$$\tilde{\rho}(t) = e^{iL_0 t}\rho$$
$$\widetilde{L}_V(t) = e^{iL_0 t}L_V e^{-iL_0 t} \tag{4.31}$$

It is again assumed that the initial density operator can be factorized into the system and the bath parts; that is, the system and bath are assumed to be uncorrelated at time $t = 0$,

$$\tilde{\rho}(0) = \tilde{\sigma}(0)\widetilde{\rho_B}(0) \tag{4.32}$$

where $\tilde{\sigma}(0)$ and $\widetilde{\rho_B}(0)$ are the system and bath density operators, respectively. Since the dynamics of the system degrees of freedom is our interest, the system density operator in time t is determined by

$$\tilde{\sigma}(t) = \langle \tilde{\rho}(t) \rangle_B \tag{4.33}$$

where $\langle\cdots\rangle_B$ denotes the trace over the bath degrees of freedom. Then the formal solution for $\tilde{\sigma}(t)$ is given by

$$\tilde{\sigma}(t) = \left\langle \exp_+\left[-i\int_0^t d\tau \widetilde{L_V}(\tau)\right]\right\rangle_B \tilde{\sigma}(0) \tag{4.34}$$

where $\exp_+\{\cdots\}$ is the time-ordered exponential operator. Equation (4.34) can be expanded in terms of the generalized moments. However, as proven by Kubo [40], one can construct an equation of motion of the form

$$\frac{d\tilde{\sigma}(t)}{dt} = \hat{K}(t)\tilde{\sigma}(t) \tag{4.35}$$

where the generalized rate kernel operator $\tilde{K}(t)$ is given by a series summation of the cumulant terms,

$$\hat{K}(t) = \sum_n \hat{K}_n(t) \tag{4.36}$$

Note that $\hat{K}(t)$ is still an operator of the system degrees of freedom, since only the trace over the bath degrees of freedom was taken. Here each cumulant term can be related to the generalized moments, for example [39],

$$\hat{K}_1 = 0 \tag{4.37a}$$

$$\hat{K}_2(t) = (-i)^2 \int_0^t d\tau_1 \, m_2(t,\tau_1) \tag{4.37b}$$

$$\hat{K}_3(t) = (-i)^3 \int_0^t d\tau_1 \int_0^{\tau_1} d\tau_2 \, m_3(t,\tau_1,\tau_2) \tag{4.37c}$$

$$\hat{K}_4(t) = (-i)^4 \int_0^t d\tau_1 \int_0^{\tau_1} d\tau_2 \int_0^{\tau_2} d\tau_3 \, [m_4(t,\tau_1,\tau_2,\tau_3) - m_2(t,\tau_1)m_2(\tau_2,\tau_3)$$

$$- m_2(t,\tau_2)m_2(\tau_1,\tau_3) - m_2(t,\tau_3)m_2(\tau_1,\tau_2)] \tag{4.37d}$$

where

$$m_n(t,\tau_1,\ldots,\tau_{n-1}) \equiv \langle \widetilde{L_V}(t)\widetilde{L_V}(\tau_1)\cdots\widetilde{L_V}(\tau_{n-1})\rangle_B \tag{4.38}$$

Here $m_n(t,\tau_1,\ldots,\tau_{n-1})$ is the nonlinear correlation function depending on n time variables. The first two terms in the fourth-order contribution to the rate kernel in Eq. (4.37d) are called the *Markovian* term, whereas the last two terms are *memory* terms [42]. Because of the second term in the integrand of $K_4(t)$, the diverging contribution from the first term cancels when t and τ_1 are well separated from τ_2 and τ_3.

Here it should be mentioned that there is a distinct difference between the two reduction schemes mentioned briefly above. If one uses the COP to calculate the fourth-order rate kernel, it turns out that the fourth-order rate kernel is determined by the Markovian terms [the first two terms in Eq. (4.37d)]. On the other hand, the POP result for the fourth-order rate kernel involves the memory terms [the third and fourth terms in Eq. (4.37d)], in addition to the Markovian terms. We next calculate the populations of the two states defined as

$$p_d(t) \equiv \text{Tr}[|D\rangle\langle D|\rho(t)] = \text{Tr}_{\text{sys}}[|D\rangle\langle D|\tilde{\sigma}(t)]$$
$$p_a(t) \equiv \text{Tr}[|A\rangle\langle A|\rho(t)] = \text{Tr}_{\text{sys}}[|A\rangle\langle A|\tilde{\sigma}(t)]$$

(4.39)

where $\text{Tr}_{\text{sys}}[\cdots]$ denotes the trace over the system eigenstates. By using the exact equation of motion for $\tilde{\sigma}(t)$ in Eq. (4.35), the time evolution of the populations is determined by the linear differential equation (*generalized master equation II*)

$$\frac{d\mathbf{P}(t)}{dt} = \mathbf{K}(t)\mathbf{P}(t)$$

(4.40)

where $\mathbf{P}(t)$ is a column vector whose elements are $\{p_d(t), p_a(t)\}$, and $\mathbf{K}(t)$ is the 2×2 matrix whose elements are given by

$$[\mathbf{K}(t)]_{ij} = \text{Tr}_{\text{sys}}[|i\rangle\langle i|\hat{K}(t)|j\rangle\langle j|]$$

(4.41)

with

$$[m_n(t, \tau_1, \ldots, \tau_{n-1})]_{ij} \equiv \text{Tr}_{\text{sys}}[|i\rangle\langle i|\langle \widetilde{L_V}(t)\widetilde{L_V}(\tau_1)\cdots\widetilde{L_V}(\tau_{n-1})\rangle_R|j\rangle\langle j|] \quad (4.42)$$

Hereafter we focus on the calculations of the off-diagonal matrix element of the generalized rate kernel $[\mathbf{K}(t)]_{ij}$ which describes the transition rate from $|j\rangle$ to $|i\rangle$ states. The diagonal rate kernels can be calculated directly by using the conservation condition,

$$[\mathbf{K}(t)]_{ii} = -\sum_{j\neq i}[\mathbf{K}(t)]_{ji}$$

(4.43)

C. Second-Order Rate Kernel: Partial Ordering Prescription

It has been found that the second-order perturbation result is often quantitatively accurate enough to describe the electron transfer reaction of a two-site system in condensed phases. As discussed above, there are two types of generalized master equations. In this section we utilize the POP result, Eq. (4.40) with Eq. (4.41). As can be seen in Eq. (4.37b), the second-order rate kernel, $[\tilde{K}_2(t)]_{21}$, representing the transition rate from the donor to acceptor states can be calculated from the two-time correlation function $m_2(t, \tau_1)$.

Inserting the definition of $\widetilde{L}_V(t)$ into Eq. (4.38), we find that the [2,1]th matrix element of $m_2(t, \tau_1)$ can be written as

$$[m_2(t, \tau_1)]_{21} = -2\,\mathrm{Re}\{\Delta_{AD}^2 \exp[-P_{AD}(t - \tau_1) - iQ_{AD}(t - \tau_1) - i\varepsilon_{AD}(t - \tau_1)]\}$$

$$(4.44)$$

where the Baker–Hausdorff theorem was used. Here the two auxiliary functions, $P(t)$ and $Q(t)$, were defined in Eq. (2.30) by replacing the spectroscopic spectral density, $\rho_S(\omega)$, with the electron transfer spectral density, $\rho_{ET}(\omega)$. Inserting $[m_2(t, \tau_1)]_{21}$ into Eq. (4.37B), one can calculate the second-order rate kernel describing the population transfer from $|D\rangle$ to $|A\rangle$.

Since the time-correlation function in Eqs. (4.44) is a rapidly decaying function in general, one can replace the upper limit of the integral in Eq. (4.37b) with an infinity. In this limit, the *rate constant* can be evaluated by invoking the stationary-phase approximation to the integral expression (4.37b) so that the Marcus result for the two-site electron transfer can be obtained [43,44]. However, since the same results were derived by numerous workers, we shall not rederive them here. Instead, we numerically calculate the rate kernel by using the spectral density obtained from the spectroscopic measurement.

Recently, Fleming and co-workers [5] measured the three-pulse photon echo peak shift of a dye, IR144, in acetonitrile as a function of the second delay time between the second and third pulses and found the corresponding spectroscopic spectral density representing three distinctive contributions: (1) the vibrational coherent part originated from the intramolecular vibrational mode of IR144, (2) ultrafast solvent inertial component, and (3) the slow diffusive motions of the acetonitrile. Consequently, the spectroscopic spectral density is given by a mixture of the three contributions. However, since the dynamical contributions from the solvent are the only relevant information one needs to extract from the spectroscopic spectral density, contributions 2 and 3 are only taken into account. The normalized spectral density thus obtained is shown in Figure 3. The spike at low frequency represents the slowly decaying component in the solvation dynamics, and

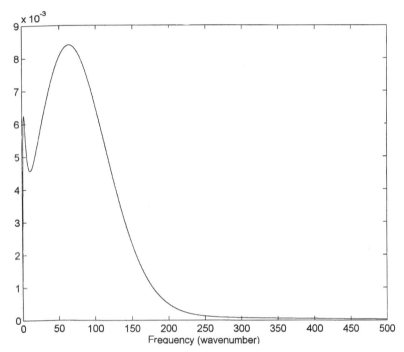

Figure 3. Normalized bare solvent spectral density $\sigma(\omega)$ of acetonitrile (refer to [5] for detailed discussion).

the broad spectrum around 80 cm^{-1} is associated with the fast inertial decaying component. Since the shape of the solvent spectral density is given in Figure 3, by scaling the magnitude of the spectral density by using the reorganization energy measured experimentally or estimated from the numerical calculation of Γ_{ET} in Eq. (3.14), the electron transfer spectral density can be obtained.

To provide a quantitative picture, we numerically calculate the second-order rate kernel by changing parameters, such as the energy gap between the two states and reorganization energies.

1. Time Dependence of the Rate Kernel

The time dependence of the rate kernel is shown in Figure 4. When the solvent (acetonitrile) spectral density in Figure 3 is used and the solvent reorganization energy is assumed to be 300 cm^{-1} (note that the unit of all energies is the angular frequency in cm^{-1} throughout this review), the second-order rate kernel reaches its asymptotic value; that is, the second-order

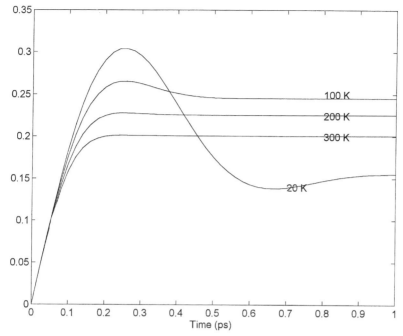

Figure 4. Time-dependent second-order rate kernel for electron transfer process of a two-state system. The electron transfer spectral density is given in Figure 3. The solvent reorganization energy is assumed to be 300 cm^{-1}, and the energy gap between the donor and acceptor states is 500 cm^{-1} exothermic. The y-axis is in arbitrary units.

rate constant, in 200 fs at room temperature. The energy gap between the two states is assumed to be 500 cm^{-1} exothermic in this calculation. As temperature decreases, the rate kernel becomes strongly time dependent. Therefore, full knowledge of the spectral density is needed to describe completely the time dependence of the rate kernel. At this point it is probably appropriate to comment on the appropriateness of using a temperature-independent spectral density (i.e., the only influence of temperature is on the occupation numbers of the harmonic modes). Nagasawa et al. have carried out a study of spectral broadening of IR144 in PMMA glass from 300 to 30 K, using the three-pulse photon echo peak shift method [6]. They find that many aspects—the absorption spectrum, the time dependence of the $\tau^*(T)$, and the stimulated echo signals themselves —of the data can be described quantitatively with a temperature-dependent spectral density, provided that both real and imaginary portions of the line-broadening function are used [6]. The one property that is not well described by this approach is the temperature dependence of the asymptotic value of the peak shift,

$\tau^*(T \to \infty)$. Whether this reflects a breakdown of the harmonic model is not yet clear.

2. Energy Gap Dependence of the Rate Kernel

For a given ET spectral density with the reorganization energy 300 cm^{-1}, the rate constants obtained by replacing the upper limit t with infinity in Eq. (4.37b), as a function of temperature as well as the energy gap between the two states, are calculated and plotted in Figure 5. At room temperature, since the frequencies of the bath oscillators are rather low in comparison to the thermal frequency $(2k_B T/\hbar)$, the classical limit of the rate constant, such as Marcus's expression obtained by applying the stationary-phase approximation to evaluate the integral in Eq. (4.37b), is acceptable. Therefore, in this high-temperature limit, the Gaussian functional form for the rate constant with respect to the effective energy gap $\varepsilon_{AD} + \lambda_{ET}$ (i.e., Marcus form) is an excellent approximation. Furthermore, in this case the solvent reorganization energy λ_{ET} is the only parameter representing the effect of the fluctuating bath degrees of freedom. However, as temperature is lowered,

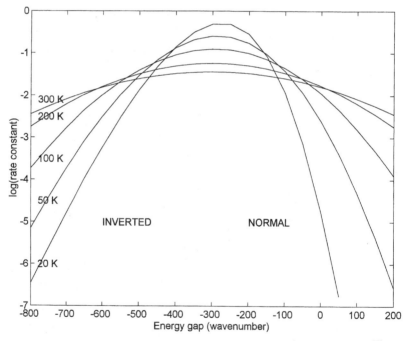

Figure 5. Energy gap dependence of the two-state electron transfer rate constant. The parameters are the same in Figure 4.

the deviation of the rate constant from the Gaussian form is clearly notice-able in Figure 5. The decrease of the rate constant in the normal region is much more drastic than that in the inverted region. In contrast, if the electron transfer is activationless (around -300 cm^{-1}), the rate constant increases as temperature decreases. Since the initial wavepacket on the donor potential surface is more localized at the curve-crossing seam at low temperature in the activationless case, it is likely that the transition from the donor to acceptor states accelerates as the temperature decreases. However, in the cases of the normal and inverted regions, the activation energy has to be input from the bath. If the temperature is low, the fluctua-tion amplitude of the bath degrees of freedom is consequently small so that the probability of the initial wavepacket to visit the curve-crossing region decreases. Hence the rate constant decreases with decreasing temperature. Also, it should be noted that the quantum tunneling process in the inverted region plays a crucial role when there exist high-frequency modes [45,46].

From the analysis presented above, one can conclude that a full knowl-edge of the ET spectral density instead of a single scalar quantity λ is needed to describe the electron transfer process in condensed phases at low tem-perature or when the solvent shows a marked deviation from the classical description, such as liquid water.

D. Transition from Nonadiabatic to Adiabatic Rate Kernel

In the electron transfer reaction in condensed phases, there are two import-ant time scales determining the reaction rate, which are the inverse of the electron exchange matrix element and the correlation time of the bath fluc-tuation. If the coupling matrix element Δ is very small, the nonadiabatic electron transfer rate (i.e., the Fermi golden-rule expression) is quantita-tively acceptable. However, the statement above is not entirely correct in general. The reason is because the time scale of the bath is another import-ant factor determining the adiabaticity of the reaction rate. If the time scale of the bath is sufficiently fast, the wavepacket created in the curve crossing (or transition state) region, where the Franck–Condon factor becomes a maximum, can be quickly relaxed into an equilibrium state on the potential energy surface of the product state. This means that the survival probability of the created wavepacket on the transition state region is very small. In this case the second-order Fermi golden rule can indeed be useful in calculating the reaction rate, and the rate is proportional to the square of the electron exchange matrix element. On the other hand, if the bath correlation time is very slow, the reaction rate is now determined by the solvation process. Consequently, the rate does not depend on the electron exchange matrix element. Perhaps Zusman [47] was the first to obtain a theoretical expression connecting between the nonadiabatic and adiabatic reaction rate constants.

Later numerous workers [41,44,48–50] generalized Zusman's theory, although the essential aspects of those results are quite similar. The overall electron transfer rate is given by the standard formula

$$1/k = 1/k_a + 1/k_{na}$$

where k_a and k_{na} are the adiabatic and nonadiabatic reaction rate constants, respectively. In contrast to the nonadiabatic reaction rate, the adiabatic rate does not depend on the magnitude of the electron exchange matrix element but instead, depends on the relaxation rate (survival time) as $1/\tau$. One can understand this crossover behavior as follows. Suppose that the initial state is in thermal equilibrium with the electron donor state. The fluctuation of the bath degrees of freedom can create a nuclear wavepacket at the curve-crossing region, where the potential energy surfaces are constructed by the solvent nuclear degrees of freedom. The second-order action of the coupling potential can induce a transition from the donor to the acceptor electronic states. If the solvation dynamics on the acceptor state is sufficiently fast (i.e., the survival time of the reactive state around the transition state region is very short), the rate is determined mainly by the second-order process (i.e., the nonadiabatic reaction rate). On the other hand, if the solvation is very slow (long survival time of the reactive state around the crossing region), the recrossing of the wavepacket back to the donor state (this recrossing process is associated with the higher-order terms with respect to Δ) becomes possible, so that the rate-determining process is now the solvation dynamics. Consequently, the adiabatic reaction rate is *solvent controlled*. Sparpaglione and Mukamel [41] presented a formal derivation of the generalized rate equation by using projection operator techniques. Expanding the generalized rate kernel perturbatively and invoking the static approximation, they were able to perform exactly the resummation of the perturbatively expanded rate kernel. This is identical to the resummation of the expansion of the rate kernel by considering the first two terms, which are the second- and fourth-order rate kernels, and using the [1,0]-Pade approximant,

$$k = k^{(2)} + k^{(4)} + \cdots \simeq \frac{k^{(2)}}{1 - k^{(4)}/k^{(2)}} \tag{4.45}$$

When the zero-frequency component is considered to be the *rate constant* and the characteristic solvent time scale is properly considered, Zusman's result can be obtained from Sparpaglione and Mukamel's result.

In the derivation above, the generalized rate kernel was expanded in terms of the even-order perturbation terms. We now discuss the resummation of

Eq. (4.27) based on the stationary variational principle. From now on, we only focus on the rate kernel $K_{da}(\omega)$ and omit the subscript da for the sake of notational simplicity.

To rewrite the perturbation expansion of the rate kernel, it is useful to define the zeroth-order state $|\phi_0(\omega)\rangle\rangle$ and a new perturbation operator $v(\omega)$ in the Liouville space as [51]

$$|\phi_0(\omega)\rangle\rangle \equiv G_0^{1/2}(\omega)K_1|\hat{D}\rho_d^{1/2}\rangle\rangle$$

$$v(\omega) \equiv \frac{G_0^{1/2}(\omega)L_1 G_0(\omega)\hat{Q}L_I G_0^{1/2}(\omega)}{|\Delta|^2} \tag{4.46}$$

The rate kernel in Eq. (4.27) can be rewritten in terms of $|\phi_0(\omega)\rangle\rangle$ and $v(\omega)$

$$K(\omega) = i\langle\langle\phi_0|\chi(\omega)\rangle\rangle \tag{4.47}$$

where

$$|\chi(\omega)\rangle\rangle \equiv \sum_{n=0}^{\infty} [|\Delta|^2 v(\omega)]^n |\phi_0(\omega)\rangle\rangle = \sum_{n=0}^{\infty} |\Delta|^{2n} |\phi_n(\omega)\rangle\rangle \tag{4.48}$$

Here the nth-order state vector is naturally defined as above; that is, $|\phi_n(\omega)\rangle\rangle \equiv [v(\omega)]^n |\phi_0(\omega)\rangle\rangle$. The physical meaning of Eq. (4.48) is that the *Liouville state vector $|\chi(\omega)\rangle\rangle$ is given by a linear combination of $|\phi_n(\omega)\rangle\rangle$'s, where the expansion coefficient of the nth term is given by $|\Delta|^{2n}$.* Therefore, a set of $\{|\phi_n(\omega)\rangle\rangle\}$ is a complete basis set in this case, and Liouville space is completely spanned by this basis set.

From the definition of $|\chi(\omega)\rangle\rangle$, one can find the relation

$$|\chi(\omega)\rangle\rangle = |\phi_0(\omega)\rangle\rangle + |\Delta|^2 v(\omega)|\chi(\omega)\rangle\rangle \tag{4.49}$$

which corresponds to the Lippman–Schwinger equation [52] extended to *Liouville space*. We next apply Schwinger's variational principle to find the variational functional $K^S(\omega)$ [51]:

$$K^S(\omega) = i\langle\langle\phi_0(\omega)|\chi^T(\omega)\rangle\rangle + i\langle\langle\chi^T(\omega)|\phi_0(\omega)\rangle\rangle - i\langle\langle\chi^T(\omega)|1 - \Delta^2 v(\omega)|\chi^T(\omega)\rangle\rangle \tag{4.50}$$

which is assumed to be *stationary* for small variations of the trial state $|\chi^T(\omega)\rangle\rangle$ about $|\chi(\omega)\rangle\rangle$. This means that the *stationary value of $K^S(\omega)$ is $k(\omega)$.* There exist several attempts using variational approaches to calculate the rate constant. Most of them are concerned with the energy of the transition state or the position of the dividing surface and utilize the minimum-energy variational principle. Unlike those approaches, here the rate kernel itself is the objective of the variational procedure instead of the energy. Thus this approach is perhaps a more direct way to calculate the rate kernel in general.

We now introduce a trial state vector, $|\chi^T(\omega)\rangle\rangle$, as

$$|\chi^T(\omega)\rangle\rangle = \sum_{n=0}^{N-1} c_n |\phi_n(\omega)\rangle\rangle \tag{4.51}$$

where c_n are the variational parameters. Note that the trial state is expressed as a linear combination of $|\phi_n(\omega)\rangle\rangle$ (for $n = 0$ to $N - 1$): that is, the trial state $|\chi^T(\omega)\rangle\rangle$ is expanded in a subspace constructed by $|\phi_n(\omega)\rangle\rangle$ (for $n = 0$ to $N - 1$). This type of trial state vector is known as the Cini–Fubini trial function [53], where the trial function is given by a linear combination of a finite set of $|\phi_n(\omega)\rangle\rangle$'s instead of infinite basis functions.

To determine the variational parameters, $\{c_i\}$, we solve the linear equation

$$\frac{\partial K^S(N, \omega)}{\partial c_i} = 0 \qquad \text{for all } i$$

and find that the stationary value $K^S(N, \omega)$ becomes identical to the $[N, N - 1]$ Padé approximants [54].

1.   One-Dimensional Case ($N = 1$): [1, 0] Padé Approximant

First consider the simplest case of all, that is, the case when $N = 1$. The trial function assumes that

$$|\chi^T(\omega)\rangle\rangle = c_0 |\phi_0(\omega)\rangle\rangle \tag{4.52}$$

where c_0 is the only variational parameter determined from the stationary variational principle. In this case the Cini–Fubini subspace is a one-dimensional space constructed by $|\phi_0(\omega)\rangle\rangle$. Then one finds that the variationally determined stationary value for the rate kernel is

$$K^S(N = 1, \omega) = \frac{[K^{(2)}(\omega)]^2}{K^{(2)}(\omega) - K^{(4)}(\omega)} \tag{4.53}$$

Now if we take the zero-frequency components of the perturbative rate kernels that are the corresponding rate constants, we recover Eq. (4.45). Since the frequency dependence of the rate kernels in Eq. (4.53) is fully retained, Eq. (4.53) should be considered as an improved version of Eq. (4.45). This result was also obtained by Mukamel and co-workers [41]; however, a justification based on the variational principle was not given previously.

2. Frequency-Dependent Transmission Coefficient

One can reinterpret the result, Eq. (4.53), as

$$K^S(N = 1, \omega) = \chi^{(1)}(\omega) K_{na}(\omega) \tag{4.54}$$

where $K_{na}(\omega)$ is the nonadiabatic rate kernel that is equal to the second-order Fermi golden-rule expression (4.29), and $\chi^{(1)}(\omega)$ denotes the frequency-dependent transmission coefficient calculated in the one-dimensional subspace and is defined as

$$\chi^{(1)}(\omega) \equiv \frac{1}{1 + K_{na}(\omega)\tau(\omega)} \tag{4.55}$$

with

$$\tau(\omega) = |K^{(4)}(\omega)|/[K_{na}(\omega)]^2 \tag{4.56}$$

Here $\tau(\omega)$ is the frequency-dependent survival time and is related, *not identical*, to the bath correlation time. Note that the transmission coefficient $\chi^{(1)}(\omega) < 1$ for all frequencies. It is also possible to interpret the factor $\chi^{(1)}(\omega)$ as describing the renormalization effect on the coupling matrix element Δ induced by the higher-order rate contribution and the bath fluctuation. The product $K_{na}(\omega)\tau(\omega)$ in Eq. (4.41) is often interpreted as the frequency-dependent *adiabaticity parameter* since if $K_{na}(\omega)\tau(\omega) \ll 1$ for all frequencies, the rate is determined completely by the second-order rate process (nonadiabatic limit), whereas if $K_{na}(\omega)\tau(\omega) \gg 1$ for all frequencies, the reaction is governed by the (survival) time $\tau(\omega)$ and does not depend on the electron exchange matrix element; that is,

$$K^S(N = 1, \omega) = \tau^{-1}(\omega) \tag{4.57}$$

This is the case of the adiabatic limit and is realized when the solvent bath correlation time is very slow, so that the probability of recrossing, more precisely multiple actions of the transition operator $v(\omega)$, becomes large.

3. Time-Dependent Transmission Coefficient and Rate Kernel

As can be seen in Eq. (4.54), the generalized frequency-dependent rate kernel is given by a product of the frequency-dependent transmission coefficient and the second-order rate kernel. Thus the time-dependent rate kernel is given by a convolution, such as

$$k_{da}^{S}(t) = \int_{-\infty}^{\infty} d\tau\, \chi^{(1)}(t - \tau) k_{na}(\tau) \qquad (4.58)$$

where $\chi^{(1)}(t)$ and $k_{na}(t)$ are the inverse Fourier transforms of $\chi(\omega)$ and $K_{na}(\omega)$, respectively. The rate kernel thus obtained is determined by the cross correlation between the time-dependent transmission coefficient and second-order rate kernel. If the time scale of the second-order rate kernel is fast compared to that of the transmission coefficient [i.e., $k_{na}(\tau) \approx k_{na}\delta(\tau)$], the time dependence of the rate kernel is determined by that of the transmission coefficient, so that $k^{S}(t) \approx \chi^{(1)}(t) k_{na}$. On the other hand, if the time scale of the transmission coefficient is much faster than that of the second-order rate kernel [i.e., $\chi^{(1)}(t - \tau) \approx \chi_{0}^{(1)}\delta(t - \tau)$], the rate kernel becomes $k^{S}(t) = \chi_{0}^{(1)} k_{na}(t)$.

4. Nonadiabatic Limit, $\chi^{(1)}(\omega) \approx 1$

Next we consider the nonadiabatic limit more in detail. This limit is the case when the transmission coefficient $\chi^{(1)}(\omega)$ is approximately equal to unity. The nonadiabatic condition, $K_{na}(\omega)\tau(\omega) \ll 1$, can be met when the electron exchange matrix element is very small regardless of the characteristic solvent time scale. Usually, the nonadiabatic limit has been assumed in such cases. However, there is another possibility satisfying this inequality [i.e., when the survival time scale, $\tau(\omega)$, is much shorter than the inverse of the second-order rate kernel]. This limit is identical in spirit to the celebrated noninteracting blip approximation [9], where there is no correlation between the two consecutive off-diagonal density matrix evolutions (blips) separated by diagonal density matrix evolutions (sojourns). This approximation is applicable when the bath correlation time is fast; that is, the memory of the bath fluctuation decays very rapidly. These two cases, (1) small Δ and (2) short bath memory (short bath correlation time), are apparently independent of each other because the former depends on the intrinsic properties of the electron donor and acceptor states, whereas the latter is determined by the dynamical aspect of the bath. However, if either condition 1 or 2 is satisfied, the rate kernel is determined by the second-order expression $K_{na}(\omega)$. Finally, we find it interesting to note that *based on Schwinger's stationary variational* *principle, the first-order approximation to the rate kernel is equal to Eq. (4.54), not to $K_{na}(\omega)$.*

To summarize this section, the two-state electron transfer rate kernel can be calculated once the corresponding spectral density is obtained from a rescaled spectroscopic spectral density measured, for example, via the photon echo technique. There appear to be two seemingly different methods for calculating the time evolution of the populations. The partial ordering prescription (generalized master equation II) is found to be useful in a time-domain analysis such as the direct calculation of the population evolution. On the other hand, the chronological ordering prescription approach (generalized master equation I) can be used in a calculation of the frequency-dependent rate kernel that is valid from nonadiabatic to adiabatic limits. Schwinger's stationary variational principle was found to be useful for providing a systematic procedure for the latter calculation.

V.   THREE-SITE ELECTRON TRANSFER

For a three-state system, the total Hamiltonian can be rewritten as

$$
H = \begin{bmatrix} 0 & -\Delta_{BD} & 0 \\ -\Delta_{BD} & \varepsilon_{BD} & -\Delta_{AB} \\ 0 & -\Delta_{AB} & \varepsilon_{AD} \end{bmatrix}
$$
$$
+ \sum_{\alpha} \left\{ \frac{p_{\alpha}^2}{2m_{\alpha}} + \frac{m_{\alpha}\omega_{\alpha}^2}{2} \begin{bmatrix} q_{\alpha}^2 & 0 & 0 \\ 0 & (q_{\alpha} - a_{\alpha}/m_{\alpha}\omega_{\alpha}^2)^2 & 0 \\ 0 & 0 & (q_{\alpha} - b_{\alpha}/m_{\alpha}\omega_{\alpha}^2)^2 \end{bmatrix} \right\}
$$
$$
(5.1)
$$

Here the three states are denoted as $|D\rangle$, $|B\rangle$, and $|A\rangle$, respectively. If the whole electron transfer system consists of three sites, the initial electron donor (final electron acceptor) state $|D\rangle$ ($|A\rangle$) is the state with the transferring electron is localized at the donor (acceptor) site. Likewise the bridge state $|B\rangle$ is a state with the electron localized at the bridge site. This model Hamiltonian is formally identical to that used in [35] to study nonequilibrium photoinduced electron transfer in condensed phases. Two neighboring states are coupled by nonzero electron exchange matrix elements, Δ_{BD} and Δ_{AB}. Note that the direct coupling between states $|D\rangle$ and $|A\rangle$ is not considered in this review. As a specific example, the model Hamiltonian given in Eq. (5.l) is appropriate to electron transfer system in the photosynthetic reaction center. The primary charge separation step has been studied extensively both experimentally and theoretically. This process involves the transfer of an electron from the photoexcited special pair, SP*, to BPL (bacteriopheophytin-b on the L-branch). Since the center distance between SP and BPL is about 17 Å, direct coupling between the two pigments is usually assumed to be negligible. However the electron transfer is very fast, occurring in roughly 10 ps or less [55]. There exists an accessory bacteriochlorophyll (BCL) between SP and BPL, whose role has been extensively debated. Thus the three sites in the photosynthetic reaction center can be identified, and the three states, $|D\rangle$, $|B\rangle$, and $|A\rangle$, are $|SP^*, BCL, BPL\rangle$, $|SP^+, BCL^-, BPL\rangle$, and $|SP^+, BCL, BPL^-\rangle$, respectively.

It turns out that it is useful to define three spectral densities as

$$\rho_{BA}(\omega) \equiv \sum_\alpha \frac{a_\alpha^2}{2m_\alpha\omega_\alpha^3} \delta(\omega - \omega_\alpha)$$

$$\rho_{CA}(\omega) \equiv \sum_\alpha \frac{b_\alpha^2}{2m_\alpha\omega_\alpha^3} \delta(\omega - \omega_\alpha) \qquad (5.2)$$

$$\rho_{CB}(\omega) \equiv \sum_\alpha \frac{(b_\alpha - a_\alpha)^2}{2m_\alpha\omega_\alpha^3} \delta(\omega - \omega_\alpha)$$

Since the three states are coupled to a common bath, the spectral distributions of the three spectral densities are likely to be very similar. Extending the procedure to obtain the electron transfer spectral density in Section III.C to the three-state system, the three spectral densities can be written in factorized forms:

$$\rho_{BA}(\omega) \equiv \Gamma_{BA}\sigma(\omega) \qquad \rho_{CA}(\omega) \equiv \Gamma_{CA}\sigma(\omega) \qquad \rho_{CB}(\omega) \equiv \Gamma_{CB}\sigma(\omega) \quad (5.3)$$

where Γ_{BA}, for example, is given by Eq. (3.l4) with replacing the two electric fields replaced with those of the $|B\rangle$ and $|A\rangle$ states. Similarly, Γ_{CA} and Γ_{BC} are defined as

$$\Gamma_{BD} \equiv (2\pi)^{-3} \int d\mathbf{k} |\mathbf{E}_B(\mathbf{k}) - \mathbf{E}_D(\mathbf{k})|^2$$

$$\Gamma_{AD} \equiv (2\pi)^{-3} \int d\mathbf{k} |\mathbf{E}_A(\mathbf{k}) - \mathbf{E}_D(\mathbf{k})|^2 \qquad (5.4)$$

$$\Gamma_{AB} \equiv (2\pi)^{-3} \int d\mathbf{k} |\mathbf{E}_A(\mathbf{k}) - \mathbf{E}_B(\mathbf{k})|^2$$

The solvent spectral density $\sigma(\omega)$ can be estimated from nonlinear spectroscopic studies as discussed in Sections II and III. Therefore, the three spectral densities are completely specifiable.

A. Two-Dimensional Solvation Coordinate System

Although the general model Hamiltonian in Eq. (5.1) shows that the potential surfaces of the three states are truly multidimensional, it is possible to reduce the description to a two-dimensional solvation coordinate system. In case of two-state problems, such as a two-state chromophore or a two-state electron transfer system in the condensed phases, a one-dimensional solvation coordinate (e.g., a single collective mode described by a Brownian oscillator) is sufficient to describe the system–bath interaction (i.e., one can find a reduced description involving a single solvation coordinate). As shown in the Appendix, this approach is exactly identical to the effective harmonic bath model used in Sections II and IV. Similarly, one can start with two collective modes (i.e., two Brownian oscillator) and show that the system–bath interaction can be rewritten in terms of an effective harmonic bath oscillators in Eq. (5.1) [2]. As a consequence, the three spectral densities are not entirely independent. More specifically, the three reorganization energies associated with the three spectral densities obey the following relationship:

$$\lambda_{AD} = \lambda_{BD} + \lambda_{AB} + 2\sqrt{\lambda_{BD}\lambda_{AB}}\cos\theta \tag{5.5}$$

where the three reorganization energies are defined by Eq. (2.10) with the proper spectral densities given in Eqs. (5.3). Equation (5.5) means that each reorganization energy is associated with the square distance of one of side of a triangle (see Figure 6). Here θ represents the dimensionality parameter that is the angle shown in Figure 6. (Refer to [35] for detailed discussion on the two-dimensional nature of the solvation coordinate system for a three-state electron transfer in condensed phase.)

B. Transformed Three-State ET Hamiltonian

As shown earlier in this review, it is convenient to transform the effective harmonic bath Hamiltonian (5.1) by using the unitary transformation operator

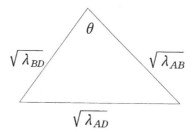

Figure 6. Triangle relationship among the three reorganization energies in a three-state electron transfer system. θ denotes the dimensionality parameter.

$$U \equiv \exp\left\{\frac{i}{\hbar}\sum_\alpha p_\alpha \begin{bmatrix} 0 & 0 & 0 \\ 0 & \dfrac{a_\alpha}{m_\alpha \omega_\alpha^2} & 0 \\ 0 & 0 & \dfrac{b_\alpha}{m_\alpha \omega_\alpha^2} \end{bmatrix}\right\} \tag{5.6}$$

Then the transformed Hamiltonian becomes

$$H' \equiv UHU^{-1} = H_0 + V \qquad \text{and} \qquad H_0 = H_S + H_B \tag{5.7}$$

where

$$H_S = \begin{bmatrix} 0 & 0 & 0 \\ 0 & \varepsilon_{BA} & 0 \\ 0 & 0 & \varepsilon_{CA} \end{bmatrix}$$

$$H_B = \sum_\alpha \left\{ \frac{p_\alpha^2}{2m_\alpha} + \frac{1}{2}m_\alpha \omega_\alpha^2 q_\alpha^2 \right\} \tag{5.8}$$

$$V = -\begin{bmatrix} 0 & \Delta_{AB}\exp(-iF) & 0 \\ \Delta_{AB}\exp(iF) & 0 & \Delta_{BC}\exp(-iG) \\ 0 & \Delta_{BC}\exp(iG) & 0 \end{bmatrix}$$

with

$$F \equiv \sum_{\alpha} \frac{a_{\alpha}}{\hbar m_{\alpha} \omega_{\alpha}^2} p_{\alpha}$$

$$G \equiv \sum_{\alpha} \frac{b_{\alpha} - a_{\alpha}}{\hbar m_{\alpha} \omega_{\alpha}^2} p_{\alpha}$$

(5.9)

In comparison to the Hamiltonian given in Eq. (5.l), the coupling matrix whose elements are Δ_{BD} and Δ_{AB} are dressed by the bath harmonic oscillators to give V in Eq. (5.8). The interaction Hamiltonian V is again considered to be a perturbational one.

C. General Fourth-Order Rate Kernel

To estimate the fourth-order contribution to the generalized rate kernel, the four-time correlation function must be evaluated. From the definition of $m_4(t, \tau_1, \tau_2, \tau_3)$ in Eq. (4.42), the $[i,j]$th matrix element of $m_4(t, \tau_1, \tau_2, \tau_3)$ can be rewritten as

$$
\begin{aligned}
[m_4(t_1, \tau_1, \tau_2, \tau_3)]_{ji} = {}& \langle \widetilde{V_{ji}^{(2)}}(\tau_3, \tau_2) \widetilde{V_{ij}^{(2)}}(t, \tau_1) \widetilde{\rho_B}(0) + \text{H.c.} \rangle \\
& + \langle \widetilde{V_{ji}^{(2)}}(\tau_3, \tau_1) \widetilde{V_{ij}^{(2)}}(t, \tau_2) \widetilde{\rho_B}(0) + \text{H.c.} \rangle \\
& + \langle \widetilde{V_{ji}^{(2)}}(\tau_2, \tau_1) \widetilde{V_{ij}^{(2)}}(t, \tau_3) \widetilde{\rho_B}(0) + \text{H.c.} \rangle \\
& - \langle \widetilde{V_{ji}^{(1)}}(\tau_3) \widetilde{V_{ij}^{(3)}}(t, \tau_1, \tau_2) \widetilde{\rho_B}(0) + \text{H.c.} \rangle \\
& - \langle \widetilde{V_{ji}^{(1)}}(\tau_2) \widetilde{V_{ij}^{(3)}}(t, \tau_1, \tau_3) \widetilde{\rho_B}(0) + \text{H.c.} \rangle \\
& - \langle \widetilde{V_{ji}^{(1)}}(\tau_1) \widetilde{V_{ij}^{(3)}}(t, \tau_2, \tau_3) \widetilde{\rho_B}(0) + \text{H.c.} \rangle \\
& - \langle \widetilde{V_{ji}^{(3)}}(\tau_3, \tau_2, \tau_1) \widetilde{V_{ij}^{(1)}}(t) \widetilde{\rho_B}(0) + \text{H.c.} \rangle \quad (5.10)
\end{aligned}
$$

where

$$\widetilde{V^{(n)}}(T_1, T_2, \ldots, T_n) \equiv \tilde{V}(T_1)\tilde{V}(T_2) \cdots \tilde{V}(T_n) \qquad (5.11)$$

Here the interaction representation was used,

$$\tilde{V}(T) \equiv \exp(iH_0 T) V \exp(-iH_0 T) \qquad (5.12)$$

Substituting the interaction Hamiltonian in the interaction representation and using Baker–Hausdorff operator equality three times, one can calculate the fourth-order nonlinear correlation functions given above. For example, to calculate the direct transfer rate kernel from $|D\rangle$ to $|A\rangle$ [i.e., the [3,1]th matrix element of $K_4(t)$], only the first three terms in Eq. (5.10) are needed, since the matrix elements such as $V_{31}^{(1)}$ and $V_{31}^{(3)}$ are all zero. In contrast, to calculate the fourth-order rate kernels, $[K_4(t)]_{ij}$ for $|i - j| = 1$, the last four terms in Eq. (5.10) have to be taken into account.

D.   Superexchange Rate Kernel, $[K_4(t)]_{31}$

Here we especially focus on the calculation of the rate kernel, $[K_4(t)]_{31}$, which describes the transfer rate from $|D\rangle$ to $|A\rangle$ directly. Although the coupling matrix element between $|D\rangle$ and $|A\rangle$ is zero, because the process has a finite rate the bridge state $|B\rangle$ acts as a virtual state for the coherent transfer from the initial electron donor state to the final acceptor state or as an intermediate state for the sequential electron transfer process. Inserting the interaction Hamiltonian into the nonlinear correlation function (5.10) gives

$$[m_4(t, \tau_1, \tau_2, \tau_3)] = [m_4(t, \tau_1, \tau_2, \tau_3)]_{31} + [m_4(t, \tau_1, \tau_2, \tau_3)]_{31} + [m_4(t, \tau_1, \tau_2, \tau_3)]_{31}$$

$$(5.13)$$

with

$$[m_4(t, \tau_1, \tau_2, \tau_3)]_{31}^{I} = \Delta_{AB}^2 \Delta_{BC}^2 \langle e^{i\varepsilon_{CB}(t-\tau_2)+i\varepsilon_{BA}(\tau_1-\tau_3)} e^{-iF(\tau_3)} e^{-iG(\tau_2)}$$

$$\times e^{iG(t)} e^{iF(\tau_1)} \rho_B(0) + \text{H.c.}\rangle$$

$$= 2\Delta_{AB}^2 \Delta_{BC}^2 \operatorname{Re}[e^{i\varepsilon_{CB}(t-\tau_2)+i\varepsilon_{BA}(\tau_1-\tau_3)} F^{I}(t, \tau_1, \tau_2, \tau_3)]$$

$$[m_4(t, \tau_1, \tau_2, \tau_3)]_{31}^{II} = \Delta_{AB}^2 \Delta_{BC}^2 \langle e^{i\varepsilon_{CB}(t-\tau_1)+i\varepsilon_{BA}(\tau_2-\tau_3)} e^{-iF(\tau_3)} e^{-iG(\tau_1)} e^{iG(t)}$$

$$\times e^{iF(\tau_2)} \rho_B(0) + \text{H.c.}\rangle$$

$$= 2\Delta_{AB}^2 \Delta_{BC}^2 \operatorname{Re}[e^{i\varepsilon_{CB}(t-\tau_1)+i\varepsilon_{BA}(\tau_2-\tau_3)} F^{II}(t, \tau_1, \tau_2, \tau_3)]$$

$$[m_4(t, \tau_1, \tau_2, \tau_3)]_{31}^{III} = \Delta_{AB}^2 \Delta_{BC}^2 \langle e^{i\varepsilon_{CB}(t-\tau_1)-i\varepsilon_{BA}(\tau_2-\tau_3)} e^{-iF(\tau_2)} e^{-iG(\tau_1)}$$

$$\times e^{iG(t)} e^{iF(\tau_3)} \rho_B(0) + \text{H.c.}\rangle$$

$$= 2\Delta_{AB}^2 \Delta_{BC}^2 \operatorname{Re}[e^{i\varepsilon_{CB}(t-\tau_1)-i\varepsilon_{BA}(\tau_2-\tau_3)} F^{III}(t, \tau_1, \tau_2, \tau_3)]$$

$$(5.14)$$

where the interaction representations of the two operators, $F(t)$ and $G(t)$, were defined in Eq. (5.9). Here the three auxiliary functions, $F^m(t, \tau_1, \tau_2, \tau_3)$, for $m =$ I, II, III, are defined as

$$
\begin{aligned}
F^m(t, \tau_1, \tau_2, \tau_3) \equiv \exp\Big\{ &\frac{1}{2} P_{CA}(T_{12}) - \frac{1}{2} P_{CA}(T_{13}) - \frac{1}{2} P_{CA}(T_{24}) + \frac{1}{2} P_{CA}(T_{34}) \\
&- \frac{1}{2} P_{BA}(T_{12}) + \frac{1}{2} P_{BA}(T_{13}) - P_{BA}(T_{14}) \\
&+ \frac{1}{2} P_{BA}(T_{24}) - \frac{1}{2} P_{BA}(T_{34}) \\
&- \frac{1}{2} P_{CB}(T_{12}) + \frac{1}{2} P_{CB}(T_{13}) - P_{CB}(T_{23}) \\
&+ \frac{1}{2} P_{CB}(T_{24}) - \frac{1}{2} P_{CB}(T_{34}) \Big\} \\
\times \exp\Big\{ &\frac{i}{2} Q_{CA}(T_{12}) - \frac{i}{2} Q_{CA}(T_{13}) \\
&- \frac{i}{2} Q_{CA}(T_{24}) + \frac{i}{2} Q_{CA}(T_{34}). \\
&- \frac{i}{2} Q_{BA}(T_{12}) + \frac{i}{2} Q_{BA}(T_{13}) - i Q_{BA}(T_{14}) \\
&+ \frac{i}{2} Q_{BA}(T_{24}) - \frac{i}{2} Q_{BA}(T_{34}) \\
&- \frac{i}{2} Q_{CB}(T_{12}) + \frac{i}{2} Q_{CB}(T_{13}) - i Q_{CB}(T_{23}) \\
&+ \frac{i}{2} Q_{CB}(T_{24}) - \frac{i}{2} Q_{CB}(T_{34}) \Big\}
\end{aligned}
\tag{5.15}
$$

Here the three sets of time arguments, such as T_{12}, are different for each nonlinear correlation function and listed in Table I.

The three contributions in Eq. (5.13) have interesting physical meanings as discussed by Hu and Mukamel. The first contribution corresponds to the case when, during the three time periods $t - \tau_1$, $\tau_1 - \tau_2$, and $\tau_2 - \tau_3$, the system evolves in electronic coherence states. In contrast, the second and third contributions involve population periods in the $|B\rangle$ state during $\tau_1 - \tau_2$. Therefore, the last two contributions, when they are integrated, as can be seen in Eq. (4.37d), diverge as t increases; more specifically, they

TABLE I
Time Arguments Associated with the Three Nonlinear Correlation Fucntions

Case	T_{12}	T_{13}	T_{14}	T_{23}	T_{24}	T_{34}
I	$\tau_3 - \tau_2$	$\tau_3 - t$	$\tau_3 - \tau_1$	$\tau_2 - t$	$\tau_2 - \tau_1$	$t - \tau_1$
II	$\tau_3 - \tau_1$	$\tau_3 - t$	$\tau_3 - \tau_2$	$\tau_1 - t$	$\tau_1 - \tau_2$	$t - \tau_2$
III	$\tau_2 - \tau_1$	$\tau_2 - t$	$\tau_2 - \tau_3$	$\tau_1 - t$	$\tau_1 - \tau_3$	$t - \tau_3$

increase linearly with respect to t asymptotically. However, because of the second term in Eq. (4.37d), which is also linearly increasing function with respect to t, the divergence is canceled.

We now make a comparison with work reported previously in this area. The coherent electron transfer mechanism, where the bridge state is merely treated as a virtual state donating a nonzero effective coupling matrix element, was discussed by McConnel [56] and Marcus [57] by effectively eliminating the $|B\rangle$-state from the Hamiltonian with Löwdin's partitioning technique. Then the electron is transferred from $|D\rangle$ to $|A\rangle$ directly with a nonzero effective matrix element determined by the energy gap between $|D\rangle$ and $|B\rangle$ as well as the two coupling matrix elements. In this case the fluctuation of the $|B\rangle$ state was neglected completely. The coherent process is the dominant mechanism for the ultrafast electron transfer in the three-state system, when the bridge state $|B\rangle$ is a very high-lying state in comparison to those of the donor and acceptor states. In this limit, one can rederive the well-known result for the coherent rate constant by (1) taking the first term in Eq. (5.13) and neglecting all the other terms in Eq. (4.37d), (2) ignoring the coupling between the $|B\rangle$ state and the bath oscillators (i.e., putting $a_\alpha = 0$ for all α), and (3) applying the stationary-phase approximation to the integrations over the two time periods $t - \tau_1$ and $\tau_2 - \tau_3$. However, as the energy of the $|B\rangle$ state is lowered and becomes comparable to the energy of the donor and acceptor states, the sequential process is also likely to contribute to the electron transfer reaction. In this case the transient population contribution, which originates from the combination of the last two terms in Eq. (5.13) and the second term in Eq. (4.37b), should be taken into account. In contrast to the common belief that the nuclear relaxation on the bridge surface is fast enough that it is possible to ignore the coupling between the bridge state and bath degrees of freedom, since the time scale of the bath is on the order of hundreds of femtoseconds, the transient population contribution timescale is of the same order. Therefore, if the electron transfer is ultrafast, it is necessary to fully include the transient population contribution in addition to the coherent contribution. The latter two terms, called the *memory terms*, in Eq. (4.37d) are rather short-lived contributions since the time orderings are not fully retained.

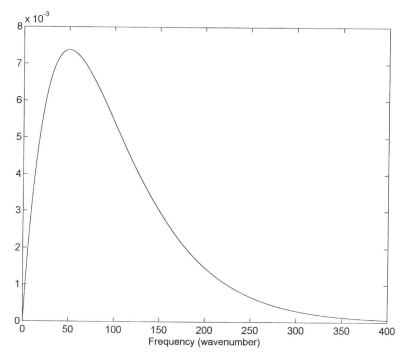

Figure 7. Model spectral density used in the calculation of the three-state electron transfer is shown. $\sigma(\omega) \propto (1/\omega) \exp(-\omega/50)$.

To provide a quantitative picture of the superexchange rate kernel, some numerical calculations for a model system are given in the following. The spectral density is assumed to be (Fig. 7)

$$\sigma(\omega) \propto (1/\omega) \exp(-\omega/50)$$

where the unit of frequency is cm^{-1} (angular frequency). In Figure 8 the time-dependent superexchange rate kernel, $[K_4(t)]_{31}$, is shown as a solid curve denoted as "total". Note that the rate kernel is strongly time dependent and does not reach its asymptotic value by 1 ps. The reason for the slow variation can be seen by examining the transient population contribution shown in Figure 9. On the other hand, the coherent contribution [the first term in Eq (5.13)] shown in Figure 9 quickly reaches its asymptotic value. The Markovian terms, obtained by combining the first two contributions in Eq. (4.37d), are drawn in both Figures 8 and 9. As discussed briefly, the memory terms are time dependent only for a short period of time and then

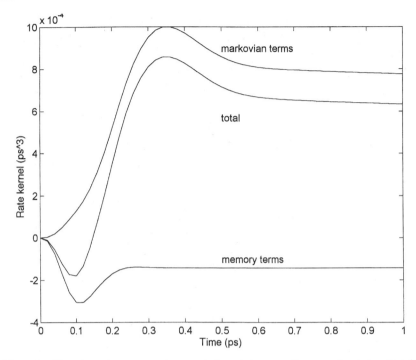

Figure 8. Time-dependent superexchange rate kernel shown as the curve denoted as "total." Markovian and memory terms are separately drawn (see the text for detailed discussion on the separation of the total rate kernel into Markovian and memory terms). The y-axis is in arbitrary unit. The three independent reorganization energies are $\lambda_{BD} = 300\,\mathrm{cm}^{-1}$, $\lambda_{AD} = 500\,\mathrm{cm}^{-1}$, and $\lambda_{AB} = 1348\,\mathrm{cm}^{-1}$. In this case the dimensionality parameter θ is equal to 25.5°. The three energy gaps are $\varepsilon_{BD} = 400\,\mathrm{cm}^{-1}$, $\varepsilon_{AD} = -700\,\mathrm{cm}^{-1}$, and $\varepsilon_{AB} = \varepsilon_{AD} - \varepsilon_{BD} = -1100\,\mathrm{cm}^{-1}$. Note that this case is not one-dimensional so that the two-dimensional potential energy surfaces should be considered.

reach their limiting values. Overall, the superexchange rate kernel is a rather slowly varying function with respect to time, in comparison to the second-order rate kernel discussed in Section IV.C. Consequently, it may be necessary when interpreting ultrafast electron transfer among three states, such as the primary charge separation in the photosynthetic reaction center, to consider this source of nonexponentiality. Since the time scale of the rate kernel is quite comparable to that of the electron transfer, the time-dependent rate kernel, instead of the asymptotic value, the superexchange rate constant, should be used to describe the population evolution. To take this work further, the next step would be to calculate all the rate kernel matrix elements up to fourth order with respect to the coupling Hamiltonian V in Eq. (5.8).

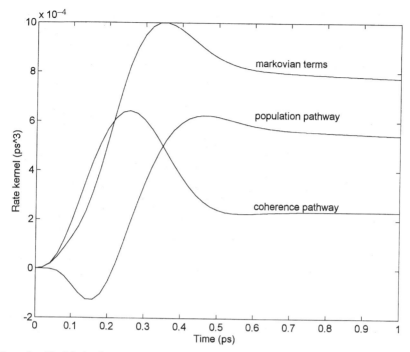

Figure 9. The Markovian terms are separated into contributions from the population pathway and coherent pathway. The former involves a population evolution period, whereas the latter involves three consecutive coherent state evolutions. The y-axis is in arbitrary unit.

VI. CONCLUSIONS

In this chapter we presented a systematic procedure for calculating the electron transfer rate kernels for two- and three-state systems by using linear and nonlinear spectroscopic measurements. We showed that the solvent spectral density plays a crucial role in the connection between spectroscopic observations and electron transfer processes in a common condensed medium. In cases when the electron transfer rate is ultrafast, when the temperature is sufficient low, or when the spectral density contains high-frequency contributions, the complete spectral density representing the dynamical aspect of the condensed phases should be used to describe the electron transfer processes properly.

APPENDIX

In this appendix we establish the equality between the multimode Brownian oscillator model and the effective harmonic bath model. In general, the two models may not be identical to each other since the generalized Langevin equation can be formally obtained from an arbitrary system–bath interaction. For example, as Cortés et al. [58] showed, the generalized Langevin equation of motion can be obtained even when the system–bath coupling is nonlinear. However, in this case, the well-known fluctuation–dissipation relation takes a different form. Therefore, the standard microscopic model to obtain a generalized Langevin equation is to assume that the system–bath coupling potential is bilinear with respect to the system and bath oscillator coordinates. This model was taken by Tanimura and Mukamel [59] to develop the multimode Brownian oscillator model to investigate the role of the system–bath interaction in spectroscopic line-broadening phenomena in condensed phases. We start with the standard model of the multimode Brownian oscillator model, and by following the normal-mode transformation procedure first used by Garg et al. [50], the equality of the Brownian oscillator model to the effective harmonic bath model will be derived.

We consider the multimode Brownian oscillator model Hamiltonian

$$H = |g\rangle H_g \langle g| + |e\rangle H_e \langle e| \tag{A.1}$$

where the two Hamiltonians associated with the ground and excited states are

$$H_g = \varepsilon_g + \sum_{n=1}^{N_S} \left(\frac{P_n^2}{2M_n} + \frac{1}{2} M_n \Omega_n^2 Q_n^2 \right) + h_{SB}$$

$$H_e = \varepsilon_e + \sum_{n=1}^{N_S} \left[\frac{P_n^2}{2M_n} + \frac{1}{2} M_n \Omega_n^2 (Q_n - Q_n^0)^2 \right] + h_{SB} \tag{A.2}$$

Here ε_g and ε_e are the electronic energies of the ground and excited states, respectively. Q_m and P_m are the optically active harmonic-mode coordinate and its conjugate momentum operators. It is assumed that there are N_s optically active Brownian oscillators. Note that the excited-state nuclear Hamiltonian is just a displaced harmonic oscillator, which is the reason why these modes are optically active. Now the system–bath interaction Hamiltonian is assumed to be

$$h_{SB} = \sum_{\alpha=1}^{N} \left[\frac{p_\alpha^2}{2m_\alpha} + \frac{1}{2} m_\alpha \omega_\alpha^2 \left(q_\alpha - \sum_n \frac{c_{n\alpha} Q_n}{m_\alpha \omega_\alpha^2} \right)^2 \right] \tag{A.3}$$

where $c_{n\alpha}$ denotes the coupling constant between the nth Brownian oscillator and the αth bath oscillator coordinates.

Hereafter we focus on the case of a single Brownian oscillator coupled to a set of bath harmonic oscillators, even though it is straightforward to proceed with an arbitrary number of Brownian oscillators. Let Q and P denote the coordinate and momentum operators of the Brownian oscillator, and Q_0 is the displacement of the BO on the excited-state potential energy surface. The classical reorganization energy is $\lambda = \frac{1}{2}M\Omega^2 Q_0^2$. Now the ground-state nuclear Hamiltonian is a coupled harmonic oscillator system, where only one harmonic oscillator is optically active and all the other ones are not. The generalized Langevin equation can be derived directly from the Hamiltonian in Eqs. (A.1) to (A.3) by obtaining the Heisenberg equations of motion for the annihilation (or creation) operators for the system and bath oscillator:

$$\dot{a} = -i\omega_0 a + \frac{i}{\hbar}\sum_\alpha \frac{\hbar}{2\sqrt{Mm_\alpha\omega_0\omega_\alpha}} c_\alpha(b_\alpha^+ + b_\alpha)$$

$$\dot{b}_\alpha = -i\omega_\alpha b_\alpha + (a^+ + a)\frac{ic_\alpha}{2\sqrt{Mm_\alpha\omega_0\omega_\alpha}}$$

(A.4)

where the renormalized frequency of the Brownian oscillator is

$$\omega_0 \equiv \sqrt{\Omega^2 + \frac{1}{M}\sum_\alpha \frac{c_\alpha^2}{m_\alpha\omega_\alpha^2}}$$

(A.5)

Calculating the formal integration of the annihilation operator of the bath oscillator and its Hermitian conjugate and using integration by parts, we find the generalized Langevin equation,

$$\dot{a} + i\omega_0 a - \frac{i}{M\omega_0}\sum_\alpha \frac{c_\alpha^2}{2m_\alpha\omega_\alpha^2}(a^+ + a) + i\int_0^t d\tau\, K(t-\tau)[\dot{a}^+(\tau) + \dot{a}(\tau)] = iR(t)$$

(A.6)

where the time-dependent kernel and forcing operator are, respectively,

$$K(t-\tau) \equiv \frac{1}{M\omega_0}\sum_\alpha \frac{c_\alpha^2}{2m_\alpha\omega_\alpha^2}\cos\omega_\alpha(t-\tau)$$

$$R(t) = \sum_\alpha \frac{c_\alpha}{2\sqrt{M\omega_0 m_\alpha\omega_\alpha}}\left\{\left[b_\alpha^+(0) - \frac{c_\alpha[a^+(0) + a(0)]}{2\omega_\alpha\sqrt{M\omega_0 m_\alpha\omega_\alpha}}\right]\exp(i\omega_\alpha t) + \text{H.c}\right\}$$

(A.7)

One can prove the quantum fluctuation–dissipation relation between the dissipative kernel and the symmetrized correlation function of $R(t)$ (see [59]). Perhaps a more familiar form of the generalized Langevin equation is one involving the time derivative of the kinematic momentum operator. From the definition of the momentum and coordinate operators in terms of the creation and annihilation operators, it is straightforward to derive the following operator differential equation;

$$\dot{P} + \int_0^t \gamma(t - \tau)\dot{Q}(\tau) + M\Omega^2 Q = f(t) \tag{A.8}$$

where the damping kernel and the fluctuating force operator are, respectively, given by

$$\gamma(t - \tau) = \int d\omega \, \frac{J_{\text{BO}}(\omega)}{\omega} \cos \omega(t - \tau)$$

$$f(t) \equiv \sqrt{\frac{\hbar M \omega_0}{2}} \{F^+(t) + F(t)\} \tag{A.9}$$

Here the spectral density associated with the damping process of a Brownian oscillator (BO) is defined as

$$J_{\text{BO}}(\omega) \equiv \sum_\alpha \frac{c_\alpha^2}{m_\alpha \omega_\alpha} \delta(\omega - \omega_\alpha) \tag{A.10}$$

Then the fluctuation–dissipation relation takes the form

$$\langle f(t)f(\tau) + f(\tau)f(t) \rangle \equiv \hbar \int d\omega \, J_{\text{BO}}(\omega) \coth \frac{\beta\hbar\omega}{2} \cos \omega(t - \tau) \tag{A.11}$$

So far we showed that the harmonic oscillator Q coupled linearly to a set of bath harmonic oscillators obeys a generalized Langevin equation and its dynamics is specified completely by the spectral density defined in Eq. (A.10). We next prove that the effective harmonic bath model is identical to the Brownian oscillator model presented above.

As shown by Garg, Ambegaokar, and Onuchic, the normal model transformation can be applied to the Hamiltonian in Eq. (A.3) since it contains quadratic terms only. Assuming that the unitary matrix U diagonalizes the Hamiltonian, one can find a set of normal coordinates $\{\chi_\alpha\}$ for $\alpha = 1$ to $N + 1$,

$$\chi_\alpha = \sum_\beta U_{\alpha\beta}^{-1} \widetilde{q}_\beta \tag{A.12}$$

where $\widetilde{q} = \{Q, q_1, q_2, \ldots, q_N\}$. Then the total nuclear Hamiltonians of the ground and excited states are

$$
\begin{aligned}
H_g &= \varepsilon_g + \sum_{\alpha=1}^{N+1} \left(\frac{\widetilde{p}_\alpha^2}{2\widetilde{m}_\alpha} + \frac{1}{2} \widetilde{m}_\alpha \widetilde{\omega}_\alpha^2 \chi_\alpha^2 \right) \\
H_e &= \varepsilon_e + \lambda + \sum_{\alpha=1}^{N+1} \left(\frac{\widetilde{p}_\alpha^2}{2\widetilde{m}_\alpha} + \frac{1}{2} \widetilde{m}_\alpha \widetilde{\omega}_\alpha^2 \chi_\alpha^2 \right) - \sum_\alpha d_\alpha \chi_\alpha
\end{aligned}
\tag{A.13}
$$

where the transformed coupling constants are $d_\alpha \equiv M\Omega^2 Q_0 U_{1\alpha}$. Here $\widetilde{m}_\alpha$ and $\widetilde{\omega}_\alpha$ are the transformed mass and frequency of the αth normal mode. Note that the excited-state Hamiltonian consists of $(N + 1)$-displaced harmonic oscillators. This can be rewritten as the effective harmonic bath model in Eq. (A.13) and proves the equality between the two models. We further consider the interconnection between the spectral density associated with the effective harmonic bath (HB) model,

$$J_{HB}(\omega) \equiv \sum_\alpha \frac{d_\alpha^2}{\widetilde{m}_\alpha \widetilde{\omega}_\alpha^2} \delta(\omega - \widetilde{\omega}_\alpha) \tag{A.14}$$

Next the linear response functions obtained from the two models presented above are considered. The fluctuating parts of the energy difference operator for the Brownian oscillator and effective harmonic oscillator bath models are, respectively,

$$\delta V_{SB}(t) = M\Omega^2 Q_0 Q(t) \tag{A.15}$$

$$\delta V_{SB}(t) = \sum_\alpha d_\alpha \chi_\alpha(t) \tag{A.16}$$

Inserting Eqs. (A.16) into the definition of the linear response function in Eq. (2.5), one can find two response functions based on the two models. Since the two models should provide the same response functions, it is possible to find the equality

$$J_{HB}(\omega) = \text{Im} \left[\frac{2\lambda}{(1 - \omega^2/\Omega^2) - \dfrac{\omega^2}{M\Omega^2} \displaystyle\int_0^\infty d\omega' \dfrac{J_{BO}(\omega')}{\omega'(\omega'^2 - \omega^2)}} \right] \tag{A.17}$$

where the fact that the classical reorganization energy λ of $\frac{1}{2}M\Omega^2 Q_0^2$ was used. Equation (A.17) provides the microscopic relationship between the two spectral densities associated with the two models. If it is assumed that the spectral density for the Brownian oscillator model, $J_{BO}(\omega)$, has the following three properties: (1) $J_{BO}(\omega)/\omega$ has no singularity, (2) $J_{BO}(\omega)$ is exponentially cut off like $J_{BO}(\omega) = g(\omega)\exp(-\omega/\Lambda)$, and (3) $g(\omega)/\omega$ is an even function with respect to ω. Then the integral in the square brackets of Eq. (A.17) can be evaluated by using the residue theorem, and we find

$$J_{HB}(\omega) = M^2 \Omega^4 Q_0^2 \, \frac{\pi J_{BO}(\omega)/2}{M^2(\Omega^2 - \omega^2)^2 + \pi^2 J_{BO}^2(\omega)/4} \qquad (A.18)$$

This equality between the two spectral densities was obtained by several workers, although they took a bit different route. Note that once the spectral density of the Brownian oscillator, $J_{BO}(\omega)$, with its mass (M), frequency (Ω), and displacement (Q_0), is specified, it is possible to calculate the spectral density of the harmonic bath model. However, the reverse procedure is not in general unique, since a single function $J_{HB}(\omega)$ can correspond to an infinite number of Brownian oscillators depending on the three parameters mass (M), frequency (Ω), and displacement (Q_0). Therefore, it is concluded that if there are no readily identifiable Brownian oscillators in the composite system, which can be identified without ambiguity, the effective harmonic bath model should be considered to be more convenient not only for the practical calculations but also for the conceptual understanding. It is worthwhile mentioning that the latter model does not need to introduce any abstract (Brownian) oscillator that does not exist in reality; instead, it assumes a collection of displaced harmonic modes with a properly defined spectral density carrying most of the dynamical information of the system–bath interaction.

REFERENCES

1. S. Mukamel, *Principles of Nonlinear Optical Spectroscopy*, Oxford University Press, New York, 1995.
2. G. R. Fleming and M. Cho, *Ann. Rev. Phys. Chem.*, **47**, 109 (1996).
3. G. R. Fleming, T. Joo, and M. Cho, *Adv. Chem. Phys.*, **101**, 141 (1996).
4. T. Joo et al., *J. Chem. Phys.*, **104**, 6089 (1996).
5. S. A. Passino, Y. Nagasawa, T. Joo, and G. R. Fleming, *J. Phys. Chem.*, A **101**, 725 (1997).
6. Y. Nagasawa, S. A. Passino, T. Joo, and G. R. Fleming, *J. Chem. Phys.*, **106**, 4840 (1997).
7. P. Vohringer, D. C. Arnett, T. S. Yang, and N. F. Scherer, *Chem. Phys. Lett.* **237**, 387 (1995).
8. M. S. Pshenichnikov, K. Duppen, and D. A. Wiersma, *Phys. Rev. Lett.* **74**, 674 (1995).

9. A. J. Leggett, S. Chakravarty, A. T. Dorsey, M. P. A. Fisher, A. Garg, and W. Zwerger, *Rev. Mod. Phys.* **59**, 1 (1987).

10. R. J. Silbey and R. A. Harris, *J. Phys. Chem.* **93**, 7062 (1989).

11. V. G. Levich and R. R. Dogonadze, *Dokl. Acad. Nauk. SSSR*, **124**, 123 (1959).

12. R. Zwanzig, *J. Stat. Phys.* **9**, 215 (1973).

13. K. Lindenberg and B. J. West, *Phys. Rev. A* **30**, 568 (1984).

14. (a) J. S. Bader, R. A. Kuharski, and D. Chandler, *J. Chem. Phys.* **93**, 230 (1990); (b) X. Song and R. A. Marcus, *J. Chem. Phys.* **99**, 7768 (1993).

15. M. Cho, J. Y. Yu, T. Joo, Y. Nagasawa, S. A. Passino, and G. R. Fleming, *J. Phys. Chem.* **100**, 11944 (1996).

16. S. J. Rosenthal, X. Xie, M. Du, and G. R. Fleming, *J. Chem. Phys.* **95**, 4715 (1992).

17. M. L. Horng, J. Gardecki, A. Papazyan, and M. Maroncelli, *J. Phys. Chem.* **99**, 17311 (1995).

18. $\text{Im}[\tilde{G}(\omega)]$ is related to $C''(\omega)$ in [11].

19. T. Joo, Y. Jia, J. Y. Yu, M. J. Lang, and G. R. Fleming, *J. Chem. Phys.* **104**, 6089 (1995).

20. M. Maroncelli and G. R. Fleming, *J. Chem. Phys.* **89**, 5044 (1988).

21. E. A. Carter and J. T. Hynes, *J. Chem. Phys.* **94**, 5961 (1991).

22. E. Neria and A. Nitzan, *J. Chem. Phys.* 96, 5433 (1992).

23. L. E. Fried and S. Mukamel, *Adv. Chem. Phys.* **84**, 435 (1993).

24. S. Roy and B. Bagchi, *J. Chem. Phys.* **99**, 9938 (1993).

25. F. O. Raineri, H. Resat, B. C. Perng, F. Hirata, and H. L. Friedman, *J. Chem. Phys.* **100**, 1477 (1994).

26. P. Madden and D. Kivelson, *Adv. Chem. Phys.* **56**, 467 (1984).

27. A. M. Stoneham, *Rev. Mod. Phys.* **41**, 82 (1969).

28. Y. Ooshika, *J. Phys. Soc. Jpn.* **9**, 594 (1954)

29. E. G. McRae, *J. Phys. Chem.* **58**, 1002 (1954).

30. Yu. T. Mazurenko and N. G. Bakshiev, *Opt Spectrosc.* **28**, 490 (1970).

31. B. Bagchi, D. W. Oxtoby, and G. R. Fleming, *Chem. Phys.* **86**, 257 (1984).

32. G. van der Zwan and J. T. Hynes, *J. Phys. Chem.* **89**, 4181 (1985).

33. R. F. Loring, Y. J. Yan, and S. Mukamel, *J. Chem. Phys.* **87**, 5840 (1987).

34. D. Chandler, *Introduction to Modern Statistical Mechanics*, Oxford University Press, New York, 1987.

35. M. Cho and R. J. Silbey, *J. Chem Phys.* **103**, 595 (1995).

36. H. Mori, *Prog. Theor. Phys.* **33**, 423 (1965).

37. R. Zwanzig, *Annu. Rev. Phys. Chem.* **16**, 67 (1965).

38. B. Yoon, J. M. Deutch, and J. H. Freed, *J. Chem. Phys.* **62**, 4687 (1975).

39. S. Mukamel, I. Oppenheim, and J. Ross, *Phys. Rev. A* **17**, 1988 (1978).

40. R. Kubo, *J. Math. Phys.* **4**, 174 (1963).

41. M. Sparpaglione and S. Mukamel, *J. Chem. Phys.* **88**, 3263 (1988).

42. R. H. Terwiel, *Physica* **74**, 248 (1974).

43. R. A. Marcus and N. Sutin, *Biochim. Biophys. Acta* **811**, 265 (1985).

44. I. Rips and J. Jortner, *J. Chem. Phys.* **87**, 6513 (1987).

45. E. Buhks, M. Bixon, J. Jortner, and G. Navon, *J. Phys. Chem.* **85**, 3759 (1981).

46. P. Siders and R. A. Marcus, *J. Am. Chem. Soc.* **103**, 741 (1981).

47. L. D. Zusman, *Chem. Phys.* **74**, 6746 (1981).

48. P. G. Wolynes, *J. Chem. Phys.* **86**, 1957 (1987).

49. J. T. Hynes, *J. Phys. Chem.* **90**, 3701 (1986).

50. A. Garg, J. N. Onuchic, and J. V. Ambegaokar, *J. Chem. Phys.* **83**, 4491 (1985).

51. M. Cho and R. Silbey, *J. Chem. Phys.*, **106**, 2654 (1997).

52. B. A. Lippman and J. Schwinger, *Phys. Rev.* **79**, 469 (1950).

53. M. Cini and S. Fubini, *Nuovo Cimento* **10**, 1695 (1953).

54. G. Baker and J. Gammer, eds., *Padé Approximants in Theoretical Physics*, Academic Press, San Diego, Calif., 1970.

55. G. R. Fleming and R. van Grondelle, *Phys. Today*, Feb. 1994, p. 48.

56. H. M. McConnel, *J. Chem. Phys.* **35**, 508 (1961).

57. R. A. Marcus, *Chem. Phys. Lett.* **133**, 471 (1987).

58. E. Cortés, B. J. West, and K. Lindenberg, *J. Chem. Phys.* **82**, 2708 (1985).

59. Y. Tanimura and S. Mukamel, *Phys. Rev. E* **47**, 118 (1993).

ULTRAFAST INTERMOLECULAR ELECTRON TRANSFER IN SOLUTION

KEITARO YOSHIHARA

Japan Advanced Institute of Science and Technology, Tatsunokuchi, Ishikawa 923-1292, Japan

CONTENTS

I. INTRODUCTION

Intermolecular electron transfer (ET) occurs in the wide range of organic and inorganic chemical reactions, polymerization reactions, photography, electrochemistry, and many other processes and the subject of extensive research for many years. Recent synthesis of a vast number of tailormade donor–acceptors and donor–insulator–acceptors gives us all possible molecular systems with varieties of the free energy difference and the distance and orientation of donors and acceptors. Nature facilitates fine molecular

Electron Transfer: From Isolated Molecules to Biomolecules, Part Two, edited by Joshua Jortner and M. Bixon. Advances in Chemical Physics Series, Volume 107, series editors I. Prigogine and Stuart A. Rice.

assemblies of electron relay systems such as photosynthetic units in plants and photosynthetic bacteria. We now chemically reconstitute cofactors in the electron transfer relay and even change protein environment by genetic engineering. All of these experimental developments brought about a new dimension in interpretation of the mechanism of ET and new possibilities of controlling ET.

The theory of ET was most successfully developed by Marcus for the classical nonadiabatic reaction [1–7] and has been widely applied to many different types of molecular and biomolecular systems. The theoretical understandings of the effects of vibrational quantum modes on ET were further developed by Levich, Jortner, and others [6–12]. These theories have been widely applied in many different types of molecular and biomolecular systems. More recently, dynamical aspects of ET received wide attention in respect to solvent and nuclear motions, and extensive theoretical studies have been made [13–28]. With recent development of ultrafast spectroscopies, new experimental evidence was obtained on dynamics of intramolecular ET [29–32] and intermolecular ET [33–47].

The purpose of this chapter is to describe recent understandings and experimental developments on the dynamical aspects of intermolecular ET in solution. Let us discuss the mechanism of ET with experimentally available parameters: namely, the rate constant of ET (k_{ET}) and solvent relaxation time (τ_s). This is actually not simple in practice, since k_{ET} is not only a function of the preexponential factor but also that of the Boltzmann factor. In the later classification we consider only the preexponential factor (i.e., barrierless ET rate constant). Among various parameters for describing solvent relaxation, we choose solvation time, which suitably describes the current experimental conditions.

1. When the rate of ET is much slower than the solvent relaxation time ($k_{ET} \ll \tau_s^{-1}$, we can safely assume that the equilibrium conditions are hold before and after the reaction and nonadiabatic transition-state theory can be applied [1–7]. The solvent is usually treated as static dielectric medium and controls the free energy difference and activation barrier by static interaction. A quantum mechanical theory including the effects of high- and low-frequency vibrational modes on ET was worked out using perturbation theory [6–12]. This theoretical description was realized to be very important to analyze real experimental systems, especially to describe the Marcus inverted region and temperature dependence of ET.

2. When the electronic matrix element for ET becomes greater, k_{ET} could become in the same order τ_s^{-1}($k_{ET} \sim \tau_s^{-1}$). In such a case the adiabaticity of the reaction increases (see the next section for a more

precise definition of adiabaticity). An important conclusion of the theory is that k_{ET} becomes a direct function of τ_s^{-1} and is inversely proportional to the solvent relaxation times. In this case ET rate is restricted by solvent relaxation time and becomes a solvent-controlled reaction.

3. If ET is faster than the solvent relaxation time ($k_{ET} > \tau_s^{-1}$), the arguments above become invalid and the effects of much faster motion such as intermolecular and intramolecular motion, have an important role in dynamical processes. In this article we describe mainly recent experimental results and their analyses in this case.

4. When ET is very much faster than the solvent relaxation time ($k_{ET} \gg \tau_s^{-1}$), the solvent motions are completely frozen and solvent dynamics have no role on ET.

A. Nonadiabatic Electron Transfer

The theory of ET in solution has been reviewed in many excellent review articles. In the present section we briefly summarize the relevant concepts and equations, which will serve as a basis for a comparison with experiments. To derive an expression for the rate constant of ET, we start from a nonadiabatic treatment and then extend the theory which is applicable to more general cases. ET is a subject of free energy surface crossing from a reactant energy surface to a product surface like any other chemical reaction. The fundamental assumption of nonadiabatic ET theory is that the reactant is kept under quasi-equilibrium with the transition state during the reaction. The rate constant of nonadiabatic ET is written as [1–7]

$$k_{NA} = \frac{2\pi}{\hbar} \frac{V_{el}^2}{\sqrt{2\pi\lambda_s k_B T}} \exp\left(-\frac{\Delta G^*}{k_B T}\right) \tag{1.1}$$

where ΔG^* is the free energy of activation, $k_B T$ the thermal energy, and λ the reorganization energy, which is defined as the value of free energy on the reactant surface at the position of the bottom of the product potential well. V_{el} is the electron exchange matrix element between the wavefunction of the reactant and product.

Whenever a large fluctuation of the solvent nuclear motion brings the reactant to the transition state, reaction takes place. Then the excess energy will be disposed into the solvent heat bath, and the entire system relaxes to a new equilibrium for the product. In some cases ET does not occur in the gas phase but occurs in a polar solvent. One of the interactions that we should consider in ET is dielectric interaction between an electric dipole moment of the reactant species and a reaction field induced by this dipole in the surrounding solvent. To describe the free energy surface for ET, a solvent coordinate is often chosen, which describes the collective motion of the solvent molecules and corresponds to the solvent polarization. The reorganization dynamics of the solvent are treated fast enough compared to other dynamics involved in the reaction. The total reorganization energy is defined as the sum of two contributions:

$$\lambda = \lambda_s + \lambda_i \tag{1.2}$$

where λ_s is the solvent reorganization energy, which is defined as the one along the solvent reaction coordinate, and λ_i is the nuclear reorganization, which is defined as the one along the nuclear reaction coordinate.

When a quadratic function is assumed for the free energy surfaces of the reactant and product with a same curvature, the activation energy can be simply described by

$$\Delta G^* = \frac{(\lambda + \Delta G)^2}{4\lambda} \tag{1.3}$$

with the free energy gap ΔG of the reaction. The energy gap dependence of ET rate constant can be separated in three regions:

1. $-\Delta G < \lambda$. The reaction becomes faster as $-\Delta G$ increases.
2. $-\Delta G = \lambda$. The reaction occurs the fastest because the activation barrier vanishes. The free energy surface of the product crosses the bottom of the reactant surface.
3. $-\Delta G > \lambda$. The reaction become slower again as $-\Delta G$ increases.

Therefore, the rate constant of ET shows a bell-shape dependence on the free energy difference. Case (1) is called the *normal region*, and case 3 is called the *inverted region*. Many experimentalists have tried to confirm the prediction of theory in vain until there were observations of ET between donor and acceptor moieties separated by a rigid spacer with steroid structure [47,48], ET between adsorbate and substrate [49–52], and the charge shift and charge recombination reaction of ion radical pairs [53–56].

B. Effects of Vibrational Modes

In Section I.A we were mainly concerned primarily with the effects of solvent motion to the reaction and ignored intramolecular degree of freedom. The effect of high-frequency intramolecular vibrational quantum modes as well as low-frequency solvent modes was worked out by generalizing the nonadiabatic expression [6–11]. The expressions of ET theory including low-frequency solvent vibrational mode (ω_s) and high-frequency intramolecular vibrational mode (ω) are given as follows [11]. At the low temperature limit, $kT_B \ll \hbar\langle\omega_s\rangle \ll \hbar\omega$, it is given by

$$k_{ET} = \frac{2\pi V_{el}^2}{\hbar^2\langle\omega_s\rangle} \exp(-S_s - S) \sum_{m=0}^{\infty} \frac{S_s^{p(m)} S^m}{[p(m)]! \, m!} \tag{1.4}$$

where $S_s = \lambda_s/\hbar\langle\omega_s\rangle$ and $S = \lambda_\nu/\hbar\omega$ are the coupling strengths for solvent low-frequency and intramolecular high-frequency modes, respectively, $p(m)$ the Bessel function, and m the vibration quantum number of the product states. λ_ν is the reorganization energy for the intramolecular high-frequency mode. The contributions of all product vibrational quanta are added. For the intermediate-temperature region, $\hbar\langle\omega_s\rangle \ll k_B T \ll \hbar\omega$, it is described as

$$k_{ET} = \frac{2\pi V_{el}^2}{\hbar\sqrt{4\pi S_s\hbar\langle\omega_s\rangle k_B T}} \exp(-S) \sum_{m=0}^{\infty} \exp\left[-\frac{(\Delta G - S_s\hbar\langle\omega_s\rangle - m\hbar\omega)^2}{4S_s\hbar\langle\omega_s\rangle k_B T}\right] \frac{S^m}{m!} \tag{1.5}$$

These equations are different expressions of Eq. (1.1) obtained by introducing the effect of high-frequency quantum states m of the product state. In this scheme the reactant molecule is in its vibrational state, with Boltzmann distribution and product in its higher vibrational excited states, and thus the effect of the high-frequency vibrational mode is explicitly included.

In some cases we can simplify the expression above only by taking into account the product vibrational state and the ET that takes place from the ground vibrational state. The overall rate constant is the sum over all the individual rate constants for different vibronic channels,

$$k_{NA} = \sum_n k_{NA}^{0\to n} \tag{1.6}$$

Here $k_{NA}^{0\to n}$ is the nonadiabatic rate constant between the ground vibrational state of the reactant and the nth excited vibrational state of the product. $k_{NA}^{0\to n}$ has a functional form similar to that of the nonadiabatic expression of Eq. (1.1) except the effective energy gap, which takes into account the quantization of energy,

$$\Delta G_0^{0\to n} = \Delta G_0 + nh\nu \tag{1.7}$$

and the effective electronic matrix element that considers the Franck–Condon overlap between the two vibronic states,

$$(V_{el}^{0 \to n})^2 = V_{el}^2 \langle 0|n \rangle^2 \tag{1.8}$$

The Franck-Condon factor is expressed as

$$|\langle 0|n \rangle|^2 = (S^n/n!) \exp(-S) \tag{1.9}$$

where the electron–vibrational coupling strength S is given by

$$S = \lambda_{hf,vib}/h\nu_{hf,vib} \tag{1.10}$$

Here $\lambda_{hf,vib}$ and $\nu_{hf,vib}$ are the reorganization energy and frequency of the quantized high-frequency mode. Consequently, the individual rate constant is expressed as

$$k_{NA}^{0 \to n} = \frac{2\pi}{\hbar} \frac{2\pi}{\sqrt{4\pi k_B T}} (V_{el}^{0 \to n})^2 \exp\left[-\frac{(\Delta G_0^{0 \to n} + \lambda_s)^2}{4\lambda_s k_B T}\right] \tag{1.11}$$

The high-frequency quantum mode is particularly important in the inverted region, as shown in Figure 1. The ET rate constant becomes greater in the inverted region compared to the normal region and makes the bell shape asymmetric. This effect has actually been observed in the long-distant ET of separated donor and acceptor by inert rigid spacer [46,48]. Most recently we have reported such an effect for ET from the reduced primary electron acceptor chlorophyll a (A_0^-) to the secondary acceptor quinone (Q) in the photosystem I reaction center, as shown in Figure 2 [57]. The free energy change of the reaction was varied for a wide range by reconstituting different quinones after extraction of intrinsic phylloquinone. The rate constants of ET decreased significantly in both the small and large ΔG regions (i.e., in the normal and inverted regions), as expected from Eq. (1.5). Data points can be fitted by a bell-shaped curve (a solid line in Fig. 2) calculated according to Eq. (1.5). The classical expression of the simple Marcus theory [Eq. (1.1)] that gives a symmetrical parabola curve for log k_{ET} versus ΔG did not give a good fit to the data points. This means the operation of vibrational coupling through the high-frequency modes, which elevates the rates in the inverted region.

The expressions of ET theory including both high- and low-frequency modes, are powerful in analyzing ET studied over a wide temperature range. In photo-induced ET from cytochrome to chlorophyll in *Chromatium* [11],

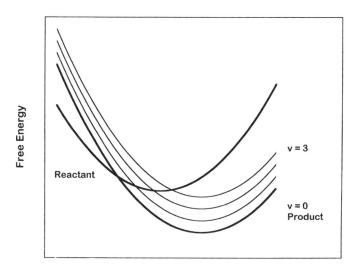

Figure 1. Free energy surface crossing at the inverted region, showing the efficient contribution of the high-frequency vibrational modes of the product surface, which creates bell shape of the free eneray dependence of electron transfer asymmetric.

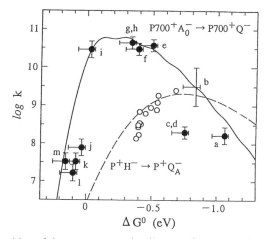

Figure 2. Logarithm of the rate constant ($\log k$) versus free energy change (ΔG^0) plot (solid circles) of the $P700^+A_0^- \to P700^+Q^-$ reaction in the photosystem I reaction center (RC) of spinach chloroplast, in which Q was replaced by different quinones (a–m). The data (open circles) for the $P^+H^- \to P^+QA^-$ reaction in the purple bacterium *Rb. sphaeroides* RCs [81], which contained different quinones, were also plotted. The solid and dashed curves were the best fits obtained using Eq. (1.5).

the rate constant of ET is only slightly temperature dependent at low temperature and strongly dependent at higher temperatures. In diffusionless donor–acceptor systems such as ET from solid organic substrate to dye adsorbate, temperature dependence was well analyzed by including a high-frequency mode [46,49].

C. Adiabatic Electron Transfer

In Sections I.A and I.B we have assumed that the solvation process is fast compared to other dynamics involved in ET and ignored the effect of the finite response of the solvent polarization relaxation. Since a time scale of the solvation dynamics is in the femto second and picosocond region for normal organic liquids, this assumption no longer holds, at least for ET, which occurs in the same time range as or faster than solvation dynamics.

Theoretical treatment that considers finite response of the solvent polarization was initiated by Zusman [13–15]. He solved a Landau–Zener ET problem in terms of the stochastic Liouville equation, where the quantum mechanical electronic transition between the reactant and product and classical diffusional motion in the potential well are explicitly included. Rips and Jortner treated the same problem in the framework of quantum mechanical propagators via path integral calculations for a two-level system coupled to a dielectric and gave the following simple equation [21]:

$$k_{ET} = k_{NA}/(1 + \kappa) \tag{1.12}$$

$$\kappa = \frac{4\pi V_{el}^2 \tau_s}{\hbar \lambda_s} \tag{1.13}$$

κ is called the adiabatic parameter. Equation (1.12) exhibits ET from the nonadiabatic limit ($\kappa \ll 1$) to the adiabatic limit ($\kappa \gg 1$) in one expression. The adiabatic rate can be written for $\kappa \gg 1$:

$$k_A = \frac{1}{\tau_s} \sqrt{\frac{\lambda_s}{16\pi k_B T}} \exp\left(-\frac{\Delta G^*}{k_B T}\right) \tag{1.14}$$

The reaction rate constant is inversely proportional to τ_s and the reaction is called a solvent-controlled adiabatic reaction [21,58]. At room temperature, $16\pi k_B T = 1.2\,\text{eV}$, and the value of λ_s is usually smaller than this. Therefore, the maximum of k_A is about $1/\tau_s$. Thus, for ultrafast ET, solvent fluctuation should be the limiting step. This kind of effect has become observable only recently with the development of ultrashort pulse lasers. Kosower and Huppert examined the excited-state intramolecular electron transfer of arylaminonaphthalene sulfonates in alcohol solutions [26]. They found a correlation between the ET rate constant and the inverse of longitudinal relaxation time, τ_L, of the solvent. τ_L is calculated using the Debye relaxation time τ_D:

$$\tau_L = \frac{\varepsilon_\infty}{\varepsilon_0} \tau_D \qquad (1.15)$$

where ε_∞ and ε_0 are the high- and low-frequency dielectric constants, respectively. τ_D is obtained by dielectric measurements. The different solvent relaxation times in relation to intramolecular ET are discussed by various authors [13–28]. Kang et al. studied the charge separation of bianthryl from locally excited state to the charge transfer state and correlated this with τ_s [29]. They found this activationless ET to be a solvent-controlled adiabatic reaction. Same kind of adiabatic process was also found for 4-(9-anthryl)-N,N-dimethylaniline (ADMA) [30,31].

In macromolecular systems such as protein, there are complex relaxation times for protein with fast atomic motions, large-amplitude motions of some protein moieties of intermediate speed, and very slow rotational diffusional motion of an entire protein. The rate of ET in a photosynthetic reaction-center protein complex varies from subpicosecond for the primary process to much slower (millisecond) final oxygen evolution. The ET reactions at different stages and rates are strongly related to the molecular dynamics of their location in protein. For example, the low-frequency large-amplitude motion that gives a large value of τ_s may still give quasiequilibrium non-adiabatic slow ET. In such a case it seems to be a problem for future researcher to identify the adiabaticity of the reactions at each step of the electron relay.

D. Electron Transfer Faster Than Solvent Dynamics, $k_{ET} > \tau_s^{-1}$

Recently, ET faster than solvation process has been discovered. The ET theory based simply on the solvation coordinate no longer holds. Ultrafast ET can be caused by both intramolecular vibrational motion (high frequency) and intermolecular vibrational motion (low frequency). Sumi and Marcus introduced two reaction coordinates explicitly: the nuclear coordinate (q) and solvent coordinate (X) as shown in Figure 3 [18]. Hereafter, we call this model two-dimensional ET theory.

$$G_r = \lambda_s X^2 + \lambda_\nu' q^2 \qquad (1.16)$$

$$G_p = \lambda_s (X - 1)^2 + \lambda_\nu' (q - 1)^2 + \Delta G \qquad (1.17)$$

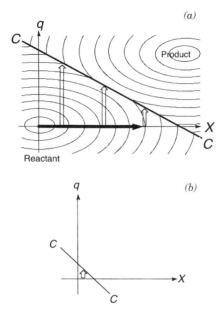

Figure 3. (*a*) Two-dimensional expression of the reactant and product free energy surface spanned by solvent coordinate X and nuclear coordinate q. The line C–C represents the transition-state curve. ET occurs along the q coordinate for every point on X as shown by the bold arrows. (*b*) simplified scheme for a free energy surface with almost no energy barrier.

Here λ'_ν is the reorganization energy for the classical coordinate q. In this scheme relaxation in nuclear coordinate is assumed to be much faster than that in the solvent coordinate. The quasiparticle on the free energy surface in Figure 3 moves along X by thermal fluctuation and transition from a reactant to a product occurs with a probability at a certain point of X. The population can be expressed by a diffusion-reaction equation,

$$\frac{\partial p(X,t)}{\partial t} = D \frac{\partial}{\partial X}\left[\frac{\partial}{\partial X} + \frac{1}{k_B T}\frac{dV(X)}{dX}\right]p(X,t) - k(X)p(X,t) \qquad (1.18)$$

where D is the diffusion coefficient, $V(X) = \lambda_s X^2$ is the free energy surface on the solvent coordinate, and $p(X,t)$ is the population at X and t. The first term on the right-hand side represents slow diffusion along the solvent coordinate X, and the last term represents the fast reaction along the nuclear coordinate q.

In the two-dimensional treatment, the vibrational nuclear motion is assumed to be infinitely fast, but the solvent motion is assumed to be diffusive. Thus, although the distribution along X may evolve in time, the distribution along q is always at equilibrium. Therefore, one can define a rate constant $k(X)$ at each X with suitable averaging over the population in the q coordinate:

$$k(X) = \nu_q \exp\left[-\frac{\Delta G^*(X)}{k_B T}\right] \qquad (1.19)$$

where $\Delta G^*(X)$ is the activation energy for the reaction observed at a point X and ν_q is the preexponential factor.

If the motion of the solvent is effectively frozen, the time-dependent full population on the reactant surface $P(t)$ will be given by the sum of each population $p(X, t)$ at each X;

$$P(t) = \int p(X, 0) \exp[-k(X)t] \, dX \qquad (1.20)$$

The main point of this theory is that the different reaction rate $k(X)$ at each X changes the distribution of the reactant during the reaction and gives rise to nonexponentiality. If the solvent is not frozen, the population dynamics can be described with a diffusion-reaction equation [Eq. (1.18)]. The time-dependent diffusion coefficient $D(t)$ is obtained from experiment in the form

$$D(t) = -\frac{k_B T}{2\lambda_s} \frac{1}{\Delta(t)} \frac{d\,\Delta(t)}{dt} \qquad (1.21)$$

where

$$\Delta(t) = a_1 \exp(-t/\tau_1) + a_2 \exp(-t/\tau_2) \qquad (1.22)$$

τ_1 and τ_2 are the time constants of the first and second components of the solvation dynamics and $a_1 + a_2 = 1$. For these parameters we use constants obtained by the dynamic, Stokes shift experiment.

The two-dimensional ET model treats both the solvent and intramolecular modes classically. However, the actual system should not only have the classical modes but also quantum mechanical high-frequency modes. To introduce the effect of vibrational frequency Walker et al. introduced a hybrid model of Sumi–Marcus and Jortner–Bixon as mentioned before [50]. This method is applied in explaining the nonexponential feature of ultrafast ET treated in this review.

Recently, ET reactions much faster than solvation processes were actually discovered for ultrafast *intermolecular* ET [33–44]. Fluorescence decays of dyes in electron-donating solvents as well as in electron–accepting solvents have been investigated. We found a fluorescence quenching as fast as about 0.1 ps for Nile Blue A perchlorate (NB) in N,N-dimethylaniline (DMA), and somewhat slower nonexponential fluorescence quenching in aniline (AN) (see below). We concluded that the fluorescence quenching is due to intermolecular ET from the solvent to the dye [35]. Later, many similar examples were found with oxazines and coumarins [28,36–44]. The main character of this reaction is that ET occurs much faster than the solvent diffusive process. For weakly polar solvents, the solvent reorganization process becomes less important, and nuclear reorganization may play a significant role. The vibrational motion is much faster than the solvent motions, and thus the reaction can precede the solvent relaxation process. Barbara et al. observed ultrafast *intramolecular* ET in betaine-30 and mixed valence compounds, and analyzed the results, including the effects of high-frequency vibrational motion [59,60]. Poellinger et al. observed intramolecular ET rate constants of porphyrin–quinone cyclophanes and found them to be on the order of $10^{12}\,s^{-1}$. It was almost independent on solvent polarity, which means that the reaction was not necessarily controlled by the solvent polarization dynamics [32].

II. SOLVENT DYNAMICS

As has been described in Section I, ET in solution is often a direct function of solvent properties. However, it was only recently that the contribution of dynamical properties of solvent to the microscopic mechanism of ET attracted researchers' attention. Thus reliable information on short-time behavior of solvent relaxation is particularly important. The solvent relaxation times were obtained historically by measurement of the dielectric loss spectrum and assume continuum dielectric media for solvent properties. In this article we use solvent relaxation times obtained from the dynamic fluorescence Stokes shift, which gives more realistic times of molecular systems under investigation than the ones obtained by dielectric loss measurement. When a probe dye molecule is photoexcited, the dipole moment of the molecule increases instantaneously. The polarization of the surrounding solvent molecules responds to this change and starts to reorganize. Therefore, the energy relaxation process, caused by the solvent reorganization, shifts the fluorescence spectrum to longer wavelengths. To obtain extract dynamical information of solvent relaxation process, the normalized spectral shift correlation function $C(t)$ is defined as [61]

$$C(t) = \frac{\nu(t) - \nu(\infty)}{\nu(0) - \nu(\infty)} \tag{2.1}$$

where $\nu(t)$, $\nu(\infty)$, and $\nu(0)$ are the fluorescence peak frequencies at times, t, ∞, and 0, respectively. For this method, a probe molecule that undergoes a large instantaneous change of dipole moment upon photoexcitation has to be chosen. Coumarin-102 (C102) is used as a probe molecule for this type of experiment, since it has a large difference in dipole moment between the excited and ground states but the difference in molecular structure is expected to be small [62–64]. It has a fluorescence lifetime of 2.8 ns in DMA and 1.4 ns in AN, and thus the effect of excited-state ET is minimal.

Figure 4 shows fluorescence decays of C102 observed at different wavelengths by a fluorescence up-conversion method with a femtosecond mode-locked laser [36,37]. At shorter wavelengths there is a fast decay, and at longer wavelengths there is a concurrent rise. This implies that the fluorescence spectrum of C102 is shifting toward red with time. One can use the spectral reconstruction method to obtain a time-resolved fluorescence spectrum [62]. Actual experimental results of $C(t)$ for the present solvents, AN and DMA, are given by biexponential fitting. The observed solvent relaxation times are 7.9 ps (19%) and 18.7 ps (81%) for DMA and 1.2 ps (28%) and 17.8 ps (72%) for AN. These are much slower than the speed of ET described in the present article, which is represented by the fluorescence lifetimes of Nile Blue in DMA (0.07 ps (95%)) [44] and of coumarin-151 in DMA [0.21 ps (100%)] [41]. Therefore, the prediction made by Kobayashi et

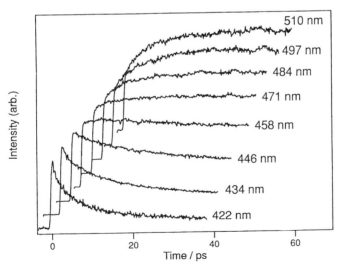

Figure 4. Fluorescenee decays of coumarin-102 (C102) in aniline measured at various wavelengths. Fast decay at shorter wavelenaths and the rise in longer wavelengths indicate a fluorescence peak shift due to the solvation of excited C102.

al. is confirmed—that the ET occurs much faster than the diffusive solvation
process by a factor of more than 50 [34].

III.   Fluorescence Spectra of Fast-Reacting Systems

Before describing the ultrafast dynamical observation of ET, let us show
that the effect of ultrafast reactions could be observed even in the steady-
state fluorescence spectra. The donor and acceptor molecules used in the
present studies are shown in Figure 5. If the reaction occurs much faster
than the solvation process, excited molecules only have a chance to fluoresce
from the non-equilibrium state before solvent relaxation. A conceptual

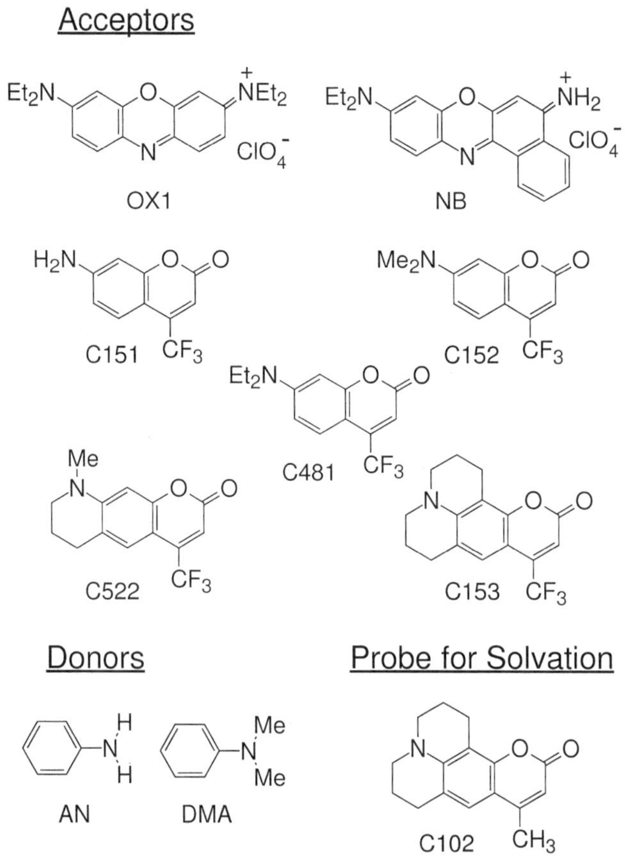

Figure 5.   Molecular structures of dyes (oxazines and coumarins) described in this article. Two
typical electron-donating solvent molecules, aniline (AN) and *N,N*-dimethylaniline (DMA), are
also shown.

drawing of these processes is shown in Figure 6. In this figure the solvation process occurs on the solvent coordinate at the same time the reaction may take place on the classical low-frequency vibrational coordinate to the reactant surface. If the initially populated nonequilibrium state is distributed above the activation energy, the reaction also occurs to higher vibrational levels of the quantum mechanical modes.

Dyes used in this article are strong fluorescent substances and are often used as laser dyes. However, when they are dissolved in electron-donating

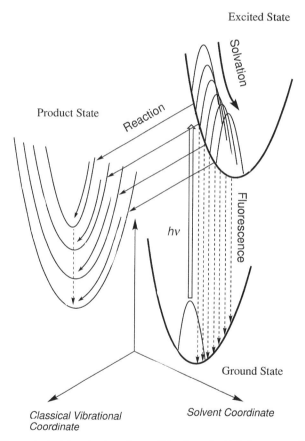

Figure 6. Conceptual picture to show the participation of a high-frequency vibrational mode of the product state in the ET mechanism. Also shown is the fact that the ET occurs along the q coordinate and competes with the solvation process along the X coordinate. The nonequilibrium initial population distribution in the reactant state (following photoexcitation) along the X coorddinate is also indicated. (From [41].)

solvents, the fluorescence is severely quenched. The fluorescence intensity of oxazine 1 (OX1) in DMA is more than 1000 times weaker than in inert solvents. In the previous studies [28] it was concluded that fluorescence quenching was actually due to intermolecular ET from solvent to dye by observing the transient product of DMA cation and neutral OX1 and its decay dynamics [35,39].

The Stokes shift obtained from the peaks of the absorption and fluorescence spectra of large organic molecules is usually greater than 4000 cm^{-1} for most of these dyes, however, for the fast-reacting dyes, it becomes as small as about 2600 cm^{-1}. The steady-state fluorescence spectra of reacting molecules are given Figure 7. A slower dye, C153, in DMA, which has a lifetime close to solvation time, has a Stokes shift of 4200 cm^{-1}; however, one of the faster dyes, C522, gives a Stokes shift of 2910 cm^{-1}. The fastest

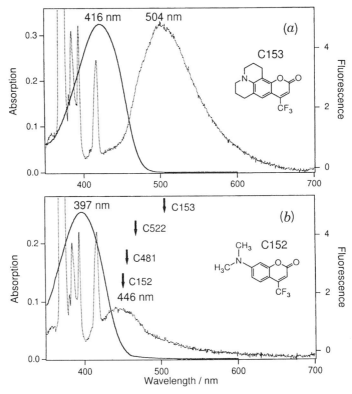

Figure 7. Steady-state fluorescence spectra of seleeted coumarins in DMA. The arrows indicate the fluoresce peak positions of different coumarins. Electron transfer is faster in the order C153 < C522 < C481 < C152. The structured spectra are Raman bands of solvent.

dye, C151, in DMA gives only $2630 \, cm^{-1}$. As the rate constant of ET increases, the amount of fluorescence Stokes shift decreases. This is an indication of the effect of ultrafast reaction to the steady-state fluorescence spectra. The effect can perhaps be called *chemical timing* or *fluorescence gating* by ultrafast reaction since the reaction controls the degree of relaxation on solvation coordinate [28,40]. Chemical timing was originally proposed in the intramolecular vibrational relaxation in the gas phase using collisional quencher. The degree of the intramolecular vibrational redistribution (IVR) process was controlled by the oxygen pressure, which determined the vibrational structure of the fluorescence spectrum of *p*-difluorobenzene [65].

IV. ELECTRON TRANSFER DYNAMICS: EXPERIMENTAL OBSERVATIONS

As we discussed in the preceding section, we now observe ET dynamics by measuring the fluorescence decays of an excited dye, which are shortened by three to four orders due to ET process. Several interesting features are observed:

1. The ET dynamics in many systems are much faster than the diffusive solvation process.
2. ET dynamics show non-single-exponential behavior.
3. With a system of OX1 dissolved in DMA, no temperature dependence on ET is found, while in AN there is a clear temperature dependence.
4. For the fast reactions, the lifetimes in DMA are shorter than in AN with the same acceptors.
5. The ET dynamics depend strongly on the substituent groups of the coumarin dyes.
6. The ET rates also depend strongly on the substituents of the amino group of donor anilines.
7. Perdeuterated AN (AN-d7) and aminodeuterated AN (AN-d2) shows a clear reduction in ET rates in comparison to those in normal AN.

The features above should be explained and analyzed using a suitable theoretical model.

The fluorescence decays of OX1 in the donor solvents are shown in Figure 8 [40,44]. The major part of the decay of OX1 in DMA shows nearly a single exponential function with a lifetime of about 0.07 ps, whereas OX1 in AN gives a nonexponential decay. The decay was tentatively analyzed by a triple exponential function with fitting parameters of 0.46 ps (40%), 1.6 ps (57%), and 18 ps (3%) at 285 K [40].

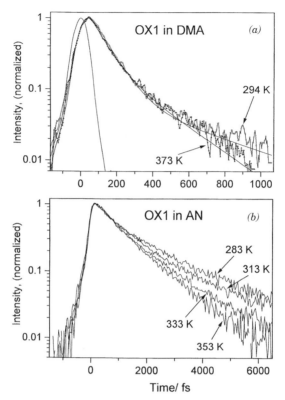

Figure 8. Fluorescence decays of OX1 (*a*) in DMA and (*b*) in AN. The former shows a fast exponential decay (ca. 70 fs) with a minor temperature-dependent component. The latter shows nonexponential decay with a clear temperature dependence for the slower decay component.

The weakly polar nature of the solvent used is one reason for the very fast ET, since it reduces the solvent reorganization energy. The dielectric constants (ε) are only 5.01 and 6.89 for DMA and AN, respectively. These values are quite small compared to polar solvents such as acetonitrile ($\varepsilon = 37.5$) and methanol ($\varepsilon = 32.6$). In weakly polar solvents, the effect of solvent reorganization is much smaller and the solvation process will not dominate the reaction.

Although oxazines are ionic and coumarins are neutral dyes, the mechanism of general scheme of excitation and reaction seems to be same. The charge shift reaction for oxazines and charge separation for coumarins both give an ultrafast ET in the present systems. The absorption spectra of OX1 shows a slight broadening in DMA compared to the weakly polar non-electron-donating solvents, which may indicate formation of a weak charge transfer complex between OX1 and DMA. However, a spectral change was not observed with coumarins. The possibility of observing intra-molecular vibrational redistribution (IVR) and solvent relaxation has been considered, and studies on excitation and observation wavelength depen-dence were carried out [25]. There were no effects, which means that such processes were not much involved in the present experiment. Back ET occurs successively from the neutral radical of the dye to the solvent cation, and the system goes back to the ground state again. The lifetime of the radical ion pair of NB and DMA, which was observed with pump-probe spectroscopy, was 4.0 ps [35], and that of OX1 and DMA was 4.7 ps [38].

V. TEMPERATURE DEPENDENCE ON ELECTRON TRANSFER

The temperature dependence of fluorescence decays of OX1 in anilines also gives us an insight to the ET mechanism. As is shown in Figure 8, the ultrafast decay of OX1 in DMA does not show any temperature dependence for the fast component, with a minor temperature dependence for the slower part [44]. In AN, however, a clear temperature dependence was observed in its non-exponential decay [40]. The first component (ca. 430 fs) of the triple exponential fittings does not show any detectable temperature dependence, but the second component (1.6 ps at 280 K) clearly becomes faster for the temperature range from 283 K   (10°C) to 353 K (80°C), which gives an activation energy of 1.0 kcal mol^{-1}. The results indicate that the slow ET is influenced by the temperature dependence of solvent motions, but this effect was negligible for the ultrafast part of ET.

Now let us analyze the ET dynamics of present systems, keeping in mind the various characteristics described above. Among several interesting dif-ferences in solvent properties (i.e., viscosity, hydrogen-bonding ability, and ionization potential), viscosity seems not to affect the ET dynamics. The high viscosity slows down the solvent dynamics and therefore may slow down the reaction. The amino group of AN can form hydrogen bonds with the solute molecule or with solvent molecules themselves, but that of DMA cannot. The hydrogen bonding between solvent molecules may be the reason for high viscosity of AN relative to DMA, which is described later.

The remaining difference which should be noted is the difference in bulk ionization potential. The ionization potential was calculated to be smaller by 0.17 eV [40] for DMA than for AN using the formula [66]

$$\Delta I_{cond} = I_{vap}^{AN} - I_{vap}^{DMA} - \left(1 - \frac{1}{\varepsilon_{AN}}\right)\frac{e^2}{2r_{AN}} + \left(1 - \frac{1}{\varepsilon_{DMA}}\right)\frac{e^2}{2r_{DMA}} \quad (5.1)$$

where I_{vap} is the ionization potential in the gas phase, ε the dielectric constant, e the electronic charge, and r the radius of the charged solvent molecule. This difference will cause a difference in the free energy gap for the reaction and thereby the free energy barrier for the reaction [40].

It seems that the faster reaction in DMA than in AN can readily be explained by the difference in energy gap. On closer examination it seems that this difference in energy gap can also be the cause of the difference in exponentiality. If there is no reaction barrier, the reactant can react very rapidly regardless of the inclination of curve C, and exponential kinetics may occur as can be expected from Figure 3(b). Thus the difference in energy gap may be the major reason for the different ET kinetics in AN and DMA.

In the present system the experiments on temperature dependence indicate that ET in AN is in the "normal region" and in DMA nearly on the top of the bell shape. To explain the observed experimental results, we apply a two-dimensional coordinate model with inclusion of a high-frequency vibrational mode. It is estimated that the minimum of the S_0 and S_1 free energy surface of OX1 is located at the same point. This means that the excited- and ground-state dipole moments of OX1 are not very different. The Stokes shift of OX1 is very small; therefore, this approximation seems quite reasonable. It is also assumed that the reaction occurs only from the lowest vibrational state of the reactant. We used 0.062 eV for the solvent reorganization energy λ_s, 0.186 eV for the reorganization energy of the classical low-frequency mode λ_ν', and 0.0105 eV for the electronic matrix element V_{el}.

The results of the simulations of temperature-dependent fluorescence decays are shown in Figure 9. The normalized population of the reactant $P(t)$ is plotted against time (ps) on a logarithmic scale. The energy gaps $-\Delta G$ are 0.248 eV for Figure 9(a) and 0.07 eV for Figure 9(b). Temperature is changed in each set, from 273 K to 373 K, with an assumption that the physical properties of the solvent (τ_s, λ_s, and dielectric constants) do not change very much with temperature. All other ET parameters are the same as we use to simulate the ET rate constants versus ΔG (see below). The simulated reaction shown in Figure 9(a) is almost exponential on the fast decay component, which coincides roughly with the experimental result of OX1 in DMA [Fig. 8(a)]. The free energy surface of the product crosses the bottom of the reactant surface. Such a situation occurs when $-\Delta G = \lambda_\nu' + \lambda_s = 0.248$ eV. Figure 9(b) perhaps corresponds to the ET of OX1 in AN [Figure 8(b)], which gives nonexponential dynamics, and the reaction becomes faster as the temperature increases. The major difference in the reaction dynamics seems to be the difference in $-\Delta G$ of 0.17 eV, as described above. In the simulation, the first component shows almost no activation barrier (0.28 kcal mol^{-1} or 0.012 eV), whereas the second component has an activation barrier of $0.77\,\text{kcal}\,\text{mol}^{-1}$ or $0.33\,\text{eV}$. This value is in good agreement with the experimental result ($1.0\,\text{kcal}\,\text{mol}^{-1}$ or $0.043\,\text{eV}$) [40].

It is worth noting that the reaction becomes slightly slower at higher temperatures [Fig. 9(a)]. When the population is located near the bottom of the reactant well, a higher temperature causes a wider distribution, which spreads the population farther from the bottom of the potential well. This makes the reaction slower. However, no such prediction of temperature

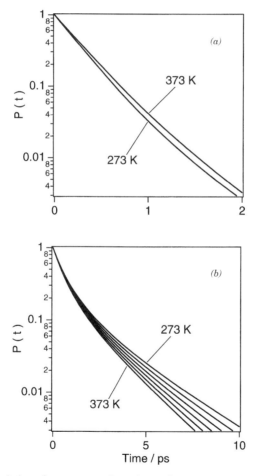

Figure 9. Simulation of temperature dependence of fluorescence decays using a two-dimensional coordinate model: (a) $-\Delta G = 0.248\,\text{eV}$, $T = 273$ and $373\,\text{K}$; (b) $-\Delta G = 0.0781\,\text{eV}$, $T = 273$, 293, 313, 333, and $373\,\text{K}$.

dependence has been observed for OX1 in DMA within the experimental error.

VI. SUBSTITUENT EFFECTS ON INTERMOLECULAR ELECTRON TRANSFER

The fluorescence decays of 4-CF_3 coumarins with five different 7-amino groups are shown in Figure 10 [41]:

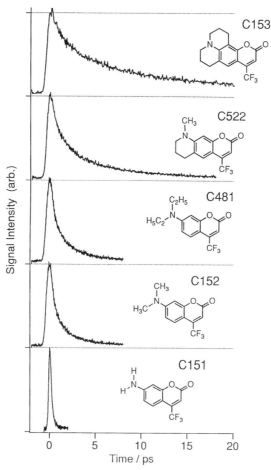

Figure 10. Fluorescence quenching of 7-amino-4-CF_3-coumarins in DMA excited at 395 nm and observed at 470 nm for C153, C522, C481, and C152 and 446 nm for C151. Concentrations are 2×10^{-3} M.

1. The ET rate depends on the substituent groups of the coumarin and it changes systematically with the substituent group on the 7-position.
2. Most of the ET of 4-CF$_3$ coumarins are faster than the diffusive solvation process.
3. Most show nonexponential fluorescence decay. The results of analysis are tentatively made with a double-exponential function.
4. For the fast reactions, the lifetimes in DMA are shorter than the ones of the same coumarins in AN.

The results, similar to observations 2 to 4 are, already discussed in Section IV with oxazines. When alkyl chain on the 7-amino group is extended, the reaction becomes slower, and when it forms a hexagonal alkyl ring with the benzene moiety, the reaction becomes the slowest. For the 7-amino group, the reaction rates increase in the order of

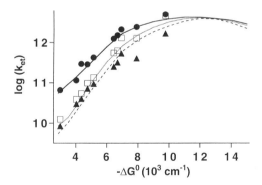

The effect of substitution could be discussed from two aspects: It can change (1) the vibrational mode of coumarin and (2) free energy difference of the reactant and product.

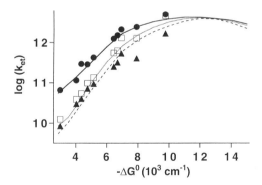

Figure 11. Plot of the experimental (Fig. 10) and simulated $\log(k_{ET})$ versus ΔG^0. The symbols are the experimental points: circles, fast component; triangles, slow component; squares, average value. The curves are the calculated values of the rate constants obtained from analysis of the simulated single wavelength decays at the finally shifted fluorescence peaks for the various ΔG^0 values: solid curve, fast component; dashed curve, slow component; dotted curve, average value. The ET and solvation parameters are $\lambda_s = 1600\,\mathrm{cm}^{-1}$, $\lambda_h = 7200\,\mathrm{cm}^{-1}$, $\lambda_l = 4800\,\mathrm{cm}^{-1}$, $v_h = 1180\,\mathrm{cm}^{-1}$, and $V_{el} = 170\,\mathrm{cm}^{-1}$, and the solvation parameters used are those of AN.

The ET rate constants obtained using a biexponential analysis are plotted in Figure 11 as a function of ΔG^0. The ET parameters are the same as in Section V. The experimentally observed rate constants are also plotted for comparison. It is seen from this figure that the correlation between the simulated and experimental rate constants is quite satisfactory. We arranged the data for different donor solvents in the same figure; however, this would not be quite right, although the agreement with theory and experimental data looks rather good. Since isotopic substitution can in principle affect different ET parameters; it is logical, however, to consider all the ET parameters separately and thus to see their effects on the simulated ET dynamics [43].

VII. DEUTERIUM ISOTOPE EFFECTS ON ELECTRON TRANSFER

In this section we wish to identify the roles of various ET parameters, such as the solvation time, vibrational frequencies, driving force, and so on, by using the deuterium isotope substitution. The ferrous-ferric (Fe^{3+}/Fe^{2+}) exchange reaction in water was predicted and observed to be slower by a factor of 2 in D_2O than in H_2O; $(k_{H_2O}/k_{D_2O} = 2)$ [67–71]. The cross reactions involving $Fe^{3+}(aq)/Fe^{2+}(aq)$, has also been studied [72]. Recently, a deuterium isotope effect on the ultrafast intramolecular ET rate in deuterated glycerol and in D_2O were explained in terms of the librational solvent motions [60,73,74]. For metal-aqueo complexes the major contribution to the isotope effect observed was considered mainly to have arisen from the solvent structural effects, which were different for D_2O and H_2O solvents [75,76].

In the present systems, deuterium-labeled solvents, which act as electron donors, are used: namely, AN-d7, aniline with all the seven hydrogens changed to deuterium, and AN-d2, with only the two hydrogens on the amino group deuterated. The isotope effect of AN-d7 is shown in Figure 12. In this figure the fluorescence decays of C151 (fast ET dye), C152, C481, C522, and C153 (slow ET dye) in AN-d7 (solid line) are compared with those in normal AN (dotted line) [42]. The decay in AN-d2 is identical to that in AN-d7 (not shown in Fig. 12) [42]. Therefore, the deuteration of amino hydrogen seems to be much more important than that of the phenyl hydrogen. The isotope effect on the fluorescence lifetimes is defined as

$$\text{isotope effect} = \frac{\langle \tau_D \rangle - \langle \tau_H \rangle}{\langle \tau_H \rangle} \qquad (7.1)$$

where $\langle \tau_H \rangle$ and $\langle \tau_D \rangle$ are the average fluorescence lifetimes of the coumarins in the normal and deuterated solvents, respectively. The isotope effects thus

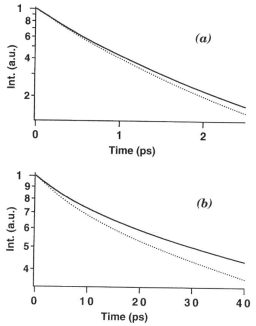

Figure 12. Deuterium isotope effect on ET dynamics. The fluorescence decays of coumarins in AN (dotted curves) and in AN-d7 (solid curves) are shown. For all the coumarins the decays are slower in AN-d7 than those in AN.

calculated are 1.1, 1.2, 1.7, 1.9, and 1.9 for C151 (fast ET dye), C152, C481, C522, and C153 (slow ET dye), respectively. It is interesting to note that the extent of the isotope effect gradually decreases as the ET process becomes faster. Thus with the fastest-reacting dye, coumarin-151 (C151), the isotope effect is smallest.

We also investigated the isotope effect on the ET dynamics using DMA and DMA-d6 as the electron donors for three coumarins (C151, C481 and C153; fast, medium, and slow dyes, respectively) in DMA and DMA-d6. The fluorescence decays for all these dyes are almost identical in both normal and deuterated DMA and thus show no isotope effect on the ET dynamics. The fluorescence decays of the deuterated coumarins, namely, C151-d2 and C152-d6, in both AN and DMA are just similar to those for normal C151 and C152, respectively. Let us now consider the effects of various ET parameters.

By considering the effect of isotopic substitution of the donor solvents on the ET dynamics it is seen that the effect is just related to the NH_2 group of AN. A similar isotope effect has also been observed on the solvation dynamics of AN (i.e., the solvation dynamics in AN-d2 and AN-d7 are slower than that in normal AN) [77]. A straightforward consideration of ET dynamics that arises due simply to the isotope effect on the solvation dynamics cannot, however, predict the observed isotope effect on the ET dynamics by our simulation results using the two-dimensional model [42]. Actually, this conclusion is quite reasonable because in the present system ET takes place quite rapidly along the q coordinate even when the solvent motions are frozen.

From our studies on the solvation dynamics of AN we have observed that the relative contribution of the inertial component to the total solvation energy is quite small ($< 10\%$) [39]. We thus expect that the effect on the ET dynamics of this component on the isotope effect observed might not be very appreciable. Developing an ET theory to include the effect of the inertial solvation component is, however, still an open question.

The isotope effect due to the solvent reorganization energy (λ_s) and reorganization energy (λ_l) of the low-frequency modes (q coordinate) will also not be affected much by isotope substitution. The ET rate is independent of the frequencies, ν_l, and depends only on the reorganization energy, λ_l, associated with these modes [69]. It is logical to assume [29] that for this low-frequency vibrational mode the force constant and the absolute displacement associated with the ET process are equal for both the normal and the deuterated solvent donors. It is thus expected that the isotope effect on ET dynamics cannot arise from the low-frequency modes involved in the ET process.

The ET parameters we consider now are the frequency (ν_h) and the reorganization energy (λ_h) of the high-frequency accepting mode. It is expected [67–70] that substitution of the amino group hydrogens of AN by deuterium can reduce the frequency of the high-frequency accepting mode (assumed to be 1180 cm^{-1} in this simulation) coupled to the ET process. The reorganization energy (λ_h) of this mode remains the same for both normal and deuterated AN. Thus to see the isotope effect of the high-frequency mode on the ET dynamics we should consider the effect of the frequency alone. We simulated the fluorescence decay (ET dynamics) tentatively using the ν_h value by 100 cm^{-1} less in AN-d7 than in AN, along with incorporation of the solvation parameters of AN-d7 but keeping the other ET parameters, including λ_h, the same as used for AN. The simulated fluorescence decays are slower in AN-d7 than those in normal AN. Thus the simulation could predict an isotope effect similar to that observed experimentally [42].

Although data on the IR active modes or the high-frequency Raman modes of AN$^{.+}$ and DMA$^{.+}$ and their deuterated analogs are not available, the transient Raman data indicate some interesting features. There are a few modes whose frequencies are changed significantly on ionization and thus become significantly reorganized when going from the neutral to the cationic state. For AN the nitrogen-ring carbon stretching mode $(N - \phi)$ is the one most affected by ionization. The other modes that are changed appreciably are the C—C stretching modes. Comparing the frequencies of the different vibrational modes of the normal and deuterated AN and AN radical cations it is expected that the effect of isotopic substitution cannot be appreciable for either of the modes, except for the nitrogen-ring carbon stretching mode of AN. The interesting point, however, is that on isotopic substitution the frequency of the nitrogen-ring carbon stretching mode of both AN and AN radical cation is increased a little instead of decreasing [42]. It is thus expected that if this mode would be contributing, it should give an opposite isotope effect to what we observe experimentally. It could be possible, however, that this negative isotope effect of the high-frequency mode is relatively weaker than another strong positive isotope effect (probably arising through ΔG^0) and that the latter dominates over the former to give the isotope effect observed for ET dynamics. With more extensive spectroscopic data on the vibrational modes of the neutral and cationic forms of aniline and its deuterated analogs it would be possible to come to a definite conclusion regarding the contribution of the high-frequency vibrational modes toward the isotope effect observed for ET dynamics.

The isotope effect on the free energy difference of ET (ΔG^0) can arise from two possible reasons: the zero-point energy effect and the other is the solvent structural effect [75,76]. The origin of the zero-point energy effect on ΔG^0 is that the isotopic substitution of a solute can reduce the frequencies of some of its vibrational modes, and this in turn can reduce the zero-point energy of the deuterated solute [67–70,75–79]. If this reduction in the zero-point energy is more in the initial state than in the final state, it will cause a reduction in the free energy difference for the rodox process of the deuterated solute. In our cyclic voltammetric measurements we observe that in a common solvent (acetonitrile) the oxidation potentials of both normal and deuterated anilines are the same within experimental error. These observations thus indicate that the contribution of the zero-point energy effect to the observed isotope effect on the ET dynamics could be negligible.

The origin of the solvent structural effect on ΔG^0 is the intermolecular hydrogen bonding of the solvent. For the metal-aqueo complexes in H_2O and D_2O, it is suggested [75,76] that for going from the state of lower electrostatic field to that of higher electrostatic field there is a higher tendency of formation of more number of intermolecular hydrogen bondings around the reacting system. This increase in the number of intermolecular hydrogen bondings results in a reduction in the total entropy of the system for going from the state of lower electrostatic field to that of higher

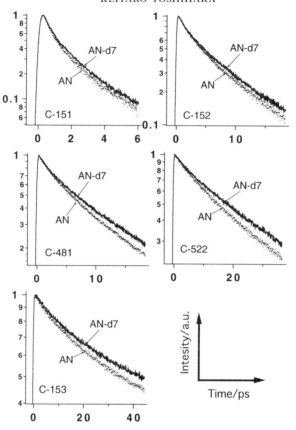

Figure 13. Effect of $-\Delta G^0$ on the simulated fluorescence decays at the fluorescence peaks in AN (dotted curves) and AN-d7 (solid curves): (*a*) fast ET ($\Delta G^0 = -6900\,\text{cm}^{-1}$ for AN-d7 and $-7000\,\text{cm}^{-1}$ for AN); (*b*) slow ET ($-\Delta G^0 = 3900\,\text{cm}^{-1}$ for AN-d7 and $4000\,\text{cm}^{-1}$ for AN). For AN-d7, ΔG^0 is assumed to be $100\,\text{cm}^1$ less than that in AN. The other ET parameters are the same as in Figure 11.

electrostatic field. A similar situation can also be expected for AN because AN is an intermolecular hydrogen-bonding solvent [80,81], and it is quite likely that AN-d2 and AN-d7 could be more strongly hydrogen bonded than normal AN. An indication in favor of this has already been observed in our solvation dynamics studies in normal and deuterated ANs [42]. Thus it is expected that the isotopic substitution of the solvent, AN, can cause a reduction in the driving force due to the solvent structural effect.

From our cyclic voltammetric measurements the reduction potentials of coumarins are observed to be about 10 mV more negative in deuterated AN than in normal AN. Similarly, the oxidation potentials of the anilines are seen to be more positive in MeOD than in MeOH. It is expected that for a particular ET process the driving force (ΔG^0) will be relatively lower with deuterated AN than with normal AN. We thus simulate the ET dynamics in AN-d7 using the driving force tentatively less by $100 \, \text{cm}^{-1}$ (12 meV) in this solvent than in normal AN but keeping the other ET parameters the same as in AN. Such a simulation qualitatively reproduces the observed isotope effect on the ET dynamics, as shown in Figure 13(a) and (b) for relatively fast ($\Delta G^0 = -6900 \, \text{cm}^{-1}$ for AN-d7 and $-7000 \, \text{cm}^{-1}$ for AN) and relatively slow ($\Delta G^0 = -3900 \, \text{cm}^{-1}$ for AN-d7 and $-4000 \, \text{cm}^{-1}$ for AN) ET cases, respectively. It is seen further from these simulation results that as the ET process becomes faster (ΔG^0 more negative) the isotope effect on the ET dynamics becomes smaller. It is exactly the same trend as we observe experimentally for the present ET systems. These observations thus suggest that ΔG^0 for a particular ET process is reduced in a deuterated solvent due to the solvent structural effect arising from the intermolecular hydrogen bondings and this is probably the main reason for the observed isotope effect on the ET dynamics in the present systems.

VIII.   CONCLUSIONS

The theories of electron transfer are described briefly with some emphasis on the dynamical aspects of ET on solvent and nuclear vibrational motions. The method of obtaining solvation time by dynamic Stokes shift measurement and some typical data relevant to this article are given. Intermolecular ET faster than solvation dynamics recently discovered is described for systems of oxazines in weakly polar electron donating solvents. The ET in the excited oxazine 1 in N,N-dimethylaniline occurs as fast as about 70 fs, which is more than an order of magnitude faster than diffusive solvation process. This reaction shows nearly single exponential kinetics with no activation energy. ET with nonexponential kinetics with an activation energy of 0.043 eV is observed with aniline as a solvent. The observations are explained by a two-dimensional reaction coordinate theory with solvent and nuclear vibrational coordinates (two-dimensional ET theory).

A systematic change in the rate of ET is observed with various substituted coumarin dyes as electron acceptors. This is explained by two-dimensional ET theory with a principal parameter of the free energy difference for ET (ΔG^0), which is obtained by optical and electrochemical measurements. It is also observed that the steady-state fluorescence spectra shifts to the blue for faster-reacting dyes and ascribed to the competition between the reaction (ET) and the solvation process of the excited dyes. One might be able to call this phenomenon fluorescence gating or chemical timing in solution.

Deuterium isotope effects are found for amino substitution of AN. From the analyses of the experimental results and considering the probable isotope effects on the different ET parameters, it is concluded that ΔG^0 for the ET reactions are probably reduced on isotopic substitution of the solvent, AN. The redox measurements indicate that the isotope effect on ΔG^0, and hence on the ET dynamics, arises primarily from the differences in the solvent structural effects of the normal and the deuterated AN, due to the differences in their intermolecular hydrogen-bonding strengths. Other possibilities for the role of vibrational motions on isotope substitution are also discussed.

ACKNOWLEDGMENTS

The author wishes to thank all who participated in the work presented here. He thanks K. Tominaga for reading the manuscript and I. V. Rubtsov and H. Shirota for preparing some of the figures. This work was supported in part by a grant-in-aid from the Grant-in-Aid for Scientific Research of New Program (07NP0301) from the Ministry of Education, Science, Sports, and Culture of Japan.

REFERENCES

1. R. A. Marcus, *J. Chem. Phys.* **24**, 976 (1956).

2. R. A. Marcus; *J. Chem. Phys.* **26**, 867 (1957).

3. R. A. Marcus, *Discuss. Faraday Soc.* **29**, 21 (1960).

4. R. A. Marcus and N. Sutin, *Biochim. Biophys. Acta* **811**, 265 (1985).

5. M. D. Newton and N. Sutin, *Annu. Rev. Phys. Chem.* **35**, 437 (1984).

6. V. G. Levich and R. R. Dogonadze, *Dokl. Acad. Nauk SSSR* **124**, 123 (1969). [*Proc. Acad. Sci. Phys. Chem. Sect.* **124**, 9 (1969)].

7. V. G. Levich, *Adv. Electrochem. Electrochem. Eng.* **4**, 249 (1966).

8. R. N. Kestner, J. Logan, and J. Jortner, *J. Phys. Chem.* **78**, 2148 (1974).

9. E. Efrima and M. Bixon, *Chem. Phys. Lett.* **25**, 34 (1974).

10. J. Ulstrup and J. Jortner, *J. Chem. Phys.* **63**, 4358 (1975).

11. J. Jortner, *J. Chem. Phys.* **64**, 4860 (1976).

12. J. Jortner and M. Bixon, *J. Chem. Phys.* **88**, 167 (1988).

13. L. D. Zusman, *Chem. Phys.* **49**, 295 (1980).

14. L. D. Zusman, *Chem. Phys.* **80**, 29 (1983).

15. L. D. Zusman, *Chem. Phys.* **119**, 51 (1988).

16. D. F. Calef and P. G. Wolynes, *J. Phys. Chem.* **87**, 3387 (1983).

17. J. T. Hynes, *J. Phys. Chem.* **90**, 3701 (1986).

18. H. Sumi and R. A. Marcus, *J. Chem. Phys.* **84**, 4894 (1986).

19. W. Nadler and R. A. Marcus, *J. Chem. Phys.* **86**, 3906 (1987).

20. J. Jortner and M. Bixon, *J. Chem. Phys.* **87**, 167. (1988).

21. I. Rips and J. Jortner, *J. Chem. Phys.* **88**, 818 (1988).

22. M. Sparpaglioni and S. Mukamel, *J. Chem. Phys.* **88**, 3265 (1988).

23. I. Rips, J. Klafter, and J. Jortner, *J. Chem. Phys.* **94**, 8557 (1990).

24. M. Bixon and J. Jortner, *Chem. Phys.* **176**, 467 (1993).

25. B. B. Smith, A. Sataib, and J. T. Hynes, *Chem. Phys.* **176**, 521 (1993).

26. E. M. Kosower and D. Huppert, *Chem. Phys. Lett.* **96**, 433 (1983).

27. H. Heitele, *Angew. Chem. Int. Ed. Engl.,* **32**, 359 (1993).

28. K. Yoshihara, K. Tominaga, and Y. Nagasawa, *Bull. Chem. Soc. Jpn.* **68**, 696 (1996).

29. T. J. Kang, W. Jarzeba, P. F. Barbara, and T. Fonseca, *Chem. Phys.* **149**, 81 (1990).

30. K. Tominaga, G. C. Walker, W. Jarzeba, and P. F. Barbara, *J. Phys. Chem.* **95**, 10475 (1991).

31. K. Tominaga, G. C. Walker, T. J. Kang, P. F. Barbara, and T. Fonseca, *J. Phys. Chem.* **95**, 10485 (1991).

32. F. Poelfinger, H. Heitele, M. E. Michel-Beyerle, C. Anders, M. Futscher, and H. A. Staab. *Chem. Phys. Lett.* **198**, 645 (1992).

33. K. Kemnitz and K. Yoshihara, *Chem. Lett.* **1991**, 645 (1991).

34. T. Kobayashi, Y. Takagi, H. Kandori, K. Kemnitz, and K. Yoshihara, *Chem. Phys. Lett.* **180**, 416 (1991).

35. H. Kandori, K. Kemnitz, and K. Yoshihara, *J. Phys. Chem.* **96**, 8042 (1992).

36. A. Yartsev, Y. Nagasawa, A. Douhal, and K. Yoshihara, *Chem. Phys. Lett.* **207**, 546 (1993).

37. Y. Nagasawa, A. P. Yartsev, K. Tominaga, A. E. Johnson, and K. Yoshihara, *J. Am. Chem. Soc.* **115**, 7922 (1993).

38. K. Yoshihara, A. Yartsev, Y. Nagasawa, A. Douhal, and K. Kemnitz, *Pure Appl. Chem.* **65**, 1671 (1993).

39. K. Yoshihara, Y. Naaasawa, A. Yartsev, S. Kumazaki, H. Kandori, A.E. Johnson, and K. Tominaga, *J. Photochem. Photobiol. A* **80**, 169 (1994).

40. Y. Nagasawa, A. P. Yartsev, K. Tominaaa, A. E. Johnson, and K. Yoshihara, *J. Chem. Phys.* **101**, 5717 (1994).

41. Y. Nagasawa, A. P. Yartsev, K. Tominaga, P.B. Bisht, A. E. Johnson, and K. Yoshihara, *J. Phys. Chem.* **99**, 653 (1995).

42. H. Pal, Y. Nagasawa, K. Tominaga, and K. Yoshihara, *J. Phys. Chem.* **100**, 11964 (1996).

43. H. Shirota, H. Pal, K. Tominaga, and K. Yoshihara, *J. Phys. Chem.* **102**, 3089 (1998).

44. I. V. Rubtsov, H. Shirota, and K. Yoshihara, to be published.

45. F. Markel, N. S. Ferns, I. R. Gould, and A. N. Myers, *J. Am. Chem. Soc.* **114**, 6208 (1992).

46. K. Wynne, C. Galli, and R. M. Hochstrasser, *J. Chem. Phys.* **100**, 4797 (1994).

47. J. R. Miller, L. T. Calcaterra, and G. L. Closs, *J. Am. Chem. Soc.* **106**, 3047 (1984).

48. G. L. Closs and J. R. Miller, *Science,* **240**, 440 (1988).

49. K. Kemnitz, N. Nakashima, and K. Yoshihara, *J. Phys. Chem.* **92**, 3915 (1988).

50. J. C. Moser and M. Gratzel, *Chem Phys.* **176**, 493 (1993).

51. J. M. Rehm, G. M. McLendon, Y. Nagasawa, K. Yoshihara, J. Moser, and M. Graetzel, *J. Phys. Chem.* **100**, 9577 (1996).

52. B. Burfeindt, T. Hannappel, W. Storck, and F. Willig, *J. Phys. Chem.* **100**, 16463 (1996).

53. N. Mataga, T. Asahi, Y. Kanda, T. Okada, and T. Kakitani, *Chem. Phys. Lett.* **127**, 249 (1988).

54. T. Asahi and N. Mataga, *J. Phys. Chem.* **93**, 6575 (1992).

55. I. R. Gould, R. H. Young, R. E. Moody, and S. Farid, *J. Chem. Phys.* **95**, 2068 (1991).

56. W.-S. Chung, N. J. Turro, I. R. Gould, and S. Farid, *J. Phys. Chem.* **95**, 7752 (1991).

57. M. Iwaki, S. Kumazaki, K. Yoshihara, T. Erabi, and S. Itoh, *J. Phys. Chem.* **100**, 10802 (1996).

58. I. Rips and E. Pollak, *J. Chem. Phys.* **103**, 7912 (1995).

59. G. C. Walker, E. Akesson, A. E. Johnson, N. E. Levinger, and P. F. Barbara, *J. Phys. Chem.* **96**, 3728 (1992).

60. K. Tominaga, D. A. V. Kliner, A. E. Johnson, N. E. Levinger, and P. F. Barbara, *J. Chem. Phys.* **98**, 1228 (1993).

61. B. Bagchi, D. W. Oxtoby, and G. R. Fleming, *Chem. Phys.* **86**, 257 (1984).

62. M. Maroncelli and G. R. Fleming, *J. Chem. Phys.* **86**, 6221 (1987).

63. M. A. Kahlow, T. J. Kang, and P. F. Barbara, *J. Chem. Phys.* **88**, 2372 (1988).

64. M. A. Kahtow, W. JarLeba, T. J. Kang, and P. F. Barbara, *J. Chem. Phys.* **90**, 151 (1989).

65. R. A. Coveleskie, D. A. Dolson, and C. S. Parmenter, *J. Chem. Phys.* **72**, 5774 (1980).

66. T. Tani and S.-I. Kikuchi, *Rep. Inst. Sci. Univ. Tokyo* **18**, 51 (1968).

67. R. A. Kuharski, J. S. Bader, D. Chandler, M. Sprik, M. L. Klein, and R. W. Impey, *Chem. Phys.* **89**, 3248 (1988).

68. J. S. Bader and D. Chandler, *Chem. Phys. Lett.* **157**, 501 (1989).

69. J. S. Bader, R. A. Kuharski, and D. Chandler, *J. Chem. Phys.* **93**, 203 (1990).

70. M. Marchi and D. Chandler, *J. Chem. Phys.* **95**, 889 (1991).

71. J. Hudis and R. W. Dodson, *J. Chem. Phys.* **78**, 911 (1956).

72. T. Guarr, E. Buhks, and G. McLendon, *J. Am. Chem. Soc.* **105**, 3763 (1983).

73. P. J. Reid, C. Silva, P.F. Barbara, L. Karki, J. T. Hupp, *J. Phys. Chem.* **99**, 2609 (1995).

74. P. J. Reid and P. F. Barbara, *J. Phys. Chem.* **99**, 3554 (1995).

75. M. J. Weaver and S. M. Nettles, *Inorg. Chem.* **19**, 1641 (1980).

76. M. J. Weaver, P. D. Tyma, and S. M. Nettles, *J. Electroanal. Chem.* **114**, 53 (1980).

77. H. Pal, Y. Nagasawa, K. Tominaga, S. Kumazaki, and K. Yoshihara, *J. Chem. Phys.* **102**, 7758 (1995).

78. E. Buhks, M. Bixon, J. Jortner, and G. Navon, *J. Phys. Chem.* **85**, 3759. (1981).

79. E. Buhks, M. Bixon, and J. Jortner, *J. Phys. Chem.* **85**, 3763 (1981).

80. H. Wolff and D. Mathias, *J. Phys. Chem.* **77**, 2081 (1973).

81. K. C. Medhi and O.S. Kastha, *Indian J. Phys.* **37**, 139, 275 (1963).

ELECTRON TRANSFER IN MOLECULES AND MOLECULAR WIRES: GEOMETRY DEPENDENCE, COHERENT TRANSFER, AND CONTROL

V. MUJICA[1], A NITZAN[2], Y. MAO[2], W. DAVIS[2], M. KEMP[2], A ROITBERG[3], and M. A. RATNER[3]

Department of Chemistry and Materials Research Center, Northwestern University, Evanston, IL 60208
[1] *Universidad Central de Venezuela, Facultad de Ciencias, Escuela de Quimica, Apartado 47120, Caracas 1020 A, Venezuela.*
[2] *School of Chemistry, The Sackler Faculty of Sciences, Tel Aviv University, Tel Aviv 69978, Israel.*
[3] *National Institute of Standards and Technology, Biotechnology Division, Bulding 222, A-353, Gaithersburg, MD 20899.*

CONTENTS

Electron Transfer: From Isolated Molecules to Biomolecules, Part Two, edited by Joshua Jortner and M. Bixon. Advances in Chemical Physics Series, Volume 107, series editors I. Prigogine and Stuart A. Rice.
ISBN 0-471-25291-3 　ⓒ 1999 John Wiley & Sons, Inc.

I. INTRODUCTION

Because a very large number of chemical processes, in molecules, biological systems and solids, correspond to transfer of electronic charge from one subregion or extended system to another, electron transfer is one of the most important processes in chemistry [1–24]. Extensive work over the past half century, beginning with Taube's measurements on metal complex systems [2,25] and with the Marcus [26,27] and Hush [28] formulation of adiabatic electron transfer rate theory [1,26,27], has increased our understanding of electron transfer systems. The current state of the art in both experiment and theory, as illustrated elsewhere in this volume, indicates a very high level of understanding of electron transfer rate phenomena and in some instances has permitted actual control of electron transfer rates by synthetic modification of molecular systems.

For a rate process to be defined, one requires irreversibility. This means that one particular subsystem, the primary system of interest, interacts with a bath that provides the source of actual irreversibility [20]. In ordinary molecular electron transfer processes, that bath arises from the vibrational, librational, and diffusive motions of the nuclei. Under these conditions, the appropriate bath density of states is weighted by Franck–Condon factors, so that the rate constant can be expressed in a simple form as [23,29,30]

$$k = \frac{2\pi}{\hbar} T_{\mathrm{DA}}^2 (\mathrm{DWFC}) \qquad (1.1)$$

Here we have assumed that the process is nonadiabatic, so that the electron transfer occurs in the vicinity of a *coincidence event* [10,13,24,31,32] (degenerate donor and acceptor states), and we have ignored dynamical processes in the bath. The matrix element TDA denotes the effective electronic mixing between donor and acceptor, and DWFC is the density-of-states weighted Franck–Condon factor, essentially an effective density of vibronic states.

Much more recently, measurements have been made on a similar set of systems, in which, again, a molecular structure acts as a bridge or linker between two sites that might be termed donor and acceptor. In these systems [33–39], which we will call *molecular wires*, the donor and acceptor are actually localized site wavefunctions on a molecule that are linked directly to metallic sites on an electrode. Examples of such systems include scanning tunneling microscopy measurements of tunneling currents along molecules chemisorbed or physisorbed on a metallic surface [33,35–39], mechanically controlled break junction measurements across molecules covalently attached on both ends to metallic surfaces [34], and measurement of currents

among gold nanodots functionalized with (α, ω) bridging molecular structures [36,40,41]. In these 4 molecular wire situations, the appropriate continuum states of the bath are no longer vibrational, but rather, are the
continuum electronic states of the electrodes. Measurement is no longer of
a rate process but of a current that flows along the molecule between electrodes, as sketched in Figure 1.

In both the intramolecular electron transfer nonadiabatic rate process
and the molecular wire tunneling current measurement, the overall Hamiltonian characterizing the dynamical process can be written

$$H = H_s + H_b + H_{sb} \tag{1.2}$$

Here the three terms on the right are, respectively, the Hamiltonian of the
system, that of the bath, and the interaction between the two. The system
Hamiltonian, in turn, can ordinarily be written as

$$H_s = |1\rangle\langle 1|\varepsilon_1 + |N\rangle\langle N|\varepsilon_N + (|1\rangle\langle N| + |N\rangle\langle 1|)T_{1N} \tag{1.3}$$

Here we have assumed that the two sites between which electron transfer
occurs can be labeled 1 and N, with respective site energies ε_1 and ε_N, and
that the effective electronic mixing between them is characterized by a
matrix element called T_{1N}.

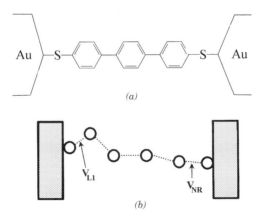

Figure 1. (a) Suggested structure of a terminal dithiol between the gold electrodes of a
mechanically controlled break junction, as suggested in [34]. (b) Model system on which conductance is calculated in molecular wires. The circles represent site orbitals within the molecular
wire, and the shaded blocks are the electrode structures. The matrix elements describe couplings
between the atoms fixed to the surface (terminal metal atoms) and the first atom in the molecular strand. In this case the molecular strand is envisioned as being four units long.

The identification of the bath and the system–bath interactions will differ in the molecular wire and nonadiabatic electronic transfer situations. With the molecular wire, the bath is simply the electronic states of the continuum electrode, and thus the bath interaction is provided by local mixing matrix elements between sites on the metallic surface and the chemisorbed atomic orbital functions (such as sulfur s and p functions for a thiol–gold inter- action). The bath Hamiltonian H_b then simply corresponds to the band structure of the electrode. In the simplest, hemispherical band model, this can be characterized [42–58] by the bandwidth (which we call 2γ) and the Fermi energy, E_f.

For the electron transfer problem, on the other hand, the system interacts with the nuclear motion states, normally referred to as the components of the outer-sphere reorganization energy [1–19]. The total reorganization energy is essentially the first moment of a spectral density function, and to some extent its breakup into inner and outer shell components is artificial [20]. From a dynamical point of view, however, the high-frequency modes, which should really be treated quantum mechanically, provide discrete levels; the lower-frequency modes, in particular those of the solvent, provide a much denser set of states and can be thought of classically as comprising a bath. In any case, the expression of Eq. (1.1) indicates a breakup between the electronic mixing within the system Hamiltonian and the effective state density that modulates that mixing, and exchanges energy to produce the rate.

The effective similarity between the nonadiabatic transfer rate and the molecular wire conductance implies that one can be informative about the other, and that the experimental interpretation of one can lead to insights about the other. In particular, from these measurements and their interpret- ation we can learn about the distance dependence of the charge transfer process, coherent versus incoherent transfer, the energetic dependence of the rate, its structural modification, and the possibility of control of the rate process for the conductance. This set of correlations comprises the subject of this chapter.

Section II briefly summarizes the experimental background for the study of nonadiabatic electron transfer and molecular wire conductance. In Sec- tion III we very briefly overview the polaron picture of nonadiabatic elec- tron transfer, in the absence of bridge localization, bridge scattering, or dephasing. Similarly, in Section IV we present a formalism for computing molecular wire conductance, in the absence of bridge localization, scattering or dephasing. In Section V, bridge localization, scattering, and dephasing are included, in a formalism appropriate for both situations. The results of this are described in Section VI, which is devoted to the length and geometry dependence, the role of coherent and incoherent transfer, and the limits under which exponential behavior will occur. Finally, some conclusions are presented in Section VII.

II. EXPERIMENTAL BACKGROUND

A. Intramolecular Electron Transfer Reactions

Intramolecular electron transfer, particularly in the nonadiabatic limit, has been of signal importance in understanding the different regimes of electron transfer reactions, ranging from simple energetic control through the inverted region to dynamical control and ultrafast rate processes [1–24]. Additionally, biological electron transfer reactions such as those between the docked complex of cytochrome c and cytochrome oxidase, those in the photosynthetic pathway, or various electron transfer enzymes, occur largely by nonadiabatic intramolecular electron transfer [11,14–21]. Biomimetic systems, and artificial photosynthetic reactions [59–61], also feature intramolecular electron transfer.

The original rate formulations, both the transition-state approaches of Marcus [26,27] and Hush [28] and the vibronic approaches based on a polaron-type [3–7,10,13,23,30,62–65] model, assumed that the initial state was in local thermal equilibrium; this set of conditions characterizes ground-state electron transfer reactions, usually prepared by electron transfer from an initiating species. Photoexcited electron transfer, on the other hand, generally begins from an initial state prepared by optical excitation, so that it is not in local thermal equilibrium. The overall rate process can still [67] be defined by averaging over initial and summing over final substates, but the initial average is no longer thermal.

Work in ultrafast electronic transfer kinetics has demonstrated a number of important issues (including solvent control, dephasing effects, coherent transfer) that were not present in the standard formulations of electron transfer theory; this has led to extensions of that theory [19,20].

From a conceptual point of view, all intramolecular electron transfer reactions become irreversible because of state density: the states whose density matters are those of the vibronic continuum or semicontinuum. The fundamental insight of Marcus [1,26] was that solvent polarization modes modulate the electron transfer reaction, satisfying the Franck–Condon principle, and can act as energy sinks or sources for the intramolecular electron transfer dynamics. Therefore, the expression of the rate constant (ignoring, for now, the issue of equilibration in the initial state) in terms of the Franck–Condon density of states, as in Eq. (1.1), arises because of the identification of the bath states for the electronic system as arising from vibrational or solvent modes.

B. Molecular Wires

Situations in which electrons flow through a molecular wire structure [33–41] between two electrodes [68] differ substantially. Now the initial state is described not by the nuclear positions and vibrational states but by the Fermi energies and initial states in the metal. The continuum that produces rate-type behavior, then, is no longer the vibronic continuum, but rather, the electronic continuum in the metal. This means that if we choose the electronic mixing as a perturbation, the bath states are those of the electronic continua in the two metals [69]. In this sense, the process of conductance is simply a scattering process, as the electronic wavefunction evolves from the anode to the cathode through the discrete states of the molecular wire. In this sense, conductance is scattering [70].

These considerations suggest that there are both deep-seated similarities and substantial differences in the comparison of molecular wire conductance with intramolecular electron transfer rate constants. Table I suggests both the similarities and differences. The two most striking differences are the nature of the continuum to which the electronic motion is coupled and the nature of the initial state. These differences will have substantial effects on the range of observable results and the possibilities for very long range transfer.

TABLE I
Similarities and Differences Between Molecular Wire Structures and Nonadiabatic
Intramolecular Electron Transfer Situations.

Aspect	Molecular Wire	Nonadiabatic Intramolecular Electron transfer
Potential energy diagram		
Observable	Current	Rate constant
Continuum	Electronic states	Nuclear motions
Initial state	Electrode	Vibronic levels
Process	Electron tunneling	Electron tunneling
Theory	$g = \dfrac{2\pi}{\hbar} e \sum f(E_1)[1 - f(E_N + eV)T_{1N}^2 \delta(E_1 - E_N)$	$k = \dfrac{2\pi}{\hbar} T_{DA}^2 (DWFC)$

The first column lists particular aspects of the system, and the second and third columns compare the behavior of the molecular wire and intramolecular electron transfer situations.

Until quite recently, these speculations on molecular wire behavior were just speculations, since no direct measurements had been made. Within the past three years, however, the use of scanning tunneling microscopy and mechanically controlled break junctions has permitted direct measurement of electron transfer through molecular wires, in the form of measured conductances [33–41,71]. In this paper we concentrate on the characterization of the conductance and comparison of very simple theoretical model for the conductance with early reported observations.

III. INTRAMOLECULAR NONADIABATIC ELECTRON TRANSFER: POLARON RATE THEORY AND CORRELATION FUNCTIONS

Current applications of the vibronic theory of intramolecular electron transfer rates can produce closed-form expressions assuming harmonic oscillator baths, and a few discrete modes, or in a more general, correlation-function form for arbitrary, anharmonic vibrations. These considerations are discussed extensively elsewhere [5–7,10,13,20,23,30,62–67,72–80]. The important point, for current purposes, is that the standard nonadiabatic electron transfer theory rate constant, in the form

$$k = \int_0^\infty dt\, e^{-it\Delta E/\hbar} \langle i| T_{DA} e^{iH_F t/\hbar} T_{DA} e^{-iH_I t/\hbar}|i\rangle \qquad (3.1)$$

requires some very specific assumptions on the nature of the transfer process. In Eq. (3.1), we denote by T_{DA} the electronic matrix element between initial and final electronic states. The correlation function is averaged over the set of nuclear degrees of freedom (solvent modes, intramolecular vibrations) that comprise the vibronic bath. The Hamiltonian operators are defined on the initial (H_I) and final (H_F) electronic states in the diabatic representation of Figure 2 ; and ΔE is the standard energy change for reaction [81].

This really rather simple and elegant form has been used to understand the time evolution of electron transfer process, contributions of different states, and the nature of dephasing and relaxation processes [66,67,72–79,81]. Perhaps most important, it is a general formulation that permits inclusion of effects such as breakdown of the Condon approximations [82–84] and preparation of an initial state that is not in local thermal equilibrium.

Nevertheless, there are some substantial limitations of this equation. First, the Condon approximation nearly always is made, permitting the rate constant to be written, instead of Eq. (3.1), by

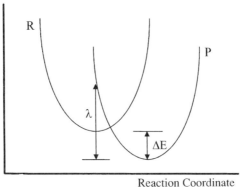

Reaction Coordinate

Figure 2. Usual (diabatic) reaction coordinate picture for nonadiabatic electron transfer situations. The reaction coordinate is best thought of as an energy difference, although it also corresponds to a geometric distortion. The reorganization energy, λ, and the overall exoergicity, ΔE, are indicated; the curves are labeled as reactant (R) and product (R)

$$k = T_{DA}^2 \int_0^\infty e^{-it\Delta E/\hbar} \langle i|e^{iH_F t/\hbar} e^{-iH_I t/\hbar}|i\rangle \, dt \qquad (3.2)$$

In this standard notation, no modulations of the electronic mixing amplitude, or of the electronic matrix element T_{DA}, are permitted. Obviously, T_{DA} is not simply the bare matrix element, but rather, an effective t-matrix element that should take into account all higher-order mixings in the electronic manifold [7,6,9,85–91]. This does not permit either dephasing or modulation by geometric effects. Therefore, this formulation does not permit localization of the electron (corresponding to geometric modification and trapping) on an intermediate site. No scattering of the electron by defects or vibrations is directly included, and dephasing effects on the bridge are absent. Despite these limitations, this formulation is completely standard in the field and has led to in-depth understanding of an enormous number of electron transfer reactions in solid, liquid, and membrane molecular situations.

IV. MOLECULAR WIRES: FORMAL ANALYSIS, SELF-ENERGIES, AND CONDUCTANCE

As suggested in Table I, for the molecular wire situation, the observable, the continuum, and the initial states differ from intramolecular electron transfer. Nevertheless, a very similar, perturbation theory analysis for the conductance is available [38,42–58]. It is based on the observation that the relevant bath is the electronic continuum in the electrodes. The phenomenon then bears a striking resemblance to chemisorption, in which (again) the discrete states of an isolated molecule interacts with the continuous electronic states of a metal.

Using the approach developed by Newns [42] and Anderson [43] for chemisorption, it is possible to rewrite the effective Hamiltonian for the molecular wire. This Hamiltonian is modified from the bare, electronic Hamiltonian of the molecule itself by self-energy terms that arise from interaction between the discrete and continuous states. In the simplest picture for the wire, such as that shown in Figure 1, in which only the terminal (1 and N) sites of the wire interact with the electrode, one can rewrite the Green's function for the electron propagation through the wire in terms of the electronic Hamiltonian H_{el} as [48–55]

$$G^{-1}(E_f) = E_f + iO^+ + \sum (E_F) - H_{el} \tag{4.1}$$

Here the self-energy terms, denoted by $\sum$, occur only at the first and last sites of the molecular wire structure; that is, they occur where the discrete molecular sites interact with the continuum structure of the electrode. The self-energy is described extensively elsewhere [42–58], the important point is that self-energy terms depend on three important parameters: the Fermi energy E_f, the bandwidth 2γ in the metal, and the coupling matrix elements V_{L1} and V_{NR}. These matrix elements characterize the strengths of the interaction between the first site on the wire with the left electrode, and the last, Nth site on the wire and the right electrode, respectively. In a very simple, extended Hückel picture they are just the analog of the Hückel β, acting between the orbital function on the terminal site of the wire and a local orbital function on the metal.

The appropriate conductance formula for the molecular wire structure of Figure 1, ignoring all nuclear motion, follows from the general Bardeen perturbation analysis [92]. Utilizing the Newns–Anderson formalism and simple hemispherical bands, the current formula is [55]

$$I = \frac{2e}{\pi\hbar} \int_{E_f - eV}^{E_F} dE \, \Delta_1(E)\Delta_N(E + eV)|G_{1N}(E; V)|^2 \tag{4.2}$$

Here I, V, and Δ_i; are, respectively, the current, the applied potential, and the imaginary part of the self-energy at site i. We note that the applied voltage V modifies the wire electronic Hamiltonian, hence the implicit dependence of G on V. The details of the relationship between the self-energy and the band structure follow from the Newns–Anderson model, as given elsewhere [55]. Here we simply state that

$$\sum_k (E) = \Lambda_k(E) - i\Delta_k(E) \qquad k = 1, N \tag{4.3}$$

$$\Delta_k(E) = \frac{\beta_k^2}{\gamma} \begin{cases} \sqrt{1 - \dfrac{E^2}{4\gamma^2}} & \left|\dfrac{E}{2\gamma}\right| < 1 \\[2ex] 0 & \left|\dfrac{E}{2\gamma}\right| > 1 \end{cases} \tag{4.4}$$

with $\beta_k = V_{NR}$ or V_{L1}, for $k = N, 1$, respectively, and where the energy E is measured from the center of the band.

Equation (4.2) gives the overall electron transfer rate for finite bias potential. If one were to assume linear response (Ohm's law), the integral simplifies, and the more approachable form

$$g = \frac{2e^2}{\pi h} \Delta_1(E_F)\Delta_N(E_F)|G_{1N}|^2 \tag{4.5}$$

appears for the conductance g. This is easily interpreted physically; the conductance consists of three terms. First is the density of donor states in the electrode on the left, second is the density of acceptor states in the electrode on the right, and third is the Green's function of the bridge that mixes these states with one another. In this sense, there is a direct analogy with Eq. (1.1): The mixing matrix element there compares to the Green's function of Eq. (4.5), and the density-of-states term in Eq. (1.1) compares to the product of self-energies in Eq. (4.5).

As in the vibronic intramolecular electron transfer rate of Eq. (3.1), however, the simple Green's function formulation in Eq. (4.2) is missing some physically critical components: It does not include the possibility of bridge localization of the electron, nor does it allow vibronic modification of the transfer amplitude described by Green's function, nor does it permit scattering of the electron or dephasing by n 'ear motions within the molecular wire.

All of these missing dynamical effects can be important. While the perturbation results of Eqs. (3.1) and (4.2) are adequate for many of the physical situations encountered in electron transfer and in molecular wires, they are, in themselves, inadequate. Despite this inadequacy, the simple conductance formula of Eqs. (4.2) and (4.5) have utility comparable to that of a golden-rule intramolecular electron transfer rate formula Eq. (3.1). Table II presents computational results [93] using a very simple extended Hückel model for the electronic structure of a molecular wire, the simple hemispherical band assumption for the electrodes (gold), and the assumption of injection at midgap for the Fermi energy. The experiments with which

TABLE II
Comparison of Calculated and Measured Resistance for Molecular Wires

System	Calculated Resistance (MΩ)	Experimental Results (MΩ)
Gold–sulfur–aryl–sulfur–gold	0.3^a	13.3–22 [34]
Gold–sulfur–aryl–sulfur–gold	9.2^a	6–9 [35]
Gold–C_{60}–tip	116.6^b	54.8 ± 13 [33]

[a] Using a value of $\beta_1 = \beta_n = 3.4$ eV in Eq. (4.5).
[b] Using a value of $\beta_1 = \beta_n = 0.47$ eV in Eq. (4.5).

these computations are compared are three of the earliest reported on single molecular wire conductances. The comparison with the two measurements made using scanning tunneling microscopy is quite satisfactory; agreement within a factor of 2 is remarkable considering the very simplified nature of the electronic structure model, the assumption (incorrect) of linear response, and the ignoring of dephasing and Coulomb blockade [39,45,55,94–97] behavior. Substantial disagreement with the mechanically controlled break junction measurement may arise because of an incorrect assumption about the geometry of the molecules bridging the break junction tip. It is assumed that a single benzenedithiol bridges between the gold electrodes, but this is not necessarily true given the preparation (evaporation from a solution of benzenedithiols) [34]. The important part about this table probably is the existence of these experiments: these are simply early indicators of the growing availability of measurements of transport through single molecular wires.

V.   BRIDGE LOCALIZATION AND DEPHASING EFFECTS

Molecular metals and conductive polymers are materials in which electrons, traveling through a molecular structure, are scattered by the vibrational and librational modes sufficiently strongly that they reach a terminal velocity and ohmic behavior is observed [98–100]. In conductive polymers, carrier motion at reasonable temperatures occurs by quasiparticle excitations corresponding to trapped electrons, either solitons (in degenerate conductors such as polyacetylene [101,102]) or polarons (in nondegenerate condutors such as polythiophene [102]). One expects, therefore, that for molecular wires or for intramolecular nonadiabatic electron transfer, when the distance between initial and final states becomes large enough, scattering and dephasing events will localize electrons, resulting in incoherent rather than coherent transfer, with substantially different length dependence.

There have been several theoretical analyses of this problem [103–105]. In the great majority of nonadiabatic intramolecular electron transfer reactions, it has been assumed that the Condon approximation holds. Use of perturbation theory then yields (for single connected, wirelike structures) exponential decay of the rate constant with distance between donor and acceptor [5–7,51,84–88,106–109]. Accordingly, one expects exponential decay in cases of coherent transfer, and indeed, exponential decay is nearly always seen in measured intramolecular electron transfer rate constants over distances of order 20 Å or less [4–9,11–15,22,110].

It is, however, anticipated that with long enough molecules, with a small enough energy gap between the injection level and the bridge orbitals, dephasings can occur, so that incoherent transfer is observed [103–105]. Indeed, under certain conditions actual electron localization on the bridge can be found. Earlier experimental results by Wasielewski suggested localization [111]. Hush, Reimers, and their collaborators have investigated extensively, from several theoretical points of view, the possibility of nonexponential decay [112]. One attractive model for such nonexponential decay generalizes the simple bridge-assisted superexchange picture of Figure 3 to include dephasing. That is, the evolution equation for the electronic density matrix is written as [105,113]

$$\dot{\rho} = \frac{1}{\hbar} i[H, \rho] + \mathscr{L}_d \rho \qquad (5.1a)$$

Here H is the electronic Hamiltonian corresponding to the levels sketched in Figure 3, ρ the electronic density matrix describing the populations and coherences in these levels, and $\mathscr{L}_d$ denotes the dissipative effect of the environment, which causes dephasing and relaxation in the electronic manifold. On the bridge levels, the dissipative time evolution from the second term is assumed to be of the form

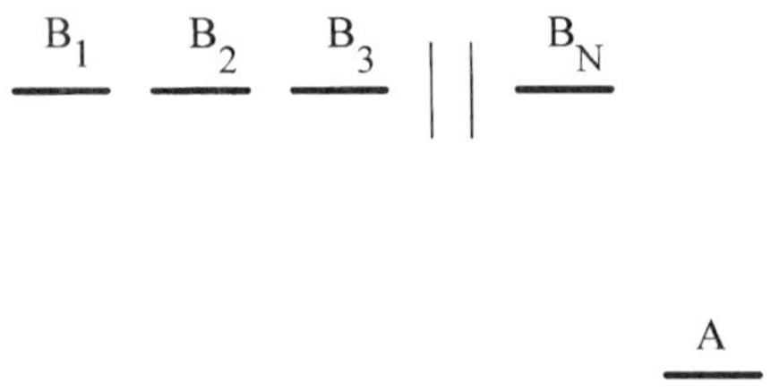

Figure 3. Electronic energy-level model for bridge-assisted electron transfer. D, A, and B_i represent, respectively, the donor level, the acceptor level, and the ith level of the bridge. The bridge is envisioned as N-site structure, assumed (for simplicity) to be degenerate.

$$\mathscr{L}_d \rho_{ij} = -\Gamma \rho_{ij} \qquad (5.1b)$$

In the absence of dephasing effects, the linear structure of Figure 3 is expected to exhibit exponential decay; that is, the decay of the initially prepared population will obey

$$-\dot{\rho}_{11} = A \, \exp(-\beta R) \qquad (5.2)$$

with the decay parameter being defined by

$$\beta = \frac{2}{R_0} \ln \frac{t_\beta}{E_g} \qquad (5.3)$$

In these equations, A is a prefactor, R the distance between initial and final sites, R_0 the intersite distance along the bridge, t_b the mixing matrix element between two adjacent bridge sites, and E_g the gap energy between the electron injected from the donor and the bridge electronic level.

Equaton (5.3) follows from the McConnell analysis [107], and can be derived in many ways [55,85–87,106,108,109]. More generally, one expects exponential decay to occur whenever there is a substantial gap between the injection level and any available delocalized molecular orbital of the bridge. Investigations of the conditions under which resonance flow (nonexponential dependence) might occur have been given, but for the most part, exponential behavior is found.

Dephasing, however, can overcome this exponential behavior. This is shown [105] directly in Figure 4, in which the intramolecular nonadiabatic rate constant is plotted as a function of the number of sites along the bridge. Although more extensive studies are available, and the problem has been studied as a function of the appropriate parameters, the physical result is quite clear and striking: For long-enough chains, the exponentially decaying coherent rate becomes smaller than the incoherent, multiple dephasing rate, which corresponds to ohmic behavior. This suggests that because of the dephasing events, connected molecular bridges really can function as wires—that is, charge can be transferred over very long distances with only minor (definitely not exponential) loss of current. This is important if molecular wires are ever actually to be used [38,58] as current carriers. The existence of conductive polymer systems [98–100], in which the transport is indeed incoherent, acts as proof of concept for this idea. Direct experimental demonstrations of the behavior of Figure 4, in isolated molecular structures, have yet to appear.

The actual plot shown in Figure 4a arises [105] from two parallel channels: One is due to coherent transfer, and indeed decays exponentially with

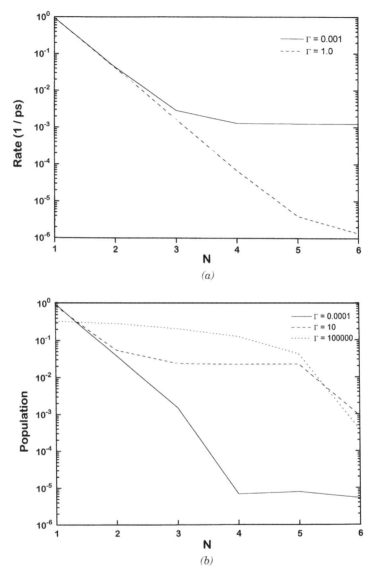

Figure 4. (a) Rate of intramolecular electron transfer for the electronic structure model of Figure 3, with the dissipative coupling described in the text. The parameters for this model are, in addition to the damping parameters, a matrix element coupling the levels equal to $300\,\mathrm{cm}^{-1}$, a gap (the energy difference between the bridge levels and the donor) of $4000\,\mathrm{cm}^{-1}$. (From [93].) (b) Population on the different sites, for the parameters of Figure 4a. Notice that with small dissipation, the population is almost entirely on the donor site, with only a very small population on the bridge sites; alternatively, with large dissipation all the sites on the bridge are essentially equally populated and have the same population as the donor. (From [105].)

distance, and the other comes from dephased, multiply scattered evolution, which decays only very weakly with distance. Figure 4(b) shows the population calculated on the sites; notice that the bridge occupation, small in the superexchange limit, increases when the incoherent pathway dominates. Calculations have also been performed with the bridge site populations represented by single vibrational oscillators [104] or by two-level systems [105], where the thermal populations on bridge levels can be found.

The idea that both coherent (superexchange) and incoherent (localized populations on the bridge) pathways contribute to an overall electron transfer rate has been suspected and demonstrated formally in the photosynthetic reaction center problem [21], where it now seems clear that the dominant contribution to the rate for charge separation from the special pair is due to an incoherent process involving intermittent population on the bridging bacteriochlorophyll.

The incoherences that we are discussing arise from thermal dephasing and relaxation effects involving the bridge levels. They should be distinguished from the usual sources of incoherence that arise from the continuum nature of the final-state manifold and the associated coupling from initially populated doorway states [75].

In the case of molecular wires, one expects similar dephasing sorts of behavior, with associated long-distance ohmic transport. Very recent reports of transport through carbon nanotube structures suggests coherent motion over very long distances [39], but incoherent motion has yet to be demonstrated in molecular wire circuits.

VI.   GEOMETRIC DEPENDENCE: LENGTH AND ANGLE MODIFICATIONS OF CONDUCTANCE

Formulas (3.1) and (4.2), for rate constant and conductance, respectively, do not directly contain dephasings. This means that one expects coherent processes through the bridge linking of either the donor and acceptor sites for the rate constant, or of the two electrodes for the molecular wire. One can then distinguish two situations: If the injection occurs into a resonance state of the bridge—that is, if the electron is injected at an energy very close to that of a bridge eigenstate—coherent, resonance, long-range transfer might be anticipated [55]. This follows directly from formal analysis of the wire structure and has apparently been observed in carbon nanotube wires [39]. In this regime, interesting effects of static scattering disorder are predicted [54,66,114], and it will be of interest to attempt measurement of these scattering and disorder effects.

Far more common, however, is injection away from resonance. Indeed, in the electron transfer reaction, injection at resonance is really not feasible;

one means is by a chemically prepared initial state, a state in which the electron is localized in some region of space. If that region of space becomes larger due to degeneracies, this simply means that the initial electronic state becomes larger, and therefore the bridge becomes shorter. This is one sense in which molecular wire experiment yields a richer data set, because the initial state is (Table I) localized in the electrode [50,115], and resonance does not cause it to leak prematurely onto the wire.

If, however, injection occurs into the gap region between discrete molecular orbitals of the bridge, the overall behavior of molecular wires and that of intramolecular electron transfer molecules becomes very similar. In particular, the length dependence of both (ignoring distance-dependent reorganizational energy changes in the molecule) will scale in the same way with distance and energetics; this is simply because the matrix element squared of Eq. (3.2) and the Green's function elements squared of Eq. (4.5) will be proportional to one another.

At this point, chemical design of the bridge becomes important. One has some simple expectations based on organic concepts of localization, delo-

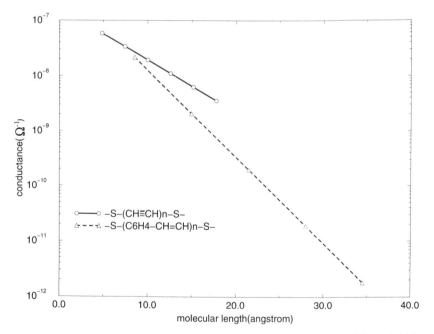

Figure 5. Conductance computed as a function of molecular length for two different bridging units (polyalkynes and poly-*p*-phenylene). Notice the exponential decay in both cases, with a more rapid decay occurring for the phenylene system. (From [93].)

calization, and mixing: One expects that π-electron systems will exhibit slower decay of distance than σ-electron systems or triple-bonded systems. One expects the stereochemistry of the bridge to be important, so that the introduction of gauche bonds will substantially reduce the rate of electron transfer. One expects twisting of the π system will reduce π mixing and therefore increase the rate of decay.

All of these anticipations have been discussed for the intramolecular electron transfer rate, and calculations show [49,93] that they also occur for molecular wire conductance. Figures 5 and 6 and Table III show, indeed, exponential decay with wire length and substantial reduction of the conductance for twisting around the interring π-bonds in (α, ω)-diphenydithiol.

The actual energy at which the Green's function in Eq. (4.5) is calculated is still a point of some confusion. We, and other workers [49] assume that the injection occurs in midgap [55] between the HOMO and LUMO levels, as illustrated in Figure 7 for a characteristic bridge. The actual point of

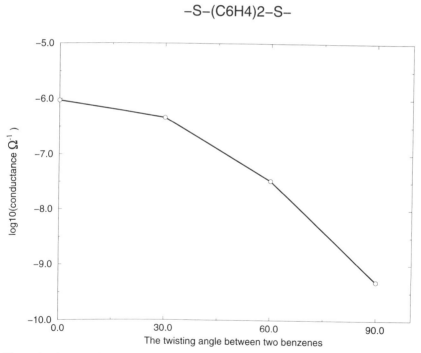

Figure 6. Computed conductance for molecular wire consisting of two phenyl rings with terminal thiols, between gold electrodes. Notice the very substantial fall-off in the conductance as a function of the twist angle. (From [93].)

TABLE III
Computed Conductances for Molecular Wires at Gold[a]

	Prefactor A	Exponent β	Conductance at 10 Å Length (Ω^{-1})	C—C bond Length (Å)[b]
—S—$(C_6H_4)_n$—S—	3.5×10^{-7}	0.3	1.9×10^{-8}	
—S—$(C_6H_4$—CH=CH$)_n$—S—	4.5×10^{-7}	0.4	1.2×10^{-8}	C=C:1.3335
—S—$(C_6H_4$—C≡C$)_n$—S—	6.3×10^{-7}	0.4	1.1×10^{-8}	C≡C:1.1934
—S—$(CH2)_n$—S— (n: even)	4.6×10^{-7}	0.8	1.8×10^{-10}	C—C:1.5500
—S—$(CH2)_n$—S— (n: odd)	2.8×10^{-7}	0.8	1.0×10^{-10}	C—C:1.5500
—S—$(CH=CH)_n$—S—	2.4×10^{-7}	0.2	3.6×10^{-8}	C=C:1.3373 =C—C=:1.4700
—S—$(C≡C)_n$—S—	1.7×10^{-7}	0.2	2.0×10^{-8}	C≡C:1.2039 ≡C—C≡:1.3783

[a] From Eq. (4.5). Conductance $g = A \exp(-\beta * L)$, where g is in Ω^{-1}, β in Å^{-1}, and L in Å (molecular length).
[b] The geometries of all the molecules optimized by Mopac program.

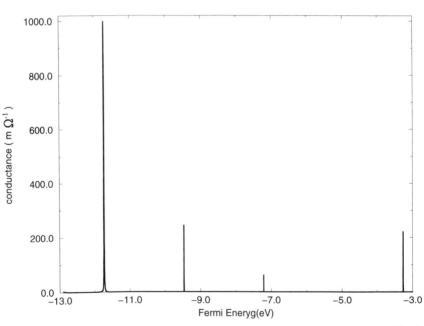

Figure 7. Conductance computed for a polyene as a function of the injection Fermi level. Resonances occur (damped by self-energy effects) whenever a π-type energy level crosses the Fermi energy. (From [93].)

injection is complicated, depending on the nature of the electrode, its Fermi level structure, and the self-consistent space charge in the molecular wire circuit. [116]. This is part of the complex behavior that one anticipates for actual molecular wires.

VII.   CONCLUSIONS

We have seen that there are substantial similarities, and substantive differences, between intramolecular nonadiabatic electron transfer and molecular wire conductance. The standard, perturbation-theoretic results of Eqs. (3.1) and (4.2) are significant for the interpretative power they have with respect to a large variety of experiments, and for permitting actual calculation and numerical data for comparison with experimental measurement. They also take into account what are probably the dominant terms in the Hamiltonian referred to in Section I.

However, as we have stressed in connection with bridge delocalization and dephasing in Section V and with the geometric dependence in Section VI, these standard descriptions do miss some important components of the bridge-assisted charge transport situation. In particular, the bath and system–bath terms, H_B and H_{SB}, are restricted in both Eqs. (3.1) and (4.2) to the donor and acceptor termini of the system. In the electron transfer case, the vibronic coupling occurs on donor and acceptor ends, while in the wire case, the electronic continuum of the metal couples only to the terminal sites of the bridge. In fact, there is a different k.... of bath, and a different system–bath interaction, that can act on the bridge itself. This could arise from non-Condon effects (vibrational motions, bath collisions, twists). These can completely change the parametric dependence of the conductance or the rate constant, as demonstrated in Section V. There are several well-known experimental situations in which this occurs, such as for the conductive molecular metals (polaron formation, phononlike scattering from vibrational modes or, simply, scattering events that cause ohmic behavior all correspond to interactions along the bridge linking initial- and final-state sites). In the electron transfer case, the chemical mechanism [2] in which charge localization actually causes reactive events on the bridge, and the intermediate ion formation discussed by Wasielewski [111], are examples of interaction of electronic states on the bridge with a bath.

Formally, the bath acting on the bridge could be treated differently depending on the nature of its effects; the first few theoretical analyses [103–105] have been done in the context of irreversible relaxation theory, in which no specific energetic account is taken of these bath interactions. This is probably appropriate for dephasing but may not be for actual trapping or energy exchange with the bath.

There is a fascinating issue of time scale involved in this question of ignoring scattering and dephasing on the bridge. Following the analysis of Landauer and Buttiker [117,118], one should be able to calculate an appropriate time scale both for tunneling through the bridge (coherent transfer) and for interaction and energy loss on the bridge (incoherent process). This problem is normally formulated for continuous potentials; its generalization to a site potential situation, such as that involved in Hückel Hamiltonian linking donor and acceptor through intermediate sites, is relatively straightforward [119].

Interelectronic interactions of the transfering electron on the bridge that go beyond the extended Hückel model can be important [55,85–87]. In electron transfer reactions, a number of extended ab initio [85–87] studies have demonstrated that electron repulsion effects can quantitatively modify the behavior expected from simple superexchange [85,107,120] models, but that, qualitatively, these effects are less important. For molecular wires the situation can be very different. In particular, if the effective conductance between the terminal site of the wire and the electrodes is small, equivalent circuit analysis suggests that one might actually observe different transport channels, corresponding to buildup of charge on the bridge, under the influence of an external field [45,49,55]. Physically, this simply means that a capacitive, self interaction potential has to be overcome in order to inject charge onto the bridge. Indeed, under these conditions the current–voltage characteristic will never exhibit a nonvanishing linear slope at the origin, because of Coulomb blockade effects (the potential must become large enough to overcome the capacitive charging energy of the wire) [55,94–97]. These effects have been very well documented in semiconductor structures [94,95], and there are some preliminary examples, both theoretical [45,49,51] and experimental [24,38,39,96], in the case of isolated molecule wires. Figure 8 shows both experimental data [39] and some calculated results of Coulomb blockade structure [computationally, of course, Eq. (4.5) cannot be used since it assumes ohmic behavior; these results were obtained using Eq. (4.2)]. Appropriate calculation of these Coulomb effects requires [49,116] self-consistent solution of the conductance of Eq. (4.2) and the Poisson equation that determines the overall potential due to electron interactions. This Poisson equation has unusual boundary conditions, because of the continuum electrode and the discrete electronic structure on the bridge, but its solution can be obtained. Qualitatively, the potential acting across the bridge should look more like a double-layer potential from electrochemistry (rapid modification of the potential near the electrodes, effectively flat potential across the bulk of the bridge); our initial calculations indeed indicate such a structure [116].

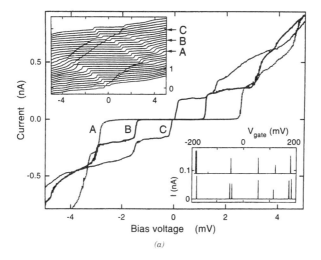

(a)

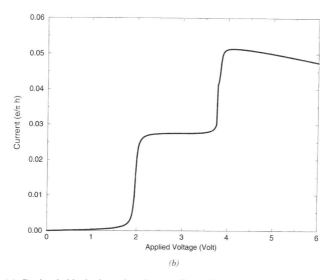

(b)

Figure 8. (*a*) Coulomb blockade and staircase effects. The experimental data are from the paper of Tans et al. on a carbon nanotube between gold electrodes; note the vanishing conductance at the origin (Coulomb blockade) and the steps in the conductances as a function of applied voltage. Curves A, B, and C have different gate voltages. The right insets show dependence on gate voltage. The steps arise from different accessible states in the voltage window. The measurements were performed at 5 mK. (From [39].) (*b*) Theory of a similar situation, based on the full conductance formula of Eq. (4.2). Notice, again, the blockade at the origin and gaps corresponding to the Coulomb staircase, in this case modified by the molecular energy levels. (From [55].)

Dynamical scattering processes due to time-dependent bath interactions can, as we have stressed above, change the transport behavior. In addition, however, static disorder (modification of local site energy levels along the bridge) can substantially modify the predicted conductance in the molecular wire situation and rate constant for intramolecular electron transfer. Calculations have shown that [51,54] static energetic disorder can completely change the expected resonant or superexchange dependence of the distance-dependent tunneling current in molecular wires; compare Figure 9. In the thermodynamic limit, this kind of disorder will lead to Anderson localization and lack of charge transport, but in finite molecular wires it simply modifies the conductance. In molecular wires, as in nonadiabatic electron transfer situations, one isolated site that is either of high or low energy can reduce the mixing, and therefore the rate constant; similar considerations should enter into any site

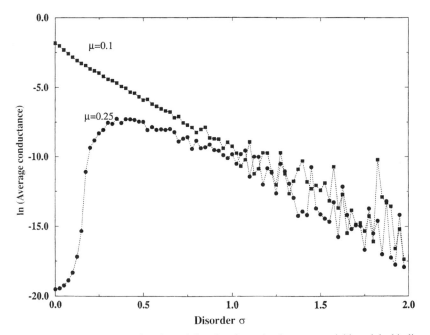

Figure 9. Conductance as a function of disorder, for a simple nearest-neighbor tight-binding model. The parameter μ is the average energy difference between the Fermi energy and the site energies. When injection occurs near resonance ($\mu = 0.1$), the increase is disorder causes substantially reduced conductance. on the other hand, for highly off-resonant ($\mu = 0.25$) situations, adventitious crossings of disorder levels with the injection level can result in increased average conductance. In general, the distribution of the conductance under these conditions is of a lognormal rather than a normal type. (From [51].)

penalty model, such as the Beratan–Onuchic approach [88], for calculating bridge-assisted currents in a site basis.

One fascinating aspect of these bridge variation phenomena (Coulomb blockade, Coulomb staircase, inelastic scattering, static disorder, charge trapping, and geometric modification) is that they permit artificial control of the rates of electron transfer or wire conductance. In this sense, these effects join such well-studied phenomena as the inverted regime, the control of reorganization energy, solvent glassing and dynamics, and the covalent and electronic design of the bridges and donor–acceptor sites as control elements [13,14,121,122] for electron transfer and electron tunneling. It is this last set of phenomena, which (to some extent) goes beyond the simple formulations of perturbation theory that are enshrined in Eqs. (3.1) and (4.2), that may constitute the next exciting chapter in the study of charge transfer and delocalization in molecules and molecular wires.

ACKNOWLEDGMENTS

We are grateful to the Chemistry Divisions of the National Science Foundation and the Office of Naval Research for funding this work. M.R. and V.M. are grateful to the joint NSF/CONICIT International Program for research support. We thank Deborah Evans, Joshua Jortner, Paul Barbara, Joe Hupp, Brian Hoffman, Richard VanDuyne, and Mark Reed for helpful comments. We are grateful to Cees Dekker for a prepublication copy of [39].

REFERENCES

1. R. A. Marcus, *Rev. Mod. Phys.* **65**, 599 (1993).

2. H. Taube, *Electron Transfer Reactions of Complex Ions in Solutions*, Academic Press, San Diego, Calif., 1970.

3. J. Ulstrup, *Charge Transfer Processes in Condensed Media*, Lecture Notes in Chemistry, Vol. 10, Springer-Verlag, New York, 1979.

4. M. E. Michel-Beyerle, G. J. Small, R. M. Hochstrasser, and G. L. Hofacker, *Chem. Phys.* **197**, 233 (1995) (special issue).

5. M. D. Newton and N. Sutin, *Annu. Rev. Phys. Chem.* **35**, 437 (1984).

6. R. A. Marcus and N. Sutin, *Biochim. Biophys Acta* **811**, 265 (1985).

7. M. D. Newton, *Chem. Rev.* **91**, 676 (1991).

8. R. D. Cannon, *Electron Transfer Reactions,* Butterworth, London,1980.

9. M.A. Fox and M. Chanon, eds., *Photoinduced Electron Transfer,* Vols. A–D, Elsevier, New York, (1988).

10. G. C. Schatz and M. A. Ratner, *Quantum Mechanics in Chemistry,* Prentice Hall, Upper Saddle River, N.J., 1994, Chap. 10.

11. J. Jortner and B. Pullman, eds., *Perspectives in Photosynthesis,* Kluwer, Dordrecht; The Netherlands 1990.

12. L. Arnaut, ed., *J. Photochem. Photobiol.* **82**, (113) (1994).

13. P. F. Barbara, T. J. Meyer, and M. A. Ratner, *J. Phys. Chem.* **100**, 13148 (1996).

14. *Inorg. Chim. Acta,* **242** (1996).

15. R. Cave and M. D. Newton, in [58].

16. G. D. Billing and K. V. Mikkelsen, *Molecular Dynamics and Chemical Kinetics,* Wiley, New York, 1996.

17. B. Chance, D. C. DeVault, H. Frauenfelder, R. A. Marcus, J. R. Schrieffer, and N. Sutin, eds., *Tunneling in Biological Systems,* Academic Press, San Diego, Calif., 1979.

18. D. DeVault, *Quantum-Mechanical Tunneling in Biological Systems,* Cambridge University Press, Cambridge, 1984.

19. P. F. Barbara and W. Jarzeba, *Adv. Photochem.* **15**, 1 (1990).

20. G. R. Fleming, M. Cho, *Annu. Rev. Phys. Chem.* **47**, 109 (1996).

21. M. E. Michel-Beyerle, ed., *Reaction Centers of Photosynthetic Bacteria.* Springer-Verlag , Berlin, 1990; J. Breton and A. Vermeglio, eds., *The Photosynthetic Reaction Center, Structure, Spectroscopy and Dynamics,* Vols. I and II., Plenum Press, New York: 1952, 1987, M. E. Michel-Beyerle, ed., *The Reaction Center of Photosynthetic Bacteria: Structure and Dynamics,* Springer-Verlag, Berlin, 1995; M. E. Michel-Beyerle, work in progress; M. E. Michel-Beyerle, Part I of this volume.

22. *Prog. Inorg. Chem.,* **30**, (1983).

23. M. Bixon and J. Jortner, *J. Phys. Chem.* **93**, 13061 (1993).

24. K. V. Mikkelsen and M. A. Ratner, *Chem. Rev.* **87**, 112 (1987).

25. R. D. Cannon, *Electron Transfer Reactions,* Butterworth, London, 1980.

26. R. A. Marcus, *Annu. Rev. Phys. Chem.* **15**, 155 (1964).

27. R. A. Marcus, *J. Chem. Phys.* **24**, 679 (1956); **43**, 679 (1965).

28. N. S. Hush, *Trans. Faraday Soc.* **57**, 155 (1961); *Electrochim. Acta* **13**, 1005 (1968).

29. G. W. Robinson and R. P. Frosch, *J. Chem. Phys.* **37**, 1962 (1962).

30. J. Jortner, *J. Chem. Phys.* **64**, 4860 (1976); M. Bixon and J. Jortner, *Chem. Phys.* **176**, 467 (1993).

31. C. Liang and M. D. Newton, *J. Phys. Chem.* **97**, 13083 (1993).

32. D. Emin, *Phys. Rev. B* **4**, 3639 (1971).

33. C. Joachim, J. K. Gimzewski, R. D. Schlitter, and C. Chavy, *Phys. Rev. Lett.* **74**, 2102 (1995); C. Joachim and J. K. Gimzewski, *Europhys. Lett.* **30**, 409 (1995).

34. M. A. Reed, C. Zhou, C. Muller, T. P. Burgin, and J. M. Tour, *Science* **278**, 252 (1997).

35. M. Dorogi, J. Gomez, R. Osifchin, R. O. Andres, and R. Reifenberger, *Phys. Rev. B* **52**, 9071 (1995); R. P. Andres, T. Bein, M. Dorogi, et al., *Science* **272**, 1323 (1996).

36. R. Reifenberger, S. Datta, and C. Kubiak, personal communication,

37. J. Rabe, in *Introduction to Molecular Electronics,* M. C. Petty, M. R. Bryce and D. Bloor, eds., Oxford University Press, New York. (1995), p. 261. J. Rabe, in [38].

38. C. Joachim, ed., *Atomic and Molecular Wires,* (Kluwer, Dordrecht, The Netherlands, 1997.

39. S. J. Tans, M. H. Devoret, H. Dai, A. Thess, R. E. Smalley, L. J. Geerligs, and C. Dekker, *Nature* **386**, 474 (1997).

40. R. H. Terrill and R. W. Murray, in [58].

41. K. C. Graber, R. G. Freeman, M. B. Hommer, and M. J. Natan, *Anal. Chem.* **67**, 735 (1995).

42. D. M. Newns, *Phys. Rev.* **178**,1123 (1969).

43. P. W. Anderson, *Phys. Rev.* **124**, 41 (1961).

44. H. Ou-Yang, R. A. Marcus, and B. Källebring, *J. Chem. Phys.* **100**, 7814 (1994).

45. M. Sumetskii, *Phys. Rev. B*, **48**, 4586 (1993); *J. Phys. Condens. Matter* **3**, 2651 (1991).

46. M. Sumetskii, *Sov. Phys. JETP* **62**, 355 (1985).

47. E. G. Petrov, I. S. Tolokh, A. A. Demidenko, and V. V. Gorbach, *Chem. Phys.* **193**, 237 (1995); I. A. Gaychuck, E. G. Petrov, and V. May, *J. Chem. Phys.* **103**, 4937 (1995).

48. V. Mujica and G. Doyen, *Int. J. Quantum Chem.* **527**, 687 (1993).

49. S. Datta, *Electronic Transport in Mesoscopic System*, Cambridge University Press, Cambridge, 1995; M. P. Samanta, W. Tian, S. Datta, J. I. Henderson, and C. P. Kubiak, *Phys. Rev. B* **53**, 7626 (1996).

50. V. Mujica, M. Kemp, and M. A. Ratner, *J. Chem. Phys.* **101**, 6849 (1994); **101**, 6856 (1994).

51. M. Kemp, V. Mujica, and M. A. Ratner, *J. Chem. Phys.* **101**, 5172 (1994); J. M. Lopez-Castillo, A. Filali-Mouhim, I. L. Plante, and J. P. Jay-Gerin, *J. Phys. Chem.* **99**, 6864 (1995); J. M. Lopez-Castillo, A. Filali-Mouhim, I. L. Plante, and J. P. Jay-Gerin, *Chem. Phys. Lett.* **239**, 223 (1995).

52. A. Cheong, A. E. Roitberg, V. Mujica, and M. A. Ratner, *J. Photochem. Photobiol.* **82**, 81 (1994).

53. M. Kemp, Y. Mao, V. Mujica, A. Roitberg, and M. A. Ratner, in [38].

54. M. Kemp, A. Roitberg, V. Mujica, T. Wanta, and M. A. Ratner, *J. Phys. Chem.* **100**, 8349 (1996).

55. V. Mujica, M. Kemp, A. Roitberg, and M. A. Ratner, *J. Chem. Phys.* **104**, 7296 (1996).

56. V. Mujica, M. Kemp, A. Roitberg, and M. A. Ratner, in *Condensed Matter Theories*, Vol. 11, E. V. Ludena, P. Vashishta, and R. F. Bishop, eds., Nova Science, Commack, N.Y., 1996.

57. R. Landauer, *IBM J. Res. Dev.* **1**, 223 (1957); in *Localization, Interaction and transport Phenomena*, B. Kramer, G. Bergmann, Y. Bruynseraede, eds., Springer-Verlag, New York, 1985, pp. 35–50; *Z. Phys.* **368**, 508. (1961).

58. J. Jortner and M. A. Ratner, eds., *Molecular Electronics*, Blackwell, London, 1997.

59. D. Gust, T. A. Moore, and A. M. Moore, *Acc. Chem. Res.* **26**, 198 (1993).

60. M. R. Wasielewski, *Chem. Rev.* **92**, 435 (1992).

61. C. A. Mirkin and M. A. Ratner, *Annu. Rev. Phys. Chem.* **43**, 719 (1992).

62. S. F. Fischer and R. P. VanDuyne, *Chem. Phys.* **26**, 9 (1977).

63. P. P. Schmidt, *Electrochem. Spec. Per. Rep.* **6**,128 (1978).

64. H. Scher and T. Holstein, *Philos. Mag. B* **44**, 343 (1981).

65. R. Kubo and Y. Toyozawa, *Prog. Theor. Phys.* **13**, 60 (1955).

66. R. D. Coalson, D. G. Evans, and A. Nitzan, *J. Chem. Phys.* 101, 436 (1994); E. Neria and A. Nitzan, *Chem. Phys.* **183**, 351 (1994).

67. M. Cho and R. Silbey, *J. Chem. Phys.* **103**, 595 (1995).

68. H. Finklea, *Electroanal. Chem.* **19**, 105 (1996).

69. C. Joachim and J. F. Vinvesa, *Europhys. Lett.* **33**, 635 (1996); M. Magoga and C. Joachim, *Phys. Rev.* **B56**, 4722 (1997).

70. L. Kouwenhoven, *Science* **271**, 1689 (1996).

71. L. A. Bumm, J. J. Arnold, M. T. Cygan, et al. *Science* **271**, 1705 (1996).

72. M. D. Todd, A. Nitzan, M. A. Ratner and J. T. Hupp, *J. Photochem. Photobiol. A* **82**, 87 (1994).

73. D. Makarov and N. Makri, *Chem. Phys. Lett.* **221**, 482 (1994).

74. R. Egger, C. H. Mak, and U. Weiss, *J. Chem. Phys.* **100**, 2651 (1994).

75. J. Cao, C. Minichino, and G. A. Voth, *J. Chem. Phys.* **103**, 1391 (1995).

76. M. Bixon and J. Jortner, *J. Chem. Phys.* **107**, 1470 (1997).

77. K. V. Mikkelsen, L. Kifkov, H. Nar, and O. Farver, *Proc. Natl. Acad. Sci. USA* **90**, 5443 (1993).

78. H. J. Kim, R. Bianco, B. J. Gertner, and J. T. Hynes, *J. Phys. Chem.* **97**, 1723 (1993).

79. T. Fonseca, H. J. Kim, and J. T. Hynes, *J. Mol. Liq.* **60**, 161 (1994).

80. J. Tang, *J. Chem. Phys.* **98**, 6263 (1993).

81. We will ignore volume and entropy changes, so that $\Delta E = \Delta G$, the free energy change.

82. P. D. Hale and M. A. Ratner, *J. Chem. Phys.* **83**, 5277 (1985).

83. M. A. Ratner and A. Madukhar, *Chem. Phys.* **30**, 701 (1978).

84. E. S. Medvedev and A. Stuchebrukhov, *J. Chem. Phys.* **107**, 3821 (1997).

85. Y.-P. Liu and M. D. Newton, *J. Phys. Chem.* **99**, 11382 (1995); M. A. Ratner, *J. Phys. Chem.* **94**, 4877 (1990); A. A. daGama, *Theor. Chim. Acta* **68**, 159 (1985); Y. Magarshak, J. Malinsky and A. D. Joran, *J. Chem. Phys.* **95**, 418 (1991); C. Goldman, *Phys. Rev. A* **43**, 4500 (1991).

86. M. J. Shepard, M. N. Paddon-Row, and K. D. Jordan, *Chem. Phys.* **176**, 289 (1993); K. Kim, K. D. Jordan and M. M. Paddon-Row, *J. Phys. Chem.* **98**, 11053 (1994); A. Broo and S. Larsson, *J. Phys. Chem.* **95**, 4925 (1990); C. A. Naleway, L. A. Curtiss, and J. R. Miller, *J. Phys. Chem.* **95**, 8434 (1991); L. A. Curtiss, C. A. Naleway, and J. R. Miller, *Chem. Phys.* **176**, 387 (1993); K. D. Jordan and M. N. Paddon-Row, *Chem. Rev.* **93**, 395 (1992); M. J. Shepard, M. N. Paddon-Row, and K. D. Jordan, *J. Am. Chem. Soc.* **116**, 5328 (1994).

87. H. Heitele and M. E. Michel-Beyerle, in *Antennas and Reaction Centers of Photosynthetic Bacteria*, M. E. Michel-Beyerle, ed., Springer-Verlag, Berlin, 1985.

88. J. N. Betts, D. N. Beratan, and J. N. Onuchic, *Science* **252**,1285 (1991); J. J. Regan, S. M. Risser, D. N. Beratan, and J. N. Onuchic, *J. Phys. Chem.* **97**, 13083 (1993).

89. A .A. Stuchebrukhov, *J. Chem. Phys.* **104**, 8424 (1996).

90. M. A. Ratner and M. J. Ondrechen, *Mol. Phys.* **32**, 1233 (1976).

91. P. O. Scherer, E. V. Knapp, and S. F. Fischer, *Chem. Phys. Lett.* **106**, 191 (1984).

92. J. Bardeen, *Phys. Rev. Lett.* **6**, 57 (1961).

93. S. Yaliraki, V. Mujica and M. A. Ratner, Manuscript in preparation.

94. D. V. Averin, A. N. Korotkov, and K. K. Likharev, *Phys. Rev. B* **44**, 6199 (1991); B. Su, V. J. Goldman, and J. E. Cunningham, *Phys. Rev B* **46**, 7644 (1990); H. Grabert and M. H. Devoret, eds., *Single Charge Tunneling: Coulomb Blockade Phenomena in Nanostructures*, Plenum Press, New York, 1992; B. L. Altshuler, P. A. Lee, and R. A. Webb, eds., *Mesoscopic Phenomena in Solids.*, Elsevier, Amsterdam, 1991; W. P. Kirk and M. A. Reed,

eds., *Nanostructures and Mesoscopic Systems,* Academic Press, San Diego, Calif., 1991; A. N. Korotkov, in [58].

95. C. W. J. Beenakker and H. vanHouten, *Solid State Phys.* **44**, 1 (1995).

96. C. M. Fischer, M. Burghard, S. Roth, and K.von Klitzing, *Europhys. Lett.* **28**, 129 (1994); C. M. Fischer, M. Burghard, and S. Roth, in [58].

97. A. N. Korotkov, in [58].

98. W. R. Salaneck, I. Lundstrom, and B. Ranby, eds., *Conjugated Polymers and Related Compounds,* Oxford University Press, Oxford, 1993.

99. J. L. Bredas and R. Silbey, eds., *Conjugated Polymers,* Kluwer, Dordrecht, 1991.

100. D. Jerome and L. G. Caron, eds., *Low-Dimensional Conductors and Superconductors,* Plenum Press, New York; 1987.

101. J. A. Pople and S. H. Walmsley, *Mol. Phys.* **5**, 15 (1962).

102. Y. Lu, ed., *Solitons and Polarons in Conducting Polymers,* World Scientific, Singapore, 1988.

103. S. Skourtis and S. Mukamel, *Chem. Phys.* **197**, 367 (1995).

104. A. K. Felts, W. T. Pollard, and R. A. Friesner, *J. Phys. Chem.* **99**, 2929 (1995); W. T. Pollard, A. K. Felts, and R. A. Friesner, *Adv. Chem. Phys.* **93**, 77 (1996).

105. W. Davis, M. Wasielewski, V. Mujica, M. A. Ratner, and A. Nitzan, *J. Phys. Chem.,* **101**, 6158 (1997).

106. J. R. Miller and J. V. Beitz, *J. Chem. Phys.* **74**, 6746 (1981).

107. H. M. McConnell, *J. Chem. Phys.* **35**, 508 (1961).

108. C. P. Hsu and R. A. Marcus, *J. Chem. Phys.* **106**, 584, (1997).

109. J. W. Evenson and M. Karplus, *J. Chem. Phys.* **96**, 5272 (1991); J. W. Evenson and M. Karplus, *Science* **262**, 1247 (1993).

110. G. L. Closs and J. R. Miller, *Science* **240**, 440 (1988).

111. M. R. Wasielewski, D. G. Johnson, W. A. Svec, K. M. Kersey, D. E. Cragg, and D. W. Minsek, in: *Photochemical Energy Conversion,* J. R. Norris and D. Meisel, eds. Elsevier, New York, 1989.

112. J .R. Reimers and N. S. Hush, *J. Photochem. Photobiol. A* **82**, 31 (1994); *Chem. Phys.* **146**, 89 (1990).

113. D. A. Weitz, S. Garoff, J. I. Gersten, and A. Nitzan, *J. Chem. Phys.* **78**, 5324 (1983).

114. J. N. Onuchic and D. Beratan, in [58].

115. Y. V. Sharvin, *Sov. Phys. JETP* **48**, 984 (1965).

116. V. Mujica, A. Roitberg, M. Kemp, and M. A. Ratner, *J. Chem. Phys.* submitted.

117. M. Buttiker and R. Landauer, *Phys. Rev. Lett.* **49**, 1739 (1982).

118. Z. Kotler and A. Nitzan, *J. Chem. Phys.* **88**, 3571 (1988).

119. J. Wilkie, A. Nitzan, J. Jortner, and M. A. Ratner, *J. Chem. Phys.,* submitted.

120. J. Halpern and L. E. Orgel, *Discuss. Faraday Soc.* **29**, 32 (1960).

121. N. Sutin, B. Brunschwig, M. K. Johnson, et al., eds., *Electron Transfer in Biology and the Solid State,* American Chemical Society, Washington, D.C., 1990; M. A. Ratner, *Nav. Res. Rev.* **49** (1994).

122. H. B. Gray and J. R. Winkler, *Annu. Rev. Biochem.* **65**, 537 (1996).

ELECTRON TRANSFER AND
EXCIPLEX CHEMISTRY

NOBORU MATAGA

*Institute for Laser Technology, 1-8-4 Utsubo-Honmachi, Nishi-ku,
Osaka 550, Japan*

HIROSHI MIYASAKA

*Department of Polymer Engineering and Science, Kyoto Institute of
Technology, Matsugasaki, Sakyo, Kyoto 606, Japan*

CONTENTS

Electron Transfer: From Isolated Molecules to Biomolecules, Part Two, edited by Joshua Jortner and M. Bixon. Advances in Chemical Physics Series, Volume 107, series editors I. Prigogine and Stuart A. Rice.
ISBN 0-471-25291-3 © 1999 John Wiley & Sons, Inc.

I. INTRODUCTION

It is well known that photoexcitation of a molecule in solutions generally induces changes in its electronic and geometric structure, which induces changes in its interactions with nearby molecules and/or its environment [1,2]. In many cases those molecular interactions are greatly enhanced in the excited electronic state, and excitation energies are used partly as driving forces of the reaction dynamics. Entire parts of photochemical reactions in solutions start from such electronic and geometrical structural changes induced by photoexcitations. Among these excited-state molecular interactions, photoinduced charge transfer (CT) and electron transfer (ET) are the most important fundamental interactions underling most problems in photophysical and photochemical processes in solutions [1–7]. In this chapter we use *exciplex phenomena* and *exciplex chemistry* in a broad sense, including various molecular interactions in the excited electronic state, such as solute–solvent interactions, hydrogen bonding and proton transfer, photo-induced CT and ET leading to the formation of excited CT complexes, exciplexes (EXs), and various ion pairs (IPs) as well as subsequent chemical reactions.

In the excited-state molecular interactions describe above, the excited dipolar solute–polar solvent interactions are not the usual typical donor (D)–acceptor (A) CT or ET interactions. As is well known, however, this type of reorientation interaction of polar solvents surrounding the dipolar solute or charged solute is very important in the ultrafast ET reaction between D and A molecules, where the reaction rate is controlled by the solvent dynamics [6,8–12]. In the case of a dipolar solute molecule composed of D and A groups in polar solvent, one can observe a large fluorescence Stokes shift due to the solvation, which was expressed first in 1955 by [13,14] (Lippert–Mataga equation):

$$h(\nu_a - \nu_f) = \text{const.} + [2(\boldsymbol{\mu}_e - \boldsymbol{\mu}_g)^2/a^3][(\varepsilon_s - 1)/(2\varepsilon_s + 1) - (\varepsilon_0 - 1)/(2\varepsilon_0 + 1)]$$

$$= \text{const.} + 2[(\boldsymbol{\mu}_e - \boldsymbol{\mu}_g)^2/a^3]F(\varepsilon_s, \varepsilon_0) \tag{1.1}$$

where $\boldsymbol{\mu}_e$ and $\boldsymbol{\mu}_g$ are solute dipole moments in the excited and ground states, respectively, ε_s and ε_0 are static and optical dielectric constants of solvent, respectively. Equation (1.1) was derived under an approximation of linear response for the solute dipole–solvent interaction, where the solute dipole was assumed to be put in a spherical Onsager cavity of radius a with unit dielectric constant, using the dielectric continuum for solvent.

The solute dipole moment changes from $\boldsymbol{\mu}_g$ to $\boldsymbol{\mu}_e$ by light absorption transition but the solvent orientation polarization remains the same at the instant of transition, which causes a nonequilibrium polarization state of solvent with respect to $\boldsymbol{\mu}_e$ producing the destabilization free energy λ (reorganization energy).

$$\lambda = [(\boldsymbol{\mu}_e - \boldsymbol{\mu}_g)^2/a^3]F(\varepsilon_s, \varepsilon_0) \tag{1.2}$$

In the course of time, the solvent polarization begins to approach the equilibrium corresponding to $\boldsymbol{\mu}_e$. If the solvent reorientation relaxation time τ_L is sufficiently shorter than the solute fluorescence lifetime τ_f, the equilibrium fluorescence state is realized before fluorescence transition and the Stokes shift due to the solvation is given approximately by Eq. (1.1). In the course of the relaxation of the solvent polarization toward the equilibrium state, the time-dependent fluorescence shift as given by

$$h[\nu_a - \nu_f(t)] = \text{const.} + [2(\boldsymbol{\mu}_e - \boldsymbol{\mu}_g)^2/a^3]F(\varepsilon_s, \varepsilon_0)[1 - \exp(-t/\tau_1)] \tag{1.3}$$

on the basis of the above-described simplified model may be observed [15,16].

The foregoing treatment of the fluorescence Stokes shift due to the polarization of solvent surrounding the large solute dipole is conceptually quite simple, and Eq. (1.1) was widely used for the studies of fluorescence solvatochromism. It should be noted here that the nonequilibrium polarization with respect to the solute–solvent interaction in the Franck–Condon state in light absorption and emission is analogous to that in the classical Marcus equation for ET reaction given in 1956, where the ET rate constant k and solvent reorganization free energy are given, respectively, by [17–19]

$$k = \nu_0 \exp[-(\lambda' + \Delta G)^2/4k_B T\lambda'] \tag{1.4}$$

$$\lambda' = e^2[(2a_1)^{-1} + (2a_2)^{-1} - r^{-1}](\varepsilon_0^{-1} - \varepsilon_s^{-1}) \tag{1.5}$$

where $\nu_0, \Delta G, a_1, a_2$, and r are the frequency factor, free energy gap for the reaction, reactants' radius (a_1, a_2), and their center-to-center distance, respectively. In this case also, linear responses for the solute–solvent interactions and dielectric continuum model for the solvent were assumed.

Although both of these simplified treatments are conceptually quite simple and useful, they may be not more than first approximations, and actual systems will show more complex behavior, indicating deviations of the solute–solvent interactions from such simple pictures. For example, electronic structures of intermolecular exciplexes, intramolecular exciplexes with nonrigid bridge between D and A groups, and excited CT complexes seem to change due to the nonlinear solute–solvent interactions, which induces further geometrical structural changes, depending on solvent polarities [2–5,7,20,21]. This problem is very important for exciplex chemistry (i.e., for elucidation of the mechanisms of photoinduced CT and subsequent chemical reactions) in solutions and is closely related to the mechanisms of the fluorescence quenching reaction in solution as discussed later in this chapter.

The ET mechanisms for the fluorescence quenching reaction in solution, including various dyes and aromatic molecules as fluorescers and various organic and inorganic molecules and ions as quenchers, was proposed in the 1930s and its studies have a long history [1]. However, one of the most important experimental investigations leading to contemporary understanding of the mechanisms of photoinduced ET reactions in solutions and exciplex chemistry in general was virtually begun by those pioneering studies on exciplex phenomena, mainly during 1950–1970 [2–5,13,14,21] and has been greatly advanced by more recent picosecond–femtosecond laser photolysis investigations [5–7,11,12,22,23].

In the following we discuss fundamental problems of exciplex chemistry, such as (1) the solvation-induced changes in electronic and geometrical structures of inter- and intramolecular exciplex systems, (2) energy gap dependences of ET in the fluorescence quenching reaction in strongly polar solutions and behaviors of product loose ion pairs, (3) mechanisms and dynamics of charge separation in the excited CT complexes and behavior of product compact ion pairs, (4) ET coupled with proton shift or transfer in some hydrogen-bonding and related systems, (5) extensions of the studies on intramolecular exciplexes to photosynthetic reaction center models, and so on and (6) interrelations among these exciplex phenomena.

II. SOLVENT EFFECTS ON ELECTRONIC AND GEOMETRICAL STRUCTURES OF INTER- AND INTRAMOLECULAR EXCIPLEX SYSTEMS AND MULTISTATE MECHANISMS

The effect of polar solvent on the excited polar solute molecule is not limited to stabilization of the solute fluorescence state with rigid dipoles due to the reorientation of the surrounding solvent dipoles, but its electronic as well as geometrical structures may change due to the interaction with polar solvent molecules, especially in the case of the molecular composite systems such as inter- and intramolecular exciplex systems as described in Section I. In relation to this problem, early fluorescence studies on some typical inter-molecular exciplexes such as pyrene (A)–DMA (N,N-dimethylaniline) (D) and anthracene (A)–DEA (N,N-diethylaniline) (D) systems showed that upon increasing solvent polarity, both fluorescence yield and decay time decreased, whereas the extent of the decrease in fluorescence yield was much larger [21,24].

1. One possible interpretation of the results above was to assume solvent-induced change in electronic and geometrical structures of the exci-plex: Namely, the electronic structure of the exciplex $(A^{-\delta}D^{+\delta})*$ ($\delta \leq 1$) will become more polar ($\delta \rightarrow 1$) with increased solvent polar-ity, leading to a decrease in the radiative transition probability k_f due to the decreased transition moment and increased radiationless transi-tion probability k_i due to the decreased energy gap between the exciplex and ground states that results from lowering the exciplex state as a result of the strong solvation. Accordingly, the fluores-cence yield $\phi_1 = k_f(k_f + k_i)^{-1}$ will decrease strongly but the lifetime $\tau_f = (k_f + k_i)^{-1}$ will not show such a large change. With further increase in the solvent polarity, the geometrical structure of the exciplex will change to a loose one due to strong solvation (increase of the $D^+ \cdots A^-$ distance in the exciplex), and k_f as well as ϕ_f will decrease further in this loose exciplex [20,21].

2. Another possible interpretation was to assume competition between nonfluorescent ion-pair formation and fluorescent exciplex formation at the encounter between A* and D, where the exciplex fluorescence yield was assumed to be given by $\phi_f = \phi_e k_f(k_f + k_i)^{-1}$ (ϕ_e: yield of the fluorescent exciplex formation at the encounter), and the electronic and geometrical structures of the exciplex were assumed to be almost independent of solvent polarity while ϕ_e decreases with increased solvent polarity [24].

To examine the validities of these proposals, nanosecond, picosecond, and femtosecond time-resolved laser spectroscopic studies on D–A systems linked by flexible methylene chains in several kinds of solvents with different polarities were conducted [3–5,7,22,25–31]. In these linked systems, the motion of the chromophores is partly restricted compared to the free D–A systems. In the following, we discuss first the results of pico-second and femtosecond time-resolved transient absorption spectral studies on p-$(CH_3)_2N$-phenyl-$(CH_2)_n$-(1-pyrenyl) ($P_n, n = 1, 2, 3$) and p-$(CH_3)_2N$-phenyl-$(CH_2)_n$-(9-anthryl) (A_n, $n = 1, 2, 3$) [5,7,22,25–31]. These resulted from one of the earliest systematic studies of intramolecular exciplexes, from which some important information and useful concepts concerning the nature of molecular interactions in the excited electronic state were obtained. Those concepts and information constituted a basis for the subsequent vast development of investigations of exciplex phenomena, including polymer photochemistry [32–34], organic photochemical reactions [35,36], and more detailed elucidations of the intramolecular exciplex phenomena in linked D–A systems [37–41].

For these D–A chromophores, their close approach assuming a sandwich structure is necessary to form exciplex (compact exciplex) in nonpolar solvents. Therefore, practically no CT fluorescence can be observed from the exciplex state in nonpolar solvents for $n = 1, 2$ compounds. In the case of $n = 3$ compounds, stationary-state fluorescence due to the exciplex state is overwhelming in nonpolar solvents because formation of a sandwich structure by means of internal rotations around —CH_2—CH_2— bonds during the S_1 state lifetime of A chromophore is possible. Actually, the exciplex formation processes of these $n = 3$ compounds in hexane at room temperature were confirmed to take a few nanoseconds, which also means that these compounds in the ground state assume an extended structure in solutions.

When solvent dielectric constant is increased the exciplex state can be realized without taking the compact configuration due to lowering of the CT state energy by solvation. For example, the time constants τ_r for the exciplex formation in acetonitrile solution for $n = 1, 2$, and 3 compounds were 1.7, 6.1, and 11 ps for P_n and 0.65, 2.1, and 2.7 ps for A_n, respectively [31], which shows that the photoinduced ET takes place without appreciable conformation changes in loose configurations (loose exciplexes). On the other hand, the effect of the solvent polarity upon the exciplex formation dynamics and also upon the exciplex structures can be demonstrated by picosecond– femtosecond time-resolved transient absorption spectral measurements, as follows in the case of P_3 [28–31]:

$$
\begin{aligned}
A* \diagup\!\diagdown\!\diagup\!\diagdown D &\xrightarrow{\ \tau_r\,=\,4\ \text{ns}\ } \text{compact exciplex in hexane} \\
A* \diagup\!\diagdown\!\diagup\!\diagdown D &\xrightarrow{\ \tau_r\,=\,45\ \text{ps}\ } \text{loose exciplex in acetone} \\
A* \diagup\!\diagdown\!\diagup\!\diagdown D &\xrightarrow{\ \tau_r\,=\,25\ \text{ps}\ } \text{loose exciplex in butyronitrile} \\
A* \diagup\!\diagdown\!\diagup\!\diagdown D &\xrightarrow{\ \tau_r\,=\,11\ \text{ps}\ } \text{loose exciplex in acetonitrile}
\end{aligned}
\tag{2.1}
$$

In hexane solution, reflecting the strong interaction between acceptor and donor in the close-contact sandwich configuration of the exciplex, the time-resolved absorption spectrum of the exciplex state is much broadened compared with the superposition of spectra of radical ions A^- and D^+. In contrast to this, the spectrum of the intramolecular exciplex in such highly polar solvent as acetonitrile is very similar to the superposition of the free-ion-radical spectra in acetonitrile, which evidently shows that the exciplex in acetonitrile solution has an intramolecular ion-pair state with an extended loose structure. In solvents of intermediate polarity, τ_r is a little longer and the absorption spectrum of the exciplex state is slightly broadened compared with those in the acetonitrile solution, respectively, indicating a slight conformation change from an extended structure toward a compact one. Nevertheless, such conformation changes in the polar solvents acetone and butylonitrile may be very small since the τ_r values in these solvents are much shorter than the time constants for the conformation changes determined by internal rotations around $-CH_2-CH_2-$ bonds (in the nanosecond range).

On the other hand, the τ_r values in polar solvents given in Eq. (2.1) are much longer than the solvent reorientation time τ_L: τ_L is approximately 0.2, 0.5 and 0.6 ps for acetonitrile, butyronitrile, and acetone. Therefore, the photoinduced charge separation processes in these systems are not controlled by solvent dynamics and are not barrierless but will be controlled mainly by the magnitude of the electronic interaction responsible for ET (tunneling matrix element) between A* and D groups and a little activation energy arising from the Franck–Condon factor (energy gap) of the reaction. Assuming a nonadiabatic ET mechanism, the rate constant for ET may be given approximately by the following equation, taking into consideration contributions to the Franck–Condon factor not only from the low-frequency solvent modes but also from the high-frequency quantum modes of intrachromophore vibrations [42–44]:

$$k_{ET} = (\pi/\hbar^2\lambda_s k_B T)^{1/2} V^2 \sum_n [e^{-S}(S^n/n!)] \exp[-(\Delta G + \lambda_s + n\hbar\langle\omega\rangle)^2/4\lambda_s k_B T]$$

$$(2.2)$$

where V is the matrix element of electronic interaction responsible for ET between A* and D, $S = \lambda_v/\hbar\langle\omega\rangle$ is the electron-vibrational coupling constant, and λ_v is the reorganization energy due to the average intrachromophore vibrational frequency $\langle\omega\rangle$, and λ_s is the solvent reorganization energy. The activation energy from the Franck–Condon factor arises in case the free energy gap $-\Delta G$ is not sufficiently large to offset the reorganization energy. When solvent polarity decreases, $-\Delta G$ decreases, while λ_v is independent of the solvent polarity, leading to increases in the activation energy and the τ_r value. On the other hand, V decreases exponentially with increased interchromophore distance, leading to the increase in the τ_r value with an increase in the number of the intervening methylene chain, as described above for the photoinduced charge separation of P_n and A_n ($n = 1, 2, 3$) in acetonitrile solutions. In any case, the electronic and geometrical structures of the intramolecular exciplex systems described above change depending on solvent polarity, and it has been demonstrated that P_3 and A_3 exciplexes show the change in geometrical structures from the compact structure in nonpolar solvent to the extended loose structure in polar solvent, and the exciplex electronic structure also becomes ion-pair-like with increase in the solvent polarity. These results support interpretation 1 described at the beginning of Section II. In this respect it should be noted here that detailed fluorescence studies on various D's and A's with different oxidation potentials and reduction potentials, respectively, linked with flexible methylene chains and semiflexible saturated hydrocarbon spacers in various solvents of different polarities, have recently confirmed the change in the intramolecular exciplex structures to a looser structure with increased solvent polarity [37–41].

Of course, when sufficiently strong electron donating and accepting chromophores are used in the intramolecular exciplex systems, photoinduced charge separation seems to be possible even in a nonpolar solvent and in extended conformations [37–41] due to the large energy gap, $-\Delta G$, in Eq. (2.2). Direct observations of the photoinduced formation of the almost completely charge-separated ion-pair state by means of the time-resolved absorption spectral measurements were made, for example, in the case of a series of fixed-distance dyads of porphyrin–quinone systems in the nondipolar solvent benzene [45]. By using such a series of fixed-distance dyads with various $-\Delta G$ values in Eq. (2.2), we can examine the validity of such equations and effects of solute–solvent interactions on the photoinduced ET when the D–A distance is fixed.

For example, the experimental results of k_{cs} (k_{ET} for the photoinduced charge separation) as a function of $-\Delta G_{cs}$ for P–Q dyads (Fig. 1) in benzene, THF (tetrahydrofurane), and BuCN (butyronitrile) are indicated in Figure 2. To reproduce the experimental results, best-fit parameters for Eq. (2.2) were determined as follows: $\lambda_s(\text{benzene}) = 0.18$ eV, $\lambda_s(\text{THF}) = 0.87$ eV, $\lambda_s(\text{BuCN}) = 1.15$ eV with common parameter values $V = 3.8$ meV, $\lambda_v = 0.6$ eV, $\hbar\langle\omega\rangle = 0.15$ eV [45]. The λ_s value for benzene solution calculated by Marcus equation (1.5) assuming a linear response for solute–solvent interactions and the dielectric continuum model for the solvent is almost zero. This result indicates that the actual interactions between the solute in the CT or ion-pair state and the solvent are more complex, including the effects of some nonlinear solute–solvent interactions and discrete molecular structures of the solution. Recent theoretical treatments on energetics of CT reactions in solutions based on such molecular

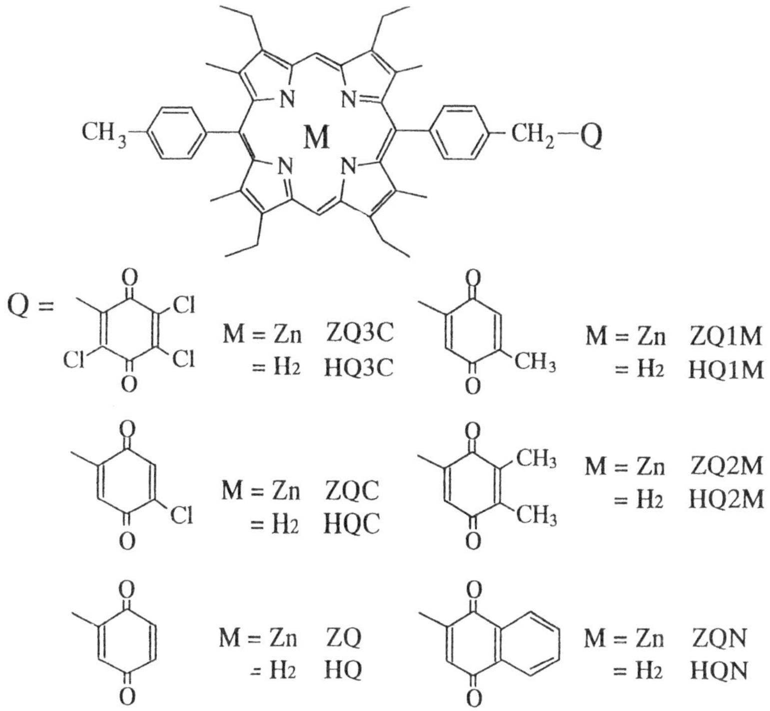

Figure 1. Structural formulas of P–Q dyads.

theories of solution structures [46] appear to agree with the experimental results of λ_s values above. Nevertheless, for qualitative or semiquantitative discussions the functional form of Eq (1.5) may be useful.

The results and discussions above, which support interpretation 1—that the electronic and geometrical structures of the exciplex change depending on the solute–solvent interactions from a compact exciplex in nonpolar solvent to a loose exciplex with increase in the solvent polarity, and the fact that in the case of the strong electron-donating and electron-accepting chromophores with favorable $-\Delta G$ values for photoinduced charge separation, the loose exciplex formation is possible not only in polar solvents but also in nonpolar solvents [37–41,45]—suggest that the photoinduced CT reactions in solutions, the formation of CT states of various electronic and geometrical structures depending on solvent polarities and strengths of donor and acceptor chromophores is possible and there may arise an ensemble of various exciplexes and ion pairs distributed over such CT states in solutions [5,7,22,47]. In relation to such a problem, results of the

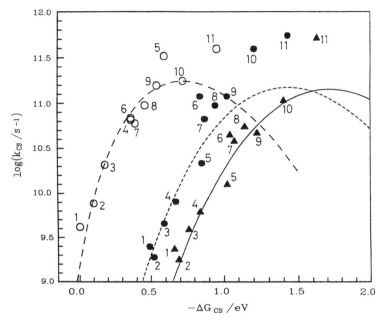

Figure 2. Solvent effect on the k_{CS} versus-ΔG_{CS} relation. Observed values in benzene (open circles): 1, HQN; 2, HQ1M; 3, HQ; 4, HQC; 5, HQ3C; 6, ZQN; 7, ZQ2M; 8, ZQ1M; 9, ZQ; 10, ZQC; 11, ZQ3C; in THF (solid circles) and in BuCN (triangles): 1, HQN; 2, HQ2M; 3, HQ1M; 4, HQ; 5, HQC; 6, ZQN; 7, ZQ2M; 8, ZQ1M; 9, ZQ; 10, ZQC; 11, ZQ3C. Curves were calculated using Eq. (2.2).

investigations on the lifetime (τ_f) of CT fluorescence (exciplex fluorescence) and on the rise time (τ_r) of picosecond laser-induced photoconductivity of some typical exciplex systems such as pyrene–DMA and pyrene–DCNB (p-dicyanobenzene) in polar solutions [47] should be discussed here.

The photoconductivity arises due to the dissociation of loose ion pairs (LIPs) into free ions. The experimental values of τ_f and τ_r in the same polar solvent were rather close, which suggested interconversion between fluorescent exciplex (EX) and LIP undergoing dissociation, corresponding to the result of an external magnetic field effect upon the exciplex fluorescence, which also indicated interconversion between the exciplex and LIP [48]. Nevertheless, because τ_f was a little longer than τ_r in all cases examined, the two-states model considering direct interconversion between EX and LIP,

$$\mathrm{EX} \rightleftharpoons \mathrm{LIP} \longrightarrow \mathrm{A}_s^- + \mathrm{D}_s^+ \tag{2.3}$$

leading to exciplex fluorescence decay exactly corresponding to the photo-conductivity rise was not valid. These results of exciplex fluorescence decay time and the picosecond laser-induced photoconductivity rise time led to the multistate model, assuming the formation of an ensemble of various exci-plexes and ion pairs in the photoinduced CT reaction as discussed above and partial interconversion between them. Such a reaction scheme gave multi-exponential functions for the exciplex fluorescence decay, $I_f(t)$, and the photoconductivity rise, $I_c(t)$:

$$I_f(t) = \sum_i A_i \exp(-k_i t) \tag{2.4}$$

$$I_c(t) = \sum_i A_i'[1 - \exp(-k_i t)] \tag{2.5}$$

where the LIPs that undergo dissociation may not be directly combined with the fluorescent exciplexes but there may be intervening ion pairs and exci-plexes, and the overwhelming term in $I_f(t)$ and $I_c(t)$ will give the approx-imate single exponential decay or rise.

It should be noted here that the above-discussed model of the solvent polarity dependence of the exciplex structures and the multistate model of exciplexes and ion pairs are closely related to the problems of energy gap dependences of ET at encounter between fluorescer and quencher in polar solutions where the inverted region predicted by classical Marcus equation (1.4) as well as Eq. (2.2) has not been observed [3,7,22,49,50]. As discussed in Section IV, ion pairs produced by ET in the fluorescence quenching reactions have a distribution of interionic distances in the pair due to the distribution of encounter distances where ET quenching of fluorescence takes place. Moreover, this distribution depends on the energy gap $-\Delta G$ for the charge separation reaction, and the average interionic distance increases with increased energy gap [7,22,23,51–55]. That is, for a higher $-\Delta G$ value, a higher solvent reorganization energy λ_s is favorable to retain a high ET rate, while λ_s becomes larger with increased interionic distance according to Marcus equation (1.5). Although the matrix element of elec-tronic interaction V in Eq. (2.2) decreases with increased distance between donor and acceptor, detailed theoretical examinations showed that a little increase in the distance was favorable to keep the large ET rate for the increase in $-\Delta G$ [52–54]. This sort of change of the structure of loose exciplexes and/or geminate ion pairs (LIPs) with increased $-\Delta G$ for charge separation lead to the result that observation of the inverted effect in the fluorescence quenching reaction due to ET in polar solutions is very difficult.

Moreover, it is well known that many photochemical reactions in solutions are induced by excited-state ET where the exciplexes and geminate radical ion pairs play important central roles. Typical examples of such reactions may be the coupled process of photoinduced ET and proton transfer from cation to anion radicals in the geminate pair, leading to hydrogen abstraction and subsequent reactions of neutral radical pairs. In Section VI we discuss such examples and some related problems that were investigated directly by means of picosecond–femtosecond laser spectroscopic methods.

Before going into a more detailed discussion of the dynamics and energy gap dependences of photoinduced charge separation and the behavior of the product geminate ion pairs, we discuss briefly some problems related to the behaviors of bichromophoric systems linked directly by a single bond, which is a special case of the intramolecular exciplex systems discussed above and also closely related to general problems of exciplex phenomena, such as solvation-induced changes in the exciplex electronic and geometrical structure from the compact structure in nonpolar or less polar solvents to the loose and more completely charge-separated structure with increased solvent polarity, as discussed earlier.

III. BEHAVIOR OF BICHROMOPHORIC SYSTEMS LINKED DIRECTLY BY A SINGLE BOND

In this case, the electronic interaction between two chromophores seems to be much stronger than in the case of weakly interacting donor–acceptor systems linked by saturated hydrocarbon spacers, as discussed in Section II. In this strongly interacting system, the geometrical freedom of the electron-donating and electron-accepting chromophores is greatly restricted compared to such intramolecular exciplex systems linked by flexible methylene chains as P_3 and A_3, and only restricted rotation around the linking single bond is possible. Nevertheless, the excited-state electronic structure of this bichromophoric system might be affected rather significantly not only by interaction with the surrounding polar solvent but also by this partial rotation around the single bond because of the strong interchromophore interactions due to their proximity to each other.

The dual fluorescence of DMABN (4-N,N-dimethylaminobenzonitrile) was discovered by Lippert et al. [56,57] in the 1960s. They suggested that the mechanism of dual fluorescence was ^1L_a ^1L_b level reversal due to the stabilization of the more polar ^1L_a state by solute–solvent interaction during the excited-state lifetime. The TICT (twisted intramolecular charge transfer) model was proposed in the 1970s to interpret the mechanism of this dual fluorescence [58–60]. It was suggested by Rotkiewicz et al. [58] that the dimethylamino group was rotated 90° in the S_1 state of DMABN in polar solvent, leading to a completely charge separated state with a large dipole moment, which underwent large stabilization due to interaction with polar solvent, giving a rather strongly red-shifted fluorescence band.

Since then, extensive experimental and theoretical studies on DMABN and similar molecules have been made in relation to the TICT mechanism [59,60] and are continuing. Nevertheless, conclusions as to the detailed quantitative mechanisms of photoinduced charge separation, including geometrical change in solute and dynamics of surrounding solvent molecules and on the generalities of the TICT mechanism through this type of bichromophoric compound have not been sufficiently clarified yet.

As noted at the end of Section II, this problem is analogous to the general mechanisms in exciplex phenomena, such as the solvation-induced changes in the exciplex electronic and geometrical structure from the compact structure with more or less strong electronic interactions between two chromophores in nonpolar solvent to a loose, almost completely charge separated structure with increased solvent polarity.

Such solvent effects on the exciplex structures, however, may depend on the oxidation and reduction potentials of electron-donating and electron-accepting chromophores, respectively, and the magnitude of the electronic interaction between them. Therefore, the electronic and geometrical structure of directly linked bichromophoric systems will change from system to system, depending on the nature of the chromophores. Systems composed of large aromatic chromophores such as ADMA [p-(9-anthryl)-N,N-dimethylaniline, A_n, $n = 0$] and a series of its derivatives [6,11,12,26,30,31,59,60–68] especially, as well as 9,9′-bianthryl and some derivatives [6,7,10,12,22,30,59,60,69–76], show complex behavior different from original TICT concept.

Time-resolved transient absorption spectra of ADMA indicated that the spectra were different from the superposition of spectra of DMA cation and anthryl anion radicals, probably due to the strong electronic interactions between the chromophores in the tilted configurations, and results of measurements of time-resolved absorption spectra in 1-butanol solution suggested strongly that there were distributions to various states with different degrees of intramolecular CT due to different degrees of tilting, and this distribution gradually changed in the course of time from the less polar Franck–Condon excited state toward the more polar equilibrium state [65]. Moreover, detailed fluorescence studies on ADMA and a series of derivatives showed that their excited-state CT degrees were not as complete as expected from the TICT but only partial, due to strong interchromophore interactions even in strongly polar solvent [67,68].

9,9′-Bianthryl is composed of identical halves but shows photoinduced intramolecular CT due to "solvation-induced broken-symmetry" [71]. It is well known that its ground-state absorption spectrum in nonpolar solvent is very similar to that of anthracene, due to its perpendicular structure, although a little red shifted as a whole, while the fluorescence spectrum in the same solvent is much broader and more red shifted than that of anthracene. This result can be understood on the basis of the fact that contrary to the perpendicular configuration in the ground state, the minima in the S_1 excited state are at considerably tilted positions in the double minimum torsional potential, which has been confirmed by the Franck–Condon and band-shape analysis of the temperature-dependent fluorescence spectra in nonpolar 2-methylbutane and slightly polar benzene solution [73,74]. The relaxation process from the perpendicular Franck–Condon excited state to the tilted equilibrium excited state was observed directly by femtosecond–picosecond time-resolved absorption spectral measurements (22,76,77], and approximate relaxation times were determined to be a few picoseconds in some aliphatic hydrocarbon solvents at room temperature.

In view of the above-described results of the photophysical primary process of 9,9′-bianthryl in nonpolar solvents, whether the solvation-induced broken symmetry takes place from the locally excited state before relaxation to the tilted configuration or from the relaxed tilted configurations should be elucidated. In a viscous polar solvent such as 1-pentanol, where the solvent reorientation relaxation is slow, the time-resolved absorption spectra showed clearly that the intramolecular CT took place from the relaxed S_1 state with tilted configuration [72]. The rise time of the CT state was obtained from analysis of the transient absorption spectra as well as the decay of the fluorescence emitted competing with the CT process and found to be 170 ps, in agreement with the τ_L value of 1-pentanol [72]. Namely, the broken symmetry of the excited electronic state was induced by solvent fluctuation in this case. On the other hand, in nonviscous aprotic polar solvents, the photoinduced intramolecular CT of 9,9′-bianthryl may be able to compete with the relaxation to the tilted configuration because the solvent τ_L may be comparable to or shorter than relaxation from the perpendicular to the tilted configuration.

Actually, measurements of dynamics of fluorescence emitted competing with the CT process of 9,9′-bianthryl in such polar solvents showed very rapid decay [6,10,12], which was interpreted on the basis of generalized Langevin equation or generalized Smoluchowsky equation as a function of only the solvent reorientation coordinate [6,10,12]. However, in view of the torsional relaxation to the tilted configuration in the S_1 state clearly observed in nonpolar solvents as well as in the viscous polar solvent, whether a slight torsional motion is necessary for the photoinduced intramolecular CT and whether the most stable configurations in the CT state are considerably tilted in addition to the stabilization by strong solvation should be elucidated. In this respect, results of some studies on this system by means of femtosecond–picosecond time-resolved absorption spectral measurements in some alkanenitriles may be interesting and useful [76].

The femtosecond–picosecond time-resolved transient absorption spectra of 9,9′-bianthryl in hexanenitrile, butyronitrile, and acetonitrile solutions were examined [76]. The observed time-dependent absorbance rise corresponding to the formation process of the CT state indicated the existence of a rather slow component in addition to the fast rise (<1 ps). From the approximately exponential rise curve of the slow component, $[CT(t)] \approx \text{const.}[1 - \exp(-t/\tau_r)]$, the rise times τ_r of the CT state in the alkanenitrile solutions were obtained to be 7.5; 3.4, and 1.8 ps for hexanenitrile, butyronitrile, and acetonitrile solutions, respectively. These τ_r values are much longer than the solvent τ_L values [6,9], as shown in Table I.

Detailed quantitative mechanisms of this CT state formation much slower than the solvent τ_L are not very clear at the present stage of investigations. Nevertheless, it seems to be possible that the torsional fluctuations are closely coupled with the CT process in addition to the solvent fluctuation. Due to the torsional fluctuation, there may be a distribution of torsional angles even in the Franck–Condon excited state. The electronic interaction between anthryl groups responsible for the intramolecular CT will depend significantly on the torsional angle and will be stronger in the more tilted configuration, leading to a distribution of the CT rate in the excited state. The CT rate in the configuration with the angle very close to 90° may be small and a slight torsional motion toward the tilted form will facilitate the CT process. In addition, those molecules undergoing very fast CT by solvent fluctuations owing to a slightly favorable configuration with respect to the initial distribution of the torsional angle will undergo further torsional relaxation toward the equilibrium CT state with a more tilted configuration. The above-discussed coupling between the torsion and the CT process may result in the observed slow rise in the CT state, which

TABLE I
Rise Time (τ_r) of the Intramolecular CT State of 9,9′-Bianthryl in Alkanenitriles (ps)

	Solvent		
	Acetonitrile	Butyronitrile	Hexanenitrile
τ_r	1.8	3.4	7.5
τ_L	0.19	0.53	0.98–1.1

indicates the importance of the torsional relaxation from the perpendicular to the tilted conformation in the photoinduced CT dynamics of 9,9'-bianthryl in very polar solvents.

In relation to the experimental results and discussions above, a recent result of molecular dynamics simulations on the torsional potential surface and dynamics of 9,9'-bianthryl in acetonitrile [75] seems to be important. According to this study [75], the torsional potential in the excited state has double minimum and the twist angle from 90° to tilted equilibrium conformation in the CT state in acetonitrile solution is approximately the same as that in the excited state where no CT takes place in nonpolar solvent. The torsional relaxation time τ_t from the perpendicular to the tilted configuration was estimated to be $\tau_t \sim 1.5$–2.0 ps in both locally excited and CT states, which was much longer than the calculated solvent fluctuation (solvation) time $\tau_{solv} \sim 200$ fs of acetonitrile for the CT state of 9,9'-bianthryl. The calculated τ_t value [75] is very close to the τ_r value obtained by femtosecond–picosecond time-resolved absorption spectral measurements in acetonitrile solution [22,76] (Table I), and τ_{solv} is almost the same as the τ_L value of acetonitrile. For the rather large difference between the τ_t and τ_{solv} values, similar reasoning to that given above concerning coupling between the torsion and CT processes may be applicable.

There is an additional experimental result that seems to favor the tilted conformation of the CT state of 9,9'-bianthryl in strongly polar solvents. In this directly linked system, electronic interaction between the anthryl groups may be strong in the tilted conformation, leading to the result that the absorption spectra of the relaxed CT state in acetonitrile solution [72] cannot be reproduced by simple superposition of the spectra of radical ions of the chromophores [78,79], while the absorption spectra of the CT state of 1,2-dianthrylethane, $(A^+ \diagup\diagdown A^-)$ with two anthryl groups linked by two methylene chains in acetonitrile solution can be well reproduced by superposition of the spectra of radical ions of anthryl chromophores, owing to the very small interchromophore interactions in the loose conformation [78,79].

In the case of the latter system, 1,2-di(1-anthryl)ethane, a relatively slow process of direct intramolecular excimer (partial overlap type) formation by structural rearrangement with a time constant of several hundred picoseconds takes place in nonpolar solvents [78,79]:

$$(A \diagup\diagdown A)^* \xrightarrow{\quad \tau \sim 450 \text{ ps} \quad} (A \diagup\diagdown A)^* \quad \text{(in hexane)} \qquad (3.1)$$

With increased solvent polarity, formation of the intramolecular CT state in a loose conformation followed by excimer formation can be observed in addition to the direct excimer formation and, in acetonitrile solution, very rapid formation of a large amount of CT state (ion-pair state) followed by a much slower excimer formation was observed. The very rapid CT state formation may not be unreasonable. The $-\Delta G$ value for this photoinduced charge separation reaction is estimated to be ca. $0.3-0.4\,\text{eV}$ and the theoretical consideration on rate for this type of reaction as discussed in Section IV indicates a k_{ET} value of $\sim 10^{11}\,\text{s}^{-1}$. The spectra of the intermediate CT state with loose conformation in the following reaction can be well reproduced by the superposition of spectra of radical ions of the anthryl chromophores [78,79]:

$$(A \diagup\diagdown\diagup A)^* \xrightarrow{\tau_r \sim 10\text{ ps}} (A^+ \diagup\diagdown\diagup A) \xrightarrow{\tau \sim 430\text{ ps}} (A \diagup \diagdown A)^* \quad (3.2)$$

As described already to some extent in this section, interesting results were obtained by ultrafast fluorescence dynamics studies on 9,9'-bianthryl and ADMA in aprotic polar solutions, for which quantitative analyses were made on the basis of the generalized Langevin equation or generalized Smoluchowsky equation, taking into consideration only the solvation coordinate [6,10–12]. However, in view of the above-described results on 9,9'-bianthryl, which indicate strongly the importance of torsional relaxation in the CT process from a perpendicular to a tilted conformation, quantitative analysis of the experimental results taking into consideration both the solvation coordinate and torsional relaxation may be highly desirable. In the case of ADMA, which has a tilted conformation in the ground state and Franck–Condon excited state, however, interchromophore interaction may already be sufficiently strong in the excited Franck–Condon state, which might lead to an ultrafast CT reaction process on the adiabatic surface along the solvation coordinate without appreciable torsional conformation change [10–12] in sufficiently polar solvent with a low τ_L value.

IV. ENERGY GAP DEPENDENCE OF ELECTRON TRANSFER ON INTERACTION BETWEEN DONOR AND ACCEPTOR IN FLUORESCENCE QUENCHING REACTIONS IN POLAR SOLUTIONS AND THE BEHAVIOR OF PRODUCT LOOSE ION PAIRS

As described in Section I, investigations of the ET mechanism of fluorescence quenching reactions in polar solutions have a history of 60–70 years [1]. Nevertheless, more or less direct and detailed experimental and theoretical studies leading to a contemporary understanding of the ET mechanism of fluorescence quenching was started mainly by pioneering studies on exciplex phenomena around 1960–1970, and such studies are still developing [2–5,21–24,51–55].

In general, ET reaction processes such as photoinduced charge separation (CS) and charge recombination (CR) of the product CT and/or geminate IP states are regulated by several factors, including the magnitude of the electronic interaction responsible for the ET reaction, the energy gap $-\Delta G$ between the initial and final states, the reorganization energy λ for the reaction [17–19,22,23,42–44], and solvent polarity effects including solvent dynamics and so on [6,8,9]. Among them, the solvent effects and the effect of $-\Delta G$ on the reaction rates as well as the structures of the CT and/or geminate IP states have been most widely examined, as discussed to some extent in Section II. It should be noted here, however, that the mechanism of the photoinduced CS or CR of the geminate IP can change depending on the magnitude of the electronic interactions and/or mutual configurations between D and A. Namely, as discussed in Section V, the CS process in the case of the photoexcitation of the CT complexes where the interaction between D and A is strong in a compact configuration and which is stable in the ground state is quite different [22,80,81a,b,83] from CS by weak interaction between a fluorescer and quencher in strongly polar solutions. Moreover, the CS process in the former case leads to the formation of compact IP (CIP), which shows quite different $-\Delta G$ dependence of the CR rate [82–85] from the bell-shaped dependence of the LIP formed in the fluorescence quenching reaction [86,87].

In this section we discuss mainly the mechanisms of the CS in the fluorescence quenching reaction in acetonitrile solutions and the CR of the product geminate LIPs, especially the remarkable difference between the bell-shaped energy gap dependence of the CR rate of geminate LIPs and the very broad energy gap law of a CS rate constant in the fluorescence quenching reaction without manifestation of the inverted region [22,23].

As discussed to some extent earlier, the usual theories of ET assuming weak D–A interaction predict symmetrical bell-shaped energy gap dependence [17–19] or asymmetrical dependence owing to contributions from high-frequency quantum modes of intramolecular vibrations [42–44]. However, the experimental results of the bimolecular rate constant of fluorescence quenching, due to the intermolecular CS obtained by stationary measurements in acetonitrile solution, showed a steep rise around a zero energy gap and became a constant of diffusion-controlled value with increases $-\Delta G_{CS}$ but did not show a decreased rate constant even at very large $-\Delta G_{CS}$ values [22,23] (Fig. 3).

Even at present, no clear-cut manifestation of the inverted region has been reported for the photoinduced CS in both cases of bimolecular fluorescence quenching reactions and intramolecular fluorescence quenching in

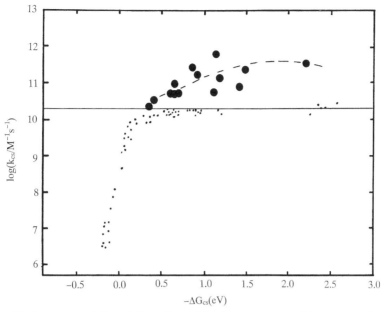

Figure 3. k_{CS} versus $-\Delta G_{CS}$ relations observed for fluorescence quenching reactions in acetonitrile solutions. Small circles, bimolecular fluorescence quenching rate constant obtained by stationary measurements [50]; large circles, "true" k_{CS} values at favorable $-\Delta G_{CS}$ values obtained by analysis of the transient effect in fluorescnence quenching processes [90].

bichromophoric systems linked by spacers [22,115] (see, e.g., Fig. 2 for the latter case). The bimolecular fluorescence quenching rate constant in the region of favorable $-\Delta G_{CS}$ values (≥ 0.5 eV) (Fig. 3) is diffusion limited, which masks the "true" rate constant of the CS process. However, there is a possibility that we can examine whether the photoinduced CS shows the bell-shaped energy gap dependence or not, by measuring the transient effect [88,89] in the fluorescence quenching process with ultrafast laser spectroscopy. Namely, approximate values of ET rate constant for CS between the fluorescer and quencher, k_{CS}, may be estimated by accurate measurements of fluorescence decay curves and by analyzing the transient effects in the quenching process for a series of fluorescer–quencher pairs covering a wide energy gap range.

Such analyses of the transient effect on the basis of the Collins–Kimball model [88,89] were made [90] for acetonitrile solutions covering an energy gap range of ca. 2.0 eV, where the diffusion-limited values of bimolecular rate constant were observed by the stationary fluorescence quenching measurements. As indicated in Figure 3, however, the k_{CS} values obtained do not

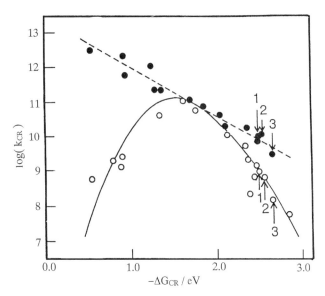

Figure 4. k_{CR} versus $-\Delta G_{CR}$ relations for geminate IPs produced by photoinduced CS in acetonitrile solutions for various D–A systems. Open circles, geminate LIPs produced by fluorescence quenching reactions [83,87]; solid circles, CIPs (compact IPs) produced by excitation of ground-state CT complexes [82,83,95–97,99]. The numbered points represent k_{CR} values for benzophenone (BP)–tertiary aromatic amine IPs [95–97]: 1, BP/N,N-diethylaniline; 2,BP/N,N-dimethylaniline; 3, BP/N-methyl-diphenylamine.

show typical bell-shaped energy gap dependence as observed in the case of the CR of geminate LIPs (Fig. 4) but a rather flat and broad shape of the energy gap dependence and no inverted region has been confirmed.

On the other hand, contrary to the photoinduced CS reaction, *only the inverted region was observed in many cases of D and A systems of organic chromophores for the energy gap dependence of the CR rate in both of the geminate LIPs formed by bimolecular fluorescence quenching reactions* [91,92] *and the intramolecular IPs formed by photoinduced CS in the bichromophoric systems linked by spacers* [93,94]. Nevertheless, systematic experimental studies on geminate LIPs by directly observing the dynamics of CR processes with ultrafast laser spectroscopy [86,87] for a wide range of the energy gap $-\Delta G_{CR} \sim 2.3$ eV in acetonitrile solutions have demonstrated a nearly bell-shaped energy gap dependence, including both normal and inverted regions (Fig. 4). On the other hand, unambiguous demonstrations of such bell-shaped energy gap dependence of the CR rate for the intramolecular IP states of linked bichromophoric systems are not yet available (see, e.g., [22] and [45]).

One of the reasons for the lack of inverted region in the results of the CS observed in the fluorescence quenching reaction might be the participation of the excited electronic states of geminate LIPs when the exothermicity of the reaction becomes too large [50,98,99]. However, since the energies of the excited electronic states of radical ions may change from compound to compound irrespective of $-\Delta G_{CS}$ value, this may not be a very general mechanism for this problem.

Another mechanism suggested more than a decade ago is a nonlinear effect in the polarization of solvent surrounding solutes undergoing an ET reaction, leading to the result that the energy gap relation of the CS reaction becomes broader, the top region becomes flat, and the energy gap law of the CR reaction becomes narrow with a prominent peak [100]. The physical interpretation of the nonlinear effect is the dielectric saturation of solvent surrounding the ionic species of reactants or products [101]. Assuming parabolic free energy curves with a larger curvature due to the dielectric saturation in the charged state of reactants or products than that in the neutral state, it was demonstrated that the observed energy gap laws of CS and CR reactions could be consistently fitted by those calculated theoreticaly [100].

After the studies above on nonlinear polarization or dielectric saturation models of solvent around charged solutes, many kinds of molecular dynamics and Monte Carlo simulation calculations were made for fluid [23,101b,c,102–106] and frozen solutions as well as some heterogeneous systems [23,107,108] to examine quantitatively whether the nonlinear effect is large or not in real molecular systems. For fluid solutions the solvent layer in the first shell was shown dielectrically saturated, but solvent outside this shell was indicated to follow the linear response, showing relatively small nonlinearity as a whole, while nonlinearity in frozen or heterogeneous systems, where translational motions of solvent are restricted, was found to be substantial. Moreover, the nonlinearity was shown to be enhanced, when solvent electronic polarizability was taken into account [23,109].

In the treatment above assuming parabolic free energy curves, curvatures of the free energy curves determined at the minima of the initial and final states were extrapolated along, the reaction coordinate in quadratic forms. However, for determining the exact form of the free energy curves as a function of the reaction coordinate, a more accurate statistical mechanical consideration of the reaction coordinate is necessary (reaction coordinate theory) [102–106,110–114], according to which the simple treatment above assuming the parabolic free energy curves with curvatures determined at the minima may be erroneous and the curvatures could change along the reaction coordinate. Nevertheless, it was demonstrated that even when we used the exact reaction coordinate theory, the energy gap law for the CS reaction was broader than the CR reaction [105]. However, it was also true that this calculated difference in the energy gap laws between the CS and CR reactions [109] was not large enough to interpret the observed difference between the energy gap laws of photoinduced CS and the CR of the product geminate LIPs in acetonitrile solutions [22,23,87,90].

Therefore, it is necessary to invoke another mechanism to explain the observed large difference between the energy gap laws of the photoinduced CS and CR of the geminate LIPs. As described briefly at the end of Section II, the most appropriate interpretation at present seems to take into account the donor–acceptor distance distribution on the energy gap law [23,51–54]. Namely, for larger $-\Delta G_{CS}$ values a higher solvent reorganization energy λ_S is favorable to keep a high ET rate according to Marcus equation (1.4), while λ_S becomes larger with increased donor–acceptor distance r according to Eq. (1.5), which makes it difficult to observe the inverted effect in the fluorescence quenching reaction due to the CS in polar solutions. In the following we discuss more or less quantitatively, the mechanisms underlying the very broad energy gap law of the CS process in the fluorescence quenching reaction and the rather typical bell-shaped energy gap law of the CR process of the product geminate LIPs.

We write Eq. (1.4) in a more explicit form:

$$W_S(-\Delta G) = \nu(k_B T/4\pi\lambda_S)^{1/2} \exp[-(\Delta G + \lambda_S)^2/4\lambda_S/k_B T] \qquad (4.1)$$

where ν is a function of the electronic matrix element $V(r)$. In the limit of the nonadiabatic mechanism with small $V(r)$, $W_s(-\Delta G)$ can be written as

$$W_S(-\Delta G) = \frac{2\pi}{\hbar}|V(r)|^2(4\pi\lambda_S k_B T)^{-1/2} \exp[-(\Delta G + \lambda_S)^2/4\lambda_S k_B T] \quad (4.2)$$

On the other hand, in the limit of the adiabatic mechanism with large $V(r)$, ν in Eq. (4.1) becomes independent of $V(r)$:

$$\nu = \nu_{ad} \qquad (4.3)$$

Combining these equations and considering that $V(r)$ is expected to decrease exponentially with increased r, r-dependent $W_S(-\Delta G(r), r)$ can be written as [23,52–54]

$$W_S(-\Delta G(r), r) = \nu(r)[k_B T/4\pi\lambda_S(r)]^{1/2} \exp\{-[\Delta G(r)$$
$$+ \lambda_S(r)]^2/4\lambda_S(r)k_B T\} \qquad (4.4)$$

$$\nu(r) = \nu_{ad}\{1 + \exp[\beta(r - r_a)]\}^{-1} \qquad (4.5)$$

where β is a constant, the nonadiabatic mechanism will predominate for $r > r_a$, and the adiabatic mechanism will predominate for $r < r_a$. The r-dependence of $-\Delta G(r)$ can be expressed as

$$-\Delta G_{CS}(r) = -\Delta G_{CS}^0 + (e^2/\varepsilon r) \tag{4.6}$$

$$-\Delta G_{CR}(r) = -\Delta G_{CR}^0 - (e^2/\varepsilon r) \tag{4.7}$$

where $-\Delta G^\circ$ is the standard free energy gap. On the other hand, Eq. (1.5) usually overestimates $\lambda_S(r)$ considerably. Therefore, use of a semiempirical formula may be appropriate, which has the same r dependence as Eq. (1.5):

$$\lambda_S(r) = c - (d/r) \tag{4.8}$$

When the $-\Delta G$ value is very large, the donor–acceptor pairs at large distances can play a significant role, where $\lambda_s(r)$ approaches a c value, while the parameter d will become important when the donor–acceptor pairs at short distances play an important role at small $-\Delta G$ values.

Although the Franck–Condon factor due to the intramolecular vibrational quantum modes is not directly r dependent, we must take into account this factor for the total ET rate as follows by convolution:

$$W(-\Delta G(r), r) = \int F_q(\varepsilon_1) W_S(-\Delta G(r) - \varepsilon_1, r) \, d\varepsilon_1 \tag{4.9}$$

where $F_q(\varepsilon_1)$ is the thermally averaged Franck–Condon factor for quantum modes and is given by

$$F_q(\varepsilon_1) = \frac{1}{\hbar\bar{\omega}} \exp[-S(2\bar{\nu}+1)] I_p[2S\sqrt{\bar{\nu}(\bar{\nu}+1)}][(\bar{\nu}+1)/\bar{\nu}]^{p/2} \tag{4.10}$$

$$\bar{\nu} = [\exp(\hbar\bar{\omega}/k_B T) - 1]^{-1} \qquad p = \varepsilon_1\sqrt{\hbar\bar{\omega}}, \quad S = \delta^2/2 \tag{4.11}$$

where I_p is the modified Bessel function, $\bar{\omega}$ the effective angular frequency of the quantum modes, $\bar{\nu}$ the averaged vibrational number, S the electron-vibrational coupling strength, and δ the shift of the normal coordinate between the initial and final states of the quantum mode.

To compare the total ET rate $W(-\Delta G(r), r)$ with the observed values of the CS as well as CR rate constants at various $-\Delta G$ values, it is necessary to average W over distance. Since the fluorescer and quencher are neutral in the photoinduced CS reaction, we can assume that the distance distribution is uniform at the initial time of the CS reaction. Therefore, $k_{CS}(-\Delta G)$ may be given by

$$k_{cs}(-\Delta G_{CS}) = \int_{r_c}^{\infty} W(-\Delta G_{CS}(r), r) 4\pi r^2 \, dr \qquad (4.12)$$

where r_c is the critical distance at which the donor and acceptor can come closest to each other. We can compare $k_{CS}(-\Delta G_{CS})$ values calculated by Eq. (4.12) with those obtained by analysis of the transient effect in the fluorescence quenching processes.

In the case of the CR reaction of the geminate IPs, the initial distance distribution will vary with the $-\Delta G_{CS}$ of the photoinduced CS reaction from which the geminate IP was produced. Therefore, the normalized initial distance distribution of the geminate IPs may be given by [23,52,53]

$$g_0(r) = 4\pi r^2 W(-\Delta G_{CS}(r), r)/k_{CS}(-\Delta G_{CS}) \qquad (4.13)$$

The initial distribution may not be stationary but may be deformed in the course of time during CR. If the CR reaction takes place quickly, the deformation might be negligible, while the deformation before the CR might be considerable if the CR takes place slowly. Therefore, the energy gap law of the CR rate observed by the time-resolved transient absorption spectral measurements on geminate IPs [22,23,87] will depend on the energy gap law of the CS reaction and diffusion-reaction dynamics of geminate IPs. Exact treatment of this problem is rather difficult and it still under investigation [116]. Nevertheless, the effect of the diffusional motions in the geminate IPs seems to be rather small, and the CR as well as the dissociation processes observed by the time-resolved transient absorption spectral measurements can be reproduced approximately by simple reaction scheme

$$D_s^+ \cdots A_s^- \xrightarrow{k_{diss}} D_s^+ + A_s^-$$
$$\downarrow k_{CR}$$
$$D \cdots A \qquad (4.14)$$

leading to the exponential decay of the absorbance $A(t)$ of the geminate IPs [87]:

$$A(t) = A(0)\Phi + A(0)(1 - \Phi) \exp(-k_d t) \qquad (4.15)$$

where $k_d = k_{CR} + k_{diss}$ and $\Phi = k_{diss}/k_d$.

The results above seem to indicate that contrary to the case of the neutral fluorescer–quencher pairs, the diffusional motions in the IP might not be easy even in very polar solvents, and essential features of its structure, including surrounding polar solvent molecules strongly interacting with ion radicals, might be approximately maintained during the CR and dissociation. A slight modification of the distance distribution may be taken into account by using the following Gaussian distribution function:

$$g(r, \sigma) = \exp[-(r - r_n)^2/(d\sigma)^2] \bigg/ \int_{r_c}^{\infty} \exp[-(r - r_n)^2/(d\sigma)^2] r^2 \, dr \qquad (4.16)$$

where r_n and d are constants that depend on the condition when the geminate IP was produced. Factor σ represents a normalized time and it is assumed that the width of the distribution will increase a little with time but the distance corresponding to the maximum distribution remains the same. We set $g(r, \sigma)$ for $\sigma = 1$ as the distribution function immediately after the formation of geminate IP [i.e., $g(r, 1) \sim g_0(r)$]. For small $-\Delta G_{CS}$, $g_0(r)$ may not necessarily be Gaussian but we approximate $g(r, 1)$ to be a Gaussian having the same width and the maximum position as $g_0(r)$. For large $-\Delta G_{CS}$, $g_0(r)$ is close to Gaussian and the correspondence between $g(r, 1)$ and $g_0(r)$ is very easy. Under these conditions, the CR rate averaged over the distance distribution which depends on σ and corresponding to the one observed can be written as

$$k_{CR}(-\Delta G_{CR}, \sigma) = \int_{r_c}^{\infty} W(-\Delta G_{CR}, r) g(r, \sigma) r^2 \, dr \qquad (4.17)$$

Calculations of k_{CS} considering the $-\Delta G_{CS}$ dependence of the donor–acceptor distance distribution in the fluorescence quenching reaction according to Eq. (4.12) and also calculations of k_{CR} taking into account the effect of a slight deformation of distance distribution before the CR decay of the geminate LIP according to Eq. (4.17) by using the common numerical values for parameter, such as β, c, d, r_c, and so on, gave a fairly satisfactory interpretation for the observed large difference between the energy gap dependences of k_{CS} (as shown in Figure 3) and the k_{CR} of the geminate LIP (as shown in Fig. 4) [23]. However, it should be noted here again that the interpretation above on the $k_{CR} \sim -\Delta G_{CR}$ relation may be no more than a semiempirical simulation of the observed energy gap dependence of k_{CR} derived from the fact that the experimental results of the time-resolved transient absorption spectra satisfy the reaction mechanism of Eqs. (4.14) and (4.15). On the other hand, it does not seem easy to reproduce the observed results of the CR and dissociation processes of geminate LIPs theoretically by straightforward applications of the diffusion-reaction equations to those processes of the LIPs together with the precedent CS processes producing LIPs with donor-acceptor distance distributions [116].

There remains also a problem to be discussed here concerning the validity of the Collins–Kimball (CK) model [88,89] used for analysis of the transient effect in the fluorescence quenching reaction [90], since the reaction is assumed to occur at a definite distance R (encounter distance) in the CK model, while actually the reaction can take place over a range of distances between fluorescer and quencher. The latter fact is due to the radial dependence of the local energy gap law and the electron tunneling matrix element, as discussed above. Namely, the overall time dependent rate constant $k(t)$ is given by

$$k(t) = \int_{r_c}^{\infty} W(-\Delta G, r) p(r, t) 4\pi r^2 \, dr \qquad (4.18)$$

The distribution function $p(r, t)$ is determined by solving the following diffusion-reaction equation under appropriate boundary conditions:

$$\frac{\partial}{\partial t} p(r, t) = D \frac{1}{r^2} \frac{d}{dr} r^2 \frac{d}{dr} p(r, t) - W(-\Delta G, r) p(r, t) \qquad (4.19)$$

where D is the mutual diffusion coefficient of the fluorescer and quencher.

As ET proceeds, the excited fluorescer concentration decreases and the fluorescence intensity $I(t)$ decays with time as follows:

$$I(t) = I(0) \exp\left[-(t/\tau_0) - [Q] \int_0^t k(t') \, dt' \right] \qquad (4.20)$$

where τ_0 is the fluorescence lifetime in the absence of the quencer and [Q] is the quencher concentration. Subsituting $W(-\Delta G_{CS}, r)$ of Eq. (4.9) into Eq. (4.19), we can solve Eq. (4.19) numerically for $p(r, t)$ by accurately taking into account the r dependence of $W(-\Delta G_{CS}, r)$ under the following initial and boundary conditions:

$$p(r, t = 0) = 1 \quad \text{for } r > r_c \qquad (4.21)$$

$$p(r = \infty, t) = 1 \qquad (4.22)$$

$$[dp/dr]_{r=r_c} = 0 \qquad (4.23)$$

We can obtain $k(t)$ by putting $p(r, t)$ obtained numerically into Eq. (4.18) and calculate numerically the fluorescence decay curve according to Eq. (4.20). The results of this numerical simulation of the fluorescence decay curve were compared with the results of analyses of transient effects in the fluoresence quenching processes by means fo the CK model [23,54].

For example, at $-\Delta G_{CS}^0 = 0.0\,eV$ and $-\Delta G_{CS}^0 = 2.8\,eV$, the results of the numerical simulation agreed very well with those of the analyses of the experimental results by means of the CK model, respectively. At $-\Delta G_{CS}^0 = 1.4$ eV, corresponding to the top region in the genergy gap dependence of k_{CS}, the ET reaction occurs very rapidly without diffusion at the most early stage, where the agreement between the result of the simulation and that of the analyses by the CK model was rather poor due to the poor experimental accuracy, while the agreement became very good at the later stage, where the diffusional motion became effective for the quenching process.

In Figure 5, $4\pi r^2 W(r)p(r, t)$, divided by the maximum value of $k(t)$ $[k(t=0)]$, is plotted against r. The distribution in Figure 5 depends not

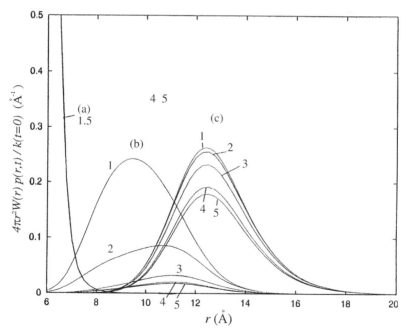

Figure 5. Time evolution of the value of $4\pi r^2 W(r)p(r, t)/k(t = 0)$ for (a) $-\Delta G_{CS}^0 = 0.0\,eV$, (b) $-\Delta G_{CS}^0 = 1.4\,eV$, and (c) $-\Delta G_{CS}^0 = 2.8\,eV$. Each curve from 1 to 5 corresponds to $t = 0$, 10^{-11}, 10^{-10}, 10^{-9}, and 10^{-8} s, respectively. All curves for case (a) appear overlapped with each other for $r > 6.5\,Å$.

only on r but also on t. Although the time-dependent change of the distribution is very small at $-\Delta G_{CS}^0 = 0.0$ eV as well as $-\Delta G_{CS}^0 = 2.8$ eV, the distribution at $-\Delta G_{CS}^0 = 1.4$ eV shows a shift toward larger r with time, owing to the rapid ET without diffusion in the early stage of the reaction, as discussed above. The peak position or the average position of the distribution (R) may be regarded as corresponding to the reaction radius or the encounter distance between the fluorescer and quencher in the CK model and has been estimated to be $R = 6$, 11.2, and 12.4 Å for $-\Delta G_{CS}^0 = 0.0$, 1.4, and 2.8 eV, respectively [23,54]. For $-\Delta G_{CS}^0 = 1.4$ eV, R was estimated to be 11.2 Å by using the curves at $t = 10^{-10}$ and 10^{-9} s because reactions in these time regions play the most significant role in the fluorescence quenching processes in this energy gap region. It should be noted here also that the results of the calculation of $k_{CS}(t)$ by numerical simulation at various $-\Delta G_{CS}$ values give the time-dependent energy gap law for the photoionduced CS reaction. The calculated results at $t = 0$ correspond to the energy gap law obtained by analyses of the $I(t)$ observed at various energy gaps with the CK model [23,54,90] and those at $t = \infty$ agree very well with the experimetal results of the steady-state measurements on fluorescence quenching reactions [23,54,49,50].

The results and discussions above give a quantitative interpretation on the fundamental mechanisms of fluorescence quenching reaction due to intermolecular ET in polar solutions and on the longstanding problem of the absence of the inverted region in the energy gap dependence of such photoinduced CS reactions in solutions. Moreover, the investigations described above give an interpretation of the conspicuous difference between the energy gap laws of the photoinduced CS in the fluorescence quenching reaction and the CR of the product geminate IP in polar solutions, where CR shows a rather typical bell-shaped energy gap law while CS shows a very broad one without the inverted region.

An important conclusion established by the investigations described above is that there is a distribution of the distance between the fluorescer and quencher where the CS takes place and that the average distance increases a little with increased energy gap $-\Delta G_{CS}$, which are essential for interpretation of the apparent lack of inverted region in the energy gap dependence of the photoinduced CS. Such a fluorescer–quencher distance distribution and change in the average distance with the energy gap for CS inevitably produces a distribution of the interionic distance in the product geminate IP and an increase in the interionic distance in the pair with increased $-\Delta G_{CS}$. This effect of the energy gap on the structure of the geminate IP together with the solvent effects on the structures of exciplex as discussed in Section II play roles of crucial importance in photochemical reactions via geminate IPs or exciplexes as discussed in Section VI.

Thus the important role of the donor–acceptor distance distribution in the interpretation of the very broad energy gap law without the inverted effect of the bimolecular CS rate constant for the fluorescence quenching reaction in polar solutions seems to be established now as discussed above. In the case of the distance-fixed donor–acceptor bichromophoric systems linked by spacers, there should not be such distance distribution effects. Nevertheless, a clear-cut result that shows the inverted effect has not been observed yet, as demonstrated, for example, by some recent investigations on the porphyrin–quinone bichromophoric systems [45,115]. Presumably, the examined energy gap for the CS might not be sufficiently large or there might be some low-lying excited electronic states of ion radicals in the pair. Anyhow, this is a quite unsatisfactory situation regarding agreement between experimental results and theoretical prediction on the energy gap dependence of the ET rate. Further efforts seem to be necessary to resolve this contradictory situation.

V. DYNAMICS OF CHARGE SEPARATION IN THE EXCITED STATE OF CHARGE TRANSFER COMPLEXES AND THE BEHAVIOR OF PRODUCT COMPACT ION-PAIR STATES

In Section IV we have discussed mainly the dynamics and mechanisms of the CS processes in the fluorescence quenching reaction by weak interaction at encounter between D* and A or D and A* in strongly polar solvent, acetonitrile, leading to the formation of LIPs (loose ion pairs). The long-standing problem of the very broad energy gap law without the inverted effect of the CS rate constant and the approximately bell-shaped energy gap law of the CR rate of the product LIPs especially, has been interpreted on the basis of the distribution of the distance between fluorescer and quencher molecules where the CS takes place and a small increase in the average distance with increased energy gap $-\Delta G_{CS}$. This change in the donor–acceptor average distance at CS in the encounter brings on a change in the interionic distance in the product LIPs with change in $-\Delta G_{CS}$, which will affect profoundly the photochemical reactions via LIPs [95–97].

On the other hand, when D and A are complexed in the ground state (CT complex formation), the D–A distance distribution effect may not be as important as in the case of the diffusional encounter complex and LIPs even in very polar solvents, although the CT complex may undergo low-frequency and small-amplitude D–A stretching and/or other intracomplex slow vibrations. It is well known owing to extensive studies on CT complexes since the proposal of Mulliken's theory of the CT complexes

[2,117,118]. Usually, the degree of CT from D to A in the ground state is very small as predicted by the Mulliken theory and confirmed by many experimental works [2,117,118], while the theory predicts almost complete one-electron transfer from D to A in the S_1 state of the complex, and this prediction was confirmed many years ago by fluorescence solvatochromic measurements on some weakly fluorescent CT complexes with the aid of Eq. (1.1) [2,119,120] and transient absorption spectral measurements on $(D^+ \cdots A^-)$ ion-pair-like spectra of an S_1-state CT complex with nanosecond laser photolysis methods [3,121,122]. However, most of the CT complexes are almost nonfluorescent, and their S_1-state lifetimes are very short. Accordingly, investigations on the S_1-state properties and dynamics of CT complexes with nanosecond spectroscopy or even with picosecond (10 ps) laser spectroscopy were very difficult despite the importance of such investigations in elucidation of the photoinduced ET mechanisms in the strongly interacting D–A systems. With the advent of femtosecond laser photolysis methods capable of femtosecond–picosecond time-resolved spectral measurements, it has become possible to observe directly excitation of the CT complex to the Franck–Condon state and its relaxation, leading to the vibrationally relaxed Compact Ion-Pair (CIP) state, which undergoes CR and ionic dissociation, and so on, which is of crucial importance for disclosing the true story of these dynamical processes. It should be noted here that because the electronic structure of the CT complex in the S_1 state is very different from that in the ground state as described above, the equilibrium geometrical structure of the S_1 state could be considerably different from that of the ground state. In such a case, the relaxation processes from the Franck–Condon to the equilibrium S_1 state will include such a change in the geometrical structure of the complex.

We discuss here the first results of picosecond–femtosecond laser photolysis studies on relatively weak CT complexes composed of TCNB (1,2,4,5-tetracyanobenzene) and methyl-substituted Bz (benzene), which seem to undergo geometrical structural change in the S_1 state [80,81] and then those of stronger CT complexes composed of various conjugate π-electronic D and A systems [82–85,123–126). There are also some recent reports on ultrafast laser spectroscopic studies concerning CT complexes of seemingly more simple composition, Bz and methyl-substituted Bz–halogen atom systems [127,128]. However, we concentrate our discussions on those CT complexes with the above-mentioned conjugate π-electronic D and A systems, and also mainly discuss the problem of $k_{CR} \sim -\Delta G_{CR}$ relations of CIPs formed by excitation of the CT complexes.

For example, results of picosecond–femtosecond laser photolysis and time-resolved transient absorption spectral measurements on a TCNB–toluene complex in toluene solution can be summarized as follows [80,81a]:

$$(\text{TCNB}^{-\delta}\text{Tol}^{+\delta}) \xrightarrow{h\nu} {}^{1}(\text{TCNB}^{-\delta'}\text{Tol}^{+\delta})* \xrightarrow[\substack{\text{structural changes} \\ \text{within 1:1 complex}}]{\tau_d \sim 1.5}$$

$$^{1}(\text{TCNB}^{-\delta''}\text{Tol}^{+\delta''})* \xrightarrow[\text{Tol}]{\tau_r \sim 30\,\text{ps}} {}^{1}(\text{TCNB}^{-}\text{Tol}_2^{+})* \qquad (5.1)$$

where δ, δ', and δ'' represent degrees of partial CT and $\delta'' > \delta' > \delta$. Immediately after excitation, a little sharpening of the TCNB anionlike band shape takes place with a time constant of 1.5 ps, which indicates an increase in the CT degree due to a slight geometrical structural change. In this system, however, this structural change within 1:1 complex does not lead to the complete CS, but further interaction with a donor toluene molecule and the formation of 1:2 complex with a time constant of 30 ps is necessary. Similar results have been observed for other solutions in benzene and mesitylene, where $\tau_d \sim 2.1$ ps, $\tau_r \sim 20$ ps and $\tau_d \sim 550$ fs, $\tau_r \sim 40$ ps, respectively, for the former and latter solutions.

On the other hand, photoinduced CS of these CT complexes is facilitated by solvation in a polar solvent such as acetonitrile. In the case of TCNB/methyl-substituted Bz complexes in acetonitrile solutions, rapid CS with a time constant shorter than 1 ps is induced by solvent reorientation considerably but not completely immediately after excitation. For the almost complete CS leading to the CIP formation, further intracomplex structural change and solvation, which take place with a time constant of a few 10 ps to a few picoseconds are necessary [80,81b]:

$$(\text{A}^{-\delta}\text{D}^{+\delta})_s \xrightarrow{h\nu} {}^{1}(\text{A}^{-\delta'}\text{D}^{+\delta'})_s* \xrightarrow[\text{solvation}]{<1\,\text{ps}}$$

$$(\text{A}^{-\delta''}\text{D}^{+\delta''})_{s'}* \xrightarrow[\substack{\text{structural changes and} \\ \text{further solvation}}]{\tau_{\text{CS}}} (\text{A}^{-}\cdots\text{D}^{+})_{s''}(\text{CIP}) \qquad (5.2)$$

The τ_{CS} value in acetonitrile solution has been confirmed to become shorter with decreased oxidation potential of D [i.e., $\tau_{\text{CS}} = 20, 13, 12, 7, 5$ ps, respectively, for the TCNB complexes with toluene, mesitylene, p-xylene, durene, and hexamethylbenzene (HMB)]. These τ_{CS} values are considerably larger than the solvent τ_L. Accordingly, the CS process seems to involve an intracomplex structural change in the 1:1 complex even in acetonitrile solution, which might be induced by solvation. This τ_{CS} value is also affected by the energy of the CT or IP state $(D^+ \cdots A^-)$ and shortened with lowering of the $(D^+ \cdots A^-)$ state due to the decreased ionization potential of D as described above for the methyl-substituted Bz series. When we use for A PMDA (pyromellitic dianhydride), which is a little stronger acceptor than TCNB, τ_{CS} for PMDA–toluene complex has been found to be 7 ps [83], compared with $\tau_{CS} = 20$ ps for the TCNB–toluene complex in acetonitrile solution. Moreover, when we use much stronger acceptor TCNE (tetracya-noethylene), τ_{CS} becomes very short. For example, it was demonstrated that τ_{CS} was shorter than 100 fs for pyrene–TCNE and perylene–TCNE complexes in acetonitrile solutions [82,83], and τ_{CS} for the former system has been determined to be 50 fs by recent measurement with higher time resolution [126].

Therefore, most of the CT complexes composed of moderate or strong D and A undergo rapid ^{1}CIP formation with rather a short time constant τ_{CS} in polar solvents and those ^{1}CIPs make rapid CR decay to the ground state as discussed in the following, leading to the formation of no dissociated free-ion radicals or only a very small number of them [81–86,99,123,124]. In the case of CT complexes composed of relatively weak D and A in moderately and strongly polar solvents, CIP can undergo ionic dissociation competing with the CR deactivation to the ground state [5,81–85,124,129,130]. In such a case it is generally believed that the CIP formed from the excited Franck–Condon state of the CT complex undergoes ionic dissociation via LIP competing with the CR decay from each IP state:

$$\text{CIP} \underset{k_{solv}}{\overset{k_{solv}}{\rightleftharpoons}} \text{LIP} \xrightarrow{k_{diss}} D_s^+ + A_s^- \qquad (5.3)$$

$$\downarrow k_{CR}^{CIP} \qquad \downarrow k_{CR}^{LIP}$$

Examinations of such dissociation processes by means of time-resolved absorption spectral measurements should give a multiexponential decay curve of transient absorbance converging to a constant value of dissociated free-ion radicals. We have observed for the first time the biphasic decay of the transient absorbance in the case of the TCNB complexes with Bz, toluene, and xylene in acetonitrile solutions [81c], and reasonable values for the rate constants in Eq. (5.3) have been obtained by analyzing the absorbance decay curves [81c]. To obtain the rate constant for the reverse process of k_{solv}, a different kind of experimental information in addition to the time-resolved absorption spectral measurements seems to be necessary because discrimination of spectra of CIP, LIP and dissociated in radicals is rather difficult. There is a report concerning the measurement of fluorescence decay curves of CIP in less polar solvent for such different kinds of information [130]. However, there is the problem that the structure of CIPs as well as LIPs and reaction mechanisms might be affected by decreased solvent polarity, complicating the interpretations.

On the other hand, transient absorbances of most of the other CIPs produced by excitation of the CT complexes in acetonitrile solutions show single exponential decay, of which the reaction scheme may be given by

$$
\text{CIP} \xrightarrow{\quad k_{\text{diss}} \quad} D_s^+ + A_s^- \tag{5.4}
$$
$$
\downarrow \begin{matrix} \text{CIP} \\ k_{\text{CR}} \end{matrix}
$$

where in most cases $k_{\text{CR}}^{\text{CIP}} \gg k_{\text{diss}}$.

These results suggest that in many excited CT complex systems undergoing ionic dissociation in strongly polar solutions, it may not necessarily be appropriate to consider one definite kind of LIP as an intermediate of the ionic dissociation from CIP, but the dissociation process from CIP may be a gradual change through multiple kinds of intermediate IP states with a different number of intervening solvent molecules between ions in the pair leading to the practical loss of pair correlation between geminate ions. This argument of ionic dissociation is closely related to the multiple-states model of geminate IP for the ionic dissociation of exciplex systems as discussed in Section II to interpret the results observed for the exciplex fluorescence decay time and photocurrent rise time of pyrene–DMA and pyrene–DCNB systems in polar solutions. This argument also corresponds to the interionic distance distribution in the LIPs produced by fluorescence quenching reaction in acetonitrile solution, which has been proposed to interpret the lack of inverted region in the observed $k_{\text{CS}} \sim -\Delta G_{\text{CS}}$ relation, as discussed in Section IV.

By means of the picosecond–femtosecond laser photolysis and time-resolved transient absorption spectral measurements on various CT complexes, CR rate constants of CIPs have been obtained [80–86,95–97,99]. For many D–A systems, comparative studies have been made on the properties of CIPs formed by excitation of the ground-state CT complexes and LIPs produced by CS at encounter in the bimolecular fluorescence quenching reaction in acetonitrile, using the same D–A system in the same polar solvent and observing their decay processes directly with picosecond–femtosecond laser spectroscopy, [82,83,86,87,95–97,99].

In Figure 4, $k_{\mathrm{CP}}^{\mathrm{CIP}}$ values are plotted against $-\Delta G_{\mathrm{CR}}$ in compaison with $k_{\mathrm{CR}}^{\mathrm{LIP}}$ for various D–A systems. The $-\Delta G_{\mathrm{CR}}$ values in Figure 4 are estimated by the conventional formula

$$-\Delta G_{\mathrm{CR}} = E_{\mathrm{ox}}(\mathrm{D}^+/\mathrm{D}) - E_{\mathrm{red}}(\mathrm{A}/\mathrm{A}^-) - e^2/\varepsilon R \qquad (5.5)$$

with $(e^2/\varepsilon R) = 0.5\,\mathrm{eV}$), which is frequently used for LIP in acetonitrile solutions.

A semiempirical estimate of $-\Delta G_{\mathrm{CR}}$ for CIP by evaluating reorganization energy $\lambda = \lambda_v + \lambda_s$ from the CT absorption and fluorescence spectral data of some CT complexes [119,120] in polar solutions and by using the approximate formula

$$-\Delta G_{\mathrm{CR}} \sim h\nu_{\mathrm{max}} - \lambda \qquad (5.6)$$

suggested a slight shift to the higher-energy side as a whole compared with the value estimated with Eq. (5.5), where ν_{max} is the frequency of the CT absorption peak of the CT complexes used to obtain the results in Figure 4 [83]. Therefore, the functional form of the energy gap dependence of $k_{\mathrm{CR}}^{\mathrm{CIP}}$ may not be affected by using the $-\Delta G_{\mathrm{CR}}$ values estimated with Eq. (5.5). The most noteworthy results in Figure 4 are as follows:

1. There is an extremely large difference between the $k_{\mathrm{CR}}^{\mathrm{CIP}}$ and $k_{\mathrm{CR}}^{\mathrm{LIP}}$ values of the same strong D–A systems, such as perylene–TCNE and pyrene–TCNE where CR of LIP is in the normal region with $k_{\mathrm{CR}}^{\mathrm{LIP}} \sim 10^9\,\mathrm{s}^{-1}$ in contrast to the fact that there is no indication of normal-region-like behavior in the CR of CIP with $k_{\mathrm{CR}}^{\mathrm{CIP}} > 10^{12}\,\mathrm{s}^{-1}$.

2. $k_{\mathrm{CR}}^{\mathrm{CIP}}$ is larger than $k_{\mathrm{CR}}^{\mathrm{LIP}}$ in general also for relatively weaker D–A systems such as pyrene–phthalic anhydride and tertiary aromatic amine–benzophenone. This fact plays an important role in the mechanism of the photoinduced hydrogen abstraction reaction via geminate ion radical pair in the case of the latter systems as discussed in Section VI.C.

3. Contrary to the approximately bell-shaped energy gap dependence of $k_{\mathrm{CR}}^{\mathrm{LIP}}$ the energy gap dependence of $k_{\mathrm{CR}}^{\mathrm{CIP}}$ is given by an exponential form of

$$k_{\mathrm{CR}}^{\mathrm{CIP}} = \alpha \, \exp[-\beta|\Delta G_{\mathrm{CR}}|] \qquad (5.7)$$

where α and β are constants independent of ΔG_{CR} and the slope β is considerably more gentle than that of $k_{\mathrm{CR}}^{\mathrm{LIP}}$ in the inverted region.

The remarkable difference in the energy gap dependence of the CR rate between two kinds of IPs as described above may be originating from the difference in the structures, including donor and acceptor configurations, in the pair where CIP has a more compact structure with stronger electronic interaction between donor and acceptor. In this respect it should be noted here that solvent polarity effects on the energy gap dependence of k_{CR}^{CIP} for aromatic hydrocarbon–acid anhydride systems have been examined by changing solvent from acetonitrile to acetone and ethylacetate [84]. Although the energy of the CIP state is affected by interaction with polar solvent, little or no solvent polarity effect on the exponential energy gap law itself has been observed [84]. This result is similar to the fact that little or no effect of solvent polarity on the $k_{CR} \sim -\Delta G_{CR}$ relation of the intramolecular IP state of porphyrin–quinone dyads in the inverted region is observed when the solvent was changed from benzene to THF [45].

On the other hand, CR decay of the CIP produced in rigid cyclophane-derived electron acceptor with a central cavity that accommodates many 1,4-disubstituted benzene derivative donors in acetonitrile solution has been studied by means of ultrafast laser spectroscopy over a wide energy gap range for CR [124]. These CIP systems kept in the rigid cyclophane seem to have more compact structure than the uncapsulated CIP in polar solvent. The results of measurements on k_{CR}^{CIP} of this cyclophane-derived CIP [124] were quite similar to those uncapsulated CIP systems in acetonitrile [82–84], and the same energy gap dependence as Eq. (5.7) with approximately the same slope β has been confirmed. Moreover, the CR decay of CIPs produced by exciting CT complexes of a series of aromatic hydrocarbon with TCNB adsorbed on porous glass without solvent has been investigated recently, results of which showed the same exponential energy gap law with practically the same slope β [131] as those observed in polar solutions [82–84].

The above-described experimental results on the CR of CIPs in various environments seem to suggest strongly the dominant effect of the intramolecular high-frequency quantum modes or various intracomplex vibrational modes and the rather minor role of solvent reorganization in the CR process. As is clear from the exponential energy gap law in Eq. (5.7) and from the schematic diagram in Figure 6, this energy gap dependence of the CR transition is analogous to that of radiationless transition [132] in the weak-coupling limit [132b,c]. In regard to this problem, some related theoretical studies [133] and femtosecond laser spectroscopic studies on excited CT complexes [125,126] have been reported.

Qualitatively speaking, k_{CR}^{CIP} is determined by electronic coupling between the CIP state and ground state and by vibrational overlaps between the low-lying initial vibrational levels of the CIP and the isoenergetic vibrational

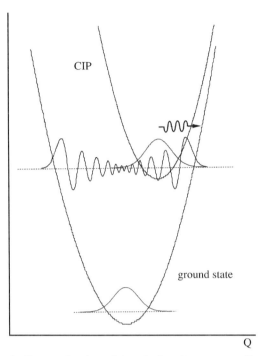

Figure 6. Schematic diagram for the $-\Delta G_{CR}$ (ordinate) versus coordinate (Q) (abiscissa) showing the CR process from the CIP state due to quantum mechanical tunneling by intra-complex vibrational modes.

levels of the ground state as indicated in Figure 6. On the basis of the previous theoretical treatment of the radiationless transition in the weak-coupling limit [132b,c], it is possible to give an approximate equation for the energy gap dependence of k_{CR}^{CIP} [134]:

$$\ln(k_{CR}^{CIP}) = \ln C - (\gamma/\hbar\omega_M)|\Delta G_{CR}| \tag{5.8a}$$

$$\gamma = \ln(|\Delta G_{CR}|/S_M\hbar\omega_M) - 1 \tag{5.8b}$$

$$S_M = \sum_j S_j \tag{5.8c}$$

$$\omega_M = \sum_j \omega_j S_j \Big/ \sum_j S_j \tag{5.8d}$$

$$S_j = \tfrac{1}{2}(M_j\omega_j/h)(\Delta Q_j)^2 \tag{5.8e}$$

where the preexponential factor C represents mainly the contribution from the electronic coupling term connecting the CIP and ground state, and also some additional term [134]. The term $-(\gamma/\hbar\omega_M)|\Delta G_{CR}|$ arises from the energy release into acceptor modes in the CIP $\rightarrow$ ground-state transition. S_M and ω_M are average quantities for electron-vibrational coupling constants and angular frequencies of the quantum modes, respectively, where contributions from individual acceptor modes j are given by Eqs. (5.8C) and (5.8d), and S_j is given by Eq. (5.8e) as a function of equilibrium displacements ΔQ_j. Although γ contains $|\Delta G_{CR}|$, its variation with change of $|\Delta G_{CR}|$ is much smaller than that for $\ln(k_{CR}^{CIP})$.

It should be noted here that, in general, the right-hand side of Eq (5.8a) should contain the term due to the solvent reorganization because both intramolecular (or intracomplex) modes and solvent modes are important. Nevertheless, as discussed above, the solvent polarity and solvent reorganization do not seem to be important for the CR of CIP because the observed $k_{CR}^{CIP} \sim -\Delta G_{CR}$ relations are not affected by the change of solvent polarity [84], and even for the CIP state adsorbed on porous glass in the absence of solvent, the same $k_{CR}^{CIP} \sim -\Delta G_{CR}$ relations as in strongly polar solutions are observed [131]. On the other hand, it seems to be possible that the low-frequency intermolecular complex vibrations contribute to the reorganization associated with the intracomplex electron transfer. Actually, recent resonance Raman and electronic absorption as well as fluorescence spectral studies on TCNE–hexamethylbenzene complex suggested the important role of such intermolecular complex vibrations in the electron transfer [135]. Moreover, vibrational coherence of the complex stretching modes in the CR process in the excited state of the same CT complex has been observed by means of femtosecond laser spectroscopic measurements [125], which suggests the importance of the complex stretching modes in the intracomplex electron transfer. Therefore, not only the intramolecular high-frequency quantum modes in donor and acceptor moieties but also intermolecular complex vibrations may contribute to the observed CR processes. At any rate, Eq. (5.7) has the same form as Eq. (5.8a), with $\alpha = C$ and $\beta = \gamma/\hbar\omega_M$).

In an extreme case of a supermolecule composed of very strongly interacting D with low ionization potential and A with high electron affinity, the main contribution to C may be the electronic matrix element of the kinetic energy operator for internal conversion [132,136] between an excited state with large CT degree and a ground state with much smaller CT degree. That is, the two electronic states are mixed by coupling to promoting vibrations that break down the orthogonality of the Born–Oppenheimer states. This is just the radiationless transition in the weak-coupling limit and seems to make difficult detection of normal-region-like behavior even at very small $-\Delta G_{CR}$ values for these CIPs.

The slope in the $\ln(k_{CR}^{CIP}) \sim |\Delta G_{CR}|$ relation is determined by $\gamma/\hbar\omega_M$. The absolute value of k_{CP}^{CIP} increases with increased factor C, which includes the electronic coupling matrix element. It was confirmed that k_{CR}^{CIP} values for the PMDA-methyl-substituted Bz series were larger than those of the acid anhydrides–polycyclic aromatic hydrocarbon series indicating the large electronic coupling matrix element for the smaller dimensions of the donor series, which seems to be quite reasonable [82,84]. The slope $\gamma/\hbar\omega_M$ is determined by S_M and ω_M and becomes more gentle with increases in these quantities. The increase in S_M is caused by an increase in the number of participating acceptor modes [134b,c,42,137] according to Eq. (5.8c), including high-frequency intramolecular modes and complex vibrational modes, which means that the rate of energy dissipation increases with the number of participating acceptor modes. Moreover, a large change in charge distribution accompanied by transition causes a considerable equilibrium displacement ΔQ_j, which also increases S_M.

The factors responsible for determination of the slope in the $\ln(k_{CR}^{CIP}) \sim |\Delta G_{CR}|$ linear relation for D–A systems may change depending on the nature of those systems. If the CT degree in the excited state is small, being practically the same as in the ground state, the slope of the radiationless transition probability versus energy gap relation may become very steep, as in the case of the usual internal conversion in the weak-coupling limits, owing to the small equilibrium diplacements and smaller number of the participating quantum modes in the transition. For those CIP systems discussed above, those factors may be much larger leading to the much more gentle slope in the $\ln(k_{CR}^{CIP}) \sim |\Delta G_{CR}|$ relation. Those factors may remain approximately the same throughout a series of similar compounds, as in the case of the systems discussed above. For the interpretation of results including a much wider ΔG_{CR} range, it may be necessary to consider changes in the participating multimodes and equilibrium shifts, leading to changes in reorganization energies with change in ΔG_{CR} as a cause for the gentle slope in the $\ln(k_{CR}^{CIP}) \sim |\Delta G_{CR}|$ relation, as indicated schematically in Figure 7 [83,84,138]. To elucidate more detailed mechanisms of the CR transitions of CIPs, assessing the importance of various vibrations that couple with the CR process, including high-frequency quantum modes of D and A moieties, low-frequency complex vibration modes and phononlike modes of environment, investigations on the temperature dependence of the CR process in the adsorbed state without solvent may be useful; the phononlike modes and low-frequency complex vibration modes seem to be frozen and only the higher-frequency quantum modes seem to remain effective to promote CR at sufficiently low temperatures [139].

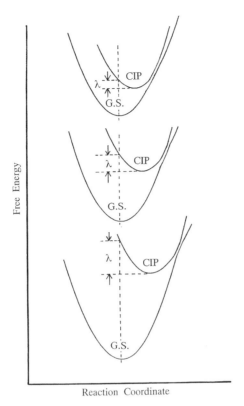

Figure 7. Free energy curves for the CIP and ground state versus reaction coordinate, where the potential minimum position of CIP depends on the $-\Delta G_{CR}$ value, illustrating that the CR of the CIP is in the inverted region even for small $-\Delta G_{CR}$ values.

VI. PHOTOINDUCED ELECTRON TRANSFER COUPLED WITH PROTON SHIFT OR TRANSFER IN SOME HYDROGEN-BONDING AND RELATED SYSTEMS

One of the most typical organic reactions of chemical product formation via a CT or ET state is the coupled process of photoinduced CS followed by proton transfer (PT) within the geminate IPs, resulting in the formation of neutral radical pairs that give stable final chemical product. The most extensively studied well-known example is the photochemical hydrogen abstraction reaction between benzophenone (A) and tertiary aromatic amines (D-H) in such polar solvents as acetonitrile:

$$A^* + H\text{-}D \rightarrow A^* \cdots H\text{-}D \rightarrow A^- \cdots H\text{-}D^+ \rightarrow A\text{-}H \cdots D \rightarrow \qquad (6.1)$$

Detailed picosecond–femtosecond laser photolysis studies on these reaction processes [95–97] have revealed that mechanisms underlying these reaction processes are closely related to the most fundamental mechanisms of photo-induced CS and CR of the product geminate IPs, and so on, in solutions discussed in preceding sections.

For example, the donor–acceptor distance distribution in the photoin-duced bimolecular CS reaction in polar solvents and the dependence of the average distance on the energy gap of the CS reaction producing the gemi-nate LIP with longer interionic distance for the larger $-\Delta G_{CS}$ value, which are discussed in detail in Section IV, affect profoundly the reaction of Eq. (6.1) because the proton transfer rate within the geminate LIP is very sensi-tive to the interionic distance in a geminate LIP [95–97].

Moreover, the reason for the well-known fact that the quantum yield of the photochemical hydrogen abstraction reaction between the benzophe-none (BP) and the tertiary aromatic amines decreases under very high con-centrations of the amine in polar solutions has been elucidated by means of ultrafast laser photolysis studies [95–97]. Namely, it is due to the fact that the CIP formed by excitation of the ground-state BP–amine CT complex formed under high amine concentration undergoes CR decay to the ground state much faster than LIP, as discussed in Section V and indicated in Figure 4, leading to the no ketyl radical formation. Presumably, in addition to the rapid CR process from CIP, the geometry of its compact structure may not be favorable for PT.

On the other hand, we showed for the first time that the inter-molecular hydrogen-bonding interaction frequently induced fluorescence quenching, especially when two conjugate π-electronic systems were directly combined by hydrogen-bonding interaction in nonpolar solvents [140]. We proposed the CT interaction between the proton donor (D-H) and acceptor (A) in the excited state of the hydrogen-bonding complex, leading to the formation of a kind of nonfluorescent exciplex $(D^+\text{-}H \cdots A^-)$, as a possible mechanism of the quenching [140] and proved it by means of picosecond–femtosecond time-resolved transient spectroscopic measurements on several typical hydrogen bonding systems [5,22,141,142].

For example, in the case of an 1-aininopyrene (AP)(D-H)–pyridine (Pyr)(A) system in hexane solution, picosecond-femtosecond time-resolved absorption spectral measurements have revealed that equilibrium between the locally excited (LE) state $(D^*\text{-}H \cdots A)$ and the nonfluorescent exciplex or IP state due to photoinduced ET $(D^+\text{-}H \cdots A^-)$ is realized within a few picoseconds and the equilibrium state undergoes decay to the ground state in the time regime of nanoseconds–100 ps via the IP state. In this system the photoinduced ET seems to be facilitated greatly by a slight shift of proton from D-H toward A in the excited state. On the other hand, in the case of a 1-pyrenol (PyOH)–Pyr system in hexane solution, a slight proton shift from D-H toward A in the excited state of the hydrogen-bonded complex seems to induce CT followed by a large-scale proton shift or PT, which induces ultrafast radiationless transition to the ground state without neutral radical pair product formation by coupled CT–PT processes.

In contrast to these typical hydrogen-bonding systems, we confirmed the neutral radical pair formation due to the coupled CT and PT processes in the case of aromatic hydrocarbon (A)–secondary aromatic amine (D-H) exciplex in nonpolar solvent, where a special hydrogen-bonding interaction seems to exist between the π-electron system of the aromatic hydrocarbon molecule and the N-H of the secondary aromatic amine in the exciplex state, $(A^- \cdots H\text{-}D^+)$ [5,28,143].

In the following, we discuss picosecond–femtosecond laser photolysis studies on excited hydrogen-bonding systems between heteroaromatic proton donor and acceptor systems [141,142], exciplexes between aromatic hydrocarbons and secondary aromatic amines [28,143], and BP–tertiary aromatic amine as well as BP–secondary aromatic amine systems [144], examining the mechanisms underlying the coupled ET and proton shift or transfer phenomena of these four types of systems.

A.   Electron Transfer Coupled with Proton Shift in Hydrogen-Bonding Systems in Nonpolar Solvents: 1-Aminopyrene/Pyridine and 1-Pyrenol/Pyridine Systems Studied by Femtosecond–Picosecond Laser Spectroscopy

To observe directly the dynamic processes of the LE → IP equilibrium state formation from the LE state, detailed femtosecond laser photolysis studies on AP–Pyr and AP–MePyr (methyl-substituted Pyr) systems in hexane solutions were made [142a]. Rapid formation of the LE → IP equilibrium mixture from the LE state has been observed clearly by means of the femtosecond time-resolved absorption spectral measurements and can be described by the reaction scheme

$$
\begin{array}{ccc}
 & \overset{k_1}{\underset{k_{-1}}{\rightleftharpoons}} & \\
\text{LE}(t) & & \text{IP}(t)
\end{array}
$$

$$hv_f + (\text{D-H}\cdots\text{A}) \quad (\text{D-H}\cdots\text{A}) \quad (\text{D-H}\cdots\text{A}) \quad (\text{D-H}\cdots\text{A}) + hv_f' \qquad (6.2)$$

$$k_1, k_{-1} \gg k_f, k_i, k_f', k_i' \qquad\qquad k_i' \gg k_f, k_i, k_f'$$

From this reaction scheme, [LE(t)] and [IP(t)] are given by

$$[\text{LE}(t)] = k_{-1}\tau_e[\text{LE}(0)][1 + (k_1/k_{-1})\exp(-t/\tau_e] \qquad (6.3)$$

$$[\text{IP}(t)] = k_1\tau_e[\text{LE}(0)[1 - \exp(-t/\tau_e)] \qquad (6.4)$$

respectively, where $\tau_e = (k_1 + k_{-1})^{-1}$, and the transient absorbance $D_\lambda(t)$ at a wavelength λ, where the molar extinction coefficients of LE and IP states are denoted by $\varepsilon_{\text{LE}}^\lambda$ and $\varepsilon_{\text{IP}}^\lambda$, respectively, is given by

$$\begin{aligned} D_\lambda(t) &= \varepsilon_{\text{LE}}^\lambda[\text{LE}(t)] + \varepsilon_{\text{IP}}^\lambda[\text{IP}(t)] \\ &= \tau_e[\text{LE}(0)][(k_{-1}\varepsilon_{\text{LE}}^\lambda + k_1\varepsilon_{\text{IP}}^\lambda) + k_1(\varepsilon_{\text{LE}}^\lambda - \varepsilon_{\text{IP}}^\lambda)\exp(-t/\tau_e)] \end{aligned} \qquad (6.5)$$

The observed time-dependent spectral change in the early stage was reproduced satisfactorily by Eq. (6.5), from which τ_e values were obtained to be approximately a few picoseconds–10 ps for these AP–Pyr and AP–MePyr systems. In contrast to this, the decay times τ_d of the equilibrium state determined by the time-resolved absorption spectral measurements as well as the LE fluorescence decay curve measurements of the hydrogen-bonded complexes were a few hundred picoseconds to 1 ns.

The energy gap for the LE $\rightarrow$ IP process, $-\Delta G_{\text{CS}}$, and that for the IP $\rightarrow$ ground state, $-\Delta G_{\text{IP}}$, have been evaluated by the conventional method using the oxidation potential of AP, E_{ox}, and the reduction potential of Pyr, E_{red}, measured in polar solvents with dielectic constant ε_r and the $S_1 \leftarrow S_0$ energy gap, ΔE_{00}, of AP as follows:

$$-\Delta G_{\text{CS}} = \Delta G_{\text{IP}} + \Delta E_{00} \qquad (6.6)$$

$$-\Delta G_{\text{IP}} = E_{\text{ox}} - E_{\text{red}} - (e^2/\varepsilon R) + \Delta G_S \qquad (6.7)$$

$$\Delta G_S = (e^2/2)(1/R^+ + 1/R^-)(1/\varepsilon - 1/\varepsilon_r) \qquad (6.8)$$

where R is the center-to-center distance between cation and anion assumed to be 7 Å and ΔG_s is the sum of the correction terms of the solvation energies calculated using the Born formula for the cation and anion with radii R^+ and R^- in a solvent with dielectric constant ε. The energy gap $-\Delta G_{\text{CS}}$ for these systems is calculated to be about -1.0 eV. Although $-\Delta G_{\text{CS}}$ becomes more negative with an increase of methyl substitutions in Pyr, τ_e shows only a very small increase with increased number of sub-stituents, probably because k_1 and k_{-1} should show the opposite dependence on $-\Delta G_{CS}$ [142a]. A slight increase in τ_e in the case of the multiple-substituted Pyr acceptor may be the effect of the steric hindrance, which decreases the interaction responsible for ET.

Contrary to the above-discussed behavior of τ_e, τ_d shows a much larger systematic increase with increased $-\Delta G_{IP}$. τ_d seems to be determined by k_i because $(k_f + k_i) < 10^8\,s^{-1}$ in the case of AP in hexane solution and k_f of the IP state is usually very small $(< 10^6\,s^{-1})$. Moreover, although the energy gap range is rather small, the exponential energy gap law, $\log(\tau_d^{-1}) \sim \log k_i' = \alpha - \beta|\Delta G_{IP}|$, has been confirmed for these systems, which is similar to that of the CR of CIP discussed in Section V. However; the β value of this hydrogen-bonding system is considerably larger than that of CIP as indicated in Figure 8.

In contrast to the large positive value of $-\Delta G_{IP}$, $-\Delta G_{CS}$ is negative, amounting to about $-1.0\,eV$. If we assume the usual ET mechanism, the ET process should be very slow. However, very rapid ET with a rate constant of about $10^{11}\,s^{-1}$ was observed. This fact means that $-\Delta G_{CS}$ should become a considerably large positive value due to the hydrogen-bonding interactions. As described above, $-\Delta G_{CS} \sim -1.0\,eV$ was estimated without taking into account the effect of the hydrogen-bonding interaction. In

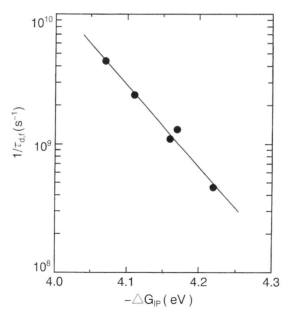

Figure 8. Exponential energy gap law, $\log(\tau_d^{-1}) \sim \alpha - \beta|\Delta G_{IP}|$, for the CR transition from the IP state of the AP–Pyr and AP–MePyr hydrogen-bonding complexes in hexane solutions.

relation to this problem, it should be noted here that a considerable lowering (0.8 eV) of the ionization potential of D-H (dibenzocarbazole) hydrogen-bonded with Pyr in condensed phase was confirmed [145]. This lowering of ionization potential of the D-H π-electron system is due to the increase in the intramolecular CT from the $\rangle$N—H group to the aromatic hydro-carbon ring by hydrogen-bonding interaction. On the other hand, hydrogen bonding will enhance the electron affinity of the π-electron system of Pyr leading to the stabilization of the IP state by more than 1 eV. Very similar mechanism of the stabilization of the IP state by hydrogen bonding inter-action will prevail also in AP–Pyr and AP–MePyr systems.

Moreover, the extent of the intramolecular CT from the substituent to the aromatic hydrocarbon ring of D-H is larger in the excited electronic state than in the ground state, which enhances the proton acidity of the substituent group, causing a proton shift toward the nitrogen of Pyr. This proton shift further lowers the ionization potential of D-H and enhances the electron affinity of Pyr, leading finally to a considerably large positive value of $-\Delta G_{CS}$. Therefore, in this mechanism, rapid proton shift in the hydrogen bond in the excited electronic state plays a crucial role in facilitating the fast ET [142]. This interpretation is supported by ab initio molecular orbital CI calculations on some heteroaromatic hydrogen-bonded model systems, which predict that one-electron transfer from D*-H to hydrogen-bonded A takes place when the D*-H proton is moved to the vicinity of the mid-point between the proton-donating and proton-accepting atoms [146].

A similar mechanism of photoinduced ET coupled with a proton shift in the hydrogen bond also appears to be working in the case of the PyOH–Pyr and PyOH–MePyr systems in nonpolar solvents. Femtosecond laser photo-lysis studies on these hydrogen-bonding systems in hexane solution [142] have clearly indicated only rapid (exponential) decay of the $S_n \leftarrow S_1$ absorp-tion band of PyOH with decay time τ_d of about 10 ps, without any other product formation. Nevertheless, we can recognize in this case a systematic relation between the τ_d observed and the $-\Delta G_{CS}$ estimated by Eqs. (6.6) to (6.8), as indicated in Table II. Namely, τ_d increases a little as $-\Delta G_{CS}$ becomes more negative, which suggests that the ET mechanism is respon-sible for decay of the LE state. Presumably, the IP state formed from the LE state may undergo much faster nonradiative CR transition to the ground state than its formation process, contrary to the case of the AP–Pyr and AP–MePyr systems.

Although the $-\Delta G_{CS}$ values of PyOH–Pyr and PyOH–MePyr systems estimated without taking into account the effect of the hydrogen-bonding interaction are more negative (-1.5 eV) than those of AP–Pyr and AP–MePyr systems, a little greater proton shift in the hydrogen bond is possible,

TABLE II

Decay Times (τ_d) of the LE State of the Hydrogen-Bonding Complexes (D*-H $\cdots$ A) and the Free Energy Gap $(-\Delta G_{CS})$ for the (D*-H $\cdots$ A) $\rightarrow$ (D$^+$-H $\cdots$ A$^-$) Reaction of PyOH–Pyr and PyOH–MePyr Systems in Hexane Solutions

Acceptor	τ_d (ps)	$-\Delta G_{CS}$ (eV)
Pyr	8.8 ± 1.0	-1.46
2MePyr	9.6 ± 1.1	-1.50
4MePyr	14.2 ± 0.8	-1.56
2,6DiMePyr	11.1 ± 1.0	-1.55

owing to the larger proton acidity of PyOH compared with AP, which will make the actual $-\Delta G_{CS}$ values approximately the same as those of the AP–Pyr and AP–MePyr systems. After CS by such a mechanism, an ultrafast and much larger scale proton shift will take place in the PyOH–Pyr system, which will induce a large destabilization of the ground state, leading to ultrafast degradation from the IP state to the ground state.

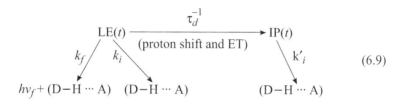

$$ \text{(6.9)} $$

where $\tau_d^{-1} \gg k_f + k_i$, $k_i' \gg \tau_d^{-1}$, and k_i' includes an ultrafast large-scale proton shift, which induces ultrafast nonradiative decay to the ground state.

It should be noted here that in both AP–Pyr and PyOH–Pyr systems, formation of neutral radicals due to complete proton transfer in the IP state has not been observed. In the case of AP–Pyr and AP–MePyr systems, we have observed a clearly long-lived (100 ps–1 ns)(LE $\leftrightarrow$ IP) equilibrium state, where only a moderate amount of proton shift seems to take place. Contrary to this, it might be possible that the large-scale ultrafast proton shift in the IP state of PyOH–Pyr and PyOH–MePyr systems leads to the ultrafast formation of pyrenoxy radicals. The fact that we could not detect such neutral radical formation when practically all PyOH molecules are hydrogen bonded with Pyr in the ground state and LE fluorescence is completely quenched suggests strongly the following reaction mechanism and dynamics. Namely, ultrafast crossover to the ground electronic state coupled with large-scale proton motion, followed by ultrafast relaxation

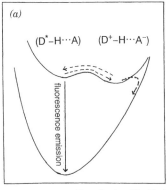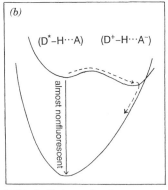

Coordinate of the Proton Shift in the Hydrogen Bond

Figure 9. Conceptual diagrams for ET coupled with the proton shift in the hydrogen bond and nonradiative decay via the ET state in hydrogen-bonded complexes in nonpolar solvent: (*a*) (AP ⋯ Pyr); (*b*) (PyOH ⋯ Pyr).

to the vibrational ground state of the hydrogen bond might take place. These suggestions are shown schematically in Figure 9. Although the theoretical details of this ultrafast nonradiative relaxation process are not very clear at the present stage of investigation, it seems to be an interesting and challenging problem.

B.   Coupled Electron Transfer and Proton Transfer in Pyrene–Secondary Aromatic Amine Exciplexes in Nonpolar Solvents

As a typical example of a special case of hydrogen-bonding interaction between the excited aromatic hydrocarbon and the secondary aromatic amine which forms exciplex in nonpolar solvent [5,28,143], we discuss here the results for the pyrene–N-ethylaniline (NEA) system, which has been studied in detail using picosecond laser photolysis [5,28,143]. Contrary to the pyrene–N,N-dimethylaniline (DMA) exciplex, which shows strong fluorescence with a long lifetime [21], the pyrene–NEA exciplex shows much weaker fluorescence with a much shorter lifetime [5,143] in hexane solution. The exciplex fluorescence yield (ϕ_f) and lifetime (τ_f), radiative and radiationless transition probabilities were obtained as follows [5,21,143]:

$$
\begin{array}{c}
{}^1A^* + H-D \longrightarrow {}^1(A^- \cdots H - D^+)^* \xrightarrow{\quad k_n \quad} \\[2mm]
\swarrow k_{\text{isc}} \qquad \searrow k_f \\[2mm]
{}^3A^* + H-D \qquad\qquad A + H-D + h\nu_f
\end{array}
\tag{6.10}
$$

Pyrene–DMA: $\phi_f = 6.6 \times 10^{-1}$, $\tau_f = 130$ ns, $k_f = 5.1 \times 10^6$ s^{-1},

$\quad\quad\quad\quad k_i = k_{\mathrm{isc}} + k_n = 2.6 \times 10^6$ s^{-1}

Pyrene–NEA: $\phi_f = 5.1 \times 10^{-3}$, $\tau_f = 3.4$ ns, $k_f = 1.5 \times 10^6$ s^{-1},

$\quad\quad\quad\quad k_i = k_{\mathrm{isc}} + k_n = 2.9 \times 10^8$ s^{-1}

It is evident from these experimental results that, although the k_f value of pyrene–NEA is a little smaller than that of pyrene–DMA, the large difference in ϕ_f as well as τ_f between the pyrene–DMA and pyrene–NEA systems can be ascribed to the fact that the k_i value of the latter is 10^2 times larger than that of the former. It has been confirmed that the most important difference in the k_n of these two systems is that there is a high feasibility of proton transfer from the NEA cation to the pyrene anion in the exciplex state, leading to the formation of hydropyrenyl radical, whereas such a process is not possible in the pyrene–DMA exciplex, and also the k_{isc} value of the pyrene–NEA system is much larger than that of the pyrene–DMA exciplex $(k_{\mathrm{isc}} \sim 10^6$ s$^{-1})$. We have also examined the effect of the deuteration of $\overset{\diagdown}{\diagup}$N—H hydrogen of NEA on the decay time τ_f and $k_n = k_{\mathrm{PT}} + k_{\mathrm{ic}}$, and obtained the following results:

$$
\begin{array}{ccc}
& \overset{k_{\mathrm{PT}}}{\longrightarrow} & (\mathrm{A\!-\!H \cdots D} \longrightarrow \mathrm{A\!-\!H} + \mathrm{D} \\
{}^1(\mathrm{A} \cdots \mathrm{H\!-\!D}^+)^* & & \\
k_{\mathrm{isc}} \diagup \quad \Big| k_f \quad \diagdown k_{\mathrm{ic}} & & \\
{}^3\mathrm{A}^* + \mathrm{H\!-\!D} \quad \Big\downarrow \quad \mathrm{A} + \mathrm{H\!-\!D} & & \\
\mathrm{A} + \mathrm{H\!-\!D} + h\nu_f & &
\end{array}
\tag{6.11}
$$

Pyrene–NEA: $\quad k_{\mathrm{PT}} + k_{\mathrm{ic}} = 1.9 \times 10^8$ s^{-1}, $k_{\mathrm{isc}} = 1 \times 10^8$ s^{-1},

$\quad\quad\quad\quad\quad \tau_f = 3.4$ ns, $k_f = 1.5 \times 10^6$ s

Pyrene–NEA-d: $k_{\mathrm{PT}} + k_{\mathrm{ic}} = 0.4 \times 10^8$ s^{-1}, $k_{\mathrm{isc}} = 1 \times 10^8$ s^{-1},

$\quad\quad\quad\quad\quad \tau_f = 6.8$ ns, $k_f = 1.5 \times 10^6$ s

The decrease in $k_{\mathrm{PT}} + k_{\mathrm{ic}}$ by deuteration is probably due to the decreased k_{PT}.

The rapid proton transfer indicates strongly the hydrogen-bonding interaction between $\overset{\diagdown}{\diagup}$N—H of the NEA cation and the π-electron system of pyrene anion at the C_1 position in view of the formation of 1-hydro-1-pyrenyl radical. In other words, this exciplex should have an oblique configuration, contrary to the plane-parallel structures usually assumed (or experimentally proved) in the case of aromatic hydrocarbon-tertiary aromatic amine exciplexes and excited CT complexes and even in nonpolar solvent where stabilization of such a CT or IP state with non-compact or loose structure by solvation with polar solvent is absent. Nevertheless, this non-plane-parallel hydrogen-bonded structure is also supported by the fact that the shape of the absorption band due to pyrene anion part in the pyrene–NEA exciplex in nonpolar solvent is very sharp and rather similar to that of pyrene anion radical in acetonitrile solution, contrary to the much broader band of pyrene/N,N-diethylaniline exciplex in hexane, which seems to have a plane parallel structure [147]. This structure is also favorable for the much larger k_{isc} value of pyrene–NEA exciplex than that of pyrene–DMA exciple, because the spin–orbit interaction matrix element between the singlet CT state and the pyrene local triplet state (^3L_a) $\langle ^1(A^- \cdots D^+)|H_{SO}|(^3A^* \cdots D)\rangle$ is enhanced in the oblique configuration, as confirmed experimetally by using the aromatic hydrocarbon–aromatic amine system, linked by one methylene chain at the amino nitrogen, which seems to have an oblique configuration [5,143,148,149].

Thus, despite the almost perpendicular oblique configuration of pyrene and NEA in the exciplex, hydrogen-bonding interaction between the pyrene π-electron system and the N—H of NEA can produce ET or IP state even in nonpolar solvent, facilitating rapid PT that results in hydropyrenyl radical formation. MO theoretical investigations on such a loose exciplex state in a nonpolar solvent will be an interesting subject of photochemistry. On the other hand, the following facts should be noted here. Contrary to the case of the pyrene–NEA system, neither neutral radical formation as well as enhanced local triplet-state formation nor CT fluorescence emission can be observed in the case of AP–Pyr or PyOH–Pyr hydrogen-bonding systems. This difference might be ascribed to the different structures of the CT states between these two types of systems. In the oblique configuration of the pyrene–NEA exciplex, the direct π-electronic overlap between donor cation and acceptor anion may be considerably larger than that in the coplanar structures of AP–Pyr and PyOH–Pyr systems. Moreover, in the latter, the hydrogen-bonding interaction is not with conjugate π-electron systems but with n-electrons of Pyr. These differences of structures might lead to the larger CT fluorescence probability and also facilitate neutral radical formation in the case of pyrene–NEA. More detailed theoretical studies on potential energy surfaces and reaction dynamics there may be of interest.

Finally, we should note here that such hydrogen-bonding interaction and proton transfer in the pyrene–NEA exciplex becomes difficult in a polar solvent because the solvation of A and D-H disturbs the hydrogen-bonding interaction and in a strongly polar solvent such as acetonitrile, solvated loose IP is formed, where rapid proton transfer and fast local triplet formation are difficult. There is, however, a well-known donor–acceptor system which can easily undergo rapid proton transfer coupled with ET, producing neutral radical pairs in the strongly polar solvent acetonitrile (i.e., benzophenone–amine systems). We discuss these systems on the basis of femtosecond–picosecond laser photolysis studies in the next section.

C. Electron and Proton Transfer Reactions in Excited Benzophenone–Tertiary Aromatic Amine Systems

In benzophenone (BP)–aromatic amine systems, ions in the geminate IP produced by photoinduced ET can interact strongly even in the solvated loose configuration in acetonitrile solutions, producing ketyl radicals by rapid intra-IP proton transfer (PT) while such neutral radical formation by PT in the IP state is possible only in the more strongly interacting exciplex state with a hydrogen bond between the acceptor anion and donor cation in nonpolar solvents in the case of aromatic hydrocarbon (A) and secondary aromatic amine (D-H).

Detailed femtosecond–picosecond laser photolysis studies on the hydrogen abstraction reactions of excited BP with such tertiary aromatic amines as DMA, DEA (N,N-diethylaniline), and MDPA (N-methyldiphenylamine) in acetonitrile solutions have revealed the following reaction mechanisms [95–97]:

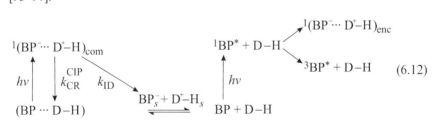

$$
\begin{array}{lll}
{}^{1}(BP^{\bar{\cdot}}\cdots D^{+}\text{-}H)_{com} & & {}^{1}BP^{*}+D\text{-}H \nearrow {}^{1}(BP^{\bar{\cdot}}\cdots D^{+}\text{-}H)_{enc} \\
& & \qquad\qquad\qquad\qquad \searrow {}^{3}BP^{*}+D\text{-}H \\
hv \uparrow \downarrow k_{CR}^{CIP} \; k_{ID} & & hv \uparrow \\
(BP\cdots D\text{-}H) \qquad BP_{s}^{-}+D^{+}\text{-}H_{s} & & BP+D\text{-}H
\end{array}
\qquad (6.12)
$$

$$
{}^{1}(BP^{\bar{\cdot}}\cdots D^{+}\text{-}H)_{enc}
\begin{array}{l}
\xrightarrow{\;k_{ID}\;} BP_{s}^{-}+D^{+}\text{-}H_{s} \\
\xrightarrow{\;k_{PT}\;} BPH+D \\
\xrightarrow[k_{CR}^{LIP}]{} BP+D\text{-}H
\end{array}
\qquad (6.13)
$$

$$^3BP^* + D-H \xrightarrow{k_{CS}} {}^1(BP^{-\cdots} D^{+}-H)_{enc} \begin{array}{c} \nearrow^{k_{ID}} BP_s^- + D^{\pm}-H_s \\ \\ \searrow_{k_{PT}} BPH + D \end{array} \tag{6.14}$$

where $^1(BP^- \cdots D^+\text{-H})_{com}$ and $^{1,3}(BP^- \cdots D^+\text{-H})_{enc}$ represent ^{1}CIP formed by excitation of the ground-state CT complex $(BP \cdots D\text{-H})$ and ^{1,3}LIP formed by CS at encounter between BP* and D-H, respectively.

^{1}CIPs undergo rapid CR transition to the ground state without forma-tion of ketyl radicals by intra-IP PT. Although a small amount of ionic dissociation from ^{1}CIP was observed in the case of DEA and MDPA donors, no ionic dissociation was detectable for the ^{1}CIP with DMA. The k_{CR}^{CIP} values are plotted in Figure 4 against $-\Delta G_{CR}$, together with those of many other ^{1}CIPs formed by excitation of ground-state CT complexes. It is evident that the ^{1}CIPs of BP–tertiary aromatic amine systems constitute the exponential energy gap law together with other ^{1}CIPs. ^{1}BP* undergoes a coupled process of ^{1}LIP formation by ET with the amine and intra-IP PT, leading to the formation of ketyl radicals. ^{1}LIP also undergoes ionic dis-sociation as well as CR deactivation to the ground state. k_{CR}^{LIP} of ^{1}LIP is considerably smaller than the k_{CR}^{CIP} of ^{1}CIP with the same amine and appears to constitute a bell-shaped energy gap law together with many other ^{1}LIPs formed by encounters between donor and acceptor in fluorescence quench-ing reactions, as shown in Figure 4. ^{3}BP* also undergoes a coupled process of ^{3}LIP formation and intra-IP PT, producing a ketyl radical. ^{3}LIP also undergoes ionic dissociation in competition with ketyl radical formation but does not show CR to the ground state, owing to the difference in spin multiplicity. Not only the k_{CR} values indicated in Figure 4 but also values of other rate constants, k_{ID} and k_{PT}, indicated in Eqs. (6.12) to (6.14), have been obtained by detailed computer simulation of femtocecond–picosecond time-resolved transient absorption spectra [95–97].

The quite different behavior of CIP from that of LIP with respect to the intra-IP PT may be ascribed to the difference in their structures. The struc-ture of CIP may not only be more compact than that of LIP, but also relative orientations of donor and acceptor in CIP should be unfavorable for PT and such a structure seems to be maintained during its lifetime of about 100 ps. The different nature of CIP from LIP is also manifested in its energy gap law for the CR process, as discussed above. Another interesting point on the behavior of BP–tertiary aromatic amine geminate IPs is the dependence of k_{PT} and k_{ID} of LIP on the energy gap for CS, $-\Delta G_{CS}$, of the IP formation. The ^{3}LIP especially, shows systematic dependences of k_{PT} and k_{ID} on $-\Delta G_{CS}$ or the oxidation potential of the amine. That is, k_{PT}

TABLE III
Dependencies of the Reaction Rate Constants of ^{3}LIP Between BP and Tertiary Aromatic Amines upon the Oxidation Potential of the Amine (E_{ox}) in Acetonitrile Solutions

	E_{ox} vs. SCE (V)	k_{PT} (s^{-1})	k_{ID} (s^{-1})
$^3(BP^- \cdots MDPA^+)_{enc}$	0.86	8.2×10^9	1.1×10^9
$^3(BP^- \cdots DMA^+)_{enc}$	0.76	5.4×10^9	1.4×10^9
$^3(BP^- \cdots DEA^+)_{enc}$	0.72	7.3×10^8	2.1×10^9
$^3(BP^- \cdots DET^+)_{enc}{}^a$	0.69	$\ll 2 \times 10^8$	(ca. 2×10^9)

^{a}DET, N,N-diethyl-p-toluidine.

decreases significantly and k_{ID} increases a little with increased $-\Delta G_{CS}$ or decreased amine oxidation potential, as indicated in Table III. This result can be understood by considering the theoretically expected donor–acceptor distance distribution in the CS reaction and its dependence on on $-\Delta G_{CS}$, as discussed in detail in Section IV and indicated in Figure 5 [22,23,52–55].

The results calculated for the theoretical CS rate constant (k_{CS}) using reasonable parameter values clearly show a distance-dependent distribution and average distance or distance where k_{CS} shows a peak at each value of the standard free energy gap $-\Delta G_{CS}^0$ increases with increase of $-\Delta G_{CS}^0$. Therefore, the interionic distance of the product LIP immediately after CS should increase with increased free energy gap for CS, in agreement with the results above for the energy gap dependence of k_{PT} and k_{ID} of the ^{3}LIP.

D. Photoinduced Electron Transfer Coupled with Ultrafast Proton Transfer in Benzophenone–Secondary Aromatic Amine Systems

As discussed above, the photochemical reaction mechanisms of BP–tertiary aromatic amine systems in relation to the hydrogen abstraction reaction in acetonitrile solutions can be described by Eqs. (6.12) to (6.14). The reaction mechanism of ^{3}BP*–tertiary aromatic amines in acetonitrile, especially, which has been studied for a long time, has now been established unambiguously by femtosecond–picosecond laser photolysis studies [95–97]. On the other hand, the reaction mechanism of the hydrogen abstraction of ^{3}BP* from secondary amines such as diphenylamine (DPA) in various kinds of solutions could be described by the following reaction scheme according to our investigations [144]:

$$\text{In isooctane: } {}^3BP^* + DH \begin{cases} \nearrow BPH + D \\ \searrow {}^3(BP^- \cdot DH^+)_{enc} \end{cases} \qquad (6.15)$$

$$\text{In acetonitrile:} \quad {}^3\text{BP}^* + \text{DH} \Big\langle {}^{\displaystyle \nearrow \text{BPH} + \text{D}}_{\displaystyle \searrow {}^3(\text{BP}^- \cdots \text{DH}^+)_{\text{enc}} \longrightarrow \text{BP}_s^- + \text{DH}_s^+} \qquad (6.16)$$

where ${}^3(\text{BP}^- \cdot \text{DH}^+)_{\text{enc}}$ is a triplet exciplex, or ${}^3\text{CIP}$, which undergoes no ketyl radical formation by intracomplex PT but only CR decay. The ketyl radical, BPH, seems to be formed very rapidly upon interaction. The reaction proceeds very similarly in acetonitrile, where, however, ${}^3(\text{BP}^- \cdots \text{DH}^+)_{\text{enc}}$ is the ${}^3\text{LIP}$, which undergoes ionic dissociation with a very high quantum yield but no ketyl radical formation by intra-IP PT.

Although the apparent reaction scheme of these systems with secondary aromatic amines as hydrogen donors is quite different from that of the BP–tertiary aromatic amines, ET or CT interaction seems to play the most important role in both cases, because the reaction yield is also very high for secondary amine donors compared with other hydrogen donors with higher oxidation potentials. In this respect, to elucidate more details of the role of ET or CT interaction in the hydrogen abstraction of ${}^3\text{BP}^*$ from secondary aromatic amines, we have performed picosecond laser photolysis studies on the intramolecular reaction process of BP and DPA linked by spacers, where mutual geometry between BP and DPA is more or less restricted in the ground-state static configuration and in the course of a dynamic process in the excited state. We have employed 1-(4-benzophe-noxy)-2-[4-(phenylamino)phenoxy]ethane, 1-(4-benzophenoxy)-3-[4-(phenyl-amino)phenoxy]propane, and 1-(4-benzophenoxy)-8-[4-(phenylamino)-phenoxy]octane (BP-O-$(\text{CH}_2)_n$-O-DPA; BOnOD: $n = 2, 3, 8$) and have examined the effects of the chain length, solvent polarity, and temperature on the reaction mechanisms [l44b,c].

In a benzene solution, the apparent mechanism of hydrogen abstraction was a "direct" process; the rise time of the ketyl radical was identical with the decay time of ${}^3\text{BP}^*$ and no absorption spectrum of the CS state was observed. The decay of ${}^3\text{BP}^*$ appeared to be regulated by dynamic rearrangement of the mutual configuration between ${}^3\text{BP}^*$ and DPA to a more compact configuration favorable for the hydrogen abstraction. Although the formation of a small amount of triplet exciplex or ${}^3\text{CIP}$ which did undergo no ketyl radical formation by intracomplex PT was detected for the unlinked system in nonpolar solvent, no such compact IP state was detectable for the linked system. Presumably, owing to the restriction of mutual configurations between ${}^3\text{BP}^*$ and DPA moieties by the spacer, formation of such compact configuration which can undergo CS but cannot produce ketyl radical may be difficult.

In polar solvents such as acetonitrile, the ET, from DPA to the ^{3}BP* moiety took place in a loose configuration producing a IP state detectable with picosecond transient absorption spectral measurements, in addition to the apparent direct hydrogen abstraction from DPA by ^{3}BP*. Two kinds of IP were formed, one of which did undergo ketyl radical formation by intra-IP PT, the other being long-lived without ketyl radical formation. The nature of the long-lived IP might be somewhat similar to the IP of the unlinked system observed in polar solvents [Eq. (6.16)]. The reaction scheme of the linked system in polar solvents can be represented by

$$^3\text{BP}^* - \text{D-H} \rightleftarrows \begin{matrix} \text{(BPH} - \text{D)} \\ \;\;\;^3(\text{BP}^- - \text{D-H}^+) \longrightarrow \text{(BPH} - \text{D)} \\ \text{long-lived IP} \end{matrix} \qquad (6.17)$$

Summarizing the results and discussions above, the photoreduction processes of ^{3}BP* by the secondary aromatic amine (DPA) moiety can be interpreted consistently from the viewpoint that the CS reaction in a loose configuration, which is more feasible for larger energy gap $-\Delta G_{CS}$ and in polar solvents, takes place in competition with dynamic rearrangement processes to a more compact conformation favorable for the hydrogen abstraction reaction and that the mutual distance and orientations between the ^{3}BP* and DPA groups immediately before the reaction are of crucial importance in determining the apparent reaction mechanisms.

Moreover, even in the case of direct hydrogen abstraction of the ^{3}BP* from DPA, it seems quite feasible that immediately after the CS in a conformation with donor and acceptor in close proximity to each other in a favorable orientation for proton transfer, ultrafast PT in the pair will take place, which makes detection of the IP state extremely difficult. This reaction mechanism is similar to that proposed for the strong fluorescence quenching induced by hydrogen-bonding interaction between PyOH* and Pyr, where immediately after photo-induced ET from PyOH* to hydrogen-bonded Pyr assisted by a slight proton shift, ultrafast large-scale proton shift induces ultrafast nonradiative decay to the ground state [Eq. (6.9)]. In the case of the ^{3}BP* $\cdots$ DPA system, however, immediately after ET assisted by the configuration rearrangement, ultrafast PT takes place, producing neutral radicals:

$$^3\text{BP}^* \cdots \text{D-H} \xrightarrow[\substack{\text{configuration rearrangement} \\ \text{and ET}}]{\tau_r^{-1}} \text{IP} \xrightarrow{h_{\text{PT}}} \text{BPH} \cdots \text{D} \qquad (6.18)$$

where $k_{\text{PT}} \gg \tau_r^{-1}$.

It should be noted here that in the interpretations above of the photo-chemical reaction mechanisms of BP–secondary aromatic amine and BP-tertiary aromatic amine systems, the concepts of "loose" and "compact" exciplexes and/or IPs, depending on solvent polarities and energy gaps for CS $-\Delta G_{CS}$, and so on, play extremely important roles. These concepts were proposed many years ago by the present author (N.M.) and proved by means of picosecond–femtosecond laser photolysis studies on exciplex systems and photoinduced CS reactions of linked and unlinked typical donor–acceptor systems in solutions [5,7,21–23,28,52–55,150,151], for which further strong support has been given by recent detailed investigations [35,37–40] (see Sections II to VII).

We have described some comparative studies of electron transfer coupled with proton shift or transfer in some excited hydrogen–bonding complexes between conjugate π-electronic heteroaromatic proton donor and acceptor systems, in aromatic hydrocarbon–aromatic amine exciplexes, and excited benzophenone–aromatic amine systems and discussed underlying reaction mechanisms. Theoretical investigations on the similarities and/or differences among the reaction mechanisms of these systems, including calculations of potential energy surfaces and reaction dynamics, should be an interesting and challenging problem.

VII. PHOTOREACTION DYNAMICS AND MECHANISMS IN PHOTOSYNTHETIC REACTION CENTER MODELS IN RELATION TO EXCIPLEX CHEMISTRY

In relation to photoinduced CS mechanisms in a biological photosynthetic reaction center (RC), investigations on model systems have been and are being made vigorously [152]. Such studies on the dynamics and mechanisms of photoinduced CS in various model systems will be very important to reveal why nature adopted the present biological RC systems. Nevertheless, the realization of the model systems, containing chromophores analogous to chlorophylls and acceptors such as quinones, which can undergo such ultra-fast and very high efficiency CS processes as those in biological systems is a difficult problem. In this respect it is of crucial importance to elucidate the mechanisms of photoinduced CS, CR of the product IP state, and CSH (charge shift) reactions in those biomimetic chromophores in order to design RC models.

Although the porphyrin (P)–quinone (Q) systems were studied quite frequently, it is not possible to produce a long-lived IP state by using such a dyad, because of the rapid CR decay of the IP state. Especially in the case of the P–Q system in polar solutions, the rate of CR decay of the IP state is comparable with or larger than that of the photoinduced CS

[45,87,153,154b], because the $-\Delta G_{CR}$ value of the P–Q system is around 1.5 eV in very polar solutions, and this value is very close to the top region of the bell-shaped energy gap dependence of k_{CR} indicated in Figure 4 [87].

Because formation of a long-lived IP state by photoinduced CS with P–Q dyads is not possible, as discussed above, nature utilized multistep ET in multichromophoric systems. In this respect we made a first examination of a two-step ET for the system P-(CH$_2$)$_4$–Q–(CH$_2$)$_4$–Q'(P4Q4Q'), where P is etioporphyrin, Q benzoquinone, and Q' trichlorobenzoquinone [154]. The lifetime of the IP state of this system (300–400 ps in dioxane and THF) was longer than that of the P4Q dyad (ca. 100 ps) due to more extensive CS by the two-step ET mechanism [154]. Nevertheless, formation of a very long-lived IP state was not possible, due to the flexibility of the methylene chain spacer. Since then, numerous studies have been made on multistep ET systems held by more rigid spacers [152]. In the following we discuss experimental results on some triad systems [155–157] which are indicated in Figures 10 to 12. In these compounds, the Im group shows a very characteristic sharp absorption band in the visible region when accepting an electron from donor porphyrins, which is very convenient for the kinetic analysis of time-resolved transient absorption spectra.

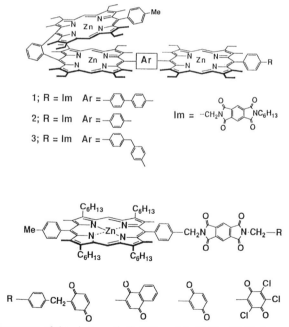

Figure 10. Structures of the photosynthetic RC models (ZnP)$_2$–(ZnP)–Im and (ZnP)–Im–Q.

Formation of the long-lived IP state with lifetime of 2.5 μs has been confirmed in the case of the a $(ZnP)_2–(ZnP)$–Im system with a biphenyl spacer (Fig. 10) in THF solution [155]. Detailed picosecond time-resolved studies on this system have revealed the path of long-lived IP state formation as follows [155]:

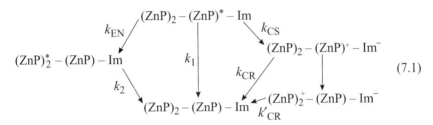

$$\tag{7.1}$$

Namely, $(ZnP)_2^*–(ZnP)$–Im produced by excitation energy transfer cannot undergo CS but is simply deactivated. In addition, k_{CR} from the $(ZnP)_2$–$(ZnP)^+$–Im$^-$ state is larger than k_{CSH}. Accordingly, the quantum yield of the final CS state $(ZnP)_2^+$–(ZnP)–Im$^-$ is small (ca. 0.1). Moreover, the mode of photoinduced multistep CS in this system is different from that of the bacterial photosynthetic RC:

$$(BChl)_2^*–(BChl)–(BPheo)–Q \rightarrow (BChl)_2{}^+–(BChl)^-–(BPheo)–Q \rightarrow$$

$$(BChl)_2{}^+–(BChl)–(BPheo)^-–Q \rightarrow (BChl)_2{}^+–(BChl)–(BPheo)–Q^-$$

where photoinduced ET takes place from a special dimer to (BPheo) via (BChl) and further shift to Q. In this respect, studies on the (ZnP)–Im–Q system (Fig. 10) seem to be important. The photoinduced CS in this system has been confirmed to take place by a reaction mode similar to that of the bacterial RC:

$$(ZnP)^*–Im–Q \xrightarrow{k_{CS}} (ZnP)^+–Im^-–Q \xrightarrow{k_{CSH}} (ZnP)^+–Im–Q^-$$

By employing substituted quinone with a high electron affinity, such as trichloroquinone, and adjusting the $k_{CSH} \sim -\Delta G_{CSH}$ relation near its peak, practically 100% quantum yield of the final CS state formation was attained in THF solution [156]. However, due to the rather short distance between $(ZnP)^+$ and the Q$^-$ group in the final CS state, its lifetime was a few nanoseconds. Moreover, the initial CS in this system is not between porphyrin chromophores, as in the case of the biological RC, but between porphyrin and Im.

On the other hand, we have succeeded recently in realizing the long-lived IP state formation with high efficiency (0.8–0.9) via the photoinduced CS between metal porphyrin dimer and free-base porphyrin followed by the shift of electron to Im:ZnDP*–HP–Im → ZnDP$^+$–HP$^-$–Im → ZnDP$^+$–HP–Im$^-$ [157a], and that between metal–methylenechlorin (ZC) and porphyrin (HP or ZP), followed by the shift of electron to Im: ZC*–HP–I → ZC$^+$–HP$^-$–I → ZC$^+$–HP–I$^-$ or ZC*–ZP–I → ZC$^+$–ZP$^-$–I → ZC$^+$–ZP–I$^-$ [157b] (Figs. 11 and 12). The reaction mode of these systems is very similar to that of bacterial RC.

By using compounds 1 and 2 and their components 3 to 7 in Figure 11, mechanisms of photoinduced CS and formation of the long-lived CS state in these systems have been elucidated [157a]. Namely, by means of picosecond laser photolysis studies on THF solutions of compound 1, it has been confirmed that immediately after excitation, photoinduced CS takes place with $k_{CS} \geq 5 \times 10^{10}\,s^{-1}$, giving the product ZnDP$^+$–HP$^-$–Im followed by CSH with $k_{CSH} = 4.1 \times 10^9\,s^{-1}$, leading to a long-lived ($\tau \sim 280\,ns$) CS state ZnDP$^+$–HP–Im$^-$ with quantum yield > 0.8. Similar reaction processes leading to a long-lived CS state also takes place in the case of compound 2 in THF solution. The distance between porphyrins in this case, however, is longer, due to the biphenyl spacer, Ar, between them, which gives a smaller tunneling matrix element, leading to a smaller initial CS rate constant, $k_{CS} = 8.3 \times 10^9\,s^{-1}$. Nevertheless, because the k_{CSH} value ($4.6 \times 10^9\,s^{-1}$) is almost the same as in compound 1, the quantum yield of the ZnDP$^+$–HP-Im$^-$ state (ca. 0.79) is almost the same as in compound 1. Moreover, because the distance between ZnDP$^+$ and Im$^-$ in the final CS state is longer in compound 2 than in compound 1, its lifetime becomes considerably longer (ca. 2.1 μs). Thus in the ZnDP–HP–Im system, not only are the component chromophores and their arrangements analogous to those of the bacterial RC, but its initial photoinduced CS and the CSH succeeding CS proceed smoothly along the steps of free energy gaps $-\Delta G_{CS} \sim 0.4\,eV$ and $-\Delta G_{CSH} \sim 0.1\,eV$, giving a long-lived CS state with very high yield.

Furthermore, by means of picosecond laser photolysis studies on the triad compounds and their component chromophore systems in Figure 12, some important results on the mechanisms of CS processes have been obtained [157b]. Because excitation energies of the S_1 state of ZC, HP, and ZP are 1.94, 1.96, and 2.13 eV, respectively, owing to the ultrafast energy transfer, excitation energy is trapped at ZC, where the initial CS starts. This is analogous to ultrafast excitation energy transfer from antenna chlorophylls to the RC special dimer, where the photoinduced CS begins. We discuss next the reaction mechanisms in THF solutions of ZC–HP–I and ZC–ZP–I in Figure 12 studied by means of ultrafast laser spectroscopy.

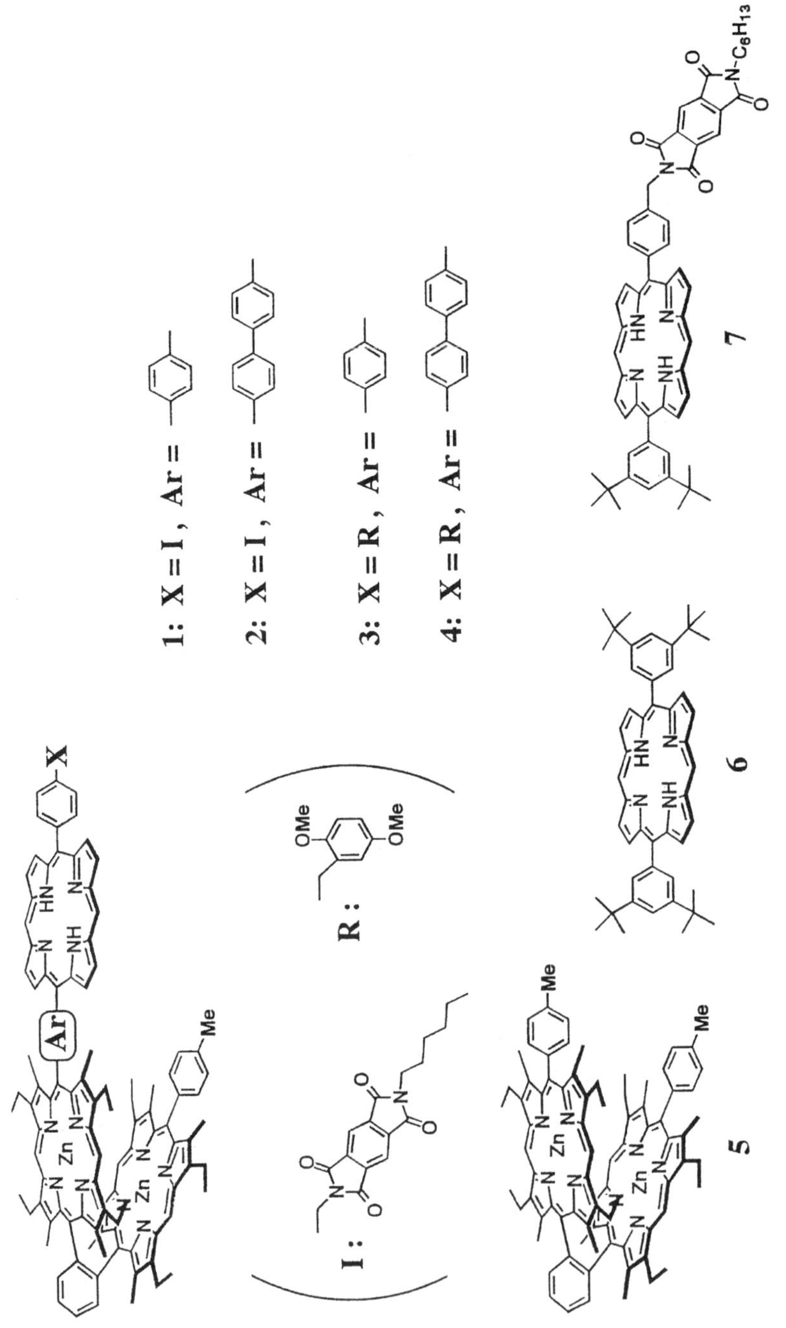

Figure 11. Structure of the photosynthetic RC models ZnDP–HP–Im and related component systems.

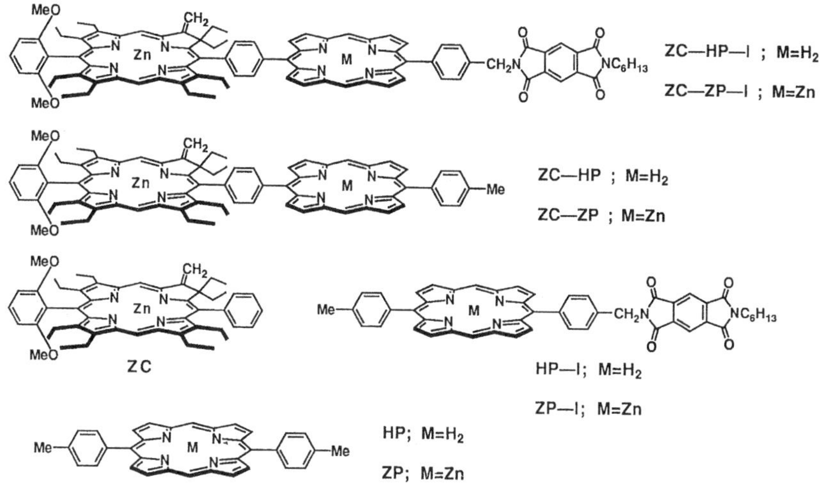

Figure 12. Structure of the photosynthetics RC models ZC–HP–I, ZC–ZP–I, and related component systems

In the case of ZC–HP–I, the reaction,

$$\text{ZC*–HP–I} \xrightarrow{k_{CS}} \text{ZC}^+\text{–HP}^-\text{–I} \xrightarrow{k_{CSH}} \text{ZC}^+\text{–HP–I}^-$$

proceeds smoothly with $-\Delta G_{CS} = 0.23\,\text{eV}$, $k_{CS} = 4.9 \times 10^{10}\,\text{s}^{-1}$ and $-\Delta G_{CSH} = 0.22\,\text{eV}$, $k_{CSH} = 5.8 \times 10^9\,\text{s}^{-1}$, leading to the final CS state (lifetime $\sim$ several 100 ns) with a quantum yield of about 0.7. On the other hand, as indicated in Figure 13, it is necessary to consider not only the k_{CS} but also the back reaction k_{ret} in the case of the ZC–ZP–I. Moreover, since k_{CSH} seems to be considerably larger than k_{CS}, detection of the intermediate state $\text{ZC}^+\text{–ZP}^-\text{–I}$ appears to be difficult. Nevertheless, a weak transient absorption band due to ZP$^-$ of the intermediate state could be detected, and on the basis of this absorption and according to the reaction scheme of Figure 3, the results observed could be interpreted consistently with the following rate constants and the energy gaps for the reaction processes: $k_{CS} = 4.5 \times 10^9\,\text{s}^{-1}$, $-\Delta G_{CS} = -0.03\,\text{eV}$, $k_{CSH} = 4.0 \times 10^{10}\,\text{s}^{-1}$, $\Delta G_{CSH} = 0.48\,\text{eV}$, $k_{ret} = 1.3 \times 10^{10}\,\text{s}^{-1}$, $-\Delta G_{ret} = 0.03\,\text{eV}$, $k_{CR1} = 2.7 \times 10^8\,\text{s}^{-1}$, $-\Delta G_{CR1} = 1.97\,\text{eV}$, from which the quantum yield of the final CS state (with lifetime of several 100 ns) has been obtained to be 0.9.

When the energy of the intermediate state of the reaction is higher than the initial state and the final state is sufficiently lower than the intermediate state with the appropriate energy gap value leading to a very fast reaction from the intermediate to the final state, detection of the transient intermedi-

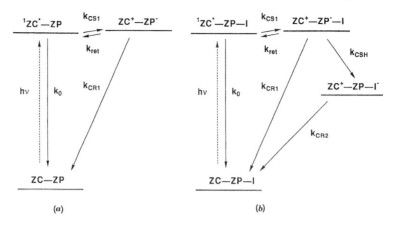

Figure 13. Reaction scheme of ZC–ZP (*a*) and ZC–ZP–I (*b*) in THF solution.

ate state becomes very difficult, even by means of ultrafast laser spectroscopy. When a long-range photoinduced ET is detected in a composite system despite the absence of a detectable intermediate state, there is a possibility that the superexchange mechanism is working, where the tunneling matrix element is given by the well-known equation

$$V_s = V_{mi} \cdot V_{nm}/\Delta E_{mi} \tag{7.2}$$

where i and n are the initial and final states and m is the virtual state, respectively, V_{mi} and V_{nm} are respective electronic interaction matrix elements between states i and m and m and n, and ΔE_{mi} is the energy difference between states i and m. In the example above of the ZC-ZP-I system in THF solution, the superexchange matrix element V_s might be fairly large because ΔE_{mi} is small (0.03 eV). Actually, however, the experimental results can be interpreted consistently with the *real* intermediate state ZC^+-ZP^--I.

For the initial CS processes in the photosynthetic RC of bacteria,

$$(BChI)_2^*-(BChl)-(BPheo)-Q \rightarrow (BChl)_2{}^+-(BChl)^--(BPheo)-Q \rightarrow$$
$$(BChl))_2^*-(BChl)-(BPheo)^--Q \rightarrow$$

vigorous studies were made on the problem of whether the (BChl)⁻ state is virtual or real [158]. Detailed subpicosecond spectroscopic studies on this problem indicate strongly that the (BChl) state is real [159]. Nevertheless, the superexchange seems to be the overwhelming important mechanisms at sufficiently low temperatures. In the superexchange mechanism, the i, m, and n states must coupled coherently. In order that the coherent coupling is not disturbed by environmental perturbation at higher temperature, the coupling must be fairly strong. At any rate, construction of artificial molecular devices using ultrafast, long-range electron transfer with coherent mechanism should be an interesting and challenging problem.

REFERENCES

1. Th. Förster, *Fluoreszenz Organischer Verbindungen*, Vandenhoeck & Ruprecht, Göttingen, Germany, 1951.

2. N. Mataga and T. Kubota, *Molecular Interactions and Electronic Spectra*, Marcel Dekker, New York, 1970.

3. N. Mataga and M. Ottolenghi, in *Molecular Association*, Vol. 2, R. Foster, ed., Academic Press, London, 1979 p. 1.

4. N. Mataga, in *Molecular Interactions*, Vol. 2, H. Ratajczak, and W. J. Orville-Thomas, eds., Wiley, New York, 1981, p. 509.

5. N. Mataga, *Pure Appl. Chem.* **56**, 1255 (1984).

6. P. F. Barbara and W. Jarzeba, *Adv. Photochem.* **15**, 1 (1990).

7. N. Mataga, *Pure Appl. Chem.* **65**, 1605 (1993).

8. M. Maroncelli, J. McInnis, and G R. Fleming, *Science*, **243**, 1674 (1989).

9. I. Rips, J. Klafter, and J. Jortner, *J. Phys. Chem.* **94**, 8557 (1990).

10. P. F. Barbara, T. J. Kang, W. Jarzeba, and T. Fonseca, in *Perspectives in Photosynthesis*, J. Jortner, and B. Pullman, eds., Kluwer, Dordrecht, The Netherlands, 1990, p. 273.

11. K. Tominaga, G. C. Walker, W. Jarzeba, and P. F. Barbara, *J. Phys. Chem.* **95**, 10475 (1991).

12. K. Tominaga, G. C. Walker, T. J. Kang, and P. F. Barbara, *J. Phys.Chem*, **95**, 10485 (1991).

13. E. Lippert, *Z. Naturforsch.* **10a**, 5411 (1955).

14. N. Mataga, Y. Kaifu, and M. Koizumi, *Bull. Chem. Soc. Jpn.* **28**, 690 (1955).

15. B. Bagchi, D. W. Oxtoby, and G. R. Fleming, *Chem. Phys.* **86**, 257 (1984).

16. G. van der Zwan and J. T. Hynes, *J. Phys. Chem.* **89**, 4181 (1985).

17. R. A. Marcus, *J. Chem. Phys.* **24**, 966 (1956).

18. R. A. Marcus, *J. Chem. Phys.* **24**, 979 (1956).

19. R.A. Marcus, *Annu. Rev. Phys. Chem.* **15**, 155 (1964).

20. N. Mataga, T. Okada, and N. Yamamoto, *Bull. Chem. Soc. Jpn.* **39**, 2562 (1966).

21. N. Mataga, T. Okada, and N. Yamamoto, *Chem. Phys. Lett.* **1**, 119 (1967).

22. N. Mataga and H. Miyasaka, *Prog. React. Kinet.* **19**, 317 (1994), and references cited thererin.

23. T. Kakitani, N. Matsuda, A. Yoshimori, and N. Mataga, *Prog. React. Kinet.* **20**, 347 (1995), and references cited thererin.

24. H. Knibbe, K. Röllig, F. P. Schäfer, and A. Weller, *J. Chem. Phys.* **47**, 1184 (1967).

25. N. Mataga, T. Okada, H. Masuhara, N. Nakashima, Y. Sakata, and S. Misumi, *J. Lumin.* **12/13**, 159 (1976).

26. S. Masaki, T. Okada, N. Mataga, Y. Sakata, and S. Misumi, *Bull. Chem. Soc. Jpn.* **49**, 1277 (1976).

27. T. Okada, T. Saito, N. Mataga, Y. Sakata, and S. Misumi, *Bull. Chem. Soc. Jpn.* **50**, 331 (1977).

28. N. Mataga, M. Migita, and T. Nishimura, *J. Mol. Struct.* **47**, 199 (1978).

29. T. Okada, M. Migita, N. Mataga, Y. Sakata, and S. Misumi, *J. Am. Chem. Soc.* **103**, 4715 (1981).

30. M. Migita, T. Okada, N. Mataga, Y. Sakata, S. Misumi, N. Nakashima, and K. Yoshihara, *Bull. Chem. Soc. Jpn.* **54**, 3304 (1981).

31. N. Mataga, S. Nishikawa, T. Asahi, and T. Okada, *J. Phys. Chem* **94**, 1443 (1990).

32. R. D. Stramel, C. Nguyen, S. E. Webber, and M. A. J. Rodgers, *J. Phys. Chem.* **92**, 2934 (1988).

33. J. A. Delaire, M. Sanquer-Barrie, and S. E. Webber, *J. Phys. Chem.* **92**, 1252 (1988).

34. Y. Morishima, Y. Tominaga, M. Kamachi, T. Okada, Y. Hirat, and N. Mataga, *J. Phys. Chem.* **95**, 6027 (1991).

35. F. D. Lewis, G. D. Reddy, D. M. Bassani, S. Schneider, and M. Gahr, *J. Am. Chem. Soc.* **116**, 597 (1994), and references cited therein.

36. F. D. Lewis and B. E. Cohen, *J. Phys. Chem.* **98**, 10591 (1994), and references cited therein.

37. J. W. Verhoeven, T. Scherer, and R J. Willemse, *Pure Appl. Chem.* **65**, 1717 (1993).

38. W. Schuddeboom, T. Scherer, J. M. Warman, and J. W.Verhoeven, *J. Phys. Chem.* **97**, 13092 (1993).

39. I. H. M. van Stokkum, T. Scherer, A. M. Brouwer, and J. W. Verhoeven, *J. Phys. Chem.* **98**, 852 (1994).

40. T. Scherer, I. H. M. van Stokkum, A. M. Brouwer, and J. W. Verhoeven, *J. Phys. Chem.* **98**, 10539 (1994).

41. J. W. Verhoeven, *Pure Appl. Chem.* **62**, 1585 (1990).

42. J. Ulstrup and J. Jortner, *J. Chem. Phys.* **63**, 4358 (1975).

43. J. Jortner, *J. Chem. Phys.* **64**, 4860 (1976).

44. M. Bixon and J. Jortner, *J. Phys. Chem.* **95**, 1941 (1991).

45. T. Asahi, M. Ohkohchi, R. Matsusaka, N. Mataga, R. P. Zhang, A. Osuka, and K. Maruyama, *J. Am. Chem. Soc.* **115**, 5665 (1993).

46. B.-C. Perng, M. D. Newton, F. O. Raineri, and H. L. Friedman, *J. Chem. Phys.* **104**, 7153, 7177 (1996).

47. Y. Hirata, Y. Kanda, and N. Mataga, *J. Phys. Chem.* **87**, 1659 (1983).

48. N. Kh. Petrov, A. I. Shushin, and E. L. Frankevich, *Chem. Phys. Lett.* **82**, 339 (1981).

49. D. Rehm and A. Weller, *Ber. Bunsenges. Phys. Chem.* **73**, 834 (1969).

50. D. Rehm and A. Weller, *Isr. J. Chem.* **8**, 259 (1970).

51. R A. Marcus and P. Siders, *J. Phys. Chem.* **86**, 622 (1982).

52. T. Kakitani, A. Yoshimori, and N. Mataga, in *Electron Transfer in Inorganic, Organic, and Biological Systems*, J. R. Bolton, N. Mataga, and G. McLendon, eds., Advances in Chemistry, Vol. 228, American Chemical Society, Washington, D.C. 1991, Chap 4.

53. T. Kakitani, A. Yoshimori, and N. Mataga, *J. Phys. Chem.* **96**, 5385 (1992).

54. N. Matsuda, T. Kakitani, T. Denda, and N. Mataga, *Chem. Phys.* **190**, 83 (1995).

55. M. Tachiya and S. Murata, *J. Phys. Chem.* **96**, 8441 (1992).

56. E. Lippert, W. Luder, and F. Moll, *Spectrochim. Acta* **10**, 858 (1959).

57. E. Lippert, W. Luder, H. Moll, W. Nagele, H. Boos, H. Prigge, and J. Seibold-Blankenstein, *Angew. Chem.* **73**, 695 (1961).

58. K. Rotkiewicz, K. H. Grellmann, and Z. R. Grabowski, *Chem. Phys. Lett.* **19**, 315 (1973).

59. Z. R. Grabowski, K. Rotkiewicz, A. Siemiarczuk, D. J. Cowley, and W. Baumann, *Nouv. J. Chim.* **3**, 443 (1979).

60. Z.R. Grabowski, *Pure Appl. Chem.* **65**, 1751 (1993), and references cited therein.

61. T. Okada, T. Fujita, and N. Mataga, *Z. Phys. Chem. NF,* **101**, 57 (1976).

62. W. Baumann, F. Petzke, and K. D. Loosen, *Z. Naturforsch.* **34a**, 1070 (1979).

63. W. Baumann and H. Bischof, *J. Mol. Struct.* **84**, 181 (1982).

64. N. Detzer, W. Baumann, B. Schwager, J. C. Fröhling, and C. Brittinger, *Z. Naturforsch.* **42a**, 395 (1987).

65. T. Okada, N. Mataga, W. Baumann, and A. Siemiarczuk, *J. Phys. Chem.* **91**, 4490 (1987).

66. W. Baumann, B. Schwager, N. Detzer, T. Okada, and N. Mataga, *J. Phys. Chem.* **92**, 3742 (1988).

67. J. Herbich and A. Kapturkiewicz, *Chem. Phys.* **158**, 143 (1991).

68. J. Herbich and A. Kapturkiewicz, *Chem. Phys.* **170**, 221 (1993).

69. F. Schneider and E. Lippert, *Ber. Bunsenges. Phys. Chem.* **72**, 1155 (1968).

70. F. Schneider and E. Lippert, *Ber. Bunsenges. Phys. Chem.* **74**, 624 (1970).

71. N. Nakashima, M. Murakawa, and N. Mataga, *Bull. Chem. Soc. Jpn* **49**, 854 (1976).

72. N. Mataga, H. Yao, T. Okada, and W. Rettig, *J. Phys. Chem.* **93**, 3383 (1989).

73. R. Wortmann, K. Elich, S. Lebus, and W. Liptay, *J. Chem. Phys.* **95**, 6371 (1991).

74. R. Wortmann, S. Lebus, E. Karsten, S. Assar, N. Detzer, and W. Liptay, *Chem. Phys. Lett.* **198**, 220 (1992).

75. M. J. Smith, K. Krogh-Jespersen, and R. M. Levy, *Chem. Phys.* **171**, 97 (1993).

76. N. Mataga, S. Nishikawa, and T. Okada, *Chem. Phys. Lett.* **257**, 327 (1996).

77. T. Okada, S. Nishikawa, K. Kanaji, and N. Mataga, in *Ultrafast Phenomena* Vol. VII, C. B. Harris, E. P. Ippen, G. A. Mourou, and A. H. Zewail, eds., Springer-Verlag, Berlin, 1990, p. 397.

78. N. Mataga, H. Yao, and T. Okada, *Tetrahedron* **45**, 4683 (1989).

79. H. Yao, T. Okada, and N. Mataga, *J. Phys. Chem.* **93**, 7388 (1989).

80. H. Miyasaka, S. Ojima, and N. Mataga, *J. Phys. Chem.* **93**, 3380 (1989).

81. (a) S. Ojima, H. Miyasaka, and N. Mataga, *J. Phys. Chem.* **94**, 4147 (1990); (b) **94**, 5834 (1990); (c) **94**, 7534 (1990).

82. T. Asahi and N. Mataga, *J. Phys. Chem.* **93**, 6575 (1989).

83. T. Asahi and N. Mataga, *J. Phys. Chem.* **95**, 1956 (1991).

84. T. Asahi, M. Ohkohchi, and N. Mataga, *J. Phys. Chem.* **97**, 13132 (1993).

85. T. Asahi, N. Mataga, Y. Takahashi, and T. Miyashi, *Chem. Phys. Lett* **171**, 309 (1990).

86. N. Mataga, Y. Kanda, and T. Okada, *J. Phys. Chem.* **90**, 3880 (1986).

87. N. Mataga, T. Asahi, Y. Kanda, T. Okada, and T. Kakitani, *Chem. Phys.* **127**, 249 (1988).

88. F. C. Collins and G. E. Kimball, *J. Colloid Sci.* **4**, 425 (1949).

89. R. M. Noyes, *Prog. React. Kinet.* **1**, 129 (1961).

90. S. Nishikawa, T. Asahi, T. Okada, N. Mataga, and T. Kakitani, *Chem. Phys. Lett.* **185**, 237 (1991).

91. I. R. Gould, D. Ege, S. I. Mattes, and S. Farid, *J. Am. Chem. Soc.* **109**, 3794 (1987).

92. I. R. Gould, J. E. Moser, D. Ege, and S. Farid, *J. Am. Chem. Soc.* **110**, 1991 (1988).

93. M. R. Wasielewski, M. P. Niemczyk, W. A. Svec, and E. B. Pewitt, *J. Am. Chem. Soc.* **107**, 1080 (1985).

94. R. J. Harrison, B. Pearce, G. S. Beddard J. A. Cowan, and J. K. M. Sanders, *Chem. Phys.* **116**, 429 (1987).

95. H. Miyasaka, K. Morita, K. Kamada, and N. Mataga, *Bull. Chem. Soc. Jpn.* **63**, 385 (1990).

96. H. Miyasaka, K Morita, T. Nagata, M. Kiri, and N. Mataga, *Bull. Chem. Soc. Jpn.* **64**, 3229 (1991).

97. H. Miyasaka, T. Nagata, M. Kiri, and N. Mataga, *J. Phys. Chem.* **96**, 8060 (1992).

98. N. Mataga, *Bull. Chem. Soc. Jpn.* **43**, 3623 (1970).

99. N. Mataga, Y. Kanda, T. Asahi, H. Miyasaka, T. Okada, and T. Kakitani, *Chem. Phys.* **127**, 239 (1988).

100. (a) T. Kakitani and N. Mataga, *Chem. Phys.* **93**, 381 (1985); (b) *J. Phys. Chem.* **89**, 8 (1985); **89**, 4752 (1985); (c) **90**, 993 (1986).

101. (a) T. Kakitani and N. Mataga, *Chem. Phys. Lett.* **124**, 437 (1986); (b) Y. Hatano, M. Saito, T. Kakitani, and N. Mataga, *J. Phys. Chem.* **92**, 1008 (1988); (c) Y. Enomoto, T. Kakitani, A. Yoshimori, and Y. Hatano, *Chem. Phys. Lett.* **186**, 366 (1991).

102. J. K. Hwang and A. Warshel, *J. Am. Chem. Soc.* **109**, 715 (1987).

103. R. A. Kuharski, J. S. Bader, D. Chandler, M. Sprik, M. L. Klein, and R. W. Impey, *J. Chem. Phys.* **89**, 3248 (1988).

104. E. A. Carter and J. Hynes, *J. Phys. Chem.* **93**, 2184 (1989).

105. A. Yoshimori, T. Kakitani, Y. Enomoto, and N. Mataga, *J. Phys. Chem.* **93**, 8316 (1989).

106. G. King and A. Warshel, *J. Chem. Phys.* **93**, 8682 (1990).

107. Y. Hatano, T. Kakitani, Y. Enomoto, and A. Yoshimori, *Mol. Simul.* **6**, 191 (1991).

108. A. Warshel, Z. T. Chu, and W. W. Parson, *Science* **246**, 112 (1989).

109. M. Saito, T. Kakitani, and Y. Hatano, in *Dynamics and Mechanisms of Photoinduced Electron Transfer and Related Phenomena*, N. Mataga, T. Okada, and H. Masuhara, eds., Elsevier, New York, 1992, p. 131.

110. R. A. Marcus, *Discuss, Faraday Soc.* **29**, 21 (1960).

111. A. Warshel, *J. Phys. Chem.* **86**, 2218 (1982).

112. D. E. Calef and P. G. Wolynes, *J. Chem, Phys.* **78**, 470 (1983).

113. A. Warshel and J. K. Hwang, *J. Chem. Phys.* **84**, 4388 (1986).

114. M. Tachiya, *J. Phys. Chem.* **93**, 7050 (1989).

115. M. R. Wasielewsky, G. L. Gaines III, M. P. O'Neil, W. A. Svec, M. P. Niemczyk, L. Prodi, and D. Gosztola, in *Dynamics and Mechanisms of Photoinduced Electron Transfer and Related Phenomena*, N. Mataga, T. Okada, and H. Masuhara, eds., Elsevier, New York, 1992, p. 87.

116. For example, (a) A. I. Burshtein, A. A.: Zharikov, and N. V. Shokhirev, *J. Phys. Chem.* **96**, 1951 (1992); (b) A. Yoshimori, K. Watanabe, and T. Kakitani, *Chem. Phys.* **201**, 35 (1995); (c) A. I. Burshtein and E. Krissinel, *J. Phys. Chem.* **100**, 3005 (1996).

117. R. S. Mulliken and W. B. Person, *Molecular Complexes*, Wiley, New York, 1969.

118. R. Foster, *Organic Charge Transfer Complexes*, Academic Press, London, 1969.

119. J. Czekalla and K O. Meyer, *Z. Phys. Chem. (Munich)* **27**, 184 (1961).

120. N. Mataga and Y. Murata, *J. Am. Chem. Soc.* **91**, 3144 (1969).

121. R. Potashnik and M. Ottolenghi, *Chem. Phys. Lett.* **6**, 525 (1970).

122. H. Masuhara and N. Mataga, *Chem. Phys. Lett.* **6**, 608 (1970).

123. H. Segawa, C. Takehara, K. Honda, T. Shimidzu, T. Asahi, and N. Mataga, *J. Phys. Chem.* **96**, 503 (1992).

124. A. C. Benniston, A. Harriman, D. Philp, and F. Stoddart, *J. Am. Chem. Soc.* **115**, 5298 (1993).

125. K. Wynne, C. Galli, and R. M. Hochstrasser, *J. Chem. Phys.* **100**, 4797 (1994).

126. K. Wynne, G. D. Reid, and R. M. Hochstrasser, *J. Chem. Phys.* **105**, 2287 (1996).

127. A. Hormann, W. Jarzeba, and P. F. Barbara, *J. Phys. Chem.* **99**, 2006 (1995).

128. W. Jarzeba, K. Thakur, A. Hormann, and P. F. Barbara, *J. Phys. Chem.* **99**, 2016 (1995).

129. I. R. Gould, D. Noukakis, L. Gomez-Jahn, R. H. Young, J. L. Goodman, and S. Farid, *Chem. Phys.* **176**, 439 (1993).

130. B. R. Arnold, D. Noukakis, S. Farid, J. L. Goodman, and I. R. Gould, *J. Am. Chem. Soc.* **117**, 4399 (1995).

131. H. Miyasaka, S. Kotani, and A. Itaya, *J. Phys. Chem.* **99**, 5757 (1995).

132. (a) R. Kubo and Y. Toyozawa, *Prog. Theor. Phys.* **13**, 160 (1955); (b) R. Englman and J. Jortner, *Mol. Phys.* **18**, 145 (1970); (c) K. F. Freed and J. Jortner, *J. Chem. Phys.* **52**, 6272 (1970).

133. (a) J. Jortner, M. Bixon, H. Heitele, and M. E. Michel-Beyerle, *Chem. Phys. Lett.* **197**, 131 (1992); (b) M. Bixon, J. Jortner, J. Cortes, H. Heitele, and M. E. Michel-Beyerle, *J. Phys. Chem.* **98**, 7289 (1994).

134. (a) T. J. Meyer *Prog. Inorg. Chem.* **30** 389 (1983); (b) P. Chen, R. Duesing, G. Tapolskyj, and T. T. Meyer, *J. Am. Chem. Soc.* **111**, 8305 (1989); (c) P. Chen, R. Duesing, D. K. Graff, and T. J. Meyer, *J. Phys. Chem.* **95**, 5850 (1991).

135. K. Kulinowski, I. R. Gould, and A. B. Myers, *J. Phys. Chem.* **99**, 9017 (1995).

136. S. H. Lin, *J. Chem. Phys.* **44**, 3759 (1966).

137. (a) D. B. McQueen and K. S. Schanze, *J. Am. Chem Soc.* **113**, 7470 (1991); (b) R. G. Alden, W. D. Chang, and S. H. Lin, *Chem. Phys. Lett.* **194**, 318 (1992); (c) R. Islammpour, R. G. Alden, G. Y. C. Wu, and S. H. Lin, *J. Phys. Chem.* **97**, 6793 (1993).

138. I. R. Gould, D. Noukakis, L. Gomez-Jahn, J. L. Goodman, and S. Farid, *J. Am. Chem. Soc.* **115**, 4405 (1993).

139. H. Miyasaka, S. Kotani, A. Itaya, G. Schweitzer, F. C. DeSchryvers, and N. Mataga, *J. Phys. Chem.* **101**, 7978 (1997).

140. (a) N. Mataga and S. Tsuno, *Naturwissenschaften* **10**, 305 (1956); (b) *Bull. Chem. Soc. Jpn.* **30**, 711 (1957).

141. (a) M. M. Martin, N. Ikeda, T. Okada, and N. Mataga, *J. Phys. Chem.* **86**, 4148 (1982); (b) N. Ikeda, H. Miyasaka, T. Okada, and N. Mataga, *J. Am. Chem. Soc.* **105**, 5206 (1983).

142. (a) H. Miyasaka, A. Tabata, K. Kamada, and N. Mataga, *J. Am. Chem. Soc.* **115**, 7335 (1993); (b) H. Miyasaka, A. Tabata, S. Ojima, N. Ikeda, and N. Mataga, *J. Phys. Chem.* **97**, 8222 (1993).

143. T. Okada, I. Karaki, and N. Mataga, *J. Am. Chem. Soc.* **104**, 7191 (1982).

144. (a) H. Miyasaka and N. Mataga, *Bull. Chem. Soc. Jpn.* **63**, 131 (1990); (b) H. Miyasaka, M. Kiri, K. Morita, N. Mataga, and Y. Tanimoto, *Chem. Phys. Lett.* **199**, 21 (1992); (c) *Bull. Chem. Soc. Jpn.* **68**, 1569 (1995).

145. M. M. Martin, D. Grand, N. Ikeda, T. Okada, and N. Mataga, *J. Phys. Chem.* **88**, 167 (1984).

146. H. Tanaka and K. Nishimoto, *J. Phys. Chem.* **88**, 1052 (1984).

147. H. Fujiwara, N. Nakashima, and N. Mataga, *Chem. Phys. Lett.* **47**, 185 (1977).

148. N. Mataga, *Radiat. Phys. Chem.* **21**, 83 (1983).

149. T. Okada, I. Karaki, E. Matsuzawa, N. Mataga, Y. Sakata, and S. Misumi, *J. Phys. Chem.* **85**, 3957 (1981).

150. N. Mataga, in *Electron Transfer in Inorganic, Organic, and Biological Systems*, J. R. Bolton, N. Mataga, and G. McLendon, eds., Advances in Chemistry, Vol. 28 American Chemical Society, Washington, D.C., 1991, Chap. 6.

151. N. Mataga, in *Dynamics and Mechanisms of Photoinduced Electron Transfer and Related Phenomena*, N. Mataga, T. Okada, and H. Masuhara, eds., Elsevier, New York, 1992, p. 3.

152. See for example; (a) M. R. Wasielewski, *Chem. Rev.* **92**, 435 (1992); (b) D. Gust, T. A. Moore, and A. L. Moore, *Acc. Chem. Res.* **26**, 198 (1993); (c) K. Maruyama, A. Osuka, and N. Mataga, *Pure Appl. Chem.* **66**, 867 (1994), and references, cited therein.

153. (a) N. Mataga, A. Karen, T. Okada, S. Nishitani, N. Kurata, Y. Sakata, and S. Misumi, *J. Am. Chem. Soc.* **106**, 2442 (1984); (b) N. Mataga, A. Karen, T. Okada, S. Nishitani, Y. Sakata, and S. Misumi, *J. Phys. Chem.* **88**, 4650 (1984).

154. (a) S. Nishitani, N. Kurata, Y. Sakata, S. Misumi, A. Karen, T. Okada, and N. Mataga, *J. Am. Chem. Soc.* **105**, 7771 (1983); (b) N. Mataga, A. Karen, T. Okada, S. Nishitani, N. Kurata, Y. Sakata, and S. Misumi, *J. Phys. Chem.* **88**, 5138 (1984).

155. (a) A. Osuka S. Nakajima, K. Maruyama, N. Mataga, and T. Asahi, *Chem. Lett.* 1003 (1991); (b) A. Osuka, S. Nakajima, K. Maruyama, N. Mataga, T. Asahi, I. Yamazaki, Y. Nishimura, T. Ohno, and K. Nozaki, *J. Am. Chem. Soc.* **115**, 4577 (1993).

156. M. Ohkohchi, A. Takahashi, N. Mataga, T. Okada, A. Osuka, Y. Yamada, and K. Maruyama, *J. Am. Chem. Soc.* **115**, 12137 (1993).

157. (a) A. Osuka, S. Nakajima, T. Okada, S. Taniguchi, K. Nozaki, T. Ohno, I. Yamazaki, Y. Nishimura, and N. Mataga, *Angew. Chem. Int. Ed. Engl.* **35**, 92 (1996); (b) A. Osuka, S. Marumo, N. Mataga, S. Taniguci, T. Okada, I. Yamazaki, Y. Nishimura, T. Ohno, and K. Nozaki, *J. Am. Chem. Soc.* **118**, 155 (1996).

158. See, for example, J. Jortner and B. Pulman, eds., *Perspectives in Photosynthesis*, Kluwer, Dordrecht, The Netherlands, 1990.

159. T. Arlt, S. Schmidt, W. Kaiser, C. Lauterwasser, M. Meyer, H. Scheer, and W. Zinth, *Proc. Natl. Acad. Sci. USA* **90**, 11759 (1993).

ELECTRON-TRANSFER TUBES

J. J. REGAN

Beckman Institute, California Institute of Technology, Pasadena, CA 91125

J. N. ONUCHIC

Department of Physics, University of California at San Diego, La Jolla, CA 92093-0319

CONTENTS

Electron Transfer: From Isolated Molecules to Biomolecules, Part Two, edited by Joshua Jortner and M. Bixon. Advances in Chemical Physics Series, Volume 107, series editors I. Prigogine and Stuart A. Rice.
ISBN 0-471-25291-3   © 1999 John Wiley & Sons, Inc.

I. INTRODUCTION

Electron transfer (ET) processes in proteins are characterized by the motion of a single electron between centers of localization (such as the chlorophyll dimer in photosynthetic reaction centers). An electronic donor state D is created by the injection of an electron or by photoexcitation, after which the system makes a radiationless transition to an acceptor state A, resulting in the effective transfer of an electron over several angstroms. The experimental and theoretical understanding of the rate of this process has been the focus of much attention in physics, chemistry, and biology. The remaining articles in this volume review this extensive subject in detail (e.g., [1]); we only review enough here and in the next section to define our terminology.

The ET process can be described be the following single-electron Hamiltonian, used extensively in the generic ET problem (see, e.g. [2–4]):

$$H_{\text{ET}} = H_{\mathbf{Q}} + T_{DA}(\mathbf{Q})\sigma_x + \tfrac{1}{2}[\alpha_D^{\text{eff}}(\mathbf{Q}) + \alpha_A^{\text{eff}}(\mathbf{Q})]\sigma_z + \tfrac{1}{2}[\alpha_D^{\text{eff}}(\mathbf{Q}) - \alpha_A^{\text{eff}}(\mathbf{Q})]\sigma_z \tag{1.1}$$

The Pauli spin-$\tfrac{1}{2}$ matrices σ_x and σ_z provide a convenient notation for any two-state system; here, the spin states *up* and *down* correspond to the D and A states, respectively. $H_{\mathbf{Q}}$ supplies the dynamics for the nuclear coordinates $\mathbf{Q}$, and $\alpha_D^{\text{eff}}(\mathbf{Q})$ [$\alpha_A^{\text{eff}}(\mathbf{Q})$] is the instantaneous energy for the reactants (products) state. T_{DA} is the coupling between the donor and acceptor states, the *tunneling matrix element*. The "effective" designation is discussed in Section II.

This description implies that D and A are electronic states computed for different frozen nuclear configurations in the Born–Oppenheimer approximation. Since electron transfer is a radiationless transition, electron tunneling is dominated by resonant nuclear configurations (nuclear configurations that have $\alpha_D = \alpha_A$). Since the nuclei are much slower than the electrons, it is reasonable to assume that they are frozen as tunneling occurs (Condon approximation). Thus ET is controlled by two main factors: a nuclear one permitting tunneling at resonant configurations (the activated complex), and an electronic one that measures the coupling between the donor and acceptor states at these resonant configurations. If the tunneling matrix element is very small, so that the transfer is in the nonadiabatic limit (as is appropriate for long-distance ET in semirigid systems such as proteins), the ET rate can be written as [5–14]

$$k_{\mathrm{ET}} = \frac{2\pi}{\hbar}\, T_{DA}^2 \rho_{\mathrm{FC}} \qquad (1.2)$$

The Franck–Condon factor ρ_{FC} accounts for the density of nuclear states for resonant D–A configurations.

In this chapter we focus on electronic effects, captured in T_{DA}. The methods described in this paper help one understand how the bridge electronically couples the donor to the acceptor; we do not attempt to compute ρ_{FC} or absolute rates. Nevertheless, even without knowledge of ρ_{FC}, one can use the methods described in this paper to attempt to predict the *ratios* of rates between different experiments where ρ_{FC} is expected to be nearly the same relative to changes in T_{DA}.

In Section II we define the object of our attention, T_{DA}. In Section III we discuss electron paths in their basic form. In Section IV the *pathways* model is reviewed to clarify fundamental points and to define the relevant terms. In Section IV the new *tube* methods are described, and in Section VI we give a real-world application of the preceding theory.

II. TUNNELING MATRIX ELEMENT

A. Partitioning

In Section I we recognized only two electronic states, D and A, coupled by the matrix element T_{DA}. But even in the single-electron picture with the simplest representation, real proteins have many more states, including, for example, thousands of bonding, antibonding, and lone-pair orbitals. We write the many-state electronic Hamiltonian as:

$$H^{\mathrm{full}} \equiv \begin{bmatrix} \begin{bmatrix} \alpha_D & \eta \\ \eta & \alpha_A \end{bmatrix} & [H_{DA,B}] \\ [H_{B,DA}] & [H_B] \end{bmatrix} \qquad (2.1)$$

Each state in this system corresponds to a single electron in some orbital like a σ-bond, or the D (or A) site itself. This definition is written to suggest a partition between two subsystems. The first is the two-state subsystem containing D (energy α_D) and A (energy α_A) and the direct coupling η between them (negligible in long-distance transfer). The second subsystem, H_B, is the *bridge*, representing all the remaining states in the protein. $H_{B,DA}$ contains the matrix elements coupling these two subsystems.

ET, however, is essentially a two-state process; an electron is either localized at the donor or it is localized at the acceptor. Therefore, we seek an *effective* two-state Hamiltonian to describe the problem in the nuclear conformation where D and A resonate $[\alpha_D^{\mathrm{eff}}(\mathbf{Q}) = \alpha_A^{\mathrm{eff}}(\mathbf{Q})]$.

The natural language for discussing transition probabilities (coupling) and effective Hamiltonians makes use of the Green's function operator G. The G for any H_B is

$$G(E) \equiv \frac{1}{E - H_B} \tag{2.2}$$

Matrix elements of this operator are proportional to the Fourier transforms of transition probabilities ($t \to E$) between states in H_B. The G for a bridge in a nonorthogonal basis (unavoidable in studies using bond-centered or otherwise "localized" orbitals) depends directly on an overlap S. A nonorthogonal basis G is computed not as a simple inversion of the $E - H_B$ matrix but rather by projecting the operator $\hat{G} = (E\hat{I} - \hat{H}_B)^{-1}$ in the desired basis. No qualitative changes occur, however, and the effect of overlaps can be included directly in an effective Hamiltonian [15]. The expression for T_{DA} in a nonorthogonal basis has recently been obtained [16] and is very similar to that given in Eq. (2.6).

Suppose that for some H_B with N states, one is only interested in transitions between the n states of some subset ($n < N$). An *effective* Hamiltonian $H^{\mathrm{eff}}(E)$ is an n-element matrix function of E which when plugged into Eq. (2.2) yields exactly the same G matrix elements that one would obtain using the full N-element H_B. The algebraic process of finding $H^{\mathrm{eff}}(E)$ in this context is known as Löwdin [17] partitioning, and was first introduced in ET by Larsson [18,19].

As we are only interested in transitions between D and A, Löwdin partitioning selecting the D–A subsystem ($n = 2$) yields the effective two-state system

$$H^{\mathrm{eff}}(E) = \begin{bmatrix} \alpha_D^{\mathrm{eff}}(E) & T_{DA}(E) \\ T_{AD}(E) & \alpha_A^{\mathrm{eff}}(E) \end{bmatrix} \tag{2.3}$$

where

$$\alpha_{D(A)}^{\text{eff}}(E) = \alpha_{D(A)} + \Delta_{D(A)}(E) \tag{2.4}$$

$$\Delta_{D(A)} = \sum_{d,a} \beta_{D(A)d} G_{da}(E) \beta_{aD(A)} \tag{2.5}$$

and

$$T_{DA} = \sum_{d,a} \beta_{Dd} G_{da}(E) \beta_{aA} \tag{2.6}$$

$$G_{da}(E) = \langle d|(E - H_B)^{-1}|a\rangle \tag{2.7}$$

A subset of bridge states with *direct* coupling to D and A appears explicitly in the sums above, indexed by d and a, respectively. These are the entrance and exit states of the bridge, and their identification and characterization is crucial to the subsequent methods. β_{Dd} and β_{aA} are elements of $H_{DA,B}$. T_{DA} is a sum of products, and each product includes three couplings; β_{Dd} is the coupling from D to a particular bridge entrance d, G_{da} covers the "cost" of propagating from that entrance to some bridge exit a, and β_{dA} is the coupling from that exit to A. Note that in this superexchange mechanism, electron propagation is a tunneling event [i.e., the occupation of the protein bridge orbitals (by either electron or hole) is virtual] [9,10,14,19–24].

The Löwdin partitioning yielding Eq. (2.3) preserves all information [from the many-state Hamiltonian in Eq. (2.1)] that is relevant to D–A transitions; the system has simply been notationally reduced from many states to two. The price paid is that the effective two-state Hamilitonian and its D–A coupling (the tunneling matrix element we seek) are both functions of a new parameter E.

B. E_{tun} and the Two-State Approximation

At this point the two-state system of Eq. (2.3) is just algebraic rearrangement. To get an effective two-state system, the parameter E must be replaced by a fixed number drawn from the energies and couplings in the real ET system, and these values in turn must meet certain constraints to warrant this replacement.

In electron (hole) transfer, the orbitals of the conduction (valence) bridge "band" (the energies of the diagonalized bridge) are closest to the energies of the D–A subsystem band. Recall that at transfer, the D and A states are resonant. The best two-level approximation is obtained when E is fixed at a value E_{tun} defined as the energy difference between this resonant D–A level and the nearest energies of the diagonalized bridge [25,26]. E_{tun} is referred to as the tunneling energy of the transferring electron [19].

If the resonant energy of the $D-A$ subsystem is closer to the valence band (as is the case in the protein work discussed here), the ET event could be referred to as a hole transfer. To simplify the subsequent discussion, we shift our energy scale to center the bridge valence band on zero, so that E_{tun} can be equated to the $D-A$ resonance energy, and it will be a positive number.

Let the number γ be indicative of the coupling between any two bonding orbitals in the bridge that share an atom. In passing we note that two such bonding orbitals can be viewed as a two-state system, whose symmetric (even) eigenstate has energy γ, and whose antisymmetric (odd) eigenstate has energy $-\gamma$. Since even combinations have lower energy than odd combinations, $\gamma < 0$. This means that γ/E_{tun} is a negative number (for either electron or hole transfer), a point of some importance in the interference issues discussed in Section V.E.

To avoid the mixing that would invalidate the two-state reduction, we require that γ/E_{tun} be small. The direct $D-A$ coupling η must also be small, but that is assured due to the distance involved. Finally, the coupling between the bridge and the $D-A$ subsystem (i.e., the β_{Dd} or β_{aA} above) must be small relative to E_{tun}, or the $D-A$ subsystem won't be sufficiently isolated. Under these conditions it is possible to replace E with E_{tun} (as defined above) and arrive at an effective two-state system. Skourtis discusses these issues in [26].

In Section V.F we define a specific H_B used in a particularly simple ET bridge with only one parameter, the dimensionless γ/E_{tun}. The $D-A$ to bridge coupling is discussed in Section VI.C.

III.   PATHS IN T_{DA}

Given a specific choice of Hamiltonian (CNDO, AM1, etc.) and adequate computing resources, one can compute T_{DA} for a given protein bridge in the context of that Hamiltonian and a particular choice of E_{tun} (assuming that the structural data provide an accurate model of the activated complex, that dynamic effects are not important, and that the assumptions underlying Eq. (1.2) are under control—see Section V.I). But even so, a single number (T_{DA}) offers no structural information, no understanding of the relative importance of covalent versus hydrogen bonds, no means to address the role of dynamic factors (referred to below as through-space jumps), no way to predict the effect of mutations—no interpretation of the rate. The Green's function formulation of T_{DA}, in conjunction with the *pathways* idea, offers a useful approach for understanding these issues and has the advantage that it works with any Hamiltonian.

A *path* is a specific sequence of bridge orbitals, starting at a site d (which is directly coupled to the donor D), and ending at a site a (directly coupled to A) that an electron (or hole) can virtually traverse from the bridge entrance to the bridge exit. For example, the $—N—C_\alpha—C—$ bond sequence of a protein backbone could be a segment of a path. The determination of the relevance of a particular path is the subject of Section IV; here paths are simply defined and discussed in general terms.

Consider a bridge with on two states d and a (with energies E_d and E_a respectively), coupled to each other by γ, as in Figure 1. Let a donor state D couple to d with coupling β_{Dd}, and let an acceptor state A couple to a via β_{aA}. In this simple system the sum in Eq. (2.6) has only one term; $T_{DA}(E) = \beta_{Dd}G_{da}(E)\beta_{aA}$. Recognizing the bridge as a two-state subsystem, for which the desired Green's function matrix element G_{da} has a particularly simple form, we have

$$G_{da}(E) = \frac{\gamma}{\alpha_E^2 - \gamma^2}$$

$$= \frac{1}{\alpha_E}\left[\frac{\gamma}{\alpha_E} + \left(\frac{\gamma}{\alpha_E}\right)^3 + \left(\frac{\gamma}{\alpha_E}\right)^5 + \cdots\right] \qquad (3.1)$$

where $\alpha_E^2 \equiv (E - E_d)(E - E_a)$. The G_{da} matrix element is written as a (Dyson) sum of terms, and each term corresponds to a specific path, as depicted in Figure 1. The first term represents the only purely *forward* path, with one factor of $(\gamma/\alpha_E)^2$. Each subsequent term represents a path with an additional d–a round trip, introducing an additional factor of $(\gamma/\alpha_E)^2$ (backscatter effects).

More complicated bridges have more complicated expansions, but the same idea applies. This "sum of paths" view underscores the potential role of interference in this model. There will be interference effects buried in the calculation of an individual G_{da} (like the backscatter in this example) and there will be interference effects in the overall T_{DA} sum [Eq. (2.6)], where

Figure 1. Simple two-state bridge, with its effects in G_{da} expanded as a sum of paths.

different G_{da} matrix elements between different bridge exit and entrance points are added.

These two classes of interference effects arise directly from the algebra involved and are not particularly noteworthy; in Section V this paper will focus on a different way of dividing interference effects into two categories which depend on the physical nature of proteins themselves rather than on the algebra.

The recognition of G_{da} as a sum of paths involving steps between various tight-binding localized orbitals is all well and good but does not immediately offer a practical approach for understanding electronic coupling. How does one make sense of an infinite number of paths? How does one use these paths in protein problems, where one would prefer to discuss the problem in terms of amino acid orientation and tertiary structure? The first attempt to apply fully some notion of electron transfer paths in proteins, including the notion of through-space jumps as expensive links in such paths, was the *pathways* model [27,28]. Although the *pathways* model has some well-understood limitations, it remains a very useful model for understanding electronic coupling in proteins, and offers an excellent framework for discussing the problem. The model is reviewed here briefly in the course of defining the problems this chapter addresses, and to facilitate making the distinction between *paths, tubes, pathway coupling,* and *tube coupling.*

IV. FIRST STEPS TOWARD PHYSICAL RELEVANCE: THE PATHWAYS MODEL AND PATHWAY COUPLING

A. Partitioning Revisited

Just as partitioning was used on the full electronic Hamiltonian in Section II.A, it can be used to manipulate the bridge Hamiltonian. In a straightforward T_{DA} calculation [Eq. (2.6)] the only G matrix elements one cares about are those *between the bridge entrance points d and exit points a.* It is sometimes useful, therefore, to reduce a bridge to an effective system that retains only these "gateway" states and/or other states of special significance.

As an example, consider the idealized linear alkane bridge of Figure 2(a). Simple systems such as this are frequent jump-off points in the ET literature [22,23,27,29–33]. The σ-bonding orbitals between the C—C and C—H pairs have been identified with energies α_c and α_h, respectively. Three classes of nearest-neighbor couplings have also been identified: γ_c, between C—C and C—C bonds; γ_h, between C—H and C—H bonds; and γ_x, between C—C and C—H bonds. Highlighted in Figure 2(a) is a four-orbital subsystem, wherein the two C—H bonding orbitals located between the two C—C

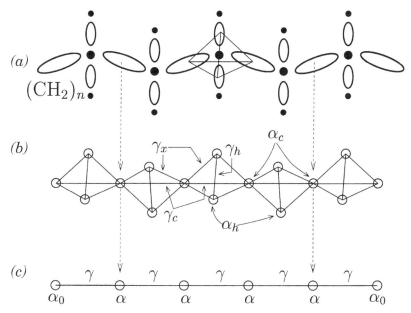

(a)

$(CH_2)_n$

(b)

γ_x γ_h α_c

γ_c α_h

(c)

γ γ γ γ γ

α_0 α α α α α_0

Figure 2. How alkane would be converted to a Hamiltonian used in this model $(a) \rightarrow (b)$, and then how this periodic Hamiltonian could be further reduced $(b) \rightarrow (c)$ to result in the "string of pearls" H below it. The new energies and couplings in system (c) have the effects of the hydrogen side groups renormalized into them.

bonding orbitals are coupled to each other and to the C—C orbitals but are otherwise isolated.

If one is only interested in G matrix elements between the C—C bonds, the C—H orbitals can be partitioned (or "renormalized") out (da Gama [31] offers a more detailed example) to obtain the effective bridge Hamiltonian

$$H_B^{\text{eff}}(E) = \begin{bmatrix} \alpha_0 & \gamma & & & & \\ \gamma & \alpha & \gamma & & & \\ & \gamma & \alpha & \gamma & & \\ & & \gamma & \ddots & \gamma & \\ & & & \gamma & \alpha & \gamma \\ & & & & \gamma & \alpha_0 \end{bmatrix} \tag{4.1}$$

whose matrix elements are functions of E:

$$\gamma(E) = \gamma_c + [2\gamma_x^2/(E - \alpha_h - \gamma_h)] \tag{4.2}$$

$$\alpha(E) = \alpha_c + [4\gamma_c^2/(E - \alpha_h - \gamma_h)] \tag{4.3}$$

$$\alpha_0(E) = \alpha_c + [2\gamma_c^2/(E - \alpha_h - \gamma_h)] \tag{4.4}$$

This H_B^{eff} corresponds to the simple "string of pearls" system (a one-dimensional chain with only nearest-neighbor interactions) shown in Figure 2(c); only the C—C bonding orbitals (and the couplings between them) have been retained. The energies, couplings, and meaning of these "effective orbitals" have changed, but the G matrix elements between them are the same as they were in the full system.

One can, via partitioning and renormalization, reduce any system to a set of "sampling points" like this (preferably stopping before one partitions away all physical relevance). If the system is periodic as in this example, then, as here, a string can be constructed that will be homogeneous (except perhaps at its endpoints).

Now, what will happen to the Green's function between the endpoints of a string of pearls if the string is extended by one unit? Consider a "growing" string as in Figure 3, where all the bridge energies and couplings are homogeneous (as above, except for the endpoints). For simplicity, we place this string on an energy scale where its degenerate orbital energies are zero ($\alpha = 0$). The following equation describes the growth of this system (the superscript k indicates the stage of growth):

$$H^k = H^{k-1} + V^k. \tag{4.5}$$

By the definition in Eq. (2.2), $(E - H^k)G^k = 1$, so

$$G^{k-1}(E - H^{k-1} - V^k)G^k = G^{k-1} \tag{4.6}$$

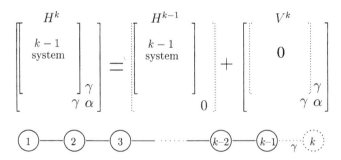

Figure 3. Growing "string of pearls" Hamiltonian.

Thus

$$G^k = G^{k-1} + G^{k-1} V^k G^k \tag{4.7}$$

Taking matrix elements between the string endpoints in this recurrence relation, one finds that for $k \geq 1$,

$$G^k_{1,k} = f_k G^{k-1}_{1,k-1} \tag{4.8}$$

where

$$f_k \equiv \frac{\gamma}{E} \frac{1}{1 - (\gamma/E)f_{k-1}} \tag{4.9}$$

Note that $G^1_{1,1} \equiv 1/(E - \alpha) = 1/E$, and we define $G^0_{1,0} \equiv 1/\gamma$ and $f_0 \equiv 0$ to extend the relation to $k = 1$. Expanding f in terms of γ/E (which will be small when $E \to E_{\text{tun}}$; see Section II.B), one finds that

$$\frac{G^k_{1,k}}{G^{k-1}_{1,k-1}} = f_k \approx \begin{cases} f_2 + O\left[\left(\dfrac{\gamma}{E}\right)^5\right] & k > 2 \\[2mm] f_3 + O\left[\left(\dfrac{\gamma}{E}\right)^7\right] & k > 3 \end{cases} \tag{4.10}$$

So if we are only interested in the *end-to-end* coupling of a homogeneous string of pearls, we see that once the string has only two or three links, the act of adding another link just *multiplies* the total G matrix element by a simple term which for all practical purposes is *independent of the string length* (it depends only on the internal structure of the periodic unit and can thus be precomputed).

This discussion started by noting that a string of pearls can represent the renormalization of an arbitrarily complicated (but approximately periodic, chainlike) system. This means that the extension of the bridge can correspond to the addition of a complicated subnetwork with an infinite number of new intersecting and backscattering paths. But all this will do to the end-to-end coupling is multiply it by a simple (nearly) constant term.

McConnell [20] was the first to notice that end-to-end superexchange coupling in a string of n pearls (orbitals) could be estimated as a product of $n - 1$ identical terms. In their original work on the pathways method, Beratan and Onuchic [27,28] arrived at this conclusion using a translational symmetry (Bloch state) approach. Many other Green's function investigations in this theme have been taken to varying levels of detail [1], but the simple argument above is sufficient for what follows.

B. Decay per Step

Consider a *forward path*, a path that never visits the same bridge state twice. Although there are an infinite number of general paths between a given bridge entrance and exit, there are only a finite number of strictly forward paths. There is only one such path in the bridge of Figure 1. The term *pathway* is associated with a *forward path*.

A protein is a complicated bridge, and there are many, many pathways between a given D and A. One might imagine throwing away everything in the protein but the orbitals on one of these paths, and, say, the nearest orbitals (like the C—H orbitals above, or general residue links C—R). This would leave only a chainlike system, reducible to a string of pearls, and the overall coupling would be just a product of nearly identical terms, one for each step in the path. But given that one is interested in making this single-path assumption, and given a sufficiently complicated molecule, a question arises—which path?

The protein itself can answer the question regardless of the validity of the single-path assumption in its particular case. A network can be created in which every *node* corresponds to a bridge orbital and every *edge* corresponds to a *decay factor* ϵ between some pair of orbitals (details of this process are covered in Section IV.E). The ϵ associated with the step between a given pair of orbitals is not the coupling between those two orbitals. Rather, ϵ is as suggested in Eq. (4.10)—it is the *cost* of taking the step between the two orbitals under the assumption that the protein has been stripped to just some chain or backbonelike substructure which contains that step.

This network can then be searched (Dijkstra's algorithm) for forward paths between D and A. Every forward path can be assigned a *pathway coupling*

$$PC = \prod_{i=1}^{n} \epsilon_i \tag{4.11}$$

The index i is just a label on the steps (orbital pairs) in the given path and n is the number of steps in the path (the number of path orbitals minus one). Path orbital counting (as defined here) does *not* include D and A; rather, counting begins and ends with the first and last bridge states, states that would be respectively labeled by d and a in Eqs. (2.6) and (2.7). The coupling into $(D \rightarrow d)$ and out of $(a \rightarrow A)$ the path is discussed in Sections IV.D and VI.C.

A given ϵ factor corresponds to the simple term f_2 (or f_3) of Eqs. (4.9) and (4.10) for a periodic unit [taken at some fixed $E = E_{tun}$ (see Sections IV.E and V.H)] or they correspond to a jump decay (see below). The value of ϵ is always less than unity, regardless of the complexity of the original system that was reduced to the string of pearls to obtain f_2 or f_3 and is therefore referred to as a *decay per step*. *It is the cost (to the coupling) of extending a chain-like bridge by one more unit.* The product form of Eq. (4.11) emphasizes that pathway coupling is developed as a product of decay factors as one proceeds down a path. Since ϵ is the cost of extending a bridge, it makes no sense to consider paths that turn back on themselves; *pathway coupling is only associated with forward paths*, and every forward path can be associated with a pathway coupling (backscatter effects are included in ϵ).

This network story needs one more element. If one only included covalent link decay factors, the ϵ network for a protein would be essentially one-dimensional, reflecting the protein backbone, and a path search would always yield a backbone path. To incorporate the *fold* of the protein into the network properly, decay factors associated with hydrogen bonds and close contacts in space (through-space jumps) must be added. These kinds of decay factors were not formally developed in the chain argument given above, but they can easily be viewed as the cost of tunneling through the vacuum between two chain endpoints that are brought close together. They should just be the same as ϵ above, but then scaled by an exponential decay factor that depends on the jump distance (this factor and the special case of H-bonds are discussed in Section IV.E). These additional edges in the network greatly enrich the complexity of the network and are the key to the pathways model.

The best path between a given D and A is defined as the path with the maximum pathway coupling, PC_{max}. The final leap in the pathways model is to say

$$T_{DA} \propto PC_{max} \qquad (4.12)$$

which is just the single-path assumption. *A primary feature of the pathway model is that it reveals which path is most important under the assumption that only one path matters.*

C. Why Has the Pathways Model Been Successful?

The single-pathway model can expect *some* success, simply because proteins have an obvious one-dimensional character; the peptide backbone. Coupling will usually be dominated by some segment of the backbone or two backbone segments linked by some jump. A protein backbone is an example of the periodic bridge discussed above; one sees only a repeated series of peptide bonds $(-N-C_\alpha-C-)_n$, with bonds pointing off to the sides $(C{=}O, N-H, C_\alpha-H, C_\alpha-R)$. If one works at a resolution that ignores atom species, extending this bridge out along a residue (away from the backbone) to reach a donor or acceptor site does not change the story. One still has a chain of bonds with side groups, an approximately repeated unit.

But the real success of the pathways model (as an improvement over a simple exponential decay model) is based on the fact that the couplings between spatially neighboring orbitals are highly heterogeneous, spanning a scale from covalent bonds to through-space jumps [34]. Covalently or hydrogen-bond-connected orbitals are *far more* strongly coupled than orbitals that may be close together in space but are not so joined.

After folding, sulfide bridges, hydrogen bonds, and/or through-space jumps provide *shortcuts* between points on the main chain that are otherwise distant (measuring distance along the backbone). These shortcuts may, of course, involve the sidechain residues, or cofactors in the protein.

A through-space jump requires tunneling through vacuum, an expensive proposition usually exceeding the cost of propagating through several bonds. But the longer the best-through bond pathway gets, the less expensive some jump will appear in terms of its effect on the overall PC. In many cases, there will be a clear choice between a simple backbone pathway or a shorter pathway that takes a jump. It is when these effects start to balance one another, when shortcuts introduce competing pathways, that the single pathway picture has trouble. This single-path approach has been successful for studying ET in chemically modified cytochrome c [35,36], but it has not been sufficient to analyse ET in certain reactions in chemically modified azurin, as discussed later.

D. Six Important Features of the Pathways Model

First, to reiterate, the two primary features of the pathways model are: (1) it tells you which path is most important given that you agree to assume that only one path (or family of similar paths) matters, and (2) its success is based on the bond versus jump competition.

Second, there will always be more than one forward path with the same PC; such paths will usually be just minor variations on some "core" path, but sometimes they will differ by major structure units. Consideration of these issues leads to the tube methods of Section V.

Third, a path's PC is *not* the same as the G_{da} one would compute for a bridge that contained only the states on the path. Pathway coupling uses an ϵ that *implicitly includes* the effects of orbitals hanging on the side of the path, whereas a calculation of G_{da} for the path states alone would *not* include these hanging orbitals. The inclusion of these side orbitals implicitly represents the inclusion of secondary paths, which interfere destructively with the core path (having one more step, they differ in sign with the core path; see Section V.E) and thus reduce the path's PC relative to the path's G_{da}.

Fourth, the PC_{max} for some D and A is *not* the full electronic coupling T_{DA}, even given that one is comfortable with the implicit pathways approximation [Eq. (4.12)] discarding the bulk of the protein, retaining only the orbitals on and near the best path. Rather, it is an estimate of the magnitude of a G_{da} computed over the severely butchered protein bridge described at the beginning of Section IV.B (a core path, plus neighboring orbitals), where the d and a are the endpoints of the best path. PC is an estimate of this G_{da}; it is only an estimate because the first few terms [$k = 1, 2$ in Eq. (4.9)] aren't equal to the f_2 or f_3 used for ϵ in Eq. (4.11). This is not a problem when identifying the *best* path, because every path suffers from the same problem, by roughly the same error. But more important, even if only one path mattered and PC_{max} matched its corresponding G_{da} exactly, one would still have to multiply PC_{max} by β_{Dd}, the coupling from the donor to the bridge entrance point, and β_{aA}, the coupling from the bridge exit into the acceptor, to get an estimate of the *magnitude* of T_{DA}. The product of these two couplings is often referred to as the electronic prefactor which appears in front of exponential decay models of electronic coupling. This prefactor and its meaning are sometimes erroneously neglected when one is only concerned with determining a β for an exponential decay model. When using measured ET rates to make deductions about β's, it is critical that one have good estimates for the ρ_{FC} factor and the electronic prefactors (the D–A to bridge coupling).

Fifth, to risk stating the obvious, pathway couplings do not correspond to terms in the G_{da} sum; it makes no sense to attempt to *add* PC's associated with the multiple paths of a network to find some sort of "total" coupling (in Section VI.E we describe an attempt to construct this sum properly).

Sixth, although pathways provides a quick and dirty way to estimate coupling and identify a best path, it is not just a stopgap measure for stripping a protein, useful only until the day when we can load *entire proteins* into quantum chemistry programs. Such attempts began at the extended Huckel level some years ago [37] and have recently gotten more daring (Section VII), if not more accurate (Section V.I). Given care and feeding, any self-respecting quantum chemistry code will output *some* result at the end of the day. But as discussed in Section V, the pathways technique, expanded with tubes, remains a centrally useful way to guide and interpret such brute-force calculations.

E. Standard Pathways Model

The pathways implementation and its standard parameters are described below.

The protein is first transformed into a set of tight-binding sites centered in σ-bonding orbitals and lone-pair orbitals. π and antibonding orbitals are neglected due to their weaker contribution relative to the σ framework [38]. A network is then constructed whose nodes correspond to these orbitals.

A constant decay factor ϵ is associated with every two orbitals (nodes) which directly share an atom (e.g., the two σ-bonds in H_2O share the O). This is the first set of edges added to the network, collectively called the *via-atom* or *through-bond* links.

Additional edges, with decay factors ϵ_{TS}, are then added between every remaining pair of sites that have a center-to-center separation r that is less than some cutoff (typically, 4 Å). These are called the through-space links, or jumps. Some of these jumps will meet certain chemical and geometrical requirements to earn the designation *hydrogen bonds* and then be assigned the decay ϵ_{HB} instead. The standard values of these parameters are as follows:

$$\epsilon = 0.6 \tag{4.13}$$

$$\epsilon_{TS} = \epsilon \exp[-1.7(r - 1.4)] \tag{4.14}$$

$$\epsilon_{HB} = \epsilon \tag{4.15}$$

The factor $\epsilon = 0.6$ corresponds to steps down a simple alkanelike structure with a specific E_{tun} (see [22, 23, 27, 29, 30, 32, 33]). Given the notion of a decay per step, the decay ϵ can be empirically determined from model compounds (e.g., a set of simple spacers with differing lengths) or simple β-strands in proteins [35, 36, 39–42]. The through-space decay factor 1.7 represents an electron tunneling out of a well [34], a much stronger decay than the 1.0–1.5 general decay factor noted for general protein media [42–44]. The distance 1.4 represents an average orbital center-to-center equilibrium distance.

The pathways method defines a hydrogen bond (H-bond) to be the link between (1) the σ-bond connecting a heavy atom to its hydrogen, and (2) the lone-pair orbital on the heavy atom acceptor, when certain chemical and geometrical constraints are satisfied [45]. In the pathways method the ϵ_{hB} link representing an H-bond is equated to ϵ, to reflect the original hypothesis [38] that H-bonds can be as important to electronic coupling as covalent links. It is simple enough (in a program implementation) to equate it to ϵ_{TS} instead, or to make both decays more complicated functions. Before making these adjustments, however, one must have at least as much faith in the hydrogen positions as one has in the functional improvements. We feel it more important simply to be aware of when H-bonds are involved than it is to make these minor adjustments to this simple model. H-bonds are discussed more in Section V.G.

In the early pathways implementation [46–48], sites were atom centered rather than being centered on bonding orbitals and lone-pair orbitals, primarily because hydrogen atoms were not explicitly included in the structural model (although their presence was accounted for in the ϵ decay, as described above). This also meant that H-bond coupling could not be directly associated with an interaction between a σ-bonding orbital and a lone pair; rather, H-bonds were associated with the decay of *two* covalent bonds (ϵ^2), imagined between the two heavy atoms involved in the H-bond.

V. FROM PATHWAYS TO TUBES: DEALING WITH INTERFERENCE

The pathways method has always had the modest goal of stepping beyond the square barrier tunneling model (simple exponential $D–A$ separation dependence) of electronic coupling and including just enough structural information to explain why rates could vary by orders of magnitude even though the distances and driving forces involved were the same (or why rates could match over very different $D–A$ separations). The *tube* methods that follow (first used in 1995 [49,50]) permit the use of more complicated Hamiltonians but retain the simple utility of the pathways model and in fact exploit its limitations to expose interference effects. The present paper serves as the full development of the tube approach.

A. What Is a Tube?

A typical path analysis first finds the best path (PC_{max}) and then finds all paths with a PC greater than or equal to $Z\%$ of PC_{max} (Z being some cutoff percentage). This search can easily generate tens of thousands of forward paths, each path unique in its sequence of states, although perhaps not unique in its PC.

This large number of paths obscures the fact that there are actually only a relatively small number of truly different routes between the D and A sites; routes that differ by a significant secondary or tertiary structural feature, like a hydrogen bond, sulfide bridge, or even a through-space contact between two residues that lie far apart in the sequence but close together in space.

The inherent order in a protein (as opposed to that in, say, the same volume of liquid water) will result in paths that can be roughly divided into different *path families*, where each path in a family exploits the same secondary and/or tertiary features. In such a family, any path will have only minor differences from other paths in the family. For example, any path that follows some section of the $—N—C_\alpha—C—$ backbone will certainly have a sibling path that is identical to it except that it takes a brief detour through a C_α—H bond before returning to the same section of backbone. On the other hand, if two paths follow the same section of a backbone, but then one jumps through a hydrogen bond to another backbone, the two paths belong in different families.

It is the lower-resolution path family view, not the high-resolution individual path view, that is most accessible to biochemists and nature through the process of mutation. Any mutation is likely to change thousands of individual pathways, but its effect on the network of path families will be far less complicated and therefore easier to characterize and understand. This is why one needs a picture of ET that treats the path family on a quantitative level.

Path families are best thought of as *tubes*; there is a central path, a path in the family with the shortest length, called a *tube core*. The remaining paths shift around the tube core, sharing common states with it for most of its length, like light beams bouncing around in a fiber optic cable. In the pathways language, if T_{DA} is dominated by a single pathway, the entire protein matrix may be reduced to a single tube and the remainder of the protein plays no important role in the electronic mediation of tunneling. Tubes are the proper quantitative way of dealing with the effects of multiple pathways.

B.   Quantitative Definition of a Tube

How to find tubes? There are a number of ways one might do so; the approach described below is based on decimating a set of paths to leave only tube cores. First, find all forward paths that provide a coupling that is at least $Z\%$ of the coupling of the best path. The lower Z is, the more paths one will find, but if Z is lower than 0.1, one will start finding paths that probably would not matter even if they all miraculously managed to interfere constructively. Thus one does not have to consider *all* forward paths. The search method is certain (resources permitting) to find all paths with a PC above a certain cutoff. This set is *complete* with respect to the cutoff. One ensures (either implicitly via the path search or via a subsequent sort) that the paths can be accessed in order of diminishing PC.

From this set of paths (which can easily number in the tens of thousands), we wish to select paths that we will use as tube cores. The first tube core is easy to select. The first path in the ordered set is also the best path, so it becomes the first tube core.

To find more tube cores, the remaining paths must be considered one by one in diminishing PC order. When a path is considered, it is compared to all the tube cores that were already selected. Each comparison notes L_i, the length of the smallest entirely *new* segment the path has with respect to known tube core i. This will be the smallest continuous section, measured in number of virtual states, that the two paths do not have in common. We compare the path to each known tube core until we have a full set of L_i. Then if $\min(L_i) > 3$, we keep the path as a new tube core, knowing that it is sufficiently different from all known tubes. Further criteria can be employed, but this method works well. The comparison scan uses minimal differences to ensure noticing a wide selection of tube cores rather than just keeping those that are radically different from each other. The final test ensures that the minimal difference is significant, not just a hanging group detour.

This decimation proceeds until all forward paths with a nonnegligible PC have been considered. The final set of tube cores comprises those paths that survive this decimation. For example, in the case of the azurin Cu–126:Ru interaction discussed in Section VI, a cutoff of $Z = 1\%$ results in 16,839 paths, a set that the scan above decimates, leaving only eight tube cores (Figure 4).

The orbitals composing the tube core form an initial set of tube orbitals. This set is expanded to include all neighboring orbitals, and then expanded again to include the neighbors of those neighbors. The resulting set will contain all the states one would be likely to find in the family of paths exploiting the same general route as the tube core.

In our simple (no π-orbitals) Hamiltonian (Section V.G), a *neighbor* N to site X is any site that (1) shares a heavy atom with X (i.e., N and X are σ-bonding orbitals associated with a common atom), or (2) shares a hydrogen with X in a hydrogen bond [i.e., X (N) is the σ-bond between the heavy donor atom and the hydrogen, while N (X) is the lone-pair on some other atom], or (3) is connected to X via a *favorable* through-space jump, a jump providing a coupling that is greater than some nonnegligble percentage of the PC associated with the best through-bond path connecting X and N.

C. Tube Structure and Overlap

The tube construction process builds tubes with radii of two orbitals, not counting the core. For example, if the core tube follows the peptide backbone exclusively, this definition catches the σ-bond of the carbonyl and the two lone pairs on the oxygen, but stops there. It will *not* cross any hydrogen bond that the lone pairs make; that is the job of a *different* tube. A more sophisticated tight-binding Hamiltonian (e.g., with π-bonds) would require an expanded nearest-neighbor definition list, but the meaning of the tubes would remain the same.

The only exception to this "radius of two" rule is that in some cases it makes sense to fatten a tube at some point along its length to completely absorb a small loop structure such as an imidazole ring, rather than generate overlapping tubes that differ only in that part of the ring they traverse. Proteins are simple enough that this is probably the only exception that

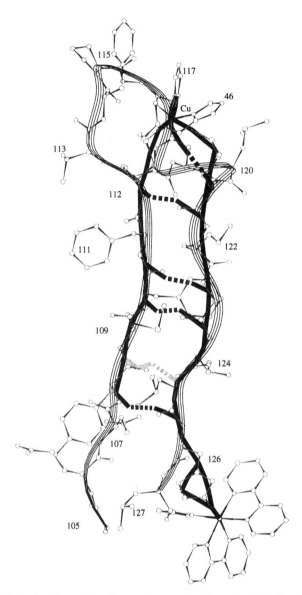

Figure 4. Sets of pathway tubes for the ET coupling from Cu to Ru(bpy)$_2$(im) in HIS126 modified azurin. The shaded lines (dashed lines are H-bonds) indicate the cores of the tubes, which together are responsible for effectively all the electronic coupling of the protein matrix. Positions 122 and 124 are like 126, but with a subset of the tubes shown here. Position 107 is like this picture, only in a mirror image, as 107 is on the left. Similarly, 109 is a mirror of 124. The coupling is dominated by the β-strand directly linked to the copper at 112. See Figures 5, 7, 9 and 10 for details.

would require a slight modification of the tube selection and construction algorithm described above.

A *branching tube* is just two tubes with an overlapping section. Tubes can overlap, but the overlap between any two tubes is of course limited—if there were continuous overlap, the two tubes would be the same. Nevertheless, the overlap can be very significant. For example, one could have two completely different tubes, associated with two different β-strands but then have a third tube that follows the first tube over half of its length and then traverses a main-chain-linking hydrogen bond to follow the second tube for the remainder of its length (see the azurin tubes depicted in Figure 4).

D. Tube Coupling Versus Pathway Coupling

Although based on a single core path, a tube contains a large number of forward paths and an infinite number of backscatter paths. The interference contained in this tubular network of paths can be renormalized away (as discussed above in the context of the Green's function formulation of *pathways*) and understood as a simple decay of coupling with distance. For this reason we call this kind of interference *trivial interference*, to distinguish it from the *tube interference* of Section V.E.

We define *tube coupling* as the actual Green's function matrix element between the tube endpoints (its particular d and a) over the states composing the tube; it includes backscatter to infinite order and will have a characteristic sign (unlike *path coupling*, which has no meaningful sign).

Will the tube coupling and path coupling of the path defining the tube core have proportional magnitudes? This was discussed to some extent in the fourth point of Section IV.D. Some tubes will have a simple interior structure (e.g., tubes that follow backbones exclusively). The tube coupling in these cases should be proportional to the PC associated with the tube's core path, because the hanging orbitals can be renormalized into a simple step decay, as in Section IV.B.

But not all tubes will be so simple. As mentioned above, a tube could enter a residue with a ring structure, and then we could either define two tubes differing only in which side of the ring they traversed, or we could expand the tube to include the entire ring. The latter is the best choice if for no other reason than we may simply *not care* how the interference breaks down in detail in the ring, since in practice the only thing a biochemist might do with the ring is remove it entirely. In some cases, the interference attributable to the ring in the expanded tube may result in a tube coupling that is significantly different from the sum of the two tube couplings associated with the two overlapping tubes one might otherwise define, and may even effect the overall sign. In any case, these deviations from a simple pathway-like result will be captured in the tube coupling.

E. Interference Between Tubes

The quantum mechanical interference between tubes, between different structural units in the protein that one might change biochemically, is what we find the most interesting. This is the *nontrivial interference* in the ET problem, the interference that arises at the level of the sum in Eq. (2.6). A given term in this sum can directly correspond to one or more tubes, the tubes that connect the particular $d-a$ pair in the term.

For example, if the electronic coupling is dominated by two tubes, with similar magnitude but different signs, the final Green's function propagator will be much weaker than the propagator for the individual tubes (destructive interference). In this case, destroying one tube could actually increase the overall coupling. This is the kind of interference that one can probe experimentally and expose via tube calculations. Tubes encapsulate trivial interference, and they expose nontrivial interference.

The phase of the wavefunction of each tube is fundamental since it determines if one has constructive or destructive interference. It was pointed out in Section II.B that in either electron or hole transfer, the relevant intrabridge coupling γ and the relevant E_{tun} will have opposite sign. This means that the central ET parameter γ/E_{tun} will always be negative. This ratio leads the expression for the cost of extending a pathway by one step [Eq. (4.9)], meaning that in a *rough* picture, two tubes that differ by one step in their length will differ in their phase.

For very extended chains, as in the β-strand, this sign oscillation in the wavefunction with every bond is robust. For a more contorted tube, the situation is not so clear. Interaction between non-nearest-neighbor bonds ("crosstalk" interactions [22,23]) may become strong enough that this simple sign alternation picture could break down. We return to this discussion at the end of Section V.G, where the pathways Hamiltonian is introduced.

F. Tubes as a Reduced Protein

The smallest collection of tubes that properly reproduce the $D-A$ coupling provided by the full protein constitute a *reduced protein* and have the advantage of directly exposing the quantum interference (if any) in the T_{DA} sum. Several reductionist approaches have been developed to identify what protein subsets dominate the tunneling matrix element. Broo and Larsson showed that most of the $D-A$ coupling was still present [51] in a particular protein fragment retaining only amino acids between the D and A. Path families were first discussed by Regan et al. in 1993 [37], and it was shown that by simply stripping away everything but states on the paths with the highest PC, one would arrive at a significantly reduced protein that still provided all the coupling of the full protein (indicating that most of a protein is irrelevant to a particular ET reaction). Siddarth and Marcus [52] used a path search with a more complicated heuristic to show this as well. Tubes were introduced in 1995 [49,50] as a means to divide interference effects into trivial (intratube) and nontrivial (intertube) categories and interpret multiple tubes in azurin.

The concept of estimating the importance to the coupling of particular part of a protein by considering the effect of a mutation was proposed by Kuki and Wolynes [53], and implemented more effectively through "importance values" by Skourtis et al. [54]. To be able to quantitatively estimate the participation of amino acids, Skourtis recently developed the concept of contact maps [55] and described multiple paths on a more quantitative level [56]. An alternative approach for estimating the importance of amino-acid participation takes the "derivative" of T_{DA} with respect to a particular amino acid. This was developed by Okada et al. [57,58] and later refined with importance values by Gehlen et al. [59]. These techniques can also serve to reduce a D–A system to its essential parts.

G. The Pathways Hamiltonian

Although the procedures described here could use any Hamiltonian, we want to keep this initial analysis simple and focus on tube effects rather than (among other things) be forced to address the effectiveness of a given Hamiltonian. For this reason we use a very simple pathways Hamiltonian, so called because its parameters are similar to those defined for the pathways model [Eqs. (4.13) to (4.15)].

The only states in the bridge Hamiltonian H_B are the σ-bonding and lone-pair orbitals in the protein; these represent the states a hole can virtually occupy in a transition from A to D, to realize an electron transfer from D to A. As with the pathways model, couplings between these states are divided into three categories.

First is covalent coupling γ. If two orbitals share an atom (like two different C—H bonds sharing the carbon in a methyl), they are "covalently coupled". Orbital-to-orbital coupling is, of course, some continuous function of local atomic elements and geometries, consistent with the basis being used. It is easy to add a simple stereochemical dependence to this function, but in the current work the pathways Hamiltonian uses a constant value of γ regardless of atomic species involved or their relative geometry.

The second class of coupling involves through-space jumps γ_{TS}, and this covers every interaction (below a distance cutoff of 4 Å) not falling into the category above. Analogous to Eq. (4.14), γ_{TS} is normally set to γ scaled by an exponential decay depending on the center-to-center separation.

The third class of coupling is γ_{HB}, representing H-bonds. H-bonds are a special subset of the through-space jumps. They are identified by scanning all jumps and selecting those that jump between (1) the σ-bond connecting a heavy atom to a hydrogen, and (2) a lone-pair orbital on some heavy-atom hydrogen acceptor. This set is further reduced by rejecting the jumps that do not meet the chemical and geometrical criteria associated with H-bonds [45]. This particular selection process presumes that the hydrogens and lone-pair placement has been made in a reasonable fashion (that, incidentally, should be done to maximize hydrogen bonding).

Although the term *hydrogen bond* covers a relatively broad spectrum of configurations, in protein work it is usually clear that when hydrogens and lone pairs are properly placed, a given jump is either a "good" hydrogen bond (where the σ-bonding and lone-pair orbitals start to resemble two covalent bonds sharing the hydrogen), or the usual "bad" (expensive) through-space jump. In proteins there is usually a strong step-function character to this H-bond selection process. We therefore elect to set $\gamma_{HB} = \gamma$ rather than equate γ_{HB} to γ_{TS}. The value of γ_{HS} is easily adjusted to test the H-bond's contribution to coupling. Equating γ_{HB} to γ reflects the growing evidence [60–62] that H-bonds are as important to ET as covalent bonds. In protein work, the determination of exactly *where* the H-bonds are and in what *relative* range their coupling should be (to support observed ET rates) is an issue that can be resolved without more detailed Hamiltonians.

In the experiments discussed here, the covalent and H-bond coupling overwhelmingly dominate the situation; the use of an exponential function [as in Eq. (4.14)] for γ_{TS} makes no difference to the conclusions of this paper (its only effect is generally to reduce all T_{DA} values but leave our ratios largely unperturbed). We therefore set $\gamma_{TS} = 0$ to achieve sparser matrices and save a little time and storage space (there are cases where jumps are certainly not negligible, e.g., the case of the metal to position 72 coupling in cytochrome c [36]).

The outcome of the preceding discussion is that the H_B used in what follows is simply γ times a sparse matrix of 1's. All results are given in energy units of γ. The coupling to the D–A subsystem is described in the next section. This simple Hamiltonian still features the coupling anisotropy (several orders of magnitude of scatter about a simple exponential decay line) that has been associated with the "hot and cold spot maps" [37,48] of the pathways model.

The number γ is negative on our energy scale, and the E_{tun} we use is positive, consistent with the discussion in Section II.B. This will result in the sign oscillation discussed in Section V.E. It would, however, be impetuous to believe that the phase of the tubes are properly retained in *all situations* by this Hamiltonian. When we use this H_B later in azurin, we are careful to point out where it seems to be working and where the results are dubious. Although the pathways Hamiltonian will have quantitative problems, it offers good qualitative results and certainly succeeds at exposing the relevant tubes and where one should pay close attention to their relative phases. This information is vital to the proper design of experiments.

We also want to reinforce that even though all the results discussed here arise from the simple pathways Hamiltonian, the qualitative features of the tube picture, the reduction of the protein to a collection of tubes, and the question of how they interfere are robust to any level of sophistication in the electronic Hamiltonian being used.

H. T_{DA}'s Dependence on E_{tun}

In the pathways model, a fixed ϵ is used [Eq. (4.13)], corresponding to a particular value of E_{tun} substituted for E in Eq. (4.9) and (4.10) to complete the two-state approximation as discussed in Section II.B. For the pathways model to be successful in different ET reactions, which presumably have different values of E_{tun}, T_{DA} must have a weak E_{tun} dependence over some range of E_{tun}, and the value implicitly used in the pathways model must fall in this range.

Recent detailed calculations for more sophisticated Hamiltonians have suggested that T_{DA} does indeed have a weak E_{tun} dependence in a biologically relevant range [15,63,64]. For the energetic region relevant to ET [i.e., between 2 and 5 eV above the "valence band" (HOMO orbitals)], the dependence of the Green's function on the tunneling energy is extremely weak in these studies, making the pathways hypothesis reasonable.

The pathways Hamiltonian, on the other hand, since it does not include orbital overlap and many-electron effects explicitly, gives the impression of a strong E_{tun} dependence (see, e.g., Fig. 11). This incorrect conclusion arises from an inappropriate interpretation of the pathways Hamiltonian. This Hamiltonian and the tunneling energy are *interdependent*; our first task in using this Hamiltonian is to calibrate the dimensionless ratio γ/E_{tun} properly (Section 6.7), by fixing γ and adjusting E_{tun} to provide the proper wavefunction decay down a simple tube (this is how the proper effective parameters are obtained). This is a simultaneous calibration of *both* the Hamiltonian and E_{tun}. For this reason the pathways Hamiltonian is not appropriate for studies of T_{DA}'s dependence on tunneling energy.

After calibration to a particular D and A, the γ/E_{tun} used for a bridge study cannot change. The same parameter has to be used to explain the rates for the entire collection of different D–A pairs in a given series of reactions with the same driving force, and so on.

I. T_{DA} **Accuracy**

The goal of any T_{DA} calculation, or the related calculations of importance values [54], contact maps [55], and tunneling currents [65], is to understand experimentally measured ET rates and design new experiments to further this understanding. The theory of these calculations has been carried to minute levels of detail on a variety of theoretical fronts, many of that may never be tested experimentally, due to the difficulty of matching theoretical assumptions with experimental conditions.

Some of these assumptions have been mentioned in the preceding development [i.e., the assumptions relating to nuclear and electronic separation in Eq. (1.2) and its implicit comment on time scales, the validity of the two-state approximation and the notion of the tunneling energy, and the conditions on the equivalence of ρ_{FC} in various experiments]. To this we add all the usual problems and trade-offs in quantum chemistry, but on an unprecedented protein scale. The choice of a method or basis involves making trade-offs between experimental agreement in electronic spectra, vibrational spectra, bond angles and lengths, charge distribution, multipole moments, and heat of formation (e.g., [66]). Optimizing for experimental agreement in one area usually means sacrificing agreement in another. Are these trade-offs (identified for small molecules) transferable to proteins? What do they mean for protein ET? Work in this field has only begun [67].

Proteins introduce other new difficulties that require at least as much attention as the choice of the Hamiltonian. Foremost is the obvious point that proteins *move*; issues relating to dynamics are discussed in Section V.J. But there are basic structural questions even in a static picture. The preparatory work for a T_{DA} calculation involves more than just a drive-through withdrawal from the protein data bank. From the point of view of a T_{DA} calculation, the atomic coordinates obtained from x-ray crystallography offer only a baseline structure; the coordinates will have areas of ambiguity (with high-temperature factors), may be far from even a local energy minimum, and will not have hydrogens. Before doing a T_{DA} calculation, one must resolve these ambiguities and "place" hydrogens. The latter is problematic in ET, often coupled to proton transfer [68,69]. In a protein of moderate size, there may often be no unique solution to the problem of hydrogen arrangement, reflecting the notion that internal hydrogen exchange is nearly activationless and ubiquitous [70,71]. These issues give rise to a range of distinct structures from that one must choose to begin a coupling calculation. Exactly *which* structure one decides to use may have more of an impact on the result than the choice of the Hamiltonian or even dynamic effects in an MD simulation (in which the σ-bonds to hydrogens in the initial structure are immutable).

The tube approach can support any Hamiltonian; why bother to use the simple pathways Hamiltonian? It certainly does not rack up impressive CPU usage times; the calculations described later take only seconds on a Gnu/ Linux-Intel-based machine. Other work has already shown that today's computing resources permit the application of more complicated Hamiltonians to full proteins (Section VII). This ability, however, does not detract from the usefulness of simple models. Simple models continue to offer a baseline reference, as in Section VI.D, where pathways results are used to look for interference effects before tube calculations begin. The goal in using the pathways Hamiltonian is not to claim high "accuracy" in a T_{DA} calculation in a protein, but rather, to motivate a method with some basic calculations that capture the essential features of the problem in a biochemically useful way through path searches and the idea of interfering tubes. The pathways Hamiltonian is too simple to depend on minute structural details; it depends only on covalent bond assignments, along with a rough picture of through-space contacts and hydrogen-bond placement, requirements that can be filled with some measure of confidence. This captures enough of the physics of the problem to pose some relevant questions.

More complicated Hamiltonians will necessarily depend more heavily on the assumptions made in the choice of coordinate set. Any use of such Hamiltonians should also be accompanied by an explanation of how the T_{DA} result depends on these structural decisions, the protein's dynamics (see below), and on the choice of the Hamiltonian.

J. Tube Dynamics

Proteins are, of course, dynamic entities fluctuating around some equilibrium structure. How does this affect ET? This is a topic of active research but very few results exist to date other than to say, yes, sure enough, the T_{DA} for a particular Hamiltonian changes if the MD snapshot structure changes [72,73]. The question is: Does it change the problem in a way that makes a static picture useless for T_{DA} coupling calculations? At what actual tunneling lengths will ET become inelastic (strongly coupled to vibrational motion) [27,73]? Or can the effects of dynamics be folded into some average cost of a through-space jump or crosstalk interaction?

It is clear that if through-space jumps are important, dynamics should be considered. Is that all, or will dynamics be important even for the covalent chain interactions? The only answer at this time is partially qualitative. When the coupling between bonds is dominated by nearest-neighbor interactions in the chain (as in a β-strand, for example), one expects the dynamical effects to be small. The situation changes if distant-neighbor interaction effects become important. Dynamics may substantially change the overlap between non-nearest-neighbor orbitals. A proper study of these effects should classify the orbital interactions into these two categories, and for the latter, it will have to account properly for dynamics. Dynamical studies that fail to analyze the effects on the individual parts of the dominant tubes will not be able to provide useful and transferable information.

VI.   EXAMPLE IN AZURIN

Azurin was discovered by Horio [74] while isolating and purifying cytochromes. It is one of a class of blue copper proteins, all of which have the following properties [75]. They are found only in bacteria and plants, they have about 130 amino acids, and although the functions of all of these proteins are not completely understood, those that are known involve electron transport systems. The proteins absorb at 600 nm (hence the blue color), and they are paramagnetic when oxidized, and both of these features are attributable to the singly occupied $d_{x^2-y^2}$ orbital on the copper, which engages in ligand-to-metal charge transfer with a π-bonding orbital between the copper and a sulfur [76–78]. The Cu has three proper ligands, CYS-112:SG, HIS-46:ND1, and HIS-117:ND1, but is also relatively close to MET-121:SD and GLY-45:C=O (Figs. 5 and 6).

The ET properties of azurin have been studied extensively. Early work examined the properties of blue copper proteins in general [79,80]. Azurin received more attention with a study of the role of its hydrophobic patch in bimolecular ET [81,82], and subsequent publication of its crystal structure by Nar, Van de Kamp, Canters, and coworkers [83,84]. Farver and Pecht have investigated many features of azurin ET [85,86], focusing on ET between a pulse-generated disulfide radical anion (CYS 3-26) and the Cu^{II} center in various single-site mutants [86–89], and recently reporting the pH dependence of the ET rates [90]. Gray et al. [91] constructed CYS-112-ASP mutants, to explore 112's role in ET [92] (121 and 46 are also under investigation [93]). Canters et al. used mutants to determine the important role of the 117 ligation in preserving ET function [94]. More recently they have reported a detailed structural analysis of the Cu neighborhood in the MET-121-GLN mutant [95].

On the theoretical side, Broo and Larsson have investigated the role of aromatic side groups in azurin ET [51]. Mikkelsen et al. have computed ET self-exchange rates [96]. Regan et al. [49,50] looked at azurin in initial tube work (the present paper is a full development of the tube approach). Gehlen et al. [59] looked at azurin ET as well, in work focusing on reducing the protein (Section V.F) to facilitate an extended Hückel analysis.

The following is a full application of the concepts described in previous sections: the identification of tubes and the exploration of how

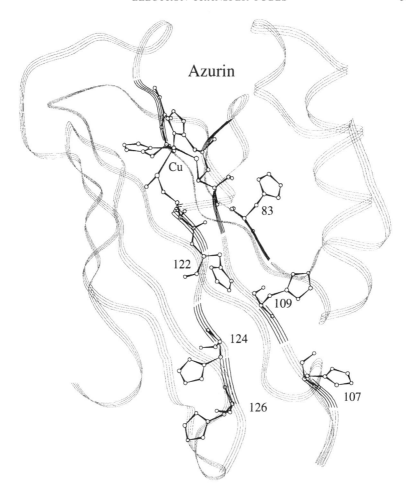

Figure 5. Model of the azurin molecule showing the copper, its ligands, and the Ru(bpy)₂(im) attachment points at positions 83, 107, 109, 122, 124, and 126 (each site was a separate experiment). The probe at 83 points into the page, and is at the back in this view. Only 83 is a native HIS; the remaining HIS sites represent mutations (double mutations, since the HIS at 83 was changed to GLN). Model based on data from PDB entry *4azu* [83,84].

they interfere. This analysis is driven at every point by experimental results and offers a detailed example of how to analyze the coupling between different points in a large molecule or protein. Unlike some of the work mentioned above, neither these experiments nor this analysis were meant to explore the *biological* role of ET in azurin, but rather, they were meant to shed light on the ET process itself.

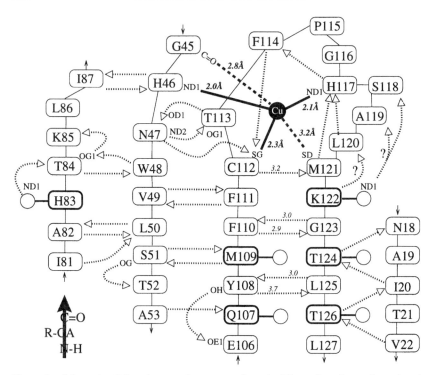

Figure 6. Schematic of the relevant substructure of azurin (Figure 5), to be used as a key for the tube schematics that follow (Figures 7, 9, and 10). It shows the backbone, the copper, and the location of the HIS Ru(bpy)₂(im) probes (circles) and the H-bonds (dashed arrows). The relationship of the copper to its nearest atomic neighbors are shown, solid lines indicating a bond, dashed lines indicating possible strong overlap. The solid lines lie in a plane, to which the dashed lines are roughly perpendicular. The dashed arrows are H-bonds, leading from backbone N to backbone O unless otherwise indicated. Some H-bond N → O distances are given.

A. Experiments

Ralf Langen and Angelo DiBilio, working with Harry Gray, Jay Winkler, Lars Skov, and others [42,49], have recently performed a series of ET experiments on azurin. Azurin has four HIS amino acids, two of which are coordinated to the copper, and of the remaining two only one is on the surface (83). Five mutations of this structure were made, in which the HIS at 83 was removed, while at the same time another HIS was placed at a different location (107, 109, 122, 124, and 126) on the protein surface (see Fig. 5). The HIS sites on these six versions of azurin (five mutants, one native) were each chemically modified by the attachment of Ru(bpy)₂(im) probes to their outermost nitrogen (NE2). ET rates were measured from the Cu to the Ru, using the flash-quench technique [97]. The rates appeared to be

TABLE I
Experimental rates (s^{-1}) and Number of Steps[hb] on the Best Path from Each Cu Exit

Position	Rate	112	121	46	117	45
83	1.1(1)e6	18[3]	24[3]	21[2]	29[4]	21[2]
109	8.0(5)e5	18[0]	22[1]	25[1]	27[2]	25[1]
107	2.4(3)e2	24[0]	28[1]	31[1]	33[2]	31[1]
122	7.1(4)e6	15[1]	13[0]	22[2]	18[1]	28[2]
124	2.2(2)e4	21[1]	19[0]	28[2]	24[1]	28[2]
126	1.3(6)e2	27[1]	25[0]	34[1]	30[1]	34[2]

temperature independent, which is consistent with the observation that $\Delta G^0 \approx \lambda \approx 0.8\,\text{eV}$ [41], indicative of a maximal rate.

The mutants at 107, 109, 122, 124, and 126 offer insight into the coupling down a β-strand. Logarithmic fitting of the rate results yield a distance decay constant of $1.09\,\text{Å}^{-1}$ with a close contact $(r_0 = 3\,\text{Å})$ rate of $10^{13}\,\text{s}^{-1}$, a number that indicates a considerably softer decay than the $1.4\,\text{Å}^{-1}$ cited by Dutton [44] as a general decay constant for proteins. All rates are listed in Table I.

B. Bridge Structure

The structure appearing in Figure 5 is based on the Brookhaven Protein Data Bank entry *4azu*, from the crystallographic work of Nar [83,84]. This model, and various structures derived from it using computer molecular replacement and energy minimization techniques [98], were used in the analysis that follows.

The schematic shown in Figure 6 is a representation of the subset of the protein that provides a bridge coupling equivalent to that provided by the full protein (Section VI.H) for the ET experiments to be discussed here. Neglecting the rest of the protein does not significantly change any results. The schematic shows the backbone, the copper, and the location of the HIS Ru(bpy)$_2$(im) probes (circles) and the H-bonds (dashed arrows). The relationship of the copper to its nearest atomic neighbors are shown, solid lines indicating a bond, dashed lines indicating relatively close contact (but presumably not enough for a bond [84]). The distances to the Cu are those for "chain A," one of four molecules in the unit cell. The distances listed in [83] (Table 9, p. 441) are averages of the four molecules and thus differ slightly from these numbers, which are for chain A only. The dashed arrows are hydrogen bonds (H-bonds, leading from backbone N to backbone O unless otherwise indicated). The H-bonding pattern of an antiparallel β-sheet is evident between 49–53 and 112–106, and between 112–106 and 121–127.

The H-bonding pattern of a parallel sheet can be seen between 123–127 and 18–22.

The schematic in Figure 6 records an H-bond between the ND1 atom of the HIS ring in 83 and the backbone carbonyl oxygen of 84. This H-bond was first noted by the original crystallographers [83,99]. It also appears reasonable in the refinement structures of Ru(bpy)$_2$(im)-HIS83-modified azurin [100], suggesting that the addition of Ru(bpy)$_2$(im) to HIS:83:NE2, replacing the proton there, does not disrupt the H-bond being made from ND1. If anything, the Ru(bpy)$_2$(im) attachment should keep NE2 of the HIS imidazole oriented away from the protein, forcing the other nitrogen (ND1) to point toward the protein surface, probing for H-bonding partners. The backbone oxygen of 84 sticks out of the surface underneath ND1, ready to make such a contact.

Such a stabilizing H-bond does not appear to be possible for the remaining sites. In all cases except 122, the probes are positioned over β-sheet structures, and all nearby O's are lying down in these sheets, already taking part in a stable H-bonding network before the probe appears on the scene. For 122, it appears that the HIS:ND1 atom might be able to make a weak H-bond to 118:O or 119:O, but with the Ru probe attached this seems unlikely. 122:N might also make a weak bond to 120:O. This is the reason for the question mark by the 122 H-bonds in Figure 6; they were tested both "on" and "off", and in the end did not appear to make a difference in the results either way, so they were left out. Also, two H-bonds involving the native residues 124 and 126 have been left out, because they are broken (or irrelevant to the coupling) in the mutants. The H-bond between 83 and 84, however, plays an important role in the computed coupling, and was retained throughout.

The Hamiltonian H_B associated with this bridge was described in Section V.G; in the next section we define the coupling to the bridge.

C. D–A to Bridge Coupling

To progress further, the D–A to bridge coupling for these experiments must be defined. Without these couplings one is unable to evaluate the absolute amplitudes of T_{DA}. Since we are interested in a series of reactions that have similar local environments at the D (Cu) and A (Ru) sites, we presume that these couplings will be similar from experiment to experiment, as will the Frank–Condon factor ρ_{FC}. Although we do compute T_{DA}'s in what follows, we do so more for calibration of parameters in the model than for direct comparison to rates. As far as comparisons to experiment are concerned, we only compare ratios of computed T_{DA}^2 to ratios of rates, with the assumption that factors common to both experiments in the ratio will cancel.

The acceptor state A will be an orbital with an electron physically localized at the Ru. We assume only one bridge exit a, only one state with direct coupling to A; the bonding orbital X(HIS):NE2-Ru, where X is one of the aforementioned sites (83, 107, 109, 122, 124, or 126). Although the Ru couples to other sites in the Ru(bpy)$_2$(im) probe, we assume that these sites do not couple to the remainder of the protein, meaning that whatever effect they would have on T_{DA} would be the same in all experiments discussed here, and thus would not matter in T_{DA} ratios. Similarly, having only one bridge exit means having only one β_{aA}, which can then be pulled out of the sum as a T_{DA} scaling factor that goes away in any T_{DA} ratio, so we do not consider it further.

The donor end of the transfer is more complicated; it will be an orbital localized on Cu, but we will model it as having a direct coupling to five other orbitals, those involved in or near the Cu ligation (see Fig. 6). These are the bridge entrance points d referenced in Eq. (2.7): the bonding orbitals 112:SG-Cu, 46:ND1-Cu, and 117:ND1-Cu, and a lone-pair orbital at 121:SD and at 45:C=O. Attempting to follow the notation of Eq. (2.6), the couplings to these orbitals will be labeled, respectively, β_{D112}, β_{D46}, β_{D117}, β_{D121}, and β_{D45}. Simply selecting these points as bridge entrance points does not mean they are all important; in the particular Ru placements discussed here, β_{D45} would have to be very large compared to β_{D112} for it to be important, and this turns out not to be the case.

In a set of detailed studies of the coordination of metals in blue copper proteins, Solomon, Lowery and co-workers [76–78,101] computed the electronic structure of the Cu and its ligands, to better understand the ligand-to-metal charge transfer band, which, among other things, gives these proteins their striking blue color. Lowery and co-workers conclude that the strongest coupling is to the SG sulfur in 112, while the SD at 121 experiences very little coupling, the carbonyl of 45 even less so, and the two nitrogens fall somewhere in the middle, toward the lower end of the scale. Additional theoretical study of blue copper spectra has recently been performed by Larsson et al. [102].

Recent experimental work by Mizoguchi et al. [92] has sought to quantify the role of the 112 ligand in azurin ET. As with the 122 work studied in this paper, a double mutation, HIS-83-GLN and LYS-122-HIS, was constructed [with a Ru(bpy)$_2$(im) redox partner on the HIS at 122], but with the additional mutation CYS-112-ASP. In the resulting structure, the Cu^I to Ru^{III} ET rate could not be measured directly, but was assigned an upper limit of 10^3 s^{-1}, almost four orders of magnitude slower that the analogous experiment studied here with a native CYS at 112 (7×10^6 s^{-1}, Table I). ET function of the Cu center was not entirely destroyed, however, as bimolecular reactions with WT azurin could still be measured. These results speak to the importance of the 112 ligation in azurin ET. The drop in ET rate to 122 cannot, of course, be attributed solely to a change in T_{DA}; after all, CYS-112-ASP lost the blue color [91] retained by the other mutants described here. Nevertheless, Marcus analysis of the bimolecular rates suggested that reorganizational energy change could not nearly account for the dramatic rate reduction, and that electronic effects must be involved.

We need only a rough estimate for the magnitude of the D–A to bridge couplings to interpret the tube results; the relevance of their relative *signs* is discussed later. From now on, the D–A to bridge couplings β_{Dd} will be referred to as tube "weights" since they set the relative importance of the tube G_{da}'s in Eq. (2.6).

Since we only use ratios, we can work with the weights on a normalized scale. If the coupling β_{D112} is 1, the coupling β_{D121} is something like 0.1. This is reasonable if only from the distances depicted in Figure 6. Any interaction with the lone-pair orbitals on 45:O is likely to be no more than this weak 121 coupling, so we also set $\beta_{D45} = 0.1$. The coupling to the N's of 46 and 117 is, on the other hand, somewhat stronger; about 0.25 on this scale. The basic rule is CYS > HIS > MET. This set of weights is called the "rational" set. The 112 coupling is clearly the strongest and will be seen to dominate determination of the coupling in the work considered here.

D. Experimental Data Versus Best Paths

The experimental data [42] from flash transient spectroscopy is listed in Table I, with the experimental error in the last decimal place listed in parentheses. Next to it are listed the number of steps in the best path from each of the five bridge entrances d to the single bridge exit a discussed above. The number of H-bonds in each path is shown as a subscript.

Our interest in ratios has already been discussed. The $\log_{10}$ of the experimental ratios are shown in Table II, with the upper and lower limits placed by the error. Next to it are corresponding PC_{max} ratios associated with the best paths. Each of these paths starts at the 112 ligation of the Cu. For 107 and 109 this is obvious from Figure 6; for 83 it is also true, but perhaps not as obvious. In the cases of 122, 124, and 126, however, the *shortest* path from the Cu starts at 121, but since the coupling from Cu to 121 is so weak $(\beta_{D112} \gg \beta_{D121})$, 112 still offers the best (albeit one H-bond longer) paths to these sites as well. We now examine what this single-path assumption tells us.

Since all couplings in the bridge are the same (including H-bonds), there is only one ϵ, so path coupling ratios are computable directly from the 112 column of Table I. They are just $\log_{10}(\epsilon^{2(N_1 - N_2)})$, where N_1 is the number of steps in the numerator path, N_2 is the number of steps in the denominator path, and the factor of 2 is because T_{DA} is squared in the rate expression.

TABLE II
Rates versus Pathway Couplings [$\log_{10}$(ratios)]

Ratio	Experimental	PC$_{max}$ (via 112)
83/109	$0.09 \leq 0.14 \leq 0.18$	$0.00(-0.14)$
83/107	$3.59 \leq 3.66 \leq 3.72$	$2.66(-1.00)$
83/122	$-0.86 \leq -0.81 \leq -0.77$	$-1.33(-0.52)$
83/124	$1.64 \leq 1.70 \leq 1.75$	$1.33(-0.37)$
83/126	$3.65 \leq 3.93 \leq 4.09$	$3.99 \ (0.07)$
109/107	$3.46 \leq 3.52 \leq 3.58$	$2.66(-0.86)$
109/122	$-0.99 \leq -0.95 \leq -0.91$	$-1.33(-0.38)$
109/124	$1.51 \leq 1.56 \leq 1.61$	$1.33(-0.23)$
109/126	$3.52 \leq 3.79 \leq 3.96$	$3.99 \ (0.20)$
107/122	$-4.54 \leq -4.47 \leq -4.42$	$-3.99 \ (0.48)$
107/124	$-2.04 \leq -1.96 \leq -1.90$	$-1.33 \ (0.63)$
107/126	$-0.02 \leq 0.27 \leq 0.44$	$1.33 \ (1.06)$
122/124	$2.46 \leq 2.51 \leq 2.55$	$2.66 \ (0.15)$
122/126	$4.47 \leq 4.74 \leq 4.90$	$5.32 \ (0.59)$
124/126	$1.95 \leq 2.23 \leq 2.40$	$2.66 \ (0.43)$

The standard $\epsilon = 0.6$ is used (Section IV.E). Pathways/experiment ratio differences appear in parentheses. Clearly, with N data points, there are only $N - 1$ unique ratios, so Table II has redundant information. However, if (for example), the ratios a/b, b/c, c/d, and d/e were all slightly too large, it might go unnoticed unless one looked at a/e directly, in that case the error would be evident. To keep it obvious, all ratios are listed.

If the conditions associated with the measured rates are consistent with the assumptions of this model, important rules to keep in mind in considering Table II are as follows. A positive ratio difference (pathways/experiment) means that the pathways assumption of a single path overestimates the experimental ratio, suggesting that in the real experimental measurement, destructive interference *hinders* the measured ET rate in the numerator (with respect to what a single path would provide) and/or constructive interference *helps* the measured ET rate in the denominator (with respect to what a single path would provide). Conversely, a negative ratio difference means the pathways assumption underestimates the experimental ratio, suggesting that destructive interference hinders the denominator and/or constructive interference helps the numerator.

Table II can be used as a guide to determine where a single-pathway assumption does well and where the assumption fails. In the former case the single-pathway picture may still be incorrect (through a lucky combination of multiple paths that gives a single path result), but in the latter case it is clear that a more detailed analysis is called for, so we concentrate on the large deviations from zero (i.e., problems) in the differences column. Examination of this column reveals the following.

The single-pathway assumption overestimates the strength of 107 versus 122 (a little), 124 (a little more), and especially 126 (a lot more). This same *ordering* of differences applies for 109 and 83 versus 122, 124, and 126; the farther one gets from the Cu down the 122–124–126 β-strand, the more positive the ratio differences involving these sites become. This suggests that (in the experiments) constructive interference is helping as we move down the 122–124–126 β-strand. Similarly, 124/126 is overestimated and 122/126 even more so. This could imply that 126 has more than one route helping it.

Conversely, the single-pathway assumption underestimates the 109/107 and 83/107 ratios. So either 109 and 83 are experimentally fast relative to pathways expectations of these ratios, or 107 is too slow (or a little bit of both). Considering 107 alone, we note that in the ratios where 107 is in the numerator, the pathways model overestimates, and when 107 is in the denominator (as with 83 and 109), pathways underestimates. It is a straightforward trip from the copper to 109 and on to 107; and there does not appear to be a possibility for interference effects to help 109 or hinder 107 significantly. On the other hand, interference could change the other sites, and change these ratios. Site 83 looks like it could easily be served by multiple paths (Figs. 7 and 8); it is not obvious how they might help or hurt 83's case.

Ordered by increasing disagreement, the pathways model does best with respect to experiment in the following ratios: 83/126, 83/109, 122/124, and

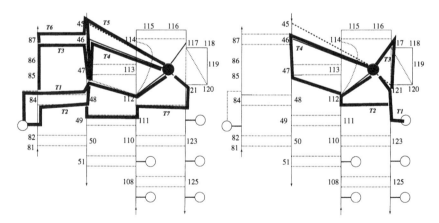

Figure 7. Schematic of tube cores to 83 and 122; see Figure 6 for details.

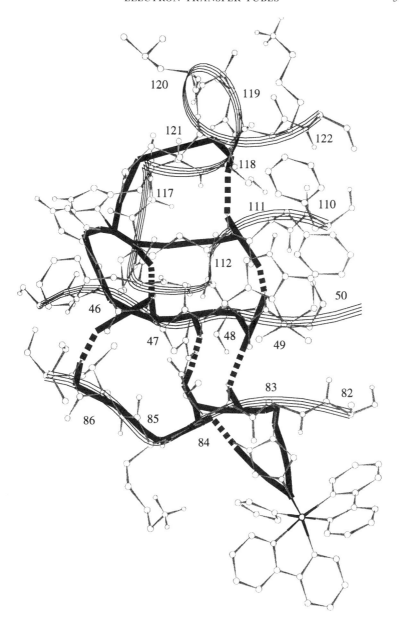

Figure 8. Sets of pathway tubes for the ET coupling from Cu to Ru(bpy)₂(im)-HIS83 in wild-type azurin. The shaded lines (dashed lines are H-bonds) indicate the cores of the tubes, which together are responsible for effectively all the electronic coupling of the protein matrix. For 83, multiple tubes traverse a β-sheet between donor and acceptor. See Figures 6 and Figure 7 for details.

109/126. In the same order, it does worst with: 109/107, 83/107, and 107/126. It is interesting that 83 and 109 appear on both these lists.

Armed with these preliminary pathways impressions, we break the problem down to a tube-by-tube analysis in the next section, and in the final part of this section see how it all adds up.

E.   Tube Breakdown

Examples of tubes for the cases of 126 and 83 are shown in Figures 4 and 8, respectively. The tubes shown in these and the remaining figures result from the process described in Section V.B. Although the tubes are drawn beginning from the Cu, they actually begin at the bridge entrances (Section VI.C). Tube results for all six possible acceptors (83, 122, 109, 124, 107, and 126) are presented in Tables III to V and Figures 7, 9, and 10. The tubes are ordered (labeled $T1$, $T2$, etc.) by the length of their cores and thus by the PC associated with their core paths. When different cores have the same length, the ordering is arbitrary. The ordering is *not* a statement of the tube's importance to the coupling, as these numbers must be weighted by the appropriate D–A to bridge coupling. Tables III to V give the tube number (corresponding to the figure), the exit taken from the copper, the number of steps (subscripted by the number of H-bonds), and finally some special G' matrix elements computed at a specific energy E_{tun} for the tubes.

The numbers in the columns labeled G'_{da} represent the G matrix element computed over the tube, *and only the tube*, associated with the pathway listed, at the E_{tun} specified as an argument to G'_{da}. The particular value of E_{tun} used is discussed in the next section. As defined previously, the tube is a subset of the bridge states, composed only of the states along the core of the pathway (between state d, inferable from column d (and the discussion in Section VI.C), and state a, which is always the final NE2-Ru σ-bond), plus their nearest neighbors, plus their nearest neighbors again. This G' is defined with respect to a submatrix of the full H_B, and the prime on G' flags this special condition. As discussed previously, this takes into account all trivial interference and local backscatter effects and is basically a pathways-like result, *for the tube being considered, at the particular energy specified as an argument to G'.*

The sum of the numbers for G'_{da} given in the table, when weighted properly by the appropriate D–A to bridge couplings for the particular copper exits involved, is expected to roughly approximate the full T_{DA} value as computed over the entire protein. The full T_{DA} can, of course, be computed "exactly" at any time (Section VI.H); what we are after in this tubular breakdown is an understanding of the factors determining the final value. It is, for example, critical to know if one is adding a set of large numbers

TABLE III
Tube Results for 83 and 122[a]

Tube	83			122		
	d	$Step_{[hb]}$	G'_{da}	d	$Step_{[hb]}$	G'_{da}
T1	112	$18_{[3]}$	7.8×10^{-5}	121	$13_{[0]}$	-5.6×10^{-4}
T2	112	$19_{[2]}$	-4.9×10^{-5}	112	$15_{[1]}$	-3.1×10^{-4}
T3	112	$21_{[3]}$	-2.6×10^{-5}	117	$18_{[1]}$	-5.5×10^{-6}
T4	46	$21_{[2]}$	-8.9×10^{-5}	46	$22_{[2]}$	7.5×10^{-6}
T5	45	$21_{[2]}$	-2.5×10^{-5}			
T6	45	$22_{[2]}$	2.0×10^{-5}			
T7	121	$24_{[3]}$	9.3×10^{-6}			

[a]See tubes in Figure 7. $E_{tun} = -2.06\gamma$.

TABLE IV
Tube results for 109 and 124[a]

Tube	109			122		
	d	$Step_{[hb]}$	G'_{da}	d	$Step_{[hb]}$	G'_{da}
T1	112	$18_{[0]}$	6.1×10^{-5}	121	$19_{[0]}$	-4.8×10^{-5}
T2	121	$22_{[1]}$	1.6×10^{-5}	112	$21_{[1]}$	-2.2×10^{-5}
T3	121	$22_{[1]}$	1.4×10^{-5}	112	$21_{[1]}$	-2.5×10^{-5}
T4	121	$22_{[1]}$	1.4×10^{-5}	112	$21_{[1]}$	-2.6×10^{-5}
T5	46	$25_{[1]}$	-1.5×10^{-6}	117	$24_{[1]}$	-4.6×10^{-7}
T6	45	$25_{[1]}$	-4.2×10^{-6}	46	$28_{[2]}$	5.3×10^{-7}
T7	117	$27_{[2]}$	1.6×10^{-7}	45	$28_{[2]}$	1.5×10^{-6}

[a]See tubes in Figure 8. $E_{tun} = -2.06\gamma$.

TABLE V
Tube results for 107 and 126[a]

Tube	107			126		
	d	$Step_{[hb]}$	G'_{da}	d	$Step_{[hb]}$	G'_{da}
T1	112	$24_{[0]}$	4.5×10^{-6}	121	$25_{[0]}$	-3.5×10^{-6}
T2	121	$28_{[1]}$	1.2×10^{-6}	112	$27_{[1]}$	-1.6×10^{-6}
T3	121	$28_{[1]}$	1.2×10^{-6}	112	$27_{[1]}$	-1.6×10^{-6}
T4	121	$28_{[1]}$	1.0×10^{-6}	112	$27_{[1]}$	-1.6×10^{-6}
T5	121	$28_{[1]}$	1.1×10^{-6}	112	$27_{[1]}$	-1.8×10^{-6}
T6	121	$28_{[1]}$	1.2×10^{-6}	112	$27_{[1]}$	-1.9×10^{-8}
T7	46	$31_{[1]}$	-1.1×10^{-7}	117	$30_{[1]}$	-3.4×10^{-8}
T8	45	$31_{[1]}$	-3.1×10^{-7}	46	$34_{[2]}$	3.9×10^{-8}
T9	117	$33_{[2]}$	1.2×10^{-8}	45	$34_{[2]}$	1.1×10^{-7}

[a]See tubes in Figure 10. $E_{tun} = -2.06\gamma$.

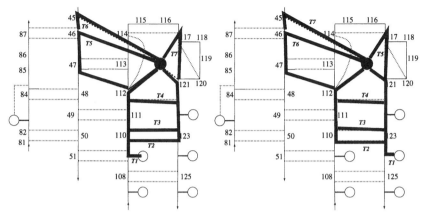

Figure 9. Schematic of tube cores to 109 and 124; see Figure 6 for details.

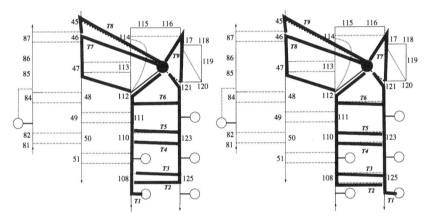

Figure 10. Schematic of tube cores to 107 and 126; see Figure 6 for details.

of *opposite* sign to arrive at a sum with a small absolute magnitude. An example of this arises later.

F. Weighting the Matrix Elements

The final step in examining the tubes is to compute T_{DA} values. This means that we must select a particular value of E_{tun} and compute G_{da} values over the full protein bridge and over individual tubes between the d and a gateway states. These elements are then weighted (D–A to bridgecoupling) and added to arrive at T_{DA} [Eq. (2.6)]. Recall, however, that we are not particularly concerned about the signs of these weights, just their relative magnitudes. Our concern remains focused on the signs and relative contributions of the tubes.

Figure 11 shows a plot of T_{DA}^2 as a function of E_{tun} for the 122, 124, and 126 couplings for two different choices of weights. The highest H_B eigenvalue is at 2 (in units of γ); an attempt to use an E_{tun} too close to this value will of course blow up the G_{da} matrix elements and violate the two-state

Tunneling Coupling vs. Energy

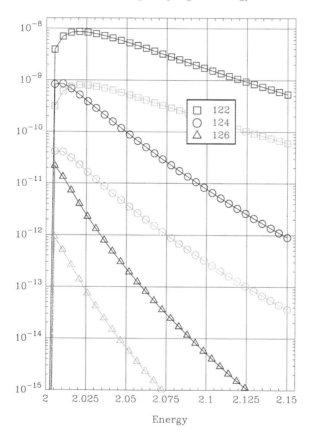

Figure 11. $T_{DA}^2(E_{tun})$ versus E_{tun} (units $-\gamma$) for the 122, 124, and 126 couplings, with two different sets of weights on the donor couplings. In the gray lines, all weights are 0.1. In the black lines, $\beta_{D112} = 1$, $\beta_{D46} = \beta_{D117} = 0.25$, $\beta_{D121} = \beta_{D45} = 0.1$. Our interest is in the *relative* shifts of these lines.

model (Section II.B). In both cases, as E_{tun} gets farther from the bridge energy, T_{DA} drops off, as one would expect from $G = 1/(E_{\text{tun}} - H)$. In this single-electron model (no shielding), G's dependence on E_{tun} is too strong (Section V.H), but the general trend should be that as expected for hole tunneling. We want to focus on a value of E_{tun} that is not only outside the "band" of bridge eigenvalues, but that yields the appropriate distance decay in a simple covalent chain.

The tubes to 126 are the longest, followed by the tubes to 124 and 122 (Fig. 4). Note that the longer the tubes involved are, the faster the fall-off as a function of E_{tun}. This is essentially a pathways-like result, since tube coupling goes like $(\gamma/E_{\text{tun}})^n$ where n is the number of steps in the tube core.

One set of T_{DA}^2 lines (gray) in Figure 11 was generated with all weights set to 0.1, the smallest weight in the rational set. The second set of lines (black) was generated using the rational weights, in which the CYS-112 coupling is favored. All lines shift upward, but the upward shift for 124 exceeds that for 122, and the upward shift of 126 exceeds that for 124 (and 122). This is because 126 has more tubes feeding it from the strong 112 coupling than 122 does, as seen in Figures 7, 9, and 10. These tubes all interfere coherently, so any increase in the 112 coupling is bound to help 126 more than it will help 122. One knows that these tubes interfere coherently by looking back at Tables III to V. The values for G_{da}' listed there were calculated using a single energy (determined below) and show the individual contribution of each route. The information for 126 in Table V, for example, shows five essentially identical tubes, all contributing the same value, with the same *sign*. These tubes traverse the ladder of H-bonds between the two parallel β-strands seen in Figure 10. The same argument applies to the 124 shift versus that of 122.

In contrast, 107 and 109 (not shown) increase by *exactly* the same amount (with the increase in 112 coupling associated with using the rational weights); 83 increases by the same amount, plus about 1%, due to its extra feeds from HI5:46 (Table III). This also suggests that the coupling for these cases is mediated solely by 112; 107 and 109 are clearly single tube reactions, but 83 may see multiple-tube involvement, although tubes leaving via 112 will be the most heavily weighted.

G. Comparing to Experiment

Figure 12 shows ratios of T_{DA}^2 plotted as a function of E_{tun}, computed over the full protein, not just individual tubes. As in Figure 11, the black lines in Figure 12 are computed from T_{DA} values computed with the rational weights, emphasizing 112, and the gray lines represent equal coupling from all Cu exits. As noted above, the increase at 112 helps 124 and 126;

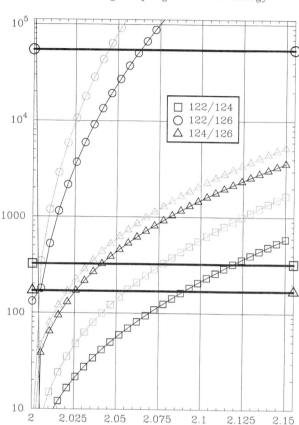

Figure 12. Ratios of T_{DA}^2 compared to ratios of experimental rates. The T_{DA}^2 ratios are functions of E_{tun}; experimental ratios are shown as straight lines. Theory lines cross experiment lines at roughly the same place, suggesting -2.06γ as an appropriate value for E_{tun}. Two different sets of donor couplings were used (Fig. 10).

hence the drop in going from gray to black, as 124 and 126 are in the denominators of these ratios.

This figure shows how experimental data can be used to calibrate E_{tun} for the pathways Hamiltonian. In each of the ET reactions considered here, D and A (and the ρ_{FC} factors) are assumed to be the same, so one anticipates that E_{tun} will also be the same for all reactions. But at what value of E_{tun}? The proper E_{tun} to use in this model is where the curved theory lines intersect the straight experimental lines in Figure 12: $E_{\text{tun}} \sim -2.06\gamma$. Recall that γ/E_{tun} is the parameter being calibrated here, for use with the simple pathways Hamiltonian. It should not be viewed as a real energy.

The tube results suggest that in the cases of 122, 124, and 126 (and the other three sites not yet considered), the principal coupling is provided by the tubes from 112 and to a lesser extent 121 (because of weaker copper coupling), and that in any event, these tubes interfere *coherently*. Because of the dominance of the 112 tubes, ratios of T_{DA}'s (for acceptors on the same β-strand) will not depend so strongly on the choice of weights that shifting the weights slightly will select a substantially different E_{tun}. For the case with all weights the same, the dominant tube leaves the Cu at 121 to feed directly into 122, and so on. For the rational set of weights, the main coupling is through 112. Since the paths down the 112 strand are just a constant three steps longer than those down the 121 strand, the ratios in one strand are similar to those in the other, so both sets of theory lines in Figure 12 (for each set of weights) cross experimental lines at roughly the same energy.

By "cross" it should be understood that whenever the theoretical ratio is within, say, half an order of magnitude of the experimental line (half of one of the major divisions on the plot), we are gratified. To get such results when using the same γ for every coupling in the protein (including that provided by H-bonds) is as much as one could hope for. Also, the lines representing experimental ratios have a confidence interval (see Table II).

We can now compare *pathways* to tubes to experiment. The multiple tubes into 126 *soften* the decay of its coupling (with respect to a single pathways result). The pathways prediction for the 122/126 ratio is 5.3 (Table II), the tube prediction can be read from Figure 12 as somewhere around 4.7 (near the -2.06γ line). The experimental ratio is about 4.7 (Table II). Similarly for 124/126, pathways ≈ 2.7, tubes ~ 2.7, experiment ~ 2.5. For 122/124, pathways ~ 2.7, tubes ≈ 1.8, experiment ~ 2.5. In the latter two cases, although it is clear that the simple pathways Hamiltonian produces the expected positive tube interference for 124, it looks like it helps 124 too much.

H. Tubes Versus the Full Protein

The full G_{da} matrix elements taken between bridge entrance and exit points for the full protein bridge (not just individual tubes as in Tables III to V) are provided for $E_{\text{tun}} = -2.06\gamma$ in Table VI. From this table and the expression for T_{DA} [Eq. (2.6)], one can immediately see that bridge entrances are important in that reactions and what effects the D–A to bridge couplings will have, since the weights multiply these numbers directly in the T_{DA} sum. The largest bridge couplings to the β-strand acceptors (107, 109, 122, 124, 126) are of course via 121 and 112 (although the better weight at the 112:SG-Cu coupling will make the 112 tubes more important). It appears 83 will be served by strong routes from 46 and 45 as well (although they will be weighted relatively weakly).

According to Table VI, the total bridge coupling from the bridge entrance at 112 to the exit at 126 is -3.1×10^{-7}, while the corresponding β-strand tubes (Table V) all contribute about -1.6×10^{-6} each to the total coupling, a factor of 5.2 better than the total. This comparison of β-strand tubes to the full protein G_{da} matrix elements for the bridge entrances 112 and 121 (with a common exit at the Ru) is laid out in Table VII. The tubes do better than the full protein (for the same entrance and exit), but sometimes they do much better.

Two effects are in play here. The first effect arises from trivial interference as defined in Section V.E. Anytime a bridge is simply expanded, without significant changes such as the introduction of new routes between the D and A, the extra side trips on the paths destructively interfere (in their

TABLE VI

$G_{da}(-2.06\gamma)$ for Full Azurin (above) and the Azurin Subset (below) Consisting of the Union of All Tubes $(-\gamma^{-1})$

d	X = 83	X = 122	X = 124	X = 126	X = 109	X = 107
112	3.0×10^{-6}	-6.1×10^{-5}	-5.3×10^{-6}	-3.1×10^{-7}	7.1×10^{-6}	3.2×10^{-7}
121	1.6×10^{-6}	-3.1×10^{-4}	-1.4×10^{-5}	-6.6×10^{-7}	7.8×10^{-6}	4.6×10^{-7}
117	1.5×10^{-8}	-3.0×10^{-6}	-1.5×10^{-7}	-7.0×10^{-9}	7.6×10^{-8}	4.7×10^{-9}
46	9.7×10^{-7}	1.1×10^{-6}	7.9×10^{-8}	4.4×10^{-9}	-7.9×10^{-8}	-3.6×10^{-9}
45	3.0×10^{-6}	2.1×10^{-6}	1.5×10^{-7}	7.8×10^{-9}	-1.5×10^{-7}	-6.8×10^{-9}
112	3.2×10^{-6}	-6.3×10^{-5}	-5.5×10^{-6}	-3.3×10^{-7}	7.5×10^{-6}	3.5×10^{-7}
121	1.8×10^{-6}	-3.2×10^{-4}	-1.4×10^{-5}	-7.0×10^{-7}	8.3×10^{-6}	4.6×10^{-7}
117	2.0×10^{-8}	-3.0×10^{-6}	-1.6×10^{-7}	-7.4×10^{-9}	7.9×10^{-8}	5.0×10^{-9}
46	9.9×10^{-7}	1.1×10^{-6}	7.2×10^{-8}	3.5×10^{-9}	-5.1×10^{-8}	-2.3×10^{-9}
45	3.1×10^{-6}	2.2×10^{-6}	1.4×10^{-7}	6.8×10^{-9}	-9.9×10^{-8}	-4.5×10^{-9}

$^a a$ is X:HIS:NE2-RU, d sites are described in Section VI.C.

TABLE VII

Absolute Ratios of Best Tubes G'_{da} to Full G_{da} for Bridge Entrances 112 and 121

d	122	124	126	107	109	83
112	5.1	4.2	5.2	14.1	8.6	26.0
121	1.8	3.4	5.3	2.6	2.1	5.8

leading-order contribution) with the couplings and energies of the states on the important paths, reducing the overall coupling between any two points. The numbers in the lower half of Table VI, representing G_{da} elements taken over the *reduced* protein (the union of the tubes); are slightly larger than those for the *entire* protein in the upper half, all by roughly the same constant factor (around 1.05). The full protein coupling between two points will generally be smaller than that provided by a single tube. Moreover, the longer the distance covered by the tube (the more surface area it has), the more it will be dragged down by this renormalization, which is why the ratios in Table VII generally increase with DA separation (note the 122, 124, 126 sequence from 121, or the 109, 107 sequence from 112). The CYS:112-107 distance is one step shorter than the MET:121-126 distance (because the Cu is coupled to SG in CYS, as opposed to SD in MET), so the 112-107 tube does better, relative to its full protein counterpart, than the 121-126 tube does (14.1>5.3).

The second effect in play is of course due to multiple tubes. If multiple tubes interfere constructively, this will help the full protein number relative to the single best tube number, decreasing the best tube to full protein ratio (and of course the opposite will happen for destructively interfering tubes). There are multiple constructively interfering tubes from 112 to 126, and these help the full protein number to keep this ratio as low as 5.2, lower than the 5.3 ratio for the *shorter*, more direct route via 121. This effect is even more pronounced for 107, where multiple tubes from 121 help the denominator in the tube/protein ratio to keep the ratio as low as 2.6, as opposed to its value of 14.1 for the single-tube direct route from 112.

I.   The Rest of the Story

Position 107 is added to the mix, using the proper weights, in Figure 13, and we observe that the lines involving 107 also cross the experiment lines at roughly the same energy $E_{tun} = -2.06\gamma$. The full Green's function treatment does much better for 107 than the single pathway results did (see Section VI.D). Looking at 107/126 (the worst pathways result in Table II), we have pathways ~ 1.3, tubes ~ 0.5 (Figure 13), experiment ~ 0.3. This improvement is directly attributable to the coherent interference in the 126 tubes, and helps to resolve some of the concerns about 107 raised at the end of Section VI.D.

The story is not so promising when we consider 109 (Figure 14). Note that 109 is two amino acids closer to the Cu than 107, as is the case for the 122 versus 124 and 124 versus 126. We would thus expect these ratios to be similar (certainly on a log scale), even including the multiple tube effects (having 3 tubes to 124 and 5 tubes to 126 will only change the ratio by $3^2/5^2 \sim 1/3$). Pathways and tubes aside, we note that the log_{10} of the

Tunneling Coupling Ratios vs. Energy

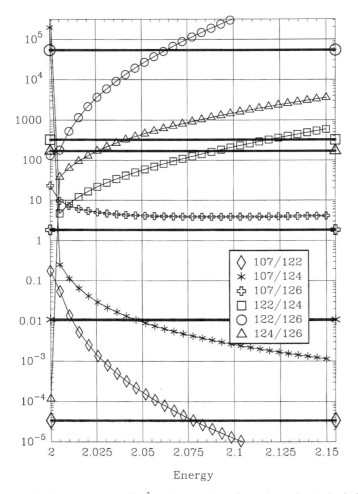

Figure 13. Another comparison of T_{DA}^2 ratios to ratios of experimental rates (as in Fig. 11), but this time including 107 ratios with 122, 124 and 126 ratios and the vertical scale is expanded. The rational weights are used on the Cu couplings for this and the remaining plots.

experimental ratios for 122/124, 124/126, and 109/107 are, respectively, 2.5, 2.2, and 3.5 (Table II). The ratio of 109 to 107 is a factor of 10 bigger than expected, and the fact that 107 appears to be in agreement with everything suggests that experimentally, 109 is just coming on too strong. The theory ratios with 109 in the numerator undershoot the experimental lines by an order of magnitude in Figure 14.

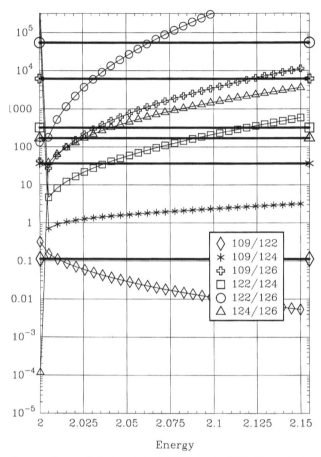

Figure 14. Same as Figure 13, but we include 109 instead of 107. Theoretical ratios involving 109 in the numerator are seen to undershoot their experimental counterparts (at $E_{\text{tun}} = -2.06\gamma$), because the theory (or its parameterization or the structure being examined by the theory) cannot account for the high rate seen at 109.

Before discussing this problem further, 83 is added to the mix. Recall that with respect to pathways results, 83 did well in ratios with 126 and 109 (end of Section VI.D), but now we have seen that for different reasons, 126 and 109 are not consistent with pathways results, leaving us without these clues about 83.

The ratios for 83 (with 83 in the numerator) appear in Figure 15, under-shooting experimental values even more than the ratios with 109 in the

Tunneling Coupling Ratios vs. Energy

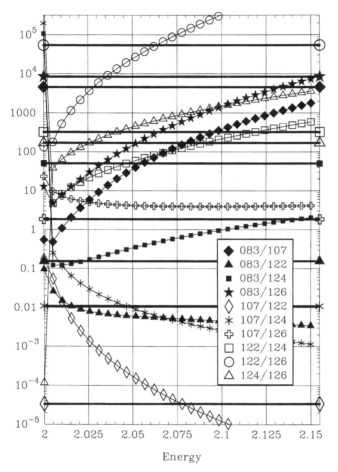

Figure 15. Same as Figure 13, but with 83 added to the mix, and 107 left in. As with 109, theoretical ratios with 83 in the numerator are seen to undershoot their experimental counterparts.

numerator. Position 83 is expected to show a rate difficult to understand, because the direct line from the Cu to the Ru site is perpendicular to the intervening backbone structure; the tube structure (Fig. 8) is not as simple as the cases considered above. Not only does the 83 acceptor have multiple tubes, the tubes take different exits from the copper. Because of this, the 83 result may be more dependent on the choice of weights than the β-strand results were. As can be seen from Table VI, one can increase the coupling to

83 by increasing the weights on 46 and 45, without having any effect on the β-strand acceptors. This suggests a possible resolution to the problem of 83: is it possible that 45 and 46 can provide a coupling comparable to that of 112's sulfer, increasing the calculated coupling to bring it in line with experiment? That is doubtful and would violate our assumptions about the Cu couplings, but in any case the real problem is more fundamental and is found in the tube breakdown.

The tube breakdown for 83 in Table III shows that tubes come from 112 with different signs, with magnitudes around $5 \pm 3 \times 10^{-5}$. Also, there are additional tubes coming from 45 and 46 with different signs, with magnitudes around 10^{-5}. The best tube to 83 goes via 112 in 18 steps; using it alone for T_{DA} yields 7.8×10^{-5}. On the other hand, the full $T_{DA}(-2.06\gamma)$ (computed from Table VI and the rational weights) is 3.7×10^{-6}. Thus, for 83, the best tube is doing better than the full protein by a factor of 20, a much larger number than the analogous numbers in Table VII for the other acceptor locations, even though the 83 tube has about the same length as the 112–109 and 121–124 routes (strictly, the numbers in Table VII show G_{da} ratios, not T_{DA} ratios, but including the weights will not change this observation).

If *only* this tube's contribution were included in the plots of T_{DA}^2 in Figure 15, the 83 coupling would improve by a factor of 20^2, bringing the theory ratio lines up by the $\log_{10}$ of that, or about two and a half orders of magnitude, where they would actually be a little above the experimental lines at -2.06γ. But the total coupling is the sum of the tubes, and for 83 one notes that small (less than an order of magnitude) changes in the coupling of one tube can cause large (more than an order of magnitude) changes in the sum, because the numbers going into the sum have opposites signs. The tube method reveals that the case of 83 will be much more sensitive to tube interference effects and thus more sensitive to both dynamical effects and the details of the Hamiltonian.

Position 109 is a different story. The shortest, best tube also happens to start at the strongest Cu ligand, 112. If one increases the 121 weight enough to help 109, the ratios for 107, 122, 124, and 126 are ruined, because they depend on 112 being stronger than 121, with the 126 reaction being helped by constructive path interference. A possible resolution to the problem of 109's fast rate is that the Ru probe itself might be making an additional contact to the protein, at some point closer to the Cu than the root of HIS-109. Such a contact is not represented in our bridge. Perhaps the best suspect for this contact would be the external hydrophylic SER-118; its OG oxygen is available for H-bonding, and 118 neighbors one of the Cu ligands (117). The stereochemistry of the static model, however, does not support this possibility. There is not enough room to get the bulky HIS-Ru(bpy)$_2$(im) into the position an H-bond would require.

Nor is it likely that 83 is taking advantage of a favorable tube we have not recognized. A large bulk of the protein sits between 83 and the copper, so merely shifting the probe around on the surface is not going to find a noticeably shorter route. Also, as already discussed in Section 6.2, 83 : HIS : ND1 appears to H-bond to 84 : O in all crystallographic data available, and this stabilizes the probe in a position that provides exactly the tubes appearing in Figures 7 and 8, no more, no less. The exact role (or lack thereof) of interference in the rate to 83 is impossible to determine with the pathways Hamiltonian. Such a simple Hamiltonian cannot possibly be accurate under these conditions, but *by showing that large numbers of opposite sign are adding up to a smaller number*, the tube method tells us where we need to be more careful. It is doubtful that any biologically important ET reaction would have evolved to function under such equivocal circumstances.

J.   Summary of Example

Azurin's structure was examined to identify the hydrogen bonds and the bridge entrance and exit points. The relative magnitudes of the couplings into and out of the bridge via these points were estimated from reports of electronic structure calculations and from simple arguments involving distances.

An ϵ network was constructed, and a pathways analysis determined that jumps would not play a role in the reactions being considered (due to high H-bond connectivity). Ratios of PC$_{max}$ were compared to experimental ratios to see how well pathways did and to look for anomalies where tube interference could play a role. A pathways Hamiltonian was constructed for the bridge, and a subset of experimental results associated with a single β-strand were used to calibrate the single bridge parameter γ/E_{tun}. This was done by plotting full protein T_{DA} values as a function of E_{tun} and looking to see where T_{DA} ratios fell into closest agreement with experimental rate ratios.

Tubes were identified using a tube selection process that decimated a full set of forward paths. The primary features of the tubes, including their endpoints, the number of steps in their cores, and their tube couplings (computed with the pathways Hamiltonian at the E_{tun} selected), were laid out in tables. Simple tube schematics were drawn. Tube couplings exhibited the expected sign oscillation and proper decay with distance. Tubes with the same length and character (i.e., same number of H-bonds, same number of turns, etc.) were found to have couplings with the same sign and magnitude.

The remaining analysis examined how each tube contributed to the overall coupling between the $D–A$ pair in each experiment. It was determined that multiple tubes could positively interfere to enhance the coupling to the 124 and 126 sites. This conclusion was predicated on (1) the strength of the CYS-112 to Cu coupling relative to the MET-121 to Cu coupling, and (2) tubes using hydrogen bonds to span the gap between the 112 and 121, β-strands.

The ET rate to the 107 site, mediated by a single tube, made sense with respect to the 122, 124, and 126 rates. The 109 rate appeared to be anomalously large. The failure to account for this rate may be related to an error in bridge construction. Modeling led us to assume that the HIS-Ru(bpy)$_2$(im) probe could not (through simple dihedral angle adjustments in the HIS) adopt a position that would give it a shortcut to the Cu (beating its backbone route). Nevertheless, without more structural data it would be impossible to rule out another contact.

The coupling to 83 was a case of large tube couplings of differing sign adding to a total coupling with a magnitude that was smaller than that of its largest tube component. The tube method identifies this as a critical situation, highly dependent on relative phases, and therefore also dependent on details in the Hamiltonian and possible dynamic factors as well.

To better understand the story of β-sheet-mediated ET, it would be enlightening to have another set of ET rate measurements, like the 122, 124, 126, 107, and 109 series discussed here, but with a different balance between the 112 and 121 couplings, diminishing 112 or enhancing 121 or both. It is unfortunate that the CYS-112-APS mutation is so deleterious to the ET rate [92]; perhaps gentler adjustments to this ligation are possible. Also, the high rate to 109 may merit further investigation. Is the HIS-Ru(bpy)$_2$(im) probe making a contact that provides a better $D–A$ route? Modeling suggests that this is unlikely. Also, the role of H-bonds in ET needs further elucidation. This study suggest that their role is as important as the role of covalent bonds.

VII.   New Hamiltonians and New Directions

Do more detailed Hamiltonians simply "improve" results, or will they to give rise to qualitatively new features? It is essential that theory be performed in a controlled manner, adding complexity in systematic and testable steps, that make sense in terms of preceding models (hence the elaborate review of the pathways model of this paper). Direct comparisons between calculations performed at an arbitrary level of complexity with experimental data will not necessarily increase our understanding in this field. It is essential to compare Hamiltonians derived at different levels of complexity for static proteins and to analyze sensitivity to geometrical uncertainty and dynamical fluctuations. As shown in the long example above, each of these sources of uncertainty may be responsible for large influences on the final prediction. Only by understanding each factor independently will we achieve reliable and responsible comparisons between theory and experiment.

Efforts are under way to improve ET Hamiltonians. In the early 1990s, several attempts were made to compute T_{DA} in proteins using molecular orbital theory. Regan et al. [37], Marcus, Siddarth, and co-workers [52,- 59], Gruschus and Kuki [103], Broo [104,105], Braga and Larsson [106,107], Evenson and Karplus [108], Goldman [109], Priyadarshy et al. [110,111], Ratner et al. [21,63], Mikkelsen et al. [96], Okada et al. [57,58] utilized simple molecular orbital models. These models employed independent-electron Hamiltonians (extended-Hückel, pathways-like Hamiltonians, hole tunneling Hamiltonians, or hybrid approaches). Intensive efforts continue toward building meaningful Hamiltonians, computing couplings in very large systems, and analyzing large matrices of orbital interactions in a productive manner.

What are the challenges associated with the development of better Hamiltonians for ET? The most popular computations beyond the pathways level use extended-Hückel or other independent-electron Hamiltonians. While extended-Hückel methods can probably be parameterized to succeed in the ET problem, in their present standard forms they are extremely limited in their predictive power because of their inconsistent treatment of the protein virtual intermediate state energetics [112]. Protein mediated tunneling is facilitated by virtually populated (i.e., energetically forbidden but very slightly allowed by the uncertainty principle) oxidized ($N - 1$ electron) and virtually reduced ($N + 1$ electron) protein states. As such, reliable protein ET rate calculations must reproduce these energies correctly, and Hartree–Fock or better electronic structure theory is needed to achieve this reliably. Yet, direct use of state-of-the-art computational strategies (employed widely for small molecules) are simply out of reach in proteins. As such, one central theoretical challenge is to establish ways to apply reliable electronic structure methods to these very large systems [64].

The tube strategy presented here provides the framework to solve this challenge with better Hamiltonians. It creates the link between the simple ideas of the pathways model and the more sophisticated Hamiltonians needed for more quantitative results. Using this framework we hope to integrate the wisdom of the simple models, the sophistication of better Hamiltonians, and the proper treatment of protein dynamics, leading to a fully quantitative model for electron transfer.

ACKNOWLEDGMENTS

Work at UCSD was performed with the support of the NSF (MCD-96-03838), the Program in Mathematics in Molecular Biology (NSF DMS-94-06348) and the NIH (GM48043). Caltech work was supported by the Department of Energy (DE-FG03-96ER62219). We thank Ralf Langen, Angelo DiBilio, Jay Winkler, and Harry Gray for fruitful discussion.

REFERENCES

1. S. S. Skourtis and D. N. Beratan, *Adv. Chem. Phys.* (1996); also Part 1 of this volume.

2. J. N. Onuchic and D. N. Beratan, in *Protein Electron Transfer,* D. S. Bendall, ed., BIOS Scientific Publishers, Oxford, 1996.

3. J. N. Onuchic, D. N. Beratan, J. R. Winkler, and H. B. Gray. *Annu. Rev. Biophys. Biomol. Struct.* **21**, 349 (1992).

4. A. Garg, J. N. Onuchic, and V. Ambegaokar, *J. Chem. Phys.* **83**(9), 4491 (1985).

5. R. A. Marcus, *J. Chem. Phys.* **24**(5), 966 (1956).

6. N. S. Hush, *Trans. Faraday Soc.* **57**, 155 (1961).

7. R. A. Marcus, *Annu. Rev. Phys. Chem.* **15**, 155 (1964).

8. N. S. Hush, *Electrochim. Acta.* **13**, 1005 (1968).

9. J. J. Hopfield, *Proc. Natl. Acad. Sci. USA.* **71**(9), 3640 (1974).

10. J. Jortner. *J. Chem. Phys.* **64**, 4860 (1976).

11. M. D. Newton and N. Sutin, *Annu. Rev. Phys. Chem.* **35**, 437 (1984).

12. R. A. Marcus and N. Sutin. *Biophys. Acta* **811**, 265 (1985).

13. J. R. Reimers and N. S. Hush, *Chem. Phys.* **134**, 323 (1989).

14. M. D. Newton, *Chem. Rev.* **91**(5), 767 (1991).

15. I. A. Balabin and J. N. Onuchic, *J. Phys. Chem.* **100**(28), 11573 (1996).

16. S. Priyadarshy, S. Skourtis, S. M. Risser, and D. N. Beratan, J. Chem. Phys. **104**, 9473 (1996).

17. P. O. Löwdin. *J. Math. Phys.* **3**, 969 (1962).

18. S. Larsson. *J. Am. Chem. Soc.* **103**, 4034 (1981).

19. S. Larsson, *J. Chem. Soc. Faraday Trans. II* **79**(P9), 1375 (1983).

20. H. M. McConnell, *J. Chem. Phys.* **35**(2), 508 (1961).

21. M. A. Ratner, *J. Phys. Chem.* **94**, 4877 (1990).

22. C. Liang and M. D. Newton, *J. Phys. Chem.* **96**, 2855 (1992).

23. K. D. Jordan and M. N. Paddon-Row, *Chem. Rev.* **92**, 395 (1992).

24. L. A. Curtiss, C. A. Naleway, and J. R. Miller, *Chem. Phys.* **176**(2/3), 387 (1993).

25. A. S. Davydov and Y. B. Gaididei, *Phys. Status Solidi B* **132**(1), 189 (1985).

26. S. S. Skourtis, D. N. Beratan, and J. N. Onuchic, *Chem. Phys.* **176**(2/3), 501 (1993).

27. D. N. Beratan, J. N. Onuchic, and J. J. Hopfield, *J. Chem. Phys.,* **86**(8), 4488 (1987).

28. D. N. Beratan and J. N. Onuchic, *Photosyn. Res.* **22**, 173 (1989).

29. D. Beratan and J. J. Hopfield, *J. Am. Chem. Soc.* **106**, 1584 (1984).

30. J. N. Onuchic and D. N. Beratan, *J. Amer. Chem. Soc.* **109**(22), 6771 (1987).

31. A. A. S. da Gama, *J. Theor. Biol.* **142**, 251 (1990).

32. C. Liang and M. D. Newton, *J. Phys. Chem.* **97**, 3199 (1993).

33. L. A. Curtiss, C. A. Naleway, and J. R. Miller, *J. Phys. Chem.* **97**(16), 4050 (1993).

34. D. N. Beratan, J. N. Onuchic, and J. J. Hopfield, *J. Chem. Phys.* **83**, 5325 (1985).

35. D. N. Beratan, J. N. Onuchic, J. R. Winkler, and H. B. Gray, *Science,* **258**(5089), 1740 (1992).

36. D. S. Wuttke, M. J. Bjerrum, J. R. Winkler, and H. B. Gray, *Science,* **256**(5059), 1007 (1992).

37. J. J. Regan, S. M. Risser, D. N. Beratan, and J. N. Onuchic, *J. Phys. Chem.* **97**(50), 13083 (1993).

38. D. N. Beratan and J. N. Onuchic, in *Electron Transfer in Inorganic, Organic, and Biological Systems,* J. R. Bolton, N. Mataga, and G. McLendon, eds., American Chemical Society, Washington, D.C., 1991, Chap. 5.

39. G. L. Closs and J.R. Miller, *Science,* **240**, 440 (1988).

40. S. S. Isied, M. Y. Ogawa, and J. F. Wishart, *Chem. Rev.* **92**, 381 (1992).

41. T. B. Karpishin, M. W. Grinstaff, S. Komar-Panicucci, G. Mclendon, and H. B. Gray, *Structure,* **2**(5), 415 (1994).

42. R. Langen, I. J. Chang, J. R. Germanas, J. H. Richards, J. R. Winkler, and H. B. Gray, *Science,* **268**(5218) 1733 (1995).

43. V. Mikkelsen and M. A. Ratner, *Chem. Rev.* **87**(1), 113 (1987).

44. C. Moser, J. M. Keske, K. Warncke, R. S. Farid, and P. L. Dutton, *Nature,* **355**(6363), 796 (1992).

45. G. A. Jeffrey and W. Saenger, *Hydrogen Bonding in Biological Structures,* Springer-Verlag, Berlin (1991).

46. J. N. Betts, D. N. Beratan, and J. N. Onuchic, *J. Am. Chem. Soc.* **114**, 4043 (1992).

47. D. N. Beratan, J. N. Betts, and J. N. Onuchic, *Science,* **252**, 1285 (1991).

48. D. N. Beratan, J. N. Betts, and J. N. Onuchic, *J. Phys. Chem.* **96**(7), 2852 (1992).

49. J. J. Regan, A. J. Di Bilio, R. Langen, L. K. Skov, J. R. Winkler, H. B. Gray, and J. N. Onuchic, *Chem. Biol.* **2**(7), 489 (1995).

50. J. J. Regan, F. K. Chang, and J. N. Onuchic, in *Photochemistry and Radiation Chemistry,* J. F. Wishart and D. G. Nocera, eds., ACS Symposium Books, American Chemical Society, Washington, D. C., 1996.

51. A. Broo and S. Larsson, *J. Phys. Chem.* **95**(13), 4925 (1991).

52. P. Siddarth and R. A. Marcus, *J. Phys. Chem.* **97**(10), 2400 (1993).

53. A. Kuki and P. G. Wolynes, *Science,* **236**, 1647 (1987).

54. S. S. Skourtis, J. J. Regan, and J. N. Onuchic, *J. Phys. Chem.* **98**(13), 3379 (1994).

55. S. S. Skourtis and D. N. Beratan, J. Phys. Chem. B. **101**(7), 1215 (1997).

56. S. S. Skourtis, J. N. Onuchic, and D. N. Beratan, *Inorg. Chim. Acta,* **243**(1/2), 167 (1996).

57. H. Nakagawa, Y. Koyama, and T. Okada, *J. Biochem.* **115**(5), 891 (1994).

58. A. Okada, T. Kakitani, and J. Inoue, *J. Phys. Chem.* **99**(10), 2946 (1995).

59. J. N. Gehlen, I. Daizadeh, A. A. Stuchebrukhov, and R. A. Marcus, *Inorg. Chim. Acta,* **243**(1/2), 271 (1996).

60. C. Turro, C. K. Chang, G. E. Leroi, R. I. Cukier, and D. G. Nocera, *J. Am. Chem. Soc.* **114**(10), 4013 (1992).

61. J. F. de Rege, S. A. Williams, and M. J. Therien, *Science,* **269**, 1409 (1995).

62. J. L. Sessler, B. Wang, and A. Harriman, *J. Am. Chem. Soc.* **117**(2), 704 (1995).

63. V. Mujica, M. Kemp, and M. A. Ratner, *J. Chem. Phys.* **101**(8), 6849 (1994).

64. I. V. Kurnikov and D. N. Beratan, *J. Chem. Phys.* **105**(21), 9561 (1996).

65. A. A. Stuchebrukhov, *J. Chem. Phys.* **104**(21), 8424 (1996).

66. T. Helgaker and P. R. Taylor, in *Modern Electronic Structure Theory*, D. R. Yarkony, ed., World Scientific, Singapore (1995).

67. M. D. Todd and K. V. Mikkelsen, *Inorg. Chem.* **226**(1/2), 237 (1994).

68. R. I. Cukier, *J. Phys. Chem.* **99**(43), 16101 (1995).

69. J. Waluk, *J. Mol. Liq.* **64**(1/2), 49 (1995).

70. D. Morikis and P. E. Wright, *Eur. J. Biochem.* **237**, 212 (1996).

71. S. W. Englander and N. R. Kallenbach, *Q. Rev. Biophys.* **16**(4), 521 (1983).

72. J. Wolfgang, S. M. Risser, S. Priyadarshy, and D. N. Beratan, *J. Phys. Chem.* B. **101**(15), 2986 (1997).

73. I. Daizadeh, E. S. Medvedev, and A. A. Stuchebrukhov, *Proc. Natl. Acad. Sci. USA,* submitted.

74. T. Horio. *J. Biochem (Tokyo),* **45**, 195 (1958).

75. E. T. Adman, in *Metalloproteins,* Part 1, *Metal Proteins with Redox Roles,* P. M. Harrison, ed., Verlag Chemie, Weinheim, Germany, 1985, pp. 1–42.

76. M. D. Lowery and E. I. Solomon, *Inorg. Chim. Acta,* **200**, 233 (1992).

77. E. I. Solomon and M. D. Lowery, *Science,* **259**(5101), 1575 (1993).

78. E. I. Solomon, K. W. Penfield, A. A. Gewirth, M. D. Lowery, S. E. Shadle, J. A. Guckert, and L. B. Lacroix, *Inorg. Chim. Acta,* **243**(1/2), 67 (1996).

79. H. B. Gray, *Chem. Soc. Rev.* **15**, 17 (1986).

80. A. G. Sykes, *Inorg. Chem.* **36**, 377 (1991).

81. M. Van de Kamp, R. Floris, F. C. Hall, and G. W. Canters, *J. Am. Chem. Soc.* **112**(2), 907 (1990).

82. M. Van de Kamp, M. C. Silvestrini, M. Brunori, J. van Beeumen, F. C. Hall, and G. W. Canters, *Eur. J. Biochem.* **194**(1), 109 (1990).

83. H. Nar, A. Messerschmidt, R. Huber, M. Van de Kamp, and G. W. Canters, *J. Mol. Biol.* **218**(2), 427 (1991).

84. H. Nar, A. Messerschmidt, R. Huber, M. Van de Kamp, and G. W. Canters, *J. Mol. Biol.* **221**(3) 765 (1991).

85. O. Farver and I. Pecht, *J. Am. Chem. Soc.* **114**, 5764 (1992).

86. O. Farver and I. Pecht, Biophys. Chem. **50**(1/2), 203 (1994).

87. O. Farver, L. K. Skov, H. Nar, M. Van de Kamp, G. W. Canters, and I. Pecht, *Eur. J. Biochem.,* **210**(2), 399 (1992).

88. O. Farver, L. K. Skov, T. Pascher, B. G. Karlsson, M. Nordling, L. G. Lundberg, T. Vänngård, and I. Pecht, *Biochemistry,* **32**, 7317 (1993).

89. O. Farver, L. K. Skov, G. Gilardi, G. Vanpouderoyen, G. W. Canters, S. Wherland, and I. Pecht, *Chem. Phys.* **204**(1/2), 271 (1996).

90. O. Farver, N. Bonander, L. K. Skov, and I. Pecht, *Inorg. Chim. Acta,* **243**(1/2), 127 (1996).

91. T. J. Mizoguchi, A. J. Dibilio, H. B. Gray, and J. H. Richards, *J. Am. Chem. Soc.* **114**(25), 10076 (1992).

92. T. J. Mizoguchi, PhD thesis, California Institute of Technology, 1996.

93. H. B. Gray, A. DiBilio, and C. Kiser, personal communication.

94. A. C. F. Gorren, T. Denblaauwen, G. W. Canters, D. J. Hopper, and J. A. Duine, *FEBS Lett.* **381**(1/2), 140 (1996).

95. J. W. A. Coremans, O. G. Poluektov, E. J. J. Groenen, G. C. M. Warmerdam, G. W. Canters, H. Nar, and A. Messerschmidt, *J. Phys. Chem.* **100**(50), 19706 (1996).

96. C. V. Mikkelsen, L. K. Skov, H. Nar, and O. Farver, *Proc. Natl. Acad. Sci. USA* **90**, 5443 (1993).

97. I. J. Chang, H. B. Gray, and J. R. Winkler, *J. Am. Chem. Soc.* **113**(18), 7056 (1991).

98. Biosym Technologies, Inc., *InsightII/Discover Software,* Biosym Technologies Inc., San Diego, Calif., 1991.

99. E. N. Baker, *J. Mol. Biol.* **203**, 1071 (1988).

100. M. Day, Ph.D. thesis, California Institute of Technology, 1995.

101. E. I. Solomon, M. J. Baldwin, and M. D. Lowery, *Chem. Rev.* **92**, 521 (1992).

102. S. Larsson, A. Broo, and L. Sjolin, *J. Phys. Chem.* **99**(13), 4860 (1995).

103. J. M. Gruschus and A. Kuki, *J. Phys. Chem.* **97**(21), 5581 (1993).

104. A. Broo, *Chem. Phys.* **169**, 135 (1993).

105. A. Broo, *Chem. Phys.* **169**, 151 (1993).

106. M. Braga and S. Larsson, *Int. J. Quant. Chem.* **44**(5), 839 (1992).

107. M. Braga and S. Larsson, *J. Phys. Chem.* **97**(35), 8929 (1993).

108. J. W. Evenson and M. Karplus, *Science,* **262**(5137), 1247 (1993).

109. C. Goldman, *Phys. Rev. A* **43**(8), 4500 (1991).

110. S. Priyadarshy, S. M. Risser, and D. N. Beratan, *J. Phys. Chem.* **100**(44), 17678 (1996).

111. S. Priyadarshy, S. M. Risser, and D. N. Beratan, *Int. J. Quantum Chem.* **60**(8), 65 (1996).

112. W. B. Curry, I. V. Kurnikov, S. S. Skourtis, D. N. Beratan, J. J. Regan, A. J. A. Acquino, P. Beroza, and J. N. Onuchic, *J. Bioenerg. Biomembr.* **27**(3), 285 (1995).

COPPER PROTEINS AS MODEL SYSTEMS FOR INVESTIGATING INTRAMOLECULAR ELECTRON TRANSFER PROCESSES

OLE FARVER

Institute of Analytical and Pharmaceutical Chemistry, Royal Danish School of Pharmacy, 2 Universitetsparken, DK-2100 Copenhagen Ø, Denmark

ISRAEL PECHT

Department of Immunology, The Weizmann Institute of Science, Rehovot 76100, Israel

CONTENTS

Electron Transfer: From Isolated Molecules to Biomolecules, Part Two, edited by Joshua Jortner and M. Bixon. Advances in Chemical Physics Series, Volume 107, series editors I. Prigogine and Stuart A. Rice.
ISBN 0-471-25291-3   © 1999 John Wiley & Sons, Inc.

I. INTRODUCTION

Two of the first-row transition-metal ions, iron and copper, serve central roles in catalyzing or mediating biological redox processes. Copper proteins have been found to be involved in a wide range of biological processes, which with the exception of the O_2-carrying hemocyanin, perform different electron transfer reactions, predominantly those of biological energy conversion cycles [1]. The last decade has seen a dramatic advance in our understanding of structural, electronic, and functional properties of copper proteins, due mainly to the determination of three-dimensional structures at high resolution of representative members of these proteins [2]. This progress has enabled meaningful analyses of their reaction mechanism. The structural information has in particular allowed for analyzing, in more quantitative terms, the rates and activation parameters determined for different electron transfer processes and relating them to the current theoretical treatments. We have focused our attention mainly on the electron transfer reactions of copper proteins for the following reasons:

1. Electron transfer reactivity is, as stated above, the main function performed by the different forms of copper ions in biological systems.
2. In all cases known so far, the copper ions shuttle between the mono- and divalent oxidation states, mediating single electron transfer only.
3. The copper ions are coordinated directly to amino acid residues (also to the peptide carbonyl or amido groups) without intervening cofactors or other ligands (e.g. porphyrin or inorganic sulfur).
4. The rather unique coordination sites that evolved for copper ions in proteins enable them to perform most efficiently specific electron transfer reactions (e.g. protein–protein electron mediation or multiple reduction steps of small molecules such as O_2 or N_2O).

Taken together, the features listed above make copper proteins important and informative for resolving fundamental principles that control electron transfer processes within and among proteins. In addition, since other parameters, such as the role of intermediary cofactors are excluded, resolution of the role of the polypeptide matrix and separation distance determination in a more direct and unambiguous fashion becomes possible. In the following we first review structural properties of different types of copper proteins involved in electron transfer processes and then proceed to discuss results of relevant studies of these systems.

The major group of copper-containing electron transfer proteins comprise the "blue" proteins, which constitute a rather large family, widespread in nature [3]. Those blue proteins containing only a single copper ion are of a relatively low molecular weight, 10–20 kD, and mediate single electrons, usually between large immobilized, membrane-bound protein assemblies. They are named cupredoxins, in analogy to a functionally similar group of iron proteins. The blue copper–containing oxidases comprise a second group. In the multicopper enzymes, the blue copper center constitutes one type of redox center, whereby sequential one-electron steps take place from the reducing substrate to a dioxygen binding trinuclear center, where the latter oxidizing substrate becomes reduced to two water molecules [4,5]. Intramolecular electron transfer is a process where specificity as well as control are exerted in both types of blue proteins [6]. Another, less colorful group of copper proteins catalyzes the oxidation of a variety of organic substrates, with the concomitant two-electron reduction O_2 to H_2O_2. These enzymes contain a single copper ion as part of their active site, and include galactose oxidase, some of the amine oxidases, and the copper–zinc superoxide dismutase [7].

II. STRUCTURE

A considerable amount of structural information is now available on different types of copper proteins, from the cupredoxins and single copper oxidases to multicentered enzymes such as ascorbate oxidase and cytochrome c oxidase as well as for the dioxygen transport protein hemocyanin. The first cupredoxins for which the three-dimensional structure has been solved were poplar tree (*Populus nigra*) plastocyanin (Pc) [8] and *Pseudomonas aeruginosa* azurin (Az) [9] (Fig. 1). High-resolution structural information is now available for both oxidized [Cu(II)] and reduced [Cu(I)] forms of these proteins and at different pH values [10], as well as for their metal-depleted apo forms [11]. More recently, the structures of other cupredoxins, such as the cucumber blue protein [12], amicyanin (Am)s [13], and stellacyanin [14] have been determined. The first and last are unusual for their lack of methionine, which constitutes one of the metal ligands in most blue copper centers. The amicyanin structure is particularly noteworthy since this cupredoxin has also been crystallized in its complexes with methylamine dehydrogenase and cytochrome c_{551i}, both assumed to be its physiological reaction partners [13,15].

Three-dimensional structures have also been determined for different metal ion–substituted cupredoxins [2] and for an increasing number of mutated azurins [16]. These yielded important insights discussed below. The folding topology of all cupredoxins studied so far is very similar, consisting of an eight-stranded β-sheet structure with a highly hydrophobic interior. The "blue" copper site classified as type 1 (T1) is found in one end of the β-sheet protein (commonly referred to as the "Northern region").

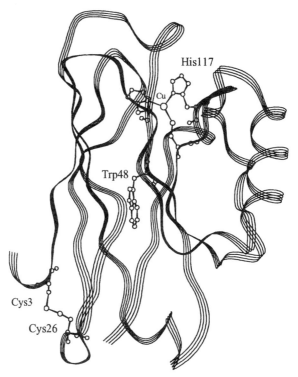

Figure 1. Three-dimensional structure of the main-chain polypeptide of *Pseudomonas aerugi-nosa* azurin. The copper center is shown together with some other amino acid residues of particular interest. (From [16]).

The metal ion is isolated from direct contact with the solvent by its ligands and a high density of surrounding hydrophobic residues. This arrangement is most probably essential in providing specificity and control of the physiological function in this group of proteins.

The T1 center seems to be coordination saturated and shows no tendency to bind additional exogenous ligands (Fig. 2). Three of the copper ligands are highly conserved: two imidazoles (His) and one thiolate group (Cys) providing a trigonal planar ligand configuration which stabilizes the Cu(I) relatively to the Cu(II) state. It is the charge transfer from the cysteine thiolate to Cu(II) that gives rise to the intense blue color of the group of proteins [17]. The observed small differences in the copper site structure of cupredoxins are seen primarily in the strength and geometry of the weaker axial ligand interactions. In contrast, the configuration of the copper site is only slightly affected by the oxidation state of the metal ion [10, 16, 18].

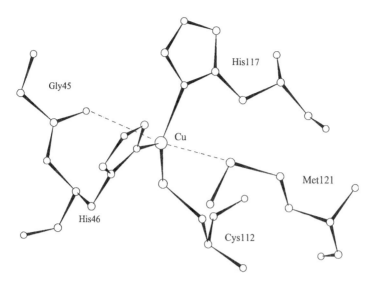

Figure 2. Blue copper center in azurin. The three close ligands are shown attached to the copper ion by solid lines, and the two distal ligands are connected to the metal center by dashed lines. (From [16].)

Even removal of the metal ion does not change the proteins' conformation to a significant extent [12]. The relation between copper site structure and their redox potentials is considered further below.

Whereas T1 sites are involved only in electron transfer activity, type 2 (T2) Cu(II) centers were identified originally in the blue oxidase laccase, and through determination of the three-dimensional structure of ascorbate oxidase (AO) assigned to be part of the trinuclear dioxygen reduction site in the blue copper oxidases. In the latter enzyme, T2 copper has two His imidazole ligands and one water or hydroxyl ligand arranged in a trigonal geometry [19]. In other enzymes, T2 centers may have quite different structures and reactivities. Thus in superoxide dismutase (SOD) the geometry of the copper site is a distorted square pyramid with four in-plane histidine imidazoles and a labile water molecule in the apical position [20]. This type of site is thus reminiscent of low-molecular-weight inorganic copper complexes.

The type 3 (T3) copper site, together with T2, constitute the trinuclear O_2 reduction site in blue oxidases. It binds a pair of copper ions which, in the Cu(II) state, is strongly antiferromagneticaly coupled [21]. In the dioxygen-carrying hemocyanin, an analogous copper pair provides the O_2 binding site. In deoxyhemocyanin the copper ions are 0.35 nm apart and each metal ion is coordinated to N_ϵ-nitrogens of two close imidazoles, while a third imidazole ligand is placed at a more distant axial position [22]. In oxyhemocyanin the two copper ions are additionally coordinated to O_2 in $\mu - \eta^2 : \eta^2$ configuration [23]. In AO the T3 copper site has quite a different geometry, being coordinated to six imidazoles, arranged in a trigonal position with an OH^- bridging the copper pair [19,24] (cf. below).

All blue oxidases have a minimum of four redox centers per molecule. In those containing only copper ions, there is one of each of the above-mentioned types: T1, T2, and T3. These enzymes catalyze oxidation of their substrates by O_2 with a concomitant four-electron reduction of dioxygen to water [3]. Ascorbate oxidase is a homo-dimer of 70-kD subunits [19,24]. Each monomer is constructed of three domains, with T1 copper bound inside one domain, while three copper ions that constitute a trinuclear T2–T3 center are bound to four histidine imidazoles from one domain and four additional imidazoles from the second one (Fig. 3). Of the three copper ions in the trinuclear center, two have coordination similar to that of the copper pair in hemocyanin and are by analogy assigned as being the T3 site, while the remaining copper ion would then by exclusion be T2. None of the imidazoles are involved in bridging the copper ions. The T2 site is the one most accessible to exogenous ligands and was also shown to bind anions such as F^-, N_3^-, and CN^-. Indeed, more recently [25] the structure of both azide and peroxide derivatives as well as of the reduced form of AO were determined. While no change was resolved in the T1 and T2 coordinations, the oxygen bridging the T3 Cu pair disappears in the reduced protein, and the distances separating the metal ions in the trinuclear center increase markedly. In the peroxide derivatives of AO, the T1 site also remains unchanged, while the oxygen atom bound to T3 disappears and the peroxide is coordinated terminally as HO_2^- to one of the T3 copper ions, which is, in addition, coordinated to histidyl imidazoles in a distorted tetrahedral configuration. A marked increase in metal ions separation at the trinuclear site takes place as well. Two azide ions bind terminally to one copper ion of the T3 pair, again releasing the oxygen (hydroxide or oxide) ligand and making it penta-coordinated with longer distances among all three metal ions in this site [25].

The function and specificity of the oxidases are provided primarily through their multiple redox centers. Since the blue oxidases are reduced in single-electron steps by the substrates, the electrons are taken up by T1 Cu(II) while O_2 is reduced by the trinuclear center [26]. Intramolecular electron transfer is an essential part of the catalytic redox cycle. It is therefore noteworthy that residues of adjacent amino acid residues constitute ligands to the T1 and T3 sites and may also provide the electron transfer pathway: In the observed His–Cys–His sequence, the cysteine thiolate is

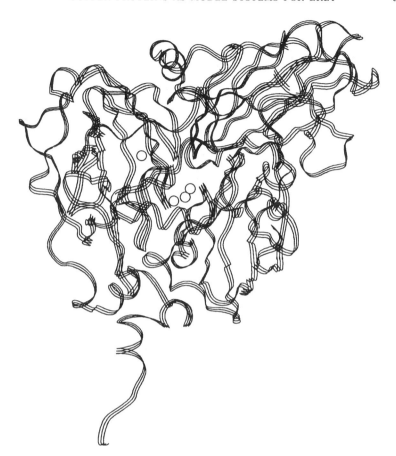

Figure 3. Ribbon drawing of ascorbate oxidase. The positions of the four copper ions are indicated by open circles. The coordinates of AO are from [19].

coordinated to T1 copper and the two histidyl imidazoles to the two T3 copper ions of the trinuclear cluster [19,24].

Undoubtedly, the most interesting and important copper-containing enzyme is cytochrome *c* oxidase (COX), which, in its various forms, serves as the terminal, dioxygen-reducing catalyst of the respiratory electron transfer chain [27]. Being an integral membrane protein, composed of 13 distinct polypeptide chains, slowed down resolution of its three-dimensional structure. Still, a range of different methodologies were combined to yield a remarkably detailed understanding of its structural properties, in particular of its redox active centers. The recent success in determining the three-dimensional structure of both a pro- [28] and eucaryotic [29] COX enabled a detailed reexamination of structure–function relations, specifically electron transfer among the active sites of this enzyme, and will naturally lead to better targeted investigations in the future. In the following only a brief overview of structural features will be presented, setting the stage for a later brief review of internal Long Range Electron Transfer (LRET) operating in this enzyme.

The Cu_A site, now shown to bind two copper ions, was used to estimate the direction of the molecule and its C and N termini relative to the cytosolic and matrix sides of the membrane [28]. The site is located in subunit II and protrudes to the cytosolic side, 0.8 nm above its surface. The heme a and the second binuclear site, constituted of heme a_3 and Cu_B are all bound to subunit I at an identical level of about 1.3 nm below the cytosolic membrane surface. The binuclear Cu_A site is constructed of eight ligands: two bridging cysteinyl thiolates, four histidyl imidazoles, one methionine and either a peptide or a glutamate carbonyl. These are arranged around the two copper ions bridged by the two thiolates, placing all four atoms in the same plane at distances of 0.27 and 0.38 nm for the Cu–Cu and S_γ–S_γ, respectively. One Cu is coordinated to a His imidazole and a methionine sulfur, while the second is bound to another imidazole and a peptide carbonyl oxygen, making both coppers tetrahedrally coordinated. This novel binuclear structure is a mixed valence $[Cu–Cu]^{3+}$ complex in its oxidized state and has its unpaired electron fully delocalized over both nuclei, which probably has important functional implications.

The heme a iron has been found to be coordinated to two His imidazoles, with the ring essentially perpendicular to the membrane plane, forming a 104° angle with heme a_3, to which another His imidazole provides one apical ligand. A phenylalanine residue lies halfway between the two planes of one of the imidazole ligands of heme a and heme a_3, at an equidistance of 0.35 nm. This arrangement is proposed to be potentially in control of electron transfer between the two hemes [28].

The Cu_B is coordinated to three His imidazoles with no other ligands, not even water, detected. It is 0.45 nm away from the heme a_3 and displaced 0.1 nm from the normal to this heme iron, where the latter is slightly displaced toward its fifth ligand. Although the short separation between heme a_3 and Cu_B suggests direct electron transfer between them, additional routes are proposed to provide multiple pathways leading to O_2 reduction catalyzed by this site [28].

III. REDOX-SITE PROPERTIES

The copper(II) ion in the T1 site is approximately coplanar with its three closest ligands, and a particularly strong π-bonding interaction with the thiolate sulfur leads to an electronic ground state that is highly covalent [30]. The structure of this site has neither the preferred geometry of low-molecular-weight Cu(II) complexes, which is tetragonal, nor that of the tetrahedral structure, typical for Cu(I) complexes. Rather, it is distorted tetrahedral (i.e., a compromise between these two configurations). This unique structure of the T1 copper site, however, is not the cause for its unusual spectroscopic properties. The intense color is due to $\pi S \rightarrow Cu(d_{x^2-y^2})$ ligand-to-metal charge transfer involving a cysteine thiolate as the electron donor [31] and with a molar extinction coefficient of $2000–6000 \, M^{-1} \, cm^{-1}$ [i.e., several 100-fold larger than that observed for regular Cu(II) complexes]. The unique electronic properties of T1 copper(II) is also expressed in its EPR spectrum, displaying a particular narrow hyperfine splitting in the $g_{\parallel}$ region (ca. $0.008 \, cm^{-1}$), due to interaction of the copper nuclear and electron spins, which is approximately 50% smaller than those of most smaller copper complexes [32]. This is attributed to delocalization of the unpaired $Cu(d_{x^2-y^2})$ electron onto the $Cys(S)p\pi$ orbital, thus reducing the nuclear–electron interaction [17].

These T1 spectroscopic properties are assumed to be related to its function. The reduction potentials of the T1 Cu(II)/Cu(I) center are higher than those generally encountered in other copper sites and complexes. As stated above, the ligand configuration changes from a square planar toward a distorted tetrahedral geometry and preferentially stabilizes the Cu(I) state relative to Cu(II). Structure provided $Cu \rightarrow L$ π-backbonding has been proposed to account further for stabilization of the cuprous state since strong π-interaction with the d_{π} orbitals results in increased ligand field strength [33]. The observed minimal difference in structure of the blue copper center upon redox change will lower the Franck–Condon barrier for electron transfer (see Sec. IV).

While T2 sites with tetragonally arranged ligands show the expected lower reduction potentials characteristic for most copper complexes, T3 copper pair generally have redox potentials as high as those of the T1 copper site, although the former ligand sphere consists solely of imidazoles and a hydroxide [25]. Distortion of the copper site toward a more tetrahedral symmetry - stabilizing Cu(I) relative to Cu(II) may again be the cause for the high potentials observed. In ascorbate oxidase the six T3 histidyl imidazole ligands (five N_{ϵ} and one N_{δ}) are arranged in a trigonal prism [19], while as noted above, in hemocyanin the arrangement of the T3 site is quite different [22,23]. This disparity in geometry is not surprising when considering the very different chemical functions of the two proteins: the irreversible four-electron reduction of dioxygen to water versus reversible O_2 binding. The reduction potentials of the T1 and T3 sites in AO are both ca. 350 mV at pH 7.0 [34]. Interestingly, the similarity of T1 and T3 potentials is also conserved in the other blue oxidases, and for example, in fungal laccase both these potentials are exceptionally high (ca. 780 mV at pH 7.0) [35]. The concordance between the T1 and T3 potentials is still an unanswered puzzle. The reversibility of the oxygen binding to hemocyanin places the Cu(II)/Cu(I) reduction potential of this protein also at around 800 mV. This again suggests that high redox potentials are determined primarily by the configuration of the ligands more than by their particular chemical nature (e.g., thiolate in T1).

As already mentioned, unlike iron–sulfur and heme proteins, the copper-containing sites in electron transfer proteins have no extended coordination sphere. Rather, the copper ion is directly coordinated to protein residues, which then define the active site structure. As noted above for the T1 site, the protein fold provides a geometry that is a compromise between the preferred tetragonal Cu(II) structure and a tetrahedral structure for Cu(I). Except for a slight lengthening of the copper-ligand bonds in the Cu(I) site, the structures are remarkably similar in both oxidation states [10, 16, 18]. As the T1 sites function as electron mediators, this structural property is expected to minimize changes occurring upon redox cycling and hence cause lower Franck–Condon barriers. Thus, using published reorganization energies of Cu(I) to Ru(III) electron transfer [36], we calculate a reorganization energy for the copper center in azurin to be $53 \, \text{kJ mol}^{-1}$ (0.55 eV), in line with the notion that this site has been evolutionarily optimized for efficient electron transfer. In addition to minimal changes in copper–ligand bond lengths and angles, the metal ion is embedded in the hydrophobic interior of the protein so that solvation contributions to outer-sphere reorganization energies also become negligible. It should, however, be noted that entropic factors caused by conformational changes and hydrophobic interactions may also be significant. Comparing the electron self-exchange rate constants for the two blue copper proteins plant plastocyanin ($k_{ESE} \leq 10^4 \, \text{M}^{-1} \, \text{s}^{-1}$) [37] and azurin ($k_{ESE} \sim 10^6 \, \text{M}^{-1} \, \text{s}^{-1}$) [38], it becomes obvious that minimal reorganization energies cannot be the only decisive factor for effectve electron transfer.

Extensive electronic overlap between the electron donor and acceptor orbitals is a prerequisite for a strong electronic coupling. Spectroscopic studies of single crystals of plastocyanin have been combined with self-consistent-field calculations and indicate an orientation of the half-filled $d_{x^2-y^2}$ orbital bisecting the Cu–Cys(S_γ) bond, which leads to a high degree of anisotropic covalency favoring electron transfer through the cysteine compared with the two imidazole ligands [31]. One may expect that a similar extent of anisotropic covalency also exists in AO, where intramolecular electron transfer is assumed to take place from T1 copper(I) via the analogous thiolate ligand (cf. Sec. VII.A). However, in azurin, with a similar very intense $S(Cys)p\pi \rightarrow Cu^{2+}$ ($d_x2_{-y}2$) transition, the thiolate is directed toward the interior of the protein and the physiological electron transfer is supposed to take place via the His117 imidazole. Therefore, it still has to be established if and how the apparently identical copper sites in these two different proteins may nonetheless have distinct orientations of their electronic structures.

IV. THEORETICAL MODELS FOR ELECTRON TRANSFER PROCESSES

Several excellent reviews have been published on electron transfer theory [39]. LRET within proteins is characterized by a weak interaction between electron donor and acceptor, and in the nonadiabatic limit the rate constant is proportional to the square of the electronic coupling between reactant and product, represented in the form of a matrix element, H_{AB}. For intramolecular electron transfer, the rate constant k is given by Fermi's golden rule [39].

$$k = \frac{2\pi}{\hbar} H_{AB}^2 (FC)$$

For relatively low vibrational frequencies, where $kT > h\nu$, the Franck–Condon factor (FC) for nuclear motions can be treated clasically, a condition that often applies to biological electron transfer. The electronic motion, however, requires a quantum mechanical treatmet, and the semi-classical Marcus equation [39] may be expressed as

$$k = \frac{2\pi}{\hbar} \frac{H_{AB}^2}{(4\pi\lambda RT)^{1/2}} e^{-(\Delta G^0 + \lambda)^2/4\lambda RT}$$

Since wavefunctions decay exponentially with distance, the tunneling matrix element H_{AB} will decrease with distance $(r - r_0)$ as

$$H_{AB} = H_{AB}^0 e^{-\beta(r-r_0)/2}$$

H_{AB}^0 is the electronic coupling at direct contact between electron donor and acceptor (where $r = r_0$) and the decay of electronic coupling with distance is determined by the coefficient, β. Electron transfer pathways can now be identified by analyzing the bonding interactions that lead to the maximal H_{AB} [40].

V. CUPREDOXINS AS MODELS FOR STUDYING
INTRAMOLECULAR LRET

A. Azurin

All known azurins contain a highly conserved disulfide bridge (between Cys 3 and 26) at one end of the β-sheet structure at a distance of 2.6 nm from the copper center present in the opposite end of the protein (cf. Fig. 1). Although the role of this disulfide is most probably structural rather than electron transfer activity, it provides azurin with a potential second redox center. This, combined with the high-resolution three-dimensional structures now available for several wild-type and single-site mutated *P. aeruginosa* azurins, yields a highly interesting system for analysis of the various factors that determine the rates of LRET within the polypeptide matrix [6]. Using sufficiently strong reductants (i.e. $E^0 \leq -1$ V), the disulfide bridge can be reduced by a single electron to produce the radical anion indicated in scheme (i), which is paralleled by the competing direct reduction of the Cu(II) ion (ii). The radical anion RSSR$^-$ exhibits an intense absorption band centered at 410 nm ($E_{410} = 10,000 \, \mathrm{M^{-1} \, cm^{-1}}$). Its transient absorption was found to decay (iii) concomitantly with a decrease in the characteristic Cu(II) absorption band at 625 nm ($E_{625} = 5700 \, \mathrm{M^{-1} \, cm^{-1}}$). Both these processes take place with identical rate constants that are independent of both the protein and RSSR$^-$ concentrations [41].

$$\mathrm{RSSR\text{-}Az[Cu(II)]} + \mathrm{CO_2^-} \rightarrow \mathrm{RSSR^-\text{-}Az[Cu(II)]} + \mathrm{CO_2} \qquad \text{(i)}$$

$$\mathrm{RSSR\text{-}Az[Cu(II)]} + \mathrm{CO_2^-} \rightarrow \mathrm{RSSR^-\text{-}Az[Cu(I)]} + \mathrm{CO_2} \qquad \text{(ii)}$$

$$\mathrm{RSSR^-\text{-}Az[Cu(II)]} \rightarrow \mathrm{RSSR\text{-}Az[Cu(I)]} \qquad \text{(iii)}$$

Process (iii) is therefore an intramolecular electron transfer from RSSR$^-$ to Cu(II). Azurins isolated from distinct bacteria share highly homologous sequences and therefore most probably, similar three-dimensional structures. Still, the different azurins were found to exhibit distinct reactivities and reduction potentials. The specific rates of the foregoing intraprotein LRET at pH 7.0 and 298 K determined for a large number of different azurins, both wild types and single-site mutants, are summarized in Table I together with the Cu(II)/Cu(I) reduction potentials and activation parameters.

    LRET pathway calculations for the process above using available high-resolution three-dimensional structures of wild-type azurins and of *P. aeruginosa* single-site mutants [16] predict that similar electron transfer routes operate in all the native and mutated proteins studied so far. Two

TABLE I

Kinetic and Thermodynamic Data for the Intramolecular Reduction of Cu(II) by RSSR⁻

Azurin	k_{298} (s^{-1})	E' (mV)	$-\Delta G^0$ (kJ mol^{-1})	ΔH^* (kJ mol^{-1})	$\Delta S^{\neq}$ (J K^{-1} mol^{-1})
Wild					
Pseudomonas aeruginosa	44 ± 7	304	68.9	47.5 ± 2.2	-56.5 ± 3.5
P. fluorescens	22 ± 3	347	73.0	36.3 ± 1.2	-97.7 ± 5.0
Alcaligenes spp	28 ± 1.5	260	64.6	16.7 ± 1.5	-171 ± 18
A. faecalis	11 ± 2	266	65.2	54.5 ± 1.4	-43.9 ± 9.5
Mutant					
D23A	15 ± 3	311	69.6	47.8 ± 1.4	-61.4 ± 6.3
F110S	38 ± 10	314	69.9	55.5 ± 5.0	-28.7 ± 4.5
F114A	72 ± 14	358	74.1	52.1 ± 1.3	-36.1 ± 8.2
H35Q	53 ± 11	268	65.4	37.3 ± 1.3	-86.5 ± 5.8
I7S	42 ± 8	301	68.6	56.6 ± 4.1	-21.5 ± 4.2
M44K	134 ± 12	370	75.3	47.2 ± 0.7	-46.4 ± 4.4
M64E	55 ± 8	278	66.4	46.3 ± 6.2	-56.2 ± 7.2
M121L	38 ± 7	412	79.3	45.2 ± 1.3	-61.5 ± 7.2
V31W	285 ± 18	301	68.6	47.2 ± 2.4	-39.7 ± 2.5
W48A	35 ± 7	301	68.6	46.3 ± 5.9	-58.3 ± 6.0
W48F	80 ± 5	304	68.9	43.7 ± 6.7	-61.9 ± 9.7
W48S	50 ± 5	314	69.9	49.8 ± 4.9	-44.0 ± 3.5
W48Y	85 ± 5	323	70.7	52.6 ± 6.9	-30.2 ± 3.6
W48L	40 ± 4	323	70.7	48.3 ± 0.9	-51.5 ± 5.7
W48M	33 ± 5	312	69.7	48.4 ± 1.3	-50.9 ± 7.4

Source: [41]

main pathways were found, differing by one order of magnitude in their electronic coupling factors (Fig. 4). One pathway proceeds along the polypeptide backbone from Cys3 to Asn10, which is connected to the copper ligating imidazoe of His46 by a H-bond between its carbonyl group and $N_\epsilon H$ of the imidazole group. The alternative pathway leads directly from Cys3 to Thr30 via a H-bond and then, from Val31 by a through-space jump across 0.4 nm to Trp48 (or the mutated residue 48). The neighboring residue, Val49, is connected by another hydrogen bond to Phe111, which may lead the electron further via the peptide bond to the copper-ligating thiolate residue of Cys112 [6].

Two characteristics of this LRET in azurin are noteworthy and instructive: electron transfer proceeds via a typical β-sheet structured medium, and the direct distance separating the two redox centers is maintained at the same value in all derivatives studied. Since the same pathway from RSSR⁻ to Cu(II) has been found to operate in all azurins studied so far, it was possible to use the extensive set of independent experimental data to

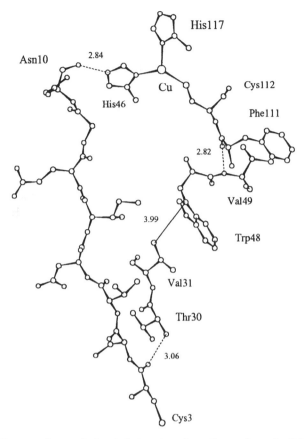

Figure 4. Two prominent calculated electron transfer pathways from disulfide to copper in azurin. Some of the interconnecting distances are shown (in Å). Hydrogen bonds are shown as dashed lines and the through-space jump is indicated by a thin line. The structure coordinates are from [16] and the pathway calculations were performed using the methodology developed by Beratan and Onuchic [40].

calculate the reorganization energy of this process, $\lambda = 99 \pm 5\,\mathrm{kJ\,mol^{-1}}$, and the electronic decay factor, $\beta = 6.7 \pm 0.3\,\mathrm{nm^{-1}}$ [42]. This β-value is in excellent agreement with both the theoretical value calculated for tunneling through a saturated $-(CH_2)_n$ chain ($6.5\,\mathrm{nm^{-1}}$) [43] and the experimentally determined decay factor for electron tunneling in different modified cytochrome c ($7.1\,\mathrm{nm^{-1}}$) [44].

Significantly, no distinction was made in these calculations between pathways involving σ- and π-orbitals. Another theoretical study, however, applying extended Hückel calculations, implies that aromatic residues may intensify the electronic coupling [45]. In several systems that can be assumed to have been selected by evolution for efficient electron transfer, aromatic residues were found in positions where they can enhance the electronic coupling between electron transfer donor and acceptor. This is illustrated by the assumed tryptophan-mediated reduction of a quinone in the photosynthetic reaction centers [46], in the tryptophan–tryptophylquinone system present in methylamine dehydrogenase [13], as well as in the [cytochrome c peroxidase:cytochrome c] complex [47]. Other examples are Tyr83 in plastocyanin and the surface-exposed aromatic Cu ligands His87, His95, and His117 in plastocyanin, amicyanin, and azurin, respectively [48]. Recent studies of an azurin mutant where an additional Trp was introduced into the wild-type sequence by substitution of valine in position 31 (V31W) shows an approximately 10-fold increase in the intramolecular electron transfer rate constant from $RSSR^-$ to Cu(II). This could not be rationalized in terms of driving force or reorganization energy and therefore probably reflects enhanced electron coupling provided by the additional aromatic residue [42].

Identification of active sites on electron transfer proteins has been attempted by affinity labeling, using Cr(II) as electron donor [49]. The electron transfer reaction between Cr(II) aqua ions and Cu(II) in several blue copper proteins (P) proceed as

$$Cr(II) + P[Cu(II)] \rightarrow Cr(III) - P[Cu(I)]$$

Cr(III) complexes are generaly substitution inert, in marked contrast to the labile Cr(II) complexes, so the essential feature of the procedure above is based on the notion that the coordination sphere of Cr(II) in the transition state is retained in the substitution inert Cr(III) product. Thus, the Cr(III) ion will remain in the very same coordination sphere as during the electron transfer from Cr(II) to the Cu(II) center, and in this way, redox active sites on the blue copper proteins, azurin [50], plastocyanin [51], and stellacyanin [52] were identified. In all three proteins the Cr(III) label was found in one particular site only, despite the fact that a number of potentially complexing chromium ligands are available. The uniqueness of the Cr(III) binding centers was demonstrated by the large effects observed on the reduction kinetics of Cr(III)-labeled proteins [53].

The issue of whether and how the highly heterogeneous polypeptide matrix of proteins may control LRET is still a subject of intense interest and discussion. There is substantial experimental evidence for through-bond superexchange electron transfer. Still, Dutton and co-workers have analyzed results of intramolecular electron transfer processes studied in a large number of biological systems and found a correlation between the exponential decrease in k_{max} (which is the free energy optimized rate constant, i.e. for

$-\Delta G^0 = \lambda$) and the edge-to-edge distance separating the electron donor and acceptor with a decay factor of $\beta = 14\,\mathrm{nm}^{-1}$ [54]. This suggested that the electron transfer rate constants are independent of the chemical nature of the separating medium and challenged ideas based on a considerable body of rate and activation parameters determined for intramolecular LRET in proteins. The former correlation relied heavily on results of electron transfer within the photosynthetic reaction centers, however, where a rather limited involvement of the polypeptide matrix in separating the electron transfer partners is present. Further, when results of LRET between the $RSSR^-$ and Cu(II) in more than a dozen different wild-type and mutated azurins are examined, it becomes even more apparent that the simple analysis above does not provide a satisfactory rationale for the experimental findings: Using a value of 2.46 nm for $(r - r_0)$ between Cys3(S_γ) and Cys112(S_γ) in *P. aeruginosa* azurin (which defines the shortest direct electron transfer distance) yields $\beta = 9\,\mathrm{nm}^{-1}$. The difference between the two calculated values (14 and $9\,\mathrm{nm}^{-1}$) for this exponential decay seems too large to be accounted for in terms of experimental errors. Also, using $\beta = 14\,\mathrm{nm}^{-1}$ yields a calculated maximum rate constant (i.e., activationless) which is almost two orders smaller than that derived from the kinetic parameters determined for azurin ($k_{max} = 120\,\mathrm{s}^{-1}$). Therefore, the experimental results of LRET in the azurins do not fit an exponential decay correlated with a direct through-space distance. Interestingly, the above $k_{max} = 120\,\mathrm{s}^{-1}$ fits perfectly into a linear plot of log k_{max} versus the through-bond distance for all results of ruthenium-substituted cytochrome c mutants drawn with a slope of $7.1\,\mathrm{nm}^{-1}$ [44].

Ruthenium ammine complexes have been widely employed as secondary redox centers coordinated to histidyl imidazole side chains of the protein surfaces [36,55,56]. Photo-induced electron transfer between ruthenium(II) and the intrinsic redox centes in c-type cytochromes and azurins for which high-resolution three-dimensional structures are available were then studied. Furthermore, by derivatizing a large number of different native and engineered surface histidines of cytochrome c and azurin with Ru complexes, internal electron transfer was studied. Thus the electron transfer distance and nature of the intervening medium were modified in a controlled and systematic manner. In parallel, the ligands of the Ru center was also varied [e.g., with 2,2′-bipyridyl (bpy) and imidazole (im) as ligands] enabling examination of the role of a wide range of driving force on the intramolecular electron transfer rates [36].

Originally, the surface-exposed His83 of *P. aeruginosa* azurin was derivatized by the pentaammineaqua–ruthenium(II) complex [36] and the intramolecular electron transfer from $[(NH_3)_5Ru(His83)]^{2+}$ to the Cu(II) center was found independent of temperature. Interestingly, the observed low rate constant ($1.9\,\mathrm{s}^{-1}$) is more than one order of magnitude lower than in

[(NH$_3$)$_5$Ru(His33)]Cyt *c* [57], although the direct distance between the redox centers is the same (1.2 nm), and the driving force is only slightly larger in azurin (23 kJ mol^{-1} versus 17 kJ mol^{-1}). The difference in rates was assigned to the longer pathway in modified azurin, which involves twice as many peptide bonds as that of Ru-modified Cyt *c*. In contrast, the intramolecular electron transfer in azurin derivatized with complexes having much higher driving forces [e.g., the (bipy)$_2$Ru(II)His83 to Cu(II) system has a driving force of 73 kJ mol^{-1}] proceeds at 10^6 s^{-1} [58]. This and other examples of the dramatic rate enhancements [36] are not unreasonable since the large driving force is sufficient to overcome the reorganization energy of the reaction. The only barrier to the electron transfer process then becomes the distance between electron donor and acceptor. Histidine residues were introduced by site-directed mutagenesis in different positions on the surface of Cyt *c* and azurin with through-bond metal-to-metal distances from 0.2 to 0.4 nm. The variation in electron transfer rate constants clearly showed the critical role of the separating medium structure in determining electronic coupling. Moreover, a detailed analysis of these data resolved interesting differences between the electronic coupling provided by distinct conformational elements of polypeptides: namely, β-strands being closer to an extended hydrocarbon chain provide separation distances that are linearly related to the number of intervening atoms. Thus the distance dependence of the electronic coupling in proteins was proposed also to be a function of the polypeptides conformation (i.e. α-helical or β-stranded elements) [59].

B.   Amicyanin

Amicyanin (Am) is another bacterial cupredoxin that mediates the electron transfer required for methylamine oxidation by methylamine dehydrogenase (MADH) producing ammonia and formaldehyde. In the bacteria, these electrons are probably further transferred via cytochrome c_{551i} to the terminal oxidase. The system of these three redox proteins provides a very interesting model for interprotein LRET. It has been studied in great detail in solution and in the crystalline heterotrimeric state [60]. Here we only briefly address the proposed electron transfer pathway whereby amicyanin mediates electrons between MADH and cytochrome c_{551i}. The tryptophan–tryptophylquinone (TTQ) redox cofactor of MADH has its unmodified Trp facing the solvent-exposed His imidazole ligand of amicyanin. The distance from the C$_{\eta2}$ atom of Trp57 where substrate oxidation is proposed to take place to the copper center of amicyanin is 1.6 nm [15]. Two electron transfer pathways were proposed. One is directly from Trp108 on MADH via a through-space jump of 0.35 nm to Pro94 on Am and further to the Cu-ligating His95. Another path leads from Trp57 to Ser56 on MADH, which is in hydrogen-bond contact via a water molecule with His95 of

Am. Thus the presence of this water molecule is apparently a prerequisite for electron transfer. Two partially overlapping pathways then lead from the copper ligand Cys92 or Met98 in Am to a common point and continue through three further residues of Am and two of Cyt c_{551i}, reaching the latter iron center. The calculated electronic coupling through these two pathways is three orders of magnitude smaller than the TTQ to copper pathways, due mainly to two through-space jumps and a longer distance [15]. Since the intramolecular electron transfer rate constant depends on the square of the electronic coupling, the Am to Cyt c_{551i} rate constant should be 10^6-fold smaller than that of the MADH to Am constant. However, both were found to be of the same order of magnitude: $10–50 \, s^{-1}$. Since the redox potential of Am drops $73 \, mV$ upon binding to MADH, the driving force of the intra-protein electron transfer would decrease accordingly. Similarly, the electron transfer between Am and Cyt c_{551i} would be more advantageous. Also, larger reorganization energies for the MADH-Am electron transfer could be expected.

C. Other Cupredoxins

Stellacyanin (St), the single blue copper protein isolated from *Rhus vernicifera*, has two surface-exposed free histidyl residues which were both derivatized with ruthenium pentaammine complexes. Apparently, the proximity and similar reactivity of both His residues exclude preparation of singly modified Ru(II) St. The intramolecular electron transfer between Ru(II) and Cu(II) has been induced by pulse-radiolytic Ru(III) reduction. The observed rate constant for the intramolecular electron transfer was $0.05 \, s^{-1}$ at $18°C$ [61]. The direct separation distance between both Ru ions and the Cu(II) is estimated to be ca. $2.0 \, nm$, but without the three-dimensional structure a more detailed pathway analysis was not feasible.

Another striking example of a rather slow electron transfer process was observed for plastocyanin modified with $(NH_3)_5Ru^{2+}$ at His59. Although the edge-to-edge distance between the two redox centers is only $1.2 \, nm$, the electron transfer rate constant is $< 0.3 \, s^{-1}$ [62]. The slow rate is probably due to a very poor electronic coupling caused by a very long through-bond electron transfer pathway. Extended Hückel calculations have been performed on five different potential electron transfer pathways in plastocyanin [63]. Interestingly, the results demonstrated that intramolecular through-bond electron transfer is electronically favorable even when compared with shorter outer-sphere electron transfer, where the wave functions decay much faster than by the superexchange through the protein matrix.

VI. ELECTRON TRANSFER IN TYPE 2 COPPER-CONTAINING PROTEINS

The T2 copper–zinc enzyme superoxide dismutase (SOD) is one of the best characterized type 2 copper proteins [64]. This type of SOD is found in all eucaryotic cells and is a homodimer of 16,000 kD subunits, each containing one copper and one zinc ion. As the name indicates, it catalyzes the dismutation of the superoxide radical anion into dioxygen and peroxide. The three-dimensional structure of this SOD has been determined [65], and the overall protein structure bears similarities to the blue copper proteins discussed in Section II [60]. The metal sites consist of a copper(II) ion coordinated to four histidines and one exchangeable water molecule in a distorted tetragonal pyramidal arrangement and a zinc(II) ligated to three histidines and one aspartate in a distorted tetrahedral geometry. One of the imidazoles bridges the two metal ions. The Cu(II) ion is found at the bottom of a crevice where anions such as cyanide and azide as well as the substrate, O_2^-, may enter and displace the water molecule, thus initiating the enzymatic dismutation reaction. In this first step, the superoxide anion coordinates to Cu(II) and reduces it to Cu(I) with concomitant O_2 liberation. Then Cu(I) becomes oxidized by another superoxide ion, producing and releasing peroxide. Thus direct (i.e. inner sphere) electron transfer takes place between substrates and the copper ion. The Zn(II) ion is solvent inaccessible and seems only to serve a structural–electronic role. A number of other metal ions, such as Ag(I), Co(II), and Cd(II), substitute for the zinc center with little or no effect on the SOD activity [66].

Another example of an inner-sphere electron transfer between a T2 copper center and a substrate ligand is found in galactose oxidase, which contains a single Cu center in a square pyramidal geometry with two histidines, a covalent cysteine–tyrosine derivative and the alcohol substrate in equatorial positions, while another tyrosine occupies the axial position [67]. The enzyme catalyzes the oxidation of primary alcohols to the corresponding aldehyde concomitantly with dioxygen reduction to hydrogen peroxide. Here the unique equatorial tyrosine ligand is apparently undergoing a redox change, which in combination with the Cu(II)/Cu(I) cycle provides the two redox equivalents involved in the reaction [67].

Organic cofactors analogous to TTQ (cf. Sec. V.B) are also found in T2 copper containing amine oxidases, where primary amines are oxidatively deaminated and two electrons are transferred to the $O_2 \rightarrow H_2O_2$ reduction. The cofactor in these oxidases is a tyrosine-derived quinone (topaquinone). Although it is not directly coordinated to the copper center [68], both redox centers are required for the amine oxidase activity [69]. Intramolecular electron transfer between the reduced topaquinone and Cu(II) has been observed and suggested to be part of the catalytic cycle [70]. The three-dimensional structure of this T2 copper enzyme has not yet been determined. Still, at least a formal analogy is apparent with the operation of methylamine dehydrogenase MADH (Section V.B).

VII.   INTRAMOLECULAR LRET IN MULTICOPPER OXIDASES

The family of blue copper oxidases shares both structural and apparently also functional properties. *Rhus vernicifera* laccase has been the most extensively investigated as the prototype member of the family. Still, as described above (Sec. II) it is zucchini ascorbate oxidase (AO) for which a high-resolution three-dimensional structure was first determined, hence attracting more attention in recent years. Three-dimensional structures are now available for both oxidized and reduced AO [19] as well as for its complexes with hydrogen peroxide and azide [25], thus providing an essential foundation for structure–function correlation.

All blue copper oxidases catalyze one-electron oxidation of specific substrates by O_2, which becomes reduced to water [35]. Their minimal catalytic unit consists of four copper ions bound to distinct sites, designated T1, T2, and T3, which have been described above. Here we review studies of internal electron transfer in several of the blue copper enzymes, such as ascorbate oxidase, ceruloplasmin, laccase, and nitrite reductase, as well as cytochrome c oxidase.

A.   Ascorbate Oxidase

AO exists as a dimer of identical 70 kD subunits, each containing a catalytic unit with one of each of T1, T2, and T3 copper sites. The reduction potentials of the T1 and T3 Cu(II)/Cu(I) couples are identical, (350 mV at pH 7.0) [34], while the T2 potential is considerably lower (< 300 mV). The catalytic cycle of this enzyme, like that of other blue oxidases, requires intramolecular electron transfer from T1[Cu(I)] to T3[Cu(II)] for the transfer of the four electrons necessary for O_2 reduction to two H_2O molecules.

Three independent groups have studied the intramolecular electron transfer from T1[Cu(I)] to T3[Cu(II)] in AO under anaerobic conditions and have reported similar rate constants [71–73]. Tollin et al. [71] employed the photochemically produced lumiflavin semiquinone to reduce T1[Cu(II)] in a fast second-order process ($k = 2.7 \times 10^7 \, M^{-1} \, s^{-1}$ at pH 7.0). This was followed by partial reoxidation of the T1[Cu(I)] site with a rate constant of $160 \, s^{-1}$. However, the flavin absorption in the near-ultraviolet region prevented monitoring changes at 330 nm, where T3[Cu(II)] absorbs. Still, the process monitored at 610 nm was interpreted to be an intramolecular T1→T3 ET. This interpretation gained support from experiments using AO from which T2 copper was specifically removed and yielded the same rate constants for the electron transfer [71]. In another study [72], pulse-radiolytically produced CO_2^- radicals served as primary electron donor, reacting in a bimolecular, diffusion-controlled process with T1[Cu(II)] $(1.2 \times 10^9 \, M^{-1} \, s^{-1}$ at pH 7.0). The ensuing processes of T1[Cu(I)] reoxidation and T3[Cu(II)] reduction were directly monitored at both 610 nm (T1) and 330 nm (T3) and proceeded at identical specific rates. These were concentration independent, confirming their assignment as an intramolecular electron transfer from T1 to T3. Interestingly, two or three distinct phases were observed in this process, with rate constants of $200 \, s^{-1}$ for the fastest phase and $2 \, s^{-1}$ for the slowest. Later, in another pulse-radiolysis study, similar electron transfer rates were determined using different organic radicals, yet monitoring only the 610 nm chromophore [73].

It is noteworthy that using pulse-radiolytically produced reducing radicals (CO_2^- and O_2^-) an intramolecular electron transfer from T1 to T3 in the related blue copper oxidase *Rhus* laccase, only one phase of intramolecular electron transfer process could be resolved, with a rate constant of $2 \, s^{-1}$ [74]. Evolutionary selection for efficient dioxygen reduction to water has probably produced the unique structural features of the blue oxidases. Hence AO provides an appropriate system for studying intraprotein electron transfer processes, specifically between T1 Cu(I) and the trinuclear center. Several questions then arise: Does the degree of AO reduction control the rate of internal electron transfer? Does the presence of substrates, electron donors or dioxygen, affect the internal electron transfer rates? How does the rate of internal electron transfer relate to the structural differences between the structures of reduced and oxidized AO? How is the intramolecular electron transfer in AO controlled so as to avoid production of reactive intermediates during four-electron reduction to O_2? These questions, as well as others, gain more significance when it is noted that steady-state kinetic measurements of AO yield turnover numbers of 12,000–15,000 s^{-1} [34,75], considerably faster than the values observed for the intramolecular T1 to T3 electron transfer.

To try to resolve the reason for this discrepancy between the internal LRET rate and the catalytic turnover rate of AO, internal electron transfer was studied using AO samples that were "activated" or "pulsed" by turning over 1 mM ascorbate in the presence of 0.25 mM O_2 prior to measuring the rates, yet resolved no difference [72]. In contrast, however, to these experiments, which were performed under strictly anaerobic conditions, when AO solutions containing low and controlled O_2 concentrations (15–65 μM) were employed, quite conspicuous differences were observed [76]. An additional faster intramolecular electron transfer phase was discovered with a rate constant that was identical, 1100 s^{-1} (293 K, pH 5.8) when monitored at either 610 nm (T1) or 330 nm (T3). This phase was maintained as long as O_2 remained in the solution. Noteworthy also were the large transient spectral changes that appeared in the near-UV region, probably reflecting reaction between the trinuclear site and dioxygen following the intramolecular electron transfer from T1[Cu(I)] to T3[Cu(II)] [77]. The formation of oxygen reduction intermediates coordinated to the trinuclear center will obviously increase the driving force of the latter electron transfer significantly, thus causing its rate enhancement. Calculations show that a 100-mV increase in the reduction potential of the T3 center would lead to the observed increase in electron transfer rate constant [76]. Similar effects of O_2 presence on the internal electron transfer rates have been reported for mammalian cytochrome oxidase during its catalytic reaction [78].

The activation enthalpy for internal electron transfer in AO under anaerobic conditions was found to be rather low ($\Delta H^{\neq} = 9.1 \text{ kJ mol}^{-1}$) while the entropy barrier was highly negative ($\Delta S^{\neq} = -170 \text{ J K}^{-1} \text{ mol}^{-1}$). The reorganization energy, λ, for this intramolecular anaerobic electron transfer process was calculated [72] from the activation parameters using the equations presented in Section IV. The through-bond distance between the T1–Cu ligating cysteine and the closest histidine imidazole coordinated to the T3–copper pair is 1.34 nm. Taken together with the previously determined electronic decay factor, $\beta = 6.7 \text{ nm}^{-1}$ for electron transfer in other β-sheet blue proteins (Sec. V.A), the reorganization energy in AO becomes $\lambda = 153 \text{ kJ mol}^{-1}$, which is significantly higher than that observed for intramolecular electron transfer in *P. aeruginosa* azurin ($\lambda = 99 \text{ kJ mol}^{-1}$; cf. Sec. V.A). Moreover, as the three-dimensional structure of different forms of AO shows, the T3 site reduction causes considerable local conformational changes at the trinuclear site. The copper–copper distance increases from 0.37 nm to 0.51 nm [25] and their antiferromagnetic coupling is disrupted.

The maximal rate constant of 1100 s^{-1} observed for intramolecular electron transfer in AO is still considerably smaller than the turnover number of about $14,000 \text{ s}^{-1}$. Thus dioxygen coordination to the trinuclear site is not sufficient to attain rates compatible with the maximal enzymatic activity. Under optimal turnover conditions the concentration of reducing substrate (e.g. ascorbate) is sufficient to maintain a steady state of fully reduced copper sites. Tollin and co-workers have therefore studied the oxidation reaction of fully reduced AO with a laser-generated triplet state of 5-deazariboflavin [71], causing a one-electron oxidation of the T2–T3 cluster. This was followed by a rapid biphasic intramolecular electron transfer from T1[Cu(I)] (and presumably) to the oxidized trinuclear center. The faster of the two observed rate constants (9500 and 1400 s^{-1}, respectively) is comparable to the turnover number for AO under steady-state conditions and renders it likely that this is the rate-limiting step in catalysis.

As mentioned in Sec. II, Messerschmidt et al. [19] proposed that the shortest electron transfer pathway from T1[Cu(I)] to T3[Cu(II)] takes place via Cys507 and either His506 or His508, (cf. Fig. 5). Both pathways consist of nine covalent bonds, yielding a total distance of 1.34 nm. The pathway calculated using the Beratan and Onuchic algoritm (cf. Sec. IV) supports the probability of this model and resolves a further alternative path via a H-bond between the carbonyl oxygen of Cys507 and His506(N_δ) (Fig. 5). The electronic coupling for the two covalent pathways is 0.010, while for the latter the coupling is 0.014. Another noteworthy point is the similarity between the structural features of AO and plastocyanin. In both proteins the T1–copper cysteine ligand is assumed to operate in electron transfer. As discussed above (Sec. III), calculations on plastocyanin suggested a high degree of anisotropic covalency in the copper center, resulting in very favorable electronic coupling via the cysteine ligand [31]. It would be interesting to learn whether the same anisotropic covalency also exists in the structural arrangement connecting the T1–T3 in AO.

One unique feature of the observed internal electron transfer process in AO is the multiplicity of its phases. Several possible rationales for this pattern were considered. One was that it may reflect reactivity of AO sub-populations differing in the degree of ligand protonation in the trinuclear center (e.g. the oxygen bridging the T3 coppers or that bound to T2). How-

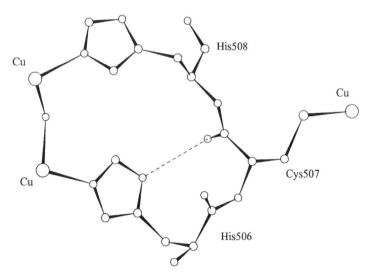

Figure 5. Suggested pathways for intramolecular electron transfer from T1 Cu(I) (right) to T3 copper(II) (left) in ascorbate oxidase. The hydrogen bond is shown as a dashed line. Pathway calculations were based on the three-dimensional structure coordinates of AO [19].

ever, our recent examination of the electron transfer process over the pH
range 5–9 resolved no differences in rates or in amplitudes of the various
phases, hence excluding this possibility (Farver and Pecht, unpublished
results). Another possible explanation could be that the different reaction
phases reflect distinct degrees of AO reduction. However, examination of
the rates and amplitudes of the various phases throughout full-reductive
titrations showed that all phases are observed in the reaction of oxidized
AO and exhibit a similar pattern until its complete reduction. Hence this
unusual pattern of reactivity is still unexplained. Moreover, as detailed
above, a new, additional phase of internal electron transfer is observed in
O_2-containing solution. This, however, is assumed to reflect an internal
transfer of a second or third electron to the trinuclear center in one equivalent reduced AO molecules with a bound O_2 molecule.

B. Laccase

Laccases from *Polyporus versicolor* and *R. vernicifera* were the first members of the blue copper oxidase family to be investigated in great detail, both
spectroscopically and kinetically [3]. They contain one each of three redox
active copper sites per catalytic subunit: T1, T2, and T3. Although the three-
dimensional structure is still unknown for any laccase, ample evidence from
spectroscopic studies and sequence homoglogy with ascorbate oxidase suggests strong similarity between these two types of oxidases [19], and therefore similar electron transfer pathways may be operative in both.

The intramolecular electron transfer from T1 Cu(I) to the T3 Cu(II)
center has been proposed to be rate determining (i.e. k_{cat}) for the catalytic
cycle [79]. Under anaerobic conditions, however, this T1 Cu(I) to T3 Cu(II)
electron transfer was found to be exceptionally slow ($k_{ET} \leq 1\,s^{-1}$) in contrast to the rapid reoxidation of T1 observed for the reaction of fully
reduced enzyme with dioxygen [80]. These observations were interpreted
as reflecting the existence of two different protein conformations, a resting
and an activated state [81]. Significantly, this discrepancy is analogous to
that encountered in AO, where the internal T1 → T3 electron transfer rate is
markedly slower than that expected on the basis of the catalytic turnover
rate.

Thus, some of the key questions concerning the catalytic mechanism of
laccases relate to formation of protein intermediates. Some early studies
addressing these questions were performed by applying the pulse radiolysis
technique to fungal laccase [82] and human ceruloplasmin [83]. The
hydrated electron, e_{aq}^-, reacted primarily with exposed aromatic residues
and disulfide groups to produce radicals, some of which decayed in subsequent intramolecular electron transfer steps to the Cu(II) center. This behavior is probably a result of the extreme reactivity of e_{aq}^-, rather than being an

indication of physiologically significant electron transfer pathways. Still, they provided early evidence that the T1 Cu(II) ion is the primary electron uptake port for these enzymes. More detailed pulse-radiolysis studies on *Polyporus* fungal laccase performed by Guissani et al. [84] showed that different optical transients are produced upon reaction with radicals such as OH, e_{aq}^-, CO_2^-, and O_2^-. Assesment of the significance of these internal electron transfer processes still awaits structure determination. Employing pulse-radiolytically produced nitroaromatic radical anions as reductants, O'Neill et al. [85] have shown that the T1 Cu(II) sites in *Rhus vernicifera* and *R. succedanea* laccase are reduced in bimolecular processes with rate constants of $2 \times 10^6 \, M^{-1} s^{-1}$ and $1.7 \times 10^7 \, M^{-1} s^{-1}$, respectively (pH 7.4).

As indicated, one of the key questions related to the catalytic cycle of the blue oxidases is the role of T2 copper. This has been addressed by Andréasson and Reinhammar [79,86] by combining rapid-freeze EPR and stopped-flow experiments of *Rhus* laccase. All results seem to involve a relatively slow transition between an active and an inactive form of the enzyme. However, with today's knowledge that the T2 and T3 sites constitute a trinuclear center, it is difficult to reconcile this fact with T2 having distinct redox behavior from T3. It therefore seems that the mechanism of T2 copper involvement in the catalytic cycle, specifically the intramolecular electron transfer, in the blue oxidases needs to be reevaluated.

Later studies tried to resolve the relationship between possible protein conformational changes and the redox changes of the copper sites by investigation of the anaerobic reduction of *Rhus* laccase [87]. The marked differences in intrinsic protein fluorescence that accompany redox changes in the enzyme indicated that two separate reduction steps may account for the reduction-linked fluorescence increase [87]. According to the proposed model, an anaerobic reduction pathway of laccase is associated with a minimum of three different fluorescence-monitored protein states which depend on redox-linked conformational changes. One proposed step in the kinetic model is a "very fast" concomitant two-electron transfer from T1 Cu(I) T2 Cu(I) to the T3 Cu(II) pair, with a rate constant independent of the concentration of the Ru(II) reductant. This intramolecular electron transfer step should follow the putative individual bimolecular reductions of T1 and T2 Cu(II), and with a rate constant 10 times faster for the latter site [87]. Concomitant two-electron transfer is in agreement with the proposal of Andréasson and Reinhammar [80] (see above), although in the earlier simulations the intramolecular electron transfer rate was proposed to be slow [79]. This mechanism where both T1 and T2 copper(I) synchronously transfer two electrons to T3 Cu(II) is in conflict with later thermodynamic and kinetic studies which have clearly established that intramolecular one-electron transfer can take place between T1 Cu(I) and T3 Cu(II) in molecules where just the former site has been reduced (see below). Finally, the issue of a direct bimolecular reduction of the T2 Cu(II) ion is still unresolved. The rate constant for intramolecular electron transfer from T1 Cu(I) to the transiently formed trinuclear center-oxygen radical complex was determined to be $3.4 \, s^{-1}$, similar to the rate-determining step in the overall catalytic cycle at pH 6.0 [87].

The notion that peroxolaccase is a bona fide distinct intermediate in the catalytic cycle of this enzyme has been established by detailed analysis of the absorption and circular dichroism spectra of *R. vernicifera* laccase occurring upon (1) addition of one equivalent of hydrogen peroxide to oxidized laccase; (2) reduction with ascorbate under aerobic conditions, (3) reaction of partial or fully reduced enzyme with O_2, (4) reduction of peroxolaccase, and (5) peroxide reoxidation of the reduced enzyme [88]. In the three-dimensional structure of the peroxo derivative of AO it is found that a hydroperoxide binds terminally end-on to one of the two copper ions constituting the T3 center [25]. In view of the close sequence similarity between laccases and AO, the notion of a T3 copper peroxide complex formation by laccase thus gained concrete structural foundation.

The T3 copper(II) pair in laccase and AO is antiferromagnetically coupled, and in the AO structure the two copper ions were shown to be bridged by OH^- or O^{2-}. Magnetic susceptibility measurements were performed on native oxidized laccase and peroxolaccase at room temperature to examine whether the exchange coupling between these two T3 Cu(II) ions is affected by interaction with peroxide [89]. Analysis of these data demonstrates that peroxide coordination to the T3 site reduces coupling between the two Cu(II) ions [89]. This is in very good agreement with the structural changes observed upon adding peroxide to AO crystals, discussed above.

In conclusion, clear evidence is now available for the reactivity of *Rhus* laccase molecules containing 1, 2, and 3 reduction equivalents, with dioxygen yielding intermediates of different stabilities. This might, however, not be the case for fungal laccase, where the properties of both T3 and T1 differ considerably compared with *Rhus* laccase, hence leading to marked differences in the catalytic mechanism.

C. Nitrite Reductase

Several blue copper proteins have also been found to participate in the denitrification cycle carried out by anaerobic bacteria, leading to the dissimilatory reduction of nitrate and nitrite, usually producing dinitrogen. Nitrate reductase (NIR) and nitrous oxide reductase (N2OR) are involved in the cascade of such anaerobic respiration processes. Dissimilatory nitrite reductases have so far been found to contain copper or hemes c and d_1 [90]. Copper-containing nitrite reductases have been characterized in several organisms, and the one isolated from *Achromobacter cycloclastes* has been crystallized and its three-dimensional structure determined [91]. It exists in the crystal and in solution as a trimer where each monomer consists of two β-sheet domains, similar to that of plastocyanin, yet containing two copper ions: one is bound to a T1 site (two His, one Cys, and one Met ligands), the other is bound to a putative T2 site with three His and one solvent water ligand. The two sites do however, share adjacent amino acid residues, His135 (T2) and Cys136 (T1), with a distance separating the two copper ions of 1.25 nm. While the T1 copper is some 0.4 nm below the protein surface, the T2 copper lies at the bottom of a 1.2 nm-deep solvent-accessible channel. The bacterial cupredoxin, pseudoazurin, also present in *A. cycloclastes* has been demonstrated to be an electron donor to NIR and might indeed be serving this role physiologically [92]. The second NIR substrate (NO_2^-) was shown to bind to the T2 copper at the bottom of the above-mentioned channel. These results suggest that the catalytic cycle of NIR involves electron uptake by T1 Cu(II) from pseudoazurin followed by intramolecular electron transfer to T2 Cu(II), which would then reduce the bound nitrite, yielding NO and water.

A recent pulse-radiolytic study of *Achromobacter cycloclastes* NIR [93] has shown that the T1 Cu(II) chromophore is reduced by *N*-methyl nicoti-neamide radicals in a bimolecular process with a rate constant of $3.4 \times 10^8 \, M^{-1} \, s^{-1}$. This was followed by a second, slower reoxidation step of T1 Cu(I), shown to be first order with a rate constant of $1.4 \times 10^3 \, s^{-1}$ and assigned to an intramolecular long-range T1 Cu(I) electron transfer to T2 Cu(II). From the crystal structure an electronic coupling for the pathway connecting the two redox centers could be calculated to be 0.010 (i.e., identical to that observed in AO).

D. Ceruloplasmin

Ceruloplasmin (CP) is an odd member of the blue copper protein family for several reasons. First and foremost, its physiological function as an oxidase is still unresolved, although its capacity to catalyze ferrous ions by dioxygen has been well established and its involvement in iron metabolism is strongly supported by considerable physiological and genetic evidence [94]. Another open question related to CP was its copper content, which has been debated for quite some time. This has now clearly been resolved by the three-dimensional crystallographic structure determination of human CP [95]. Finally, CP is the main, if not the only mammalian member of the blue copper protein family. Human CP consists of a single polypeptide with a molecular weight of 132 kD folded in six domains arranged in a triangular array. Each domain comprises a β-sheet, constructed in a manner typical for the cupre-doxins. Three of the six copper ions are bound in T1 sites present in domains

2, 4, and 6, whereas the other three copper ions form a trinuclear cluster, bound at the interface between domains 1 and 6. The spatial relation between the trinuclear center and the nearest T1 (in domain 6) closely resembles that found in AO. This was taken to further support the belief that CP has an oxidase function [95]. The three T1 sites are separated from each others by a distance of 1.8 nm. This was again related to the CP function, since this distance still allows for internal electron transfer at reasonable rates and also increases the probability for electron uptake. The trinuclear coordination site consists of four histidine pairs, two pairs from domain 1 and two pairs from domain 6. As in AO, two of the copper ions are bound to six histidines and assigned as T3, while the third copper (most distant from T1) is coordinated to two histidines only and is designated as the T2 site. In all cases histidine pairs bridge two copper ions, and by analogy to the AO structure, an oxygen atom apparently bridges the two T3 coppers and another one is bound to the T2 copper ion [95].

An additional relevant structural feature is that the cystein of domain 6, which provides its thiolate as ligand to T1, is placed between the two sequential His residues, which are coordinated to the T3 copper pair. This structural element was first observed in AO and proposed to serve as the electron transfer path between the single T1 and the trinuclear center. Naturally, the functional significance of this CP feature now calls for investigation.

A considerable body of results from activity studies of CP has accumulated during earlier decades and is now awaiting more meaningful analysis and interpretation using the three-dimensional structure available. Catalysis of amine oxidation by CP, in particular biogenic amines present in plasma, cerebral, spinal, and intestinal fluids, as well as of ferrous ions, is probably physiologically relevant and has been studied extensively. The mechanism of dioxygen reduction by CP at the trinuclear center is of particular interest, as the presence of three distinct T1 sites raises the question of which sites are involved in the internal electron transfer to the single O_2 reduction site.

As mentioned above, pulse radiolysis was first applied to studying electron transfer within this multicentered blue oxidase more than two decades ago. An internal electron transfer to the T1 Cu(II) has been observed from the disulfide radical ions produced upon CP reaction with hydrated electrons [83]. Both the transient absorption of the radicals and the 605-nm band of the T1 Cu(II) site were found to decay in a unimolecular process at an identical rate of about $900 \, s^{-1}$. As five different disulfide bonds were resolved in the human CP structure, future work will have to try and address the implications of this process.

E. Cytochrome *c* Oxidase

One of the most complex and challenging copper-containing redox enzymes is cytochrome *c* oxidase (COX). As described above, the three-dimensional structure for the enzyme isolated from two different species provided important direct insights into the nature of its redox centers, possible intra-molecular electron transfer pathways, and proton pumping mechanism. Upon reducing dioxygen to two water molecules, four scalar and four vectorial protons are pumped across the mitochondrial membrane, leading to ATP synthesis, thus converting part of the redox energy into another chemical form. The electron transfer processes catalysed by COX are carried out by three metal-containing redox active centers.

1. A copper binding site (Cu_A) is the primary electron uptake site from Cyt *c*. The x-ray crystallographic structure [28,29] has also confirmed the proposal [96] that the Cu_A site is actually a binuclear copper center with 0.25 nm separating the two ions. Significantly, the first spectroscopic evidence for the existence of such a binuclear site in COX has in fact emerged from studies of N2OR, containing a similar binuclear copper site [96].

2. Heme *a*, which is coordinated axially by two histidine imidazoles from the same subunit.

3. A bimetallic center consisting of heme a_3 and Cu_B. The fifth ligand of the heme a_3 is an imidazole, whereas Cu_B is coordinated by three imidazoles. The distance between the iron and the copper ions is 0.45 nm, yet no bridging ligand has been resolved despite the strong antiferromagnetic coupling between the two metal ions [28].

The long-standing controversy regarding the identity of the primary electron acceptor site has been settled mainly by kinetic studies such as those of Kobayashi [97] and Hill [98], who showed it to be the Cu_A site. The structural rationale for this assignment has now been given by the three-dimensional structure determination showing the other redox centers to be relatively less exposed to the water-soluble electron donor cytochrome *c* (in the mammalian system). Cu_A resides in the globular domain, which protrudes to the cytosolic side 0.8 nm above the membrane surface, and the metal is shielded only by two residues (Trp and Asp). Moreover, very close to this area a putative docking site has been proposed for cytochrome *c* with 10 negatively charged side chains available for interaction with this positively charged electron donor. The distances separating Cu_A from heme *a* and heme a_3 are 1.95 and 2.21 nm, respectively (center to center) [29]. The widely accepted reaction scheme assumes electron transfer to proceed from Cu_A to the closer heme *a* and from there over a 0.47-nm separation to heme a_3 which is part of the binuclear dioxygen reduction site. Two structural features are noteworthy in the context of electron pathways connecting these redox centers:

1. Cu_A is closer to heme a than to heme a_3, and a Mg(II) binding site is found placed in between them with a glutamate residue that is coordinated to Mg^{2+} via its carboxylate side chain and to Cu_A via its peptide carbonyl oxygen.

2. Although the proximity between hemes a and a_3 suggests direct (i.e. through-space) electron transfer, a covalent pathway is also available through the single residue present between the two His ligands, each coordinated to one of the hemes. In the beef heart structure an additional potential pathway is observed via Phe 377, which is stacked between the heme a_3 porphyrin and the His 378 ligand of the heme a [28].

Extensive studies of electron transfer to and within cytochrome c oxidase performed over decades since the pioneering work of Warburg [99] and Keilin [100] have now received novel and important insights through the foregoing structure determination. Yet it raised further questions, related primarily to the vectorial proton transfer that is coupled to the electron transfer reaction. Electron transfer from Cu_A to heme a has been shown (for a recent review, see Winkler et al. [101]) to be relatively fast $(10^4 \, s^{-1})$, especially when considering the 1.9-nm separation and 50-mV driving force. A direct electron transfer pathway linking the two centers has been proposed by Ramirez et al. [102] based on the structure above. It consists of 14 covalent and two hydrogen bonds. Using coupling values deduced from experimental work on model proteins, an estimate for the driving force optimized electron transfer rate of $4 \times 10^4 – 8 \times 10^5 \, s^{-1}$ was calculated, implying a reorganization energy between 0.15 and 0.5 eV for this step. The surprising structure of the binuclear Cu_A site, having its unpaired electron potentially delocalized over a large space, has been brought up as one cause for the features of this electron transfer step.

The relatively small difference in distances separating Cu_A from hemes a and a_3 is in contrast to the experimentally observed preferential electron transfer to the former site. This has been observed consistently in experiments where electron injection into Cu_A was attained either by radicals, by cytochrome c [97,103] or by internal reequilibration. The latter process is induced by a protocol using flash photolysis of the half-reduced COX complex with CO. The photo-induced CO dissociation yields heme a_3 with a lower reduction potential and initiates electron transfer to heme a and

Cu_A with observed rate constants $3 \times 10^5 \, s^{-1}$ and $2 \times 10^4 \, s^{-1}$, respectively [98,103].

VIII. CONCLUSIONS

Essentially four types of copper binding sites have evolved for performing electron transfer reactions at the crossroads of energy assimilation and conversion systems: The blue T1 site performs the task of electron mediation among distinct proteins or between donor substrates and internal sites, and the rather limited structural changes accompanying its redox cycle lead to low Franck–Condon barriers. The recently resolved structure of the binuclear copper site present in COX (Cu_A) and in N2OR added a distinct and surprising member to the family of copper-containing redox centers. Although a number of theoretical studies as well as structure–reactivity analysis have already focused on this site, the functional reasons that led to the evolutionary selection of this unusual site are still widely debated. The site performs electron mediation reactions similar to those performed by the T1 copper (i.e., from a soluble donor to internal sites of the respective enzyme). So far, two possible rationales have been proposed for its functional selection. One suggests that an advantage in the proviso for electron uptake and delivery via two distinct pathways; the second that of being a site with a reorganization energy even lower than that of T1. Hopefully, future studies will resolve the two alternatives or support both. The trinuclear copper center composed of T2 and T3 provides an additional challenge, as the mechanism by which dioxygen reduction to water takes place at this center is still only partially resolved.

ACKNOWLEDGMENTS

The authors wish to thank Lars K. Skov for producing the figures. Part of this work was supported by grants from the German–Israeli Foundation, the Volkswagen Foundation, and by the Danish Natural Science Research Council.

REFERENCES

1. K. D. Karlin and Z. Tyeklar, eds. *Bioinorganic Chemistry of Copper*, Chapman & Hall, New York, 1993.

2. E. T. Adman, *Adv. Protein Chem.* **42**, 145 (1991).

3. O. Farver and I. Pecht, in *Copper Proteins and Copper Enzymes*, Vol. 1, R. Lontie, e.d., CRC Press, Boca Raton, Fla., 1984, p. 183; S. K. Chapman, *Perspect. Bioinorg. Chem.* **1**, 95 (1991).

4. C. Ostermeier, S. Iwata and H. Michel, *Curr. Opin. Struct. Biol.* **6**, 460 (1996).

5. A. Messerschmidt, in *Bioinorganic Chemistry of Copper*, K. D. Karlin and Z. Tyeklar, eds., Chapman & Hall, New York, 1993.

6. O. Farver and I. Pecht, *FASEB J.* **5**, 2554 (1991).

7. M. K. Johnson et al., eds. *Adv. Chem. Ser.*, **226** (1990).

8. P. M. Colman, H. C. Freeman, J. M. Guss, M. Murata, V. A. Norris, J. A. M. Ramshaw, and M. P. Venkatappa, *Nature (London)* **272**, 319 (1978).

9. E. T. Adman, R. E. Stenkamp, L. C. Sieker, and L. H. Jensen, *J. Mol. Biol.* **123**, 35 (1978).

10. J. M. Guss and H. C. Freeman, *J. Mol. Biol.* **169**, 521 (1983); E. N. Baker, *J. Mol. Biol.* **203**, 1071 (1983); W. E. B. Shephard, B. F. Anderson, D. A. Lewandoski, G. E. Norris, and E. N. Baker, *J. Am. Chem. Soc.* **112**, 7817 (1990); H. Nar, A. Messerschmidt, R. Huber, M. v. d. Kamp, and G. W. Canters, *J. Mol. Biol.* **221**, 765 (1991).

11. T. P. J. Garret, D. J. Clingeleffer, J. M. Guss, S. J. Rogers, and H. C. Freeman, *J. Biol. Chem.* **259**, 2822 (1984); H. Nar, A. Messerschmidt, R. Huber, M. v. d. Kamp and G. W. Canters, *FEBS Lett.* **306**, 119 (1992).

12. J. M. Guss, E. A. Merritt, R. P. Phizackerley, B. Hedman, M. Murata, K. O. Hodgson, and H. C. Freeman, *Science* **241**, 806 (1988).

13. L. Chen, R. Durley, B. J. Poliks, K. Hamada, Z. Chen, F. S. Matthews, V. L. Davidson, Y. Satow, E. Huizinga, F. M. D. Vellieux, and W. G. J. Hol, *Biochemistry* **31**, 4959 (1992); A. P. Kalverda, S. S. Wymenga, A. Lommen, F. J. M. v. d. Ven, C. W. Hilbers, and G. W. Canters, *J. Mol. Biol.* **240**, 358 (1994).

14. P. J. Hart, A. M. Nersissian, R. G. Hermann, R. N. Nalbandyan, J. S. Valentine, and D. Eisenbergd, *Protein Sci.* **5**, 2175 (1996).

15. L. Chen, R. C. E. Durley, F. S. Matthews, and V. L. Davidson, *Science* **264**, 86 (1994).

16. H. Nar, A. Messerschmidt, R. Huber, M. v. d. Kamp, and G. W. Canters, *J. Mol. Biol.* **218**, 427 (1991); L. M. Murphy, R. W. Strange, B. G. Karlsson, L. G. Lundberg, T. Pascher, B. Reinhammar, and S. S. Hasnain, *Biochemistry* **32**, 1965 (1993); A. Romero, C. W. G. Hoitink, H. Nar, R. Huber, A. Messerschmidt, and G. W. Canters, *J. Mol. Biol.* **229**, 1007 (1993); L. -C. Tsai, L. Sjölin, V. Langer, T. Pascher, and H. Nar, *Acta Crystallogr. D* **51**, 168 (1993).

17. H. B. Gray and E. I. Solomon, in *Copper Proteins*, Spiro, T. G., ed. Wiley, New York, 1981, p. 1.

18. J. M. Guss, P. R. Harrowell, M. Murata, V. A. Norris, and H. C. Freeman, *J. Mol. Biol.* **192**, 361 (1986).

19. A. Messerschmidt, A. Rossi, R. Ladenstein, R. Huber, M. Bolognesi, G. Gatti, A. Marchesini, R. Petruzzelli, and A. Finazzi-Agró, *J. Mol. Biol.* **206**, 513 (1989); A. Messerschmidt and R. Huber, *Eur. J. Biochem.* **187**, 341 (1990).

20. J. A. Tainer, E. D. Getzoff, J. S. Richardson, and D. C. Richardson, *Nature (London)* **306**, 284 (1983).

21. E. I. Solomon, U. M. Sundaram, and T. E. Machonkin, *Chem. Rev.* **96**, 2563 (1996).

22. A. Volbeda and W. G. S. Hol, *J. Mol. Biol.* **209**, 249 (1989).

23. K. A. Magnus, H. Ton-That, and J. E. Carpenter, in *Bioinorganic Chemistry of Copper*, K. D. Karlin and Z. Tyeklar, eds., Chapman & Hall, New York, 1993, p. 143.

24. A. Messerschmidt, R. Ladenstein, R. Huber, M. Bolognesi, L. Avigliano, A. Marchesini, R. Petruzzelli, A. Rossi, and A. Finazzi-Agró, *J. Mol. Biol.* **224**, 179 (1992); A. Messersch-

midt, W. Steigemann, R. Huber, G. Lang, and P. M. H. Kroneck, *Eur. J. Biochem.* **209**, 597 (1992).

25. A. Messerschmidt, H. Luecke, and R. Huber, *J. Mol. Biol.* **230**, 997 (1993).

26. B. Reinhammar, in *The Coordination Chemistry of Metalloenzymes*, I. Bertini et al. eds., D. Reidel, Norwell, Mass., 1983, p. 177.

27. G. T. Babcock and M. Wikström, *Nature* **356**, 301 (1992).

28. S. Iwata, C. Ostermeier, B. Ludwig, and H. Michel, *Nature* **376**, 660 (1995).

29. T. Tsukihara, H. Aoyama, E. Yamashita, T. Tomizaki, H. Yamaguchi, K. Shinzawa-Itoh, R. Nakashima, R. Yaono, and S. Yoshikawa, *Science* **269**, 1669 (1995).

30. K. W. Penfield, A. A. Gewirth and E. I. Solomon, *J. Am. Chem. Soc.* **107**, 4519 (1985); H. E. M. Christensen, L. S. Conrad, K. V. Mikkelsen, M. K. Nielsen, and J. Ulstrup, *Inorg. Chem.* **29**, 2808 (1990); M. D. Lowery, J. A. Guckert, M. S. Gebhard, and E. I. Solomon, *J. Am. Chem. Soc.* **115**, 3012 (1993).

31. M. D. Lowery and E. I. Solomon, *Inorg. Chim. Acta* **198–200**, 233 (1992).

32. J. F. Boas, in *Copper Proteins and Copper Enzymes*, Vol. 1, R. Lontie, ed., CRC Press, Boca Raton, Fla., 1984, p. 5.

33. H. B. Gray and B. G. Malmström, *Comm. Inorg. Chem.* **2**, 203 (1983).

34. A. Marchesini and P. M. H. Kroneck, *Eur. J. Biochem.* **101** (1979) 65.

35. J. A. Fee and B. G. Malmström, *Biochim. Biophys. Acta* **153**, 299 (1968).

36. M. J. Bjerrum et al., *J. Bioenerg. Biomembr.* **27**, 295 (1995).

37. D. G. A. H. daSilva, D. Beoku-Betts, P. Kyritsis, K. Govindaraju, R. Powls, N. P. Tomkinson, and A. G. Sykes, *J. Chem. Soc. Dalton Trans.* **1992**, 2145 (1992).

38. C. M. Groeneveld and G. W. Canters *J. Biol. Chem.* **263**, 167 (1988).

39. R. A. Marcus and N. Sutin, *Biochim. Biophys. Acta* **811**, 265 (1985); A. Broo and S. Larsson, *J. Phys. Chem.* **95**, 4925 (1991); P. Siddarth and R. A. Marcus, *J. Phys. Chem.* **97**, 13078 (1993); H. B. Gray and J. R. Winkler, *Annu. Rev. Biochem.* **65**, 537 (1996).

40. D. N. Beratan, J. N. Betts, and J. N. Onuchic, *Science* **252**, 1285 (1991).

41. O. Farver and I. Pecht, *Proc. Natl. Acad. Sci. USA* **86**, 6968 (1989); O. Farver and I. Pecht, *J. Am. Chem. Soc.* **114**, 5764 (1992); O. Farver, L. K. Skov, M. v. d. Kamp, G. W. Canters, and I. Pecht, *Eur. J. Biochem.* **210**, 399 (1992); O. Farver, L. K. Skov, T. Pascher, B. G. Karlsson, M. Nordling, L. G. Lundberg, T. Vänngård, and I. Pecht, *Biochemistry* **32**, 7317 (1993); O. Farver, L. K. Skov, G. Gilardi, G. van Pouderoyen, G. W. Canters, S. Wherland, and I. Pecht, *Chem. Phys.* **204**, 271 (1996); O. Farver, N. Bonander, L. K. Skov, and I. Pecht, *Inorg. Chim. Acta* **243**, 127 (1996).

42. O. Farver, L. K. Skov, S. Young, N. Bonander, B. G. Karlsson, T. Vänngård, and I. Pecht, *J. Am. Chem. Soc.* **119**, 5453 (1997).

43. A. Broo and S. Larsson, *Chem. Phys.* **148**, 103 (1990).

44. D. S. Wuttke, M. J. Bjerrum, J. R. Winkler, and H. B. Gray, *Science* **256**, 1007 (1992).

45. A. Broo and S. Larsson, *J. Phys. Chem.* **95**, 4925 (1991).

46. M. Plato, M. E. Michel-Beyerle, M. Bixon, and J. Jortner, *FEBS Lett.* **249**, 70 (1989).

47. H. Pelletier and J. Kraut, *Science* **258**, 1748 (1992).

48. O. Farver, in *Protein Electron Transfer*, D. Bendal, ed., 1996, BIOS Scientific Publishers, Chap. 7.

49. O. Farver and I. Pecht, *Coord. Chem. Rev.* **94**, 17 (1989).

50. O. Farver and I. Pecht, *Isr. J. Chem.* **21**, 13 (1981).

51. O. Farver and I. Pecht, *Proc. Natl. Acad. Sci. USA* **78**, 4190 (1981).

52. G. Morpurgo and I. Pecth, *Biochem. Biophys. Res. Commun.* **104**, 1592 (1982); O. Farver, A. Licht, and I. Pecht, *Biochemistry* **26**, 7317 (1987).

53. L. K. Skov, U. Christensen, K. Olsen, and O. Farver, *Inorg. Chem.* **32**, 4762 (1993).

54. C. C. Moser, J. M. Keske, K. Warncke, R. S. Farid, and P. L. Dutton, *Nature* **355**, 796 (1992); R. S. Farid, C. C. Moser, and P. L. Dutton, *Curr. Opin. Struct. Biol.* **3**, 225 (1993).

55. H. B. Gray, *Chem. Soc. Rev.* **15**, 17 (1986).

56. R. Margalit, N. M. Kostic, C. -M. Che, D. F. Blair, H. -J. Chaing, I. Pecht, J. B. Shelton, J. R. Shelton, W. A. Schroeder, and H. B. Gray, *Proc. Natl. Acad. Sci. USA* **81**, 6554 (1984).

57. N. M. Kostic, R. Margalit, C.-M. Che, and H. B. Gray, *J. Am. Chem. Soc.* **105**, 7765 (1983); D. G. Nocera, J. R. Winkler, K. M. Yocom, E. Bordignon, and H. B. Gray, *J. Am. Chem. Soc.* **206**, 5145 (1984).

58. D. S. Wuttke and H. B. Gray, *Curr. Opin. Struct. Biol.* **3**, 555 (1993).

59. R. Langen, J. L. Colón, D. R. Casimiro, T. B. Karpisin, J. R. Winkler, and H. B. Gray, *J. Biol. Inorg. Chem.* **1**, 221 (1996).

60. V. L. Davidson and L. H. Jones, *Biochemistry* **35**, 8120 (1996).

61. O. Farver and I. Pecht, *Inorg. Chem.* **29**, 4855 (1990).

62. M. P. Jackson, J. McGinnis, R. Powls, G. A. Salmon, and A. G. Sykes, *J. Am. Chem. Soc.* **110**, 5880 (1988).

63. H. E. M. Christensen, L. S. Conrad, K. V. Mikkelsen, M. K. Nielsen, and J. Ulstrup, *Inorg. Chem.* **29**, 2808 (1990); H. E. M. Christensen, L. s. Conrad, J. M. Hammerstad-Pedersen, and J. Ulstrup, *FEBS Lett.* **296**, 141 (1992).

64. I. Fridovich, *J. Biol. Chem.* **264**, 7761 (1989).

65. J. A. Tainer, E. D. Getzoff, K. M. Beem, J. S. Richardson, and D. C. Richardson, *J. Mol. Biol.* **160**, 181 (1982).

66. J. S. Valentine and M. W. Pantoliano, in *Copper Proteins*, T. G. Spiro, ed., Wiley, New York, 1981, p. 291; J. S. Valentine and D.M.D. Freitas, *J. Chem. Educ.* **62**, 990 (1985).

67. N. Ito, S. E. V. Phillips, C. Stevens, Z. B. Ogel, M. J. McPherson, J. N. Keen, K. D. S. Yadev, and P. F. Knowles, *Nature* **350**, 87 (1991).

68. J. A. Duine, *Eur. J. Biochem.* **200**, 271 (1991).

69. D. Mu, S. M. Janes, A. J. Smith, D. E. Brown, D. M. Dooley, and J. P. Klinman, *J. Biol. Chem.* **267**, 7979 (1992).

70. D. M. Dooley, M. A. McGuirl, D. E. Brown, P. N. Turowski, W. S. McIntire, and P. F. Knowles, *Nature*, **349**, 262 (1991).

71. T. E. Meyer, A. Marchesini, M. A. Cusanovich, and G. Tollin, *Biochemistry* **30**, 4619 (1991); G. Tollin, T. E. Meyer, M. A. Cusanovich, P. Curir, and A. Marchesini, *Biochim. Biophys. Acta* **1183**, 309 (1993); J. T. Hazzard, A. Marchesini, P. Curir, and G. Tollin, *Biochim. Biophys. Acta* **1208**, 166 (1994).

72. O. Farver and I. Pecht, *Proc. Natl. Acad. Sci. USA* **89**, 8283 (1992).

73. P. Kyritsis, A. Messerschmidt, R. Huber, G. A. Salmon, and A. G. Sykes, *J. Chem. Soc. Dalton Trans.* **1993**, 731 (1993).

74. O. Farver and I. Pecht, *Biophys. Chem.* **50**, 203 (1994).

75. P. M. H. Kroneck, F. A. Armstrong, H. Merkle, and A. Marchesini, *Adv. Chem. Ser.* **220**, 223 (1982).

76. O. Farver, S. Wherland, and I. Pecht, *J. Biol. Chem.* **269**, 22933 (1994).

77. M. Goldberg, O. Farver, and I. Pecht, *J. Biol. Chem.* **255**, 7353 (1980); O. Farver, P. Frank, and I. Pecht, *Biochem. Biophys. Res. Commun.* **108**, 273 (1982).

78. M. Brunori, G. Antonini, F. Malatesta, P. Sarti, and M. T. Wilson, *FEBS Lett.* **314**, 191 (1992).

79. L. -E. Andréasson and B. Reinhammar, *Biochim. Biophys. Acta* **445**, 579 (1976).

80. L. -E. Andréasson, R. Brändén, and B. Reinhammar, *Biochim. Biophys. Acta* **483**, 370 (1976).

81. B. Reinhammar, *Chem. Scr.* **25**, 172 (1985).

82. M. Faraggi and I. Pecht, *Nature, New Biol.* **233**, 116 (1971).

83. M. Faraggi and I. Pecht, *J. Biol. Chem.* **248**, 3146 (1973).

84. A. Guissani, Y. Henry, and L. Gilles, *Biophys. Chem.* **15**, 177 (1982).

85. P. O'Neill, E. M. Fielden, L. Morpurgo, and E. Agostineli, *Biochem. J.* **222**, 71 (1984).

86. L. -E. Andréasson and B. Reinhammar, *Biochim. Biophys. Acta* **568**, 145 (1979).

87. F. B. Hansen, R. W. Noble, and M. J. Ettinger, *Biochemistry* **23**, 2049 (1984); F. B. Hansen, G. B. Koudelka, R. W. Noble, and M. J. Ettinger, *Biochemistry* **23**, 2057 (1984); G. B. Koudelka, F. B. Hansen, and M. J. Ettinger, *J. Biol. Chem.* **260**, 15561 (1985); G. B. Koudelka and M. J. Ettinger, *J. Biol. Chem.* **263**, 3698 (1988).

88. M. Goldberg and I. Pecht, *Proc. Natl. Acad. Sci. USA* **71**, 4684 (1974); O. Farver, M. Goldberg, D. Lancet, and I. Precht, *Biochem. Biophys. Res. Commun.* **73**, 494 (1976); O. Farver, M. Goldberg, and I. Precht, *FEBS Lett.* **94**, 383 (1978); O. Farver, M. Goldberg, and I. Precht, *Eur. J. Biochem.* **104**, 71 (1980).

89. O. Farver and I. Pecht, *FEBS Lett.* **108**, 436 (1979).

90. W. G. Zumft, in *The Procaryotes*, Vol. 1, A. Balows et al. eds., Springer-Verlag, Berlin, 1992, p. 554.

91. J. W. Godden, S. Turley, D. C. Teller, E. T. Adman, M. -Y. Liu, W. J. Payne, and J. LeGall, *Science* **253**, 438 (1991).

92. M.-Y. Liu, M.-C. Liu, W. J. Payne, and J. LeGall, *J. Bacteriol.* **166**, 604 (1986).

93. S. Suzuki, T. Kohzuma, K. Yamaguchi, N. Nakamura, S. Shidara, K. Kobayashi, and S. Tagawa, *J. Am. Chem. Soc.* **116**, 11145 (1994).

94. E. D. Harris, *Nutr. Rev.* **53**, 170 (1996).

95. V. Saitzev, I. Saitzeva, G. Card, A. Ralph, B. Bax, and P. Lindley, *J. Inorg. Biochem.* **59**, 719 (1995); I. Saitzeva, V. Saitzev, G. Card, K. Moshkov, B. Bax, A. Ralph, and P. Lindley, *J. Biol. Inorg. Chem.* **1**, 15 (1996).

96. W. E. Antholine, D. H. W. Kastrau, G. C. M. Steffens, G. Buse, W. G. Zumft, and P. M. H. Kroneck, *Eur. J. Biochem.* **209**, 875 (1992).

97. K. Kobayashi, H. Une, and K. Hayashi, *J. Biol. Chem.* **264**, 7976 (1989).

98. B. C. Hill, *J. Biol. Chem.* **266**, 2219 (1991).

99. O. Warburg, *Biochem. Z.* **152**, 479 (1924).

100. D. Keilin, *Proc. R. Soc. London* **98**, 312 (1925).

101. J. R. Winkler, B. G. Malmström, and H. B. Gray, *Biophys. Chem.* **54**, 199 (1995).

102. B. E. Ramirez, B. G. Malmström, and H. B. Gray, *Proc. Natl. Acad. Sci. USA* **92**, 11949 (1995).

103. S. Hallén and T. Nilsson, *Biochemistry* **31**, 11853 (1992).

APPLYING MARCUS'S THEORY TO ELECTRON TRANSFER IN VIVO

GEORGE McLENDON

Department of Chemistry, Princeton University, Princeton, NJ 08544

SONJA KOMAR-PANICUCCI

Cambridge, MA 02139 Peptide, Inc.

SHELBY HATCH

Department of Chemistry, Princeton University, Princeton, NJ 08544

CONTENTS

I. INTRODUCTION

The insights of radiationless transition theory applied to single-electron transfer reactions have revolutionized our understanding of this chemistry, as this volume amply demonstrates. The unique level of quantitative understanding of structure-dependent reactivity for electron transfer also provides opportunities to explore structure–function relationships in biological systems. Electron transfer reactions play many roles in biology; key

Electron Transfer: From Isolated Molecules to Biomolecules, Part Two, edited by Joshua Jortner and M. Bixon. Advances in Chemical Physics Series, Volume 107, series editors I. Prigogine and Stuart A. Rice.
ISBN 0-471-25291-3   © 1999 John Wiley & Sons, Inc.

among these is the use of redox reactions to control the release of metabolic energy via the electron transport system of mitochondria. Laboratory reactions of the isolated, purified protein components of biological redox systems show that all the hallmark features of biological electron transfer shown by small molecules are retained in protein electron transfer, although additional features are introduced by the complex structures of the biomolecules. For example, rates scale strongly (essentially exponentially) with distance for both small molecule electron transfer and for protein-to-protein electron transfer [1]. However, for proteins, additional complexity is produced, since many paths may exist between the donor and acceptor. Some are shorter "through space" but involve weak (nonbonded) electronic couplings, while other paths, although longer, involve direct covalent bonding connections and thus produce higher electronic couplings. These pathway issues are discussed in detail elsewhere in this volume in the chapter "Electron Transfer Tables". Similarly, the complex protein structures can provide multiple binding sites for the redox partner with different characteristic couplings and reorganization energies. These multiple structures may themselves interconvert on the time scale of net electron transfer. Such dynamic effects have also been discussed previously [2].

Despite such caveats, it is clear that the basic dependence of biological redox reactions on such key parameters as distance and reaction free energy precisely mirror the reactions of simple small molecules. Thus the detailed understanding of electron transfer chemistry reflected in this volume can provide especially powerful windows into how biomolecules are designed to couple structure to reactivity. In this chapter we attempt to go one level beyond the in vitro exploration of biological electron transfer to inquire what insights from fundamental studies of electron transfer can be applied to understanding the complex metabolism of a living organism, as reflected in its redox-dependent energy metabolism. As for any mechanistic study, the key to such an investigation is the ability to isolate variables that can be studied systematically.

For any electron transfer reaction, one key variable is reaction free energy. As exemplified by Marcus's theory [3] (and subsequent more detailed analyses outlined elsewhere in this volume), electron transfer rates depend quadratically on the free energy of the redox reaction (holding other variables, such as electronic coupling and reorganization energy constant). Thus

$$k_{et} = A \exp - \Delta G^{\ddagger}/kT$$

where A depends on electronic coupling and $\Delta G^{\ddagger} = (\Delta G^{\circ} - \lambda)^2/4\lambda$ in the classical limit [λ is the reorganization energy (discussed extensively in this volume) and $\Delta G°$ the reaction free energy, k the boltzmann constant, T the absolute temperature, and 4 is a well-known integer].

Alternatively, expanding the quadratic in terms of directly accessible experimental variables gives

$$k_{et} = (k_{11}k_{22}K_{12})^{1/2}$$

where k_{11} and k_{22} are self-exchange rates for each couple and K_{12} is the equilibrium constant for the reaction. Thus for biological electron transfer, $k_{et} \propto (K_{12})^{1/2}$.

To probe the free energy dependence of electron transfer on the metabolism of a living cell directly, at least two requirements need to be met. First, methods must be developed to vary the free energy of a biological couple systematically. Second, conditions must be found in which the cell's metabolism is sensitive to this free energy and associated electron transfer rates. In this chapter we outline methods for approaching these goals and provide a progress report on the ability of Marcus's theory to predict not only reaction rates but also the associated metabolism, and ultimately, the relative stability and growth of an organism.

II.   MODIFICATION OF ΔG IN VITRO AND IN VIVO

The features that control the redox potential of proteins are understood in broad outline, if not in detail [4]. Those residues that provide an increase in net polarity (i.e., the sum of all dipoles) around the heme will favor the oxidized state. In practice, it remains difficult to calculate such effects properly, but they can be measured easily. For cytochrome c, many amino acid substitutions have been isolated, which result in small (20–50 mV) changes in heme potential while the global three-dimensional structure remains constant. In our work [5] we sought synergistic effects by combining several such replacements in cytochrome c within a single protein, expressed in a corresponding strain of yeast (the simplest eucaryote). We find that multiple substitutions can indeed give synergistic (although not purely additive) effects on potential, as summarized in Table I.

Note that for several such variants the normal downhill electron flow from cytochrome c_1 to cytochrome c must reverse, and the net electron transfer is endergonic. In this way we can directly modify the free energy of this metabolic process. (Note that this is unusual. Since ΔG depends only on the structures of the reactants and products, the net ΔG is unaffected by protein (enzyme) catalysts. Thus mutagenesis can only affect ΔG if the protein itself is a reactant and/or product. This condition is met for electron transport proteins such as cytochrome c.)

TABLE I
Reduction Potential of Substituted Iso-1-Cytochrome c

Protein	$E^0 \pm 2$ (mV vs. SHE)	ΔE^0 (mV)	ΔG^0 (kcal mol^{-1})
C102A	285		-1.27
F82S C102A	247	-38	-0.39
R38A C102A	239	-46	-0.21
N521 C102A	231	-54	-0.02
R38A N521 C102A	212	-73	0.42
R38A F82S C102A	203	-82	0.62
N521 F82S C102A	189	-96	0.95
R38A N521 F82S C102A	162	-123	1.57

III. REACTIVITY OF MODIFIED CYTOCHROMES

Given the availability of structurally conserved cytochromes with different redox potentials, an obvious question is how such systems react with their partner proteins. In short, we found that the reaction rate for the highly exothermic reaction with ($\Delta G \sim 1$ V) cytochrome c peroxidase (Ccp) was minimally affected by changes in Cc potential (Table II). Similarly, the reactions of cytochrome c with cytochrome oxidase or lactate dehydrogenase are not affected by cytochrome c redox potential. By contrast, the reaction with cytochrome c_1 depends strongly on Cc potential; the net turnover rate for c_1 enzymatic oxidation of Cc scales constant quadratically with the Cc/C$_1$ equilibrium constant, *precisely* as predicted by Marcus's theory (Table III).

TABLE II
Kinetic Parameters from Numerical Analysis of Cytochrome Oxidase Reactions with Redox-Altered Cytochromes c

Iso-1-cytochrome c	k_{cat} (mol cyt c/mol cyt aa_3)	K_m (μm)
C102A	1230 ± 82	15.1 ± 2.6
F82S C102A	740 ± 41	10.0 ± 1.6
R38A C102A	918 ± 145^a	8.3 ± 4.7^a
N521 C102A	1053 ± 61	10.1 ± 1.9
R38A N521 C102A	949 ± 55^a	9.0 ± 1.7^a
R38A F82S C102A	1117 ± 34	24.2 ± 1.6
N521 F82S C102A	605 ± 10	9.9 ± 0.5
R38A N521 F82S C102A	833 ± 31	14.2 ± 1.4

^{a} Normalized to account for variations in enzymatic activity among different preparations.

TABLE III

Kinetic Parameters from Numerical Analysis of the Cytochrome bc_1 Complex Reactions with redox-Altered Cytochromes c

Iso-1-cytochrome c	$k_{12}^{measured}$ (mol cyt c/mol cyt c_1) (s)	$k_{12}^{normalized}$ (mol cyt c/mol cyt c_1) (s)	K_{12}
C102A	84 ± 9	84	8.54
F82S C102A	53 ± 1	44	1.93
R38A C102A	20 ± 3	30	1.43
N521 C102A	59 ± 6	34	1.03
R38A N521 C102A	19 ± 2	23	0.49
R38A F82S C102A	18 ± 2	16	0.35
N521 F825 C102A	7 ± 5	11	0.20
R38A N521 F82S C102A	5 ± 5	6.6	0.07

IV. MARCUS'S THEORY IN VIVO: PREDICTION OF METABOLIC RATES?

The foregoing shows that it is possible to alter ΔG systematically for an important physiological reaction and to express the genes that code for this altered reactivity within individual yeast strains. At least two more challenging questions remain. First, to what extent, if any, can the theories outlined in this volume predict variations in electron transfer reactions of these protein variants, as their reactions occur within the cell? Second, what are the observable physiological consequences, if any, of such altered reactions? To address these questions requires defining metabolic conditions that are particularly sensitive to electron transport rates. For yeast it has been found that growth on lactate confers special sensitivity to redox reactions that involve cytochrome c [6]. For the yeast to survive with lactate as the sole carbon source requires obligate aerobic respiration.

Under these conditions it has been observed that overall yeast growth depends on the intracellular concentration (as measured spectrophotometrically) of cytochrome c. In biological terms, cytochrome c is an autogenous regulator of growth on lactate. In chemical terms, cytochrome c is apparently involved in a step (or steps) which are at least partially rate determining in the overall complex reactions of growth. This observation sets the critical stage for measurement of the effects of redox potential changes on growth, since such changes may be able to modulate the rates of reactions that involve cytochrome c.

Therefore, genes that code for altered cytochrome c were used to replace the wild-type gene, and the effects on cell growth measured. Genes were replaced as a single copy and integrated in the normal genome position to eliminate any effects of differences in gene regulation on the observed physiology.

The bottom line is that for those variants with highly modified redox potential, the relative rates of overall cell growth quantitatively follow the order predicted by Marcus's theory (Fig. 1). The simplest explanation for

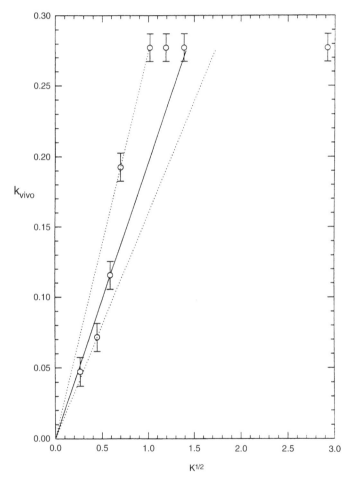

Figure 1. In vivo reactivity of diploid yeast strains containing redox altered cytochrome c as related to Marcus's theory of electron transfer.

this predictive correlation is that under the conditions used, the (uphill) electron transfer from cytochrome c_1 to cytochrome c can become net rate determining for overall cell growth. By comparison, it seems less likely that a simple equilibrium occurs. For the largest potential shift (ca. 120 mV), the *equilibrium* concentration of *reduced* Cc that would be available for reaction with cytochrome oxidase, as calculated by the Nernst equation, would be only 1% of the corresponding concentration for wild-type Cc. The aforementioned autogenous regulation studies show that when Cc is present at only a 1% level, growth on lactate cannot be sustained. These observations argue against an equilibrium mechanism, in favor of a mechanism in which the rate-determining step is slow transfer from Cc_1 to Cc, followed by rapid trapping of the reduced cytochrome c by cytochrome c oxidase (CcO), in comparison with unproductive reequilibration by electron transfer from Cc to Cc_1. The reaction in Cc is effectively irreversible, being coupled to dioxygen reduction to water.

A prediction of this mechanism is that undesired equilibration can be minimized by maximizing the CcO trapping pathway. There is an intriguing suggestion that the yeast system does precisely such optimization. When spectra of the various cytochromes are examined *within* the cells of the wild-type and variant strains, the relative absorbance of Cc, Cc_1, and cytochrome b remain relatively constant. However, for those uphill reactions that require efficient CcO trapping of reduced Cc, we find that the intracellular CcO concentration is substantially increased (Fig. 2). Such an increase may ensure that the trapping reaction efficiently competes with the unproductive $Cc \rightarrow Cc_1$ back transfer. A question left unanswered for now is how the yeast senses the need to produce additional CcO in these variant strains.

V.   EXTRAGENIC SUPPRESSION: A CRITICAL TEST IN PROGRESS

Extragenic suppression is the phenomenon by which the effect of a mutation or series of mutations in one gene is suppressed by a mutation or series of mutations in a different gene or genes. By modifying (lowering) the redox potential of cytochrome c_1, an extragenic suppressor of the redox mutants of cytochrome c should result. Additionally, this further tests the applicability of Marcus's theory in vivo.

The initial task is to make a mutation (or mutations) in cytochrome c_1 which significantly lowers the redox potential, to the point of being lower than the redox mutants of cytochrome c. These changes must not alter the overall structure of the protein. Mature cytochrome c_1 is a 248-residue heme protein for which a high-resolution structure is only starting to become available. This makes it difficult, at present, to predict which changes

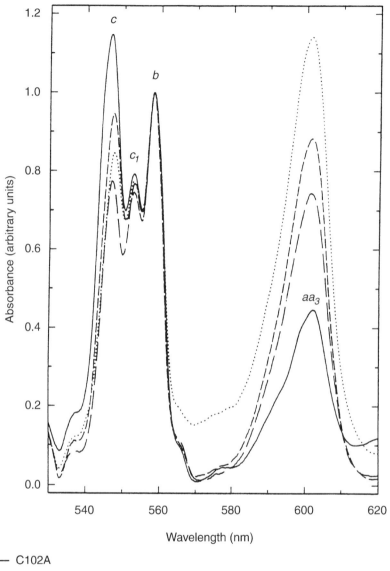

Figure 2.   Low-temperature spectra of cytochromes in intact yeast cells with increasing ratio of cytochrome aa_3 to cytochrome b.

would produce the desired results. However, it is known where the heme binds, and the secondary structure has been predicted. Therefore, three approaches are being taken.

First, the nonpolar residues in the heme environment are being altered to isosteric polar residues. This provides an increase in net polarity around the heme, favoring the oxidized state and thus lowering the redox potential. Thus far, none of these single mutations have been shown to affect growth on nonfermentable substrates, either in strains containing wild-type cytochrome c or in strains containing redox-altered cytochromes c.

Second, again in an attempt to alter the heme environment, a methionine residue, one of the axial ligands of the heme iron, has been altered via in vivo mutagenesis to all other naturally occurring amino acids. This method selects only for proteins that have function (i.e., growth on nonfermentable substrates). The only replacement that has shown any function is phenylalanine. It is unlikely that phenylalanine is acting as a sixth ligand; it is more probable that the sixth ligand has been replaced by a water molecule (or by some other side chain), which could lower the redox potential significantly. When this mutant is present in a yeast strain containing wild-type cytochrome c, it shows diminished growth on nonfermentable substrates at $30°C$. Furthermore, the spectral band for cytochrome c_1 was not visible in the low-temperature spectra of intact yeast cells. When $CYT1$ (the gene that codes for cytochrome c_1) is deleted, the spectral band for cytchrome b (another heme protein in the bc_1 complex) is also decreased significantly. However, in the phenylalanine mutant, the cytochrome b band remains at a normal level. This, along with diminished growth on nonfermentable substrates, suggests that the cytochrome c_1 absorption band may have shifted, indicating a possible change in axial ligation and an associated change in redox potential. In strains containing redox-altered cytochromes c, this mutant has no growth on nonfermentable substrates. It is currently being purified so that the redox potential can be measured.

The third method being used to construct an extragenic suppressor is random mutagenesis of cytochrome c_1. The mutagenesis is performed by using PCR. This method is used routinely to make exact copies of a gene, but in this case, the conditions are altered (by changing metal ion concentrations and lowering the concentration of one nucleotide in relation to the other three) such that an average of one "mistake" (base-pair change) is made per gene. These PCR products are then co-transferred along with a "gapped" plasmid into the desired yeast strain. The yeast strain has a deletion of $CYT1$ and redox-altered cytochrome c. The "gapped" plasmid is a plasmid in which the majority of $CYT1$ gene has been excised, but leaves regions of homology with the PCR product on both ends such that a cross-over event can occur, and a "repaired" plasmid is then present in each yeast cell — a different mutation in each one.

The cells are then grown on selective media, individual colonies are picked, and their growth on nonfermentable substrates is observed. The control strain contains redox-altered cytochrome c and wild-type cytochrome c_1. Since an extragenic suppressor is being pursued, the candidates we are interested in are those which grow *better than* the control. Thus far we have two that grow slightly better and two that grow significantly better than the control. They are currently being sequenced to discover the changes that occurred in the *CYT1* gene.

REFERENCES

1. C. C. Moser, J. M. Keske, K. Warncke, R. S. Farid, and P. L. Dutton, *Nature* **355**, 796 (1992).

2. K. Pardue, P. Bak, and G. McLendon, *J. Am. Chem. Soc.* **109**, 7540 (1987).

3. R. A. Marcus, and N. Sutin, *Biochim. Biophys. Acta* **811**, 265 (1985).

4. (a) A. L. Raphael, and H. B. Gray, *Proteins Struct. Funct. Genet.* **6**, 338 (1989); (b) D. C. Rees, *Proc. Natl. Acad. Sci. USA* **82**, 3082 (1985); (c) G. I. Williams, G. Moore, and R. J. P. Williams, Comm. Inorg. Chem. 4, 55 (1985).

5. S. Komar-Panicucci, J. Bixler, G. Bakker, F. Sherman, and G. McLendon, *J. Am. Chem. Soc.* **114**, 5443 (1992).

6. F. Sherman, H. Taber, and W. Campbell, *J. Mol. Biol.* **13**, 21 (1965).

SOLVENT-FLUCTUATION CONTROL OF SOLUTION REACTIONS AND ITS MANIFESTATION IN PROTEIN FUNCTIONS

HITOSHI SUMI

Institute of Materials Science, University of Tsukuba, Tsukuba, 305-8573, Japan

CONTENTS

OUTLINE

Dynamics in solution reactions is determined by solvent fluctuations. In second-order reactions, reacting two species of molecules approach sufficiently closely by diffusion which becomes possible due to thermal fluctuations of solvents. First-order reactions take place within a solute molecule, but in general when its solvation structure takes specific molecular arrangements most favorable for reactions. These specific conformations can be taken as a result of solvent fluctuations. Important roles are also played by intramolecular vibrational fluctuations in solute molecules, in

Electron Transfer: From Isolated Molecules to Biomolecules, Part Two, edited by Joshua Jortner and M. Bixon. Advances in Chemical Physics Series, Volume 107, series editors I. Prigogine and Stuart A. Rice.
ISBN 0-471-25291-3   © 1999 John Wiley & Sons, Inc.

conjunction with the fact that atomic arrangements in solute molecules reorganize after reaction. The former fluctuations are much slower than the latter ones. In this situation the rate constant takes a general form $1/(k_{\text{TST}}^{-1} + k_f^{-1})$ with $k_f > 0$, where k_{TST} represents the rate constant expected from the transition-state theory (TST). k_f represents the rate constant with which the molecular arrangements most favorable for reactions are attained in the solute–solvent system as a result of solvent fluctuations. k_{TST} does not depend on the speed of solvent fluctuations measured by the inverse of their relaxation time τ, but k_f is proportional to $\tau^{-\alpha}$ with $0 < \alpha \leq 1$. Usually, τ is proportional to the solvent viscosity η. In rapid solvents with small τ and η such that $k_{\text{TST}} \ll k_f$, the formula shows that the rate constant k approaches k_{TST}, recovering the TST. In slow solvents with large τ and η such that $k_f \ll k_{\text{TST}}$, on the other hand, k approaches k_f, which decreases with η. This formula covering the both regimes was confirmed for the thermally recovering Z/E isomerization of substituted azobenzenes and N-benzylideneanilines in various solvents in a variation range of η as wide as 10^8-fold under pressure. In the latter TST-invalid regime, the reaction is controlled by slow speeds of solvent fluctuations. Enzymes have often been found in this regime. In other words, they perform their functions utilizing conformational fluctuations of their large flexible body in solvents.

I. INTRODUCTION

The standard theory on the rate of chemical reactions is the transition-state theory (TST) [1]. It has recently been disclosed, however, that the fundamental assumption of TST concerning rapid thermalization in the reactant state is not satisfied in many solution reactions [2]. They cover various solution reactions, extending from elementary reactions such as electron- [3], proton- [4], excitation- [5], and atom-group- [6] transfer reactions and isomerization reactions [7] to composite reactions such as enzymatic ones mediated by a biological supramolecule [8–16]. Study of solution reactions has been regarded as one of the most traditional and important subjects in chemistry. It was disclosed, in this situation, that TST cannot describe solution reactions. This fact was therefore regarded as meaning that the general expression on rates of chemical reactions has not yet been established, and many works have been devoted to clarifying the unknown general expression [2,17–19].

It is a priori assumed in TST that thermal fluctuations are sufficiently fast in the reactant state. To be more exact, it is assumed that before reaction occurs, thermal equilibration has already been established among substates in the reactant state and also between the reactant and the transition states, because of very fast thermal fluctuations. On this assumption, the calculation of the rate constant is greatly simplified since we need only to calculate the transmission coefficient which describes how fast a reactant at the transition state changes to the product state. We also need to know the reactant population at the transition state, but it has been determined only by the free energy of the transition state relative to the reactant state since they are in thermal equilibrium by assumption. Under this assumption, the rate constant should not depend on how fast thermal equilibration in the reactant state is accomplished, that is, on the thermalization time in the reactant state. In solution reactions, the thermalization time in the reactant state increases as the viscosity of solvents increases, as explained in the next paragraph. The thermalization time is usually proportional to the solvent viscosity. In many solution reactions, however, it seems that the rate constant depends on the thermalization time in the reactant state, decreasing with an increase in the solvent viscosity. It seems, therefore, that the fundamental assumption of TST is not satisfied in these solution reactions, as mentioned earlier.

In solution reactions, solvent fluctuations play important roles in thermalization in the reactant state. This can easily be understood in second-order reactions, where reacting two species of molecules are distributed with various mutual distances in the reactant state. Thermalization of the distribution takes place by molecular diffusion, which is made possible as a result of solvent fluctuations. In first-order reactions, a solute molecule in which reaction takes place constitutes a solvated structure together with solvent molecules surrounding it. The solvated structure is composed of continuously distributed substates differing little by little in molecular arrangements in the solute–solvent system. Thermalization of the distribution of these molecular arrangements is made possible as a result of solvent fluctuations here. In solution reactions, therefore, the thermalization time in the reactant state can be estimated by the relaxation time of solvent fluctuations, which is usually proportional to the viscosity of solvents. In solution reactions mentioned earlier, the rate constant was observed to decrease with an increase in the solvent viscosity, that is, with an increase in the thermalization time in the reactant state. It should be remembered that the viscosity works to decelerate the speed of matters moving in liquid, and does not work after they stop. Therefore, it does not influence static properties such as the potential energy surface for reaction from which the rate constant can be calculated in TST. The viscosity is a quantity concerned only with dynamics. With the viscosity dependence mentioned above, therefore, reactions enter the non-TST regime, controlled by the slow speed of solvent fluctuations through viscosity.

To describe the solution reactions mentioned earlier, we need a theory which at least does not a priori assume rapid thermalization in the reactant state in the course of reaction, in contradistinction to TST. Such a theory had been presented by Kramers [20] in 1940 when TST was still in the process of establishment [21]. In his model, chemical reactions are regarded as surmount over a potential barrier from the reactant to the product well in the course of diffusive (Brownian) motions of reactants along a one-dimensional reaction coordinate. In this case, not only the reaction described above but also the thermalization in the reactant well take place as a result of diffusive motions of reactants. However, his theory had not attracted much attention for a long time. It can be said that his theory was presented too early [22]. It is a natural order in the development of theoretical descriptions that an event (such as reaction) which causes a deviation from the thermal equilibrium is regarded as a small perturbation on a thermalized system (as in TST, and more generally) as in the linear-response theory for kinetic coefficients [23]. It was only beginning in the 1980s that his theory had attracted many people, in conjunction with their attention to solution reactions whose rate decreases with an increase in the solvent viscosity.

In applying the Kramers theory to observations, however, we met a new problem. The solvent viscosity decelerates diffusive motions of reactants in the reactant well. When the viscosity is sufficiently small, the rate constant given by his theory reduces to that expected from TST, becoming independent of the viscosity. (Kramers predicted that the rate constant would increase with an increase in the friction in the small-friction limit where the reaction is limited by slow energy thermalization into the product state. This regime gives rise to the so called Kramers turnover in the rate constant [24], since it is followed by the regime mentioned above and the one in the high-friction limit mentioned later, where the rate constant decreases with the friction. In solution reactions, however, the solvent viscosity causing the friction has already reached an appreciable value not realized in gas-phase reactions. It has been considered, then, that the viscosity dependence in solution reactions starts from the TST regime mentioned earlier without experiencing the energy-relaxation-controlled regime.) In the high-viscosity region, on the other hand, his theory gives a rate constant that decreases in proportion to the inverse of the solvent viscosity, entering a regime that cannot be described by TST. This viscosity dependence of the rate constant in the non-TST regime is, however, a little different from the one observed. It decreased in general in proportion to a fractional (less-than-unity) power of the inverse of the solvent viscosity [3–16,19], as

$$k_{\text{obs}} \propto \eta^{-\alpha} \qquad \text{with} \quad 0 < \alpha \leq 1 \qquad (1.1)$$

Many people tried to understand this experimental fact by extending the Kramers model. The effort can be classified into two streams of theories. One was initiated by Grote and Hynes [25] and another by Sumi and Marcus [26]. As reviewed (in Section IV), the first unambiguous check of these two streams became possible by recent experimental data that covered both the TST and non-TST regimes in a viscosity change as wide as 10^8-fold under pressure. The data cannot be described by the former stream of theories but by the latter one. The viscosity dependence of Eq. (1.1) has often been observed also in biological reactions such as enzymatic reactions [6,8–16]. Therefore, they work in the non-TST regime where Eq. (1.1) is applicable. In this regime, however, the rate constant decreases from the TST-expected value k_{TST}, being limited by the slow speed of solvent fluctuations. Living organisms have developed their functions adjusted so as to be most convenient to their living conditions as a result of evolution over a long period. It is interesting, then, to clarify why living organisms have chosen, for their better living, a regime with rate constant smaller than k_{TST}. As shown also in the present article, it seems appropriate to assume that enzymes can accomplish their functions only by utilizing solvent fluctuations, not to assume that their activities are limited by a slow speed of solvent fluctuations. Let us begin by surveying essential characteristics of the two streams of theories mentioned earlier.

II. THE KRAMERS MODEL AND ITS EXTENSION BY GROTE AND HYNES

The Kramers model can be described by a double-well potential $W(X)$ composed of a reactant and a product well along a one-dimensional reaction coordinate X, as shown in Figure 1. Chemical reactions are accomplished as a result of diffusive surmounting, represented by a zigzag trajectory in

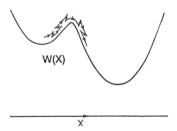

Figure 1.   Double-well potential for reaction $W(X)$ along a one-dimensional reaction coordinate X in the Kramers model, and a reactive diffusive trajectory represented by a zigzag line surmounting the potential barrier from the reactant to the product well.

Figure 1, over the potential barrier between these wells. To describe the shift from the TST-valid to the TST-invalid regime with an increase in the solvent viscosity, it is sufficient to take into account only diffusive motions of the coordinate value X under the assumption that its velocity has been thermalized at each value of X [20]. Then the distribution function $P(X;t)$ for a reactant to be found at the coordinate value X at time t in the double-well potential $W(X)$ satisfies a diffusion equation in the force field $dW(X)/dX$ as

$$\frac{\partial}{\partial t} P(X;t) = D \frac{\partial}{X} \left[\frac{\partial}{\partial X} + \frac{1}{k_B T} \frac{dW(X)}{dX} \right] P(X;t) \qquad (2.1)$$

where D and T represent, respectively, the diffusion constant and the absolute temperature. Around the bottom of the reactant well at $X = 0$, the $W(X)$ is quadratic in X as

$$W(X) \approx \tfrac{1}{2}\omega_0^2 X^2 \qquad (2.2)$$

with a curvature ω_0^2, where both the origin of energy and that of the coordinate X are put at the bottom of $W(X)$. The thermalization time in the reactant state corresponds to the relaxation time of the average motion of X in the potential of (2.2). As shown in Appendix A, it is given by

$$\tau = k_B T/(\omega_0^2 D) \qquad (2.3)$$

It is usually proportional to the solvent viscosity η in solution reactions since $D \propto \eta^{-1}$ on the basis of (macroscopic) hydrodynamics, as mentioned earlier. Around the top of the potential barrier located at $X = X_b$ we can write $W(X)$ as

$$W(X) \approx W(X_b) - \tfrac{1}{2}\omega_b^2 (X - X_b)^2 \qquad (2.4)$$

with a negative curvature $-\omega_b^2$. In the high-viscosity region of $\tau\omega_0^2/\omega_b \gg 1$, the Kramers model described by (2.1) gives the rate constant

$$k_{\mathrm{KR}} \approx \frac{1}{\tau} \frac{\omega_b}{2\pi\omega_0} \exp\left[-\frac{W(X_b)}{k_B T} \right] \qquad (2.5)$$

which is inversely proportional to the solvent viscosity η from the relation (2.3).

As mentioned earlier, the η dependence given by (2.5) cannot describe the observation that with an increase in η the rate constant in general decreased more slowly than η^{-1}, as expressed in Eq. (1.1). First, let us see the extension of the Kramers model by Grote and Hynes (GH) [25] in conjunction with this fact. They regarded it as important that when a reactant passes through the potential-barrier region around $X \approx X_b$ much faster than microscopic solvent motions, the friction decelerating its speed becomes smaller than that expected from the solvent viscosity which appears in macroscopic hydrodynamics. This occurs since solvents cannot completely follow the reactant, being called the frequency-dependent friction. In this situation, τ in Eq. (2.5) proportional to the viscosity η should be replaced by an effective τ proportional to the effective friction experienced by the reactant passing through the potential-barrier region. Then the rate constant given by Eq. (2.5) should not become as small as given by the original Kramers rate constant $k_{KR} \propto \eta^{-1}$, and hence it is expected that the rate constant does not decrease as rapidly as η^{-1} with an increase in η, as has in fact been observed.

This slow viscosity dependence of the rate constant in the GH model, however, must be obtained only in the intermediate-viscosity region since it should not last until the high-viscosity limit: In this limit, even the reactant motion passing through the potential-barrier region should become much slower than microscopic solvent motions, with speed at least on the order of $(1\,\text{ps})^{-1}$, as long as the rate constant continues to decrease. Then the rate constant should approach the original $k_{kR} \propto \eta^{-1}$ in the high-viscosity limit. It is rather expected, therefore, that toward this limit the rate constant should decrease more rapidly than η^{-1}, in contradistinction to the observation, if the rate constant is larger than the original k_{KR} due to the effect of frequency-dependent friction in the intermediate-viscosity region. This difficulty manifests itself again, but in a little different form, when we try to describe experimental data with the GH theory in Section IV.

Mathematically, the frequency dependence of the friction is equivalent to the non-Markoffian effect, in the time domain, that deceleration of the reactant motion at a certain time is determined not only by the velocity at the same time but also by its velocities at earlier times. This effect is incorporated [25] by a generalized Langevin equation for the motion $X(t)$ of the reaction coordinate X at time t in the double-well potential $W(X)$ as

$$\frac{d^2 X(t)}{dt^2} = -\frac{dW[X(t)]}{dX(t)} - \int_{-\infty}^{t} \hat{\zeta}(t-t') \frac{dX(t')}{dt'} dt' + R(t) \qquad (2.6)$$

where the second term on the right-hand side describes the non-Markoffian effect of the friction, mentioned above, with the friction kernel $\hat{\zeta}(t-t')$ ($\neq 0$, only for $t > t'$) while $R(t)$ represents the random force arising from small pulling and pushing by microscopic motions of solvent molecules. In contrast to this $R(t)$, the second term is called the systematic part due to microscopic solvent motions, giving dissipation. Since both come from the same origin, they should be related to each other. The relation is called the *fluctuation–dissipation theorem* [27] and is written as

$$\langle R(t)R(t')\rangle = k_B T \hat{\zeta}(t - t') \tag{2.7}$$

where $\langle \cdots \rangle$ represents the statistical average. The frequency-dependent friction is defined as the Laplace transform of the frictional memory function $\hat{\zeta}(t - t')$

$$\zeta(\mu) = \int_0^\infty \hat{\zeta}(t)e^{-\mu t}\, dt \tag{2.8}$$

which is called the friction at a frequency μ. In the Markoffian limit, $\hat{\zeta}(t - t')$ is regarded as decaying much more rapidly than the reactant velocity represented by $dX(t)/dt$. Then the second term on the right-hand side of Eq. (2.6) can be approximated as

$$\int_{-\infty}^t \hat{\zeta}(t - t')\,\frac{dX(t')}{dt'}\, dt' \approx \zeta(0)\,\frac{dX(t)}{dt} \tag{2.9}$$

where $\zeta(0)$ represents the friction at zero frequency, defined by Eq. (2.8) for $\mu = 0$. When the approximation of Eq. (2.9) is applied, Eq. (2.6) reduces to a Langevin equation, for which it can be shown [27] that the distribution function $P(X; t)$ for the coordinate value $X(t)$ satisfies the diffusion equation (2.1) used by Kramers, with the Einstein relation of

$$D = k_B T / \zeta(0) \tag{2.10}$$

Based on the generalized Langevin equation (2.6), GH showed that the rate constant is given in terms of μ/ω_b, which they called the transmission coefficient, by

$$k_{GH} = (\mu/\omega_b)k_{TST} \tag{2.11}$$

with

$$k_{TST} = (\omega_0/2\pi)\,\exp[-W(X_b)/k_B T] \tag{2.12}$$

where ω_0 and ω_b are angular frequencies appearing respectively in Eqs. (2.2) and (2.4). Here, k_{TST} of Eq. (2.12) represents the rate constant expected from TST in a situation that the potential energy surface for reaction is given by Figure 1. This μ corresponds to the frequency μ appearing in the frequency-dependent friction $\zeta(\mu)$ of Eq. (2.8) and is determined by solving

$$\mu/\omega_b = \omega_b/[\mu + \zeta(\mu)] \tag{2.13}$$

When $\zeta(\mu) \ll \mu$, Eq. (2.13) gives $\mu \approx \omega_b$ with which Eq. (2.11) reduces to $k \approx k_{TST}$, recovering TST. When $\zeta(\mu) \gg \mu$, on the other hand, Eq. (2.13) gives $\mu\zeta(\mu) \approx \omega_b^2$, with which Eq. (2.11) reduces to $k \approx [\omega_b/\zeta(\mu)]k_{TST}$. This rate constant agrees with that obtained from Eq. (2.5) by replacing τ therein by an effective τ which is obtained from Eq. (2.3) with Eq. (2.10) for D by replacing $\zeta(0)$ therein by $\zeta(\mu)$. This is just in accordance with the motivation of the GH model mentioned earlier. In this non-TST regime, we get the rate constant much larger than k_{KR} of Eq. (2.5) if $\zeta(\mu) \ll \zeta(0)$, and then the viscosity dependence of Eq. (1.1) can be expected, as mentioned earlier. Here, μ must become much smaller than ω_b, since $\zeta(\mu) \gg \mu$ and $\mu\zeta(\mu) \approx \omega_b^2$ must be satisfied simultaneously. As mentioned earlier, in order for $\zeta(\mu)$ to be much smaller than $\zeta(0)$, the frequency μ must be much larger than the typical speed of microscopic solvent motions which cause friction to reactant motions. [Otherwise, $\zeta(\mu)$ approaches $\zeta(0)$ proportional to the solvent viscosity η by Eq. (2.10), and k reduces to $k_{KR} \propto \eta^{-1}$.] Then μ, which equals $(k/k_{TST})\omega_b$ from Eq. (2.11), must be much smaller than ω_b, where ω_b should be of the same order of magnitude as ω_0 on the order of $10^{13}\,s^{-1}$ [28–30]. It should also be noted that the typical speed of microscopic solvent motions should be at least on the order of $(1\,ps)^{-1}$ $(= 10^{12}\,s^{-1})$, as explained in some detail in Section IV. It is therefore when $10^{12}\,s^{-1} \ll (k/k_{TST})10^{13}\,s^{-1}$ that the idea of the frequency-dependent friction of the GH model becomes effective. This causes a problem if the rate constant k is observed to become about 0.1 times as small as k_{TST} and simultaneously to deviate much from k_{KR} as Eq. (1.1). The observation could not be described by the frequency-dependent friction without assuming unrealistically small values for the typical speed of microscopic solvent motions. The same difficulty also arises when the viscosity dependence of Eq. (1.1) is observed over a rate-constant variation broader than about 10-fold. This difficulty occurs as described in Section IV.

The frequency dependence of friction has been calculated with various models [31]. As long as the rate constant is calculated by Eq. (2.11) with Eq. (2.13), however, all these calculations cannot avoid the above-mentioned criticism in describing the observation of Eq. (1.1) over a wide range of viscosity variation.

III. EXTENSION OF THE KRAMERS MODEL BY SUMI AND MARCUS

In extending the Kramers model, Sumi and Marcus (SM), on the other hand, regarded it as important that not only solvent fluctuations but also much faster intramolecular vibrational fluctuations in solute molecules take part in solution reactions. This fact can be detected, for example, in an observation that when a sudden change in electron cloud is induced by photoexcitation in a solute molecule in solvents, it gives rise to reorganization due to these two kinds of fluctuations with discriminated temporal variation in energy and width of optical spectra [32]. It is well known that reorganization induced by electron transfer reactions has these two components which correspond to the outersphere component due to solvent fluctuations and the inner-sphere component due to intrasolute vibrational fluctuations.

Let us begin with first-order reactions in solution. In this case, solvent fluctuations are diffusive molecular-arrangement fluctuations in the solvated structure of a solute molecule, while intrasolute vibrational fluctuations are ballistic atomic-arrangement fluctuations. The characteristic time of the latter can be estimated by oscillation periods of intramolecular vibrations, which are shorter than about 0.1 ps. The characteristic time of the former is on the order of 1 ps or longer for a small solute molecule, as estimated by time-resolved absorption and fluorescence spectra for dimethyl-s-tetrazine and MnO_4^- in solvents [32]. Small moieties protruding from the surface of large molecules such as proteins perform rotational diffusive fluctuations in solvents. The relaxation time of these motions is on the order of 10 ps to 1 ns, measured by Doppler broadening of the Rayleigh scattering of Mössbauer radiation [33]. Comparatively rigid large domains around a cleft in the protein structure perform hinge-bending fluctuations in tune with the solvated-structure fluctuations. The relaxation time of these motions is on the order of 10–100 ns, measured by the same method as above [33]. In association with electron transfer reactions in polar solvents, orientational reorganization of polar solvent molecules take place around solute molecules. In this case, the solvated structure fluctuations are excited and/or damped by microscopic solvent motions not only through viscosity, but also through electric fields generated by orientational fluctuations of polar solvent molecules [34,35]. The latter gives rise to a spherical Coulombic field fluctuating around a charged solute molecule, contributing to energy fluctuations of the solute–solvent system. The relaxation time of this field is given by $\tau_D \varepsilon_o / \varepsilon_s$, where τ_D denotes the Debye relaxation time, when polar solvents are represented by a dielectric continuum with the optical and the static dielectric constants ε_o and ε_s, respectively [36]. This longitudinal dielectric-relaxation time has a magnitude of the order of 1–100 ps with an exceptional case on the order of 0.1 ps for water. We thus see that solvent fluctuations are much slower than intrasolute vibrational fluctuations except in this exceptional case.

Both fluctuations contribute to promote reactions. Although slow solvent fluctuations are decelerated by an increase in the solvent viscosity, fast intrasolute vibrational motions are not influenced by the solvent viscosity. It is considered, then, that reactions are driven by intrasolute vibrational motions at each coordinate value of much slower solvent fluctuations. In this situation, we can envisage an intrinsic reactivity, given by a rate constant of reactions induced by intrasolute vibrational motions, at each coordinate value of solvent fluctuations. Then, taking the simplest one-dimensional model, we can introduce a single coordinate value X for describing molecular arrangements in the solvated structure of a solute molecule in the course of solvent fluctuations. Reactant populations at each coordinate value X at time t are written as $P(X; t)$. The intrinsic reactivity at each coordinate value X is described by a rate constant $k_i(X)$ with different values for different X's. It should be determined by TST since it is considered that thermal equilibration in the reactant state has always been accomplished due to fast intrasolute vibrational fluctuations at each coordinate value of slow solvent-induced solvated-structure fluctuations. They are also much slower than the typical speed of microscopic solvent motions at least on the order of $10^{12} \, \text{s}^{-1}$, except in the exceptional case of electric fluctuations in water. Then deceleration of the X motion by friction due to these microscopic solvent motions can be regarded as Markoffian, not influenced by the frequency dependence of the friction. Thus $P(X; t)$ should satisfy a diffusion-reaction equation

$$\frac{\partial P(X; t)}{\partial t} = D \frac{\partial}{\partial X} \left[\frac{\partial}{\partial X} + \frac{1}{k_B T} \frac{dV(X)}{dX} \right] P(X; t) - k_i(X) P(X; t) \qquad (3.1)$$

which is obtained by supplementing the right-hand side of the diffusion equation (2.1) with the second term, describing a decrease in reactant populations at each X by reaction. Since the reaction is taken into account by the second term, the potential $V(X)$ in the first term on the right-hand side can be a single-well potential. The potential is quadratic as

$$V(X) \approx \tfrac{1}{2} \omega_0^2 X^2 \qquad \text{around the bottom of } V(X) \qquad (3.2)$$

with the same form as Eq. (2.2). The relaxation time of fluctuations in X is given by Eq. (2.3).

Equation (3.1) is the fundamental starting point of the SM model, and it can also be interpreted as a generalization of the model by Agmon and Hopfield [37] for ligand binding to hemo proteins. The fundamental equation describing second-order reactions in solution is essentially the same as Eq. (3.1), as explained below. In second-order reactions, two species of molecules, say A and B, react during diffusion in solvents in the field of a mutual interaction potential $U(r)$, which depends on the mutual distance r. It is usually assumed that reaction takes place at each mutual distance r with an intrinsic reactivity $k_i(r)$ which depends on r. Then the distribution function $Q(r; t)$ for finding A at a distance r from B at time t should satisfy a three-dimensional diffusion-reaction equation

$$\frac{\partial Q(r; t)}{\partial t} = D\nabla \cdot \left(\nabla Q + \frac{Q}{k_B T} \nabla U \right) - k_i(r)Q \tag{3.3}$$

where ∇ represents the three-dimensional gradient operator and the diffusion constant D is given by a sum of diffusion constants of A and B molecules. Since $k_i(r)$ depends on space variables only through the mutual distance r, so does $Q(r; t)$. Then Eq. (3.3) is essentially a one-dimensional diffusion-reaction equation such as Eq. (3.1). Only differences are that r must be larger than the encounter distance a given in a situation of closet contact between the two molecules, and also that a multiplication factor of $4\pi r^2$ must be taken into account in the r integration from a to ∞. The distribution in the mutual distance r thermalizes in the field of the potential $U(r)$ in second-order reactions as long as reaction between A and B molecules is sufficiently slow. The thermalization is attained as a result of solvent fluctuations, since they enable these solute molecules to diffuse. The thermalization time is inversely proportional to the diffusion constant D, which is inversely proportional to the solvent viscosity η from the hydrodynamic Stokes–Einstein–Debye relation.

The general form of the rate constant obtained from the SM model, under the condition of initial thermalization in the reactant state, has been clarified by the present author [38,39]. The rate constant of solution reactions, described by Eq. (3.1) for first-order reactions and by Eq. (3.3) for second-order reactions, has a unified form

$$k = 1/(k_{\text{TST}}^{-1} + k_f^{-1}) \qquad \text{with } k_f > 0 \tag{3.4}$$

where k_{TST} represents the rate constant expected from TST and k_f represents a term causing deviation of the rate constant k from k_{TST}. As noted in Section I, k_{TST} does not depend on the thermalization time in the reactant state, while k_f does. To be more exact, k_f depends on the relaxation time τ of solvent fluctuations, as

$$k_f \propto \tau^{-\alpha} \propto \eta^{-\alpha} \qquad \text{with} \quad 0 < \alpha \leq 1 \tag{3.5}$$

where it was taken into account that τ is usually proportional to the solvent viscosity η. When $k_f \gg k_{TST}$ in low-viscosity solvents, Eq. (3.4) reduces to $k \approx k_{TST}$, recovering TST. When $k_f \ll k_{TST}$ in high-viscosity solvents, on the other hand, Eq. (3.4) reduces to $k \approx k_f$, invalidating TST. The observation of Eq. (1.1) is covered in this non-TST regime.

The explicit form of k_{TST} is given by

$$k_{TST} = \int_{-\infty}^{\infty} k_i(X) e^{-V(X)/k_B T} \, dX \Big/ \int_{-\infty}^{\infty} e^{-V(X)/k_B T} \, dX$$

$$\text{for first-order reactions} \tag{3.6}$$

while

$$k_{TST} = 4\pi \int_{a}^{\infty} r^2 k_i(r) e^{-U(r)/k_B T} \, dr \qquad \text{for second-order reactions} \tag{3.7}$$

Let us note here that k_{TST} has a dimension of inverse time in Eq. (3.6) while that of inverse time multipled by volume in Eq. (3.7).

To calculate k_f, we must first solve, for a function $p(X)$ in first-order reactions and for $\bar{p}(r)$ in second-order reactions, Fredholm's integral equation of the second kind as

$$p(X) + \int_{-\infty}^{\infty} G(X, X') k_i(X') p(X') \, dX' = g(X) \tag{3.8}$$

or

$$\bar{p}(r) + \int_{a}^{\infty} A(r, r') k_i(r') \bar{p}(r') \, dr' = \bar{g}(r) \tag{3.9}$$

with

$$g(X) = e^{-V(X)/2k_B T} \Big/ \left[\int_{-\infty}^{\infty} e^{-V(X)/k_B T} \, dX \right]^{1/2} \tag{3.10}$$

and

$$\bar{g}(r) = (4\pi)^{1/2} r e^{-U(r)/2k_B T} \tag{3.11}$$

where $G(X, X')$ and $A(r, r')$ are defined as

$$G(X, X') = D^{-1} \int_{-\infty}^{\infty} dZ \left[\int_{Z}^{\infty} f(X, X') g(X') \, dX' \right]$$

$$\times g(Z)^{-2} \left[\int_{Z}^{\infty} g(Y') f(Y', Y) \, dY' \right] \qquad (3.12)$$

with

$$f(X, X') = \theta(X - X') - g(X)g(X') \qquad (3.13)$$

where $\theta(X - X')$ represents the step function equal to unity for $X > X'$ but zero for $X < X'$, while

$$A(r, r') = \frac{rr'}{D} \exp\left[-\frac{U(r) + U(r')}{2k_B T} \right] \int_{L(r,r')}^{\infty} y^{-2} \exp\left[\frac{U(y)}{k_B T} \right] dy \qquad (3.14)$$

where $L(r, r')$ represents the larger one between r and r'. [When Eq. (3.2) is applicable in the entire X region, Eq. (3.12) can be further simplified with Eq. (6.9) of [38] since $\int_0^{\infty} \langle X | \exp(-Ht) L | X' \rangle \, dt$ in the notation there corresponds to $G(X, X')$ of Eq. (3.12) in the present text.] It has been shown [38,39] that Fredholm's integral equations of the second kind, (3.8) and 3.9), have a unique solution. Moreover, they correspond to a continuous-variable version of a set of coupled linear equations, which reduce to a set of coupled linear equations when the continuous variable X or r is approximated by digitized points for numerical calculation. The calculation can be executed without being annoyed by boundary conditions. Therefore, we can easily solve Eq. (3.8) or (3.9) for $p(X)$ or $\bar{p}(r)$ by computers. Then k_f is given by

For first-order reactions:

$$k_f = \frac{k_{TST} \int_{-\infty}^{\infty} g(X) k_i(X) p(X) \, dX}{\int_{-\infty}^{\infty} dX \int_{-\infty}^{\infty} dX' g(X) k_i(X) G(X, X') k_i(X') p(X')} \qquad (3.15)$$

For second-order reactions:

$$k_f = \frac{k_{TST} \int_{a}^{\infty} \bar{g}(r) k_i(r) \bar{p}(r) \, dr}{\int_{a}^{\infty} dr \int_{a}^{\infty} dr' \bar{g}(r) k_i(r) A(r, r') k_i(r') \bar{p}(r')} \qquad (3.16)$$

When the thermal average $\langle \cdots \rangle$ is defined by $\int \cdots g(X)^2 \, dX$ with $\int g(X)^2 \, dX = 1$ for first-order reactions, $\alpha < 1$ in the viscosity dependence of k_f of Eq. (3.5) is realized under a condition that

For first-order reactions: $\quad \langle [dk_i(X)^{-1}/dX]^2 \rangle / \langle k_i(X)^{-1} \rangle^2$ $\qquad$ (3.17)

or

For second-order reactions: $\quad \lim_{b \to \infty} b^3 \int_a^b r^2 \left[\frac{dk_i(r)^{-1}}{dr} \right]^2 dr \Big/ \left[\int_a^b r^2 k_i(r)^{-1} \, dr \right]^2$

$$\tag{3.18}$$

diverges [40]. To be more exact, this condition guarantees that k_f decreases more slowly than τ^{-1} as the thermalization time $\tau (\propto \eta)$ in the reactant state increases. Therefore, it does not exclude a possibility that k_f decreases as $\tau^{-1}[\log \tau]^n$ with $n > 0$, which decreases only very slightly more slowly than τ^{-1}, as an exceptional case, although in most cases investigated so far, the dependence of k_f of Eq. (3.5) for $\alpha < 1$ is realized [40]. It should be noted, moreover, that when $k_i(X)$ or $k_i(r)$ is a delta function working only at a single value of X or r, we get only $k_f \propto \tau^{-1}$, irrespective of the condition in Eq. (3.17) or (3.18) [38–40].

It is important that the intrinsic reactivity $k_i(X)$ or $k_i(r)$ must have its value distributed continuously over X or r values, under the condition of Eq. (3.17) or (3.18), in order for the viscosity dependence of k_f of Eq. (3.5) with $\alpha < 1$ to be realized. The analytical expression of α has been obtained in special cases. In intramolecular (first-order) electron transfer reactions with which the inner- and the outer-sphere reorganization occur, respectively, by energies λ_i and λ_o, it can be shown [38] that $k_i(X)$ is distributed with a nonvanishing width when $\lambda_i/\lambda_o \neq 0$. In this case, the quantity in Eq. (3.17) diverges when $\lambda_i/\lambda_o < 2$. In the neighborhood of the boundaries in this λ_i/λ_o region, the value of α is given by

$$\alpha \approx |1 - \lambda_i/\lambda_o| \qquad \text{for } \lambda_i/\lambda_o \ll 1 \quad \text{or} \quad 1 \gg 2 - \lambda_i/\lambda_o \, (> 0) \qquad (3.19)$$

Investigated in second-order reactions is a case that the intrinsic reactivity $k_i(r)$ in Eq. (3.3) decreases with r in proportion to r^{-n} as

$$k_i(r) = (a/r)^n f \qquad \text{for } r > a \quad \text{with } n > 3 \qquad (3.20)$$

where the last condition should be imposed from a requirement that the TST-expected rate constant k_{TST} is determined with a finite magnitude from Eq. (3.6). This case can be realized as excitation transfer reactions: When

$n = 6$, for example, the transfer is caused by Förster's interaction between transition dipole moments of reacting two molecules. Similarly, it is caused by the dipole–quadrupole interaction for $n = 8$ and by the quadrupole–quadrupole interaction for $n = 10$. In this case of Eq. (3.20), irrespective of the functional form of the interaction potential $U(r)$ in Eq. (3.3), the value of α is given by [39]

$$\alpha = (n - 3)/(n - 2) \qquad \text{with } n > 3 \qquad (3.21)$$

which approaches 0 as n approaches 3. Especially for $U(r) = 0$ in this case, the analytic form of k_f can also be obtained [39], which is given, at least for $n = 6, 8,$ or 10, by

$$k_f \approx (8\pi a^3/3) f^{1/(n-2)} (D/a^2)^{(n-3)/(n-2)} \qquad (3.22)$$

where a^2/D gives a measure for the relaxation time τ of solvent fluctuations. This expression can be applied in the non-TST regime, although k_f becomes proportional to τ^{-1} ($\propto D/a^2$ in the present case) in the TST regime as a general property [38–40].

As an example of the calculation of the rate constant by Eqs. (3.6) to (3.16), we take the second-order reaction for $k_i(r) = (a/r)^6 f$ and $U(r) = 0$ in Eq. (3.3), where reaction takes place through Förster's mechanism during free diffusion of reactants. Figure 2 shows, with the log-log plot, the

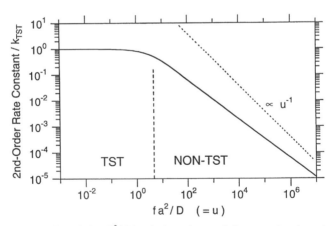

Figure 2. Log-log plot of the $fa^2/D (= u)$ dependence of the second-order rate constant, divided by k_{TST}, obtained for $k_i(r) = (a/r)^6 f$ and $U(r) = 0$ in Eq. (3.3). Shown by the dotted line is the dependence proportional to u^{-1} ($\propto \tau^{-1}$) for comparison. The dashed line shows the boundary between the TST and the non-TST regime given by $u = 14/3$.

fa^2/D ($\equiv u$) dependence of the rate constant divided by the TST-expected rate constant k_{TST} ($= f \cdot 4\pi a^3/3$), where u is proportional to the relaxation time τ of solvent fluctuations. We see therein that in the small u, (i.e., the small τ) region, the rate constant becomes independent of u with a value nearly equal to k_{TST}. In the large u (i.e. the large τ) region, the rate constant decreases as u increases, where the u dependence is weaker than that proportional to u^{-1} ($\propto \tau^{-1}$) shown by the dotted line. The former corresponds to the TST regime, and the latter to the non-TST regime. The τ dependence of the rate constant shown in Figure 2 comes from that of k_f in Eq. (3.4). The $fa^2/D (= u)$ dependence of k_f/k_{TST} is shown in Figure 3 with the log-log plot. Shown also therein by a dotted line is k_C/k_{TST} with $k_C = (56\pi/9)aD$, which can be obtained by the lowest-order perturbation in τ ($\propto D^{-1}$) for k_f, that is, by using $\bar{g}(r)$ of Eq. (3.11) instead of $\bar{p}(r)$ in the expression of k_f in Eq. (3.16). The boundary between the TST and the non-TST regime can be estimated from $k_C = k_{TST}$ which gives $fa^2/D(= u) = 14/3$, and it is shown by the vertical dashed line in Figures 2 and 3. In the TST regime of $fa^2/D < 14/3$, we see in Figure 3 that k_f practically coincides with k_C, in agreement with the note mentioned earlier. In the non-TST regime of

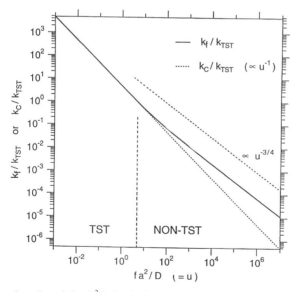

Figure 3. Log-log plot of the $fa^2/D(= u)$ dependence of k_f/k_{TST} derived from the second-order rate constant of Figure 2, together with that of k_C/k_{TST} ($= 14u^{-1}/3$), to which k_f/k_{TST} merges in the small u region, shown by a dotted line. Another dotted line shows a dependence proportional to $u^{-3/4}$ ($\propto \tau^{-3/4}$) added for comparison. The boundary between the TST and non-TST regimes, given by $u = 14/3$, is shown by the vertical dashed line.

$fa^2/D > 14/3$, on the other hand, it departs from k_C, decreasing more slowly than k_C ($\propto u^{-1} \propto \tau^{-1}$) as u increases. The dependence can be regarded as proportional to $u^{-3/4}$ ($\propto \tau^{-3/4}$), as seen in comparison with a dotted line exactly proportional to $u^{-3/4}$, in agreement with the analytical expression of k_f given by Eq. (3.22) for $n = 6$.

In the SM model, solution reactions take place, activated by two kinds of fluctuations. They are molecular-arrangement fluctuations in the solute–solvent system and intrasolute vibrational fluctuations, described, respectively, by the first and the second terms on the right-hand side of Eq. (3.1) or (3.3). Reactions are, therefore, described in a two-dimensional potential surface, as shown in Figure 4, spanned by these two components in the reaction coordinate. Here the molecular-arrangement fluctuations described by a coordinate X on the abscissa are driven by solvent fluctuations and are much slower than the intrasolute vibrational fluctuations described by a coordinate q on the ordinate. Then a reactive trajectory in this model should be the one shown in Figure 4, where a surmount over the transition-state barrier on the line C is induced by fast intrasolute vibrational motions, represented by a vertical straight line, after molecular arrangements appropriate for the surmount are prepared by their diffusive motions, represented

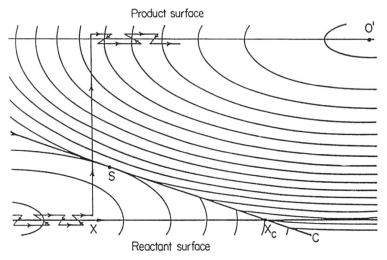

Figure 4. Two-dimensional double-well potential for reaction in the Sumi–Marcus model, spanned by slow (solvent-driven) molecular-arrangement fluctuations in the solute–solvent system on the abscissa and fast (ballistic) intrasolute vibrational fluctuations on the ordinate. Also shown is a reactive trajectory surmounting the transition-state barrier on the line C from the reactant to the product surface.

by a zigzag line along the abscissa passing through the bottom of the reactant surface.

Although the molecular arrangements appropriate for reaction are distributed continuously along this abscissa, it is instructive to approximate them as located on a single X value for a single conformation. In this approximation, reaction can be described by

$$R \underset{k_{-1}}{\overset{k_1}{\rightleftharpoons}} M \overset{k_2}{\longrightarrow} P \qquad (3.23)$$

where M represents the intermediate state with the single molecular arrangement most favorable for reaction. M is induced by solvent fluctuations with a rate constant k_1 from the reactant state R, with a reverse rate constant k_{-1} and M changes to the product state P with a rate constant k_2 as a result of intrasolute vibrational fluctuations. Since k_{-1} represents the relaxation speed of the molecular arrangement of M along the abscissa passing through the bottom of the reactant surface in Figure 4, it should be inversely proportional to the relaxation time τ of solvent fluctuations in the reactant state. Since k_1/k_{-1} should equal the equilibrium constant K_M at M relative to R from the principle of detailed balance, it is independent of τ, and we get $k_1 \propto \tau^{-1}$. Since k_2 is brought about by intrasolute vibrational motions, k_2 should be independent of τ. In the approximation of Eq. (3.23), the rate constant under the steady-state condition is given by the form of Eq. (3.4) with $k_f = k_1$ and $k_{TST} = K_M k_2$. In this approximation, we get only a case of $\alpha = 1$ in Eq. (3.5), consistent with the situation, mentioned earlier, that only $\alpha = 1$ is obtained when the intrinsic reactivity $k_i(X)$ or $k_i(r)$ is a delta function working only at a single value of X or r. The result obtained by this approximation discloses, however, a very important feature of k_f in Eq. (3.4). It represents the rate constant with which molecular arrangements most favorable for reactions induced by intrasolute vibrational fluctuations are prepared prior to the reactions as a result of solvent fluctuations in the solute–solvent system.

When TST is invalidated, the solvent viscosity appears explicitly in the expression of the rate constant, and it depends strongly on temperature. In this case, the true thermal activation energy of the rate constant, concerned with the static potential energy surface for reaction, should be derived from the isoviscosity Arrhenius plot under the condition that the solvent viscosity is fixed while the temperature is varied. In the reaction scheme mentioned in the preceding paragraph, this activation energy for k_f should be given by the energy necessary for preparing the molecular arrangements most favorable for reactions. This activation energy should be smaller than that of k_{TST}, which is given by the height of the saddle point S with the lowest energy on the transition-state line C in Figure 4. This result gives rise to an important characteristic of the rate constant in the SM model as mentioned below. Since Eq. (3.4) reduces to $k \approx k_{TST}$ in the low-viscosity TST regime for $k_f \gg k_{TST}$, the thermal activation energy we derive from the observed rate constant k is of course that of k_{TST}. In the high-viscosity non-TST regime for $k_f \ll k_{TST}$ where Eq. (3.4) reduces to $k \approx k_f$, on the other hand, the activation energy we observe should be that of k_f. Therefore the thermal activation energy should become smaller in the latter regime than in the former regime. This occurs neither in the Kramers model nor in the GH model since both these models are described by the one-dimensional double-well potential for reaction in Figure 1. In this case, the same transition-state barrier must always be surmounted in any situation, always giving the same value for the activation energy.

IV. EXPERIMENTAL CHECK FOR THESE TWO STREAMS OF THEORIES

To determine experimentally which stream of theories is appropriate, it is essential to determine experimentally not only the rate constant k deviating from k_{TST} in the non-TST regime but also the TST-expected rate constant k_{TST} there. To this end, the rate constant must be measured over a wide viscosity range covering not only the non-TST regime but also the TST regime. Such data had not been presented until very recently. As reactions in the non-TST regime, many data had been piled up on the photoinduced E/Z isomerization of stilbenes and its related molecules in various solvents [7,41]. These data showed, however, only the non-TST regime, disclosing the viscosity dependence of the rate constant in Eq. (1.1). Therefore, we could not determine unambiguously k/k_{TST} values in this regime. In these data, the photoisomerization takes place on the excited-state (S_1) potential surface where the potential barrier is not hight, with a height about $15 \, kJ/mol^{-1}$. Then the isomerization is so fast, with a reaction time on the order of 100 ps, that the reaction has always been cast into the non-TST regime with its rate controlled by solvent fluctuations that are slow relative to the reaction.

A breakthrough of this situation was given by Asano et al. [42–44], who measured the thermal Z/E isomerization of substituted azobezenes and N-benzylideneanilines, shown in Figure 5, under pressure. Although these molecules are similar in structure to stilbenes, the reaction investigated with these molecules, after photoexcitation of their E isomer, was not the fast photoinduced E/Z isomerization on the excited-state (S_1) potential surface (process 1 in Figures 5, investigated with stilbenes), but the slow thermally recovering Z/E isomerization on the ground-state (S_0) potential surface (process 2 in Figure 5). In this case the potential barrier for isomerization

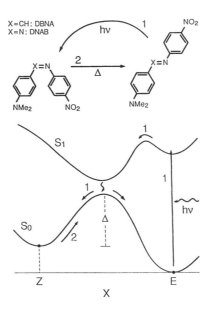

Figure 5. Photocycle of isomerization of DBNA and DNAB, composed of the photoinduced E/Z isomerization on the excited-state (S_1) potential surface (process 1) followed by the thermally recovering Z/E isomerization on the ground-state (S_0) potential surface (process 2) after photoexcitation of the E isomer from the S_0 to the S_1 surface.

(Δ in Figure 5) is so high as about $50\,kJ/mol^{-1}$, much higher than the thermal energy of about $2.5\,kJ/mol^{-1}$ at room temperature that it takes for the reaction at least about 1 ms. Then, using an apparatus with time resolution of the order of 1 μs, we can see only process 2 in Figure 5 since it is observed as if process 1 took place instantaneously after photoexcitation. Instead, we can increase the solvent viscosity up to about 10^8-fold by applying pressures of several hundreds MPa ($\approx$ several thousands atm) on a solvent. In this case the reaction was so slow in the TST regime at ambient pressure, and it was cast into the non-TST regime at high pressures on the order of 100 MPa.

Figure 6 shows the pressure dependence of the rate constant k_{obs} observed for N-[4-(dimethylamino)benzylidene]-4-nitroaniline (abbreviated as DBNA) in a solvent, Traction Fluid B (TFB), which is a commercial product (of Nippen Oil Co.) whose main consituent is 2,4-dicyclohexyl-2-methylpentane. The k_{obs} is nearly independent of pressure when it is lower than about 100 MPa. This behavior can well be understood in the framework of TST as a situation that the volume of the reactant is unchanged between the reactant and the transition state.

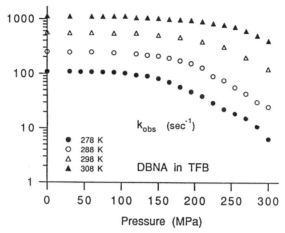

Figure 6. Pressure dependence of the rate constant k_{obs} of the thermally recovering Z/E isomerization of DBNA in TFB at 278, 288, 298, and 308 K.

This interpretation is consistent with the known fact that the isomerization of DBNA occurs by nitrogen inversion where neither charge separation nor bond scission takes place during the formation of the transition state [45]. In this case the volume change of the reactant should be very small, even in polar solvents, although the solvent TFB is nonpolar. When pressure exceeds about 150 MPa, k_{obs} begins to decrease with increasing pressure. More appropriately, it is with increasing solvent viscosity that k_{obs} begins to decrease. Therefore, we entered the non-TST regime. Since $k_{obs} = k_{TST}$ in the low-pressure region, the TST-expected rate constant k_{TST} in the non-TST regime at high pressures can be obtained by smoothly extrapolating k_{obs} in the low-pressure region into the high-pressure region, as shown by solid circles in Figure 7 for DBNA in TFB at 278 K taken as an example.

Let us note here that $k_{obs} \leq k_{TST}$ is satisfied in the entire pressure region in Figure 7. This means that the form (3.4) of the rate constant resulted from the SM model is satisfied. Then, equating k_{obs} to $1/(k_{TST}^{-1} + k_f^{-1})$, we can derive k_f values in the non-TST regime. Figure 8 shows k_f values for DBNA in TFB at 278 K as a function of the solvent viscosity that was obtained at each pressure. We see therein that the relation of Eq. (3.5) for $\alpha \approx 0.46$ is satisfied for viscosity change ranging over 2×10^4-fold. Thus the rate constant formula (3.4) with (3.5) that resulted from the SM model could experimentally be confirmed for viscosity change ranging over about 10^8-fold, including the TST regime in the low-pressure (low-viscosity) region. Similar results can be obtained for other substituted azobezenes and

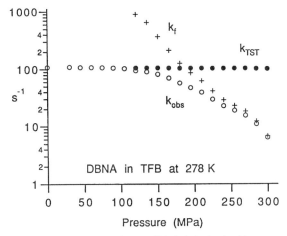

Figure 7. TST-expected rate constant k_{TST} (solid circles) obtained by extrapolating k_{obs} (open circles) in the low-pressure region into the high-pressure region for DBNA in TFB at 278 K. Also shown (by + marks) are k_f values derived by equating k_{obs} to $1/(k_{TST}^{-1} + k_f^{-1})$.

N-benzylideneanilines in various solvents, including not only nonpolar ones but also polar, protic, and aprotic ones [42–44].

Since both k_{TST} and k_{obs} were obtained in the non-TST regime, we can also check the applicability of the GH model explained in Section II. We note first that k_{TST} derived by extrapolation in Figure 7 is consistent with Eq. (2.12), where $\omega_0/(2\pi)$ takes a value from about 10^{12} to $10^{13}\,\text{s}^{-1}$ for DBNA in TFB, with $W(X_b)$ taking about $54\,\text{kJ/mol}^{-1}$ as Δ in Figure 5.

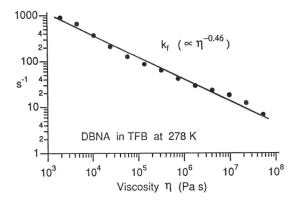

Figure 8. Viscosity (η) dependence of the solvent-fluctuation-controlled part k_f of the rate constant k_{obs} for DBNA in TFB at 278 K in the non-TST regime at high pressures.

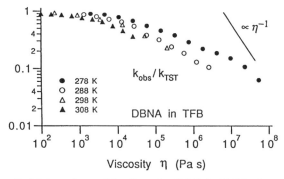

Figure 9. Viscosity (η) dependence of k_{obs}/k_{TST} for DBNA in TFB in the non-TST regime at high pressures. Shown by the solid line is the η^{-1} dependence for comparison.

Figure 9 shows the viscosity (η) dependence of k_{obs}/k_{TST} for DBNA in TFB in the non-TST regime at high pressures. With increasing η in the high-viscosity region, k_{obs}/k_{TST} decreases more slowly than the solid line, which shows the η^{-1} dependence for comparison. Therefore, the Kramers model giving the η^{-1} dependence in this region cannot be applied. In the GH model, this weak η dependence is attributed to a situation in which the frequency-dependent friction $\zeta(\mu)$ becomes much smaller than its static value $\zeta(0)$ at a frequency μ much larger than the typical speed of microscopic solvent motions which cause friction to reactant motions, as explained in Section II. Here μ is given by $\omega_b k_{obs}/k_{TST}$ from Eq. (2.11), with ω_b on the order of $10^{13}\,\mathrm{s}^{-1}$, as noted in Section II, while k_{obs}/k_{TST} decreases to a value smaller than 0.1 in Figure 7. Therefore, the situation of $\zeta(\mu) \ll \zeta(0)$ must be realized at $\mu \sim 10^{12}\,\mathrm{s}^{-1}$. This means that the typical speed of microscopic solvent motions must be much slower than $10^{12}\,\mathrm{s}^{-1}$ in order for k_{obs} in Figures 6 and 9 to be described by the frequency-dependent friction. This requirement does not seem to be realized since the typical speed of microscopic solvent motions is at least on the order of $10^{12}\,\mathrm{s}^{-1}$, as explained below. The speed can be estimated by the inverse of the width of frequency distribution of microscopic solvent motions interacting with isomerizing motions of the solute molecule. The wavelength of these micro-scopic motions should be comparable to or smaller than the dimension of the isomerizing moiety at most on the order of several angstroms. Inelastic neutron-scattering studies on the energy spectrum in molecular liquids [46] shows that even if the wavelength is fixed at a single value of about several angstroms, corresponding to a single wavevector with a magnitude of about $2\pi/$ (several angstroms) $\sim 1\,\mathring{\mathrm{A}}^{-1}$, microscopic solvent motions have a width of about several THz in their angular frequencies, that is, a width of about

1 THz in their frequencies, or a width of about $30\,cm^{-1}$ in their energies. This shows that frequencies of microscopic solvent motions exerting friction to reactant motions are distributed over a width at least larger than 1 THz. Therefore, the typical speed of these microscopic solvent motions is at least on the order of $10^{12}\,s^{-1}$. This speed should also manifest itself as that of a rapid component in energy and width variation of transient optical spectra of solute molecules in solution. The component has been observed to relax in less than 1 ps, also here [32].

The inverse of the typical speed of microscopic solvent motions mentioned above is considered to give the correlation time τ_{sc} among random forces $R(t)$ in the generalized Langevin equation (2.6). Mathematically, τ_{sc} equals the decay time of the statistically averaged correlation function $\langle R(t)R(t')\rangle$ of random forces and can be estimated by

$$\tau_{sc} \approx \int_0^\infty \langle R(t)R(0)\rangle\,dt/\langle R(0)^2\rangle = \int_0^\infty \hat{\zeta}(t)\,dt/\hat{\zeta}(0) = \zeta(0)/\hat{\zeta}(0) \qquad (4.1)$$

where the second and third equalities are obtained from the fluctuation dissipation theorem in Eq. (2.7) and the definition of the frequency-dependent friction $\zeta(\mu)$ in Eq. (2.8), respectively. It can be shown [47] by using Eqs. (2.11) and (2.13) that values of τ_{sc} of Eq. (4.1) required to describe the experimental data in Figures 6 and 9 by the GH model can be derived from these data themselves without any adjustable parameter. These values obtained for each experimental point in Figure 9 are shown in Figure 10 in a form multiplied by a number b^{-1} of order unity and ω_b, where ω_b is on the order of $10^{13}\,s^{-1}$, as noted in Section II. We see in Figure 10 that $\omega_b\tau_{sc}$ is about 5×10^5 at the viscosity $\eta \approx 10^2\,Pa \cdot s$ at 308 K at the low-pressure edge in the non-TST regime, and reaches about 5×10^{10} at $\eta \approx 5 \times 10^7\,Pa \cdot s$ at 278 K. In order that the non-TST regime realized for DBNA in TFB is describable by the GH model, therefore, the correlation time τ_{sc} among random forces in the generalized Langevin equation must be at least about 50 ns at the low-pressure low-viscosity edge in this regime, and as large as about 5 ms at the highest pressure applied. These requirements cannot be satisfied in real systems, since τ_{sc} must be at most on the order of 1 ps as the inverse of the typical speed of microscopic solvent motions. This shows, with explicit values of τ_{sc}, how unphysically slow the microscopic solvent motions must be in order to describe the experimental data in Figure 6 by the GH model.

Experimentally, the typical speed of microscopic solvent motions can be seen directly as that of a rapid component in energy and width variation of transient optical spectra of solute molecules in solution, as mentioned earlier. It has been observed [32] that the speed of the component, with a

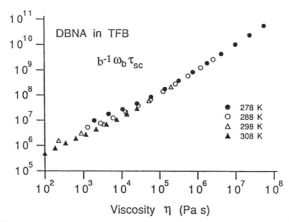

Figure 10. Viscosity dependence of the correlation time τ_{sc} among random forces shown in a form multiplied by a number b^{-1} of order unity and ω_b (ca. 10^{13} s^{-1}) in the non-TST regime for DBNA in TFB, determined under an assumption that this regime is describable by the Grote–Hynes theory.

magnitude larger than 10^{12} s^{-1}, is kept almost unchanged even if solvents change drastically from liquids to glasses beyond the glass-transition temperature. In contrast, it has also been observed [32] that the speed of a slow component coming from relaxation of solvated-structure fluctuations of the solute molecule increases almost stepwise when the temperature is increased beyond the glass-transition temperature. This reflects that the speed of these structural fluctuations are controlled by the solvent viscosity η, which increases almost stepwise in this temperature change. We see therefore that τ_{sc}, given by the inverse of the typical speed of microscopic solvent motions, is almost independent of η.

On the right-hand side of Eq. (4.1), the static value $\zeta(0)$ of the frequency-dependent friction is proportional to η from the Einstein relation (2.10) since $D \propto \eta^{-1}$, while τ_{sc} on the left-hand side is almost independent of η. This means that the initial value $\hat{\zeta}(0)$ of the time-dependent friction kernel $\hat{\zeta}(t)$, appearing in the generalized Langevin equation (2.6), is proportional to η on the right-hand of Eq. (4.1). In evaluating the η dependence of the rate constant derived from the GH model, van der Zwan and Hynes [48] assumed that it is not $\hat{\zeta}(0)$ but τ_{sc} that increases in proportion to η. Under this assumption, they obtained that the rate constant can approach a constant value with increasing η in this model. Both $[\hat{\zeta}(0) \propto \eta$ and $\tau_{sc} =$ constant] and $[\tau_{sc} \propto \eta$ and $\hat{\zeta}(0) =$ constant] satisfy Eq. (4.1), but the latter that they assumed does not seem reasonable on the basis of the observation mentioned above. Theoretically, too, it does not seem reasonable since τ_{sc}^{-1}

should have a magnitude on the order of the width of frequency distribution of microscopic solvent motions interacting with reactant motions [49]. The width does not change so much even if solvents change from liquids to glasses, remaining always at least on the order of $10^{12}\,\mathrm{s}^{-1}$.

Pressure dependence of the thermally recovering Z/E isomerization was also measured for substituted azobenzenes, with a result similar to that for substituted N-benzylideneanilines shown above for DBNA in TFB in Figure 6 [42–44]. The only difference for the former in polar solvents is that the rate constant k_{obs} initially increases with pressure. This behavior can well be understood in the framework of TST as a situation that the volume of the reactant is smaller in the transition state than in the reactant state, since the transition state in the isomerization of these molecules has a polar structure in polar solvents [50]. The pressure dependence of k_{obs} can be converted into the viscosity dependence. An example for 4-(dimethylamino)-4′-nitroazo-benzene (abbreviated as DNAB in Figure 5) in a polar solvent, glycerol triacetate (GTA), is shown in Figure 11 at various temperatures. When

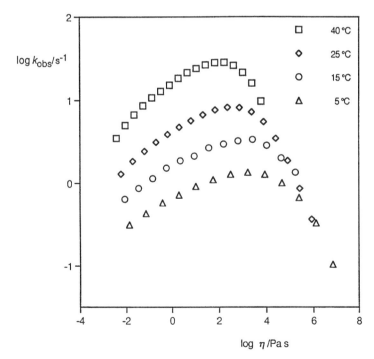

Figure 11.   Rate constant k_{obs} of the thermally recovering Z/E isomerization of DNAB in GTA at various temperatures plotted as a function of the solvent viscosity η.

we trace therein k_{obs} along a vertical line corresponding to a fixed value of the solvent viscosity η, we are just looking at its isoviscosity plot. In Figure 11, the increase in k_{obs} with η in the low-viscosity region reflects its increase with pressure. In this TST regime, vertical separation of k_{obs} for different temperatures in Figure 11 is due to the large value of the thermal activation energy, which amounts to about $50\,kJ/mol^{-1}$, as mentioned earlier. In the high-viscosity region where k_{obs} decreases with η in Figure 11, the reaction enters the non-TST regime. In this region k_{obs} tends to converge into a single line in Figure 11. This means that the thermal activation energy of k_{obs} is very small in the isoviscosity plot. In general, it has been observed that the thermal activation energy of k_{obs} in the isoviscosity plot is considerably smaller in the non-TST regime than in the TST regime [43,44]. This observation can be understood within the framework of neither the Kramers model nor the GH model, since in these one-dimensional reaction-coordinate models the same barrier must always be surmounted in any situation, as noted in Section III. In contrast, this observation can easily be rationalized within the framework of the SM model, where the rate constant is given in the form of Eq. (3.4): The thermal activation energy of k_{obs} reduces to that of k_{TST} in Eq. (3.4) in the TST regime, while to that of k_f in Eq. (3.4) in the non-TST regime. The thermal activation energy of k_f comes from the energy of the intermediate state M in the two-step approximation in Eq. (3.23). It should in general be considerably smaller than the thermal activation energy of k_{TST}, which comes from the energy of the saddle point S in the two-dimensional potential surface for reaction in Figure 4, as also noted in Section III.

The analyses in the third and fourth paragraphs of this section within the framework of the SM model can be performed quite similarly also for the thermally recovering Z/E isomerization of DNAB in GTA. The rate constant can nicely be fitted by formulas (3.4) and (3.5) with $\alpha \approx 0.7$ for viscosity change ranging over 10^8-fold on the abscissa in Figure 11 [43,44]. The viscosity (η) dependences of the solvent-fluctuation controlled part k_f of k_{obs} determined at various temperatures are collected into Figure 12. We note there that they come approximately on a single straight line irrespective of differences in temperature. The near straightness of the line confirms the η dependence of k_f in Eq. (3.5), while the near convergence into the single line represents the fact that the thermal activation energy of k_f is very small in the isoviscosity plot. The latter exemplifies the property of k_f mentioned in the preceding paragraph. It discloses a reaction dynamics that molecular arrangements most favorable for reactions driven by intrasolute vibrational fluctuations can be prepared in the solute–solvent system without appreciable loss of energy during molecular-arrangement fluctuations for DNAB in GTA.

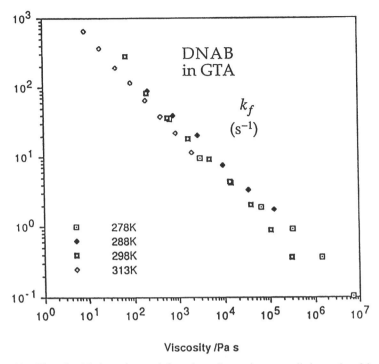

Figure 12. Viscosity (η) dependence of the solvent-fluctuation-controlled part k_f of k_{obs} for DNAB in GTA at various temperatures in the non-TST regime at high pressures.

Recently, the frequency dependence of the friction in the GH model was recalculated by Biswas and Bagchi on the basis of a mode-coupling theory [31a]. The model was refined by this recalculation, but it does not remove the fundamental difficulty of the model, which fails in describing that the isoviscosity thermal activation energy of k_{obs} is considerably smaller in the non-TST regime at high viscosities than in the TST regime at low viscosities in Figure 11. The recalculation does not modify the fundamental feature in k_{GH} of Eq. (2.11) that the transmission coefficient μ/ω_b does not change appreciably as long as the speed of solvent fluctuations is kept constant by fixing the solvent viscosity η at a constant value. This difficulty persists in any extension of the Kramers model as long as it keeps the character of a single-step model as described by Figure 1.

Their recalculation showed an η dependence of Eq. (1.1) over an η variation of at most 20-fold in a region much lower than 1 Pa $\cdot$ s. When η exceeds about 1 Pa $\cdot$ s, the mode coupling theory breaks down, as they noted, and the hard-sphere model becomes realistic for solvent friction. They predicted that

k_{obs} should approach an η dependence proportional to $1/\log \eta$ as its asymptotic form in this high-η limit. As mentioned, k_{obs} obeys Eq. (1.1) at high η's in the thermal Z/E isomerization in Figure 5. In contradiction to its validity range of $\eta \ll 1\,Pa \cdot s$ in their calculation, k_{obs} obeys Eq. (1.1) only at much higher viscosities $\gtrsim 100\,Pa{\cdot}s$, where their calculation based on the mode coupling theory breaks down, as apparent in Figures 9 and 11. Therefore, Eq. (1.1) obtained by their calculation contradicts the observation of Eq. (1.1) in the real system in Figure 5. Simultaneously, the observation disproves their prediction on the asymptotic form in the high-viscosity limit of $\eta \gg 1\,Pa \cdot s$, since k_{obs} approaches zero there not with the form proportional to $1/\log \eta$, but with the form (1.1) [31b]. In this limit k_{obs} describable in general as Eq. (3.4) can be approximated by k_f, whose η dependence can be expressed as Eq. (3.5) with $0 < \alpha < 1$ ranging over several orders of magnitude in η variations, as shown by Figures 7 and 8.

The GH model has theoretically been tested by its comparison with simulation results for a certain class of solution reactions and found to be reasonable (e.g., [31c]). In all the reactions simultated (e.g., electron transfer reactions within a diatomic molecule in solvents with a fixed molecular structure [31c]), however, intramolecular vibrations, if they exist, have energies so high that they do not contribute to drive reactions. In this case, reactions must be driven only by solvent-fluctuation-induced Brownian motions of the solvated structure of a solute molecule. The solute molecule is so small here that the solvated structure is very thin, and its fluctuation speed becomes fast, becoming comparable to the speed of individual microscopic motions of solvent molecules. In this situation, frictional forces exerted by these motions must be treated not by hydrodynamics, but by non-Markoffian dynamics with memory effects, that is, by taking into account the frequency dependence of friction as in the starting equation (2.6) of the GH model. The reactions used for the test, however, seem to lack the usual features of solution reactions. In usual cases, solute molecules are not so small as diatomic moleculs that some intrasolute vibrational motions have energies low enough as to be excitable at room temperature. Moreover, their solvated structures becoming thick enough fluctuate much more slowly than microscopic motions of solvent molecules, as mentioned in the second paragraph of Section III. As usual features of solution reactions in this situation, therefore, reactions are driven by the aid of both thermal intrasolute vibrational fluctuations and solvent-fluctuation-induced solvated-structure fluctuations in the solute–solvent system. After reaction, reorganization takes place along coordinates of these two kinds of fluctuations, as, respectively, inner- and outer-sphere reorganization, well known in electron transfer reactions in solution.

The SM model takes into account correctly both these two kinds of fluctuations in the solute–solvent system. To be more exact, reactions take place driven by thermal intrasolute vibrational fluctuations at each of solvated structures prepared by solvent-fluctuation-induced solvated-structure fluctuations, since the former is much faster than the latter. In other words, solvent fluctuations prepare in advance solvated structures [at state M in the two-step approximation in Eq. (3.23)] most favorable for reactions driven by intrasolute vibrational fluctuations. It is with the rate constant k_f in Eq. (3.4) that these solvated structures are prepared in the course of solvent fluctuations. This situation has been noted as *conformational gating* (or *control*) in electron transfer reactions in solution [3b,51]. Recently, the same term has been used to describe atom-group transfer reactions in solution [6]. Formula (3.4) gives the general expression of the rate constant in this situation, although it is more generally applicable to other solution reactions, too.

Thus Eq. (3.4) with (3.5) derived from the SM model has remained as a candidate of the general expression for the rate constant of solution reactions, which covers not only the TST regime in the low-viscosity region but also the non-TST regime in the high-viscosity region. It is shown in Appendix B that this formula is also applicable to more general cases where a reaction is accomplished through a series of intermediate products.

V. SOLVENT VISCOSITY DEPENDENCE OF ENZYMATIC REACTIONS AND ITS MEANING

Also in biological reactions mediated by proteins, rate constants have often been observed to decrease with an increase in the solvent viscosity as Eq. (1.1) [6,8–16]. This means that these biological reactions are in the non-TST regime, where the rate constant given generally by Eq. (3.4) with (3.5) reduces to $k \approx k_f$ due to $k_f \ll k_{TST}$. In this regime the rate constant k becomes smaller than k_{TST}, being controlled by the slow speed of solvent fluctuations. Then it is interesting to ask why biological reactions have chosen the non-TST regime with a small rate constant after natural selection in evolution for a long period. We would like, rather, to consider that proteins utilize solvent fluctuations (to be more exact, solvent-fluctuation-induced conformational fluctuations in proteins) to accomplish their functions. This corresponds to positively utilizing the situation of *conformational gating* (or *control*) mentioned in the last second paragraph in Section IV.

Let us consider enzymatic reactions as a typical example of biological reactions mediated by proteins. An enzymatic reaction can be decomposed into three main steps, illustrated in Figure 13: In the first step a substrate (drawn by thick lines) is captured by an enzyme (drawn by thin lines) to

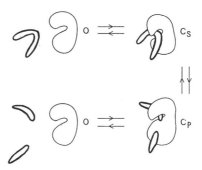

Figure 13. Three main steps, from O to C_S, from C_S to C_P, and from C_P to O, comprising the catalytic activity of an enzyme (drawn by thin lines) on a substrate (drawn by thick lines).

make a complex, and in the second step the captured substrate is transformed into a product (in the example here, the substrate is cut into two pieces and one of them is released from the enzyme), while in the third step the product is fully released from the enzyme and the enzyme recovers its original free form. Let us denote by O the state where an enzyme and substrates are free from each other in solvents, by C_S the state where one of substrates makes a complex with the enzyme, and by C_P the state where the substrate in the complex has been transformed into a product. The first step mentioned above corresponds to a process from O to C_S, the second step corresponds to that from C_S to C_P, and the third step to that from C_P to O. These three steps are quite different from each other, and an enzyme must mediate these quite different steps by a single protein body. [The rate of catalysis by enzymes can be formulated in a general form by taking into account Eq. (3.4) with (3.5) for elementary reactions. Readers not interested in the process of this formulation can skip from here to the paragraph following Eq. (5.12).]

Since the first step mentioned above is a second-order reaction, its forward rate for decay of free substrates due to capture by enzymes is proportional to the free-enzyme concentration written as [E]. Then the reaction in which substrate S is transformed into product P as a result of catalysis by enzyme E can be expressed as

$$
\begin{array}{ccccc}
& E & & E & \\
& \downarrow & & \uparrow & \\
& {\scriptstyle k(C_S,O)[E]} & & {\scriptstyle k(C_P,C_S)} & {\scriptstyle k(O,C_P)} \\
S & \underset{g(O,C_S)}{\overset{}{\rightleftharpoons}} & ES & \underset{g(C_S,C_P)}{\overset{}{\rightleftharpoons}} & EP \xrightarrow{\quad} P \\
& \downarrow & & & \\
& E & & &
\end{array}
\qquad (5.1)
$$

where $k(C_S,O)$ represents the rate constant of the second-order reaction from state O to C_S mentioned above, with its backward rate constant written as $g(O,C_S)$, and $k(C_P,C_S)$ represents the rate constant for transformation of the substrate to the product from state C_S to C_P in the complex formed with the enzyme, with its backward rate constant written as $g(C_S,C_P)$, while $k(O,C_P)$ represents the rate constant for release of the product from the state C_P of its complex with the enzyme. The ratio of $k(C_S,O)$ to $g(O,C_S)$ of

$$K(C_S,O) \equiv k(C_S,O)/g(O,C_S) \tag{5.2}$$

gives the equilibrium constant for the state C_S relative to the state O, determined only by the free energy difference $\Delta G(C_S,O)$ between them, as $K(C_S,O) = \exp[-\Delta G(C_S,O)/k_B T]$. For later use we also introduce the equilibrium constant $K(C_P,O)$ of

$$K(C_P,O) \equiv [k(C_P,C_S)k(C_S,O)]/[g(O,C_S)g(C_S,C_P)] \tag{5.3}$$

for the state C_P relative to the state O determined by the free energy difference $\Delta G(C_P,O)$ between them, as $K(C_P,O) = \exp[-\Delta G(C_P,O)/k_B T]$.

When we regard the first step from S to ES in Eq. (5.1) as a first-order reaction with a forward rate constant $k(C_S,O)[E]$, we can apply Appendix B to obtain the rate constant of the reaction, regarded as first-order, from S to P in Eq. (5.1). It is given by

$$1/\{[k(O,C_P)K(C_P,O)[E]]^{-1} + [k(C_P,C_S)K(C_S,O)[E]]^{-1}$$
$$+ [k(C_S,O)[E]]^{-1}\} \equiv b[E] \tag{5.4}$$

with

$$b = 1/[k(O,C_P)^{-1}K(C_P,O)^{-1} + k(C_P,C_S)^{-1}K(C_S,O)^{-1} + k(C_S,O)^{-1}] \tag{5.5}$$

where b represents the rate constant as a second-order reaction for transformation of substrates into products in the catalysis by enzymes. With the concentration [S] of substrates, the transformation speed of substrates into products is given by b [E] [S]. Usually, this speed is expressed in terms of the total concentration of enzymes $[E]_0$, not in terms of the concentration of free enzymes [E]. As shown in Appendix C, they are related by

$$[E] = [E]_0/\{1 + (b/a)[S]\} \tag{5.6}$$

with

$$a = 1/\{k(C_P, C_S)^{-1} + k(O, C_P)^{-1}[1 + K(C_P, C_S)^{-1}]\} \qquad (5.7)$$

where $K(C_P, C_S)$ represents the equilibrium constant for the state C_P relative to the state C_S, determined only by the free energy difference $\Delta G(C_P, C_S)$ between them, as $K(C_P, C_S) = \exp[-\Delta G(C_P, C_S)/k_B T]$. Introducing Eq. (5.6) into b [E] [S], we get the transformation speed of substrates into products in the catalysis by enzymes in the standard form.

$$a \frac{[E]_0 [S]}{K_M + [S]} \qquad (5.8)$$

with

$$K_M \equiv a/b \qquad (5.9)$$

where K_M represents the Michaelis–Menten constant.

Since all of $k(O, C_P)$, $k(C_P, C_S)$ and $k(C_S, O)$ are rate constants of solution reactions, although the former two are those of a first-order reaction while the last one is that of a second-order reaction, all of them should be expressed in the general form (3.4) with (3.5). It is usual in reactions mediated by enzymes that each of these three steps of reactions is accomplished through a series of intermediate products. Even in this case, the rate constant formula (3.4) with (3.5) can be applied without essential change, as shown in Appendix B. Since all of $k(O, C_P)$, $k(C_P, C_S)$, and $k(C_S, O)$ can be expressed in a form of Eq. (3.4), both a and b in Eqs. (5.5) and (5.7) can be expressed in the same form, as

$$a = 1/(a_{TST}^{-1} + a_f^{-1}) \qquad \text{and} \qquad b = 1/(b_{TST}^{-1} + b_f^{-1}) \qquad (5.10)$$

where a_{TST} and b_{TST} represent the TST-expected form of a and b, which can be obtained by inserting the TST-expected forms $k_{TST}(O, C_P)$, $k_{TST}(C_P, C_S)$ and $k_{TST}(C_S, O)$ respectively, for $k(O, C_P)$, $k(C_P, C_S)$, and $k(C_S, O)$ in Eqs. (5.5) and (5.7), while a_f and b_f represent the part controlled by solvent-fluctuation-induced conformational fluctuations of proteins, being obtained by inserting the fluctuation-controlled parts $k_f(O, C_P)$, $k_f(C_P, C_S)$, and $k_f(C_S, O)$, respectively, for $k(O, C_P)$, $k(C_P, C_S)$, and $k(C_S, O)$ in Eqs. (5.5) and (5.7). Especially,

$$a_f = 1/\{ k_f(C_P, C_S)^{-1} + k_f(O, C_P)^{-1}[1 + K(C_P, C_S)^{-1}]\} \qquad (5.11)$$

and

$$b_f = 1/[k_f(O, C_P)^{-1}K(C_P, O)^{-1} + k_f(C_P, C_S)^{-1}K(C_S, O)^{-1} + k_f(C_S, O)^{-1}]$$
(5.12)

We see thus that the rate constant formula (3.4) with (3.5) for solution reactions can also be applied to describe the enzymatic reaction in such a way that both the rate constants a and b appearing in the Michaelis–Menten form (5.8) with (5.9) for the transformation speed of substrates into products have the same form (5.10) as (3.4), where [E]$_0$ and [S] represents, respectively, the total concentration of enzymes and that of free substrates.

As noted in the second paragraph in this section, the three main steps comprising an enzymatic reaction, as illustrated in Figure 13, are quite different from each other. Then, protein conformations most favorable for them are in general very different from each other, too. In this situation, a conformation most favorable for one of the three steps is very unfavorable for the other two. Since the enzyme has only a single protein body, its conformation cannot be optimized to three different steps simultaneously. Then the enzyme must be fallen into a state of frustration. It can be considered that the enzyme has reached a compromise of the frustration, described below, as a result of evolution for a long period. First, let us denote the protein conformation most favorable for the first step from O to C_S by $M(C_S, O)$, that for the second step from C_S to C_P by $M(C_P, C_S)$, and that for the third step from C_P to O by $M(O, C_P)$. They can be described as three points, $M(C_S, O)$, $M(C_P, C_S)$, and $M(O, C_P)$, apart from each other in a multidimensional configuration-coordinate space, as shown in Figure 14. The protein conformation at the initial state O for the first step has been set to be different from the conformation at point $M(C_S, O)$, as described by a point O apart from point $M(C_S, O)$ in Figure 14. Simultaneously, the protein has been set to have a flexible tertiary structure able to fluctuate in solvents. Then the structure at point O can deform to that at point $M(C_S, O)$ as a result of conformational fluctuations induced by solvent fluctuations. After surmounting point $M(C_S, O)$ as the first step, the protein relaxes into a conformation different from that at point $M(C_S, O)$. This conformation constitutes the initial state C_S for the second step. This conformation has been set, again, to be different from the conformation at point $M(C_P, C_S)$, as described by a point C_S apart from point $M(C_P, C_S)$ in Figure 14, but the structure at point C_S has been arranged, again, so as to be able to deform to that at point $M(C_P, C_S)$ as a result of conformational fluctuations induced by solvent fluctuations. After surmounting point $M(C_P, C_S)$ as the second step, the same mechanism as mentioned above for the second step

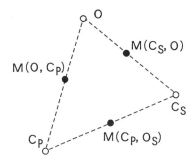

Figure 14. Arrangement, in a multidimensional configuration-coordinate space, of the configurations most favorable for the first step (from point O to C_S) at point $M(C_S, O)$, for the second step (from C_S to C_P) at point $M(C_P, C_S)$, and for the third step (from C_P to O) at point $M(O, C_P)$.

will take place for the third step, from a point C_P back to the starting point O over point $M(O, C_P)$ in Figure 14.

In this scenario, since all the protein conformations at the initial states of the three steps comprising an enzymatic reaction are not optimized to these steps, they must be accomplished as a result of conformational fluctuations induced by solvent fluctuations. In this situation, these steps cannot help being executed with a speed smaller than that expected from TST, but all of them can be accomplished at any rate. This is in accordance with the final aim of enzymes that they achieve their functions to facilitate our lives even if their speeds become a little slow. Their speeds are, in fact, at most on the order of $10^3 \, s^{-1}$ and are much slower than speeds of conformational fluctuations, which are the order of 10 to 100 ns [33] as mentioned in Section III.

To confirm this scenario, we must check that at least two of the three steps are controlled by conformational fluctuations. In the formula (5.8) for the transformation speed of substrates into products in the enzyme catalysis, a can be obtained by measuring the speed $a\,[E]_0$ in the situation of $[S] \gg K_M$ under dense substrate concentrations. It has often been called the catalytic rate constant by enzymes. It has often been observed that a has a solvent-viscosity dependence as shown on the right-hand side of Eq. (1.1) [8–16]. This means that a is regulated by a_f, reducing to $a \approx a_f$ for $a_f \ll a_{TST}$ in the first equation in (5.10) (i.e., that enzymatic reactions are in the non-TST regime controlled by slow solvent fluctuations). Let us note here that a_f^{-1} is given in Eq. (5.11) by a linear combination of the inverse of the fluctuation-controlled parts $k_f(C_P, C_S)$ and $k_f(O, C_P)$ of the rate constants $k(C_P, C_S)$ and $k(O, C_P)$, respectively, of the second and third steps. Then $a \approx a_f$ means only that at least one of the second and third steps is controlled by conformational fluctuations. On the other hand, the Michaelis–Menten constant shows a solvent-viscosity dependence reflecting at least two of the three steps. To be more concrete, K_M is written as Eq. (5.9), with b given by Eq. (5.10). In the situation that at least one of the second and third steps is controlled by conformational fluctuations, $b \approx b_f$ as well as $a \approx a_f$ should be realized, since b_f^{-1} is given by a linear combination of the inverse of all the fluctuation-controlled parts $k_f(C_S, O)$, $k_f(C_P, C_S)$, and $k_f(O, C_P)$ of $k(C_S, O)$, $k(C_P, C_S)$, and $k(O, C_P)$ as in Eq. (5.12), where $k(C_S, O)$ represents the rate constant of the first step. As the first possibility, if only one of the second and third steps in the three steps (e.g., the second step) is controlled by solvent fluctuations, both a_f and b_f should be proportional to $k_f(C_P, C_S)$. Then K_M given by a_f/b_f becomes independent of the solvent viscosity, although $k_f(C_P, C_S)$ has a viscosity dependence on the right-hand side of Eq. (3.5). As the second possibility, if both the second and third steps in the three steps are controlled by solvent fluctuations, both a_f^{-1} and b_f^{-1} should be a linear combination of only $k_f(O, C_P)^{-1}$ and $k_f(C_P, C_S)^{-1}$. When $k_f(O, C_P) \propto \eta^{-\alpha(O, C_P)}$ and $k_f(C_P, C_S) \propto \eta^{-\alpha(C_P, C_S)}$ with constants $\alpha(O, C_P)$ and $\alpha(C_P, C_S)$ in the viscosity (η) dependence on the right-hand side of Eq. (3.5), K_M ($\approx b_f^{-1}/a_f^{-1}$) is given by

$$K_M \approx [c + d\eta^{|\alpha(O, C_P) - \alpha(C_P, C_S)|}]/[e + f\eta^{|\alpha(O, C_P) - \alpha(C_P, C_S)|}] \tag{5.13}$$

with constants c, d, e, and f. Then K_M has two saturation values: c/e in the low-viscosity region and d/f in the high-viscosity region. Therefore, its viscosity dependence connects these two saturation values, increasingly for $c/e < d/f$ or decreasingly for $c/e > d/f$ as the viscosity is increased in the intermediate-viscosity region. When all three steps are controlled by conformational fluctuations, too, K_M has a viscosity dependence similar to Eq. (5.13). As the third possibility, when the first step and only one of the second and third steps (e.g. the second step) are controlled by conformational fluctuations, a_f^{-1} should be proportional to $k_f(C_P, C_S)^{-1}$, while b_f^{-1} should be a linear combination of $k_f(C_S, O)^{-1}$ and $k_f(C_P, C_S)^{-1}$. Since $k(C_S, O) \propto \eta^{-\alpha(C_S, O)}$ with a constant $\alpha(C_S, O)$ in this case, we get

$$K_M \propto 1 + c\eta^{\alpha(C_S, O) - \alpha(C_P, C_S)} \tag{5.14}$$

with a constant c. When $\alpha(C_S, O) > \alpha(C_P, C_S)$, then K_M becomes saturated in the low-viscosity region but increases without saturation in the high-viscosity region, while it becomes saturated in the high-viscosity region but increases without saturation as η decreases, when $\alpha(C_S, O) < \alpha(C_P, C_S)$.

Experimental data on the viscosity dependence of K_M are not abundant. It has been observed that K_M was nearly independent of the solvent viscosity for hydrolysis of artificial peptides catalyzed by carboxypeptidase A [9] and also for CO_2 hydration and HCO_3^- dehydration catalyzed by carbonic anhydrase [11]. This near constancy of K_M can be expected not only in the first possibility mentioned above, but also in the second and third possibilities in the viscosity region where the viscosity dependence of K_M becomes saturated. Therefore, these data alone are not decisive. For hydrolysis of adenosine catalyzed by adenosine deaminase [15], however, it has been observed that K_M increased as the viscosity η increased. These data can be rationalized only in the second or third possibilities mentioned above. For hydrolysis of ATP catalyzed by myosin ATPase [16], moreover, it has been observed that K_M had a viscosity dependence connecting, with η-increasing dependence, two saturation values, expected only in the second possibility mentioned above. It seems thus that the scenario of enzymatic reactions inferred in this section has been confirmed by experimental data, although more detailed data are needed for further confirmation.

CONCLUSIONS

In describing the viscosity dependence of the rate constant of solution reactions, the Kramers model has been extended into two streams of theories initiated by Grote and Hynes and by Sumi and Marcus. Recent experimental data on the thermally recovering Z/E isomerization of substituted azobezenes and N-benzylideneaniline under pressures seem to favor the latter. Therefore, the latter remains as a candidate of the general expression for the rate constant of solution reactions, which should cover not only the TST regime in the low-viscosity region but also the non-TST regime in the high-viscosity region. The rate-constant formula given by the latter has the form Eq. (3.4) with (3.5). Here, k_{TST} represents the TST-expected rate constant and does not depend on the thermalization time τ in the reactant state, that is, on the solvent viscosity η, which is usually proportional to τ. On the other hand, k_f in Eq. (3.4) represents the solvent-fluctuation-controlled part of the rate constant, in general decreasing in proportion to a fractional (less-than-unity) power of τ^{-1} as τ increases, as shown in Eq. (3.5). It can be calculated by Eq. (3.15) for first-order reactions in solution and by Eq. (3.16) for second-order reactions in solution, after solving, respectively, Eqs. (3.8) and (3.9). Both Eqs. (3.8) and (3.9) are Fredholm's integral equations of the second kind, with a unique solution, which are a continuous-variable version of a set of coupled linear equations. Therefore, they can easily be solved by computers with a digitization approximation for the continuous variable, without being annoyed by boundary conditions.

The rate-constant formula (3.4) with (3.5) can be applied not only to elementary solution reactions, but also to a reaction accomplished through a series of intermediate products in solution. Therefore, it has a wide generality, for example, applicable also to protein-mediated biological reactions, such as enzymatic ones, which usually proceed with a series of intermediate products. Enzymatic reactions seem to be in the non-TST regime with $k \approx k_f$ for $k_f \ll k_{TST}$ in Eq. (3.4), where reactions are controlled by solvent-fluctuation-induced slow conformational fluctuations of proteins. Stated more appropriately, enzymes execute the three main reaction steps in the catalysis by utilizing solvent-fluctuation-induced conformational fluctuations of proteins. To facilitate this reaction mechanism, enzymes must have a large flexible body made of proteins, which can easily fluctuate in tune with solvent fluctuations at ambient temperature.

In the rate constant formula (3.4), the solvent viscosity (η) dependence is primarily through the dependence of k_f on the thermalization time τ in the reactant state, as shown in the first proportionality in (3.5). The τ can be regarded as measured by the relaxation time of solvated-structure (i.e., conformational) flutuations in the solute–solvent system. In the present work, it has been assumed that τ is proportional to η. The actual relationship between τ and η is an important problem, but it is out of the scope of the present work. To be more appropriate, therefore, analysis with Eqs. (3.4) and (3.5) should be made by using the η dependence of τ measured experimentally, for example as in [52].

ACKNOWLEDGMENT

The author wishes to thank T. Asano of Oita University for helpful discussions on experimental aspects of isomerization reactions in solution.

APPENDIX A: RELAXATION TIME OF DIFFUSIVE MOTIONS

We can determine the time development of the average value of the coordinate X, given by $\bar{X}(t) \equiv \int XP(X;t)\,dX$, from Eq. (2.1). Around the bottom of the reactant well, Eq. (2.2) is applicable. Then Eq. (2.1) enables us to get $d\bar{X}(t)/dt = -\bar{X}(t)/\tau$ with τ given by Eq. (2.3), by using $\int X(d^2P/dX^2)\,dX = 0$ and $\int X[d(XP)/dX]\,dX = -\bar{X}(t)$ obtained by partial integration.

APPENDIX B: FIRST-ORDER REACTION THROUGH A SERIES OF INTERMEDIATE PRODUCTS

In general, a first-order reaction (from a reactant state R to a product state P) can be accomplished through a series of intermediate product states $(P_1, P_2, P_3, \ldots, P_N)$, as

$$R \underset{k(R,1)}{\overset{k(1,R)}{\rightleftharpoons}} P_1 \underset{k(1,2)}{\overset{k(2,1)}{\rightleftharpoons}} P_2 \underset{k(2,3)}{\overset{k(3,2)}{\rightleftharpoons}} \cdots \underset{k(N-1,N)}{\overset{k(N,N-1)}{\rightleftharpoons}} P_N \overset{k(P,N)}{\longrightarrow} P \quad (B.1)$$

where $k(m + 1, m)$, with $k(1, 0) = k(1, R)$ and $k(N + 1, N) = k(P, N)$, represents the rate constant of reactions from the mth to the $(m + 1)$th product state when the 0th and $(N + 1)$th product state are regarded as the reactant and final-product state, respectively. When Eq. (B.1) proceeds in solution, all of the elementary rate constants have a general form given by Eq. (3.4) with (3.5), for example, as

$$k(m + 1, m)^{-1} = k_{TST}(m + 1, m)^{-1} + k_f(m + 1, m)^{-1} \quad (B.2)$$

where $k_{TST}(m + 1, m)$ represents the TST-expected rate constant of reactions from the mth to the $(m + 1)$th product state and $k_f(m + 1, m)$ represents the solvent-fluctuation-controlled part of $k(m + 1, m)$. since $k(m + 1, m)$ and its backward-reaction rate constant $k(m, m + 1)$ are related by the principle of detailed balance, the same relation should also hold in their components in Eq. (B.2) [38] as

$$K(m + 1, m) = \frac{k(m + 1, m)}{k(m, m + 1)} = \frac{k_{TST}(m + 1, m)}{k_{TST}(m, m + 1)} = \frac{k_f(m + 1, m)}{k_f(m, m + 1)} \quad (B.3)$$

where $K(m + 1, m)$ represents the equilibrium constant of the $(m + 1)$th product state relative to the mth states, being given by $\exp[-\Delta G(m + 1, m)/k_B T]$ in terms of the free energy difference $\Delta G(m + 1, m)$ between these two states. Generalizing $K(m + 1, m)$, we can introduce $K(m, R)$ for $m = 1, 2, \ldots, N$ representing the equilibrium constant of the mth product state relative to the reactant state. It is given in terms of $K(m + 1, m)$ by

$$K(m, R) = K(m, m - 1)K(m - 1, m - 2) \cdots K(2, 1)K(1, R) \quad (B.4)$$

Let us investigate what form the rate constant k of reactions from R to P in Eq. (B.1) has when each elementary-reaction rate constant in Eq. (B.1) is written as Eq. (B.2).

Generalizing $k(m + 1, m)$, too, we can introduce the rate constant $k(m, \mathrm{R})$ for $m = 1, 2, \ldots, N$ of reactions from the reactant state to the mth product state. Then Eq. (B.1) is equivalent to

$$\mathrm{R} \underset{k(\mathrm{R},N)}{\overset{k(N,\mathrm{R})}{\rightleftarrows}} \mathrm{P}_N \xrightarrow{k(\mathrm{P},N)} \mathrm{P} \tag{B.5}$$

Since Eq. (B.5) is isomorphic to Eq. (3.23), the rate constant k of reactions from R to P is given by

$$k^{-1} = [k(\mathrm{P}, N)K(N, \mathrm{R})]^{-1} + k(N, \mathrm{R})^{-1} \tag{B.6}$$

with $K(m, \mathrm{R})$ for $m = N$ in Eq. (B.4). Since $k(N, \mathrm{R})$ is the rate constant of a reaction system

$$\mathrm{R} \underset{k(\mathrm{R},N-1)}{\overset{k(N-1,\mathrm{R})}{\rightleftarrows}} \mathrm{P}_{(N-1)} \xrightarrow{k(N,N-1)} \mathrm{P}_N \tag{B.7}$$

it is given by

$$k(N, \mathrm{R})^{-1} = [k(N, N - 1)K(N - 1, \mathrm{R})]^{-1} + k(N - 1, \mathrm{R})^{-1} \tag{B.8}$$

We can continue this reduction process in $k(m, \mathrm{R})$ until $k(2, \mathrm{R})$, which is given by

$$k(2, \mathrm{R})^{-1} = [k(2, 1)K(1, \mathrm{R})]^{-1} + k(1, \mathrm{R})^{-1} \tag{B.9}$$

Summing Eqs. (B.6), (B.8), and (B.9), together with intermediate equations, side by side, we get

$$k^{-1} = k(\mathrm{P}, N)^{-1}K(N, \mathrm{R})^{-1} + \sum_{m=1}^{N-1} k(m + 1, m)^{-1}K(m, \mathrm{R})^{-1} + k(1, \mathrm{R})^{-1} \tag{B.10}$$

Introducing Eq. (B.2) for $k(\mathrm{P}, N)^{-1}$, $k(m + 1, m)^{-1}$, and $k(1, \mathrm{R})^{-1}$ in Eq. (B.10), we see that the rate constant k can be expressed in the same form as Eq. (3.4), with k_{TST} and k_f given here by

$$k_{\mathrm{TST}}^{-1} = k_{\mathrm{TST}}(\mathrm{P}, N)^{-1}K(N, \mathrm{R})^{-1} + \sum_{m=1}^{N-1} k_{\mathrm{TST}}(m + 1, m)^{-1}K(m, \mathrm{R})^{-1}$$

$$+ k_{\mathrm{TST}}(1, \mathrm{R})^{-1} \tag{B.11}$$

and

$$k_f^{-1} = k_f(\mathrm{P}, N)^{-1} K(N, \mathrm{R})^{-1} + \sum_{m=1}^{N-1} k_f(m+1, m)^{-1} K(m, \mathrm{R})^{-1} + k_f(1, \mathrm{R})^{-1}$$

$$(\mathrm{B}.12)$$

In fact, k_{TST} of Eq. (B.11) does not depend on the relaxation time τ (proportional to the solvent viscosity η) of solvent-fluctuation-induced molecular-arrangement fluctuations in the solute–solvent system, while k_f of Eq. (B.12) is a decreasing function of τ, with k_f^{-1} given by the sum of terms proportional to the inverse of the right-hand side of Eq. (3.5). When only a single term in the sum gives a dominant contribution, especially, k_f of Eq. (B.12) behaves as Eq. (3.5) in the τ dependence. We see thus that the general form (3.4) of the rate constant of solution reactions is applicable also when the reaction takes place through a series of intermediate product states as shown in Eq. (B.1).

APPENDIX C: RELATION BETWEEN THE TOTAL AND THE FREE ENZYME CONCENTRATION

The reaction scheme in Eq. (5.1) can also be expressed as

$$
\begin{array}{ccc}
\mathrm{E} & & \mathrm{E} \\
\downarrow & & \uparrow \\
 & \xrightarrow{\ k(\mathrm{C_S},\mathrm{O})[\mathrm{E}]\ } & \quad k(\mathrm{O},\mathrm{C_S}) \\
\mathrm{S} & \underset{g(\mathrm{O},\mathrm{C_S})}{\rightleftharpoons} \mathrm{ES} & \xrightarrow{\qquad} \mathrm{P} \\
 & \downarrow & \\
 & \mathrm{E} &
\end{array}
\qquad (\mathrm{C}.1)
$$

As we derived b of Eq. (5.5) from the reaction scheme (5.1) through (5.4), we can derive another form of b equivalent to Eq. (5.5) from the reaction scheme (C.1) equivalent to Eq. (5.1), as

$$b = 1/\{[k(\mathrm{O}, \mathrm{C_S})K(\mathrm{C_S}, \mathrm{O})]^{-1} + k(\mathrm{C_S}, \mathrm{O})^{-1}\} \qquad (\mathrm{C}.2)$$

with the equilibrium constant $K(\mathrm{C_S}, \mathrm{O})$ of Eq. (5.2). The rate constant is justified under the steady-state condition where the concentration at the intermediate state [i.e., that [ES] at the state $\mathrm{C_S}$ in Eq. (C.1)] is kept nearly constant during reaction. Under the steady-state condition, therefore, a balance must be maintained between the incoming and outgoing current of reactants, given, respectively, by $k(C_S, O)$ [E] [S] and $[g(O, C_S) + k(O, C_S)][ES]$, at the state C_S, where [E] and [S] represent, respectively, the concentration of free enzymes and that of free substrates. Then we get $[ES] = [E][S]k(C_S, O)/[g(O, C_S) + k(O, C_S)]$. By using Eqs. (5.2) and (C.2), this relation can be rewritten as

$$[ES] = [E][S]b/k(O, C_S) = [E][S]b\{[k(O, C_P)K(C_P, C_S)]^{-1} + k(C_P, C_S)^{-1}\}$$
$$(C.3)$$

where the second equality is ensured from the expression of the rate constant $k(O, C_S)$, obtained similarly to Eq. (C.2), of the reaction from the first intermediate state C_S to the last state O in Eq. (5.1). Here $K(C_P, C_S)$ represents the equilibrium constant for the state C_P relative to the state C_S, equal to $k(C_P, C_S)/g(C_S, C_P)$.

The reaction scheme in Eq. (5.1) can be expressed in a form different from Eq. (C.1), too, as

$$
\begin{array}{ccc}
E & & E \\
\downarrow & & \uparrow \\
\underset{g(O,C_P)}{\overset{k(C_P,O)[E]}{\rightleftharpoons}} EP & \overset{k(O,C_P)}{\longrightarrow} P \\
\end{array}
$$

$$S \qquad\qquad EP \qquad\qquad P \qquad (C.1)$$

$$
\begin{array}{c}
\downarrow \\
E
\end{array}
$$

Quite similarly to the procedure described above for obtaining Eq. (C.3), we can get

$$[EP] = [E][S]b/k(O, C_P) \qquad (C.5)$$

as the concentration [EP] at the second intermediate state C_P in Eq. (5.1), where the equilibrium constant concerned is $K(C_P, O)$ of Eq. (5.3), equal to $k(C_P, O)/g(O, C_P)$, instead of $K(C_S, O)$.

The total concentration of enzymes $[E]_0$ is given by the sum of that of free enzymes [E] and that of enzymes in the state of complex with substrates or products [ES] and [EP] as

$$[E]_0 = [E] + [ES] + [EP] \qquad (C.6)$$

Inserting Eqs. (C.3) and (C.5) into Eq. (C.6), we get Eqs. (5.6) with (5.7) in the text.

REFERENCES

1. For example, k. J. Laidler, *Theories of Chemical Reaction Rates*, McGraw-Hill, New York, 1969.

2. H. Sumi, in *Dynamics and Mechanisms of Photoinduced Electron Transfer and Related Phenomena*, N. Mataga, T. Okada, and H. Masuhara, eds., Elsevier, Amsterdam, 1992, p. 177; *J. Mol. Liq.* **65/66**, 65 (1995).

3. (a)P. Finckh, H. Heitele, and M. E. Michel-Beyerle, *Chem. Phys.* **138**, 1 (1989); M. J. Weaver and G. E. McManis III, *Acc. Chem. Res.* **23**, 294 (1990); (b) L Qin and N. M. Kostic, *Biochemistry*, **33**, 12592 (1994); T. Scherer, I. H. M. van Stokkum, A. M. Brouwer, and J. M. Verhoeven, *J. Phys. Chem.* **98**, 10539 (1994).

4. A. Mordzinski, L. Lipkowski, and T. E. Granzyna, *Chem. Phys.* **140**, 167 (1990).

5. G. K. Malik, T. K. Pal, P. K. Malik, T. Ganguly, and S. B. Banerjee, *Spectrochim. Acta A*, **45A**, 1273 (1989).

6. For example, S. Yadgar, C. Tatreau, B. Gavish, and D. Lavalette, *Biophys. J.* **68**, 665 (1995); P. J. Steinbach, A. Ansari, J. Berendzen, D. Braunstein, K. Chu, B. R. Cowen, D. Ehrenstein, H. Frauenfelder, J. B. Johnson, D. C. Lamb, S. Luck, J. R. Mourant, G. U. Nienhaus, P. Ormos, R. Philipp, A. Xie, and R. D. Young, *Biochemistry*, **30**, 3988 (1991).

7. D. H. Waldeck, *Chem. Res.* **91**, 415 (1991); G. Ponterini and M. Caselli, *Ber. Bunsenges. Phys. Chem.* **96**, 564 (1992); Y. Onganer, M. Yin, D. R. Bessire, and E. L. Quitevis, *J. Phys. Chem.* **97**, 2344 (1993); T. O. Harju, J. Erostyák, Y. L. Chow, and J. E. I. Korppi-Tommola, *Chem. Phys.* **181**, 259 (1994); F. Aramendía, R. M. Negri, and E. S. Román, *J. Phys. Chem.* **98**, 3165 (1994); R. Mohrschladt, J. Schroeder, J. Schwarzer, J. Troe, and P. Vöhringer, *J. Chem. Phys.* **101**, 7566 (1994); K. Hara, N. Kometani, and O. Kajimoto, *J. Phys. Chem.* **100**, 1488 (1996); and their earlier references cited therein.

8. B. Gavish, in *The Fluctuating Enzyme*, G. R. Welch, ed. Wiley, New York, 1986, p. 263; B. Gavish and S. Yedgar, in *Protein–Solvent Interaction*, R. B. Gregory, ed., Dekker, New York, 1995, p. 343.

9. N. G. Goguadze, J. M. Hammerstad-Pedersen, D. E. Khoshtariya, and J. Ulstrup, *Eur. J. Biochem.* **200**, 423 (1991).

10. B. Somogyi, J. A. Norman, L. Zempel, and A. Rosenberg, *Biophys. Chem.* **32**, 1 (1988); A. Rosenberg, K. Ng, and M. Punyiczki, *J. Mol. Liq.* **42**, 31 (1989); K. Ng and A. Rosenberg, *Biophys. Chem.* **39**, 57 (1991).

11. Y. Pocker and N. Janjic, *Biochemistry*, **26**, 2597 (1987); **27**, 4114 (1988); *J. Am. Chem. Soc.* **111**, 731 (1989).

12. A. P. Demchenko, O. I. Rusyn, and E. A. Saburova, *Biochim. Biophys. Acta* **998**, 196 (1989); E. A. Saburova, N. B. Simonova, N. A. Pronina, T. N. Fol'kovich, and V. E. Semenenko, *Dokl. Akad. Nauk* **325**, 1259 (1992).

13. G. W. Rayfield, *Photochem. Photobiol.* **43**, 171 (1986).

14. H. Kyushiki and A. Ikai, *Proteins: Struct. Funct. Genet.* **8**, 287 (1990).

15. L. C. Kurz, E. Weitkamp, and C. Frieden, *Biochemistry*, **26**, 3027 (1987).

16. T. Ando and H. Asai, *J. Bioenerg. Biomembr.* **9**, 285 (1977).

17. J. T. Hynes, in *Theory of Chemical Reaction Dynamics*, Vol. IV, M. Baer, ed., CRC Press, Boca Raton, Fla., 1985, p. 171.

18. A special volume, *Ber. Bunsenges. Phys. Chem.* **95**, (3), (1991).

19. M. Cho, Y. Hu, S. J. Rosenthal, D. C. Todd, M. Du, and G. R. Fleming, in *Activated Barrier Crossing*, G. R. Fleming and P. Hänggi, eds., World Scientific, Singapore, 1993, p. 143.

20. H. A. Kramers, *Physica* (Leipzig) **7**, 284 (1940).

21. S. Glasstone, K. J. Laidler, and H. Eyring, *The Theory of Rate Processes*, McGraw-Hill, New York, 1941.

22. M. Dresden, *Phys. Today*, Sept. 1988, p. 26.

23. R. Kubo, *J. Phys. Soc. Jpn.* **12**, 570 (1957).

24. G. R. Fleming and P. G. Wolynes, *Phys. Today*, May 1990, p. 36.

25. R. F. Grote and J. T. Hynes, *J. Chem. Phys.* **73**, 2715 (1980); **74**, 4465 (1981); see also [17].

26. H. Sumi and R. A. Marcus, *J. Chem. Phys.* **84**, 4894 (1986).

27. For example, H. Risken, *The Fokker–Planck Equations*, Springer-Verlag, Berlin, 1984.

28. G. Rothenberger, D. K. Negus, and R. M. Hochstrasser, *J. Chem. Phys.* **79**, 5360 (1983).

29. B. Bagchi, *Int. Rev. Phys. Chem.* **6**, 1 (1987).

30. S. K. Kim and G. R. Fleming, *J. Phys. Chem.* **92**, 2168 (1988).

31. As recent works, (a) R. Biswas and B. Bagchi, *J. Chem. Phys.* **105**, 7543 (1996); (b) T. Asano and H. Sumi, *Chem. Phys. Lett.* (1998), *in press*; (c) B. B. Smith, A. Staib, and J. T. Hynes, *Chem. Phys.* **176**, 521 (1993); and references cited therein.

32. J. Yu and M. Berg, *J. Chem. Phys.* **96**, 8741, 8750 (1992); *J. Phys. Chem.* **97**, 1758 (1993); J. T. Fourkas and M. Berg, *J. Chem. Phys.* **98**, 7773 (1993), **99**, 8552 (1993); J. Ma, D. V. Bout, and M. Berg, *J. Chem. Phys.*. **103**, 9146 (1995).

33. V. I. Gol'danskii and Yu. F. Krupyanskii, *Q. Rev. Biophys.* **22**, 39 (1989); in *Protein–Solvent Interaction*, R. B. Gregory, ed., Marcel Dekker, New York, 1995, p. 289.

34. R. Zwanzig, *J. Chem. Phys.* **38**, 1605 (1963); T. S. Nee and R. Zwanzig, *J. Chem. Phys.* **52**, 6353 (1970).

35. T. H. Tjia, P. Bordewijk, and C. J. F. Böttcher, *Adv. Mol. Relax. Proc.* **6**, 19 (1974).

36. H. Fröhlich, *Theory of Dielectrics*, Oxford University Press, New York, 1949.

37. N. Agmon and J. J. Hopfield, *J. Chem. Phys.* **78**, 6947 (1983).

38. H. Sumi, *J. Phys. Chem.* **95**, 3334 (1991).

39. H. Sumi, *J. Chem. Phys.* **100**, 8825 (1994); *J. Phys. Chem.* **100**, 4831 (1996).

40. H. Sumi, *Chem. Phys.* **212**, 9 (1996).

41. See also, J. Schroeder, J. Troe, and P. Vöhringer, *Chem. Phys. Lett.* **203**, 255 (1993); J. Schroeder and J. Troe, in *Reaction Dynamics in Clusters and Condensed Phases,* J. Jortner et al., eds., Kluwer, Dordrecht, The Netherlands, 1994, p. 361.

42. K. Cosstick, T. Asano, and N. Ohno, *High Pressure Res.* **11**, 37 (1992).

43. T. Asano, H. Furuta, and H. Sumi, *J. Am. Chem. Soc.* **116**, 5545 (1994).

44. T. Asano, K. Cosstick, H. Furuta, K. Matsuo, and H. Sumi, *Bull. Chem. Soc. Jpn.* **69**, 551 (1996); T. Asano, K. Matsuo, and H. Sumi, *Bull. Chem. Soc. Jpn.*, **70**, 239 (1997).

45. T. Asano, H. Furuta, H.-J. Hofmann, R. Cimiraglia, Y. Tsuno, and M. Fujio, *J. Org. Chem.* **58**, 4418 (1993), and earlier references cited them.

46. F. J. Bermejo, F. Batallan, J. L. Martinez, M. Garcia-Hernandez, and E. Enciso, *J. Phys. Condens. Matter* **2**, 6659 (1990); E. G. D. Cohen, P. Westerhuijs, and I. M. de Schepper, *Phys. Rev. Lett.* **59**, 2872 (1987).

47. (a) H. Sumi and T. Asano, *Chem. Phys. Lett.* **240**, 125 (1995); (b) *J. Chem. Phys.* **102**, 9565 (1995).

48. G. van der Zwan and J. T. Hynes, *J. Chem. Phys.* **76**, 2993 (1982).

49. R. Zwanzig, *J. Stat. Phys.* **9**, 215 (1973); E. Cortes, B. J. West, and K. Lindenberg, *J. Chem. Phys.* **82**, 2708 (1985); B. J. Gertner, K. R. Wilson, and J. T. Hynes, *J. Chem. Phys.* **90**, 3537 (1989); see also the appendix in [47b].

50. T. Asano and T. Okada, *J. Org. Chem.* **51**, 4454 (1986), and earlier references cited therein.

51. For example, S. A. Wallin, E. D. A. Stemp, A. M. Everest, J. M. Nocek, T. L. Netzel, and B. M. Hoffman, *J. Am. Chem. Soc.* **113**, 1842 (1991); J. Feitelson and G. McLendon, *Biochemistry* **30**, 5051 (1991); M. C. Walker and G. Tollin, *Biochemistry*, **30**, 5546 (1991); C. A. Salhi, Q. Yu, M. J. Heeg, N. M. Villeneuve, K. L. Juntunen, R. R. Schroeder, L. A. Ochrymowycz, and D. B. Rorabacher, *Inorg. Chem.* **34**, 6053 (1995); and earlier references cited therein.

52. D. H. Waldeck, W. T. Lotshaw, D. B. McDonald, and G. R. Fleming, *Chem. Phys. Lett.* **88**, 297 (1982); M. van der Auveraer, M. van den Zegel, N. Boens, F. C. de Schryver, and F. Willig, *J. Phys. Chem.* **90**, 1169 (1986); M. Lee, A. J. Bain, P. J. McCarthy, C. H. Han, J. N. Haseltine, A. B. Smith III, and R. M. Hochstrasser, *J. Chem. Phys.* **85**, 4341 (1986); A. Ansari, C. M. Jones, E. R. Henry, J. Hofrichter, and W. A. Eaton, *Science* **256**, 1796 (1992); E. Laitinen, P. Ruuskanen-Järvinen, U. Rempel, V. Helenius, and J. E. I. Korppi-Tommola, *Chem. Phys. Lett.* **218**, 73 (1994); W. Mikosch, Th. Dorfmüller, and W. Eimer, *J. Chem. Phys.* **101**, 11044 (1994); M. Levitus, R. M. Negri, and F. Aramendía, *J. Phys. Chem.* **99**, 14231 (1995).

EXPERIMENTAL ELECTRON TRANSFER KINETICS IN A DNA ENVIRONMENT

PAUL F. BARBARA AND ERIC J. C. OLSON

Department of Chemistry, University of Minnesota, Minneapolis, MN 55455

CONTENTS

I. INTRODUCTION

A. Background

Electron transfer kinetics between donors and acceptors in a DNA environment has been an area of intense interest and controversy. Generally speaking, electron transfer and other charge transfer processes in DNA are relevant to the radiation damage of DNA and certain physiological processes [1–6]. The major focus of DNA electron transfer research, however, has been the fundamental question: Is conventional electron transfer theory capable of modeling electron transfer rates accurately in a DNA environment? Conventional electron transfer theory in the last two decades has had

Electron Transfer: From Isolated Molecules to Biomolecules, Part Two, edited by Joshua Jortner and M. Bixon. Advances in Chemical Physics Series, Volume 107, series editors I. Prigogine and Stuart A. Rice.
ISBN 0-471-25291-3   © 1999 John Wiley & Sons, Inc.

enormous success in quantitatively modeling electron transfer reactions of small and medium-sized molecules in a broad range of non-DNA environments, ranging from liquid and frozen solutions to more highly organized systems, including proteins [7–19]. It is important to establish whether the characteristic features of the DNA environment might necessitate a modification of conventional electron transfer theory. The unique structural features of a DNA environment are reviewed in Figure 1, which portrays the common B-form of DNA [20]. The double helix of B-DNA consists of two antiparallel polynucleotide chains. Highly charged sugar–phosphate

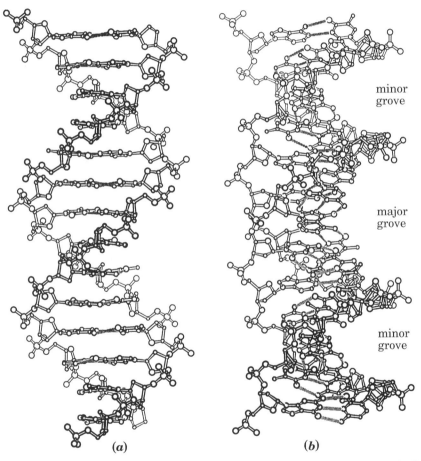

(a)   *(b)*

Figure 1. Molecular structure of the most common form of DNA: (*a*) side view of B-DNA; (*b*) the helix tilted 32° from the viewer to show the minor and major grooves. (Reproduced with permission from [20].)

backbones spiral around a well-defined stack of nucleotide bases, comprised of hydrogen-bond-paired planar aromatic rings. Many simple aromatic molecules, and aromatic regions of more complex molecules, demonstrate a high propensity for intercalation between the base pairs of DNA. Molecules that share such an intimate interaction with the base pairs of DNA have been utilized in numerous studies of DNA-mediated electron transfer. In many cases, metal–ligand complexes are employed which possess an extended-planar ligand that intercalates between base pairs, with the remainder of the metal–ligand complex in either the major or minor groove.

For electron transfer between donors and acceptors at moderate to long distances (weak coupling) the electron transfer rate constant, k_{ET} can be approximated by [12]

$$k_{ET} = \frac{2\pi^{3/2} H_{DA}^2}{h} \left(\frac{1}{\lambda k T}\right)^{1/2} \exp\left[-\frac{(\Delta G + \lambda)^2}{4\lambda k T}\right] \qquad (1.1)$$

where h is Planck's constant, k the Boltzmann constant, T the temperature, H_{DA} the electronic coupling matrix element, ΔG the free energy change of the reaction, and λ the nuclear reorganization energy (of the reactants and solvent) that is required to transform reactants into products. In favorable cases experimentalists can measure k_{ET} and independently estimate ΔG and λ. This allows for an experimental determination of H_{DA}. For many systems Eq. (1.1) is insufficient since it treats the nuclear system as a single effective classical degree of freedom, and instead, theoretical models with one or more additional high-frequency quantum mechanical degrees of freedom must be employed. The latter models have met with great success for describing electron kinetics in the inverted regime, where quantum mechanical nuclear degrees of freedom are especially important [7,14,15,21–23].

The dominate source of the distance dependence of electron transfer rates in a wide variety of environments has been demonstrated to be the distance dependence of H_{DA}. For insulating environments, such as proteins, water, organic solvents, and aliphatic organic molecules, theory [7,10,17] and experiment [24–33) show that H_{DA} is well approximated by an exponential dependence as given in

$$|H_{DA}|(r) \approx |H_{DA}^0| \exp[-(\beta/2)(r - r^0)] \qquad (1.2)$$

H_{DA}^0 is the electronic coupling when the donor and acceptor are in contact, and β reflects the distance dependence of the electronic coupling.

For electron transfer between donors and acceptors in proteins and other insulating environments, there exists a large energy difference between the energy of the donor–acceptor "activated complex" and the energy of the oxidized and reduced states of the bridging medium. Consequently, coupling of the donor to the acceptor involves quantum mechanical tunneling of charge between a localized orbital or orbitals on the donor and a localized orbital or orbitals on the acceptor. The distance dependence of electron transfer rates are especially well characterized in proteins, for which donor and acceptor distances can be fixed at long distances. The exponential fall-off of H_{DA} and k_{ET} with distance has been observed to hold for proteins for over 10 powers of 10 [25]. Reported values for β in proteins typically fall in the range 0.9–1.6 Å^{-1}. The shaded region Figure 2 represents the range of experimentally observed maximal rates for electron transfer in proteins for various donors and acceptors [25]. Electron transfer over edge-to-edge dis-

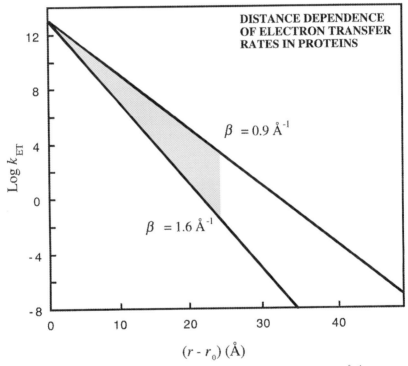

Figure 2. Theory and experiment show that β is typically in the range 0.9–1.6 Å^{-1} for electron transfer in proteins. The shaded region represents the range of experimentally observed maximal rates for electron transfer in proteins for various donors and acceptors [25]. Electron transfer over edge-to-edge distances as large as 25 Å has been observed in protein systems, although these rates are relatively slow, $< 10^2\,\mathrm{s}^{-1}$.

tances as large as 25 Å has been observed in protein systems, although the rates are relatively slow, $< 10^2 \, s^{-1}$.

B. General Issues

For electron transfer between donors and acceptors in a DNA double helix, the intervening medium is partially comprised of a highly ordered stacked arrangement of aromatic bases. A number of theoretical and experimental papers have attempted to examine whether the stacked arrangement of aromatic bases, the π-stack, is an important bridging medium for electron transfer between donors and acceptors in DNA. In particular, this work has focused on whether the π-stack dramatically alters the distance dependence of electron transfer, especially for long-range electron transfer where DNA might strongly accelerate electron transfer rates.

Many of the measurements of the distance dependence of electron transfer rates in DNA observe a proteinlike dependence ($\beta \approx 1 \, \text{Å}^{-1}$) (see Section II). However, there are a few reports of electron transfer rates that exceed proteinlike values by many orders of magnitude. The most striking example is the report of Murphy et al. of electron transfer between two metal complexes at a distance ≈ 40 Å at a rate of $> 10^9 \, s^{-1}$ [34]. This surprising result is over nine decades faster than that expected based on the protein measurements. Based on this measurement, an extraordinarily mild distance dependence ($\beta < 0.2 \, \text{Å}^{-1}$) is inferred [34]. These authors implicate the π-stack in promoting ultrafast long-range electron transfer (i.e., the "molecular wire" effect).

From a theoretical standpoint one of the key parameters that determine whether the π-stack should play an important role in promoting long-range electron transfer is the energy gap between the "activated complex" of the donor–acceptor system and the oxidized and/or reduced states of the π-stack [35]. If this gap is extraordinarily small, delocalized electronic states of the bridge might be expected to couple donors and acceptors efficiently at a distance. However, electronic structure calculations by Priyadarshy et al. predict that this energy gap is sufficiently large for ordinary donors and acceptors in DNA to ensure conventional tunneling behavior [35]. While the bases are found to dominate electronic coupling between donor and acceptors at opposite ends of the helix, the bases are not found to be extraordinarily strong mediators for the coupling. Thus a rather typical β value is predicted for electron transfer in DNA [36].

Experimentally, it is known from the redox chemistry of DNA that oxidation and reduction potentials of nucleotides are affected by many complex factors, such as pH and ionic strength. Therefore, at least on experimental grounds, there is some uncertainty in the magnitude of the energy gap between donor–acceptor states and oxidized or reduced DNA. Felts et al.

have made theoretical calculations that consider the kinetic consequences of an energy gap that is sufficiently small to allow for thermally induced virtual excitations of the bridging π-stack [37]. Even though the oxidized and reduced states of the bridge are not thermally directly accessible according to this model, thermally induced virtual excitation of such states can play a role in enhancing the electron tunneling process. These calculations, which are beyond the simple electron transfer theory (Eq. (1.1)), suggest that the distance dependence of electron transfer in DNA may be highly nonexponential [37]. In principle, virtual excitations are capable of promoting long-range electron transfer between the donor and acceptor. For certain parameters such a mechanism was shown to exhibit a typical protein-like distance dependence at short distances but a much more shallow distance dependence for longer distances. Thus no single β value would apply.

One serious potential complication when interpreting the distance dependence of electron transfer rates in DNA is the competition between direct electron transfer reactions and various slow nondirect electron transfer processes. For example large-amplitude rare fluctuations (i.e., unwinding, etc.) of the donor–acceptor–DNA system can in principle bring faraway donors and acceptors into contact and thereby induce electron transfer. Bimolecular electron transfer between separate donor- and acceptor-modified duplexes is another probable mechanism competing with slow, long-range electron transfer. Finally, rare direct oxidation or reduction of the bridge, and subsequent charge hopping on the DNA, may be another indirect slow electron transfer reaction with an apparent mild distance dependence. An approximately biexponential distance dependence is expected when slow nondirect electron transfer processes are present in competition with a strongly distant dependent direct electron transfer process, as outlined in Figure 3. For example, for very slow rates (on the order of $1\,s^{-1}$) it is potentially difficult to distinguish direct electron transfer from a variety of competitive slow processes. Thus the apparent distance dependence of extremely slow electron transfer reactions in DNA must be interpreted with great caution.

In this chapter we make a critical analysis of all available electron transfer rate measurements in DNA to identify potential strengths and weaknesses of the various measurements. In particular, apparent inconsistencies between different measurements on identical or closely related systems are considered in detail. Considering the tremendous controversy in the DNA electron transfer field and the tremendous variability of the reported electron transfer rates, a critical analysis of the various results is highly justified. The present chapter is narrowly focused on measurements of electron transfer kinetics in DNA. A longer, more in-depth, and broadly based analysis of DNA electron transfer studies has been written by Netzel [38]. Netzel's

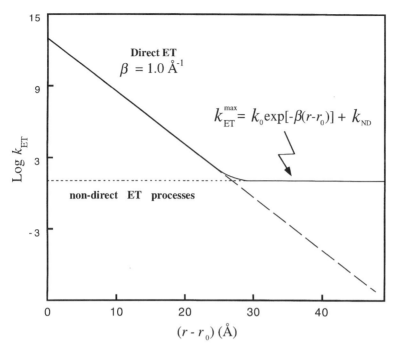

Figure 3. An approximately biexponential distance dependence for the electron transfer rate (solid line) is expected when slow nondirect electron transfer processes (represented by a short-dashed line and assumed constant, $k_{NR} = 1\,s^{-1}$) are present in competition with a strongly distance-dependent direct electron transfer process (long-dashed line).

review, which is the most exhaustive analysis of experimental results in this field to date, identifies severe limitations in many of the reported results on electron transfer rates and electron transfer product distributions. The reader is referred to Netzel's paper for a more in-depth discussion of the systems reviewed in the present chapter. The reader is also referred to review articles published by several of the active groups in the field [39–42].

Topics involving electron transfer in nucleic acid systems that are beyond the main focus of this review but may be of interest to the reader include several interesting studies on the migration and trapping of excess charges in DNA, which are relevant to understanding hopping mechanisms of charge transfer [43–46]. Additionally, charge transfer reactions are beginning to be studied in peptide nucleic acid–DNA duplexes [47], and electron transfer quenching of pyrene excited states by nucleosides, oligonucleotides, and duplexes have been investigated, as part of a larger investigation of the structure and function of ribozymes [48–52].

II. EXPERIMENTAL ELECTRON TRANSFER KINETICS

A. Overview

Table I summarizes the published measurements on electron transfer rates between donors and acceptors in DNA. A broad range of electron donors and acceptors have been utilized. Minimum details on the chemical structures of the various donors and acceptors are indicated in Figures 4 and 5, which define the structural abbreviations contained in Table I. The reader is referred to the original publications for a more complete description of the chemical structures of the donors, acceptors, and DNA environments. Some of the examples involve DNA duplexes that incorporate donor and/or acceptor groups that are covalently linked or tethered to the DNA framework. Other examples involve non-covalently-bound intercalated donors and acceptors. In many cases, a sufficiently long tether is utilized between DNA and the donor and/or acceptor groups such that intercalation is feasible.

Electron donors and acceptors include metal complexes, especially involving ruthenium and rhodium. Organic electron donors and acceptors, such as ethidium bromide, EB^+, have also been utilized. In a few examples, the DNA bases themselves participate as an electron donor in the electron transfer reactions. The control and characterization of the structure of the donor–acceptor–DNA assembly are critical factors in the analysis of electron transfer kinetics. Unfortunately, there are a number of complex issues regarding the structural aspects of intercalated systems. The reader is referred to the Netzel review [38] and the original publications for an in-depth discussion of the structure of each of the donor–acceptor–DNA systems that are described in Table I. The distances that are listed in Table I are those reported in the literature. Except where indicated, the distances referred to correspond to center-to-center distances measured along the π-stack. Edge-to-edge distances between nonintercalated portions of the donors and acceptors may be considerably shorter, depending on the example.

It should be emphasized that experimental measurements for k_{ET} in DNA environments are exceedingly complex. The synthetic chemistry is more complex than that for protein systems, for a number of reasons. The location of the donor or acceptor in a DNA environment is difficult to control or even characterize. In many studies, donors and acceptors are not covalently linked to the DNA duplex but rather are allowed to freely intercalate between base pairs. Therefore, there may exist a wide distribution of donor–acceptor separations. In addition, donors and acceptors may be intercalated in the major groove, the minor groove, and/or weakly associated with the DNA. There may even exist more than one mode of intercalation in the major groove itself. Somewhat better-defined situations are those experiments that use covalently linked (tethered) donor and acceptor groups which may or may not be intercalated. Tethered experiments, however, have an additional complication since the covalent link to the DNA may itself influence the electron transfer kinetics and associated photophysical processes that are involved in measurements that employ photoinduced electron transfer to initiate the reaction. A broad range of photoinduced electron transfer processes have been used to initiate the electron transfer reaction. In some cases the excited state participates as an electron donor [34,56,64–74]; in others, the excited state is a electron acceptor [53,54,61,62,75,76]. For some of the electron transfer examples the acceptor is an intermediate ground-state species which is prepared by photo-oxidation [55,77].

The electron transfer rates, $k_{ET}(r)$, listed in Table I in most cases are those that were reported by the authors in the primary publication. Note that many of the rates are reported as inequalities since in many examples the rates were not actually measured but were inferred to be either too fast or too slow to be measured with the experimental technique that was employed. The rates between brackets were not reported in the original publication but have been estimated from data in the literature by procedures outlined in the text below. To focus on the question of whether DNA is indeed capable of promoting ultrafast electron transfer over long distances, we have included a column in Table I of electron transfer rates at a separation of ~ 40 Å. In several of the examples, these were estimated by extrapolation using the k_{ET} values at shorter distances and the published β value. It should be emphasized that for many of the examples reported, a distribution of electron transfer rates are observed. In some cases the distribution of rates is attributed to a distribution of donor–acceptor separations due to intercalation at various base-pair positions in the DNA duplex. In other cases the source of the distribution of rates is more complex and in certain cases even controversial. A distribution of intercalation modes, a distribution of conformations of the intervening bridge, and diasteromeric intercalation have all been postulated as potential sources of the distribution of rates.

Rates have been measured by a variety of techniques, ranging from simple fluorescence yield measurements to combined transient emission and absorption studies. In some cases, only the rate of disappearance of reactants is monitored, while in the better characterized examples, both the disappearance of reactants and the appearance of products is observed. Since some of these measurements are particularly susceptible to experimental error, we indicate in detail the potential complications with the kinetic measurement for each example.

TABLE I
Summary of Experiments on Electron transfer Kinetics in DNA

Case	Donor	Acceptor	r^a (Å^{-1})	$k_{ET}(t)^{b,c}$ (s^{-1})	β^c (Å^{-1})	$k_{ET}(40)^{b-d}$ (s^{-1})	Ref.	Notes
I	G	Stilbenedicarboximide S_1	~4 7.4 10.8 14.8	1×10^{12} 2×10^{11} 3×10^9 5×10^8	~0.64	~10^1	[53]	DNA hairpin structure. k_{ET} values were measured directly from transient emission and transient absorption spectra.
II	[Ru(phen')$_2$dppz]$^{2+}$* MLCT Ru(II)*	[Rh(phi)$_2$(phen')]$^{3+}$ Rh(III)	≥41	≥10^9	≤0.2	≥10^9	[34]	Only static fluorescence yields measurements.
III	[Ru(phen)$_2$dppz]$^{2+}$* MLCT Ru(II)*	[Rh(phi)$_2$(bpy')]$^{3+}$ Rh(III)	≤44	{≤10^7}	N/A	{~10^7}	[54]	Acceptor tethered, donor nontethered. Report ~50% emission quenching. Large fraction of slow ET.
IV	[Ru(NH$_3$)$_4$(pyr)(am)]$^{2+}$ Ru(II)	[Ru(bpy)$_2$(imi)(am)]$^{3+}$ Ru(III)	21	2×10^6	N/A	≤10^6	[55]	Transient measurements. k_{ET} similar to that observed in proteins at comparable distances.
V	Et' S_1	[Rh(phi)$_2$(bpy')]$^{3+}$ Rh(III)	~20 ~32	≪10^9-70% ≥10^{10}-29% ≪10^9-92% ≥10^{10}-8%	N/A	≪10^9-92% ≥10^{10}-8%	[56]	Dynamics of Et'(S_1) fast, complex, and exhibit similar decays with or without Rh(III), k_{ET} unresolved.
VI	EB$^+$ or AOH$^+$	DAP^{2+}	10.2 13.6 17.0	4.3×10^9 2.3×10^8 1.3×10^7	~0.9	~10^{-3}	[57]	Noncovalently bound donors and acceptors. Distribution of ET rates.

	Donor / Acceptor	r				Ref.	Comments
VII	[Ru(phen)$_2$dppz]$^{2+}$* MLCT Ru(II)* / [Rh(phi)$_2$bpy]$^{3+}$ Rh(III)	10.2	$>10^{10}$	~ 1	$\leq 10^{-3}$	[58]	Efficient, fast, k_{ET} values attributd to clustering of donor–acceptor pairs.
VIII	[Ru(phen)$_2$dppz]$^{2+}$* MLCT Ru(II)* / [Rh(phi)$_2$bpy]$^{3+}$} Rh(III)	10.2	$\geq 10^{9}$	N/A	$\ll 10^{7}$	[59]	Spectroscopic data consistent with donor–acceptor clustering.
IX	[Ru(phen)$_2$dppz]$^{2+}$* MLCT Ru(II)* / [Rh(phi)$_2$bpy]$^{3+}$ Rh(III)	≥ 10	Fast ($\geq 10^{10}$) Slow ($\ll 10^{7}$)	N/A	$\geq 10^{10}$	[60]^e	Donor–acceptor pairs within "sphere of action" undergo ultrafast ET. Others blocked by disruption in π-stack
X	GC / [Rh(phi)$_2$(bpy′)]$^{3+}$* Rh(III)*	17–37	$\{10^{5} \geq k_{ET} \geq 10^{-1}\}$	N/A	$\{10^{5} \geq k_{ET} \geq 10^{-1}\}$	[61,62]	Small, apparently distance independent, quantum yield of conversion detected by DNA cleavage analysis.
XI	GG / [Ru(phen)(bpy′)dppzX]$^{3+}$ Ru(III)	37	$\{10^{5} \geq k_{ET} \geq 10^{-1}\}$	N/A	$\{10^{5} \geq k_{ET} \geq 10^{-1}\}$	[63]	Increased quantum yield of conversion. Ru(III) decays of μs time scale in mixed-sequence DNA
XII	⟨TT⟩ / [Rh(phi)$_2$(bpy′)]$^{3+}$* Rh(III)*	26	N/A	N/A	N/A	[54]	Small quantum yield of conversion. More efficient with nontethered Rh(III).

[a] Donor–acceptor distance r correspond to edge-to-edge separations. The abbreviations are defined in Figs 4 and 5.

[b] Rates in braces are estimates made for the first time in this paper using various yield and kinetic data; see the text for further details.

[c] N/A, not applicable.

[d] $k_{ET}(40)$ in examples were calculated by extrapolation using the β and k_{ET} values measured at shorter distances. For example, $k_{ET}(40)$ was measured for a sample with an average-acceptor separation close to 40 Å

[e] Barton and co-workers interpret the ultrafast quenching of the donor emission to long-range electron transfer over a distance of > 40 Å with a rate of $\geq 10^{10}$; see [61]

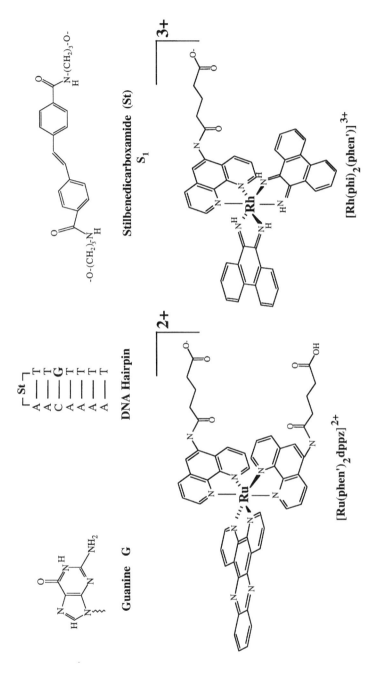

Figure 4. Chemical structures for donors, acceptors, and other species that are referred to in Table I and in the text.

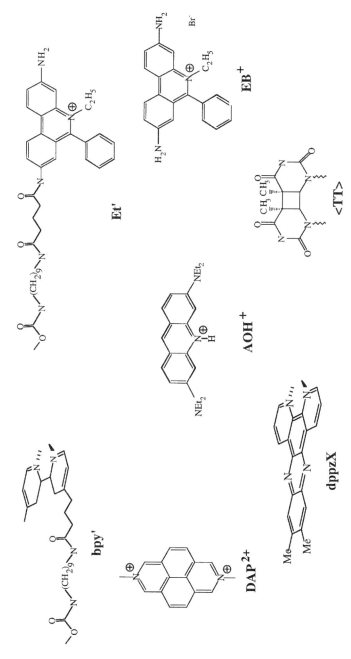

Figure 5. Chemical structures for donors, acceptors, and other species that are referred to in Table I and the text.

B.   Spectroscopic Studies of Electron Transfer

1.   Covalently Attached Donors and Acceptors

The first example in Table I is arguably the best characterized distance dependence of electron transfer rate measurement made to date [53]. It involves electron transfer in a DNA hairpin, with the electron acceptor donor being an excited state of a stilbenedicarboxamide derivative. Previous studies had demonstrated that stilbene is strongly quenched by guanine, G, and only weakly by thymine, T. No other nucleotide exhibits a quenching effect on the stilbene excited state. The kinetics of electron transfer from G to stilbene was monitored by both transient emission and transient absorption measurements. Different hairpin structures were studied having a controllable spacing between G and the electron acceptor, stilbenedicarboxamide. The observed electron transfer rates ranged from 10^{12} for stilbene–G in closest contact to nearly 10^8 for G removed approximately 18 Å. Electron transfer at farther distances was too slow to measure with this technique. The rate data exhibit an exponential distance dependence with a β value of about 0.64 Å^{-1}. An extrapolation of these results to a 40-Å separation predicts a rate $\approx 10^1\,s^{-1}$. While the results suggest that the DNA π-stack may mediate long-range electron transfer somewhat better than proteins, the results do not indicate DNA can act as a promoting medium for ultrafast long-range electron transfer.

   The Lewis et al. results [53] are in striking contrast to the report of ultrafast electron transfer over a distance of ≈ 40 Å at a rate $> 10^9\,s^{-1}$ by Murphy et al. (case II in Table I) [34]. Here the electron transfer involved an excited ruthenium polypyridyl MLCT state as an electron donor and a Rh(III) species as an acceptor, with the donor and acceptor tethered to complementing strands of DNA. DNA strand cleavage results, along with molecular modeling and other evidence, suggested that the tethered donor and acceptor metal complexes were intercalated in the DNA duplex at a distance of approximately 40 Å, corresponding to 11 intervening base pairs. Chemical structures of the donor and acceptor complexes are shown in Figure 4. The duplex with both donor and acceptor exhibited no detectable emission from the MLCT state of the donor, while a duplex with donor in the absence of acceptor showed the typical emission of the MLCT state of $[Ru(phen)_2 dppz]^{2+}$. Thus efficient emission quenching of the MLCT emission was taken as evidence of electron transfer at a rate of $> 10^9\,s^{-1}$, which is many orders of magnitude faster than the expected maximal rate in proteins at this distance (see Figure 2).

   The Murphy et al. results, which were published in 1993, are the strongest evidence for *ultrafast* electron transfer over 40-Å distances in

DNA. Unfortunately, these results are limited to static fluorescence yield measurements, and in particular, the existence of ultrafast long-range electron transfer is merely inferred from the absence of fluorescence intensity. Such measurements may be in error due to other processes that may reduce the fluorescence intensity, such as destruction of the material in the sample by some other process and rearrangement of the DNA structure which can in principle force the $[Ru(phen)_2dppz]^{2+}$ species out of intercalation, which itself would also cause emission quenching [78].

Case III in Table I is a partially analogous example to the Murphy et al. experiment (case II). This result from Dandliker et al. [54] involves a DNA duplex with one strand that contains a tethered Rh(III) acceptor. The electron donor, on the other hand, is a non-covalently bound, intercalated, photoexcited $[Ru(phen)_2dppz]^{2+}$. For case III, only about 50% emission quenching is reported, in contrast to the case II result, where 100% quenching is reported. Note in case III that the distance between the donor and acceptor is not fixed, and there is a distribution of donor–acceptor distances ranging from about 7 to 40 Å. An explanation for the about 50% quenching has been given by Dandliker et al. They attribute the partial quenching to incomplete intercalation of the Rh(III) acceptor and assume that the 50% quenching reflects the fraction of acceptor well stracked in the DNA. Thus experimental data and interpretations for cases II and III seem to be inconsistent, although the expalantion for this discrepancy is now known.

We suggest here that a reasonable interpretation for only partial quenching in case III is that the quenching rate is in fact strongly distance dependent, with only nearby donor–acceptor pairs undergoing fast electron transfer (emission quenching). It should be emphasized that this type of distance-dependent electron transfer is in clear contrast to the Murphy et al. result and interpretation of case II.

The fourth example in Table I is the study of Meade and Kayyem on the eight-base-pair duplex with distinct Ru–metal complexes covalently bound in nonintercalating geometries [55]. The relevant edge-to-edge distance for the electron transfer is ≈ 12.5 Å [38]. Electron transfer rates were measured by transient spectroscopy yielding a rate that is nearly identical to that observed for electron transfer between related donors and acceptors in protein environments. The Meade and Kayyem results are consistent with the conclusion that electron transfer in a DNA duplex is similar to proteins.

Another example of an electron transfer study between tethered donors and acceptors on a DNA duplex is that of Kelley et al., case V in Table I [56]. These authors synthesized a series of DNA duplexes containing a tethered-intercalating ethidium bromide donor with a tethered-intercalating Rh(III) acceptor, similar to the acceptor used in cases II and III. In this case the electron donor was the excited state of the ethidium moiety. Very little emission quenching was observed in this system, even at distances of approximately 20 Å. In fact, transient spectroscopy of the emission dynamics with about 150-ps time resolution exhibited very similar dynamics with or without the presence of the acceptor. However, static fluorescence measurements indicated that there was a small fraction of emission quenching, which was assumed to have occurred at a time scale that is much shorter than the time resolution of the emission apparatus. A serious complication with these measurements is apparent in a comparison of the emission decays observed from untethered and tethered donor species. The emission dynamics of the tethered ethidium moiety in the absence of acceptor is exceedingly complex and much more rapid than the emission dynamics of non-covalently bound (untethered) intercalated ethidium. Thus the results in this system are extremely complicated, due to efficient radiationless decay of the excited-state donor. These various complications make it extremely difficult to come to any conclusion about this system. While the authors have interpreted the presence of the small fraction of emission quenching as being due to electron transfer kinetics occurring at a rate of $> 10^{10} \, \text{s}^{-1}$, certainly it is difficult to establish this without more investigation.

The authors explain the presence of both very fast apparent quenching *and* unresolvably slow quenching as being due to a distribution of π-stacking environments in the DNA duplex. With regard to the evidence that these results present for ultrafast electron transfer at a long distance, one should consider the extrapolation of their results to long distances. Since the fraction of fast quenching decreases as a function of separation between donor and acceptor, at distances of approximately 40 Å their results suggest that the fast component of electron transfer possesses a vanishing amplitude. Thus a crude comparison of the Kelley et al. results to the Murphy et al. results would indicate that while ultrafast electron transfer is observed in the latter case for 40 Å separations, the Kelley et al. results show little or no evidence of ultrafast electron transfer at long range. In fact, the emission dynamics measurements taken alone would indicate that the electron transfer rate at all of the distances studied is unresolvably slow.

2. *Noncovalently Attached Donors and Acceptors*

One of the earliest measurements on the distance dependence of electron transfer rates in a DNA environment is that of Brun and Harriman [57]. These authors studied electron transfer between the excited state of EB^+ or AOH^+ and the acceptor DAP^+. The results involved non-covalently bound samples and a distribution of fluorescence lifetimes were interpreted as being due to a distribution of donor–acceptor separations along the DNA helix. While such an analysis is uncertain with respect to the true distribution of separations, the data do have the advantage that the photophysics of the system appear to be well defined. In particular, the ethidium bromide donor has a relatively long lifetime and does not exhibit fast components of decay in the absence of acceptor. Brun and Harriman have analyzed their data to extract a distance dependence for the electron transfer rate by assuming that the distinct kinetic components they observe correspond to distinct donor–acceptor distances. The results are in reasonable agreement with the conventional distance dependence of electron transfer rates. A β value of about 0.9 Å^{-1} is able to describe the apparent distance dependence of the rates that Brun and Harriman observe. It is interesting to compare the Brun and Harriman EB^+ quenching studies with the tethered Et' studies of Kelley et al. An electron transfer rate at distances longer than 20 Å in DNA is $< 10^6 \text{ s}^{-1}$ in both studies if the comparison is restricted to time-resolved measurements. Brun and Harriman do not refer to evidence of the *static* quenching mechanism of Kelley et al., even at high concentration where the average donor–acceptor separations are relatively short. The hypothetical static quenching mechanism of Kelley et al. is expected to contribute to the Brun and Harriman results if it were to operate in an analogous fashion in tethered and untethered DNA systems.

Cases VII to IX in Table I are concerned with the electron transfer kinetics of non-covalantly bound metallointercalators in DNA. Early reports on this system from Barton, Turro, and co-workers had identified very efficient luminescence quenching over a broad concentration range of electron acceptors [e.g., Rh(III)] [60,64,68–71,74,79–81]. In many cases, the efficient luminescence quenching was attributed to very fast long-range electron transfer. This interpretation is consistent with the report mentioned above of efficient electron transfer in the tethered metallointercalator–DNA sample of Murphy et al. [34].

In collaboration with Barton's group, our group has made extensive time-resolved emission and absorption measurements on untethered Ru–DNA–Rh samples. Initial attempts on tethered samples were unsuccessful due to photolysis of the samples and limited available quantities. We therefore have focused our research on DNA-mediated electron transfer on untethered samples. Untethered Ru–DNA–Rh samples possess a complex distribution of donor–acceptor separations, analogous to the measurements of Brun and Harriman. In collaboration with Barton and co-workers, our group published an extensive description of the kinetics of the excited-state electron transfer of the Ru–NA–Rh system [71]. Efficient and rapid electron transfer was observed for photoexcited donors in the presence of acceptors even at extremely low loading of the DNA duplex. Both the forward and reverse electron transfer reactions were studied. The kinetics were well described by

$$Ru^{II}(phen)_2dppz(MLCT) + Rh^{III}(phi)_2bpy \xrightarrow[k_{ESET}]{}$$

$$Ru^{III}(phen)_2dppz + Rh^{II}(phi)_2bpy \quad (2.1)$$

$$Ru^{III}(phen)_2dppz + Rh^{II}(phi)_2bpy \xrightarrow[k_{RECET}]{}$$

$$Ru^{II}(phen)_2dppz + Rh^{III}(phi)_2bpy \quad (2.2)$$

Here k_{ESET} and k_{RECET}, are, respectively, the electron transfer rate contacts for the excited state and recombination reactions. Both processes, forward and reverse electron transfer, exhibit a distribution of electron transfer rates.

In our earlier publication on this work [71], we pointed out that the kinetic results by themselves were consistent with donor–acceptor clustering along DNA, allowing for the observed fast component of electron transfer. Within this model, slower components of electron transfer are due to donor–acceptor pairs that exist at a greater separation. In other words, the donor–acceptor clustering model is based on the assumption that electron transfer in DNA does exhibit a substantial distance dependence and clustering causes a large fraction of donor–acceptor pairs to exist at a very short distance. It was pointed out that other than the kinetic data, there was no evidence for clustering. Furthermore, the proposed clustering model was believed to be inconsistent with results on the distribution of Rh(III) complexes inferred from electrophoresis measurements on the photocleavage of duplex DNA.

In a more recent paper we reexamined these issues and came to a very different conclusion with regard to the photocleavage evidence against clustering. We extended the analysis of the time-resolved measurements on the DNA–metallointercalator system and the electrophoresis photocleavage measurements by constructing a quantitative modeling of the photodynamics [58]. An equilibrium distribution of donor–acceptor distances on the DNA was simulated from a rudimentary one-dimensional lattice model for intercalation which allowed for variable donor–acceptor clustering and sequence selectivity (in some cases). The simulated distribution of donor and acceptor positions was used as an input for calculation of the photodynamics invoking a phenomenological kinetic scheme and an assumed exponential distance dependence of the excited-state electron transfer rate and the recombination rate. Comparison of the simulation results with a broad distribution of experimental data strongly indicated that a fairly typical β value of about $0.8 \, \text{Å}^{-1}$ in the presence of donor–acceptor clustering is able to account for all the features of the transient data (Figs. 6 to 8) and additionally for the related photocleavage studies (Fig. 9). On the

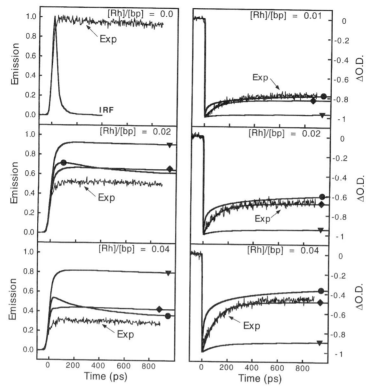

Figure 6. Picosecond experiment [71] and simulation on photoexcited Δ-Ru(phen)₂dppz bound to calf thymus DNA in the presence of various concentrations of Δ-Rh(phi)₂bpy. Simulation parameters are collected in Table II. Triangle, short-range ET, random-binding model; circle, long-range ET, random-binding model; diamond, short-range ET, clustering model. The left-hand column represents the picosecond emission measurement and simulation obtained at 10 μM Ru, 500 μM bp calf thymus DNA. The upper panel displays the instrument-response function (IRF) and Ru–DNA emission data in the absence of Rh. The middle panel shows emission data and simulation for Ru/DNA in the presence of 10 μM Rh and the lower panel for 20 μM Rh. The right-hand column contains transient-absorption experiment and simulation obtained for 20 μM Ru,1000 μM bp calf thymus DNA. The three concentrations of Rh shown are 10 μM, 20 μM, and 40 μM. (Reproduced with permission from [58].)

other hand, an attempt to model the electron transfers with a shallow distance dependence (e.g., β value $< 0.7 \text{Å}^{-1}$, corresponding to long-range ultrafast electron transfer) is not able to account for the data.

The comparison of the experimental data and the predictions (Table II) of a quantitative photodynamic model for the picosecond, nanosecond, and steady-state emission data of the Ru–DNA–Rh system are shown in Figures 6 to 8. Long-range electron transfer disagrees with each of the various types

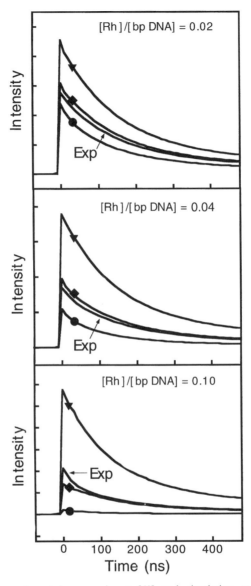

Figure 7. Nanosecond emission experiment [60] and simulation on photoexcited Δ-Ru(phen)$_2$dppz bound to calf thymus DNA in the presence of various concentrations of Δ-Rh(phi)$_2$bpy. Simulation parameters are collected in Table I. Triangle, short-range ET, random binding model; circle, long-range ET, random binding model; diamond, short-range ET, clustering model. Emission quenching data on 10 μM Ru, 500 μM bp calf thymus DNA in the presence of 10 μM (top), 20 μM (middle), and 50 μM (bottom) Rh. (Reproduced with permission from [58].)

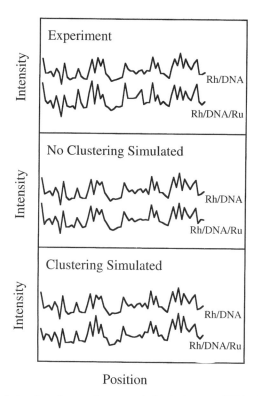

Figure 9. Gel electrophoresis experiments and simulations on DNA photocleavage by Δ-Rh(phi)₂bpy. Shown are a portion of phosphorimager scans on a DNA restriction fragment. The upper panel contains experimental results [71] on photocleavage of 500 μM bp DNA with 10 μM Δ-Rh(phi)₂bpy in the presence (bottom) and absence (top) of 10 μM Δ-Ru(phen)₂dppz. Monte Carlo simulation results at equivalent concentrations with no preferential intercalator–intercalator interactions are displayed in the middle panel. The lower panel displays Monte Carlo simulation results with clustering, at identical metallointercalator concentrations. (Reproduced with permission from [58].)

of data. For example, on the picosecond time scale it predicts resolvable emission dynamics that are not observed in the experiment. An ordinary or typical electron transfer distance dependence (β value of about $1\,\text{Å}^{-1}$) without preference for clustering of metallointercalators is also in poor agreement with experiment. However, the electron transfer model with ordinary distance dependence coupled to donor–acceptor clustering is in very good agreement with the dynamic and steady-state measurements.

These simulations strongly demonstrate that a proteinlike distance dependence of electron transfer coupled with donor–acceptor clustering is

TABLE II
Electron transfer Kinetic Models

Model[a]	Symbol	β (Å^{-1})	P	$k_{0(\text{ESET})}$ (ps^{-1})	$k_{0(\text{RECET})}$ (ps^{-1})
Long-range	●	0.16^b	1.0^c	0.3^d	0.3^e
Short-range	▼	1.0	1.0	0.3	0.01
Clustering	◆	0.8^f	13^g	0.3	0.01

Source: Reproduced with permission from [58].

[a] Three kinetic limits are presented, to simplify comparison with experimental data.

[b] β is defined by Eq. (1.2). k_0 is defined by Eqs. (1.1) and (1.2) given $r = r_0$. With a lower limit of $k_0 \geq 3 \times 10^{11}\,\text{s}^{-1}$ for the rate of excited-state electron transfer established by TCSPC measurement, $\beta < 0.16\,\text{Å}^{-1}$ must be assumed to account for the remarkable luminescence quenching reported for the tethered Ru–DNA–Rh sample where donor–acceptor separation was fixed at $> 40\,\text{Å}$.

[c] Preference for clustering, P, is defined in the Monte Carlo simulation by given a randomly selected intercalation site a certain chance that it would be rejected (and a new site selected). Physically, the preference term relates the binding constant of a donor in the nearest-available intercalation site adjacent to an intercalated acceptor to the binding constant of a donor in any other allowed intercalation site (e.g., nonnearest neighbor).

[d] Simulated quenching kinetics convolved with the TCSPC instrument-response-function indicate that k_{ESET} must be $\geq 3 \times 10^{11}\,\text{s}^{-1}$ to account for the absence of decay dynamics in the TCSPC measurements.

[e] k_{RECET} was selected to fit the observed fraction of subnanosecond MLCT bleach recovery at the lowest metallointercalator concentration.

[f] β values ranging from 0.6 through $1.4\,\text{Å}^{-1}$ were investigated. β values of less than 0.7 introduce dynamics in the predicted subnanosecond emission decays and transient-absorption recoveries not observed experimentally. β values greater than 1.0 predict insignificant nanosecond quenching components and cannot account for the overall emission quenching observed by steady-state luminescence spectroscopy.

[g] The magnitude of P in our simulations translates into a free energy change $\approx -1.5\,\text{kcal mol}^{-1}$ for transferring a donor from any nonnearest intercalation site to a nearest possible intercalation site adjacent to an acceptor.

able to account completely for a broad range of kinetic and static measurements on electron transfer between metallointercalators in DNA. It is important to emphasize that our analysis of the kinetic data predicts an electron transfer rate at 40 Å that is over 12 decades slower than the apparent rate of electron transfer reported by Murphy et al. [34]. These results are apparently incompatible with the Murphy et al. measurement, although it should be emphasized that the experiments are quite different.

The quantitative modeling is in strong disagreement with the previous claim that a photocleavage assay method is capable of ruling out donor–acceptor clustering [71]. This is demonstrated in Figure 9 where simulated and observed photocleavage measurements in the absence and presence of $[\text{Ru(phen)}_2\text{dppz}]^{2+}$ are compared [58]. These results clearly show that the photocleavage pattern is not sufficiently sensitive to Ru–Rh clustering over the range of concentrations that have been explored.

Lincoln et al. have also investigated the luminescence quenching of [Ru(phen)$_2$dppz]$^{2+}$ by [Rh(phi)$_2$bpy]$^{3+}$ in a DNA environment [59]. They interpret their results in a way similar to Olson et al. [58]. All of their data, spanning a wide range of [Ru]/[bp DNA] ratios, can be accounted for by assuming the rhodium complex preferentially binds next to the ruthenium complexes. An extended cooperative binding model is able to demonstrate that the rhodium complex is about 50 times more likely to bind adjacent to a ruthenium complex than at a site away from it, on pure poly-(dA-dT) duplex. The preference for clustering within this model is greater than the factor of 13 reported by Olson et al. for mixed sequence DNA [58], consistent with the report of very efficient fast electron transfer in the (dA-dT)-duplex environment [71]. Additionally, Lincoln et al. conclude that their data generally provide strong spectroscopic evidence that the rhodium and ruthenium species bind cooperatively in a DNA duplex.

It is important to compare the clustering model (Olson et al. [58]; Lincoln et al. [59]) with the interpretation of the Ru–DNA–Rh nontethered emission quenching data given by Barton and co-workers. In a recent paper, Barton and co-workers attribute the fast component of emission quenching to electron transfer over a distance of about 40 Å at a rate $> 10^{10}\,\mathrm{s}^{-1}$ [61]. This is, of course, consistent with their previous interpretation cited in case II. The absence of complete quenching, and indeed the presence of very slow quenching in the Ru–DNA–Rh untethered system is attributed to donor–acceptor pairs that are beyond a "sphere-of-action" distance. While this is essentially a type of distance dependence, the authors do not claim that the electron transfer rate is in fact distance dependent. Rather, they attribute the slow component of emission quenching to destacking (i.e., some disturbance of the hypothetical π-stack coupling path between donors and acceptors). The discrepancy would be resolved easily if an independent measure of the donor–acceptor separation were available. This, of course, underscores the importance of covalently bound samples where donor–acceptor separations are well defined. On the other hand, it should be emphasized that the tethered experiments of Murphy et al. have the disadvantage that only steady-state data are available, while picosecond, nanosecond, and steady-state measurements are available for the nontethered samples.

C. Chemical Measurements on Long-Range Electron Transfer

Recently, Barton and co-workers have reported a number of interesting experiments on permanent chemical changes induced by long-range electron transfer in DNA [54,61–63]. Unfortunately, these measurements are not time-resolved measurements, and it is not possible to determine kinetic information definitively from these results. The chemical experiments involve two different acceptors: a Rh(III)* species that has a lifetime of < 200 ns in the absence of intermolecular electron transfer; and a Ru(III) species generated by the flash-quench technique which possesses an apparent lifetime of hundreds of microseconds when bound to mixed sequence DNA [77]. The donors in these studies include G and a thymine dimer (⟨TT⟩). In the case of the donor G (cases X and XI in Table I), G is oxidized by electron transfer to either the Rh(III)* or Ru(III) acceptor. The initial oxidation step is chemically developed by treatment with hot piperdine, which ultimately causes release of the damaged base and strand cleavage. The location of strand cleavage is analyzed by gel electrophoresis and visualized by phosphorimaging techniques. The actual cleavage mechanism is complex and involves dioxygen trapping of a guanine cation producing the modified nucleotide 8-oxo-2′-deoxyguanosine, which is converted to the cleaved situation subsequent to treatment with piperdine. In case XII of Table I, electron transfer from ⟨TT⟩ to Rh(III)* is used to "repair" the thymine dimer (i.e., return TT to an undimerized pair of nucleotides). This dimer is a damage site in DNA induced by UV irradiation.

The main observation of these measurements is the ability to produce photo-oxidation at sites distant from the site of the *primary* photo-oxidant. The presence of damage at distant sites is presented as evidence for long-range electron transfer. Importantly, base damage is shown to occur at sites as far away 11 base pairs removed from the primary photo-oxidant. In certain situations the amount of base damage is approximately equally distributed about regions of the duplex where the DNA sequence possesses similarly oxidizable groups, independent of separation from the primary photo-oxidant. Indeed, Hall et al. conclude that long-range oxidative damage at 5′-GG-3′ doublet sites does not depend on distance [61]. Although the precise mechanism for the remote damage is unknown, evidence is presented which indicates that the damage is indeed mediated by the DNA duplex [61]. It has been observed that the amount of damage is attenuated by intervening "bulged" regions of the DNA, which the authors have interpreted as an interruption of the π-stack [62].

The quantum yields of the various photo-oxidation processes are $< 10^{-6}$. It may be that slow, nondirect electron transfer reactions such as hopping or bimolecular electron transfer may contribute to these results. Also, as mentioned above, slow dynamical fluctuations of the DNA environment may lead to extremely short donor–acceptor separations. Since the quantum-yield measurements of the chemical yield indeed are consistent with very slow rates, it is difficult to rule out these potential contributions.

Although time-resolved measurements on the electron transfer reactions in cases X to XII are not available, it is possible to estimate the rate constants for these processes roughly by considering a variety of data. For example, Stemp et al. have shown that oxidation by Ru(III) using the flash-quench technique in a pure poly-(dG-dC) duplex occurs in ≪200 ns [77]. In this experiment the non-covalently bound, intercalated Ru(III) is always adjacent to a guanine base, and rapid electron transfer is not surprising. On the other hand, the apparent rate of G oxidation is considerably slower for other DNA sequences [77]. Some of the variations in the apparent rate of reaction may reflect the variation of the oxidation potential of G as a function of surrounding bases. Calculations indicate that $5'$-G sites increase in ease of oxidation in the series GT ≈ GC ≪ GA < GG < GGG [82]. In light of these observations it is interesting to consider the kinetic measurements of Stemp et al. on Ru(III) oxidation of "mixed-sequence DNA." Presumably, mixed-sequence DNA contains a significant fraction of easily oxidized regions (e.g., GG, GA). Considering the very small quantum yield of photo-oxidation in this system, combined with the observation of a significant long-lived (100 μs) Ru(III) signal which the authors attribute to unreactive Ru(III) [77], suggests that the rate of electron transfer to many of these easily oxidized sites must be ≪(100 μs)$^{-1}$. This observation is consistent with a significant distance dependence for the rate of electron transfer from G to Ru(III).

A second indication that the rate of electron transfer from an easily oxidized G to Ru(III) is relatively slow is the observation that the yield of damage sites is about 10^3 times greater for Ru(III) than for reactions with the photo-oxidant Rh(III)* [63]. The efficiency of electron transfer should be much greater with Ru(III) (ca. 100 μs lifetime) as an electron acceptor compared to Rh(III)* (< 200 ns lifetime) for tethered samples if k_{ET} is much slower than the lifetime of the oxidant. While these arguments suggest that k_{ET} between G and Ru(III) must be ≪ 10^6 s^{-1}, a lower limit to k_{ET} can be fixed by a consideration of the quantum yield of permanent photocleaveage. If piperdine-induced photocleavage were 100% efficient for every oxidized G site, the observed yield of photocleavage would simply reflect the quantum yield of photo-oxidation. Using the 100-μs lifetime of Ru(III) in a mixed-sequence DNA environment and the largest observed yield of photocleavage observed from Arkin et al. [63], a lower limit of about 10^1 s^{-1} for the rate of electron transfer can be estimated. Finally, these rough estimates can be used to "bracket" the rates of long-range electron transfer photo-oxidation as summarized in Table I. It is significant to note that the bracketed range of electron transfer rates for photo-oxidation is many orders of magnitude slower than the long-range ultrafast rates of excited-state electron transfer reported for electron transfer between Ru(II)* and Rh(III) in case II [34]. In fact, the bracketed rates are in the range expected for electron transfer in proteins. It is also important to note that the bracketed rates are sufficiently slow that complicating factors such as nondirect electron transfer due to hopping, bimolecular mechanisms, or rearrangement of the DNA framework, may be a factor in the apparent mild distance dependence observed in the photo-oxidation yields.

The last example in Table I involves the oxidative repair of thymine dimers $\langle TT \rangle$ by [Rh(phi)$_2$bpy$'$]$^{3+}$ excited state, Rh(III)*. Oxidative repair was observed with Rh(III)* at long distances (16–26 Å) covalently bound to either end of a 16-base-pair duplex. Repair efficiencies did not decrease with increasing distance between the tethered rhodium complex and the $\langle TT \rangle$ defect, but repair was attenuated with disruptions of the intervening π-stack [54]. Dandliker et al. used these various observations as evidence for long-range electron transfer with a distance-independent rate. Assuming that the reported quantum yield for repair (ca. 10^{-6}) reflects the quantum yield for the oxidation process itself, the implied rate of electron transfer from $\langle TT \rangle$ to Rh(III)* is about 10^1 s^{-1} [assuming a typical lifetime for Rh(III)* complexes in a DNA environment]. This is slow enough to be concerned with competitive non-direct electron transfer processes.

It should also be noted that the quantum yield of $\langle TT \rangle$ repair by Rh(III)* is observed to be 30 times greater when non-covalently bound Rh(III)* is employed. We suggest that the explanation for the increased yield is an increased fraction of Rh(III)* photo-oxidants existing in much closer proximity to the $\langle TT \rangle$ sites than in the tethered samples. If the electron transfer rate is in fact highly distance dependent, the potential for binding in closer proximity in the nontethered sample would lead to the observed higher yield. Thus, while the pattern of products suggests a distance-independent rate, the extremely small quantum yield and the increased yield with non-tethered Rh(III) suggest the opposite. This discrepancy may be due to a competition of direct, distance-dependent electron transfer rates, and non-direct, very slow electron transfer due to bimolecular reaction, DNA unwinding, or other slow processes, as portrayed in Figure 3.

III. CONCLUSIONS

The various measurements on electron transfer rates between ordinary donors and acceptors in DNA reveal apparently contradictory behavior. Many of the measurements exhibit kinetics that are similar to that observed for protein environments, namely an exponential fall-off, with β values close to 1 Å^{-1}. Consequently, at short distances ultrafast rates are observed, while at longer distance the rates are much slower. In dramatic contrast, some measurements in DNA show evidence for extraordinary long-range electron transfer at ultrafast rates. Still other examples involving chemical-product yields show long-range electron transfer with an apparently slow rate and small dependence on distance. These discrepancies may be due partly to the complexity and challenge of the experiments, which are subject to several complicating factors that can be important in the interpretation the experimental results. The best-defined kinetic evidence, which involves measurements that combine various types of transient spectroscopy to measure electron transfer rates directly, support the conclusion that electron transfer between ordinary donors and acceptors in DNA is not extraordinarily rapid, nor does it have an extraordinarily shallow distance dependence. Nevertheless, more experiments will be necessary to understand the nature of the discrepancies.

ACKNOWLEDGMENTS

We acknowledge preprints and reprints from the following people: Tom Netzel, Jackie Barton, John Warman, Rich Friesner, Dave Beratan, Fred Lewis, Bengt Nordén, Harry Gray, and Tom Meade. Our research on DNA electron transfer has been supported by the NSF and by a NIH training grant to E.J.C.O. Helpful discussions with Jackie Barton and co-workers, Rich Friesner, and Tom Netzel are greatly acknowledged.

Note Added in Proof

Tanaka and Fukui recently reported [*Angew. Chem. Int. Ed. Engl.* **37**, 158 (1998)] a study on the distance dependence of the rate of electron transfer between guanine and 9-amino-6-chloro-2-methoxyacridine inserted into specific sites on small pieces of DNA. Their results, and the β value reported in their study ($\beta = 1.4 \, \text{Å}^{-1}$), significantly add to the growing body of evidence against the existence of ultrafast electron transfer in DNA.

REFERENCES

1. P. O'Neill and E. M. Fielden, *Adv. Radiat. Biol.* **17**, 53 (1993).

2. V. Michalik, *Int. J. Radiat. Biol.* **62**, 9 (1992).

3. D. Schulte-Frohlinde and E. Bothe, *NATO ASI Ser. H* **54**, 317 (1992).

4. J. H. Miller, *NATO ASI Ser. H* **54**, 157 (1992).

5. M. C. R. Symons, *NATO ASI Ser. H* **54**, 111 (1992).

6. C. Knapp, J.-P. Lecomte, A. Kirsch-De Mesmaeker, and G. Orellana, *J. Photochem. Photobiol. B: Biol.* **36**, 67 (1996).

7. P. F. Barbara, T. J. Meyer, and M. A. Ratner, *J. Phys. Chem.* **100**, 13148 (1996).

8. D. DeVault, *Quantum Mechanical Tunneling in Biological Systems*, Cambridge University Press, Cambridge, 1984.

9. M. D. Newton and N. Sutin, *Annu. Rev. Phys. Chem.* **35**, 437 (1984).

10. M. D. Newton, *Chem. Rev.* **91**, 767 (1991).

11. G. McLendon, *Acc. Chem. Res.* **21**, 160 (1988).

12. R. A. Marcus and N. Sutin, *Biochim. Biophys. Acta* **811**, 265 (1985).

13. R. A. Marcus, *Rev. Mod. Phys.* **65**, 599 (1993).

14. J. Jortner, *J. Chem. Phys.* **64**, 4860 (1976).

15. J. Jortner and M. Bixon, *Ber. Bunsenges. Phys. Chem.* **99**, 296 (1995).

16. J. W. Evenson and M. Karplus, *Science* **262**, 1247 (1993).

17. D. N. Beratan and J. N. Onuchic, *Adv. Chem. Ser.* **228**, 71 (1991).

18. D. N. Beratan, J. N. Betts, and J. N. Onuchic, *Science* **252**,1285 (1991).

19. D. N. Beratan, J. N. Onuchic, J. R. Winkler, and H. B. Gray, *Science* **258**, 1740 (1992).

20. W. Saenger, *Principles of Nucleic Acid Structure*, Springer-Verlag, New York, 1984.

21. K. V. Mikkelsen and M. A. Ratner, *Chem. Rev.* **87**, 113 (1987).

22. N. R. Kestner, J. Logan, and J. Jortner, *J. Phys. Chem.* **78**, 2148 (1974).

23. M. Bixon and J. Jortner, *Chem. Phys.* **176**, 476 (1993).

24. M. Y. Ogawa, J. F. Wishart, Z. Young, J. R. Miller, and S. S. Isied, *J. Phys. Chem.* **97**, 11456 (1993).

25. R. Langen, J. L. Colón, D. R. Casimiro, T. B. Karpishin, J. R. Winkler, and H. B. Gray, *J. Biol. Inorg. Chem.* **1**, 221 (1996).

26. R. Langen, I.-J. Chang, J. P. Germanas, J. H. Richards, J. R. Winkler, and H. B. Gray, *Science* **268**, 1733 (1995).

27. H. B. Gray and J. R. Winkler, *Pure Appl. Chem.* **64**, 1257 (1992).

28. D. S. Wuttke, M. J. Bjerrum, I.-J. Chang, J. R. Winkler, and H. B. Gray, *Biochim. Biophys. Acta* **1101**, 168 (1992).

29. D. R. Casimiro, J. H. Richards, J. R. Winkler, and H. B. Gray, *J. Phys. Chem.* **97**, 13073 (1993).

30. C. C. Moser, J. M. Keske, K. Warncke, R. S. Farid, and P. L. Dutton, *Nature* **355**, 796 (1992).

31. M. R. Wasielewski, *Chem. Rev.* **92**, 435 (1992).

32. G. L. Closs and J. R. Miller, *Science* **240**, 440 (1988).

33. M. N. Paddon-Row and J. W. Verhoeven, *New J. Chem.* **15**, 107 (1991).

34. C. J. Murphy, M. R. Arkin, Y. Jenkins, N. D. Ghatlia, S. Bossmann, N. J. Turro, and J. K. Barton, *Science* **262**, 1025 (1993).

35. S. Priyadarshy, S. M. Risser, and D. N. Beratan, *J. Phys. Chem.* **100**, 17678 (1996).

36. S. M. Risser, D. N. Beratan, and T. J. Meade, *J. Am. Chem. Soc.* **115**, 2508 (1993).

37. A. K. Felts, W. T. Pollard, and R. A. Friesner, *J. Phys. Chem.* **99**, 2929 (1995).

38. T. L. Netzel, "A Comparison of Experimental and Theoretical Studies of Electron Transfer Within DNA Duplexes" in *Organic and Inorganic Photochemistry*; Molecular and Supramolecular Photochemistry Series. Vol. 2. V. Ramamurthy and K. S. Schanze, eds., pp. 1–54. Marcel Dekker, New York, 19XX.

39. T. L. Netzel, *J. Chem. Educ.* **74**, 646 (1997).

40. R. E. Holmlin, P. J. Dandliker, and J. K. Barton, *Angew. Chem. Int. Ed. Engl.* **36**, 2714 (1997).

41. T. J. Meade, in *Probing of Nucleic Acids by Metal Ion Complexes of Small Molecules*, Vol. 33, A. Sigel and H. Sigel, Eds., Marcel Dekker, New York,1995, pp. 453–478.

42. M. Faraggi, C. Ferradini, and J.-P. Jay-Gerin, *New J. Chem.* **19**, 1203 (1995).

43. J. M. Warman, M. P. d. Haas, and A. Rupprecht, *Chem. Phys. Lett.* **249**, 319 (1996).

44. C. Beach, A. F. Fuciarelli, and J. D. Zimbrick, *Radiat. Res.* **137**, 385 (1994).

45. A. F. Fuciarelli, E. C. Sisk, J. H. Miller, and J. D. Zimbrick, *Int. J. Radiat. Biol.* **66**, 505 (1994).

46. A. F. Fuciarelli, E. C. Sisk, and J. D. Zimbrick, *Int. J. Radiat. Biol.* **65**, 409 (1994).

47. B. Armitage, T. Koch, H. Frydenlund, H. Orum, H. G. Batz, and G. B. Schuster, *Nuc. Acids. Res.* **25**, 4674 (1997).

48. T. L. Netzel, K. Nafisi, J. Headrick, and B. E. Eaton, *J. Phys. Chem.* **99**, 17948 (1995).

49. T. L. Netzel, M. Zhao, K. Nafisi, J. Headrick, M. S. Sigman, and B. E. Eaton, *J. Am. Chem. Soc.* **117**, 9119 (1995).

50. V. Y. Shafirovich, P. P. Levin, V. A. Kuzmin, T. E. Thorgeirsson, D. S. Kliger, and N. E. Geacintov, *J. Am. Chem. Soc.* **116**, 63 (1994).

51. D. O'Connor, V. Y. Shafirovich, and N. E. Geacintov, *J. Phys. Chem.* **98**, 9831 (1994).

52. V. Y. Shafirovich, S. H. Courtney, N. Q. Ya, and N. E. Geacintov, *J. Am. Chem. Soc.* **117**, 4920 (1995).

53. F. D. Lewis, T. Wu, Y. Zhang, R. L. Letsinger, S. R. Greenfield, and M. R. Wasielewski, *Science* **277**, 673 (1997).

54. P. J. Dandliker, R. E. Holmlin, and J. K. Barton, *Science* **275**, 1465 (1997).

55. T. J. Meade and J. F. Kayyem, *Angew. Chem. Int. Ed. Engl.* **34**, 352 (1995).

56. S. O. Kelley, R. E. Holmlin, E. D. A. Stemp, and J. K. Barton, *J. Am. Chem. Soc.* **119**, 9861 (1997).

57. A. M. Brun and A. Harriman, *J. Am. Chem. Soc.* **114**, 3656 (1992).

58. E. J. C. Olson, D. Hu, A. Hörmann, and P. F. Barbara, *J. Phys. Chem. B* **101**, 299 (1997).

59. P. Lincoln, E. Tuite, and B. Nordén, *J. Am. Chem. Soc.* **119**, 1454 (1997).

60. M. R. Arkin, E. D. A. Stemp, C. Turro, N. J. Turro, and J. K. Barton, *J. Am. Chem. Soc.* **118**, 2267 (1996).

61. D. B. Hall, R. E. Holmlin, and J. K. Barton, *Nature* **382**, 731 (1996).

62. D. B. Hall and J. K. Barton, *J. Am. Chem. Soc.* **119**, 5045 (1997).

63. M. R. Arkin, E. D. A. Stemp, S. C. Pulver, and J. K. Barton, *Chem. Biol.* **4**, 389 (1997).

64. J. K. Barton, C. V. Kumar, and N. J. Turro, *J. Am. Chem. Soc.* **108**, 6391 (1986).

65. P. Fromherz and B. Rieger, *J. Am. Chem. Soc.* **108**, 5361 (1986).

66. S. J. Atherton and P. C. Beaumont, *J. Phys. Chem.* **90**, 2252 (1986).

67. S. J. Atherton and P. C. Beaumont, *J. Phys. Chem.* **91**, 3993 (1987).

68. M. D. Purugganan, C. V. Kumar, N. J. Turro, and J. K. Barton, *Science* **241**, 1645 (1988).

69. G. Orellana, A. Kirsch-De Mesmaeker, J. K. Barton, and N. J. Turro, *Photochem. Photobiol.* **54**, 499 (1991).

70. C. J. Murphy, M. R. Arkin, N. D. Ghatlia, S. Bossmann, N. J. Turro, and J. K. Barton, *Proc. Natl. Acad. Sci. USA* **91**, 5315 (1994).

71. M. R. Arkin, E. D. A. Stemp, R. E. Holmlin, J. K. Barton, A. Hörmann, E. J. C. Olson, and P. F. Barbara, *Science* **273**, 475 (1996).

72. E. D. A. Stemp, M. R. Arkin, and J. K. Barton, *J. Am. Chem. Soc.* **117**, 2375 (1995).

73. A. Hörmann, E. J. C. Olson, P. F. Barbara, M. R. Arkin, E. D. A. Stemp, R. E. Holmlin, and J. K. Barton, in *Ultrafast Phenomena X*, Vol. 62 (J. G. Fujimoto, P. F. Barbara, W. H. Knox, and W. Zinth, eds.), Springer-Verlag, New York, 1996, pp. 359–360.

74. R. E. Holmlin, E. D. A. Stemp, and J. K. Barton, *J. Am. Chem. Soc.* **118**, 5236 (1996).

75. L. M. Davis, J. D. Harvey, and B. C. Baguley, *Chem.–Biol. Interact.* **62**, 45 (1987).

76. B. C. Baguley and M. Le Bret, *Biochemistry* **23**, 937 (1984).

77. E. D. A. Stemp, M. R. Arkin, and J. K. Barton, *J. Am. Chem. Soc.* **119**, 2921 (1997).

78. A. E. Friedman, J.-C. Chambron, J.-P. Sauvage, N. J. Turro, and J. K. Barton, *J. Am. Chem. Soc.* **112**, 4960 (1990).

79. M. R. Arkin, Y. Jenkins, C. J. Murphy, N. J. Turro, and J. K. Barton, *Adv. Chem. Ser.* **246**, 449 (1995).

80. M. R. Arkin, E. D. A. Stemp, Y. Jenkins, P. F. Barbara, N. J. Turro, and J. K. Barton, *Polym. Mater. Sci. Eng. Washington* **71**, 598 (1994).

81. L. S. Schulman, S. H. Bossmann, and N. J. Turro, *J. Phys. Chem.* **99**, 9283 (1995).

82. H. Sugiyama and I. Saito, *J. Am. Chem. Soc.* **118**, 7063 (1996).

AUTHOR INDEX

Numbers in parentheses are reference numbers and indicate that the author's work is referred to although his name is not mentioned in the text. Numbers in *italic* show the pages on which the complete references are listed.

SUBJECT INDEX

709